ANNALS OF THE NEW YORK ACADEMY OF SCIENCES

Volume 338

The New York Academy of Sciences
2 East 63rd Street
New York, New York 10021

AEROSOLS:
ANTHROPOGENIC AND NATURAL, SOURCES AND TRANSPORT

ANNALS OF THE NEW YORK ACADEMY OF SCIENCES
Volume 338

AEROSOLS: ANTHROPOGENIC AND NATURAL, SOURCES AND TRANSPORT

Edited by Theo. J. Kneip and Paul J. Lioy

The New York Academy of Sciences
New York, New York
1980

Library of Congress Cataloging in Publication Data

Main entry under title:

Aerosols, anthropogenic and natural, sources and transport.

(Annals of the New York Academy of Sciences; v. 338)
Papers of a conference sponsored by the New York Academy of Sciences and the Air Pollution Control Association, and held Jan. 9–12, 1979.
Includes index.
1. Aerosols—Environmental aspects—Congresses.
2. Air—Pollution—Congresses. I. Kneip, Theodore Joseph, 1926– II. Lioy, Paul J. III. New York Academy of Sciences. IV. Air Pollution Control Association. V. Series: New York Academy of Sciences. Annals; v. 338.
Q11.N5 vol. 338 [TD884.5] 500s [628.5'3]
80-12891

SP
Printed in the United States of America
ISBN 0–89766–064–1 (Cloth)
ISBN 0–89766–065–X (Paper)

ANNALS OF THE NEW YORK ACADEMY OF SCIENCES
VOLUME 338
May 15, 1980

AEROSOLS: ANTHROPOGENIC AND NATURAL, SOURCES AND TRANSPORT*

Editors and Conference Chairmen
THEO. J. KNEIP AND PAUL J. LIOY

CONTENTS

Introduction. *By* PAUL J. LIOY and THEODORE J. KNEIP ix

Part I. Urban Aerosol Characteristics
Morton Lippmann, *Chairman*

Size Distributions in Urban Aerosols. *By* MORTON LIPPMANN 1

Trace Element Abundances and Chemistry of Atmospheric Aerosols: Current Techniques and Future Possibilities. *By* ROBERT W. SHAW and ROBERT K. STEVENS 13

Secondary Aerosol: Production Mechanisms of Sulfate Compounds in the Atmosphere. *By* TIMOTHY V. LARSON 26

Sulfur and Nitrogen Compounds in Urban Aerosols. *By* ROGER L. TANNER 39

Organic Compounds in Urban Aerosols. *By* JOAN M. DAISEY 50

Optical Properties of the Urban Aerosol and Their Relation to Chemical Composition. *By* BRIAN P. LEADERER and JAN A. J. STOLWIJK 70

General Discussion 86

Part II. Source Apportionment for Urban Aerosols
Glen E. Gordon, *Chairman*

Techniques for Treating Multielement Particulate Data to Obtain Information on Sources: Overview. *By* GLEN E. GORDON 93

Source Identification and Resolution through Application of Factor and Cluster Analysis. *By* PHILIP K. HOPKE 103

The Use of Long-Term and Seasonal Trends of Elemental Compounds as an Aid to the Identification of Sources of Airborne Pollutants. *By* MICHAEL T. KLEINMAN 116

Chemical Element Balance Method Applied to Dichotomous Sampler Data. *By* THOMAS G. DZUBAY 126

*This volume is the result of a conference entitled Aerosols: Anthropogenic and Natural, Sources and Transport, held on January 9–12, 1979 by The New York Academy of Sciences and cosponsored by the Air Pollution Control Association.

Ambient Sulfate Measurements on Allegheny Mountain and the Question of Atmospheric Sulfate in the Northeastern United States. *By* WILLIAM R. PIERSON, WANDA W. BRACHACZEK, TIMOTHY J. TRUEX, JAMES W. BUTLER, and THOMAS J. KORNISKI 145

Modeling of Aerosol Dynamics: Aerosol Size and Composition. *By* RICHARD A. BEYAK and THOMAS W. PETERSON 174

General Discussion 190

Part III. The Urban Plume
Francis A. Schiermeier, *Chairman*

A Review of Urban Plume Studies. *By* FRANCIS A. SCHIERMEIER 198

Aerosol Composition of Urban Plumes Passing over a Rural Monitoring Site. *By* THOMAS G. ELLESTAD 202

Aerosol Formation in Urban Plumes over Lake Michigan. *By* DAVID F. MILLER and ABDUL J. ALKEZWEENY 219

Size and Composition of Visibility-Reducing Aerosols in Southwestern Plumes. *By* EDWARD S. MACIAS, DONALD L. BLUMENTHAL, JERRY A. ANDERSON, and BRUCE K. CANTRELL 233

Aerosol Formation in Urban Plumes. *By* KENNETH T. WHITBY 258

Formation and Transport of Ozone and Aerosols in Power Plant Plumes. *By* NOOR V. GILLANI and WILLIAM E. WILSON 276

Sulfate Formation in Urban Plumes. *By* JOHN W. WINCHESTER 297

General Discussion 309

Part IV. Natural Aerosols
Ruprecht Jaenicke, *Chairman*

Natural Aerosols. *By* RUPRECHT JAENICKE 317

The Production, Concentration, and Vertical Distribution of the Sea-Salt Aerosol. *By* DUNCAN C. BLANCHARD and ALFRED H. WOODCOCK 330

Major Contributions of Natural Primary Continental Aerosols: Source Mechanisms. *By* DALE GILLETTE 348

Organic Constituents of Natural Aerosols. *By* JÜRGEN HAHN 359

General Discussion 377

Part V. Mesoscale and Synoptic Scale Transport
George T. Wolff, *Chairman*

Mesoscale and Synoptic Scale Transport of Aerosols. *By* GEORGE T. WOLFF 379

Types of Transport Episodes in Northern Europe. *By* CYRILL BROSSET 389

Regional Scale Air Pollution: Sources and Effects. *By* R. B. HUSAR and D. E. PATTERSON 399

Evidence of Transport of Hazy Air Masses from Satellite Imagery. *By* WALTER A. LYONS 418

The Fate of Particulate Emissions from an Isolated Power Plant in the Oil Sands Area of Western Canada. *By* LEONARD A. BARRIE 434

OECD Program on Long Range Transport of Air Pollutants—Measurements from Aircraft. *By* Y. GOTAAS ... 453

The Occurrence of Atmospheric Aerosols in the Northeastern United States. *By* P. K. MUELLER, G. M. HIDY, K. WARREN, T. F. LAVERY, and R. L. BASKETT ... 463

General Discussion ... 483

Part VI. Long Range Transport
Kenneth A. Rahn, *Chairman*

On the Origin and Transport of the Winter Arctic Aerosol. *By* KENNETH A. RAHN and RICHARD J. MCCAFFREY ... 486

The Long Range Transport of Sulfurous Aerosol to Scandinavia. *By* BRYNJULF OTTAR ... 504

Long Range Transport of Desert Dust with Special Emphasis on the Sahara. *By* LOTHAR SCHÜTZ ... 515

Arctic Haze: Perturbation of the Polar Radiation Budget. *By* GLENN E. SHAW and KNUT STAMNES ... 533

Acid Precipitation and Its Effects on Terrestrial and Aquatic Ecosystems. *By* ELLIS B. COWLING ... 540

General Discussion ... 556

Part VII. Some Future Problems Associated with Energy
Merril Eisenbud, *Chairman*

Some Future Problems Associated with Energy: Introductory Remarks. *By* THOMAS R. GLENN ... 561

Potential Effects of the Projected Increase in Coal Use. *By* ARTHUR M. SQUIRES ... 563

Meteorological Effects of Rejected Heat. *By* H. E. LANDSBERG ... 569

Climatic Effects of Atmospheric Aerosols. *By* JAMES E. HANSEN, ANDREW A. LACIS, PAUTHON LEE, and WEI-CHYUNG WANG ... 575

Future Aerosols of the Southwest: Implications for Fundamental Aerosol Research. *By* S. K. FRIEDLANDER ... 588

Overview of Aerosol Research. *By* MERRIL EISENBUD ... 599

Panel Discussion. *Moderator:* THEODORE J. KNEIP ... 605

Interpretive Discussion. *By* PAUL J. LIOY and THEO. J. KNEIP ... 613

Index of Contributors ... 617

Financial assistance was received from:

- AIR POLLUTION CONTROL ASSOCIATION (APCA)
- MESA NEW YORK BIGHT PROJECT
- MID-ATLANTIC STATES SECTION OF APCA
- NATIONAL INSTITUTE OF ENVIRONMENTAL HEALTH SCIENCES
- OFFICE OF NAVAL RESEARCH
- U.S. ENVIRONMENTAL PROTECTION AGENCY

INTRODUCTION

Paul J. Lioy and Theodore J. Kneip

Institute of Environmental Medicine
New York University Medical Center
New York, New York 10016

During the past eight years, there has been a significant growth in our knowledge and understanding of many aspects of the atmospheric aerosol. The continuing concern over effects on man and the environment has resulted in studies of health effects, climatic modification, stresses in the biosphere (dry and wet deposition), and the degradation of visibility (optical effects). This conference was intended to examine the physical and chemical information that has been obtained from previous and continuing studies on aerosols, and to subsequently interrelate and/or interpret the results and conclusions to obtain a broader understanding of present knowledge and research needs.

The conference was structured to examine the urban and natural sources of aerosols and the distribution of the aerosols within the local, regional, and global atmosphere. A number of questions were raised regarding the impact of present and anticipated emissions upon various population centers and rural-agricultural areas.

The chemical and physical nature of inorganic and organic materials in the urban aerosol was examined, along with the trends that have been observed in recent years. It became apparent that there is a need for a great deal more knowledge about organic compounds and inorganic chemical species as a function of particle size, and about the reactions among these materials.

We also focused on the questions associated with source apportionment: (1) the form of the models; and (2) how well the present models can determine which sources contribute to the urban aerosol. Questions on the compatibility and applicability of the chemical element balance and the principal factor analysis techniques to various geographic locations were addressed. Questions were raised on the need for improvement of data sets used in the sophisticated mathematical treatments.

At times, locally observed high concentrations of secondary aerosols are principally due to the downwind contribution of urban and industrial sources. Papers relating to this topic discussed not only the primary emissions within individual and area plumes, but also the secondary aerosols generated through a number of heterogeneous and homogeneous reaction schemes. Particular emphasis was placed on the reaction rates determined by recent plume studies for SO_2 and NO_2 oxidation, and their applicability in other locations. Differences in the transformation rates are due to a combination of the unique characteristics of (individual) sources studied, the geographic location, and prevailing meteorology. Even at a single location, such as New York City, the extremes in weather and coincident alterations in source distribution appear to play an important role in determining the dominant reaction mechanisms.

A series of presentations dealt with the composition of natural aerosols. The types of sources and/or mass percentages that will influence only specific geographic locations were discussed. The composition of the "true" background aerosol in the atmosphere was examined, and wider-scale effects considered.

The effects of transport of the ambient aerosol were considered for both regional and synoptic scales. Since the early 1970s, a number of studies have been directed towards investigating the nature and significance of mesoscale and synoptic scale pollutant transformation and transport. Through these investigations, we have become cognizant of the magnitude of the regional accumulation and transport of aerosols, and of the quantity of materials such as $SO_4^{=}$ contributed to locations as much as 1000 kilometers downwind. The main foci of these investigations have been the northeastern U.S., Canada, and northern Europe. The presentations provided analysis of considerable information generated by these studies on trends, chemical composition, and meteorological conditions conducive to transport of the atmospheric aerosol.

To go one step further, the possibility of what could be classified as intercontinental (long range) transport was described. Particles (which can be expected to accumulate from man-made and natural emissions) and their contribution to remote areas such as the Arctic were followed from air mass trajectory analyses. Visibility reduction observed in the Arctic was attributed to transported aerosols.

The conference ended with an interactive session on potential aerosol problems associated with several major energy problems. For instance, the changes in our national energy policies and the probable increases in the use of coal and diesel fuels necessitate evaluation of the anticipated emissions and of the long-term effects that may result.

Since the production and transport of particulate pollution transcend local, state, and national boundaries, we must develop a better understanding of the processes involved. The proceedings of this conference should help to provide a substantial amount of information to guide future research on the problems. To the degree that we are successful in utilizing the assembled information with other compatible data, pertinent epidemiological and environmental studies can be done and responsible decisions can be made.

This conference was a rewarding experience for us, and we thank the authors and the attendees for the hours of valuable discussion.

We would like to acknowledge the efforts of Mrs. Ellen Marks of the New York Academy of Sciences in directing the conference, the work done by the session chairmen in assisting in the organization of the conference, and Dr. William E. Wilson of the U.S. Environmental Protection Agency who provided assistance and encouragement throughout the development process.

SIZE DISTRIBUTIONS IN URBAN AEROSOLS

Morton Lippmann

Institute of Environmental Medicine
New York University Medical Center
New York, New York 10016

The Urban Aerosol

The urban aerosol is a continuously varying suspension of liquid and solid particles in the ambient air. The particles vary considerably in size, shape, composition, and concentration because of (1) variations in the nature and strength of the local sources (primary particles, vapors that condense or react chemically to form particles, and/or vapors that react or condense on the surfaces of particles); (2) variations in the mixing characteristics and dilution capacity of the air; (3) the past history of the air mass, and hence its background burden of particles and vapors attributable to upwind sources; and (4) scavenging processes, such as sedimentation and rainfall, which act to remove particles from the air. The particles of interest range from about 0.002 μm to about 50 μm in diameter. Number concentrations may vary from 10^2 to $10^7/cm^3$. Mass concentrations may vary from 5 $\mu g/m^3$ to 2,000 $\mu g/m^3$.

Considering the diversity of the particles in the urban aerosol, it is necessary to describe the aerosols in terms of their various distributions. For each aerosol, there may be a number of different distributions of interest, such as the number, surface, volume, or mass of particles per unit volume of air. Each of these parameters may need to be characterized in terms of amount per particle size interval, because many environmental effects of interest are particle-size dependent. For example, most visibility reduction is attributable to particles in the 0.1 to 1 μm range, while most cloud nucleation is performed by particles smaller than 0.1 μm, and most surface soiling is attributable to particles larger than about 2 μm.

The discussion will be limited to the physical characteristics of the particles. Distributions of size, without respect to chemical composition, can readily and reliably be made in most cases now with commercial instrumentation, and a great deal of knowledge has been accumulated in the past decade on these functions at many urban locations under a variety of conditions. The state of current knowledge is much less satisfactory with respect to the size distribution of chemical components within the overall aerosol. However, important work is currently under way that is beginning to provide a framework for better understanding of the sources and effects of specific chemical constituents of interest.

Papers on the trace elements, sulfur and nitrogen compounds, and organics that follow in this first session of the conference will demonstrate that recent work on the characterization of the concentration of chemical species within appropriate size intervals has already provided very valuable insights. With further applications of the new and innovative techniques that they describe, it should be possible to develop a much better understanding of the origin, persistence, and fate of aerosols found in the urban air.

0077-8923/80/0338-0001 $01.75/1 © 1980 NYAS

Types of Size Distributions

The necessity of characterizing heterodisperse aerosols by relatively simple distribution functions has long been recognized. In fact, we are close to the fiftieth anniversary of the publication of a landmark paper by Hatch and Choate. Their 1929 paper in the *Journal of the Franklin Institute*[1] demonstrated the utility of the log-normal distribution for describing industrial powders and aerosols. A log-normal distribution of particles forms a skewed, bell-shaped curve when plotted on linear coordinates, as in Figure 1. The same data form a normal, or Gaussian, curve (symmetrical and bell-shaped) when the fraction is plotted against the log of diameter. Furthermore, experimental data taken from this type of distribution will plot as a straight line on log-probability coordinates; the distribution can be fully described in terms of two parameters; i.e., the median size and the geometric standard deviation and a number distribution can readily be transformed to a surface or volume distribution, or vice versa, since the slope of the line is the same for all.

Industrial hygienists have gratefully used the log-normal distribution to describe the aerosols in the workplace ever since. In those cases where it fails to fit experimental data, it is usually because there is a mixture of two or more different aerosols. If each is log-normally distributed, and if the data are sufficiently detailed and precise, the parameters of each of the component distributions can be established.

The natural desire to adopt a standard distribution function, together with a lack of knowledge about the nature of the size distributions, led many early air pollution

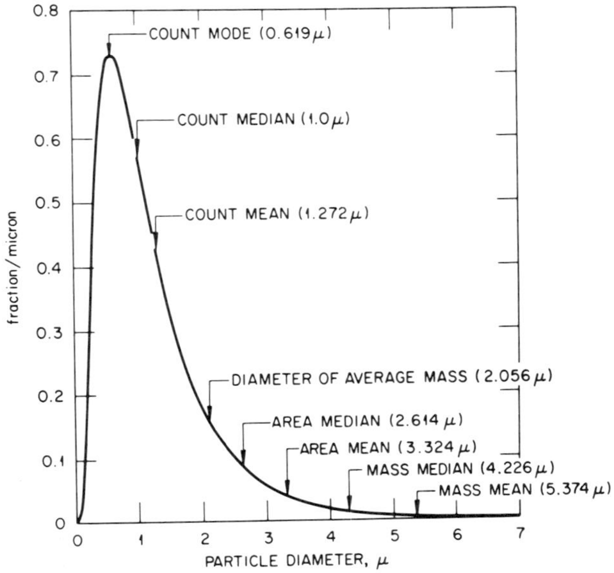

Figure 1. A log-normal distribution plotted on linear coordinates for an aerosol with a median diameter of 1.0 μm and a geometric standard deviation (σ_g) of 2.0. (Reprinted from Handbook on Aerosols.[8])

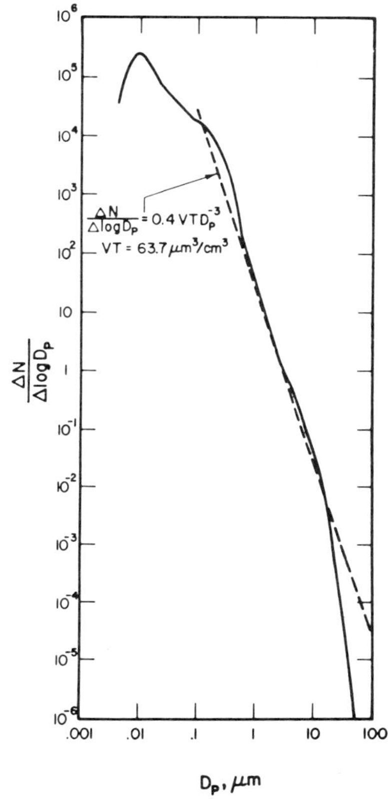

FIGURE 2. Grand average number aerosol size distribution from the 1969 Los Angeles smog experiment, compared with a Junge power law distribution calculated with the constants of Clark and Whitby. Agreement is reasonably good from 0.1 to 10 μm diameter but not above or below these sizes. (Reprinted from Airborne Particles,[9] by permission of University Park Press.)

investigators to attempt to utilize the log-normal distribution to describe ambient aerosol size distribution data. The early data were usually based on cascade impactor sampling, which lacked adequate precision or reliability, or on optical microscopy, where the limits of resolution led to truncated distributions. Hence, these earlier data, collected more than ~ 10 years ago, were not good enough to show that there were two or more distinctly different log-normal distributions in the ambient air, rather than one; and the confusion that arose over the general misinterpretation of the data led to the premature conclusion that the log-normal distribution was not applicable to air pollution aerosols.

For a period covering most of the 1960s, the distribution that appeared to best fit the available data was described by Junge[2] as

$$\frac{dN}{d\log D} = AD^{-k}$$

where dN is the number of particles per cm^3 in the size range dlog$D$, and where A and k are constants. Friedlander[3] suggested that such a distribution resulted from a balance of gains and losses of aerosols, and it was frequently designated the self-preserving distribution. The very large range of very small particles rapidly coagulate to form a smaller number of larger particles, and this process continues at a rate that diminishes with particle size. When the particles grow large enough, they can be eliminated efficiently from the atmosphere by precipitation and sedimentation. Since the processes of particle formation, coagulation, and removal occur continuously, a pseudoequilibrium condition tends to be established and can be considered as "self-preserving." As shown in FIGURE 2, large bodies of averaged data fit such a distribution reasonably well, at least over the size range of 0.1 to 10 μm, which is frequently the size range of greatest interest.

Unfortunately, such a generalization was too superficial and proved to be a dead end in terms of developing an increased understanding of the underlying processes.

With the development and refinement of continuous instruments capable of measuring number-size distribution from ~ 0.01 to 20 μm, Whitby and coworkers[4,5] were able to show that the particles in the ambient air tend to be found in three distinct size modes. The presence of these modes can best be seen when the size-number data are transformed and presented as distributions by surface and volume as well as by number, as in FIGURE 3. On these coordinates, the number distribution is usually unimodal. The surface distribution will often show peaks in the Aitken nuclei mode (<0.1 μm) and/or the coarse particle mode (>2 μm), as well as a large peak centered at ~0.2 to 0.3 μm. The volume distribution will usually exhibit two modes, one centered at about 0.3 to 0.5 μm, and another of similar magnitude in the coarse

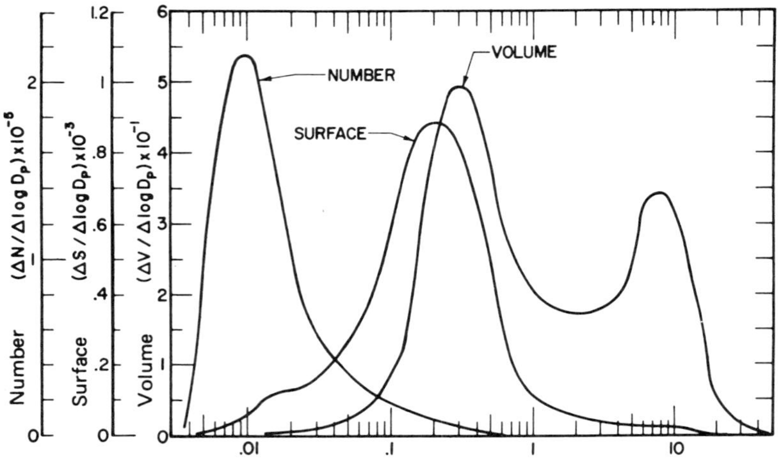

FIGURE 3. Normalized frequency plots of number, surface, and volume (particle volume times particle density) distributions for the grand average 1969 Pasadena smog aerosol. Note the bimodal distribution of mass. Each weighting shows features of the distribution not shown by the other plots. (Reprinted from Airborne Particles,[9] by permission of University Park Press.)

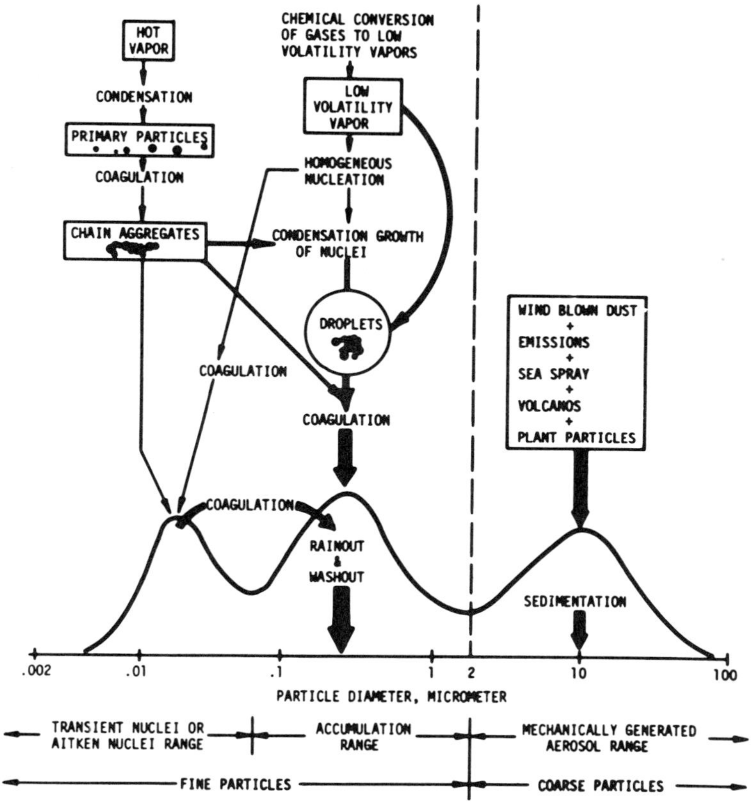

FIGURE 4. Schematic of an atmospheric aerosol surface area distribution showing principal modes, main sources of mass for each mode, the principal processes involved in inserting mass in each mode, and the principal removal mechanisms. (Reprinted from Airborne Particles,[9] by permission of University Park Press.)

particle mode. Each of the modes generally follows a log-normal distribution of diameters.

These various features of the characteristic modes are illustrated schematically in FIGURE 4. The Aitken nuclei mode is transient and can be detected only when there are active aerosol-generation processes. Its particles are continually coagulating with each other and with particles in the intermediate accumulation mode and, thereby, transferring to that mode. As the particles grow toward 0.5 μm, they become too few in number concentration to coagulate effectively, yet they are still too small to have any appreciable settling rate. Thus, they tend to accumulate in the atmosphere and do not transfer to any significant degree into the coarse particle mode.

Most of the particles in the coarse mode are primary particles generated by mechanical processes. Since the larger particles in this mode have appreciable settling velocities, they tend to settle out rapidly. High concentrations in this mode, therefore, require high rates of aerosol generation.

The relative independence of the fine and coarse mode particles is illustrated in FIGURES 5 and 6. In FIGURE 5, the arrival of aged smog aerosol from Los Angeles at a Mojave Desert sampling station greatly increased the accumulation mode volume and had much less effect on the coarse particle volume. On the other hand, FIGURE 6 demonstrates that local wind gustiness can greatly increase the windborne coarse mode dust concentration and median size without affecting the accumulation mode volume. The Aitken nuclei mode is not usually apparent on a volume plot of urban aerosol size distribution. It shows up clearly on FIGURE 7, due to a large source of nuclei and a relatively low background of particles from other sources under the conditions at the sampling site on the day of the particular test being conducted.

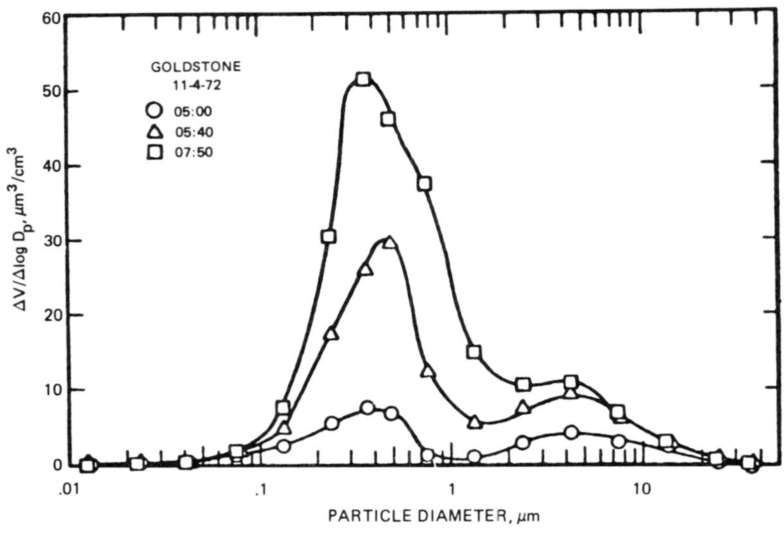

FIGURE 5. Incursion of aged smog aerosol from Los Angeles at the Goldstone tracking station in the Mojave Desert in California. Note the buildup in the accumulation mode. (Reprinted from Airborne Particles,[9] by permission of University Park Press.)

RELIABILITY OF SIZE DISTRIBUTION DATA

The validity of all size distribution measurements depends upon the reliability of the calibrations of the sampling and analytical instruments. Most of the multimodal distribution data have been generated using a series of analyzers with overlapping size ranges, as indicated in FIGURE 7. For large particle measurements using light-scattering-type aerosol spectrometers, such as the Royco 245, there are variations in response according to the surface properties, refractive indices, and sizes of the particles, as well as uncertainties in the efficiency with which the particles sampled from the ambient air are guided into the sensing zone. Also, it is important to recognize that the specific gravity of the particle does not affect the measurement. Hence, the calculated particle volume cannot be directly related to the particle's mass or aerodynamic size.

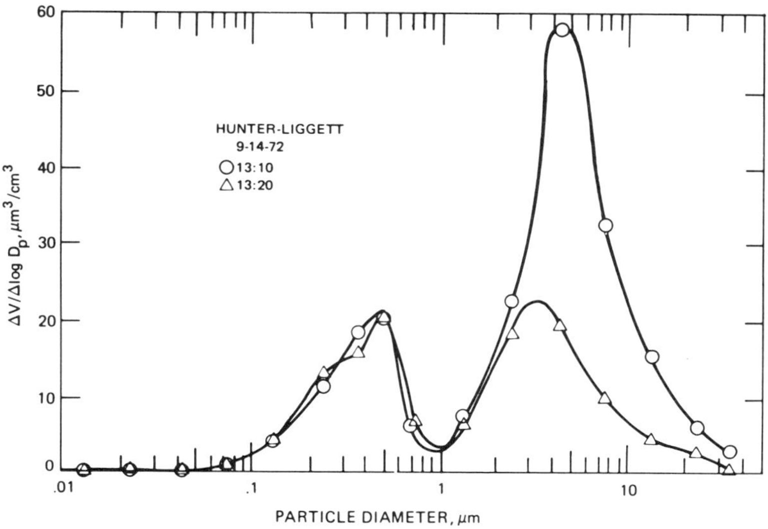

FIGURE 6. Sudden growth of the coarse particle mode due to local dust sources measured at the Hunter-Liggett Military Reservation in California. This shows the independence of the accumulation and coarse particle modes. (Reprinted from Airborne Particles,[9] by permission of University Park Press.)

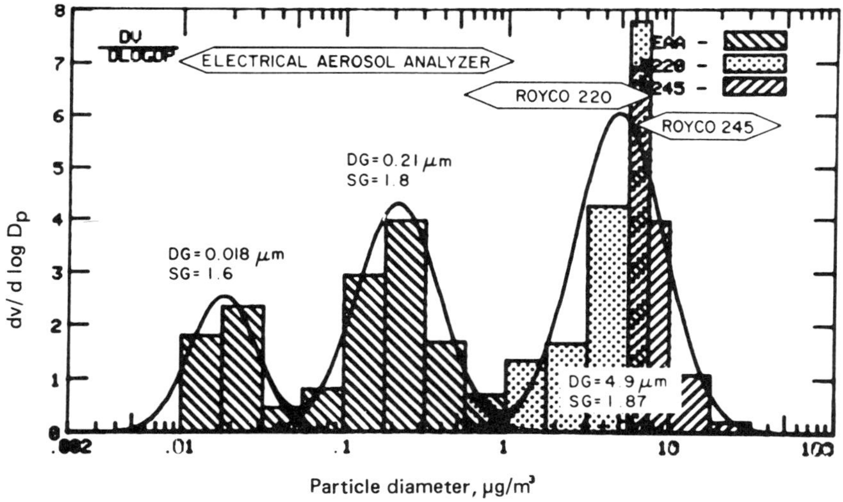

FIGURE 7. Computer-prepared volume size distribution measured during the General Motors Sulfate Study in Milford, Michigan, in the fall of 1975. The range of measurement of the three instruments used to obtain the size distribution from which the distribution was calculated is shown. The distinct trimodal nature of this distribution is the result of the low aerosol concentration (<6 $\mu g/m^3$) and the strong nuclei-mode-sized emissions from catalyst-equipped cars. (Reprinted from Airborne Particles,[9] by permission of University Park Press.)

For the small particles measured by the electrical aerosol analyzer (EAA), the particles are actually measured according to their mobilities in a voltage gradient. The accuracy of the measurements depends upon the efficiency with which the particles are charged, as well as on the precision with which they can be sorted according to their mobilities. Since there is no independent calibration standard for particles in part of the size range covered by the EAA, there has been some continuing controversy about the accuracy of the measurements made using it.

Despite the continuing uncertainties about the absolute accuracy of the EAA and optical analyzers, they are generally far superior to most other measurement systems. The diffusion battery–continuous flow nuclei counter of Sinclair[6] appears to be a suitable alternate to the EAA for the small particles, as shown in FIGURE 8, but there is no current alternative to the optical-type spectrometer for the continuous measurement of size distribution of the larger particles.

Time-averaged size distribution data have been generated by the measurement of the masses collected on the stages of cascade impactors. However, the sampling biases and particle-bounce tendencies of field samplers generally limit the reliability of the results obtained. While reliable impactor data are costly and difficult to collect, they

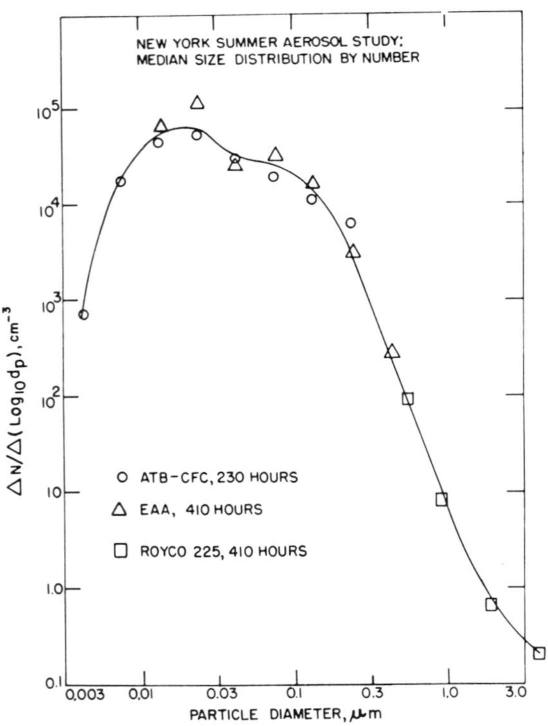

FIGURE 8. Median size distribution by number for the New York Summer Aerosol Study, 1976, showing data from the Sinclair diffusion battery and nucleus counter (ATB-CFC), electrical aerosol analyzer (EAA), and Royco Model 225 optical single-particle counter. (Reprinted from Knutson, Sinclair, & Leaderer.[6])

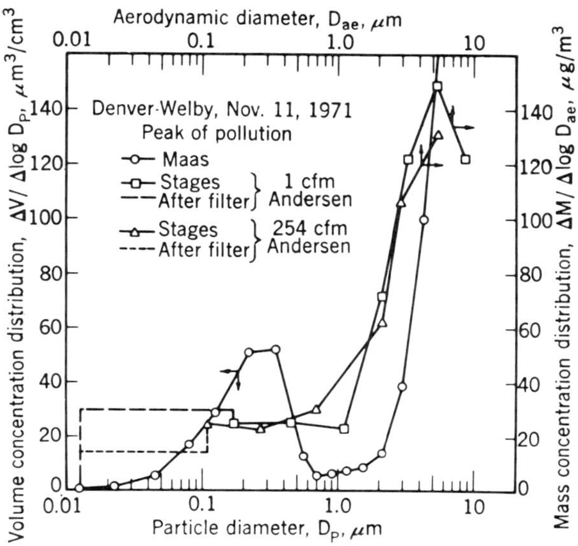

FIGURE 9. Comparison of calculated volume concentration distributions with mass concentration distributions from 8-stage Andersen samplers used in the Denver 1971 study. Data for the two cascade impactors, operated at 1 and 2.54 cfm, are plotted as a function of aerodynamic diameter, whereas the volume data are a function of particle diameter. The impactors were operated from midnight to midnight, while the EAA and Royco instruments were operated from 00:00 to 18:40. (Reprinted from Willeke & Whitby,[5] by permission of the Air Pollution Control Association.)

may be worth the effort at times, especially when it is important to determine the distribution according to aerodynamic size. Cascade-impactor-derived aerodynamic size–mass distribution data can be compared to calculated volume distribution data, as in FIGURE 9. The volume distribution generally provides a more clear-cut representation of the bimodal nature of the distribution.

FACTORS AFFECTING URBAN AEROSOL SIZE DISTRIBUTIONS

It is relatively easy to cite the major factors affecting the size distributions of urban aerosols but virtually impossible, at this time, to present any realistic summaries of distributions to be expected at any given time and place. The best data available were most frequently generated during intensive sampling campaigns and, therefore, are relatively site and condition specific. They also tend to be limited to particular times of year, especially to the summer, when it is most convenient for academic researchers to undertake sampling campaigns. The composition and size distributions in the summer tend to be different than those present at other times of the year. The amount and influence of the photochemically formed aerosols are greatest in the summer. Also, there is less atmospheric turbulence in the summer in most of the eastern U.S., and the urban aerosols are enriched in accumulation mode particles, which are transported over long distances. As a result, there is less diurnal variation in

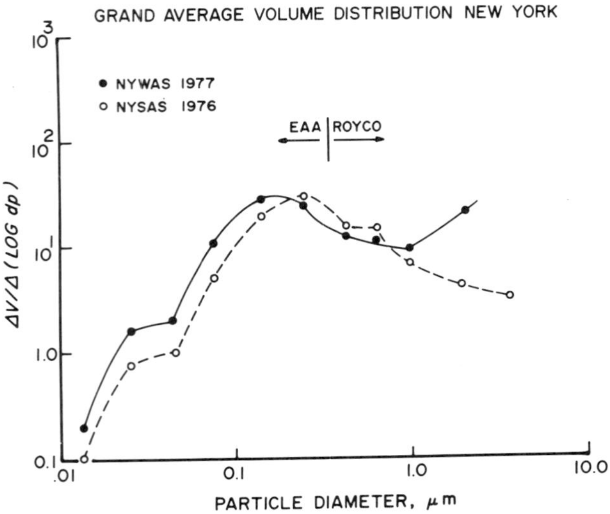

FIGURE 10. Grand average volume distributions calculated for the New York Summer Aerosol Study (1976) and the New York Winter Aerosol Study (1977).

the concentration of such particles in the summer than in the winter. The diurnal patterns of particle number concentration and accumulation mode mass are quite different in New York and Los Angeles in the summer. In Los Angeles, there is a substantial midday peak in accumulation mode volume, while there is little change throughout the day in New York. On the other hand, there is a much greater diurnal

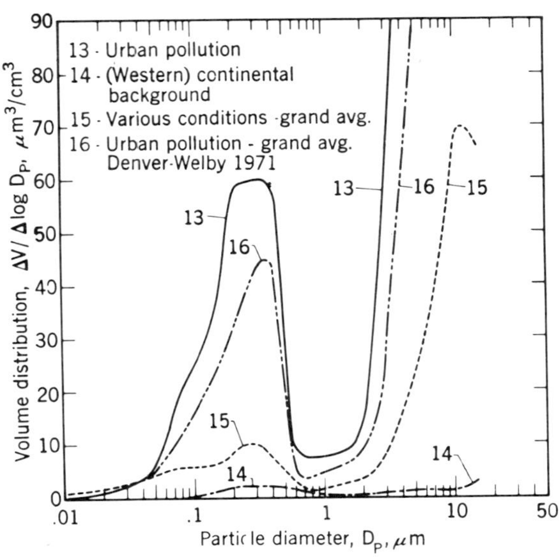

FIGURE 11. Volume concentration distributions for Denver-area aerosols. (Reprinted from Willeke & Whitby,[5] by permission of the Air Pollution Control Association.)

variation in the total numbers of particles in New York than in Los Angeles. In New York, the highest number concentration occurs in the late morning, while in Los Angeles, it is seen late in the afternoon.

A comparison of the grand average volume distributions for August 1976 and February 1977 at the New York University (NYU) Medical Center in Manhattan is shown in FIGURE 10. The total volumes are comparable, but it is clear that there is considerably more volume of <0.1-μm aerosol in the winter period, and more 0.2- to 0.7-μm aerosol volume in the summer period. The data on coarse mode particles are incomplete, but it appears that the coarse mode volume is larger in the winter when there is more atmospheric turbulence.

Another interesting feature of the New York volume distributions in both winter and summer is the clear indication of an Aitken nuclei volume mode. This is probably

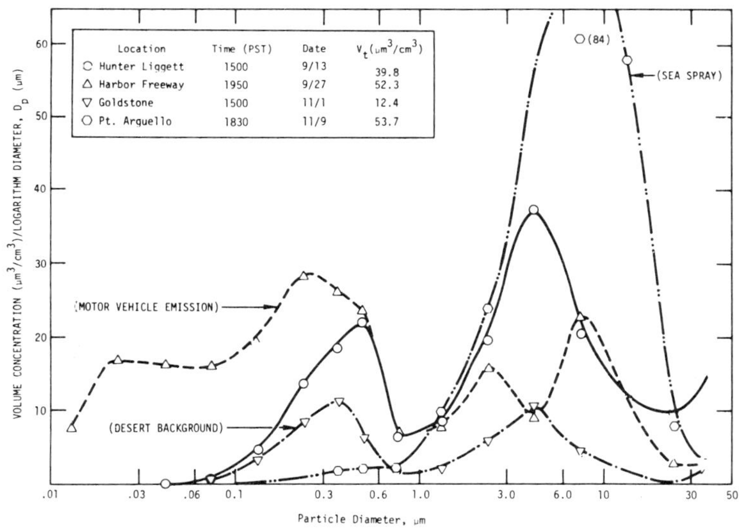

FIGURE 12. Comparison of surface and volume distributions for background and motor-vehicle-source-enriched (near Harbor Freeway) sites. (Reprinted from Hidy,[7] by permission of Rockwell International.)

accounted for by the close proximity of the NYU sampling site to the East River Drive, which is ~50 meters to the east and ~30 meters lower.

Regional and local differences in sources and ventilation also play important roles in the temporal variation in the concentration of accumulation mode aerosols. The substantial differences in size distribution that can occur within a brief period in one region (Denver) are illustrated in FIGURE 11. A variety of size distributions, seen in southern California during the ACHEX study,[7] are shown in FIGURE 12.

RESEARCH NEEDS

Quantitative characterization of the various size distributions of the ambient atmospheric aerosol can now be made routinely and reliably. However, such measurements, at present, can be made only by a limited number of groups that have acquired the considerable skills, sophistication, and relatively expensive instrumentation that are needed. Consequently, our knowledge of atmospheric aerosol size distributions is currently limited in terms of numbers of locations, times, and seasons.

More systematic size distributions need to be made, in order to clarify our understanding of the size distributions themselves at various times and places, and in order to establish more firmly the causal relationships, i.e., the nature and distribution of the sources, the temporal aspects of particle formation, growth, and removal, and the extent of long range transport. Some of the possibilities inherent in studies involving such measurements will become more readily apparent during the course of the next several days, as some of the exciting research findings are presented by the distinguished speakers gathered here for this conference. The considerable progress now being made and reported at this conference may make today's estimate of research needs obsolete by the end of the week. However, at this time, it is likely that our new knowledge will raise more questions than it settles. Also, the development of new insights will most likely continue to come from multidisciplinary studies that bring together investigators with different, but complementary, tools, backgrounds, and approaches.

References

1. Hatch, T. & S. P. Choate. 1929. Statistical description of the size properties of non-uniform particulate substances. J. Franklin Inst. **207:** 369.
2. Junge, C. 1953. Die Rolle der Aerosole und der gasförmigen Beimengungen der Luft in Spurenstoffhaushalt der Troposphare. Tellus **5:** 1.
3. Friedlander, S. K. 1961. Theoretical considerations for the particle size spectrum of the atmospheric aerosol. J. Meteorol. **18:** 753.
4. Whitby, K. T., R. B. Husar & B. Y. H. Liu. 1972. The aerosol size distribution of Los Angeles smog. J. Colloid Interface Sci. **39**(1): 177.
5. Willeke, K. & K. T. Whitby. 1975. Atmospheric aerosols: size distribution interpretation. J. Air Pollut. Control Assoc. **25**(5): 529.
6. Knutson, E. D., D. Sinclair & B. Leaderer. 1979. New York Summer Aerosol Study: number concentration and size distribution of atmospheric particles. Ann. N.Y. Acad. Sci. **322:** 11.
7. Hidy, G. M. 1975. Characterization of aerosols in California (ACHEX). Final Report No. SC524·25FR. (Prepared for California Air Resources Board.) Rockwell International. Thousand Oaks, Calif.
8. Dennis, R., Ed. 1976. Handbook on Aerosols. NTIS Pub. No. TID-26608. Springfield, Va.
9. National Research Council. 1979. Airborne Particles. University Park Press. Baltimore, Md.

TRACE ELEMENT ABUNDANCES AND CHEMISTRY OF ATMOSPHERIC AEROSOLS: CURRENT TECHNIQUES AND FUTURE POSSIBILITIES

Robert W. Shaw and Robert K. Stevens

Environmental Sciences Research Laboratory
U.S. Environmental Protection Agency
Research Triangle Park, North Carolina 27711

INTRODUCTION

The atmosphere is an enormously complex, more or less effective, mixing vessel. And yet, in spite of that complexity (numerous sources, sinks, transport processes, and transformations), the aerosol particles found there retain distinctive distributions with respect to size and composition. The bimodal size distribution for aerosols has been found in various regions by many investigators, and analysis of the aerosol size fractions has shown that the two modes differ considerably in their elemental constituents and in their chemistry. For example, the fine particle fraction is often acidic; and the coarse particle fraction, basic. This interesting dichotomy is consistent with the view that the fine particle fraction is due largely to particle formation and accretion from atmospheric gases, and that among these gases are the acidic oxides of sulfur and nitrogen.* On the other hand, the coarse particle fraction is mainly due to the breaking up of larger particles, among which are metals (Ca, K), the oxides of which are basic. These properties of atmospheric particles are schematically represented in FIGURE 1.

In general, we find that the elements sulfur and lead appear mostly in the fine fraction, and calcium, aluminum, and silicon in the coarse fraction. These elements, which usually appear in quantities measured in micrograms per cubic meter, are not thought of as trace elements.

In practice, our primary interest in measuring materials in the atmosphere is in providing information for pollution control. Two complementary orientations to this problem exist. In using "source apportionment," one begins with the experimental measurements of elemental abundances and works back to a model determination of the sources. In using "air quality simulation models,"[2] one takes a continuity (conservation) equation describing material in the atmosphere, and various input and output data (source emission rates, meteorological parameters, rates of transport and transformation, etc.), and calculates the air quality. The test of the model calculation is the agreement with experimental observation of air quality. Thus, we see that experimental determinations of elements in atmospheric particles are necessary in both approaches to the assignment of atmospheric sources: one approach working from the observations back to the source, and the other working from the source to the observations. These remarks are illustrated in FIGURE 2.

In the following sections of this paper, we will discuss a number of requirements

*The relative contribution of nitrogen oxides to particle formation is now a controversial subject, and evidence exists[1] that previous estimates were considerably too high.

0077-8923/80/0338-0013 $01.75/1 © 1980 NYAS

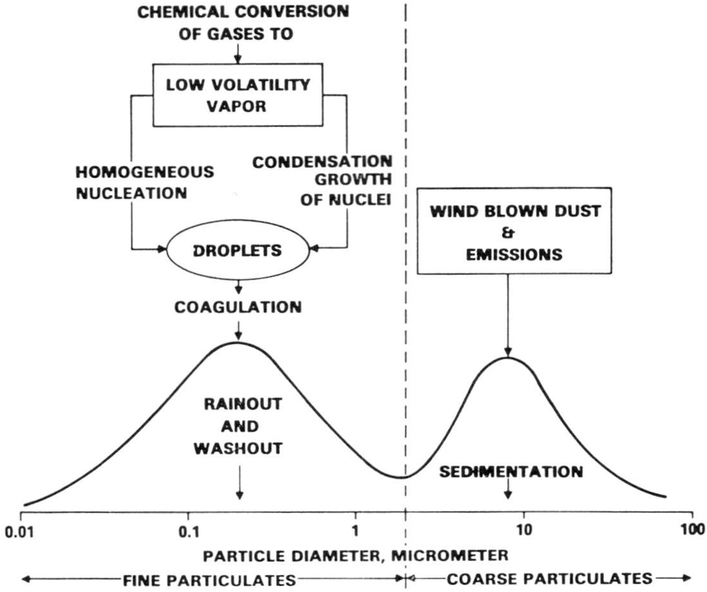

FIGURE 1. The fine and coarse particle fractions of atmospheric particles. (Courtesy of J. Hodgeson.)

for trace element analysis of atmospheric particles and describe several multielement analysis techniques that are especially suited to particular sampling instruments.

MULTIELEMENT TRACE ANALYSIS

We have observed that relative amounts of trace elements vary considerably from place to place, more so than do relative amounts of the abundant elements. It follows

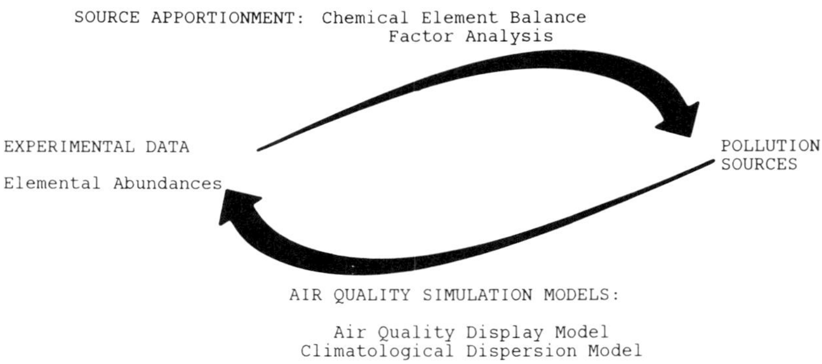

FIGURE 2. Model paths connecting experimental measurements of elements in atmospheric particles and the sources of those particles.

that analysis techniques should have the capability of detecting as many elements as possible, because at the beginning of an experiment, one generally cannot anticipate which trace elements will be present. The more one's instrument restricts the search, the more likely it is that one will miss something interesting.

This state of affairs, the desirability of detecting as many elements as possible, will probably last for some time until the data that describe the elemental composition of urban aerosols are more complete. On the other hand, from our experience, one always purchases analytical sensitivity to more elements at the cost of less analytical convenience of some other kind. For this reason, we believe that an experimental study of elemental abundances in atmospheric particles should be carefully planned to focus on the elements of greatest importance to the motivating questions behind that study. The need for planning mentioned above will be apparent in the following papers on "source apportionment."

The frequent changes in composition of atmospheric particles place a heavy burden on the atmospheric analyst. The high cost of atmospheric sampling usually precludes the use of replicate sample collectors so that, unlike the industrial analyst who can take numerous replicate samples from a product batch, the atmospheric analyst deals with single, unique samples. If the atmospheric sample is spoiled or the analytical results appear peculiar, the analyst cannot go back for a duplicate sample. Rather, the analyst must have faith in the analytical results, and this faith must be established in a firm program of frequent calibrations and checks of all aspects of the collection and analysis procedures.

Instrument Sensitivity and Elemental "Noise"

The central issue in the design or the choice of an instrument is that of sensitivity; this raises the problem of determining how much sensitivity is required. We have no complete answers to this problem. It seems likely, however, that a certain detection level exists below which it is relatively profitless to go. Because of the complexity and constantly changing nature of atmospheric samples, there may be levels and their associated variances that are characteristic of each element and that constitute a trace element source "noise." The problem here is to determine at what levels this "noise" occurs so that we can decide whether our instruments are detector noise limited and hence should be developed for increased sensitivity, or source noise limited and hence do not require increased sensitivity.

Examples of Aerosol Collection and Analysis

A principal theme of this paper is simply that a scientific problem is best solved by using the method best suited to that problem. We illustrate this theme by discussing the elemental analysis of atmospheric particles using three collection schemes: (1) the high volume (Hi-vol) sampler; (2) the dichotomous sampler; and (3) the cascade impactor. We shall see that, in each case, the method of collection imposes limitations on the method of analysis, and that investigators have solved the analysis problem by choosing an analytical tool particularly suited to the characteristics of the sample.

The Hi-vol Sampler and Optical Emission Spectroscopy

The Hi-vol sampler causes air to pass through a glass fiber filter of area 500 cm^2. A schematic diagram of the Hi-vol is shown in FIGURE 3. The design is very simple; a heavy duty turbine blower draws air through the filter at a constant flow ranging from 1.1 to 1.7 m^3/minute. Hi-vol filters are used primarily for determinations of aerosol mass concentrations but may also be used for elemental analyses. Because the aerosol particles are distributed through the glass fiber mat, they must be extracted.

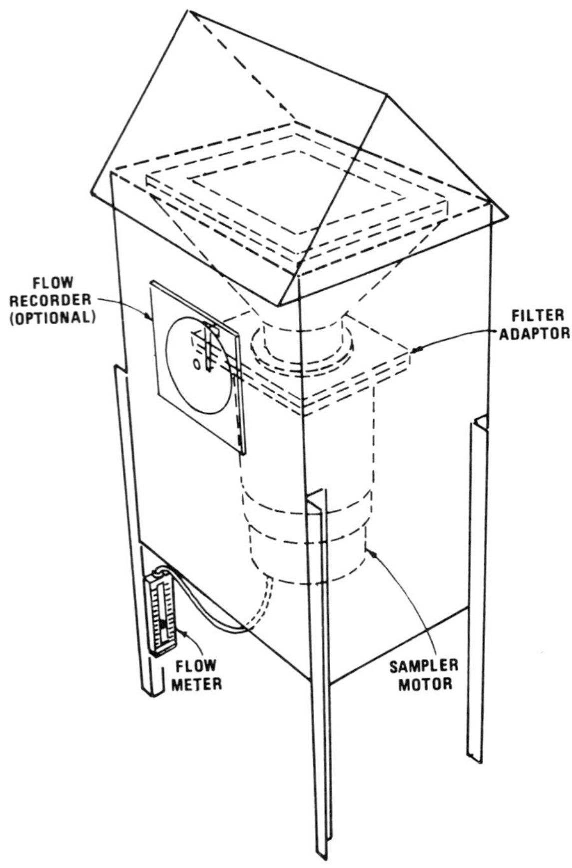

FIGURE 3. Sketch of the Hi-vol sampler.

Scott *et al.*[3] have reported the multielement analysis of about one thousand Hi-vol filters by low temperature oxygen ashing and subsequent acid extraction of the filter and analysis of the extract by optical emission spectroscopy (OES). Their paper also provides considerable information on detection limits and confidence levels determined by extensive replicate and standard sample analysis. The absolute sensitivity of

TABLE 1

FREQUENCY OF DETECTION BY OES OF ELEMENTS IN NASN SAMPLES

Number of Samples: 898
Sampling Period: 1970
Sampler and Flow Rate: Hi-vol, 1.7 m^3/min

Element	Frequency of Detection
Cu, Fe, Mn, Pb, Ti	80–100%
Al, Be, Ca, Cd, Cr, Mg, Ni, V	10–80%
Mo	<10%
As, Bi, Co, Sb, Sn, Te, Tl	<1%

OES is not so high as that of some other techniques. The Hi-vol sampler, however, collects so much sample that OES was sufficiently sensitive to have provided quantitative determinations of Al, Be, Ca, Cd, Cr, Cu, Fe, Mg, Mn, Mo, Ni, Pb, Ti, and V in a set of 898 samples from 248 sampling sites in the National Air Surveillance Network (NASN). Determinations were also made of As, Bi, Co, Sb, Sn, Te, and Tl, but the levels of these elements were such that they were above the detection limit in

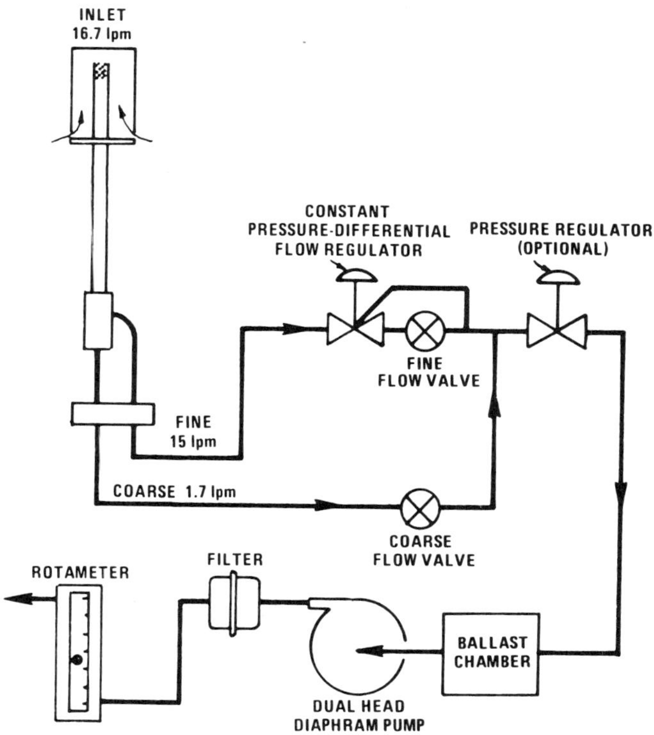

FIGURE 4. Sketch of the dichotomous sampler. (From R. K. Stevens & T. G. Dzubay. 1978. EPA Report 600/2-78-112.)

fewer than 1% of the samples. TABLE 1 shows the frequency of detection, which is defined as the percentage of samples in which the element is detected at a level above the detection limit. Scott *et al.*[3] define the detection limit as the concentration corresponding to a signal that was three times the standard deviation above the average signal due to the extraction solution. The extraction solution contained internal standards for normalization of measured values.

It is clear from TABLE 1 that OES analysis of the NASN samples has provided sufficient data to follow trends in abundances of Cu, Fe, Mn, Pb, and Ti, but certainly

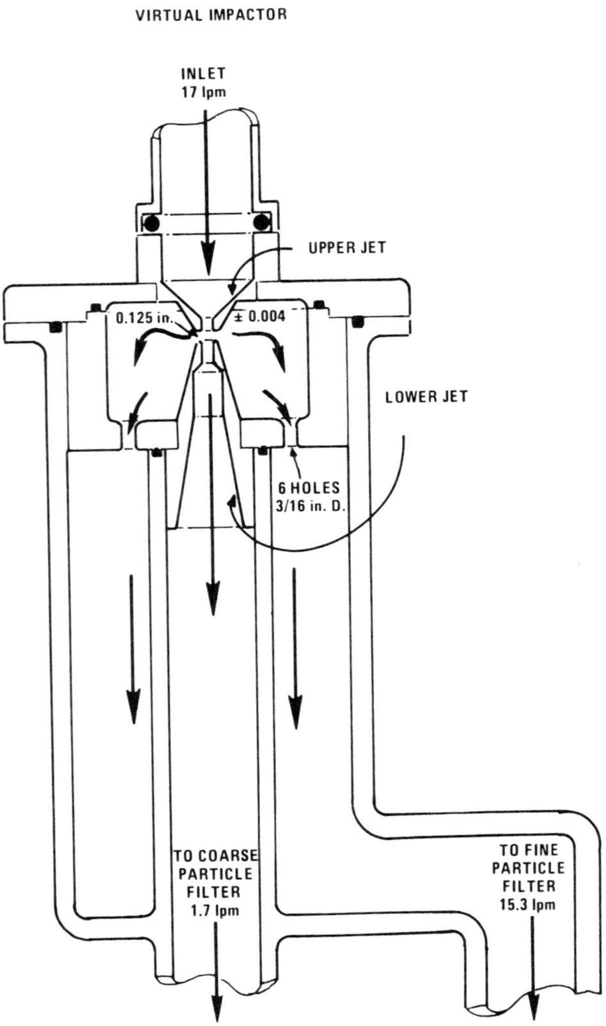

FIGURE 5. Detailed sketch of the fine–coarse particle separation region of the dichotomous sampler. (From R. K. Stevens & T. G. Dzubay. 1978. EPA Report 600/2-78-112.)

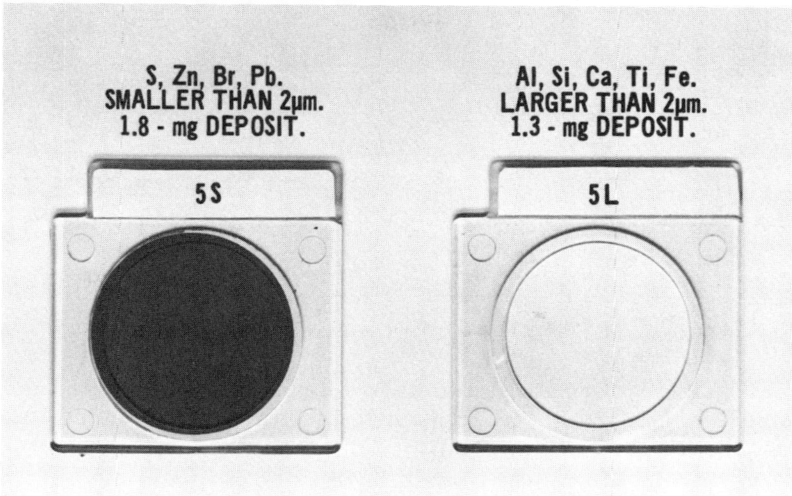

FIGURE 6. A fine fraction filter (left) and coarse fraction filter (right) from a dichotomous sampler. (Courtesy of T. Dzubay.)

not for As, Bi, Co, Sb, Sn, Te, or Tl, and probably not for Mo. Whether sufficient data exist for Al, Be, Ca, Cd, Cr, Mg, Ni, and V depends on more detailed considerations—comparisons of element levels with detection limits, interelement interferences, the requirements of the statistical treatment of the data, etc. Obviously, these results refer only to the 1970 NASN samples; samples from different sites will probably have different frequencies of detection.

The Dichotomous Sampler and X-Ray Fluorescence (XRF)

The dichotomous sampler causes the aerodynamic separation of particles into two fractions corresponding to the fine and coarse modes of the size distribution of atmospheric particles. A schematic diagram of the dichotomous sampler is shown in FIGURE 4, and a more detailed diagram of the inertial separation stage is shown in FIGURE 5. The particles are collected on thin membrane filters. The filters currently used are teflon because of their chemical inertness, good weight stability, and low trace element contamination. The sampler is designed in such a way that the particles are collected into a uniform and well-defined deposit. The sampling rate is 1 m^3/hr for most samplers in current use, so only a thin deposit of particles is formed. In situations where particle levels in the atmosphere are very high, the loading of the particles on the filter will cause a perceptible change in pressure drop across the filter, and the sampler (if appropriately controlled) will terminate its sampling at approximately 200 $\mu g/cm^2$ deposited on the filter.

The two filters from a dichotomous sampler are shown in FIGURE 6. These filters have the characteristic "dirty" and "clean" appearance of aerosol deposits from the fine and coarse particle modes, respectively. One explanation for the darkness of the

TABLE 2

FREQUENCY OF DETECTION* BY XRF OF ELEMENTS IN SAMPLES FROM NEW YORK CITY

Number of Samples: 39
Sampling Period: 2/1977
Sampler and Flow Rate: Dichotomous sampler, 1.0 m^3/hr

Element	Frequency of Detection
(Si), P, S, Cl, K, Ca, (V), (Mn), Fe, Ni, Cu, Zn, Br, Rb, Pb	100%
(Al), Sr, Hg	80–100%
Ti, Cr, Co, Ga, Ge, (As), Se	10–80%

*In fine and/or coarse fraction.

fine deposit could be the much larger fraction of the filter surface that is obscured by particles in the fine fraction when compared to the filter area obscured by the coarse fraction. This observation follows from the relations: $M \sim d^3$ and $A \sim d^2$, where M, A, and d are particle mass, area, and diameter, respectively; and from measurements of the ratios of fine to coarse mass in urban aerosols, which are typically 1–2. Alternatively, the fine fraction may contain more optically absorbing materials than the coarse fraction. Some elements commonly found in the two modes are shown in FIGURE 6.

The thin, uniform particle deposit on a flat, uncontaminated filter surface is ideally suited for x-ray emission analysis. An important advantage of x-ray emission analysis of these filters is that the sample requires no preparation whatsoever and can be analyzed straight from the sampler. Furthermore, the analysis does not change the sample in any perceptible way, and the sample remains in its original form for other analyses, such as microscopy, chemical extraction, etc.

Stevens *et al.*[4] have reported elemental abundances from dichotomous samplers situated in six cities. TABLES 2 and 3 show the frequency of detection for New York City and for Charleston, West Virginia. The elements shown in parentheses are subject to interelement interferences. For example, if Br is very high, the overlap of Br L x-rays with Al K x-rays will make the Al determination less certain. Corrections are made for interelement interferences in the computer fitting of the x-ray spectra. As shown in TABLES 2 and 3, samples from New York and Charleston were collected

TABLE 3

FREQUENCY OF DETECTION* BY XRF OF ELEMENTS IN SAMPLES FROM CHARLESTON, WEST VIRGINIA

Number of Samples: 106
Sampling Period: 4–8/1976, 1/1977
Sampler and Flow Rate: Dichotomous sampler, 1.0 m^3/hr

Element	Frequency of Detection
(Al), Si, S, Ca, (Ti), Fe, Zn, Br, Pb	100%
P, Cl, Cu, (As), Se	80–100%
(V), (Mn), Ni, Ga, Ge, Rb, Sr	10–80%
Cr, Co, Hg	<10%

*In fine and/or coarse fraction.

using dichotomous samplers, and numerous elements were detected using XRF analysis of those samples. A group of elements remain, including Cr and Co, which are not suitable for determination by XRF. This is due to the fact that their concentrations rarely exceed a few nanogram/m^3, which, even after 24 hours at sampling rates of 1 m^3/hr, does not provide enough material for XRF analysis. The minimum detectable concentration is determined using an expression by Currie[5] that is based on an analysis of x-ray counting experiments.

The Cascade Impactor and Neutron Activation Analysis

The cascade impactor causes the aerodynamic separation of particles into a series of size ranges. A schematic diagram of a typical cascade impactor design is shown in FIGURE 7. In contrast with the dichotomous sampler, which is a virtual impactor (particles impact into a volume), the cascade impactor causes impaction onto real

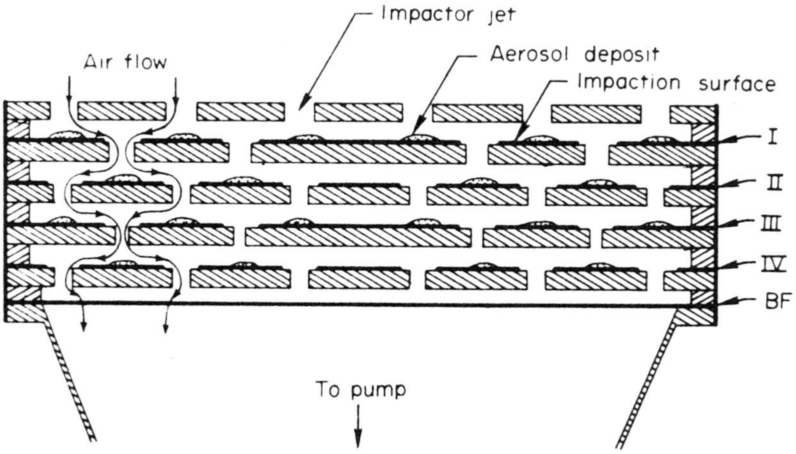

FIGURE 7. Sketch of a cascade impactor. (Courtesy of T. Dzubay.)

surfaces. The collisions of particles with real surfaces can lead to artifacts due to particle bounce or break-up unless the surfaces have the property of causing particles to stick; and a variety of greases have been used to provide stickiness. The resulting samples are not generally suitable for chemical extraction because of the presence of the grease, and are not convenient for x-ray emission analysis because of x-ray absorption in the thick sample layer. In addition, because the sampled aerosol is divided into a number (say five) of different size ranges, and because cascade impactors are mostly low volume samplers, the sensitivity of the analysis should be especially high.

Nuclear activation analysis has extremely high sensitivity for many elements, and most radiative transitions used in nuclear activation analyses emit very penetrating radiation, which passes through thick impactor samples with negligible attenuation.

Consequently, this technique is very suitable for analysis of aerosols collected by cascade impactors. At the University of Maryland, Gordon and his collaborators have made numerous elemental analyses of atmospheric particles using instrumental neutron activation analysis (INAA) and instrumental photon activation analysis (IPAA). For precise, quantitative work, the sample must be rendered into a standard shape and size to allow corrections for geometric nonuniformities in the activation source field or in the detector. Gordon *et al.* have accomplished this by dissolving the sample in a standard volume of liquid or by making the sample into a pellet.

The techniques of IPAA and INAA are complementary in that each is sensitive to particular, different elements. In the remaining remarks, we will discuss only INAA. Although INAA using thermal neutrons is relatively insensitive to Pb, Ni, and Si and has low sensitivity to S, it has extremely high sensitivity to many other elements of interest, much higher than OES or XRF. Among the elements for which INAA is very sensitive are: Sc, V, Cr, Mn, Co, Se, Ag, Sb, and Cs. This technique is also very sensitive to a number of the lanthanides, which, although not of apparent importance as an air pollution problem, may be useful in tracing sources of pollution.

Composition and size distributions of atmospheric particles in Boston, using a 6-stage cascade impactor and INAA, have been reported by Gladney *et al.*[6] These authors report analysis for Na, Al, Cl, Sc, V, Mn, Fe, Co, Zn, Se, Sb, Ce, Yb, Lu, Hf, Ta, and Th. Some of the heavier elements were present at levels lower than 0.1 ng/m^3, and only the extremely high sensitivity of INAA to these particular elements made possible their detection. Gladney *et al.* do not report the results of individual

TABLE 4

AVERAGES OF ELEMENTAL MASS CONCENTRATIONS AT MIT*

Number of Samples: 6
Sampling Period: 2/1970
Sampler and Flow Rate: Cascade impactor, 0.75 m^3/hr

Element	Concentration (ng/m^3)
Na	2340
Al	1320
Cl	2300
Sc	0.25
V	860
Mn	30
Fe	1370
Co	2.3
Zn	340
Se	4.4
Sb	7.4
Ce	4.1
Yb	0.03
Lu	0.003
Hf	0.05
Ta	0.06
Th	0.11

*Reference 6, Table 3.

TABLE 5

AEROSOL COLLECTION MEDIA FOR VARIOUS SAMPLERS

Medium Properties	Hi-vol	Dichotomous	Cascade Impactor
Composition	glass fiber	teflon membrane	adhesive surface
Density (mg/cm^2)	8	1	—
Thickness (μm)	500	7	—
pH	<10.2	neutral	—
Total sample* (mg)	150	2	2
Elemental analysis	yes	yes	yes
Chemical analysis	yes	yes	—

*Assuming aerosol concentration of 75 $\mu g/m^3$, 24-hour sampling time.

measurements, and so we cannot give a frequency of detection table. We reproduce one set of average concentrations reported by them in TABLE 4.

We summarize this section by referring to TABLE 5, which shows several characteristics of the aerosol collection media discussed above. A glass fiber Hi-vol filter is rather thick, so the sample is most conveniently prepared for analysis by extraction. The collected sample is generally rather large, so high sensitivity is not necessary for many elements. In this case, OES is a suitable analytical tool. The teflon membrane filters used with the dichotomous sampler provide a uniform, thin support that is ideal for XRF. The filter is neutral and chemically nonreactive; hence, chemical information can be gleaned from the sample following elemental analysis by XRF, which is nondestructive. The cascade impactor collects a relatively small sample and spreads it over several stages in a matrix of adhesive. These samples are well suited for INAA.

We hope that our brief discussions of OES, XRF, and INAA have illustrated the suitability of each for particular analytical problems. Each technique also has its peculiar disadvantages and requirements. OES requires considerable sample preparation under contamination-free conditions but has relatively small interelement interference corrections. XRF requires no sample preparation and is truly nondestructive; but the complex spectrum and sometimes large interelement interferences require considerable computer fitting of the data. INAA requires a well-organized schedule of two or more activation periods, followed by a sequence of counting periods ranging from seconds to months after irradiation. One could fill many pages with comparisons of the complicated characteristics of these complex techniques. The most important considerations, however, are the demands of the sample; and each of the techniques is especially suited for a general class of samples of atmospheric particulates.

Other methods of multielement analysis have provided considerable information about the constituents of atmospheric particles. Among these are particle induced x-ray emission analysis (PIXE)—practiced by groups at Florida State University, the University of California at Davis, and Brigham Young University—which has high sensitivity for many elements. A particularly interesting use of PIXE has been reported by Courtney *et al.*,[7] who have measured elements in particles near a roadway over two-hour intervals.[7] Atomic absorption spectroscopy using multiple element excitation sources is a promising tool for multielement analysis of atmospheric particles.

Our intent in this paper has not been to review multielement analysis techniques. Rather, we have endeavored to point out a few examples showing how the character of the sample dictates the method of analysis. We have not mentioned all possible means of multielement analysis, not because we have judged other methods to be generally unsuitable, but in order to be brief.

We believe that no matter what collection system or trace element analysis technique is used, an investigator can achieve good accuracy and precision only by using experienced and capable workers and only with constant vigilance. Multielement analysis systems can deliver an enormous amount of data in a relatively short time. In order to ensure the validity of those data, the operator must pay constant attention to all phases of the analytical procedure. A natural conflict may develop here. Collection of samples is relatively straightforward and does not require a high level of experience and preparation. On the other hand, analysis of these samples is quite demanding. Many laboratories, however, have neither the personnel nor the resources to establish and maintain properly a multielement analysis lab. We believe that one answer to this problem is the use of high quality, centralized analytical laboratories that are devoted to analysis of samples from standardized air samplers, such as the Hi-vol and dichotomous samplers.

Future Possibilities

Two important problems concerning trace elements in particles are: (1) How are the elements distributed in the particles?; and (2) What is the chemistry of the particles? The question about distribution can be answered only by probe analysis—using a focused beam of particles or photons to excite constituents of individual particles. Probe analyses of aerosol particles have been carried out using ion microprobe mass spectrometry,[8] Auger electron spectroscopy,[9] electron microscopy with x-ray fluorescence,[10] and laser Raman microprobe spectroscopy.[11]

These techniques may be used in a scanning mode so that a map of the elemental constituents is generated. This map indicates in detail which elements are spatially associated. All of these techniques are primarily sensitive to the surface of the target particle, not to its bulk. However, electron microscopy, coupled with detection of emitted x-rays, provides the greatest depth of penetration. Of the techniques mentioned here, laser Raman spectroscopy is the only one that provides chemical information in a relatively straightforward fashion;† it is, however, least sensitive to trace substances. Ion microprobe mass spectrometry is least likely to provide chemical information but has the greatest sensitivity to elemental constituents.

Interestingly, the use of probe techniques (especially those associated with the scanning electron microscope) is receiving increasing attention from pathologists who are interested in toxic effects of inspired particles.[10, 12, 13] These workers have employed x-ray emission analysis and electron backscattering to map the elemental composition

†Analysis depends on a perturbation of the material and an observation of the resulting change. The more severe the perturbation, the more drastic the change and, usually, the less one can infer about the initial state. For example, in ion microprobe mass spectrometry, the sample surface is torn apart into its constituents, while in laser Raman microprobe spectroscopy, the observed spectra are due to transitions within the various chemical functional groups.

of particles collected in the human lung—the most important of all aerosol samplers.

ACKNOWLEDGMENTS

We thank Dr. William Courtney for his tabulations of urban aerosol analyses from Charleston and New York, and Dr. Courtney and Dr. Charles Lewis for their helpful suggestions.

REFERENCES

1. STEVENS, R. K., T. G. DZUBAY, R. BURTON, G. RUSSWURM & E. TEW. Paper No. 98. Prepr. Pap. Nat. Meet. Div. Environ. Chem. Am. Chem. Soc. September, 1978.
2. SAUTER, G. D. 1976. EPA Report No. 600/9-76-016: 30–34. U.S. Environmental Protection Agency. Research Triangle Park, N.C.
3. SCOTT, D. R., W. A. LOSEKE, L. E. HOLBROKE & R. J. THOMPSON. 1976. Appl. Spectrosc. **30**(4): 392.
4. STEVENS, R. K., T. G. DZUBAY, G. RUSSWURM & D. RICKEL. 1978. Atmos. Environ. **12:** 55.
5. CURRIE, L. A. 1977. *In* X-Ray Fluorescence Analysis of Environmental Samples. T. Dzubay, Ed.: 289–306. Ann Arbor Science Publishers, Inc. Ann Arbor, Mich.
6. GLADNEY, E. S., W. H. ZOLLER, A. G. JONES & G. E. GORDON. 1974. Environ. Sci. Technol. **8:** 551.
7. COURTNEY, W. J., S. RHEINGROVER, J. PILOTTE, H. C. KAUFMANN, T. A. CAHILL & J. W. NELSON. 1978. J. Air Pollut. Control Assoc. **28**(3): 225.
8. MCHUGH, J. A. & J. STEVENS. 1972. Anal. Chem. **44:** 2187.
9. LINTON, R. W., P. WILLIAMS, C. EVANS & D. NATUSCH. 1977. Anal. Chem. **49:** 1514.
10. SHELBURNE, J. D., P. INGRAM, H. HAWKINS & M. WATERS. 1976. *In* Scanning Electron Microscopy/1976. Part III: 485–500. IIT Research Institute. Chicago, Illinois.
11. BLAHA, J., G. ROSASCO & E. ETZ. 1978. Appl. Spectrosc. **32**(3):292.
12. DENEE, P. B. 1972. J. Am. Ind. Hyg. October: 654.
13. BRODY, A. R., N. VALLYATHAN & J. CRAIGHEAD. 1976. *In* Scanning Electron Microscopy/1976. Part III: 477–484. IIT Research Institute. Chicago, Illinois.

SECONDARY AEROSOL: PRODUCTION MECHANISMS OF SULFATE COMPOUNDS IN THE ATMOSPHERE

Timothy V. Larson

Departments of Atmospheric Sciences and Environmental Health University of Washington Seattle, Washington 98195

"Secondary aerosol" is that portion of atmospheric particulate matter produced from gaseous molecules by chemical reactions in the atmosphere. In contrast, "primary aerosol" is that portion either directly emitted into the atmosphere or produced by condensation of supersaturated vapors that are directly emitted. Whereas the chemistry of the primary aerosol is characterized by a wide variety of chemical elements, the chemistry of the secondary aerosol is dominated by only a few elements: carbon, primarily as alkanes, alkenes, phenols, and dicarboxylic acids; nitrogen, as nitrate and ammonium compounds; and sulfur, primarily as sulfate compounds.

In this paper, we will focus on the formation mechanisms of sulfate compounds, for several reasons. First, sulfate compounds are a major fraction of the mass of the submicrometric-diameter particles in the atmosphere.[1,2] Second, there are relatively few chemical species that need to be considered: the reactant gas, sulfur dioxide (SO_2), and the major oxidation product, sulfuric acid and several of its salts. Third, there is only one main atmospheric reaction pathway: oxidation of gaseous SO_2 to a particulate sulfate compound. In contrast, some gases have several potential reaction products. For example, the oxidation of gaseous terpenes is not only a source of organic particulate compounds, but may also be a significant global source of another gas, carbon monoxide.[3] Fourth, atmospheric sulfate compounds are physically stable in particulate form, unlike the more volatile nitric acid whose higher vapor pressure does not permit formation of a condensed phase at tropospheric concentrations.[4] Finally, in contrast to the few estimates for both carbon and nitrogen compounds, there are numerous estimates of the rates of oxidation of SO_2 to sulfate compounds in the atmosphere, ranging over a spatial scale of from 10 to 1000 km.

Oxidation of SO_2 can occur not only in the gas phase (homogeneous reactions), but also in liquids or on the surface of solids (heterogeneous reactions). The gas phase processes also involve liquid and solid phases, as the products of reaction (SO_3, H_2SO_4) are expected to condense on the surface of existing particles.[5] However, the rate of oxidation of SO_2 by gas phase reactions is independent of the chemical form of the particles produced. By comparison, the rates of oxidation of SO_2 by solid and liquid phase reactions depend on the chemistry of the particles, which can be altered by the oxidation products.

An excellent, recent review by Calvert *et al.*[6] discusses the important gas phase reactions. In the photochemically active atmosphere, they include reaction with hydroxyl (OH), hydroperoxyl (HO_2), and organic peroxy (RO, RO_2) radicals. These radicals are produced in sunlight by the interaction of NO_x (nitrogen oxides) and

0077–8923/80/0338–0026 $01.75/1 © 1980 NYAS

reactive hydrocarbons. The important oxidation reactions are:

$$SO_2 + OH\ (+\ M) \rightarrow HOSO_2\ (+\ M) \quad (1)$$

$$SO_2 + HO_2\ (+\ M) \rightarrow HO_2SO_2\ (+\ M) \quad (2)$$

$$CH_3O_2 + SO_2 \rightarrow CH_3O + SO_3 \quad (3a)$$

$$CH_3O_2 + SO_2 \rightarrow CH_3O_2SO_2 \quad (3b)$$

Calvert's[6] estimates of the respective rate constants are: $k_1 = 1.1 \times 10^{-12}$ cm^3 mol^{-1} s^{-1}, based on the resonant fluorescence measurements by Davis[6,7] and Atkinson *et al.*,[8] and on the corrected data of Cox[9] and Castelman and Tang;[10] $k_2 \geqq 8.7 \pm 1.3 \times 10^{-16}$ cm^3 mol^{-1} s^{-1}, based on the $^{18}O_2$-labeling technique of Payne *et al.*,[11] with a possible upper limit of $\sim 1.5 \times 10^{-16}$ cm^3 mol^{-1} s^{-1}; and $k_{3a+3b} \simeq (5.3 \pm 2.5) \times 10^{-15}$ cm^3 mol^{-1} s^{-1}, from the measurements of Whitbeck *et al.*[12]

The product of Reaction 1, HSO_3, is itself a radical and must undergo further reactions to form sulfate. HSO_3 will add to O_2 to form the HSO_5 radical. Calvert[6] proposes a series of reactions involving HSO_5 and HSO_4 that produce nitrosylsulfuric acid (HSO_3NO_2), peroxy monosulfuric acid (H_2SO_5), and eventually sulfuric acid and sulfate-containing aerosols. I am unaware of any of these proposed intermediates having been measured in the atmosphere. Their presence could provide evidence for these reaction pathways.

In the less photochemically active atmosphere, radicals produced from O_3-alkene interactions are important. Cox and Penkett[13] and Calvert[6] have suggested that the O_3-alkene oxidation of SO_2 occurs via the Crigee intermediate, $CH_3\dot{C}HOO\cdot$. This radical could rearrange to acetaldehyde in the oxidation step:

$$CH_3\dot{C}HOO\cdot + SO_2 \rightarrow CH_3CHO + SO_3 \quad (4)$$

In gases that exit stacks, where O_2 is depleted and NO and SO_2 are enriched, the reaction:

$$O + SO_2 + M \rightarrow SO_3 + M \quad (5)$$

may also be significant.[6] Similarly, the formation of HONO at high NO_x concentrations may occur, with subsequent decomposition to OH and oxidation through Reaction 1.[6] The rates of SO_2 oxidation by both of these mechanisms will diminish as the stack gases are diluted.

Based on these considerations, Calvert[6] has estimated the oxidation rates expected to occur in polluted air and clean air, as well as in stack gases. These estimates are summarized in TABLE 1. The reaction rate of the Crigee intermediate is lower than that for HO, HO_2, and CH_3O_2 but is independent of sunlight, and this reaction should be the dominant homogeneous mechanism at night.

Heterogeneous reactions in the atmosphere include reactions of SO_2 on dry particle surfaces and in aqueous solution. Reactions on dry surfaces have been observed in the laboratory, including oxidation on carbon[14] and on fly ash particles.[15]

TABLE 1

THEORETICAL HOMOGENEOUS OXIDATION RATES IN SUNLIGHT-IRRADIATED ATMOSPHERES* ($\% \ hr^{-1}$)

	Polluted Atmosphere†	Clean Atmosphere‡	Stack Gas§
$OH + SO_2 + M$	≦2.7	0.4	≦1.2
$HO_2 + SO_2$	1.9	~0.2	—
$CH_3O_2 + SO_2$	≦2	~0.01	—
$CH_3\dot{C}HOO \cdot + SO_2$	0.2¶	—	—
$O + SO_2$	—	—	≦1.4

*Estimates from Calvert *et al.*[6]

†$[OH] = 7 \times 10^6$; $[HO_2] = 6 \times 10^9$; $[CH_3O_2] = 10^9$.

‡$[OH] = 1 \times 10^6$; $[HO_2] \simeq 10^5$; $[CH_3O_2] \simeq 10^7$.

§$(SO_2) = 500$; $(NO) = 500$; $(NO_2) = 20$; $(CO) = 1$; $(O_2) = 10^5$. [ ] ≡ molecules cm^{-3}; ( ) ≡ ppmv.

¶Value is independent of sunlight; estimate at 50% RH, $(O_3) = 0.15$, $(SO_2) = 0.05$, (total alkenes) = 0.1.

These reactions should occur in stack gases, where relative humidities can be low (≲30%) and concentrations of both carbon and fly ash particles are high.

In the atmosphere, sulfate-containing particles with diameters less than 1 μm are usually associated with water.[16] SO_2 dissolves in water to form sulfite ($SO_3^{=}$) and bisulfite (HSO_3^-) ions, according to the following equilibria:

$$H_2O + SO_2 \leftrightarrow SO_2 \cdot H_2O \qquad (6)$$

$$SO_2 \cdot H_2O \leftrightarrow HSO_3^- + H^+ \qquad (7)$$

$$HSO_3 \leftrightarrow SO_3^{=} + H^+ \qquad (8)$$

which have the following values of the equilibrium constants at 5°C:

$$K_{HS} = \frac{[SO_2 \cdot H_2O]}{SO_2} = 2.64 \text{ M atm}^{-1} \qquad \text{(References 17, 18)}$$

$$K_{15} = \frac{[HSO_3^-][H^+]}{[SO_2 \cdot H_2O]} = 2.07 \times 10^{-2} \text{ M} \qquad \text{(References 18, 19)}$$

$$K_{25} = \frac{[SO_3^{=}][H^+]}{[HSO_3^-]} = 8.1 \times 10^{-8} \text{ M} \qquad \text{(References 18, 19).}$$

We shall refer to the total amount of SO_2 dissolved as $[SO_2]_\ell = [HSO_3^-] + [SO_3^{=}] + [SO_2 \cdot H_2O]$. In water associated with acid sulfate particles ($pH \lesssim 2$), $[SO_2]_\ell \approx [SO_2 \cdot H_2O]$; in fog and cloud water ($3 \leqq pH \leqq 7$), $[SO_2]_\ell \approx [HSO_3^-]$.

The net rate of SO_2 oxidation by aqueous reactions in the atmosphere (moles of SO_2 per volume of air per unit time) is proportional to the product of the liquid water concentration (volume of water per volume of air) and the oxidation rate in solution (moles of SO_2 per volume of water per unit time). Since atmospheric liquid water concentrations are relatively low (~10^{-9} to 10^{-6} ℓ_{H_2O}/ℓ_{air}), the oxidation rates in solution must be relatively rapid in order to compete with gas phase processes.

A number of aqueous mechanisms have been proposed, including oxidation by oxygen (O_2) in the presence of metal ion catalysts[20,21] or carbon (soot),[22] by O_2 alone,[23,24] by O_3,[25–28] and by H_2O_2.[29] With the exception of the carbon-catalyzed mechanism, these reactions have been discussed in recent reviews by Beilke and Gravenhorst[30] and by Hegg and Hobbs.[31]

The significance of the metal-ion-catalyzed oxidation by O_2 depends on a number of factors. Not only are several metal catalysts present in droplets (transition 2 series: Fe^{+3}, Mn^{+2}, Ni, etc.), but the presence of one metal ion may enhance the catalytic activity of another. Barrie and Georgii[32] report that Fe^{+3} and Mn^{+2} display this synergistic behavior. Further complicating matters is the fact that manganese, and probably other metals, is unevenly distributed throughout the particle size spectrum,[31,33] suggesting that the net oxidation rate is not simply related to the solution oxidation rate. Finally, the solubility of the metals varies, depending on the particular molecular form (e.g., oxide or salt) and the chemistry of the aqueous phase (e.g., acidic or neutral).[34] There is general agreement on several points: at constant pH, the oxidation reaction in solution is first-order with respect to both $[SO_2]_\ell$ and the concentration of metal ion catalyst;[31] and the reaction rate is strongly dependent on temperature ($\Delta H \gtrsim 25$ kcal/mole).[30,31] The reader is referred to the reviews mentioned earlier for further details.

Carbon can also catalyze SO_2 oxidation by O_2 in aqueous solution. Chang *et al.*[22] studied the oxidation of SO_2 by activated carbon, acetylene soot, and natural gas soot in aqueous solution. The rate was independent of both $[SO_2]_\ell$ and pH for 1.5×10^{-4} M $\leqq [SO_2]_\ell \leqq 1 \times 10^{-3}$ M, $1.45 \leqq \text{pH} \leqq 7.50$, and $0.005\% \leqq$ carbon concentration $\leqq 0.32\%$ (by weight). These experimental conditions correspond to equilibrium SO_2 concentrations in the gas phase as high as 650 ppmv. At lower SO_2 concentrations and correspondingly lower ratios of $[SO_2]_\ell$ to dissolved carbon, the rate may be dependent on the SO_2 concentration. This possibility needs to be examined.

The uncatalyzed oxidation of SO_2 by O_2 probably involves oxidation of sulfite ion.[35] This is reflected in the fact that the rate increases with increasing pH up to pH $\cong$ (pK_{2S}). However, estimates of the rate of reaction vary as much as two orders of magnitude at pH 4.[30,31] Additionally, there are uncertainties in the pH dependence in acidic solution (see reviews of Beilke and Gravenhorst[30] and Hegg and Hobbs[31] for further details). This large discrepancy is due in part to the sensitivity of the oxidation rate to low concentrations of both positive and negative catalysts. While metal ions will accelerate the rate at concentrations as low as 10^{-7} M,[31] certain organic compounds (such as dimethylacetamide) will depress the rate by as much as a factor of 5 at 10^{-6} M.[36] This depression of the rate may be due to free radical scavenging of reactive intermediates that are involved in the oxidation. $SO_3^-\cdot$ free radicals are thought to be involved in the reaction[37] and have been identified by flash photolysis and pulsed radiolysis in oxidizing sulfite solutions.[38]

Some recent rate measurements of the SO_2–O_3 reaction in solution are summarized in TABLE 2. As with the O_2 reaction, there is disagreement over the reaction rate. The studies of Penkett[25] and of Erickson *et al.*[26] were performed at relatively high molar ratios of dissolved O_3 to $[SO_2]_\ell$; those of Barrie[27] and Larson *et al.*[28] were carried out at much lower molar ratios and absolute O_3 concentrations. The molar ratio of O_3

TABLE 2

STUDIES OF THE SO_2–O_3 REACTION IN AQUEOUS SOLUTION

Investigator	Rate Expression	Experimental pH	Molar Ratio of Reactants ($[O_3]/[SO_2]_\ell$)*	Reaction Rate† (moles SO_4^{2-}/ ℓ − s)	Comments
Penkett[25] (1972)	$k_1\,[O_3][HSO_3^-]$ where $k_1 = 3.3 \times 10^5\ M^{-1}\ s^{-1}$ at 283°K	4.65	0.03 to 0.5	1.5×10^{-10}	Measurement of O_3 depletion at 283°K in a stopped flow reactor; O_3 equivalent to ~2500 ppmv.
Barrie[27]‡ (1975)	—	4.0	10^{-6} to 5×10^{-5}	5×10^{-12}	Measurement of SO_4^{2-} formation in 1 mm drops exposed to 0.01 to 0.5 ppmv O_3 and 0.4 ppmv SO_2.
Erickson *et al.*[26] (1978)	$k_2\,[O_3][HSO_3^-] + k_3\,[O_3][SO_3^{2-}]$ where $k_2 = 3.1 \times 10^5\ M^{-1}\ s^{-1}$ and $k_3 = 2.2 \times 10^9\ M^{-1}\ s^{-1}$ at 298°K	−1.3 to 4.02	5 to 50	3×10^{-9}	Measurement similar to Penkett at 298°K and 289°K; O_3 equivalent to 2500 to 25,000 ppmv.
Larson *et al.*[28] (1978)	$k_4\,[O_3]\,[HSO_3^-][H^+]^{-0.1}$ where $k_4 = 4.4 \times 10^4\ M^{-0.9}\ s^{-1}$ at 298°K	4.0 to 6.2	6×10^{-4} to 2×10^{-3}	4×10^{-11}	Measurement of $[SO_2]_\ell$ depletion at 278°K and 298°K; O_3 concentrations from 0.28 to 0.95 ppmv.

*$[SO_2]_\ell$ = concentration of $SO_2 \cdot H_2O$, HSO_3^-, and SO_3^{2-}.

†In atmospheric droplet at pH = 4 and 10°C; SO_2 = 1 ppbv, O_3 = 40 ppbv.

‡As reported by Beilke and Gravenhorst.[30]

to SO_2 dissolved in an atmospheric droplet is approximately 10^{-4}, assuming a droplet pH of 4, a gaseous O_3 concentration of 50 ppbv, and an SO_2 concentration of 10 ppbv. At the low O_3 concentrations used in the latter two experiments, the O_3 may also have reacted with trace metal impurities, thus slowing the SO_2 oxidation rate. Future studies should incorporate not only higher O_3 levels than those used by Barrie[27] and by Larson *et al.*,[28] but also more realistic O_3 to SO_2 molar ratios than those used by Penkett[25] and by Erickson *et al.*[26]

SO_2 may also be oxidized by hydrogen peroxide (H_2O_2) in aqueous solution. The oxidation rate is expected to be fast in the pH range 3 to 5,[29] and H_2O_2 is predicted to exist in photochemically active atmospheres at concentrations of about 1 ppbv.[39] If 1 ppbv of both H_2O_2 and SO_2 were to dissolve in, and completely react in, cloud droplets,

TABLE 3

SELECTED RATE EXPRESSIONS FOR THE OXIDATION OF SO_2 IN AQUEOUS SOLUTION

Reactant	Rate Expression	Investigator
	$\left(= \dfrac{\text{moles } SO_4^{2-}}{\ell - \text{sec}} \text{ at } 278°K\right)$	
Metal ion (Manganous)	$k_M\,[Mn^{+2}][H^+]^{1/2}[SO_3^{2-}]$ where $k_M = 2.3 \times 10^6\ M^{-3/2}\ s^{-1}$	Barrie & Georgii[32] (1976)
Carbon	$k_C\,[C][O_2]^{0.7}$ where $k_C = 8.6 \times 10^{-5}\ M^{-0.7}\ s^{-1}$, $[C]$ = moles carbon/ℓ	Chang & Novakov[22] (1978)
Ozone	$k_{O_3}K_{HO_3}P_{O_3}\,[HSO_3^-]$ where $k_{O_3} = 3.3 \times 10^5\ s^{-1}\ M^{-1}$, $K_{HO_3} = 2.2 \times 10^{-2}$ M atm^{-1}, P_{O_3} = partial pressure O_3 (atm)	Penkett[25] (1972)
Oxygen (pH $\lesssim$ 7)	$k_{O_2}\,[H^+]^{1/2}[SO_3^{2-}]$ where $k_{O_2} = 7.2\ M^{-1/2}\ s^{-1}$	Larson *et al.*[28] (1978)
	or $k'_{O_2}\,[SO_3^{2-}]$ where $k'_{O_2} = 1.6 \times 10^{-3}\ s^{-1}$	Scott & Hobbs[23] (1967)

about 4 $\mu g/m^3$ of H_2SO_4 could be produced. However, competing reactions with other dissolved species (NO_x, organics) would probably occur, reducing this estimate by an unknown amount. Measurements of ambient H_2O_2 concentrations are needed to assess the importance of this oxidation mechanism.

A simple model of atmospheric liquid water chemistry will be used to compare the aqueous SO_2 oxidation reactions. The model is based on the following assumptions:

1. The aqueous concentration of catalysts is assumed to be the same in each droplet at a given atmospheric liquid water concentration;

2. The catalyst concentration in solution is directly proportional to the dry mass concentration and inversely proportional to the liquid water concentration;

3. The net SO_2 oxidation rate is proportional to both the liquid water concentration and the solution oxidation rate;

4. The aqueous rate expressions listed in TABLE 3 are correct.

Using this very simple model, we can determine that value of droplet pH and/or liquid water concentration where the rates of any pair of aqueous reactions are the same. The results are shown in FIGURE 1, where those regions of atmospheric liquid water content and droplet pH in which each of the oxidation mechanisms is most important are outlined. At low pH ($\leqq 5$) and high liquid water concentration ($W \simeq 1$ g/m^3, characteristic of clouds over polluted areas), the O_3 reaction dominates. At low pH and low liquid water content ($W \lesssim 10^{-3}$ g/m^3, characteristic of acid aerosols), the carbon-catalyzed reaction is most important. For more neutralized aerosols at higher pH, the metal-catalyzed reaction becomes important. Finally, at higher liquid water concentrations and pH values, characteristic of clouds in clean atmospheres, the O_2 reaction may be important. Uncertainties in the rate of oxidation by O_2 are represented in FIGURE 1 by the region where either O_2 or O_3 dominates.

Lines of constant overall oxidation rate (dotted lines) were plotted assuming a first-order reaction in SO_2. Since the carbon-catalyzed rate is independent of the SO_2 concentration (and the pH), this representation is not entirely appropriate. The actual carbon rate for the given conditions is 0.02 ppbv per hour, equivalent to 0.2% per hour at 10 ppbv SO_2. Further, this rate is the same over the entire carbon region. This somewhat surprising result is due to the fact that an increase in the liquid water concentration in our model implies a proportionate decrease in both the carbon concentration and the solution oxidation rate. Since the overall oxidation rate is proportional to both the solution oxidation rate and the liquid water content, the overall rate is independent of the liquid water concentration.

We can combine the results of our model with measurements of droplet pH in the atmosphere. Excluding rainwater measurements, these data are scarce. TABLE 4 summarizes the results of five studies, four studies of clouds and one of ground fogs.

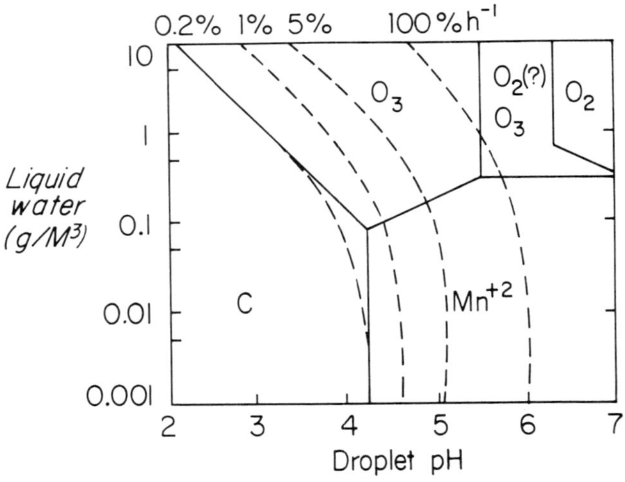

FIGURE 1. The pH and liquid water concentrations at which the O_2 reaction, the O_3 reaction, the carbon-catalyzed reaction, and the metal-catalyzed reaction are most important. [The calculations assume SO_2 = 10 ppbv, O_3 = 50 ppbv, particulate carbon = 10 μg/m^3, and soluble particulate Mn^{+2} = 1 nanomole/m^3. The dotted lines refer to the overall SO_2 oxidation rate (% hr^{-1}).]

TABLE 4
SUMMARY OF pH MEASUREMENTS IN CLOUDS

Investigator	pH Range	Comments
Oddie[40] (1962)	4.4–7.2	Cloud measurements taken over England at ~1500–2000 m; pH increased with increasing altitude.
Petrenchuk & Drozdova[41] (1966)	4.5–5.3	Cloud measurements taken over western Soviet Union; lower pH values in stratus than in nimbostratus clouds.
Mrose[42] (1966)	3.8–5.1	Fog measurements taken at surface in GDR* (46 to 1152 m); pH increased at higher elevations.
Petrenchuk[43] (1975)	3.1–4.8	Subinversion stratiform cloud measurements taken over western Soviet Union at 500 to 1770 m. pH increased with increasing height.
	5.0–5.6	Frontal precipitating cloud measurements taken over western Soviet Union; mean values of 145 measurements from 5 geographical regions.
Scott[44] (1978)	4.6–6.6	Stratocumulus cloud measurements taken over eastern Australia at 1500 to 1800 m; pH values lower downwind of Sydney than upwind.

*German Democratic Republic.

Of note are both the wide range of pH values and the observation that pH increases in cleaner air, either above or upwind of pollution. The high pH values do not exclude the presence of acid droplets. Gravenhorst[45] has shown that acid submicrometer-diameter particles can coexist with larger sea-salt particles over the ocean. For the cloud pH measurements we have summarized, the chemistry of the smaller droplets may have been overshadowed by larger, more alkaline droplets.

Since liquid water concentrations were not reported in these studies, we have estimated the range of liquid water concentrations for each study from reports of cloud type (including fog). These estimates are shown in FIGURE 2, where we have also outlined the important oxidation regions discussed earlier (dotted lines). Based on these cloud chemistry measurements, the O_3 and O_2 reactions appear to be the most important. In contrast, the catalyzed reactions should dominate the aqueous aerosol chemistry ($W \lesssim 10^{-3}$ g/m^3). For our model atmosphere, these data represent overall oxidation rates ranging from 0.4% hr^{-1} to over 100% hr^{-1}, encompassing the range of gaseous reaction rates listed in TABLE 1. The overall oxidation rate in clouds is expected to decrease with time, as the droplets are acidified by the H_2SO_4 produced. This decrease in the overall rate will be less rapid at higher liquid water concentrations, where the solutions are more dilute.

In contrast to the scarcity of cloud chemistry measurements, at least 14 studies of the rate of SO_2 oxidation in the atmosphere have been published in the last three years. These studies are summarized in TABLE 5. Most have dealt with the rate of SO_2 oxidation in the plumes of coal-fired power plants, although some have studied the oxidation over larger areas. The observed oxidation rates, reported as the equivalent first-order rate in % hr^{-1}, agree reasonably well with the rates predicted to result from both the gas phase and liquid phase reactions we have discussed.

Several investigators, including Forrest and Newman,[52] Meagher *et al.*,[53] and Smith and Jeffrey,[56] have reported that the oxidation is greater than first-order in SO_2, that is, that the apparent first-order rate decreases with plume travel time. These observations have been rationalized in terms of heterogeneous oxidation processes,[61] although they might also be explained by the stack gas mechanisms proposed by Calvert[6] that we discussed earlier. Cantrell and Whitby[49] observed a rate less than first-order. This effect can be explained by the buildup of oxidizing radicals as the plume mixes with photochemically active air. It can also be explained in terms of the carbon-catalyzed aqueous reaction. If this reaction is independent of SO_2 concentration, then a plume of SO_2 that is dispersing into, and diluted by, carbon-containing air

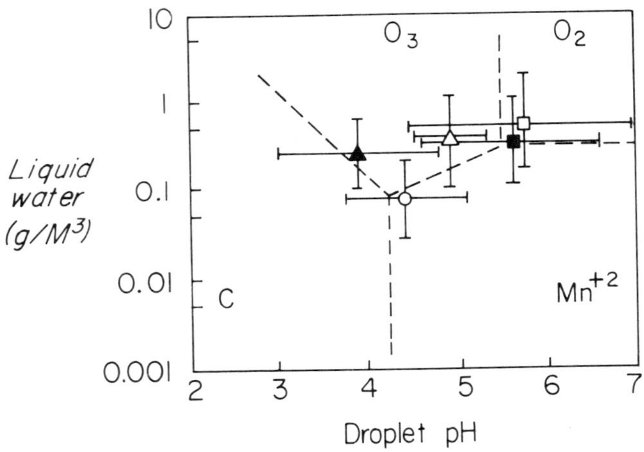

FIGURE 2. Droplet pH measurements as a function of the estimated atmospheric liquid water content. {Data are from Petrenchuk and Drosdova[41] [closed triangle], Mrose[42] [open circle], Petrenchuk[43] [open triangle], Scott[44] [closed square], and Oddie[40] [open square]. The dotted lines outline regions of differing SO_2 oxidation chemistry (see FIGURE 1).}

would have a first-order oxidation rate that increased with time, that is, a rate less than first-order in SO_2.

The most convincing argument for the importance of gas phase versus liquid phase reactions was recently described by both Gillani *et al.*[51] and Husar *et al.*[50] Gillani *et al.*[51] found that the extent of oxidation was dependent on the amount of solar radiation reaching the plume; Husar *et al.*[50] found rates up to 10 times greater during the day than at night. These observations were explained by gas phase oxidation chemistry described in the plume model of Isaksen *et al.*[62]

Based on the results of both these field studies and the laboratory rate studies, the following specific research topics are recommended:

- Examination of the products of the OH reaction with SO_2, particularly the fate of the HSO_3 radical.
- Clarification of the rates and rate expressions for the uncatalyzed oxidation of

TABLE 5
SOME RECENT SO_2 OXIDATION RATE MEASUREMENTS IN THE ATMOSPHERE

Investigator	SO_2 Source(s)	Observed Rate (% hr^{-1})	Spatial Scale (km)	Comments
Hegg *et al.*[46] (1976)	coal-fired power plant (Four Corners, N.M.)	0.3	1–6	SO_4^{2-} inferred from CN measurements; 40% ≦ RH ≦80%.
Pueschel & Van Valin[47] (1978)	coal-fired power plant (Four Corners, N.M.)	2 to 8	8–80	SO_4^{2-} inferred from CCN measurements; RH ≦20%.
Ursenbach *et al.*[48] (1977)	coal-fired power plant (Four Corners, N.M.)	0.3 to 0.7	2–65	SO_4^{2-} measured in both fine and coarse modes; oxidation rate based on fine SO_4^{2-}.
Cantrell & Whitby[49] (1978)	coal-fired power plant (Labadie, Mo.)	0.4 to 5	8–50	SO_4^{2-} inferred from particle volume measurements; rate increased with downwind distance.
Husar *et al.*[50] (1978)	coal-fired power plant (Labadie, Mo.)	0.1 to 4	8–320	Rate 1 to 4% hr^{-1} during day, ≦0.5% hr^{-1} at night.
Gillani *et al.*[51] (1978)	coal-fired power plant (Labadie, Mo.)	≦3	30–320	Cumulative SO_2/SO_4 ratio related to amount of sunlight.
Forrest & Newman[52] (1977)	four different coal-fired power plants	<1	2–70	Apparent rapid conversion near source (1 to 2% in 2 km), no conversion downwind; 32% ≦ RH ≦ 85%.
Meagher *et al.*[53] (1978)	coal-fired power plant (Cumberland)	<0.2	5–100	Apparent rapid conversion near source (0.9% in 5 km), no conversion downwind; 22% ≦ RH ≦ 71%.
Lusis & Wiebe[54] (1976)	nickel smelter (Sudbury, Ont.)	1	5–100	Maximum rate ≦3% hr^{-1}; data fit to second-order dependence by Forrest & Newman[55] (1977).
Smith & Jeffrey[56] (1975)	general emissions from east coast of U.K.	1–10	100–200	Study over ocean; most conversion occurred near source; initial rate increased with increasing RH; 70% ≦ RH ≦ 90%.
Smith & Hunt[57] (1978)	general emissions from east coast of U.K.	1	100–700	Measurements at 2700 m over North Sea.
Prahm[58] (1976)	U.K.	1.4	1000	Measurements over ocean.

(continued)

TABLE 5 (continued)

Investigator	SO_2 Source(s)	Observed Rate (% hr^{-1})	Spatial Scale (km)	Comments
Eliassen & Saltbones[59] (1975)	north central Europe	0.3–1.7	~500	Best fit of long term ground-level measurements to a model using first-order kinetics for oxidation and deposition.
OECD[60] (1977)	north central Europe	1.3	~1000	Best fit of first-order model; ~5% of SO_2 transformed to SO_4^{2-}.

SO_2 by O_2 and by O_3; the O_3 reaction rates should be remeasured at O_3 concentrations higher than those used by Barrie[27] and by Larson *et al.*,[28] and with lower ratios of dissolved O_3 to SO_2 than those used by Penkett[25] and by Erickson *et al.*[26]

- Measurement of the carbon-catalyzed rate at lower SO_2 concentrations and/or lower ratios of dissolved SO_2 to carbon.
- Investigation of the rates and rate expressions of the metal-catalyzed reactions, including the synergistic effects reported by Barrie and Georgii.[32]
- Atmospheric measurements of H_2O_2 and carbon (soot).
- Simultaneous measurements of cloud liquid water content, droplet pH, and droplet buffering capacity (alkalinity).
- Size segregated cloud droplet chemistry.

Accompanied by more refined measurements of the atmospheric conversion rate, these suggested studies should help to elucidate the mechanisms of SO_2 oxidation in the atmosphere.

SUMMARY

A brief review of current concepts regarding the homogeneous and heterogeneous processes that contribute to particle production in the atmosphere is presented, with special emphasis on the formation mechanisms of sulfate compounds. In the photochemically active urban atmosphere, SO_2 oxidation by the reactive radicals OH, HO_2, and CH_3O can account for the observed production rate of sulfate compounds (~1–5% hr^{-1}). In less photochemically active atmospheres, liquid phase oxidation of SO_2 by O_2 in the presence of catalysts, and by dissolved O_3 (and possibly H_2O_2), appears to be important. The factors that determine sulfate production rate by liquid phase processes are examined, including cloud type, atmospheric liquid water content, droplet pH and catalyst concentration, and O_3 concentration.

REFERENCES

1. CHARLSON, R. J., A. H. VANDERPOL, D. S. COVERT, A. P. WAGGONER & N. C. AHLQUIST. 1974. Science **184:** 156.

2. BROSSET, C., K. ANDREASSON & M. FERM. 1974. Atmos. Environ. **9:** 631.
3. ZIMMERMAN, P. R., R. B. CHATFIELD, J. FISHMAN, P. J. CRUTZEN & P. L. HANST. 1978. Geophys. Res. Lett. **5**(8): 679.
4. TOON, O. B. & J. B. POLLACK. 1973. J. Geophys. Res. **78:** 7051.
5. WHITBY, K. T. 1978. Atmos. Environ. **12**(1–3): 135.
6. CALVERT, J. G., F. SU, J. W. BOTTENHEIM & O. P. STRAUSZ. 1978. Atmos. Environ. **12**(1–3): 197.
7. DAVIS, D. D. 1974. Report #1, Gas Kinetics Laboratory, University of Maryland. (Quoted in Reference 6.)
8. ATKINSON, R., R. A. PERRY & J. N. PITTS, JR. 1976. J. Chem. Phys. **65:** 306.
9. COX, R. A. 1975. Int. J. Chem. Kinet. Symp. **1:** 379.
10. CASTLEMAN, A. W., JR. & I. N. TANG. 1977. J. Photochem. **6:** 349.
11. PAYNE, W. A., L. J. STIEF & D. D. DAVIS. 1973. J. Am. Chem. Soc. **95:** 7614.
12. WHITBECK, M. R., J. W. BOTTENHEIM, S. Z. LEVINE & J. G. CALVERT. 1976. Abstracts, 12th Informal Conference on Photochemistry: Kl-1 to Kl-5. U.S. National Bureau of Standards. Gaithersberg, Md.
13. COX, R. A. & S. A. PENKETT. 1972. J. Chem. Soc. Faraday Trans. 1 **68:** 1735.
14. NOVAKOV, T., S. G. CHANG & A. B. HARKER. Science **186:** 259.
15. HAURY, G., S. JORDAN & C. HOFMANN. 1978. Atmos. Environ. **12:** 281.
16. CHARLSON, R. J., D. S. COVERT, T. V. LARSON & A. P. WAGGONER. 1978. Atmos. Environ. **12**(1–3): 39.
17. JOHNSTONE, H. F. & P. W. LEPPLA. 1934. J. Am. Chem. Soc. **56:** 2233.
18. LOWELL, D. S. 1970. A Theoretical Description of the Limestone Injection Wet Scrubbing Process. **1.** NTIS Pub. No. PB-193-029.
19. YUI, T. 1940. Tokyo Inst. Phys. Chem. Res. Bull. **19:** 1229.
20. JUNGE, C. & T. G. RYAN. 1958. Q. J. R. Meteorol. Soc. **84:** 46.
21. BRIMBLECOMBE, P. & D. J. SPEDDING. 1974. Atmos. Environ. **8:** 937.
22. CHANG, S. G., R. BRODZINSKY, R. TOOSSI, S. S. MARKOWITZ & T. NOVAKOV. 1978. Catalytic oxidation of SO_2 on carbon in aqueous suspensions. Paper presented at Conference on Carbonaceous Particles in the Atmosphere, Berkeley, California, March 20–22.
23. SCOTT, W. D. & P. V. HOBBS. 1967. J. Atmos. Sci. **24:** 54.
24. MCKAY, H. A. C. 1971. Atmos. Environ. **5:** 7.
25. PENKETT, S. A. 1972. Nature (London) Phys. Sci. **240:** 105.
26. ERICKSON, R. E., L. M. YATES, R. L. CLARK & D. MCEWEN. 1977. Atmos. Environ. **11**(9): 813.
27. BARRIE, L. 1975. Berichte des Institutes für Meteorologie der Universität Frankfurt a. M., Nr. 28. (Reported in Reference 30.)
28. LARSON, T. V., N. R. HORIKE & H. HARRISON. 1978. Atmos. Environ. **12**(1–3): 1597.
29. PENKETT, S. A., K. A. BRICE & A. E. J. EGGLETON. 1976. Paper presented at European Sulfur Symposium, Ispra, Italy. (Reported in Reference 30.)
30. BEILKE, S. & G. GRAVENHORST. 1978. Atmos. Environ. **12**(1–3): 231.
31. HEGG, D. A. & P. V. HOBBS. 1978. Atmos. Environ. **12**(1–3): 241.
32. BARRIE, L. A. & H. W. GEORGII. 1976. Atmos. Environ. **10:** 743.
33. LEE, R. E. & D. J. VON LEHMDEN. 1973. J. Air Pollut. Control Assoc. **23:** 853.
34. GORDON, G. E., D. D. DAVIS, G. W. ISRAEL, H. E. LANDSBERG, T. C. O'HAVER, S. W. STALEY & W. H. ZOLLER. 1975. Study of Emissions from Major Air Pollution Sources and Their Atmospheric Interactions. NTIS Pub. No. PB-242-581-AS.
35. WINKELMANN, D. 1955. Z. Elektrochem. **59:** 891.
36. SCHROETER, L. C. 1963. J. Pharm. Sci. **52:** 559.
37. BÄCKSTROM, H. 1934. Z. Physik. Chem. **B25:** 99.
38. HAYON, E., A. TREININ & J. WILF. 1972. J. Am. Chem. Soc. **94:** 47.
39. LEVY, H. 1973. Planet. Space Sci. **21:** 575.
40. ODDIE, B. C. V. 1962. Q. J. R. Meteorol. Soc. **88**(378): 535.
41. PETRENCHUK, O. P. & V. M. DROZDOVA. 1966. Tellus **18:** 280.
42. MROSE, H. 1966. Tellus **18:** 266.
43. PETRENCHUK, O. P. 1975. *In* Proceedings of the 8th International Conference on Ice Nucleation, Leningrad, U.S.S.R.: 445–449.

44. Scott, W. D. 1978. Atmos. Environ. **12**(4): 917.
45. Gravenhorst, G. 1978. Atmos. Environ. **12**(1–3): 707.
46. Hegg, D. A., P. V. Hobbs & L. F. Radke. 1976. Electric Power Research Institute Report No. EPRI EA-270.
47. Pueschel, R. F. & C. C. Van Valin. 1978. Atmos. Environ. **12**(1–3): 307.
48. Ursenbach, W. O., A. C. Hill, W. H. Edwards & S. M. Kunen. 1977. Paper No. 77–39.1. *In* Proceedings of the 70th Annual Meeting of the Air Pollution Control Association: 1–10.
49. Cantrell, B. K. & K. T. Whitby. 1978. Atmos. Environ. **12**(1–3): 323.
50. Husar, R. B., D. E. Patterson, J. D. Husar, N. V. Gillani & W. E. Wilson, Jr. 1978. Atmos. Environ. **12**(1–3): 549.
51. Gillani, N. V., R. B. Husar, J. D. Husar, D. E. Patterson & W. E. Wilson, Jr. 1978. Atmos. Environ. **12**(1–3): 589.
52. Forrest, J. & L. Newman. 1977. Atmos. Environ. **11:** 465.
53. Meagher, J. F., L. Stockburger, E. M. Bailey & O. Huff. 1978. Atmos. Environ. **12**(11): 2197.
54. Lusis, M. A. & H. A. Wiebe. 1976. Atmos. Environ. **10:** 793.
55. Forrest, J. & L. Newman. 1977. Atmos. Environ. **11:** 517.
56. Smith, F. B. & G. H. Jeffrey. 1975. Atmos. Environ. **9:** 643.
57. Smith, F. B. & R. D. Hunt. 1978. Atmos. Environ. **12**(1–3): 461.
58. Prahm, L. P., U. Torp & R. M. Stern. 1976. Tellus **18:** 355.
59. Eliassen, A. & J. Saltbones. 1975. Atmos. Environ. **9:** 425.
60. OECD. 1977. The OECD Programme on Long Range Transport of Air Pollutants. Measurements and findings. Organization for Economic Cooperation and Development. Paris, France.
61. Freiberg, J. 1978. Atmos. Environ. **12**(1–3): 339.
62. Isaksen, I. S. A., E. Hesstvedt & Ö. Hov. 1978. Atmos. Environ. **12**(1–3): 599.

SULFUR AND NITROGEN COMPOUNDS IN URBAN AEROSOLS*

Roger L. Tanner

Environmental Chemistry Division
Brookhaven National Laboratory
Upton, New York 11973

INTRODUCTION

The need for additional knowledge concerning the spatial and temporal distribution of sulfur- and nitrogen-containing compounds, principally secondary aerosols, which result from the primary emission of sulfur dioxide, nitrogen oxides, and particulate matter, has stimulated several major field studies in the northeastern United States. The keys to the utility of such studies appear to be at least three in number: (1) the development of improved analytical techniques for ground and airborne sampling and analysis (using both filter-based and real-time methods) of trace aerosol and gaseous constituents; (2) the inclusion of emission, transport, transformation, and removal phenomena for S- and N-compounds within the scope of the studies; and (3) the application of sophisticated, potentially predictive models to this new, detailed data base.

We report in this paper some of the results for sulfate, nitrate, and related ionic species taken from a detailed chemical and meteorological data base that has been accumulated for the New York subregion since 1976 and that is generally referred to as the New York Summer and Winter Aerosol Studies.[1] We then describe some of the experimental advances in sampling and analysis of aerosol sulfur and nitrogen compounds and related gases that offer promise for measurement of all components of the atmospheric sulfate–nitrate–SO_2–NO_x–NH_3 system with time resolution of a few minutes. Additional analytical methodologies for reduced sulfur compounds in aerosols, and for H_2S and other sulfur-containing gases of biogenic origin, have been developed but are outside the scope of this paper.

NEW YORK AEROSOL COMPOSITION

Aerosol sampling during the August 1976 and February 1977 sampling periods was done only in an urban New York site and in a background site at High Point, New Jersey. The sampling program was expanded to Brookhaven (Long Island) and New Haven, Connecticut, sites during summer 1977 and winter 1978 sampling. Time resolution for aerosol filter samples was 6 hr, with some 3-hr sampling for the latter three periods. Improved analytical methods for sulfur- and nitrogen-containing constituents were used, as well as more conventional gaseous-chemical, meteorological, and particle-size-classifying instrumentation, as shown in TABLE 1.

Parameters measured included: chemical constituents—strong acid (quartz filters only), ammonium, sulfate, and nitrate, and sulfuric acid (limited data); physical

*This work was performed under the auspices of the United States Department of Energy under Contract No. EY-76-C-02-0016.

0077–8923/80/0338–0039 $01.75/1 © 1980 NYAS

TABLE 1

NEW YORK AEROSOL STUDY MEASUREMENTS*

Measured Parameter	Sampling Method	Time Resolution (hr)	Sampling Period	Sampling Method or Sensor Principle
Aerosol Sulfate	Hi-vol Filter	24 hr	S76,W77	Turbidimetry
	Hi-vol Filter	6 hr	W77,S77	Turbidimetry
	Hi-vol Filter	6 hr	W77–S78	MTB Colorimetry
	Hi-vol Filter	6 hr	W77	Ion Chromatography
	Hi-vol Filter	3 hr	S77,S78†	MTB Colorimetry
	Diffusion Sampler	12 hr	S76,W77	MTB or Turbidimetry
H_2SO_4	Hi-vol Filter	6 hr, 3 hr	S77–S78	Benzaldehyde Extraction
Aerosol Nitrate	Hi-vol Filter	24 hr	S76,W77	Reduction-Colorimetry
	Hi-vol Filter	6 hr	W77–S78	Reduction-Colorimetry
	Diffusion Sampler	12 hr	S76,W77	Reduction-Colorimetry
Strong Acid	Hi-vol Filter‡	6 hr	S77–S78	Gran Titration
	Diffusion Sampler	12 hr	S76,W77	Gran Titration
Ammonium	Hi-vol Filter	24 hr	S76	Indophenol Colorimetry
	Hi-vol Filter	6 hr	W77–S78	Indophenol Colorimetry
	Diffusion Sampler	12 hr	S76,W77	Indophenol Colorimetry
Aerosol Mass	Continuous	1 hr	S76–S78	Nephelometer
	Hi-vol Filter	24 hr	S76,W77	Gravimetry
	Hi-vol Filter	6 hr	W77–S78	Gravimetry
Size Distribution	1–6 min Ave/hr	1 hr	S76,W77	EAA
	1–6 min Ave/hr	1 hr	S76,W77	Diffusion Battery-CNC
	1–6 min Ave/hr	1 hr	S76–S78	Optical Counter
Trace Metals	Hi-vol Filter	24 hr, 1 wk	S76,W77	Atomic Absorption
	Cyclone Sampler	24 hr, 1 wk	S76–S77	Atomic Absorption
	Hi-vol Filter	6 hr	W77–S78	Atomic Absorption
Halogen Compounds	Hi-vol Filter	1 wk	S76,W77	NAA
Light Scattering	Continuous	1 hr	S76,W77	Laser Backscatter
	Continuous	1 hr	S76,S78	Nephelometer§
RH	Continuous	1 hr	S76–S78	Wet/Dry Bulb
Dew Point	Continuous	1 hr	S76–S78	
Temperature	Continuous	1 hr	S76–S78	
Wind Speed	Continuous	1 hr	S76–S78	
Wind Direction	Continuous	1 hr	S76–S78	

*Reprinted from Reference 7 with permission of *Environmental Science and Technology*. Copyright by the American Chemical Society.

†BNL, NYU site only.

‡Quartz filters only; no valid NYU data.

§See also Aerosol Mass/Nephelometer.

parameters—aerosol size distributions by cascade impactor, cyclone sampler, electrical aerosol analyzer (EAA), an optical counter and a special diffusion battery–condensation nucleus counter (CNC) apparatus, light scattering nephelometer, and other instruments; chemically speciated size classification by diffusion sampler; trace metals by atomic absorption; halogen compounds by neutron activation analysis (NAA); meteorological measurements of relative humidity (RH), temperature, and wind speed and direction; gaseous measurements of SO_2, ozone, NO_x (nitrogen oxides), and hydrocarbons at some locations for some sampling periods. Data analysis is nearing completion only for the summer 1976 sampling period, but the following trends and conclusions relative to S- and N-compounds are substantiated by the data.

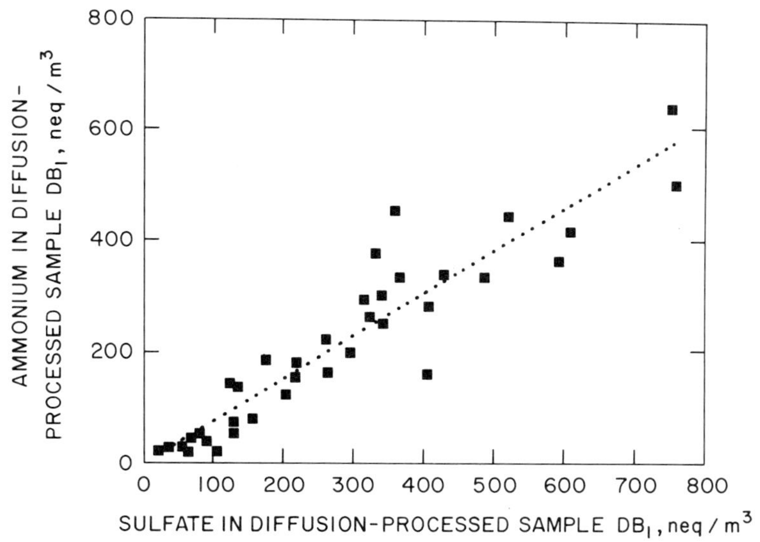

FIGURE 1. Linear regression plot of ammonium vs. sulfate in BNL diffusion-processed samples, DB_I, New York Aerosol Study, August 1976: regression line, $NH_4^+ = 0.76\ [SO_4^{2-}] - 4$; correlation coefficient, $r = 0.941$. (Reprinted from Reference 7 with permission of *Environmental Science and Technology*. Copyright by the American Chemical Society.)

The existence of aerosol sulfate in the ambient environment predominantly in the chemical form of sulfuric acid, mostly neutralized by ammonia, is now well documented.[2,3] The average composition of fine particle sulfate in summer 1976 aerosols in New York was approximately that of letovicite $[(NH_4)_3H(SO_4)_2]$, as shown by FIGURE 1 and the regression data from Reference 4, Table 3. Based on the impactor data, about 85% of the aerosol sulfate mass was in the fine particle fraction. About 50% of this fine particle sulfate was deduced to be in the suboptical size regime ($<0.25\ \mu m$) from diffusion processor data. The $H^+/SO_4^=$ ratio in suboptical aerosols did not significantly differ from that in fine fraction aerosol. The coarse particle sulfate was not associated with H^+ or NH_4^+ and comprised about 15% of the sulfate mass.

Thus, it was found that aerosol sulfate concentrations in summer were remarkably similar to those in an upwind rural location.[5] In addition, the size classification data from the diffusion sampler were in close agreement with those found in rural Illinois during July 1975. This similarity in composition and size distributions suggests that a major portion of the urban sulfate aerosol may be transported through New York City in summer, rather than formed there.

In contrast to aerosol sulfate size distribution, 58% of particulate nitrate was found to be in coarse particles (>3.5 μm), and only about 25% in the <0.25 μm size range.[6] Fine particulate nitrate levels were correlated with gaseous NO_2 at the 95% level and vanadium at the 90% level, but were only weakly related to total particulate nitrate.

The sulfate data from the winter 1977 sampling period were substantively different than those observed in summer.[7] Sulfate concentrations were lower, fine fraction sulfate was more nearly neutralized by NH_4^+ ($H^+/SO_4^= \leq 0.10$), and, based on diffusion sampler data, aerosol sulfate may have been in somewhat larger particles on the average than in the summer. Statistical analysis of the relationship of fine particulate sulfate with EAA-based volume distribution data, however, suggests the opposite. Analysis of diurnal trends from 6-hr Hi-vol sampling indicates maximum sulfate concentrations during the 06–12 hr period, but closest association of sulfate with ammonium was during the 00–06 hr period ($r > 0.9$) (see FIGURE 2). Stepwise

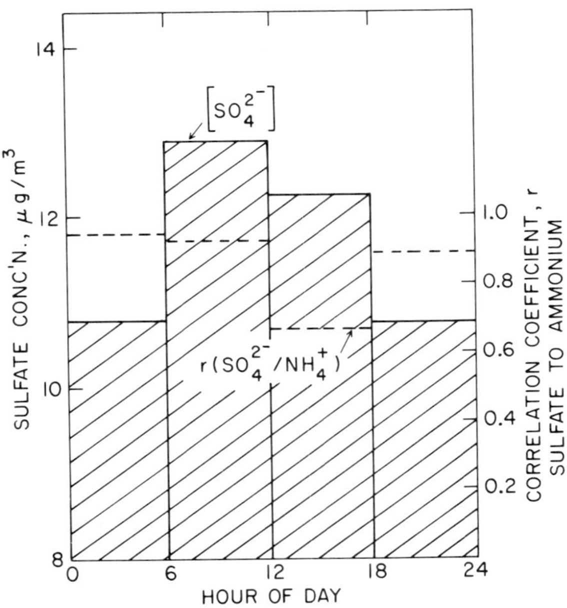

FIGURE 2. Diurnal variations of sulfate concentrations and sulfate-to-ammonium correlation, average Hi-vol data from New York, N.Y., February 1977: - = [SO_4^{2-}], $\mu g/m^3$; --- = sulfate-to-ammonium correlation coefficient, *r*. Reprinted from Reference 7 with permission of *Environmental Science and Technology*. Copyright by the American Chemical Society.)

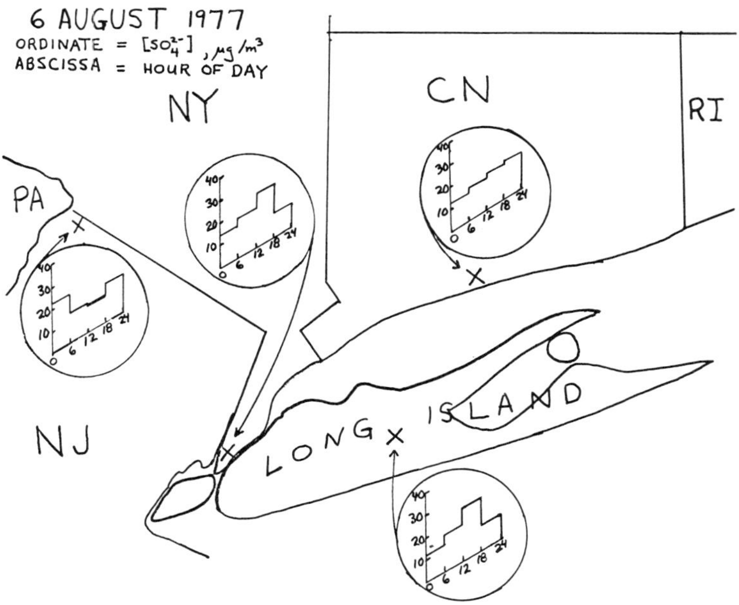

FIGURE 3. New York subregional variation in sulfate concentrations, μg/m^3, for 6 August 1977. (Reprinted from Reference 7 with permission of *Environmental Science and Technology*. Copyright by the American Chemical Society.)

regression indicates significant contribution to sulfate variability from SO_2 levels in New York, with the r^2 contribution substantively higher in winter.

Particulate nitrate levels in New York were higher in winter aerosol samples. Multiple regression of numerical sums of sulfate and nitrate (on an equivalent basis) with strong acid and ammonium indicated that particulate nitrate was significantly associated with ammonium in winter aerosol but not in summer aerosol.[7]

Investigation of the spatial variability of sulfate concentrations in the New York subregion, with inclusion of summer 1977 and winter 1978 data, is still in progress. It does appear that sulfate concentrations were elevated in New York relative to an upwind rural site by from 3–6 μg/m^3 during the winter 1977 sampling period, in contrast to summer 1976 when elevated urban values were not observed.[5] Sulfate values in the subregion were remarkably constant during July–August 1977 at the four sites, as shown in FIGURE 3, except where attenuated by local meteorological variations, such as wind direction and precipitation, affecting air mass trajectories or removal processes.

Diurnal variations were suggested by the Brookhaven National Laboratory (BNL) ionic composition data of FIGURE 4. Preliminary evaluation of diurnal variations, sorted for ozone levels >80 ppb at three of the four sites, is shown in FIGURE 5. Most noteworthy are the virtual lack of variation in sulfate levels (in contrast to mass levels) and the unique afternoon strong acid peak at BNL. Similar data for High Point show higher average $H^+/SO_4^=$ ratios but no strong diurnal trends. The intersite diurnal

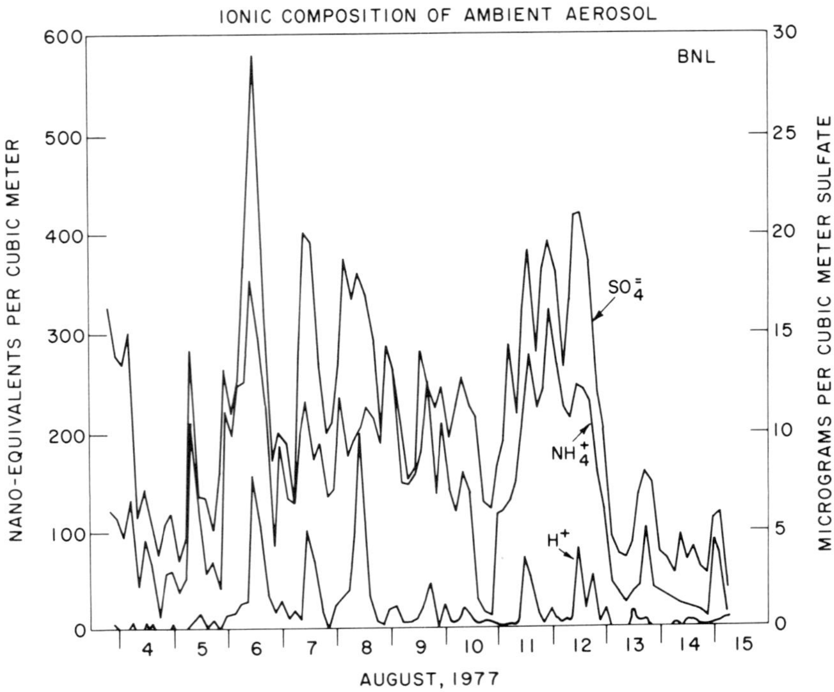

FIGURE 4. Ionic composition of ambient aerosol, Brookhaven National Laboratory site, August 1977: top line = sulfate; middle line = ammonium; bottom line = strong acid. (Reprinted from Reference 7 with permission of *Environmental Science and Technology*. Copyright by the American Chemical Society.)

variations in $H^+/SO_4^=$ for the two days, August 5th and 8th, shown in FIGURE 6 are representative of the trends observed. Nitrate concentrations were elevated at New Haven with nighttime maxima, and ammonium/sulfate ratios were higher (and $H^+/SO_4^=$ lower) at both New Haven and BNL sites as compared to the High Point, N.J., site. The relevance of the interposition of the urban area between these sites during prevailing wind directions is not yet clear.

ADVANCES IN MEASUREMENT OF AEROSOL SULFUR AND NITROGEN COMPOUNDS

The New York Aerosol Study and other urban aerosol studies have been largely dependent on laboratory analysis of filter samples collected in the field for aerosol characterization. Filter sampling methods have well-known defects (especially with regard to sulfur and nitrogen compounds): artifact sulfate formation from oxidation of adsorbed sulfur dioxide,[8] interparticle interaction both pre- and postcollection, conversions during extraction of reduced sulfur compounds,[9] and collection of nitric acid as nitrate on some filters.[10] Even the more specific filter-based techniques, such as the benzaldehyde extraction–ion chromatographic method[2,11] and the low temperature volatilization–flame photometric detection (FPD) method[11] for sulfuric acid aerosol, are marginally specific under realistic field conditions. Strong acid titrations,[2] although useful, are not species-specific and are quite prone to negative interferences.

Methods for fixing aerosol sulfur compounds with respect to strong acid-to-sulfate ratio or sulfur oxidation state continue to be active research areas. Examples are perimidylammonium bromide–coated filters for fixing sulfuric acid, pioneered by West's group,[12] and diazomethane methylation compounds of varying oxidation state with subsequent gas chromatography (GC)–FPD determination.[13] However, it is recognized that *in situ* methods for analysis of aerosols and their gaseous precursors are to be preferred, due to finer time resolution and the avoidance of filter sampling errors. Real-time methods for *in situ* analysis have not been widely applied for sulfur and nitrogen compounds or gaseous precursors in urban or rural aerosols (with the exception of SO_2 and NO, NO_2). Progress in the area of *in situ* or semi-real-time methods for aerosol sulfur, sulfuric acid, and ammonia are detailed below, as well as the application of a specific, extremely sensitive, interference-free method for nitric acid.

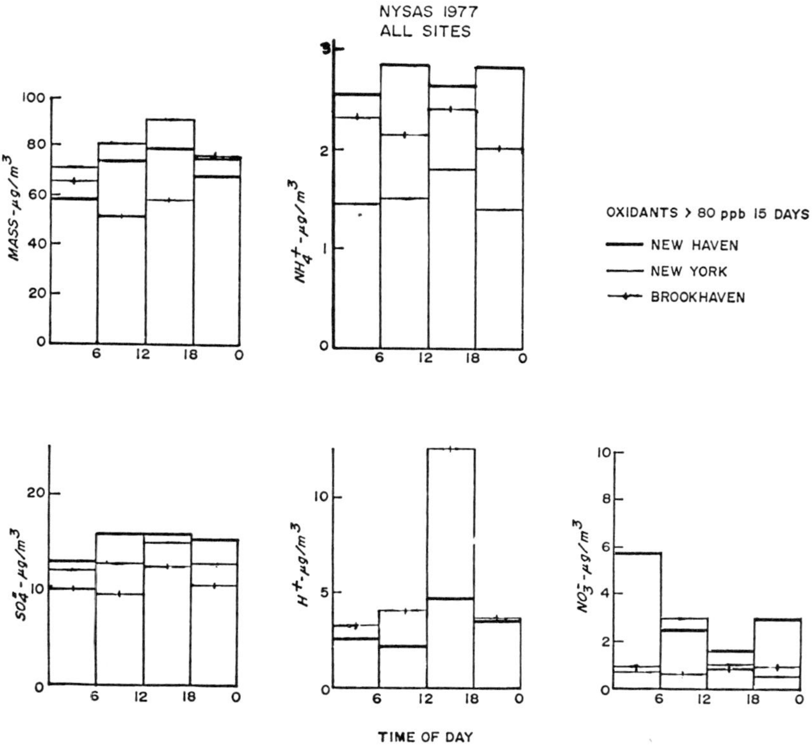

FIGURE 5. Average New York subregional diurnal variations in aerosol mass and ionic composition sorted for days of maximum $[O_3] > 80$ ppb; all concentrations in $\mu g/m^3$.

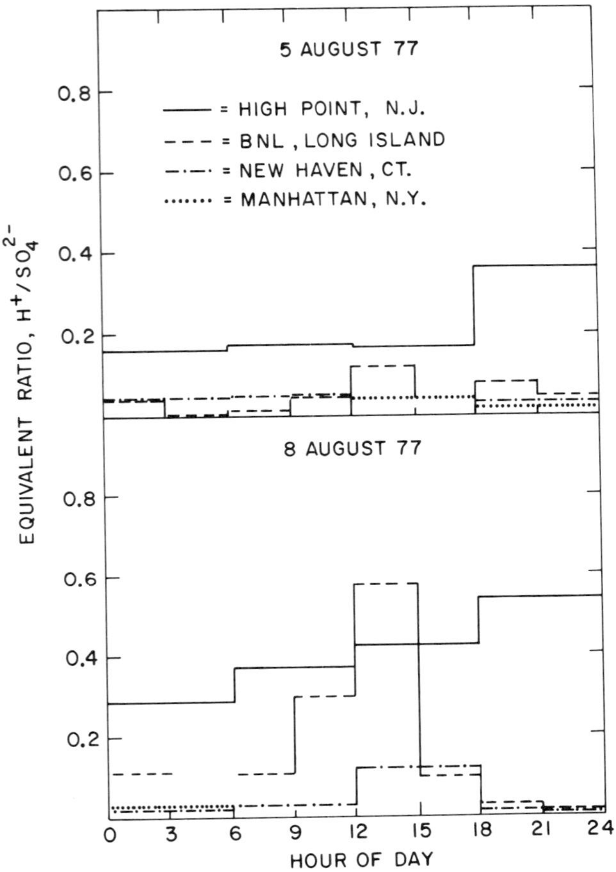

FIGURE 6. Diurnal variation in aerosol strong-acid–to–sulfate ratios in the New York subregion: top figure, 5 August 1977; bottom figure, 8 August 1977.

Aerosol Sulfur by Semi-Real-Time FPD

A system has been developed[14] for measurement of ambient levels of aerosol sulfur (principally sulfate) using a flame photometric sulfur analyzer with a heated denuder to remove sulfur dioxide. Cyclic addition of ammonia permits discrimination between sulfuric acid and other sulfate aerosols, as well as eliminating the effects of reduced instrumental sensitivity to sulfuric acid. The effects of water vapor concentration and barometric pressure on the sulfur response of the FPD system have been studied in detail. It has been shown that other environmental effects on sulfur response may be eliminated by careful and frequent zeroing of the system. Calibration plots and ambient data from July 1978 have demonstrated that the system illustrated in FIGURE 7 is selective for, and sensitive to, as little as 0.5 ppb aerosol sulfur if proper adjustment is made for electronic aberrations of commercial linearized amplification systems.

Gaseous Nitric Acid Sampling and Analysis

It is now realized that conventional glass fiber filters and other glass-based filters may collect nitric acid irreversibly, with efficiencies depending on external variables, e.g., relative humidity.[10] Hence, all historical particulate nitrate values may be presumed to be high by, in some cases, as much as an order of magnitude. Since nitrates/nitric acid are end products of atmospheric photochemical processes involving nitrogen oxides, and since nitric acid has markedly different and presumably more rapid sink processes, it is important to separate and measure independently both nitric acid and particulate nitrate in the ambient atmosphere.

Nitric acid methodologies include a real-time chemiluminescence technique involving conversion to NO, as reported by Spicer.[15] This technique is promising but not widely used, since expected HNO_3 levels are frequently less than the limit of detection of chemiluminescent NO_x analyzers (1–2 ppb). It is also subject to interference from gaseous peroxyacetyl nitrate (PAN) and possibly from other oxidized nitrogen compounds.

Nitric acid may be determined in Hi-vol samplers, using NaCl-impregnated cellulose filters downstream from a treated quartz prefilter.[16] The cellulose filters are extracted after collection and analyzed for nitrate by a conventional autoanalyzer technique—reduction with hydrazine, diazotization, and colorimetric determination. This method suffers from reversible adsorption of HNO_3 by quartz at relative humidities above about 75%, which requires desorption by postcollection heating in a

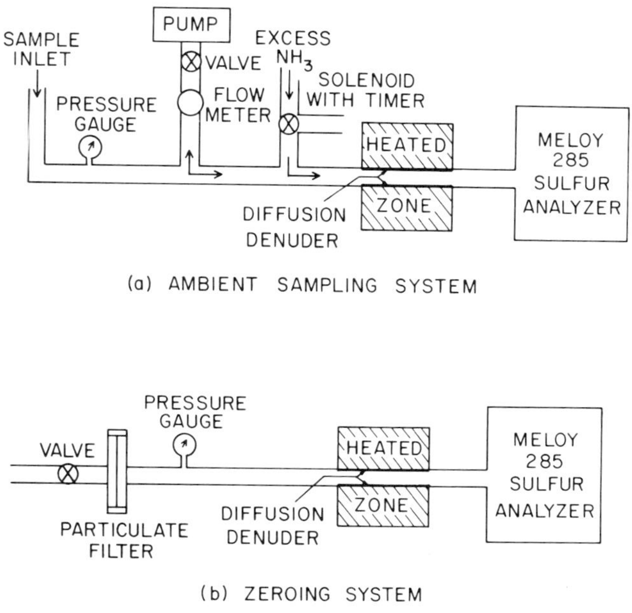

FIGURE 7. Real-time aerosol sulfate/sulfuric acid monitor: (a) ambient sampling system; (b) zeroing system.

stream of dry, clean air; but the method can be used for ambient HNO_3 levels with 1–2 hr sampling times.

Nitric acid may also be collected on nylon wool plugs located downstream from a nonadsorbent Teflon prefilter in a low-volume sampling apparatus.[17,18] The sample is extracted, then reacted with a benzene analogue in strong acid to form a nitro derivative, which is quantitatively analyzed by gas chromatography with electron capture detection (GC-ECD). Since the GC-ECD method can easily detect picogram (10^{-12} g) quantities of the derivative (especially if an electron-capture-sensitive benzene analogue is used[19]), low-volume sampling with Teflon filters may be used with sampling time/nitric acid concentration combinations of 15 min and 0.2 ppb, respectively.

Continuous Ammonia Analysis

One further consideration in the measurement of oxidized sulfur- and nitrogen-containing aerosols is the measurement of ambient levels of ammonia. A new method is being developed[20] for semi-real-time measurement of NH_3, based on collection and continuous derivatization of ammonia (and primary amines) by *o*-phthalaldehyde in the presence of a reducing agent and pH 10 borate buffer, with fluorometric determination of the derivative with a limit of detection of <1 ng/ml. This method shows promise for ambient NH_3 determination at the <1 ppb level with time resolution of 10 min or less.

The ability to determine all components of the gaseous SO_2, HNO_3, NH_3-particulate strong acid, ammonium, sulfuric acid, bisulfate, sulfate, nitrate system at ambient levels with a few minutes time resolution is an important development. For example, it allows atmospheric scientists to ask the important questions: Under what conditions is the atmospheric system containing SO_2, HNO_3, NH_3, H_2O, and aerosol particles (droplets or mixed solids containing H^+, NH_4^+, $SO_4^=$, HSO_4^-, H_2SO_4, and NO_3^-) at thermodynamic equilibrium?; and Over what spatial scale does this equilibrium extend? The answers to these and related questions will extend many fold our ability to predict the presence of acidic sulfate and nitric acid in polluted air masses.

Acknowledgments

I acknowledge the extensive contributions made by the coworkers listed below (BNL affiliation except where noted) to the work herein reported: Dr. Brian Leaderer (Pierce Foundation, Yale University) and Dr. Paul Lioy (New York University Medical Center) for contributions to the New York Aerosol Study summary, Dr. Robert Garber and Ted D'Ottavio (real-time sulfate monitor), Mary Phillips and Joseph Forrest (analytical work), Dr. Joseph Lepore (ammonia derivatization), Ruby Fajer (nitrate derivatization), and Dr. Leonard Newman, Head, Environmental Chemistry Division, BNL, for his inspiration and support of this work.

REFERENCES

1. LEADERER, B. P., D. M. BERNSTEIN, J. M. DAISEY, M. T. KLEINMAN, T. J. KNEIP, E. O. KNUTSON, M. LIPPMANN, P. J. LIOY, K. A. RAHN, D. SINCLAIR, R. L. TANNER & G. T. WOLFF. 1978. Summary of the New York summer aerosol study (NYSAS). J. Air Pollut. Control Assoc. **28:** 321.
2. TANNER, R. L., R. CEDERWALL, R. GARBER, D. LEAHY, W. MARLOW, R. MEYERS, M. PHILLIPS & L. NEWMAN. 1977. Separation and analysis of aerosol sulfate species at ambient concentrations. Atmos. Environ. **11:** 955.
3. STEVENS, R. K., T. G. DZUBAY, G. RUSSWURM & D. RICKEL. 1978. Sampling and analysis of atmospheric sulfates and related species. Atmos. Environ. **12:** 55.
4. TANNER, R. L., R. GARBER, W. MARLOW, B. LEADERER & M. A. LEYKO. 1979. Chemical composition of sulfate as a function of particle size in New York summer aerosol. Ann. N.Y. Acad Sci. **322:** 99.
5. LIOY, P. J., G. T. WOLFF, K. A. RAHN, D. M. BERNSTEIN & M. T. KLEINMAN. 1979. Characterization of aerosols upwind of New York City. II. Aerosol composition. Ann. N.Y. Acad Sci. **322:** 73.
6. KLEINMAN, M. T., B. LEADERER, C. TOMCZYK & R. TANNER. 1979. Inorganic nitrogen compounds in New York City air. Ann. N.Y. Acad. Sci. **322:** 115.
7. TANNER, R. L., W. H. MARLOW & L. NEWMAN. 1979. Chemical composition correlations of size-fractionated sulfate in New York City aerosol. Environ. Sci. Technol. **13:** 75.
8. PIERSON, W. R., R. H. HAMMERLE & W. W. BRACHACZEK. 1976. Sulfate formed by interaction of sulfur dioxide with filters and aerosol deposits. Anal. Chem. **48:** 1579.
9. EATOUGH, D. J., T. MAJOR, J. RYDER, M. HILL, N. F. MANGELSON, N. L. EATOUGH, L. D. HANSEN, R. G. MEISENHEIMER & J. W. FISCHER. 1978. The formation and stability of sulfite species in aerosols. Atmos. Environ. **12:** 263.
10. SPICER, C. W. 1977. Photochemical atmospheric pollutants derived from nitrogen oxides. Atmos. Environ. **11:** 1089.
11. RICHARDS, L. W., K. R. JOHNSON & L. S. SHEPHERD. 1978. Sulfate Aerosol Study. Report AMC8000.13FR, Air Monitoring Center, Rockwell International. Newbury Park, Calif.
12. THOMAS, R. L., V. DHARMARAJAN, G. L. LUNDQUIST & P. W. WEST. 1976. Measurement of sulfuric acid aerosol, sulfur trioxide, and the total sulfate content of the ambient air. Anal. Chem. **48:** 639.
13. PENZHORN, R. D. & W. G. FILBY. 1976. Eine methode zur spezifischen Bestimmung von schwefelhaltigen Säuren im atmosphärischen Aerosol. Staub **36:** 205.
14. TANNER, R. L., T. D'OTTAVIO, R. GARBER & L. NEWMAN. 1978. Continuous monitoring of ambient levels of aerosol sulfate using a flame photometric system. Paper No. 416, Euroanalysis III Conference, Dublin, Ireland, August 20–25.
15. JOSEPH, D. W. & C. W. SPICER. 1978. Chemiluminescence method for atmospheric monitoring of nitric acid and nitrogen oxides. Anal. Chem. **50:** 1400.
16. FORREST, J., R. L. TANNER, D. SPANDAU, T. D'OTTAVIO & L. NEWMAN. 1979. Determination of atmospheric nitric acid with NaCl-impregnated filters at high volume flow rates. American Chemical Society/Chemical Society of Japan Chemical Congress, Honolulu, Hawaii, April 1–6.
17. HARE, R. J., M. T. WININGER, W. D. ROSS & R. E. SIEVERS. 1978. Gas chromatographic measurement of nitric acid. Workshop on Measurement of Atmospheric Nitrates, Southern Pines, N.C., October 2–4.
18. TESCH, J. & R. E. SIEVERS. 1978. Selective collection and measurement of particulate nitrate, gaseous nitric acid, and nitrogen dioxide in ambient air. Workshop on Measurement of Atmospheric Nitrates, Southern Pines, N.C., October 2–4.
19. TANNER, R. L., R. FAJER & J. GAFFNEY. 1979. Determination of parts-per-billion concentrations of aqueous nitrate by derivatization gas chromatography with electron capture detection. Anal. Chem. **51:** 865.
20. TANNER, R. L. & J. LEPORE. 1979. Continuous determination of ammonia at ambient atmospheric concentrations using fluorescent derivatization. American Chemical Society/Chemical Society of Japan Chemical Congress, Honolulu, Hawaii, April 1-6.

ORGANIC COMPOUNDS IN URBAN AEROSOLS*

Joan M. Daisey

New York University Medical Center
New York, New York 10016

INTRODUCTION

The relationship between the burning of fossil fuel and occupational disease was first recognized in 1775 when Sir Percival Pott associated scrotal cancers in chimney sweeps with exposure to soot. The first carcinogenic polycyclic aromatic hydrocarbon (PAH) was isolated from soot about 150 years later.[1] Investigations of PAH compounds in urban aerosols are much more recent phenomena, due to both the complexity of particulate organic matter (POM) and the small amounts of material available for analysis. Benzo[a]pyrene was first reported in POM in the early 1950s.[2,3] Much of the pioneering work of Sawicki and coworkers[4] in the separation, identification, and quantitation of PAH compounds in particulate matter was done as recently as the 1960s.

Although application of newer, more sophisticated analytical techniques[4] has expanded our knowledge of the nature of the organic fraction of particulate matter, many of the questions posed by the early investigations still remain unanswered. These are the broader and more complex questions: What are the biological effects of POM? What are the specific sources of these compounds? How much is formed and how much is degraded in the atmosphere? To what extent and under what circumstances do these reactions occur? Does long range transport contribute significantly to urban aerosol concentrations of POM? Answers to these questions are of increasing importance as we try to assess the environmental and health consequences of increasing use of energy and of the projected conversion to coal and coal-derived fuels in the near future.

This paper will review, in PART 1, the basic physical, chemical, and biological properties of POM as determined by studies of the urban aerosol and of model systems. PART 2 will discuss some of the methods that have been, and continue to be, used to gain a better understanding of the primary and secondary sources of POM.

PART 1. PROPERTIES OF PARTICULATE ORGANIC MATTER

Biological Properties

A major reason for the study of POM has been the possibility that such compounds pose a long-term health hazard. Animal studies of the carcinogenicity of various fractions of POM, as well as the high incidence of cancer in certain occupations, are

*This work was supported by Grant No. RP-1222-1 of the Electric Power Research Institute and the American Petroleum Institute and is part of a center program supported by the National Institute of Environmental Health Sciences, Grant No. ES00260, and the National Cancer Institute, Grant No. CA 13343.

0077-8923/80/0338-0050 $01.75/1 © 1980, NYAS

the basis of this concern.[5] Hoffmann and Wynder[6] have reviewed a number of studies in which fractions of POM have been shown to be carcinogenic in mice. They concluded that fractions containing PAH are principally responsible for tumorigenicity, but that basic and oxidized neutral hydrocarbons also contain tumorigenic agents. Aliphatic, acidic, and oxidized hydrocarbon fractions can also act as tumor promoters.[6] The subfractions that have been tested on animals have been generally from the nonpolar (benzene or ether soluble) fraction of particulate matter.

More recent studies of the mutagenic character of POM, as determined by the Ames assay,[7] indicate the presence of biologically active compounds in polar as well as nonpolar fractions.[8–12] Some results of our investigations of the New York City aerosol are shown in FIGURE 1. Direct-acting mutagens, i.e., compounds that do not require metabolic activation for their effect as do the PAH compounds, are present in all three organic particulate fractions.

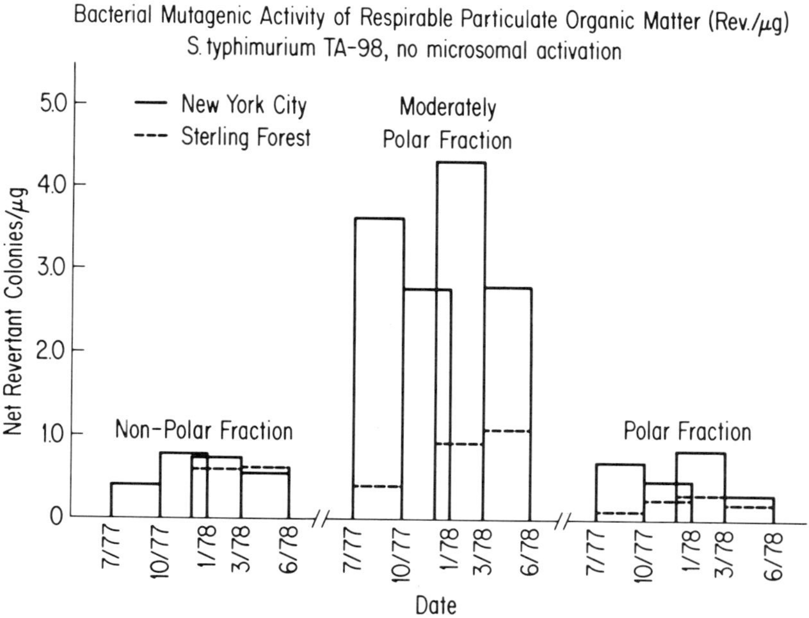

FIGURE 1. Bacterial mutagenicity of three organic fractions of respirable particulate matter collected in New York City and Sterling Forest. (The latter site is in Tuxedo, N.Y., about 70 km northwest of Manhattan.)

Physical Properties

Physical properties are factors in determining health effects and aerosol dynamics.[5] Particle size distributions have been determined for few classes of organic compounds other than the polycyclic aromatic hydrocarbons. In general, the greatest proportion, 70–90%, of these PAH compounds have been found in respirable (≦ 3.5 μm) particles,[13–17] although there are seasonal and intersite differences. Van Vaeck and Van Cauwenberge[17] very recently reported particle size distributions for several other classes of organic compounds. Most aliphatic hydrocarbons (85–90%) and carboxylic acids (90–95%) were found in particles less than 3 μm in diameter for samples collected with a cascade impactor.[17]

DeWiest[16] investigated total POM, as well as PAH, particle size distributions in Belgium. While 90% of the total mass of benzo[a]pyrene was found in particles ≦ 2 μm aerodynamic diameter, total organic matter was more uniformly distributed over particles 0.1 to 10.0 μm aerodynamic diameter, with only 55–70% of this in particles ≦2 μm. In contrast, for samples collected with a multicyclone sampler[18] in New York City in November 1977, both polar and nonpolar fractions of organic particulate matter were almost wholly (≧ 90%) in particles ≦ 3.5 μm aerodynamic diameter (unpublished data). Los Angeles investigators have found organic compounds[19] and carbon[20,21] concentrated in the submicron particles. The Los Angeles samples were generally collected during smog episodes. These data should be interpreted with some caution, however, as movement of organic compounds from upper to lower stages of the impactor samplers may occur due to the volatility of organic compounds.

Organic compounds of low molecular weight and high vapor pressure are distributed between vapor and particulate in the atmosphere. At the higher temperatures found in combustion sources, larger proportions of organic compounds will be present in the vapor phase. Natusch and coworkers[22] have shown that the polycyclic aromatic hydrocarbons emitted as vapor from power plant stacks condense on the surface of particulate matter as they cool. Surface enrichment of organic compounds in particulate matter will affect chemical reactivity in the atmosphere as well as the biological impact of these compounds.

Chemical Properties

Classes of Compounds in POM

Particulate organic matter is a complex mixture of many classes of compounds and many different individual compounds. A summary of the classes of compounds that have been identified in urban aerosols and the concentrations at which they have been reported is presented in TABLE 1. Differences in the physical and chemical properties of these classes will affect sampling and analysis. As our knowledge of these effects is still very limited, the data should be interpreted with some caution. Unstable compounds that may exist in the atmosphere could be degraded readily upon collection on a filter and thus contribute to observed levels of more stable oxidized compounds. Loss of nonpolar hydrocarbons during sampling has been well documented.[39–41] Appel and coworkers[42] have also reported evidence of positive filter artifacts.

Hydrocarbons. The hydrocarbon class includes alkanes, alkenes, aromatics, and polycyclic aromatic compounds. Such compounds are probably primary in nature, i.e., they are emitted directly into the atmosphere rather than formed by secondary reactions. The alkanes, alkenes, and some aromatics are found in what is often termed the aliphatic hydrocarbon fraction, one of the largest by mass in POM. The alkanes found in particulate matter generally range from C_{17} to C_{36}.[4] The high molecular weight alkanes are probably formed during combustion, although there are some

TABLE 1

CLASSES OF ORGANIC COMPOUNDS FOUND IN URBAN AEROSOLS

Compound Class	Example	Concentration in Urban Air* (ng/m^3)	Location, Date of Measurement(s)
Alkanes (C_{18}–C_{50})	n-$C_{22}H_{46}$	1000–4000†	217 U.S. urban stations, 1966–67 (23)‡
Alkenes	n-$C_{22}H_{44}$	2000†	217 U.S. urban stations (23)
Alkylbenzenes	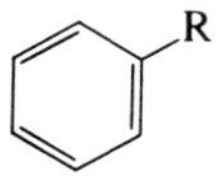	80–680†	West Covina, California July 24, 1973 (24)
Napthalenes	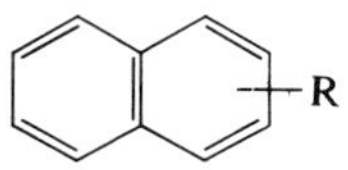	40–500†	Pasadena, California September 1972 (19)
Polycyclic aromatic hydrocarbons	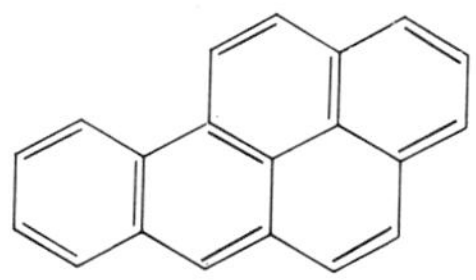	6.6	100 U.S. urban stations, 1958–59 (25)
		3.2	32 U.S. urban stations, 1966–67 (26)
		2.1	32 U.S. urban stations, 1970 (26)
Aromatic acids	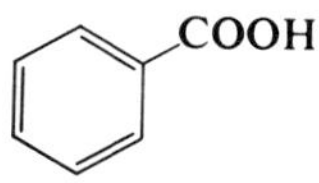	90–380	Pasadena, California September 1972 (19)
		8	Average U.S. urban prior to 1965 (28)
Cyclic ketones	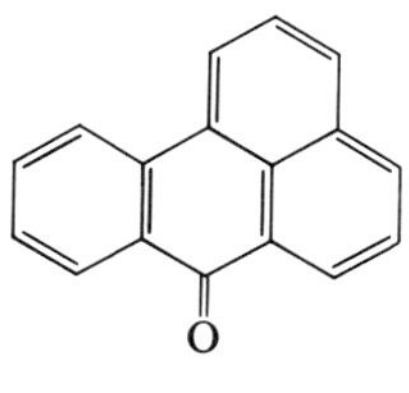	2–48	U.S. urban site January 1968 (29)
Quinones	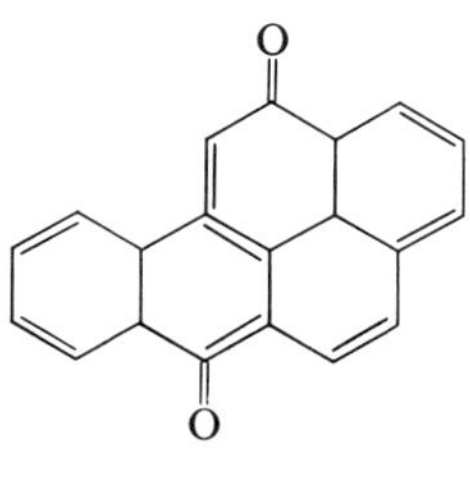	0.04–0.12§	Toronto, Ontario 1972–1973 (30)

(continued)

TABLE 1 (continued)

Compound Class	Example	Concentration in Urban Air* (ng/m^3)	Location, Date of Measurement(s)
Phenols	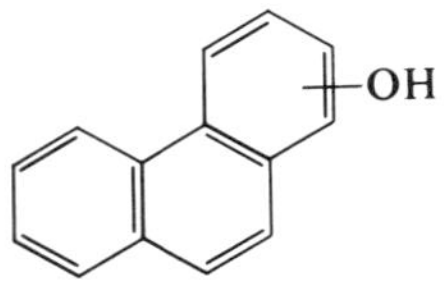	~0.3¶	Antwerp, Belgium 1975 (34)
Esters	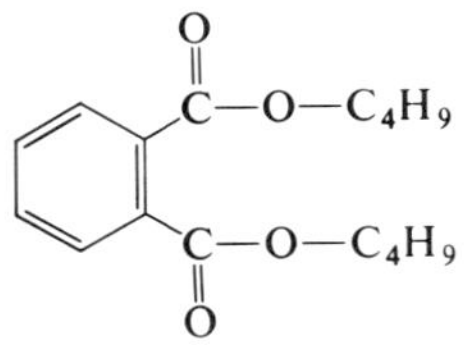	29–132	Antwerp, Belgium 1976 (27)
		2–11	New York City 1975 (31)
Aldehydes	$CHO(CH_2)_nCHO$	30–540	Pasadena, California September 1972 (19)
Aliphatic carboxylic acid	$C_{15}H_{31}COOH$	220	New York City February 1964 (6)
		36.5	Antwerp, Belgium January 1976 (27)
Aliphatic dicarboxylic acids	$HOOC(CH_2)_nCOOH$	40–1350	Pasadena, California September 1972 (19)
Aza-arenes	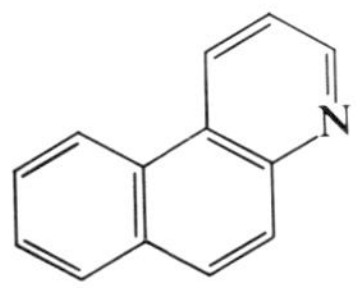	0.2	Composite 100 U.S. urban sites, 1963 (32)
		0.01	New York City 1976 (33)
		~0.5¶	Antwerp, Belgium (34)
N-Nitrosamines	$(CH_3)_2NNO$	≦0.03–0.96	Baltimore, Maryland August 1975 (35)
		15.6	New York City July 1976 (36)
Nitro compounds	$CHO(CH_2)_nCH_2ONO_2$	40–1010	Pasadena, California September 1972 (19)

TABLE 1 (continued)

Compound Class	Example	Concentration in Urban Air* (ng/m^3)	Location, Date of Measurement(s)
Nitro compounds	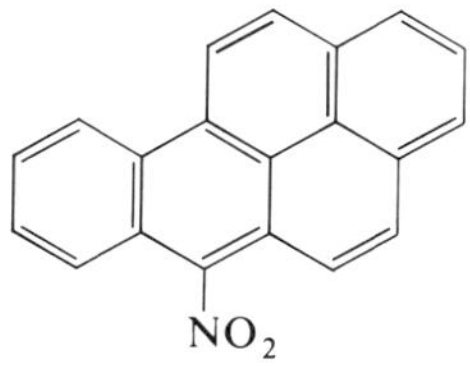	identified	Prague, Czechoslovakia (37)
Sulfur heterocyclic compounds	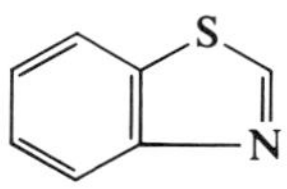	0.014–0.02	New York City 1976 (33)
	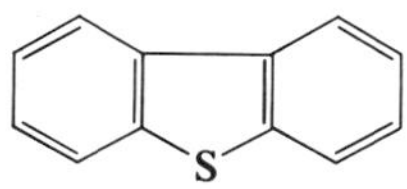	identified	Indianapolis and Gary, Indiana (38)
SO_2-adducts	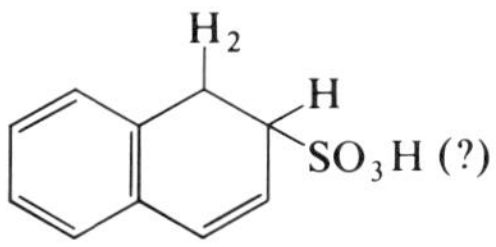	2–18 $nmoles/m^3$‖	New York City 1976 (60)
Alkylhalides	$C_{18}H_{37}Cl$	~20–320	Pasadena, California September 1972 (19)
Arylhalides	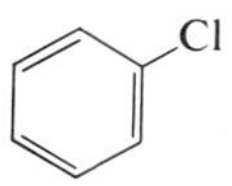	0.5–3	Pasadena, California 1972 (19)
Chlorophenols	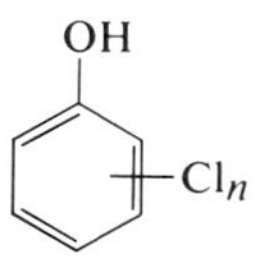	5.7–7.8	Antwerp, Belgium 1976 (27)

*Concentration of compound given as example.
†Total concentration for class of compounds.
‡References are given in parentheses.
§All isomers of given compound.
¶Estimated from data given in reference.
‖Speculative.

experimental data that indicate that they can be formed in the atmosphere from lower molecular weight alkanes.[43]

Polycyclic Aromatic Hydrocarbons. Polycyclic aromatic hydrocarbons are the most extensively studied compounds found in the ambient aerosol. Although they comprise only a small fraction by mass of urban particulate matter, some of the individual compounds are potent carcinogens. The alkylated and hydro derivatives are generally found at less than one-tenth of the concentrations of the parent compounds.[34,44] Seasonal differences in ambient levels of PAH have been well documented[5] and probably arise from seasonal differences in emission sources, photochemical degradation, and sample collection temperature.

Oxidized Hydrocarbons. The term oxidized hydrocarbon refers broadly to any compound containing carbon, hydrogen, and oxygen. It includes a wide variety of compound classes that have been detected in the aerosol (cf., TABLE 1), such as acids, aldehydes, ketones, quinones, phenols, and esters. Less stable oxidized hydrocarbons, such as epoxides and peroxides, may also be present in particulate matter.[45] Oxidized hydrocarbons may be emitted directly from combustion sources or may be produced by oxidation reactions occurring in the atmosphere. Photochemical production of oxidized hydrocarbons in the aerosol has been the subject of considerable investigation in southern California.[19,21,24] The extent to which such reactions occur and the factors affecting such conversions are not well understood. A better understanding of the primary and secondary sources of oxidized hydrocarbons is needed to develop source allocation models and to estimate the contribution of transport to the organic particulate matter in an urban area.

Organonitrogen Compounds. Aza-arenes are the only nitrogenous organic compounds in POM that have been investigated to any great extent. Urban levels of aza-arenes are generally 1–2 orders of magnitude lower than those of the polycyclic aromatic hydrocarbons.[4,33]

Nitro derivatives of organic compounds have been found in the Los Angeles aerosol during smog episodes, suggesting a photochemical origin.[19] In addition, Jäger[37] recently detected 6-nitrobenzo[a]pyrene in particulate collected in Prague. Pitts and coworkers[10] have postulated the existence of this compound in the aerosol as a result of reactions between benzo[a]pyrene and NO_2. They[10] found the compound to be highly mutagenic in the Ames assay. Whether such compounds in the aerosol present a human health hazard has not been determined.

N-Nitrosamines of low molecular weight have been detected in the ambient atmosphere.[35,36] As these compounds are potent carcinogens in animals, much research is under way to determine their occurrence, levels, and possible sources. Chamber studies[46] have suggested that these compounds, as well as nitramines, may be produced by both dark and photochemical reactions. Current work in our laboratories indicates that higher molecular weight *N*-nitrosamines may be present in POM.[47]

Organosulfur Compounds. Heterocyclic sulfur compounds have only recently been reported as present in POM.[33,38] Dong and coworkers[33] reported finding benzothiazole in the New York City aerosol at levels similar to those found for nitrogen heterocyclic compounds, i.e., 0.01–0.02 ng/m^3. Several other sulfur heterocyclics have been identified in samples collected in Indianapolis and Gary, Indiana.[38]

Reactions of POM

With the exception of PAH compounds, reactions of particulate organic matter have received little attention. Polycyclic aromatic hydrocarbons, in general, undergo electrophilic substitutions, additions, and oxidations. Atmospheric species, such as O_2, O_3, OH, HNO_3, HONO, H_2SO_4, etc., could be involved in such reactions. In the atmosphere, photochemical oxidation is probably the most important type of reaction.

Photooxidation of adsorbed PAH has been investigated in a number of model systems.[48–57] The data, in general, indicate that benzo[a]pyrene, benz[a]anthracene, perylene, and pyrene are among the most reactive PAH compounds, while triphenylene, chrysene, and coronene are relatively unreactive. Some photochemical reactions of adsorbed PAH in model systems are presented in FIGURE 2. The products

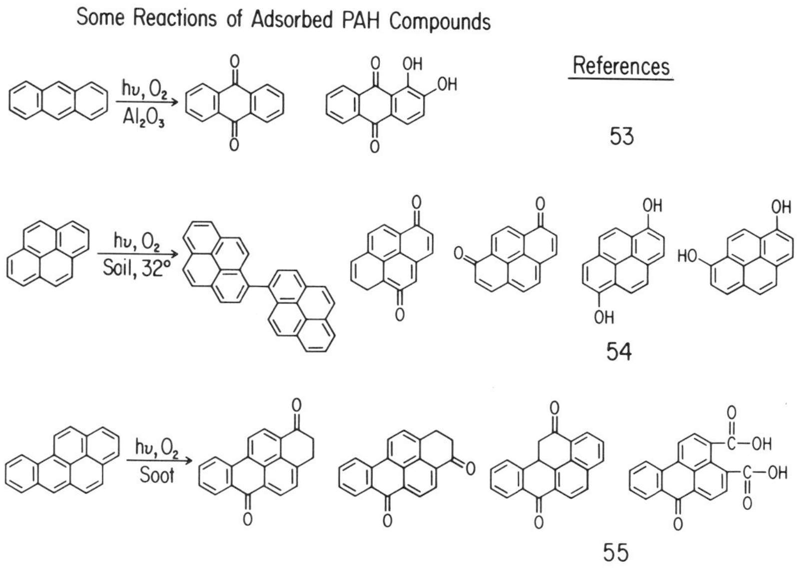

FIGURE 2. Some reactions of adsorbed PAH compounds.

formed are quinones, acids, and phenols. Studies of the oxidation of PAH compounds in solution suggest that peroxides, ozonides, and epoxides may be formed as intermediates.[45,58] However, the presence of these compounds in the aerosol has not been demonstrated as yet. Tebbens and coworkers,[52] in their studies of PAH adsorbed on soot, noted that not all of the major products formed had been identified and quantified.

There have been few measurements of the rates of photodegradation of PAH compounds.[48,50,52,55,56] A summary of the available data for benzo[a]pyrene is presented in TABLE 2. For purposes of comparison, I have estimated the half-life from the percent degradation reported for a single stated time, assuming first-order kinetics. Katz and coworkers[56] reported observing first-order kinetics for degradation of PAH adsorbed on cellulose. As can be seen from TABLE 2, there is a wide variation in the estimated half-lives (and rates of degradation). Differences in substrate, light intensity, and reaction conditions contribute to this variation. Addition of ozone or smog to the system dramatically enhances the rate of degradation. These studies suggest that benzo[a]pyrene is relatively short-lived in the atmosphere.

However, Korfmacher and coworkers[57] recently reported that photochemical decomposition of some PAH compounds (including benzo[a]pyrene) proceeds much more slowly when the compounds are adsorbed on coal fly ash than when adsorbed on other substrates, such as silica gel, or than when in solution. Maximum degradation observed for benzo[a]pyrene was about 20%, even after 100 hours of irradiation. No dependence on irradiation time was observed. The implications of such stability with

TABLE 2

RATES OF DEGRADATION OF BENZO[a]PYRENE IN MODEL SYSTEMS

System	Reactants	Substrate	Light Intensity	Estimated Half-life* (hr)	Reference
Static System	$h\nu$, air	soot	†	316	48
	smog	soot	†	3.5	
Flow Reactor	$h\nu$, air	soot	~30 watts/m^2	0.6	50–52
Carbon Needle	$h\nu$, air	carbon	2 watts/m^2	16.4	55
Semistatic System	$h\nu$, air	cellulose	500 watt lamp,	5.3	56
	$h\nu$, 0.2 ppm O_3, air	cellulose	46 cm above	0.6	
	0.2 ppm O_3, air	cellulose		0.6	

*Estimated by the author from percent degradation at time stated assuming first-order kinetics.

†Not given.

respect to human health are of concern in view of the anticipated increase in coal utilization in this country.

Korfmacher *et al.*[57] also noted very rapid oxidation of certain PAH that contain benzylic carbons, in the dark when adsorbed on fly ash. These results are of particular interest in view of the general absence of such PAH compounds in ambient particulate matter.

There is only one report of ambient measurements that relates to photooxidation of PAH compounds. Pierce and Katz[30] determined ratios of some quinone derivatives to their PAH precursors in the ambient aerosol over a period of a year. Ratios increased to a maximum during the summer months but were generally low, ranging from about 0.04 to 0.09. As benzene was used to extract the quinones,[30] and as the data of Stanley and coworkers[59] suggest that this solvent may be only about 50% efficient in extracting such compounds, the actual ratios may be higher by a factor of about two. Assuming that the quinones are the principal degradation products of PAH (which is by no means known) and that the quinones are not formed by thermal reactions during combustion, the data suggest a 10–20% degradation in the atmosphere. This is consistent with the model studies of Korfmacher *et al.*[57] Clearly, much further work is needed to characterize the products of photooxidation of hydrocarbons, to determine whether these products are present in source emissions, and to determine their rates of reaction.

Polycyclic aromatic hydrocarbons may also undergo substitution reactions with species such as NO_2, HNO_3, H_2SO_4, SO_2, etc., in the atmosphere. As previously noted, the 6-nitro derivative of benzo[a]pyrene was recently identified in ambient particulate matter.[37] It is not known at present whether this species is primary or secondary in nature or possibly even a filter artifact. Sulfonic acid derivatives have not been identified to date, although Izatt and coworkers[60] have reported the presence of organosulfur compound(s), which can be hydrolyzed to yield sulfite, in samples collected at New York University.

PART 2. DETERMINING SOURCES OF PARTICULATE ORGANIC COMPOUNDS

The ultimate source of most particulate organic matter in the urban aerosol appears to be fossil fuel. However, our knowledge of the contributions of various types of urban sources with respect to classes of compounds and amounts of POM is very limited. As with inorganic compounds, estimates of source contributions have been based largely on emission inventories,[5] which have focused on polycyclic aromatic hydrocarbons. Because of the limitations and/or lack of emission inventories, various other methods have been used to identify sources of POM and to estimate the contributions of these sources.

Long-Term Trends/Interurban Differences

Studies of long-term trends in ambient concentrations of POM can provide indications of sources and their relative significance. As coal consumption for space heating declined in the United States, average annual total suspended particulate matter (TSP) declined in urban areas.[61] Faoro[26] recently analyzed organic data for 32 urban and 19 nonurban National Air Surveillance Network (NASN) stations. On the average, benzene-soluble organics (BSO) had declined by 55% for the urban stations, while the nonurban stations showed no significant change between 1960 and 1970.[26] Benzo[a]pyrene showed similar trends. Faoro[26] concluded that the reduction in residential coal consumption during this period was probably the single greatest factor leading to the decline.

This overall trend, however, does not necessarily hold for all cities. While declines in BSO appear to have paralleled those in TSP in Los Angeles in more recent years, particulate organic matter in New York City has become proportionally more important, as shown in FIGURE 3. Annual averages for the New York University station, which have been dispersion-normalized to take into account year-to-year meteorological variations in dispersion,[65] are reported in TABLE 3. The dispersion-normalized concentration is that which would have been observed if the dispersion factor (mixing height × wind speed) had remained constant at 4200 m^2/sec, and is thus more directly proportional to emitted masses.

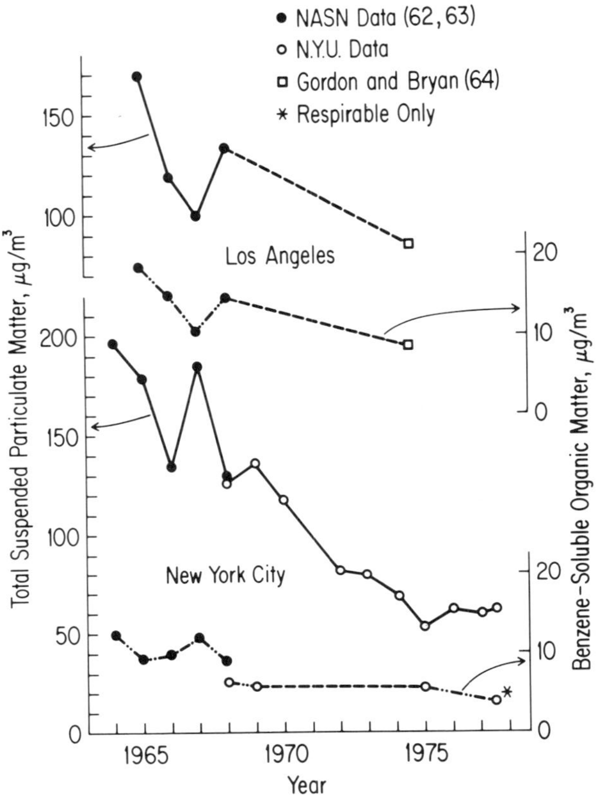

FIGURE 3. Comparison of trends in TSP and BSO in New York City and Los Angeles.

Although TSP in New York City declined by about 40% between 1968 and 1978, the decline in POM has been only about 16%. The decline in TSP in New York City is due principally to the burning of cleaner, low-ash, low-sulfur oil.[66,67] As can be seen from TABLE 3, the use of low-sulfur fuels has not led to a proportional reduction in POM. In contrast, in Los Angeles, regulations over the past ten years have been directed toward control of hydrocarbon emissions from automobiles. As might be expected, this appears to have resulted in parallel decreases in TSP and organics.

Seasonal Variations

Seasonal differences in sources of organic compounds should be reflected in the composition of the aerosol. Estimates of the contribution of space heating, for example, can be made from winter–summer differences. Seasonal differences in dispersion can be taken into account by dispersion-factor normalization.[65] Dispersion-normalized data for organic fractions of New York City particulate matter for 1969 and 1977 are shown in TABLE 4.

TABLE 3

ANNUAL AVERAGES OF ORGANIC FRACTIONS IN TOTAL SUSPENDED PARTICULATE MATTER, NEW YORK CITY,* DISPERSION-NORMALIZED

Year	TSP ($\mu g/m^3$)	Organic Fraction† ($\mu g/m^3$)	Percent Organics in TSP
1968	95.7	10.2	10.6
1969	128	10.8	8.4
1977–1978	59.8	8.8‡	14.7

*NYU Medical Center Station.
†Total of nonpolar (benzene-soluble) and polar (acetone-soluble) organics.
‡Respirable (≤3.5μ) organics only.

There are several points of interest in these data. The summer–winter differences should be largely due to space heating. Reduced levels of human activities in New York City during the summer months may also contribute to reduced emissions from automobiles, incinerators, and possibly small industrial sources. Based on an estimated 15% reduction in auto traffic during the summer months,[67] this contribution to the winter–summer difference should be of the order of approximately 1 $\mu g/m^3$, suggesting 8–11 $\mu g/m^3$ (normalized) from space heating. Seasonal differences in temperature and vapor pressures may also lead to greater condensation of organic materials and thus contribute significantly to this 8–11 $\mu g/m^3$ seasonal difference, as the average temperature difference for these two seasons is about 20°C.

Kleinman[67] estimated an annual average TSP contribution from fuel oil burning of about 30 $\mu g/m^3$ (20%) for 1969 and less than 10 $\mu g/m^3$ (~10%) after 1972, using a multiple regression model of source apportionment. The above estimate of the POM contribution from this source is consistent with Kleinman's estimates.

Comparing 1969 to 1977, winter emissions appear to have changed little since

TABLE 4

SUMMER–WINTER COMPARISONS OF ORGANIC PARTICULATE MATTER IN NEW YORK CITY*

		Dispersion-Normalized Concentrations ($\mu g/m^3$)		
Year	Fraction	Summer†	Winter†	Difference
1969‡	Nonpolar	3.6	8.3	4.7
	Polar	3.4	7.6	4.2
	Total	7.0	15.9	8.9
1977§	Nonpolar	1.4	6.7	5.3
	Polar	1.5	8.1	6.6
	Total	2.9	14.8	11.8

*Weekly samples.
†July–August 1969 and 1977; January–February 1969; December 1977; January–February 1978.
‡TSP samples.
§Respirable suspended particulate matter samples.

1969, although summer emissions have been reduced by about 60%. This latter probably reflects the more stringent emission controls and regulations.

Diurnal Patterns

Diurnal variations in POM have been used to gain insight into the sources of several classes of organics in Los Angeles during smog episodes. Cronn and coworkers[24] found midmorning and midafternoon maxima in ambient levels of nonpolar materials, such as alkanes, in samples collected in West Covina in July 1973. In contrast, carboxylic acids, phenols, amides, and alkylnitrites (determined by mass spectrometric thermal analyses) peaked between 1400 and 1600 PST, coincident with the ozone maximum, suggesting a photochemical origin. Similar results were obtained in another experiment by Schuetzle *et al.*[19]

In New York City, we have compared 12-hour daylight TSP samples with samples collected from 6 P.M. to 6 A.M. Three organic fractions—nonpolar, moderately polar, and polar—were examined. Only the nonpolar fraction showed a significant difference for night versus day. As discussed in PART 1, nonpolar organics are generally primary in origin. The higher daylight levels of nonpolar organics observed in New York City during this period support the hypothesis of a primary origin for these materials. These results are also suggestive of a local origin for nonpolar POM, rather than transport.

Mathematical Models

Because of the limitations of emission inventories, a number of investigators have begun to develop mathematical and statistical models for identifying sources of particulate matter and for estimating the contributions of these sources.

Preliminary investigations in our laboratory[68,69] indicate that it may be possible to develop such models for particulate organic matter, with appropriate modifications. The methods for source identification and allocation of polycyclic aromatic hydrocarbons that were examined recently[68] should be applicable to organic fractions of particulate matter and other classes of organic compounds. These methods fall broadly into four categories.

Ratio methods involve the use of measured emission source ratios of various organic compounds to tracer elements, in combination with ambient aerosol measurements of tracer elements. The contribution of a particular type of source to the aerosol concentration of an organic compound can be estimated if a given tracer is from a single type of source; for example, if all lead in an urban area is derived from automobiles. This method, applied recently to the New York City aerosol for a sample period in August 1976, suggested that 19–28% of the benzo[a]pyrene was automotive in origin.[68]

The chemical element balance method, developed by Friedlander and coworkers,[70,71] estimates primary source contributions to the aerosol by linearly relating source contributions of various tracer elements to ambient measurements of the elements. This method, when applied for source allocation of PAH in New York City, indicated a 64% contribution of benzo[a]pyrene from automotive sources,[68] as compared to the 19–28% estimated using the ratio method. Although this method and the ratio method have the advantage of requiring relatively little ambient data, the present lack of representative and extensive organic compound source emission data limits the applicability of these models for organic compounds.

Statistical models generally begin with a large data base of ambient aerosol measurements of TSP, tracer elements, meteorological data, and other variables. Pairwise correlations and factor analysis are used to identify sources and to select independent source tracers.[65,67] This approach has been further developed in our laboratory by Kleinman,[67] and multiple regression analysis has been used to obtain coefficients relating TSP to the source tracers. The coefficients obtained can be used to apportion TSP and/or trace elements among sources.

TABLE 5

CORRELATIONS OF VARIOUS PARTICULATE ORGANIC FRACTIONS WITH OTHER AEROSOL VARIABLES IN NEW YORK CITY*

Organic Fraction	Correlated Variable	Sampling Period: August 1976	Sampling Period: February 1977	Sources
Nonpolar	Fe, Mn	†	†	Soil, autos
Moderately polar	Pb	‡		Automobiles
	O_3	‡		Photochemistry
	Traffic	‡	§	Automobiles
	Zn	‡		Incinerators, industry
	V	†	‡	Oil burning
Polar	Pb	‡	‡	Automobiles
	Zn	‡		Incinerators, industry
	Traffic	‡	§	Automobiles
	Sulfate	†	‡	Oil burning, transport
Polycyclic aromatic hydrocarbons	Pb	‡	§	Automobiles
	Traffic	‡	§	Automobiles
	Zn	†	§	Incinerators, industry

*24-hr TSP samples.
†Correlation significant at $p < 0.05$ level.
‡Correlation significant at $p < 0.01$ level.
§Data unavailable.

We have begun to develop this type of model for particulate organic matter in New York City. The first condition that must be met for the development of such a model is that correlations be found between source tracers for a given urban area and the organic compounds or fractions of interest.

In 1963, Colucci and Begeman[72] investigated correlations of PAH compounds with various tracers in New York City. Benzo[a]pyrene and benz[a]anthracene were significantly correlated with tracers for automobiles (Pb, CO) and oil burning (V). Gordon and Bryan[64] have observed a correlation of traffic density with ambient levels of coronene in several areas in the Los Angeles basin. A summary of the correlations that have been observed recently in New York City is presented in TABLE 5. Different fractions correlate with tracers for different sources and also show seasonal differences. Thus, it should be possible to obtain equations relating the organic fractions to the tracers and to use the coefficients to estimate the contributions of various sources to those fractions or classes.

All of these results suggest that it is possible to develop source allocation models for organic compounds, although more complex models, possibly involving pattern recognition, will probably be required than those for inorganic compounds. Hoffmann and Wynder[6] have suggested using the overall pattern of alkanes in particulate to define major sources, as alkanes are chemically unreactive. Such pattern recognition has been used to advantage to identify sources of petroleum in water.[73] If different sources emit sufficiently different proportions of alkanes to allow pattern recognition, it may be possible to develop quantitative estimates of the contributions of each type of source, using methods similar to those developed by Kleinman *et al.*[67,74]

As we develop source identification and allocation models, there undoubtedly will be discrepancies between models. From such discrepancies, however, we will learn to improve the models. More important, we may also gain valuable insight into the nature of the aerosol.

TABLE 6

RATIOS OF BENZO[a]PYRENE TO BENZO[ghi]PERYLENE FOR NEW YORK CITY (AUGUST 1976)

Date	BaP/BghiP	Date	BaP/BghiP
8/10	1.38	8/17	1.24
8/11	9.00	8/23	5.77
8/12	8.80	8/24	2.64
8/13	1.85	8/27	0.49
8/14	1.03	8/28	2.39
8/16	1.08	8/29	0.42

Transport

Long range transport may be a significant source of organic particulate matter under certain meteorological conditions. Evidence for the transport of PAH compounds has been reported by several European investigators.[16,75] Transport of sulfate and ozone has been demonstrated for the northeastern section of the United States.[76,77]

Sawicki[78] and Hangebrauck and coworkers[79] suggested that ratios of certain PAH compounds might be indicative of sources. Recent work in our laboratory[68] indicates that changes in these ratios in a given city may be of use in identification of sources. Sudden increases in the ratio of benzo[a]pyrene to benzo[ghi]perylene observed in New York City in August 1976, as shown in TABLE 6, appeared to be highly correlated with aerosol transport episodes.[68] This suggests that PAH compounds may be transported from the coal-burning Midwest to the East Coast under certain meteorological conditions.

Natural sources of organic compounds may also contribute to urban aerosol burdens via transport. Photochemical reactions of volatile terpenes emitted by vegetation[80,81] have been shown to result in the formation of POM.[82] Under appropriate meteorological conditions, such aerosols might be transported into urban areas. The extent to which this may occur is not known at present.

Summary and Conclusions

Over the past 25 years, a great deal has been learned of the biological, physical, and chemical properties of POM. Organic fractions of urban particulate matter have been found to be tumorigenic in mice and mutagenic in certain strains of bacteria, indicating that such materials may be significant in the etiology of environmental diseases.

With respect to the physical properties of POM, differences in particle size distributions between classes, between cities, and between seasons are becoming apparent. Such differences are of significance to human health and to the dynamic properties of aerosols.

Hundreds of organic compounds have been identified in the urban aerosol. As sampling and analytical methods become more sophisticated, we will probably identify hundreds more and begin to see more clearly the significance of composition differences between sources, between sampling locations, and between seasons. Such differences can contribute to our understanding of both sources and chemical reactions of POM.

Organic and inorganic, vapor and particulate phase chemical reactions are not independent of each other. This interdependence is probably one of the most difficult aspects of aerosol science. Future studies must attempt to better integrate our knowledge of chemical reactions and must seek to understand the multiple factors that influence rates of reactions.

Changes in the sources of heat, light, and power have occurred repeatedly over the course of history. Conversion to coal and coal-derived fuels is anticipated in the United States in the near future. The development of methods and models for identifying sources, determining the magnitudes of source contributions, predicting products and rates of reactions, and describing transport of aerosol components is of increasing importance as we try to assess the impacts of such changes.

Acknowledgments

The comments and suggestions of Drs. T. J. Kneip, Beverly S. Cohen, M. Eisenbud, P. Lioy, and A. Agarwal during the preparation of this manuscript are gratefully acknowledged.

References

1. Kennaway, E. L. & I. Hieger. 1930. Carcinogenic substances and their fluorescence spectra. Br. Med. J. No. 3622: 1044.
2. Waller, R. E. 1952. The benzpyrene content of town air. Br. J. Cancer **6:** 8.

3. KOTIN, P., H. L. FALK, P. MADER & M. THOMAS. 1954. Aromatic hydrocarbons. I. Presence in the Los Angeles atmosphere and the carcinogenicity of atmospheric extracts. Arch. Ind. Hyg. Occup. Med. **9:** 153.
4. SAWICKI, E. 1976. Analysis of atmospheric carcinogens and their co-factors. *In* Environmental Pollution and Carcinogenic Risks. INSERM Symposia Series **52:** 297.
5. Committee on Biologic Effects of Atmospheric Pollutants. 1972. Particulate Polycylic Organic Matter. Printing and Publishing Office, National Academy of Science, Washington, D.C.
6. HOFFMANN, D. & E. L. WYNDER. 1977. Organic particulate pollutants—chemical analyses and bioassays for carcinogenicity. *In* Air Pollution. A. C. Stern, Ed. 3rd edit. **2:** 361–455. Academic Press, Inc. New York, N.Y.
7. AMES, B. N., J. MCCANN & E. YAMASAKI. 1975. Methods for detecting carcinogens and mutagens with the Salmonella/mammalian microsome mutagenicity test. Mutat. Res. **31:** 347.
8. DAISEY, J. M., I. HAWRYLUK, T. J. KNEIP & F. MUKAI. 1978. Mutagenic activity in organic fractions of airborne particulate matter. Paper presented at the Conference on Carbonaceous Particles, Berkeley, Calif., March 20–22.
9. TALCOTT, R. & E. WEI. 1977. Brief communication: airborne mutagens bioassayed in *Salmonella typhimurium*. J. Nat. Cancer Inst. **58:** 449.
10. PITTS, J. N., JR., D. GROSJEAN, & T. M. MISCHKE. 1977. Mutagenic activity of airborne particulate organic pollutants. Toxicol. Lett. **1:** 65.
11. TERANISHI, K., K. HAMADA & H. WATANABE. 1978. Mutagenicity in Salmonella typhimurium mutants of the benzene-soluble organic matter derived from airborne particulate matter and its five fractions. Mutat. Res. **56:** 273.
12. COMMONER, B., P. MADYASTHA, A. BRONSDON & A. J. VITHAYATHIL. 1978. Environmental mutagens in urban air particulates. J. Toxicol. Environ. Health **4:** 59.
13. DEMAIO, L. & M. CORN. 1966. Polynuclear aromatic hydrocarbons associated with particulates in Pittsburgh air. J. Air Pollut. Control Assoc. **16:** 67.
14. KERTESZ-SARINGER, M., E. MESZAROS & T. VARKONYI. 1971. On the size distribution of benzo[a]pyrene containing particles in urban air. Atmos. Environ. **5:** 429.
15. PIERCE, R. C. & M. KATZ. 1975. Dependency of polynuclear aromatic hydrocarbon content on size distribution of atmospheric aerosols. Environ. Sci. Technol. **9:** 347.
16. DEWIEST, F. 1978. Any factors influencing the dispersion and the transport of heavy hydrocarbons associated with airborne particles. Atmos. Environ. **12:** 1705.
17. VAN VAECK, L. & K. VAN CAUWENBERGHE. 1978. Cascade impactor measurements of the size distribution of the major classes of organic pollutants in atmospheric particulate matter. Atmos. Environ. **12:** 2229.
18. BERNSTEIN, D. M. 1977. The Influence of Trace Metals in Disperse Aerosols on the Human Body Burden of Trace Metals. Doctoral dissertation. New York University. New York, N.Y.
19. SCHUETZLE, D., D. CRONN, A. L. CRITTENDEN & R. J. CHARLSON. 1975. Molecular composition of secondary aerosol and its possible origin. Environ. Sci. Technol. **9:** 838.
20. MUELLER, P. K., R. W. MOSLEY & L. B. PIERCE. 1972. Chemical composition of Pasadena aerosol by particle size and time of day. IV. Carbonate and noncarbonate carbon content. J. Colloid Interface Sci. **39:** 235.
21. HIDY, G. *et al.* 1975. Characterization of Aerosols in California. Report SC524.25 FR, Vol. 4. California Air Resources Board, Contract 358, Rockwell Science Center. Thousand Oaks, Calif.
22. NATUSCH, D. F. S. 1976. Characterization of atmospheric pollutants from power plants. *In* Proceedings of the Second Federal Conference on the Great Lakes, 1975: 114. Great Lakes Basin Commission. Ann Arbor, Mich.
23. MCMULLEN, T. B., R. B. FAORO & G. B. MORGAN. 1970. Profile of pollutant fractions in non-urban suspended particulate matter. J. Air Pollut. Control Assoc. **20:** 369.
24. CRONN, D. R., R. J. CHARLSON, R. L. KNIGHTS, A. L. CRITTENDEN & B. R. APPEL. 1972. Survey of the molecular nature of primary and secondary components of particles in urban air by high-resolution mass spectrometry. Atmos. Environ. **11:** 929.
25. SAWICKI, E., W. C. ELBERT, T. R. HAUSER, F. T. FOX & T. W. STANLEY. 1960.

Benzo[a]pyrene content of the air of American communities. Am. Ind. Hyg. Assoc. J. **21:** 443.

26. FAORO, R. B. 1975. Trends in concentrations of benzene soluble suspended particulate fraction and benzo[a]pyrene. J. Air Pollut. Control Assoc. **25:** 638.
27. CAUTREELS, W., K. VAN CAUWENBERGHE & L. A. GUZMAN. 1977. Comparison between the organic fraction of suspended matter at a background and an urban station. Sci. Total Environ. **8:** 79.
28. SAWICKI, E. 1967. Airborne carcinogens and allied compounds. Arch. Environ. Health **14:** 46.
29. STANLEY, T. W., M. J. MORGAN & J. E. MEEKER. 1969. Rapid estimation of 7H-Benz(de)-anthracen-7-one and Phenalen-1-one in organic extracts of airborne particulates from 3-hour sequential air samples. Environ. Sci. Technol. **3:** 1198.
30. PIERCE, R. C. & M. KATZ. 1976. Chromatographic isolation and spectral analysis of polycyclic quinones. Application to air pollution analysis. Environ. Sci. Technol. **10:** 45.
31. BOVE, J. L., P. DALVEN & V. P. KUKREJA. 1978. Airborne di-butyl and di-(2-ethylhexyl)-phthalate at three New York City air sampling stations. Int. J. Environ. Anal. Chem. **5:** 189.
32. SAWICKI, E., S. P. MCPHERSON, T. W. STANLEY, J. MEEKER & W. C. ELBERT. 1965. Quantitative composition of the urban atmosphere in terms of polynuclear aza heterocyclic compounds and aliphatic and polynuclear aromatic hydrocarbons. Int. J. Air Water Pollut. **9:** 515.
33. DONG, M. W., D. C. LOCKE & D. HOFFMANN. 1977. Characterization of aza-arenes in basic organic portion of suspended particulate matter. Environ. Sci. Technol. **11:** 612.
34. CAUTREELS, W. & K. VAN CAUWENBERGHE. 1976. Determination of organic compounds in airborne particulate matter by gas chromatography–mass spectrometry. Atmos. Environ. **10:** 447.
35. FINE, D. H., D. P. ROUNBEHLER, N. M. BELCHER & S. S. EPSTEIN. 1976. *N*-Nitroso compounds: detection in ambient air. Science **192:** 1328.
36. FINE, D. H. 1976. Final Report of Monitoring for *N*-Nitroso Compounds in the States of West Virginia, New York, New Jersey and Massachusetts. Final Report on EPA Contract No. 68-02-2363. Prepared for U.S. Environmental Protection Agency, Research Triangle Park, N.C.
37. JÄGER, J. 1978. Detection and characterization of nitro derivatives of some polycyclic aromatic hydrocarbons by fluorescence quenching after thin-layer chromatography: application to air pollution analysis. J. Chromatogr. **152:** 575.
38. LEE, M. L., M. NOVOTNY & K. D. BARTLE. 1976. Gas chromatography/mass spectrometric and nuclear magnetic resonance determination of polynuclear aromatic hydrocarbons in airborne particulates. Anal. Chem. **48:** 1566.
39. KRSTVLOVIC, A. M., D. M. ROSIE & P. R. BROWN. 1977. Distribution of some atmospheric polynuclear aromatic hydrocarbons. Am. Lab.: 11.
40. DEWIEST, F. & D. RONDIA. 1976. Sur la validite des determinations du benzo[a]pyrene atmospheric pendant les mois d'ete. Atmos. Environ. **10:** 487.
41. CAUTREELS, W. & K. VAN CAUWENBERGHE. 1978. Experiments on the distribution of organic pollutants between airborne particulate matter and the corresponding gas phase. Atmos. Environ. **12:** 1133.
42. APPEL, B. R., E. M. HOFFER, M. HAIK, S. M. WALL, E. L. KOTHNY, R. L. KNIGHTS & J. J. WESOLOWSKI. 1977. Characterization of organic particulate matter. Final Report, Contract No. ARB5-682. Prepared for California Air Resources Board, Research Section, Sacramento, Calif.
43. HAUSER, T. R. & J. N. PATTISON. 1972. Analysis of aliphatic fraction of air particulate matter. Environ. Sci. Technol. **6:** 549.
44. LAO, R. C., R. S. THOMAS, H. OJA & L. DUBOIS. 1973. Application of a gas chromatograph–mass spectrometer–data processor combination to the analysis of the polycyclic aromatic hydrocarbon content of airborne pollutants. Anal. Chem. **45:** 908.
45. VAN DUUREN, B. L. 1972. Epoxides, hydroperoxides and peroxides in air pollution. Int. J. Environ. Anal. Chem. **1:** 233.

46. Pitts, J. N., Jr., D. Grosjean, K. Van Cauwenberghe, J. P. Schmid & D. R. Fitz. 1978. Photooxidation of aliphatic amines under simulated atmospheric conditions: formation of nitrosamines, nitramines, amides, and photochemical oxidant. Environ. Sci. Technol. **12:** 946.
47. Hershman, R. Unpublished data.
48. Falk, H. L., I. Markul & P. Kotin. 1956. Aromatic hydrocarbons. IV. Their fate following emission into the atmosphere and experimental exposure to washed air and synthetic smog. AMA. Arch. Ind. Health **13:** 13.
49. Inscoe, M. N. 1964. Photochemical changes in thin-layer chromatograms of polycyclic aromatic hydrocarbons. Anal. Chem. **36:** 2505.
50. Tebbens, B. D., J. F. Thomas & M. Mukai. 1966. Fate of arenes incorporated with airborne soot. Am. Ind. Hyg. Assoc. J. **27:** 415.
51. Thomas, J. F., M. Mukai & B. D. Tebbens. 1968. Fate of airborne benzo[a]pyrene. Environ. Sci. Technol. **2:** 33.
52. Tebbens, B. D., M. Mukai & J. F. Thomas. 1971. Fate of arene incorporated with airborne soot: effect of irradiation. Am. Ind. Hyg. Assoc. J. **32:** 365.
53. Voyatzakis, E., D. Jannakoudakis, T. Dorfmuller & C. Sipitanos. 1960. Photochemical conversion of anthracene into alizarin. C. R. Acad. Sci. Ser. C **250:** 112.
54. Fatiadi, A. J. 1967. Effect of temperature and ultraviolet radiation on pyrene adsorbed on garden soil. Environ. Sci. Technol. **1:** 570.
55. Barofsky, D. F. & E. J. Baum. 1976. Exploratory field desorption mass analysis of the photoconversion of adsorbed polycyclic aromatic hydrocarbons. J. Am. Chem. Soc. **98:** 8286.
56. Katz, M., C. Chan, H. Tosine & T. Sakuma. 1978. Relationship between relative rates of photochemical and biological oxidation of polynuclear aromatic hydrocarbons and their carcinogenic potential. Paper presented at the Third International Symposium on Polynuclear Aromatic Hydrocarbons, Batelle-Columbus Laboratories, Columbus, Ohio, October 25–27.
57. Korfmacher, W. A., D. F. S. Natusch & E. Wehyr. 1978. Thermal and photochemical decomposition of particulate PAH. Paper presented at the Third International Symposium on Polynuclear Aromatic Hydrocarbons, Batelle-Columbus Laboratories, Columbus, Ohio, October 25–27.
58. Tipson, R. S. 1964. Review of Oxidation of Polycyclic Aromatic Hydrocarbons. National Bureau of Standards Report 8363. U.S. Government Printing Office. Washington, D.C.
59. Stanley, T. W., J. E. Meeker & M. J. Morgan. 1967. Extraction of organics from airborne particulates. Effects of various solvents and conditions on the recovery of benzo[a]pyrene, benz[c]acridine and 7H-benz[de]anthracen-7-one. Environ. Sci. Technol. **1:** 927.
60. Izatt, R. M., J. Eatough, M. L. Lee, T. Major, B. E. Richter, L. D. Hanson, R. G. Meisenheimer & J. W. Fisher. 1978. The formation of inorganic and organic S(IV) species in aerosols. *In* Proceedings, 4th Joint Conference on Sensing of Environmental Pollutants, New Orleans, La., November 6–11, 1977: 821–824. American Chemical Society. Washington, D.C.
61. U.S. Environmental Protection Agency. 1973. The National Air Monitoring Program: Air Quality and Emissions Trends Annual Report, Vol. 1. EPA Report No. 450/1-73-001a. U.S. Environmental Protection Agency. Research Triangle Park, N.C.
62. Public Health Service. 1966 & 1968. Air Quality Data from the National Air Sampling Networks and Contributing State and Local Networks for 1966 and 1967. U.S. Department of Health, Education, and Welfare. Washington, D.C.
63. U.S. Environmental Protection Agency. 1971 & 1972. Air Quality data for 1967 and 1968 from the Air Surveillance Networks and Contributing State and Local Networks. U.S. Environmental Protection Agency. Research Triangle Park, N.C.
64. Gordon, R. J. & R. J. Bryan. 1973. Patterns in airborne polynuclear hydrocarbon concentration in four Los Angeles sites. Environ. Sci. Technol. **7:** 1050.
65. Kleinman, M. T., T. J. Kneip & M. Eisenbud. 1976. Seasonal patterns of airborne particulate concentrations in New York City. Atmos. Environ. **10:** 9.

66. KNEIP, T. J., M. A. LEYKO, M. T. KLEINMAN, M. LIPPMANN & J. M. DAISEY. 1978. Organic matter in New York City TSP: comparisons of seasonal variations and relationships to source tracers. Paper presented at the Conference on Carbonaceous Particles, Berkeley, Calif., March 20–22.
67. KLEINMAN, M. T. 1977. The Apportionment of Sources of Airborne Particulate Matter. Doctoral dissertation. New York University. New York, N.Y.
68. DAISEY, J. M., M. A. LEYKO & T. J. KNEIP. 1979. Source identification and allocation of PAH compounds in the New York City aerosol: methods and applications. *In* Carcinogenesis—A Comprehensive Survey. Polynuclear Aromatic Hydrocarbons: Third International Symposium. P. W. Jones & R. I. Freudenthal, Eds. **4:** 201. Raven Press. New York, N.Y.
69. DAISEY, J. M., M. A. LEYKO, M. T. KLEINMAN & E. HOFFMAN. 1979. The nature of the organic fraction of the New York City Summer aerosol. Ann. N.Y. Acad. Sci. **322:** 125.
70. MILLER, M. S., S. K. FRIEDLANDER, & G. M. HIDY. 1972. A chemical element balance for the Pasadena aerosol. J. Colloid Interface Sci. **39:** 165.
71. FRIEDLANDER, S. L. 1973. Chemical element balances and identification of air pollution sources. Environ. Sci. Technol. **7:** 235.
72. COLUCCI, J. M. & C. R. BEGEMAN. 1976. Carcinogenic air pollutants in relation to automotive traffic in New York. Environ. Sci. Technol. **5:** 145.
73. National Academy of Sciences. 1975. Petroleum in the Marine Environment. Workshop on Inputs, Fates and the Effects of Petroleum in the Marine Environment, Airlie House, Airlie, Va., May 21–25, 1973. Printing and Publishing Office. National Academy of Sciences. Washington, D.C.
74. KLEINMAN, M. T. 1980. The use of long term and seasonal trends of elemental compounds as an aid to the identification of sources of airborne pollutants. Ann. N.Y. Acad. Sci. (This volume.)
75. LUNDE, G. & A. BJORSETH. 1977. Polycyclic aromatic hydrocarbons in long-range transported aerosols. Nature **268:** 518.
76. LIOY, P. J., G. T. WOLFF, J. C. CZACHOR, P. E. COFFEY, W. N. STASIUK & D. ROMANO. 1977. Evidence of high atmospheric sulfates in the northeast. J. Environ. Sci. Health **A12:** 1.
77. WOLFF, G. T., P. J. LIOY, G. D. WIGHT, R. E. MYERS & R. T. CEDERWALL. 1977. An investigation of long range transport of ozone across the midwestern and eastern United States. Atmos. Environ. **11:** 797.
78. SAWICKI, E. 1962. Analysis for airborne particulate hydrocarbons: their relative proportions as affected by different types of pollution. Nat. Cancer Inst. Monogr. **9:** 201–220.
79. HANGEBRAUCK, R. P., D. J. VON LEHMDEN & J. E. MEEKER. 1967. Sources of polynuclear hydrocarbons in the atmosphere. Public Health Service Publication No. 999-AP-33. Cincinnati, Ohio.
80. RASMUSSEN, R. A. & F. W. WENT. 1965. Volatile organic material of plant origin in the atmosphere. Proc. Nat. Acad. Sci. USA **53:** 215.
81. ROLAND, L., V. MOHNEN & P. COFFEY. 1976. Measurements of background hydrocarbons in remote areas. Paper presented at the Symposium on the Non-urban Tropospheric Composition, Hollywood, Fla., November 10–12.
82. SCHUETZLE, D. & R. A. RASMUSSEN. 1978. The molecular composition of secondary aerosol particles formed from terpenes. J. Air Pollut. Control Assoc. **28:** 236.

OPTICAL PROPERTIES OF THE URBAN AEROSOL AND THEIR RELATION TO CHEMICAL COMPOSITION

Brian P. Leaderer and Jan A. J. Stolwijk

John B. Pierce Foundation Laboratory
New Haven, Connecticut 06519
and
Department of Epidemiology and Public Health
Yale University School of Medicine
New Haven, Connecticut 06520

Introduction

One of the most obvious and objectionable adverse effects of air pollution is visibility degradation. Visibility reduction in an unsaturated, precipitation-free atmosphere is caused by a complex mix of natural components and anthropogenic pollutant emissions. The basis for atmospheric visibility, or visual range, and its relationship to light scattering and mass of suspended particulate matter have been covered in detail by a number of other authors[1–7] and will only be touched upon here. This paper will deal principally with recent information relating light scattering and visibility observations with the chemical composition of the atmospheric aerosol, in an effort to examine the contribution to visibility degradation by specific aerosol species.

Optical Properties

Visibility

The presence of anthropogenic pollutants and natural components in the atmosphere clearly affects atmospheric optics. The transmission of light through the atmosphere is given by the Beer-Lambert Law:

$$dI/I = -\, b_{ext}\, dX \qquad (1)$$

where I is the intensity, b_{ext} is the extinction coefficient due to particulates and gases, and X is the distance traveled. The extinction coefficient can further be defined as:

$$b_{ext} = b_{scat} + b_{abs} \qquad (2)$$

where b_{scat} is the portion of the extinction coefficient attributable to light scattering by particulates (b_{sp}) and gas molecules (b_{sg}, Rayleigh scattering), and b_{abs} is the absorption component of the extinction coefficient attributable to light absorption by particulates (b_{ap}) and gases (b_{ag}). Of course, b_{ext} and its components are dependent on wavelength.

Visibility is dependent upon the transmission of visible light through the atmosphere and on the ability of the eye to distinguish an object due to its contrast

0077–8923/80/0338–0070 $01.75/1 © 1980, NYAS

with the background. The relationship between the extinction coefficient and visibility (visual range) V is given by:[1]

$$V = -\ln 0.02/b_{ext} = 3.9/b_{ext} \quad (3)$$

Visual range is defined as the distance, under daylight conditions, at which a typical observer can detect a specific type of target on the horizon with a visual contrast of 2%, the lower theoretical contrast limit for the human eye. Equation 3 assumes homogeneity of aerosol properties and illumination. The light scattering component, b_{scat}, of the extinction coefficient for the transmission of visible light in an urban atmosphere has been shown to be the dominant factor of atmospheric optical properties, principally visibility.[5,6] However, it has been shown that b_{ag}, principally light extinction due to absorption by NO_2, could be a significant contributor to atmospheric extinction at 550 nm when NO_2 levels are high and particulate light scattering is low.[8,9] In addition, particulate absorption may account for an appreciable portion of the total light extinction. Measurements taken near Denver have indicated that particulate absorption, most probably due to graphitic carbon, can account for as much as 40% of the total extinction coefficient for visible light.[6] Generally, however, it is assumed that b_{scat} is dominant over b_{ext} in the visible light range, so that Equation 2 becomes:

$$V = 3.9/b_{scat} \quad (4)$$

Because the molecular component (Rayleigh scattering) of the light scatter coefficient in an urban atmosphere is generally much lower than the particulate light scattering component, b_{scat} is usually assumed to be equal to b_{sp}, or particulate scattering. It should be noted, however, that Rayleigh scattering from gases in a relatively clean environment, particularly in rural areas, can account for a substantial portion of total extinction (approximately 50%).

In an unsaturated, precipitation-free urban atmosphere, b_{scat} is principally a function of wavelength of the light, size distribution, shape, and refractive index of the particles, density of the particles, and relative humidity. Atmospheric light scattering measurements of b_{scat}, utilizing an integrating nephelometer, have shown b_{scat} and visibility to be strongly related and in good agreement with Equation 4 for relative humidities less than 65%.[4,10,11] It has been indicated, however, that an empirical coefficient of approximately 2.9, rather than 3.9, in Equation 4 would result in a contrast limit of 5% and would be more suitable under conditions of less than ideal target contrast.[10–12]

Light Scattering versus Volume and Mass of Aerosol

The scattering due to particulate air pollutants in the atmosphere is usually attributed to particles of a size comparable to the wavelength of the incident light.[1] Thus, particulate light scattering and the resulting visibility would be expected to be controlled by particles in the 0.1 μm to 1.0 μm size range (the accumulation mode), the range that is dominated by secondary aerosols generated from primary pollutant emissions. Extensive size distribution measurements of urban aerosols have been used to calculate the volume of aerosol in the 0.1 μm to 1.0 μm size range for comparison with light scattering measurements.[6,11,13] These data have shown that variations in the volume of aerosol in the accumulation mode size range account for up to 90% of the variation in light scattering and that particles in this range do, in fact, dominate light scattering in the visible spectrum.

The mass size distribution of particulates in an urban atmosphere is known to be primarily bimodal in nature, with one mode between 0.1 μm and 1.0 μm and the other greater than 2 or 3 μm (the coarse mode). The accumulation mode mass typically accounts for approximately 50% of the total mass. With the light scattering sensitive size range accounting for approximately 50% of the total aerosol mass, one would not normally expect light scattering measurements to be particularly well correlated with total mass measurements, which are dominated by larger particles. Correlation studies between b_{scat}, measured by an integrating nephelometer, and total particulate mass have, somewhat surprisingly, shown a wide range of correlation coefficients. Many studies have produced consistently strong relationships.[2,10,14,15] The correlation coefficient values ranged from a low of 0.44 in a recent study in New York City[14] to a high of 0.95 in Sacramento, California.[10] In most instances, the observed strong proportional relationships between b_{scat} and total mass suggest either that variations in total mass are due principally to variations in the accumulation mode mass, or that the accumulation mode mass and coarse mode mass vary in a highly correlated manner. The observed correlation between b_{scat} and mass has led to the use of the integrating nephelometer for the continuous monitoring of total atmospheric mass, as well as for prevailing visibility readings.

Studies relating total particulate mass concentration to visibility observations have met with varying success.[3,5,10,11] Some studies have suggested that prevailing visibility is inversely proportional to total particulate mass for relative humidities less than 70%. The observed relationship between prevailing visibility and total particulate mass has resulted in a number of studies utilizing airport visibility records as an index to trace past and present total mass particulate pollutant trends. The visibility–mass relationship, as with the b_{scat}–mass relationship, suggests some consistent relationship between the accumulation and coarse mode masses.

As already discussed, it is the mass or volume of particulate matter in the 0.1 μm to 1.0 μm size range that dominates light scattering and hence visibility. The particles in this range are of various chemical compositions. Each particle can consist of a mixture of substances or principally of a single substance. A knowledge of the chemical composition of the accumulation mode particle is important to an understanding of the optical effects observed in the urban atmosphere.

Light Scattering versus Chemical Composition

In recent years, there have been a number of extensive aerosol sampling programs directed towards gathering information on the chemical and physical nature of the urban aerosol. Extensive chemical analyses of aerosol samples collected in conjunction with intensive sampling programs in California,[16,17] New York,[18,19] and St. Louis[20] have shown that sulfates and nitrates constitute a major portion of the total aerosol mass and that the dominant form of the sulfate is ammonium sulfate. Size distribution measurements taken during these studies have indicated that more than 85% of the sulfate, and approximately 50% of the nitrate, can be found in the size range less than 2.0 μm.[17,21-23] These measurements place most of the sulfate aerosol, and a large portion of the nitrate aerosol, in the light scattering size range. The studies have indicated that, at times, the sulfate aerosol can account for more than 50% of the submicron aerosol mass.[21,24,30]

Light scattering measurements (b_{scat}), taken in conjunction with the aerosol sampling programs in California and New York, have suggested that sulfates and, to a lesser extent, nitrates are principally responsible for a large portion of the light scattering variation effects observed in urban areas.[11,17,18,25,26] Light scattering measurements taken in aircraft flights over areas of continental Europe (principally over southern Scandinavia, and taken as part of the Long Range Transport of Air Pollutants project in Europe) have strongly suggested that the sulfate aerosol may dominate atmospheric optics.[27]

b_{scat} *versus Chemical Composition*

Measurements of the light scattering coefficient, as a function of relative humidity (humidograph) and with the addition of NH_3, have been made in a number of rural locations to examine the molecular or crystalline character of particles as they exist in their airborne state.[24,28,29] These measurements have indicated that sulfate aerosols—in the form of H_2SO_4 and its neutralization products with NH_3—are the dominant submicrometer-sized aerosol species. Hence, they are largely responsible for light scattering in rural Missouri and in rural locations in the midwestern and southern United States.[30]

In an effort to examine the contribution of the various chemical components of the urban aerosol to light scattering as measured by an integrating nephelometer, data gathered in Los Angeles[25] and New York[11] on the chemical composition of the ambient aerosol were regressed against the light scattering coefficient b_{scat}. Each analysis utilized a stepwise linear regression model of the form:

$$Y = B_0 + B_1 X_1 + B_2 X_2 \ldots \pm E \tag{5}$$

where Y is the dependent variable $b_{scat} \times 10^{-4}\ m^{-1}$; B_0 is the intercept; B_1, B_2, etc., are the coefficients associated with the independent variables X_1, X_2, etc., which in this analysis were sulfate mass, nitrate mass, organic mass, and the remainder of the mass (total mass minus sulfate and nitrate and organic); and E is the sum of the error terms. Only those individual independent variables that explained a significant (5% level) portion of the variation in b_{scat} were entered into the final equation in each analysis. The coefficients in this regression can be thought of as the light scattering coefficient per unit mass for each of the chemical species considered. The elemental and ionic sulfate, nitrate, and organic aerosol mass components in both analyses were adjusted to account for the probable compounds found in the air (sulfates/$SO_4^{=}$ = 1.3, nitrates/NO_3^- and organics/C = 1.4).[25] The important effect of relative humidity was considered both independently and combined with each chemical species considered. The data used in each analysis were different in that the Los Angeles data represented observations during only high pollutant episodes, while the New York data represented observations during both high and low pollutant levels in winter and summer periods. The analysis showed the calculated average contribution of each of the significant chemical components of the aerosol. The light scattering budgets, determined by the predictive equation for Los Angeles and New York, indicated that sulfates were the single most important chemical component in accounting for light scattering at both locations. In Los Angeles, sulfates generally accounted for about 33% of the scattering, while in New York, sulfate aerosol accounted for 62%. In Los Angeles, the nitrate and the remainder of the mass contributed more than 50% to the light scattering (the organic component was estimated at less than 10%), while in New York, the nitrate contribution was considerably smaller. The contribution of organics to light scattering in New York did not prove to be significant.

The computed coefficients for each aerosol chemical constituent considered in the New York and Los Angeles studies are shown in TABLE 1. In addition, TABLE 1 gives the light scattering coefficients per unit mass for sulfates, determined from measurements taken in New Jersey[31] and in southern Sweden,[27] and the light scattering coefficient per unit mass for the remainder of the suspended particulate mass, determined in the New Jersey study. Ninety percent of the observed variance in the scattering coefficients in Los Angeles was accounted for by variations in sulfate and in the remainder of the mass. Variations in nitrate and organics explained little additional variation in light scattering, although they were found to be significant. In New York, 75% of the variation in light scattering was accounted for by variations in sulfate aerosol, with variations in nitrate and organic aerosol accounting for an

TABLE 1

SUMMARY OF NEPHELOMETRIC LIGHT SCATTERING* AND CHEMICAL COMPOSITION

Site	Number of Observations	Coefficients 10^{-4} $m^{-1}/(\mu g/m^3)$					References	Notes
		r^2†	SO_4‡	NO_3‡	Organics‡	ΔMass§		
Los Angeles	58	0.92	0.07	0.05	n‖	0.025	White[25]	Measurement during high pollution levels, RH** considered.
New York	87	0.78	0.06	0.05	n	−0.002††	Leaderer[11]	Measurement made over winter and summer periods, RH considered.
New Jersey	39	0.84	0.08	—¶	—	0.01	1978[31]	Measurement made over summer period, RH considered.
Southern Sweden	77	0.53	0.05	—	—	—	Waggoner[27]	RH considered.

*Integrating nephelometer − $b_{scat} \times 10^{-4}$ m^{-1}.

†r^2 = Coefficient of determination − amount of variation in extinction explained by variations in SO_4, NO_3, etc.

‡SO_4 = 1.3 (SO_4^{2-}), NO_3 = 1.3 (NO_3^-), and Organics = 1.4 (C).

§ΔMass = total mass − (SO_4 + NO_3 + organics).

¶—, not calculated or unavailable.

‖n, not significant.

**RH, relative humidity.

††Negative coefficient due to pooling of winter and summer data—low SO_4 and high mass in winter, high SO_4 and low mass in summer.

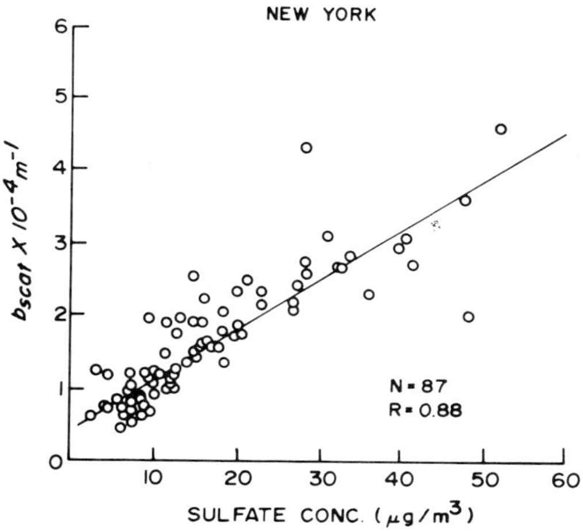

FIGURE 1. b_{scat} vs. sulfate for New York City.[12] (Sulfates = 1.3 × $SO_4^=$, and b_{scat} was measured with an integrating nephelometer equipped with a drier.)

additional 3% of the variation. The New Jersey study showed 81% of the observed variance in light scattering to be accounted for by variations in sulfate aerosol mass, with an additional 3% accounted for by the remainder of the mass. The regression developed from data taken in airplane flights over southern Sweden indicated that 53% of the observed variance in the scattering coefficient was accounted for by sulfate aerosol mass.

The regression coefficients determined for sulfates in the Los Angeles, New York, and New Jersey studies were remarkably close, while the sulfate coefficient determined for southern Sweden was somewhat lower. The modal mass ratios of sulfate/b_{scat} for all locations were in the 0.1 to 0.2 range and were consistent with ratios reported[27] for aerosols dominated by sulfates. Interestingly, a smaller than expected dependency of the scattering coefficient for sulfate on relative humidity was observed. FIGURE 1 shows the observed b_{scat}-to-sulfate relationship in the New York study and illustrates the strong relationship recorded between sulfates and light scattering. The light scattering coefficients for the nitrate aerosol in Los Angeles and in New York were also observed to be the same. The nitrate coefficients must be viewed with a good deal of skepticism, however, due to the potential for serious artifact nitrate formation on the filter media used to collect the total mass samples.[32] Conversely, there were substantial variations observed in the coefficient for the remainder of the mass, which was substantially lower than that for sulfates.

Visibility versus Chemical Composition

The observed relationship between atmospheric particulate light scattering and the sulfate aerosol mass concentration in urban and rural atmospheric environments, along with the predominance of sulfate in the submicron aerosol, has led investigators to suggest that the increased particulate light scattering observed in hazy air masses extending over much of the eastern United States and western and northern Europe may be principally due to sulfate aerosol.[27,30] The observed link between hazy air masses, increased light scattering, and elevated sulfates has led a number of investigators to use visibility isopleths to indirectly trace the genesis and movement of air masses containing high sulfates over large areas of the United States.[33–36]

In an effort to examine the suggested link between prevailing visibility observations and sulfate levels, three recent studies[37–39] compared prevailing airport visibility observations, taken at airports throughout the United States, with corresponding total particulate mass, sulfate particulate mass, nitrate particulate mass, and the remainder of the particulate mass determinations made near the airports. A stepwise linear regression model similar to Equation 5 was used to analyze the data at each site. The regression model took the following form:

$$B = (B_0 - b_{sg}) + B_1\,(\text{Sulfate}) + B_2\,(\text{Nitrate}) + B_3\,(\text{Total mass} - \text{sulfate–nitrate}) \qquad \textbf{(6)}$$

In the analysis, airport visibility observations in miles, the dependent variable (Equation 6), were converted to the extinction coefficient B (10^{-4} m^{-1}) by assuming a threshold contrast level of 2% (Equation 3) and then converting to meters. The term b_{sg} represented the contribution due to scattering by molecules, assumed to be equal to 0.15×10^{-4} m^{-1}. Relative humidity was also considered as a linear and a nonlinear function, independently and combined with each individual aerosol chemical species. The predictor equations for the significant independent variables were developed for each site. The average extinction, which could not be explained, was assumed to be the intercept, B_0, minus the contribution due to scattering by molecules (Rayleigh scattering). The other coefficients (B_1, etc.) were the extinction coefficients per unit mass for the aerosol chemical species considered. From the regression models for total extinction developed for each site, a light extinction budget was calculated. TABLE 2 gives the average light extinction budget calculated for each of the sites considered in the three studies.

As can be seen in TABLE 2, sulfate aerosol generally was found to account for a major portion of the average light extinction at almost all sites. The effect of sulfates on the average light extinction was strongest in the locations in the Northeast and in Los Angeles, where sulfates accounted for more than 50% of light extinction at most locations, and over 70% in New York, New Haven, and Los Angeles. Chicago appeared to be different from other northeastern sites in that sulfates were found to account for only 26% of light extinction, with over 69% unaccounted for. This apparently low contribution for sulfates and the high contribution unaccounted for were attributed to possible differences in air masses between the airport, where visibility observations were recorded, and the actual air pollutant sampling site in Chicago, where the particulate measurements were made.[38]

Sulfates at locations other than the Northeast (i.e., southwestern locations) typically accounted for less light extinction (about 30%). Nitrates were observed to account for very little of the average light extinction, particularly in the Northeast where the portion of visibility accounted for by sulfates was high. Nitrates did account for appreciable amounts of visibility loss in some locations in the Southwest (Phoenix and Salt Lake City), where the contribution due to sulfates was lower than in the

TABLE 2

AVERAGE EXTINCTION BUDGETS FOR VARIOUS LOCATIONS IN THE UNITED STATES UTILIZING AIRPORT VISIBILITY DATA

Location	Number of Observations	Blue-Sky Scatter	Sulfates*	Nitrates†	ΔMass‡	Unaccounted For	Reference	Notes
Rural								
Average for National Park Sites—Grand Canyon, Mesa Verde, and White Pine		45%	38%	8%	9%	n§	Trijonis[37]	Estimated
Suburban								
Lexington	86	5%	30%	n	42%	23%	Trijonis[38]	RH¶ considered
Charlotte	131	5%	56%	n	n	39%	″	″
Columbus	130	4%	65%	8%	n	23%	″	″
Urban								
Phoenix								
Maricopa County	106	17%	44%	31%	n	8%	″	″
NASN Data	202	20%	25%	19%	n	36%		″
Salt Lake City								
NASN Data	130	13%	30%	27%	30%	n	″	″
Other sites in Southwest U.S.								
Denver			marginal or poor data				″	″
Tucson, etc.								
Chicago	131	5%	26%	n	n	69%	″	″
Cleveland	102	5%	40%	n	n	43%	″	″
Newark	134	5%	52%	n	36%	19%	″	″
New York	72	5%	70%	n	n	25%	Leaderer[39]	″
New Haven	42	5%	77%	n	n	17%	″	″
St. Louis	84	5%	48%	11%	n	36%	″	″
Los Angeles	293	5%	76%	10%	n	9%	″	″

*Sulfates = 1.3 SO_4^{2-}.
†Nitrates = 1.3 NO_3^-.
‡ΔMass = total mass (sulfate and nitrate).
§n, not significant.
¶RH, relative humidity.

Northeast. In the rural areas considered, and in those in the Southwest in general, a much higher portion of the average visibility was attributable to molecular or blue-sky scattering, due to the typically better visibility and to the generally lower levels of sulfate, nitrate, and total mass.

The remainder of the suspended particulate mass appeared to contribute a significant portion to light extinction only at a few locations considered, including Lexington, Salt Lake City, and Newark. The remainder of the mass represents those nonsulfate and nonnitrate portions of the suspended particulate mass (principally organic aerosol) in the urban areas and, to some extent, the aerosol mass greater than 1.0 μm. In one study conducted in the Southwest, silicon (as SiO_2) in the submicron size range was found to account for a substantial portion of light extinction.[40]

All but two sites showed some portion of light extinction unaccounted for, and for a number of sites, the portion unaccounted for was substantial. The unaccounted-for portion in the analysis represents errors in the data base as they relate to visibility observations and to pollutant sampling and analysis. The unaccounted-for portion of light extinction associated with measurement errors may actually be related to levels of sulfate, nitrate, and the remainder of the mass.

The coefficients that were developed in the multivariate regression models relating light extinction (calculated from visibility observations) to aerosol chemical composition, and that were used to develop the extinction budgets shown in TABLE 2, are presented in TABLE 3. In addition, the coefficients calculated in other studies for which extinction budgets were not developed are presented in TABLE 3. At all sites, variations in sulfate concentrations were found to be significantly ($P \leqq 0.05$) related to variations in light extinction and accounted for the major portion of the variance in light extinction. Chicago, as discussed earlier, appeared to be an unusual case. At a number of sites, sulfates alone accounted for over 70% of the variance in light extinction, i.e., Columbus, 81%; New Jersey, 85%; New Haven, 71%; etc. At some locations, variations in nitrates and the remainder of the mass were significant in accounting for additional variations in light extinction. Generally, when it was considered in a nonlinear form, relative humidity was found to produce a better fit of the light extinction versus aerosol chemical composition models. While relative humidity was found to exert an effect on the sulfate coefficients at most sites, the effect on the sulfate coefficients at those sites was small, with one exception.[42] The effect of relative humidity on the coefficients for nitrate and remainder of the mass (at those sites where nitrate and the remainder of the mass were significant) was found to be more pronounced.

The coefficients for sulfates (in terms of light extinction per unit mass) listed in TABLE 3 are highly variable, ranging from a low of about 0.03 for Phoenix to a high of 0.16 for Los Angeles and New Haven. The sulfate coefficients are generally lower for the drier areas, the Southwest, and higher for the locations in the Northeast and in Los Angeles where the average relative humidity is higher. The range of sulfate coefficients recorded was generally within the range of coefficients calculated theoretically for secondary aerosols in the light sensitive size range,[25,27] although the sulfate coefficients for Los Angeles and New Haven are high. The high coefficients may be due to underestimates of sulfate mass because of water lost during the sampling and measurement procedure (water associated with the sulfate mass).

The observed relationship between light extinction and total particulate mass was

TABLE 3

COEFFICIENTS FOR CONTRIBUTION OF CHEMICAL CONSTITUTENTS OF AMBIENT AEROSOL LIGHT EXTINCTION IN REGRESSION MODELS

Site	Measure of Extinction	Number of Observations	Coefficients 10^{-4} $m^{-1}/(\mu g/m^3)$					Reference	Notes
			r^2†	SO_4‡	NO_3‡	Organics‡	ΔMass§		
Average for remote areas of Southwest, National Park Sites	Visibility*	—¶	—	0.04	0.05	—	0.002	Trijonis[37]	Estimated
Rural areas in Southwest	″	—	—	0.10	—	—	0.005	Ursenbach[44]	Estimated
Lexington	″	86	0.52	0.06	n‖	—	0.019	Trijonis[38]	RH** considered
Charlotte	″	131	0.53	0.11	n	—	n	″	RH considered
Columbus	″	130	0.81	0.13		—	n	″	RH considered
Phoenix									
Maricopa County	″	106	0.76	0.04	0.05	n	n	″	″
NASN Data	″	202	0.46	0.03	0.03	—	n	″	″
Salt Lake City	″	130	0.66	0.04	0.10	—	0.004	″	″
Chicago	″	131	0.27	0.03	n	—	n	″	″
Cleveland	″	102	0.52	0.07	n	—	n	″	″
Newark	″	134	0.50	0.06	n	—	0.014	″	″
New Jersey	″	39	0.81	0.10	—	—	0.01	1978[31]	Summer measurements, RH considered
New York	″	72	0.58	0.10	n	n	n	Leaderer[39]	RH considered
New Haven	″	42	0.81	0.16	n	—	n	″	″
St. Louis	″	84	0.69	0.077	n	—	n	″	″
Los Angeles	″	293	0.83	0.16	0.03	—	n	″	″
Los Angeles	″	413	—	0.09	0.05	—	n	Cass[41]	″
London	″	117	strong relationship observed between SO_4 and visibility, no coefficient given					Barnes[44]	″
—	″	16	0.49	0.11	—	—	—	Husar[33]	Uncorrected for RH

*Airport visibility readings where b_{ext} $(10^{-4}\ m^{-1}) = 24.3/\text{visibility (miles)}$.
†r^2 = coefficient of determination – amount of variation in extinction explained by variations in SO_4, NO_3, etc.
‡$SO_4 = 1.3\ (SO_4^{2-})$, $NO_3 = 1.3\ (NO_3^-)$, and Organics = 1.4 (organics).
§ΔMass = total mass – (SO_4 + NO_3 + Organics).
¶—, not calculated or unavailable.
‖n, not significant.
**RH, relative humidity.

found to be significant at most sites, but the observed correlation coefficients were found to be variable. FIGURE 2 shows the relationship of the light extinction coefficient, b_{ext}, versus total particulate mass, as observed at the Los Angeles site.[39] It is obvious from FIGURE 2 that light extinction, calculated from airport visibility observations, and total mass were poorly related. The relationship, observed at the same site, for light extinction versus sulfate is shown in FIGURE 3. The strong relationship between visibility and sulfate is obvious and demonstrates the dominant influence of sulfate on light extinction.

The observed strong relationship between light scattering light extinction (calculated from prevailing visibility observations) and sulfate aerosol mass led to one study[39] that investigated whether prevailing visibility data routinely collected at airports could be used to predict ambient sulfate concentrations, even in the presence of varying concentrations of nitrate mass and the remainder of the mass and excluding periods of precipitation or fog. Data from New York, New Haven, St. Louis, and Los Angeles were examined. The results indicated that a log sulfate vs. log visibility fitted line produced the best predictor equation. A test for equality of slopes showed that there was no significant difference in slopes between sites, although there were significant differences in the intercept. The study suggests that airport visibility observations may be useful in predicting sulfate concentrations. There was enough variability in the data from site to site, however, to suggest that the predictor equation should be developed and tested for each site considered.

There are, of course, a host of problems associated with the multivariate analyses discussed here. Some of these problems, dealing with sampling and analysis of the

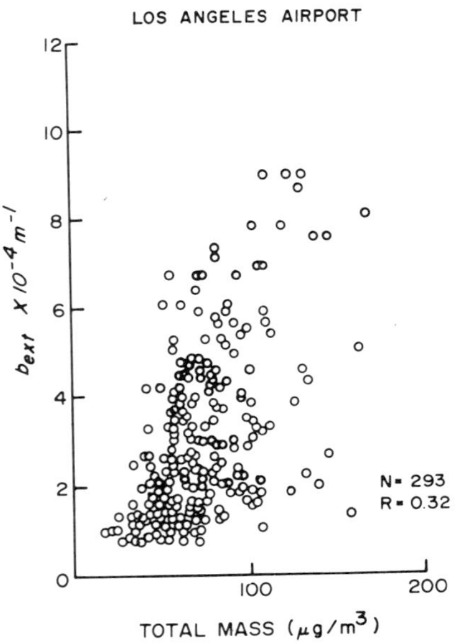

FIGURE 2. b_{scat} vs. total suspended particulate mass for Los Angeles.[37] [Total suspended particulate mass was measured by the standard high volume sampler method, and b_{scat} was calculated from airport visibility data where b_{scat} ($\times 10^{-4}$ m^{-1}) = 24.3/visibility (miles).]

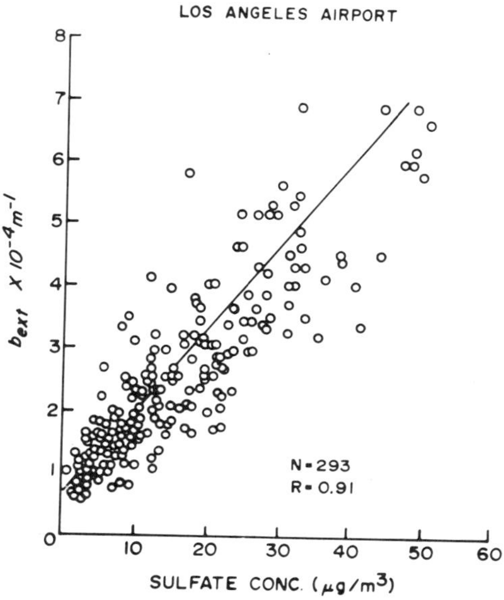

FIGURE 3. b_{scat} vs. sulfate for Los Angeles.[37] [Sulfates = $1.3 \times SO_4^{=}$, and b_{scat} was calculated from airport visibility data where b_{scat} ($\times 10^{-4}$ m^{-1}) = 24.3/visibility (miles).]

aerosol chemical species, covariance, visibility observations, etc., will be briefly discussed.

One of the major problems is artifact formation, which could contribute significant bias to chemical species measurement. Nitrate determinations obtained at each site would have to be considered suspect at this point. Recent work has shown that the filter media used to collect the total mass concentrations from which nitrate determinations were made may result in serious artifact nitrate formation, thus rendering the effect of nitrates on the average visibility for the sites appearing in TABLES 1, 2, and 3 difficult to interpret.[32] Sulfate determinations must also be interpreted with some caution, due to the potential artifact sulfate formation on the type of filtering medium used.[42,43] This effect was thought to be minimal or nonexistent for data gathered at some of the sites shown in TABLES 1, 2, and 3, particularly New York and New Haven.[39]

Water associated with the ambient aerosol (particularly the sulfate aerosol) is lost in the particulate sample due to filter equilibration, causing an additional sampling problem. This water would tend to produce larger than actual extinction coefficients per unit mass in the predictor equations but would produce no effect in the extinction budgets. This effect would be more noticeable in those areas with higher relative humidities, i.e., the Northeast.

An additional limitation in the correlational studies discussed stems from the potential for colinearity among the pollutants and relative humidity when considered as independent variables. Colinearity among the pollutant variables should not be substantial, however, due to the nature of the multivariate analysis employed in the various studies discussed and due to the generally poor intercorrelation observed between pollutant variables. Another problem is the lack of information on the chemical nature and size distribution of the submicron aerosol other than sulfate and nitrate. Additional information on the nonsulfate, nonnitrate portion of the submicron aerosol mass would help to explain more of the variation in light extinction, particularly the aerosol absorption component of the extinction coefficient. More detailed information on the size distribution of the ambient sulfate (particularly the submicron-size sulfate) would allow for a more accurate assessment of the contribution of sulfate aerosol to light scattering and light extinction.

The visibility observations recorded at airports and used to calculate total extinction have intervening factors associated with them, such as: variability in observers, nighttime vs. daytime readings, clustering of observations, maximum possible visibility, etc. The clustering effect may result from visibility observations generally being made for the greater visual ranges but not for the smaller intervals (i.e., 10, 15, 20, 25 miles, rather than 10, 11, 12, 25 miles).[33,44–46] In clean areas, where prevailing visibility typically is very good (such as the Southwest), the clustering effect could influence the analysis considerably.

The distances between the point or range over which the airport visibility readings are taken and the locations of the air pollutant monitoring sites may introduce additional errors into the analysis. This problem was thought to account for the poor light extinction vs. aerosol chemical composition equation and the poor light extinction budgets presented for Chicago in TABLES 2 and 3. If the airports and sampling sites are relatively close and are generally affected by the same air masses simultaneously, this error would be minimized for a regionally well-mixed, secondary submicron aerosol, such as sulfate. The problem would be greatest for larger aerosols and for poorly mixed, locally generated, fresh submicron aerosols.

In light of the host of problems associated with the correlation studies relating light scattering measurements and visibility observations (light extinction) with the chemical constituents of atmospheric aerosol mass, caution should be exercised in interpreting the results.

SUMMARY

Light extinction and resulting visibility degradation in an unsaturated, precipitation-free atmosphere are determined by light absorption and scattering and are caused by a complex mix of natural components and anthropogenic pollutant emissions. Correlational studies relating light scattering and light extinction (visibility) to concentrations of sulfate mass, nitrate mass, the remainder of the mass, and in some cases organic aerosol mass, while accounting for variations in relative humidity, have been reviewed. These studies indicate that sulfate aerosol mass is the dominant chemical aerosol species affecting light scattering and light extinction. The dominant effect of sulfate mass on light scattering and light extinction, even when sulfates account for a relatively small fraction of the total mass (on average, approximately 15 to 20%), stems from the fact that sulfates are secondary aerosols and are mostly found in the 0.1 μm to 1.0 μm size range (the particulate size range that scatters visible light very efficiently).

Correlational studies relating light scattering (b_{scat}) to sulfate concentrations at four different locations produced light scattering per unit sulfate mass coefficients in agreement with calculated coefficients for secondary aerosol in the 0.1 μm to 1.0 μm size range, and the coefficients were similar in value (range from 0.05 to $\approx$ 0.08 $\times 10^{-4}$ $m^{-1}/\mu g/m^3$).

Correlational studies relating airport visibility observations (light extinction) to the chemical constituents of the aerosol mass showed variations in sulfate mass over a wide range of locations to be strongly related to variations in light extinction. The light extinction coefficients per unit sulfate mass, however, did show a considerable degree of variation from location to location (0.03 to 0.16 $\times 10^{-4}$ $m^{-1}/\mu g/m^3$), but generally agreed with theory. The light extinction coefficients per unit mass for nitrate and the remainder of the mass were found to vary considerably between locations and were found statistically insignificant at many sites.

Light extinction budgets indicated that sulfate aerosol accounted for a large portion of the average light extinction; generally greater than ⅓ and, at many locations, greater than ½ (particularly in the Northeast). Nitrates and the remainder of the mass accounted for a highly variable portion of the average light extinction at those sites for which their contribution was found statistically significant. At all but one location for which measurements were made, a substantial portion of the average light extinction was unaccounted for.

The results of the multivariate analyses employed in the studies reviewed here must be interpreted with caution, due to a large number of problems associated with the data bases used. There is a need to conduct additional studies at a variety of locations to examine the relationship between light scattering and light extinction and the chemical constituents of the ambient aerosol. The studies should be designed to overcome the problems associated with the studies reviewed here.

References

1. Middleton, W. E. K. 1963. Vision Through the Atmosphere. University of Toronto Press. Toronto, Canada.
2. Charlson, R. J., N. C. Ahlquist & Helmuth Horvath. 1968. On the generality of correlation of atmospheric aerosol mass concentration and light scattering. Atmos. Environ. **12:** 455.
3. Noll, K. E., P. K. Mueller & M. Imada. 1968. Visibility and aerosol concentration in urban air. Atmos. Environ. **12:** 465.
4. Horvath, H. & K. E. Noll. 1969. The relationship between atmospheric light scattering coefficient and visibility. Atmos. Environ. **13:** 543.
5. Charlson, R. J. 1969. Atmospheric visibility related to aerosol mass concentration. Atmos. Environ. **13:** 913.
6. Waggoner, A. P. & R. J. Charlson. 1976. Measurements of aerosol optical parameters. *In* Fine Particles. B. Y. H. Lin, Ed.: 511–533. Academic Press, Inc. New York, N.Y.
7. Robinson, E. 1977. Effects on the physical properties of the atmosphere. *In* Air Pollution. A. C. Stern, Ed. **2:** 1–61. Academic Press, Inc. New York, N.Y.
8. Charlson, R. J., D. S. Covert, Y. Tomkima & P. K. Mueller. 1972. Multiwavelength nephelometer measurements in Los Angeles smog aerosol. III. Comparison to light extinction by NO_2. J. Colloid Interface Sci. **39:** 260.
9. Charlson, R. J. & N. C. Ahlquist. 1969. Atmos. Environ. **3:** 653.
10. Samuels, H. J., S. Twiss & E. W. Wong. 1973. Visibility, light scattering and mass concentration of particulate matter. Report of the California Tri-City Aerosol Sampling Project, State of California Air Resources Board.

11. LEADERER, B. P., R. L. TANNER & J. A. J. STOLWIJK. 1978. Light scattering, visibility and chemical species of the aerosol mass in New York City. In preparation.
12. STEFFENS, C. 1956. *In* Air Pollution Handbook. P. L. Magill, F. Holden & C. Ackley, Eds. Section 6: 1–43. McGraw-Hill Book Co. New York, N.Y.
13. THIELKE, J. F., R. J. CHARLSON, J. W. WINTER, N. C. AHLQUIST, K. T. WHITBY, R. B. HUSAR & B. Y. H. LIN. 1972. Multiwavelength nephelometer measurements in Los Angeles smog aerosol. II. Correlation with size distribution, volume concentrations. J. Colloid Interface Sci. **39:** 252.
14. LIPPMANN, M., M. T. KLEINMAN, D. M. BERNSTEIN, G. T. WOLFF & B. P. LEADERER. 1979. Size-mass distributions of the New York summer aerosol. Ann. N.Y. Acad. Sci. **322:** 29.
15. SIMMONS, W. A. & W. YOUNG. 1970. Correlation of the Integrating Nephelometer to the High Volume Air Sampler. Department of Public Health, Bureau of Air Use Management, Commonwealth of Massachusetts. Boston, Mass.
16. HIDY, G. M. *et al.* 1975. Summary of the California Aerosol Characterization Experiment. J. Air Pollut. Control Assoc. **25:** 1106.
17. HIDY, G. M. *et al.* 1974. Characterization of Aerosols in California (ACHEX). Vol. 1, 2, 3 and 4. (Prepared for California Air Resources Board). Rockwell International. Thousand Oaks, Calif.
18. LEADERER, B. P. *et al.* 1978. Summary of the New York Summer Aerosol Study (NYSAS). J. Air Pollut. Control Assoc. **28:** 321.
19. KNEIP, T. & M. LIPPMANN, Eds. 1979. The New York Summer Aerosol Study, 1976. Annals, Vol. 322. New York Academy of Sciences. New York, N.Y.
20. U.S. Environmental Protection Agency. 1975. Regional Air Pollution Study (RAPS). Conducted in the St. Louis region, 1975. U.S. Environmental Protection Agency. Research Triangle Park, N.C.
21. TANNER, R. L., R. GARBER, W. MARLOW, B. P. LEADERER & M. A. LEYKO. 1979. Chemical composition of sulfate as a function of particle size in the New York summer aerosol. Ann. N.Y. Acad. Sci. **322:** 99.
22. KLEINMAN, M. T., C. TOMCZYK, B. P. LEADERER & R. L. TANNER. 1979. Inorganic nitrogen compounds in New York City air. Ann. N.Y. Acad. Sci. **322:** 115.
23. MUELLER, P. K., R. W. MOSLEY & L. B. PIERCE. 1972. Chemical composition of Pasadena aerosol by particle size and time of day. *In* Aerosols and Atmospheric Chemistry. G. M. Hidy, Ed. Academic Press, Inc. New York, N.Y.
24. CHARLSON, R. J., A. H. VANDERPOL, D. S. COVERT, A. P. WAGGONER & N. C. AHLQUIST. 1974. Sulfuric acid–ammonium sulfate aerosol: optical detection in the St. Louis region. Science **184:** 156.
25. WHITE, W. H. & P. T. ROBERTS. 1977. On the nature of visibility-reducing aerosols in the Los Angeles air basin. Atmos. Environ. **11:** 803.
26. LEADERER, B. P., D. ROMANO & J. A. J. STOLWIJK. 1979. Light-scattering measurements of the New York summer aerosol. Ann. N.Y. Acad. Sci. **322:** 45.
27. WAGGONER, A. P., A. J. VANDERPOL, R. J. CHARLSON, S. LARSEN, L. GRANAT & C. TRAGARDH. 1976. Sulfate–light scattering ratio as an index of the role of sulfur in tropospheric optics. Nature **261:** 121.
28. COVERT, D. S., R. J. CHARLSON & N. C. AHLQUIST. 1977. A study of the relationships of chemical composition and humidity to light scattering by aerosols. J. Appl. Meteorol. **13:** 968.
29. CHARLSON, R. J., A. H. VANDERPOL, D. S. COVERT, A. P. WAGGONER & N. C. AHLQUIST. 1974. $H_2SO_4/(NH_4)_2SO_4$ background aerosol: optical detection in the St. Louis region. Atmos. Environ. **8:** 1257.
30. WEISS, R. E., A. P. WAGGONER, R. J. CHARLSON & N. C. AHLQUIST. 1977. Sulfate aerosol: its geographical extent in the midwestern and southern United States. Science **195:** 979.
31. Calculated from data collected in Linden, N.J., at the Exxon Research Air Monitoring Station, July 26 through September 4, 1978, and supplied in a personal communication from R. S. Brief, November 20, 1978.

32. Spicer, C. W., P. M. Schumacher, J. A. Kouyoumjian & D. W. Joseph. 1978. Sampling and Analytical Methodology for Atmospheric Particulate Nitrates. Battelle-Columbus Laboratories Final Report to the U.S. Environmental Protection Agency. Research Triangle Park, N.C.
33. Husar, R. B., N. V. Gillani, J. D. Husar, C. C. Paley & P. N. Turcu. 1976. Long range transport of pollutants observed through visibility contour maps, weather maps and trajectory analysis. *In* Preprints, Third Symposium on Atmospheric Turbulence, Diffusion and Air Quality: 344–347. American Meteorological Society. Boston, Mass.
34. Husar, R. B., D. E. Patterson, C. C. Paley & N. V. Gillani. 1977. Ozone in hazy air masses. Paper No. 6–5. *In* Proceedings of the International Conference on Photochemical Pollution. U.S. EPA Pub. No. EPA-600/3-77-0016. Research Triangle Park, N.C.
35. Lyons, W. A. & J. C. Dooley. 1978. Satellite detection of long-range pollution transport and sulfate aerosol hazes. Atmos. Environ. **12:** 621.
36. Wolff, G. T., P. J. Lioy, B. P. Leaderer, D. M. Bernstein & M. T. Kleinman. 1979. Characterization of aerosols upwind of New York City: I. Transport. Ann. N.Y. Acad. Sci. **322:** 57.
37. Trijonis, J. & K. Yun. 1977. Visibility in the Southwest. An Explanation of the Historical Data Base. Technology Services Corporation. Prepared for the U.S. Environmental Protection Agency. Research Triangle Park, N.C.
38. Trijonis, J. & K. Yun. 1978. Visibility in the Northeast. Long-term Visibility Trends and Visibility/Pollutant Relationships. Technology Services Corporation. Prepared for the U.S. Environmental Protection Agency. Research Triangle Park, N.C.
39. Leaderer, B. P., T. R. Holeford & J. A. J. Stolwijk. 1979. Relationship between sulfate aerosol and visibility. J. Air Pollut. Control Assoc. **29:** 154.
40. Macias, E. S., D. L. Blumenthal, J. A. Anderson & B. K. Cantrell. 1980. Size and composition of visibility-reducing aerosols in southwestern plumes. Ann. N.Y. Acad. Sci. (This volume.)
41. Cass, G. R. 1976. The Relationship Between Air Quality and Visibility of Los Angeles. Environmental Quality Laboratory Memorandum #18, California Institute of Technology. Pasadena, Calif.
42. Forest, J. & L. Newman. 1973. Ambient monitoring for sulfur compounds: a critical review. J. Air Pollut. Control Assoc. **23:** 761.
43. Stevens, R. K., T. G. Dzubay, G. Russwurm & D. Rickel. 1978. Sampling and analysis of atmospheric sulfates and related species. Atmos. Environ. **12:** 955.
44. Ursenbach, W. O., A. C. Hill, W. H. Edwards & S. M. Kunen. Atmospheric particulate sulfate in the western United States. Paper No. 76–7.5. Presented at the 69th Annual Meeting of the Air Pollution Control Association, Portland, Oregon, June, 1976.
45. Barnes, R. A. & D. O. Lee. 1978. Visibility in London and long range distance transport of atmospheric sulfur. Atmos. Environ. **12:** 791.
46. Reiss, N. M. & R. A. Eversole. 1978. Rectification of prevailing visibility statistics. Atmos. Environ. **12:** 945.

URBAN AEROSOL CHARACTERISTICS: GENERAL DISCUSSION*

Morton Lippmann, *Moderator*

Institute of Environmental Medicine
New York University Medical Center
New York, New York 10016

S. K. FRIEDLANDER (*University of California, Los Angeles, Calif.*): Dr. Lippmann, you alluded to the self-preserving size distributions, and I thought that it would be in order to say a few words about their relevance, or irrelevance as I think you implied.

The self-preserving concept is basically a theoretical one and, under the circumstances of its derivation, the theoretical work-up, and the laboratory tests, I think is beyond doubt correct as a fundamental concept of behavior of fine particles in the coagulation range. The question is really one of application, both to atmospheric problems and to industrial problems. I think that the applications are perhaps twofold. I still think that the basic transformation can be useful in correlating atmospheric experimental data and, in particular, that grand average that you showed can be useful in terms of establishing a general link between light scattering and total aerosol mass.

The other application, I think, is to the separate modes of the size distribution on some occasions, because you are still faced with the necessity of explaining why the size distributions are sometimes log-normal in the separate modes, which can be shown in a number of cases. The question arises, Why should they be log-normal? Well, as a good approximation, the self-preserving distribution (which is fundamentally based) is not too different from a log-normal distribution. That may be one explanation for the very-fine-particle self-preserving log-normal–type distributions.

M. LIPPMANN: Thank you. I didn't mean to imply that there was anything wrong with the self-preserving concept.

K. T. WHITBY (*University of Minnesota, Minneapolis, Minn.*): I'd like to make a comment about the question as to what might be the differences between summer and winter.

We've done quite a bit of stratification and classification analysis of the vast amount of data that has accumulated, and, in trying to identify where the differences come from, I would like to caution physicists and chemists about ignoring meteorology. The conclusion that I've come to from looking at a lot of these data is that when you take an average, such as the one that was shown for New York, the most likely explanation seems to be the differences in the contribution due to the differences in summer–winter meteorology, rather than to physics and chemistry. I think we must continually keep this in mind in analyzing these data. This is the thing we have the hardest time measuring, but it may turn out to be the most important.

R. JAENICKE (*Max-Planck-Institut für Chemie, Mainz, Federal Republic of Germany*): Dr. Lippmann, this is not a question. I just want to emphasize one point you made early in your talk.

You showed us very clearly how many measures we have with which to

*This is a compilation of the discussions held after the presentation of each of the papers in this section.

0077–8923/80/0338–0086 $01.75/1 © 1980, NYAS

characterize the aerosol, and I think we should stress the point that the aerosol can't be characterized by one measure, or even two. The aerosol is different from the gases and the gaseous air pollutants we have in the atmosphere. These can be characterized by one measure, as a function of time and volume of space, for instance; but for the aerosol, this is impossible.

G. E. GORDON (*University of Maryland, College Park, Md.*): Dr. Stevens, I was just going to say, in regard to your linking of various analysis techniques with collection methods, that neutron activation would go fine with dichotomous samplers; no incompatibility there. We haven't had them available to use with neutron activation, but there is no basic incompatibility.

E. SEARS (*California Institute of Technology, Pasadena, Calif.*): Dr. Stevens, you showed, in an early slide, a mass balance and you mentioned something about carbon, but I think it might be useful to point out that some of the organic compounds that you mentioned are also present in quantities very much the same as those of the trace elements.

R. K. STEVENS (*U.S. Environmental Protection Agency, Research Triangle Park, N.C.*): Yes, I mentioned that approximately 5% of that fraction can be identified as extractable organics. We've taken these samples and extracted, and we measure the total carbon on that. Now I'm assuming that the materials that are extractable could be identified via mass spectroscopy, but that may be an unwarranted assumption.

S. DATTNER (*Texas Air Control Board, Austin, Tex.*): Dr. Stevens, we have been using cellulose filters from Hi-vols for XRF [x-ray fluorescence analysis] and getting very good results.

R. K. STEVENS: Unfortunately, cellulose filters produce a large nitrate artifact, but go ahead.

S. DATTNER: Well, in XRF you don't pick up nitrate anyway. The one problem we've found with almost all of our equipment is that copper is usually an artifact of the motors. We have terrible problems in getting copper from the air but not from the motors.

R. K. STEVENS: If you use the high volume sampler, that is true.

P. K. MUELLER (*Environmental Research & Technology, Inc., Westlake Village, Calif.*): I think that when we talk about extractable carbon, we need to talk about the solvent. I believe, Dr. Stevens, that you are intimating that 95% of the carbon you see is graphitic.

R. K. STEVENS: In the West Virginia samples, yes.

P. K. MUELLER: We have looked at a number of samples around the United States, and we find, for instance, that a large portion of the organics—as much as 50%—is water extractable.

R. K. STEVENS: My only comment to that, Dr. Mueller, is that we are using a methanol/methylene chloride mixture, and I think that those materials that are soluble in water would also be soluble in methanol.

R. JAENICKE: Dr. Larson, you mentioned that we have one major reaction pathway from the gas to the particles. What do we know today about the opposite pathway from the particles to the gas?

T. V. LARSON (*University of Washington, Seattle, Wash.*): I believe the vapor pressures, certainly of sulfate compounds, are very low. Certain bacteria on the ground can reduce sulfates, but it is hard to see how they could be produced in the aerosol phase.

It is an interesting point, though, since sulfate is the final oxidation step for sulfur, and nitrates, essentially, for nitrogen. But for carbon, it is carbon dioxide, and so presumably there could be further reactions of the carbon aerosol with oxidation.

G. M. KLAUBER (*Environmental Measurements, Inc., Arnold Md.*): Dr. Larson, you mentioned that Calvert hypothesized an HSO_5 species in the gas phase. Did he mention, or consider, the possibility of this HSO_5 radical reacting with NO, forming NO_2, and eventually generating ozone; perhaps in a power plant plume?

T. V. LARSON: I don't know. Davis has also proposed a step with HSO_5 reaction, and I would refer you to that paper by Calvert where those rates were discussed. He disagrees with Davis.

J. ALLEN (*Aerochem Research Labs, Princeton, N.J.*): Dr. Larson, most of your talk about rates and processes was on the heterogeneous reaction. How well are the rate constants for the homogeneous gas phase reactions known?

T. V. LARSON: Four hydroxyl reaction studies agree fairly well, if you correct for the pressure dependence of the two studies that used carbon monoxide oxidation of OH as a yardstick reaction. Doing that, you get fairly good agreement for that rate. In contrast, there isn't much known about some of the organic peroxy radical reactions. There just haven't been many measurements made so far.

R. S. FEIN (*Texaco, Inc., Beacon, N.Y.*): Dr. Larson, in your discussion of liquid phase oxidation of the sulfur, you, along with most of the reviews that I've read, omitted the possibility of the sulfur being oxidized by nitrogen oxides, even though this was at one time a commercial sulfuric acid manufacturing process. I assume that this has been looked at, but are the rates low enough that you can discount them?

T. V. LARSON: I will venture to say that it hasn't been looked at as well as it might be, possibly because of the complicating chemistries of nitrosyl sulfuric acid and decomposition of that NO. I think people have shied away from these complications.

W. R. PIERSON (*Ford Motor Company, Dearborn, Mich.*): Dr. Tanner, your High Point, New Jersey, site showed the highest hydrogen to sulfate ion ratio. Do you think that a lot of the sulfate passing through High Point may have been generated from sulfur that was emitted in the Ohio Valley region? Would it be safe to expect, then, if you were farther upwind from High Point—between High Point and the Ohio Valley—that you might find still higher hydrogen to sulfate ratios?

I'm putting you on the spot, because I was operating a site there at the same time you were making your measurements.

R. L. TANNER (*Brookhaven National Laboratory, Upton, N.Y.*): I'm not sure. I really think the jury is out on the question of where one would go to find the maximum acid to sulfate ratio in the aerosol during this period.

It's quite clear, from various wind trajectories, that the trajectories under the conditions in which we got higher results at High Point than in New York, Brookhaven, and New Haven fit more exactly the precondition of the air having been in the Ohio Valley under stagnant conditions on the order of 1½ days before.

Where the trajectories were not different, the sulfate concentrations were not different. I'm not sure why the acid to sulfate ratios were higher at High Point than in the urban areas or near the ocean. The higher concentrations of ammonia in the urban areas and off the coast may account for that.

R. K. STEVENS: Dr. Daisey, that was a very interesting paper on the organic phase. Have you any comments about the recent work of Pitts, and some work in our laboratory, indicating the possibility of organic artifact on glass fiber filters due to gas phase, acidic organics?

J. M. DAISEY (*New York University Medical Center, New York, N.Y.*): The Pitts work, I believe, was done with a fiber glass filter, and he put the polycyclic aromatic hydrocarbons right on the filter. There was no particulate phase, is that right?

R. K. STEVENS: That is right.

J. M. DAISEY: So there may be differences when that material is absorbed on particulate matter.

R. K. STEVENS: I'm more concerned about the gas phase organics, like formaldehyde, acetic acid, and propionic acid, reacting with the highly basic filter itself, which I think can lead to enormous confusion.

J. M. DAISEY: Well, we are presently taking a look at sampling artifacts. I think that in collecting particulate matter and pulling air through it that contains a variety of materials in the vapor phase, there is a very good possibility of chemical reaction. I think anybody who works in organics recognizes that. It is probably going to be a very long time before we fully understand what those artifacts are and the magnitude of their effects.

D. L. SWIFT (*Johns Hopkins University, Baltimore, Md.*): Dr. Daisey, I wonder if some of the differences you observe in the summer–winter studies might be due to the blending of gasolines, and I wonder further if you'd like to hazard a guess as to the effect of the increasing use of unleaded motor fuel, which I have a feeling contains a higher proportion of aromatic compounds. I wonder if you would comment on what you think might be the outcome of that.

J. M. DAISEY: Some reports from laboratory studies indicate higher PNA [polynuclear aromatic hydrocarbon] emissions with higher concentrations of aromatics in fuels.

R. JAENICKE: Dr. Daisey, we know that the aerosol is in the mixed phase. That means that we have all the substances, either mixed in the individual particle, or mixed on different particles. Since you mentioned size distribution, couldn't you give us some ideas about the mix of aerosol with regard to organics? Could all the organic materials be in pure organic particles with no inorganic components?

J. M. DAISEY: We don't know that as yet. I suspect that some of them are probably pure organic, some of them are probably coated on surfaces, and some of them are probably mixed in with inorganic materials. It isn't an area in which we have much information at present.

T. V. LARSON: Dr. Daisey, could you make just a brief comment on the interaction of these compounds with atmospheric water vapor?

J. M. DAISEY: I did come across a paper in which someone had looked at photooxidation of organic compounds in water, and the rates seemed to be quite high. Whether this is significant in the atmosphere, I couldn't say. I think it is something that we should take a look at. It is rather difficult to do from a sampling point of view, but it may be significant.

K. T. WHITBY: Dr. Leaderer, lest the impression be made that the sulfate out in the country is always so closely related to the aerosol volume and to light scattering, I would like to mention some results we got last summer that were somewhat contradictory to this.

We did quite a bit of haze measurement work in Tennessee and Alabama last summer, as part of the STATE program. We had a continuous *in situ* sulfate analyzer running and a couple of ways of calibrating it. We found, in the power plant plume, that we could account for the nonaqueous part of the aerosol very cleanly and neatly by sulfate and, hence, the light scattering. However, much to our surprise, out in the haze areas, we never got above about one-third of the aerosol being accounted for by sulfate.

I don't know what the answer to this is, but I think that we have to be suspicious of the fact that practically all of the correlations that have been developed so far, such as yours, have been correlations of filter sulfate and nitrate with light scattering. I think we must reexamine this question, using methods that rule out some of the artifact formation filters.

B. P. LEADERER (*Yale University School of Medicine, New Haven, Conn.*): The data that came from New Haven and from New York City were collected on quartz filters from the Brookhaven laboratories. I think we feel fairly confident that the sulfate concentrations there were free of artifacts. However, in St. Louis and Los Angeles, it's much more difficult to be sure what the absolute levels were.

G. T. WOLFF (*General Motors Research Laboratories, Warren, Mich.*): Dr. Leaderer, I'd like to caution you on your assumption that the light scattering is the sole component of the extinction and that the absorption due to aerosols is negligible.

B. P. LEADERER: No, I didn't say that, and I don't want to leave that impression. The light scattering balances show other components to be quite important. However, the variations in light scattering seem to be more closely related to variations in sulfate. I think that, in many locations, absorption can be quite important.

G. T. WOLFF: We have been investigating the Denver situation for the past couple of months, and the absorption appears to be a significant proportion of the extinction.

B. P. LEADERER: About 50%, I think, in that location.

G. T. WOLFF: Now I should say that I think this has to be looked at in various parts of the country. We looked at it in some urban areas, but I think rural locations could be somewhat different.

B. P. LEADERER: Have you done any absorption measurements in New York City?

G. T. WOLFF: No, we haven't.

UNIDENTIFIED SPEAKER: Thank you, Dr. Leaderer, I enjoyed your paper and I agree with your general conclusions. There are two cautions I'd like to raise. One is that when you speak of visibility, you use the term 2% horizon visibility, because there are many other kinds of visibility measures. These are increasingly being recognized, as there are several types of visibility interference that are important to people.

The second caution is that you seemed to lump molecular scattering together with very-fine-particle scattering.

B. P. LEADERER: Not necessarily.

S. K. FRIEDLANDER: Dr. Leaderer, there are two factors that have to be considered in the case of ammonium nitrate, and I think it is important to identify them clearly. One is what I'm going to call "artifact," and that is, as I interpret it, the trapping or conversion of essentially gaseous material by the filter substrate by some kind of an interaction—absorption or chemical reaction.

The other factor—and that's a more fundamental one, in a sense—is the basic vapor–liquid and vapor–solid equilibria between ammonium nitrate and ammonia and nitric acid in the gas phase. We have done some work on that, primarily theoretical at this point. Art Stelson in our lab has used both laboratory data and data taken from the literature and developed the equilibrium relationship between the gas and solid phases. He finds that over the temperature range from 20 to 30°C, the equilibrium mass concentration of ammonia and nitric acid, that is, NH_3 and HNO_3, in the gas phase increases from 11 mg per cubic meter to 38.4 mg per cubic meter for a 20° to a 30°C change. This is what you would expect; a very sharp increase for a vapor-solid-type equilibrium system.

Now that, of course, would have very important implications for atmospheric measurements, because that is right in the range of reported values of ammonium nitrate in the gas phase; the order of 10–15–20 mg per cubic meter.

So if there is a pure solid phase, it is going to shift back and forth very markedly with temperature during the course of the day. You will never know which form you will have, the gas phase or the solid phase. And it is going to affect sampling, as well. So it will have a feedback effect on the sampling that will introduce another artifact. In other words, it also depends on the temperature at which you store your filters. In fact, this was noted recently in a paper in the *Journal of the Air Pollution Control Association* by Grosjeans and a group in Pitts' laboratory. The paper on the equilibrium relationship is going to be published as a short note from our group in *Atmospheric Environment.*

P. K. MUELLER: Dr. Leaderer, I'd like to make a couple of comments on your presentation, which is very thorough, like many of the review presentations that have been made. What continues to bother me in this business are the correlations that one can get between sulfate and other properties of the aerosol. I think the reason this kind of indication is so widespread is that the sulfate is very readily measured and quite accurately measured, possibly even more accurately than the weight. Sulfate is also more precisely measured, notwithstanding the artifact situation.

Furthermore, we have demonstrated that most of the sulfate is in the accumulation mode, and we have demonstrated that the accumulation mode is responsible for light scattering. So naturally you are going to have a correlation of sulfate with the accumulation mode mass, especially in Los Angeles where most of the particles' mass is in that size range. This doesn't mean that most light scattering is necessarily due to sulfate. Let me, therefore, continue with the litany.

The litany for the Los Angeles aerosol is that there are four major components that contribute to the mass, and we have accounted for 85% of the mass for three of these components, the fourth one being very difficult to measure. The fourth one is the liquid water component of the aerosol, which we don't measure accurately. The other three components are (1) the sum of ammonium sulfates and ammonium nitrates; (2) graphitic carbon and organic carbon; and (3) the sum all of the other elements. Thank you.

B. P. LEADERER: Those studies—the New York Summer Aerosol Study, and Whitby's work in Los Angeles—that did correlate light scattering with the measured organics, nitrates, and sulfates seemed to indicate that it was sulfates that were by far the predominant factor related to scattering.

P. K. MUELLER: Excuse me, I believe that in those studies, including ACHEX, these other components weren't adequately measured. It is true that, at times, the sum of ammonium sulfate and ammonium nitrate could be 20 or 30%. I never recall having seen it to be as much as 50% of the total mass.

B. P. LEADERER: There are some data that show 50% of mass.

P. K. MUELLER: Well, the more recent data, I think, are more accurate than ACHEX. But it can fluctuate a great deal. There are occasions when the sum of ammonium sulfate and ammonium nitrate is less than 10% of the total mass. The relative concentration of the three major chemical entities can fluctuate a great deal from day to day.

B. P. LEADERER: We have seen fluctuations in New York of from 1 or 2 micrograms to well over 40 in the course of one day.

UNIDENTIFIED SPEAKER: I'd like to comment a little bit about what Dr. Friedlander had to say with regard to what we call "negative artifact formation"; that is, the release of nitrates after they have been collected on a filter or by the equilibrium gas-phase reaction.

We have performed an experiment and have designed an experimental protocol that can account for this artifact formation. We performed that experiment in a 10-day study in the Smoky Mountains. The results of that experiment showed that there was no evidence of a negative artifact, although the ozone concentrations were between 80 to 100 ppb, both at ground level and aloft, while the nitrate concentrations at ground level were actually below our detection limit. This meant that the nitrates were below 0.1 microgram per cubic meter, or right at that level.

We also took some samples two years ago in roadway experiments, and have been continuing roadway experiments, with samplings of 20 minutes in duration, for the past year, under a contract with Dr. Will Richards at Rockwell. At no time have we seen any nitrate concentrations in either the fine or the coarse fraction that were in excess of 1 microgram per cubic meter. However, simultaneous measures taken at the CHESS site, depending upon which filter medium they were using, showed values of nitrates up to 20 or 30 micrograms per cubic meter.

I just cannot, at this time, support the hypothesis that there is a significant gas phase release of nitrates from the filter. Hopefully, an experiment that is being conducted this summer will resolve this, but experimentally, at least in the field and in some laboratory experiments, I just can't see that happening.

T. W. PETERSON (*University of Arizona, Tucson, Ariz.*): Just one comment about an alternative approach that might get around some of this difficulty with correlations. With the improvement in sampling devices—such as the low-pressure impactor, the type that has been developed by Friedlander and others—it is possible to get a rather good size distribution for sulfur and other species that are abundant in urban and rural aerosols. By obtaining a good size distribution for individual elemental or chemical species, it's possible to make rather accurate Mie scattering calculations and to actually synthesize the light scattering coefficient.

If there are also measurements of such things as graphitic carbon (soot), and if one can estimate absorption, or obtain an independent measurement, then one can actually synthesize the total extinction coefficient. This gets around the problem that Mueller and others discussed—that the presence of a correlation doesn't necessarily explain the percentage contribution by any given species. I'll show such a light scattering budget for the rural Southwest in my talk tomorrow morning.

TECHNIQUES FOR TREATING MULTIELEMENT PARTICULATE DATA TO OBTAIN INFORMATION ON SOURCES: OVERVIEW*

Glen E. Gordon

Department of Chemistry
University of Maryland
College Park, Maryland 20742

INTRODUCTION

At present, the only air pollution standards affecting aerosols in most areas are those placed on the total suspended particulate matter (TSP). However, as instruments for routine measurements of more detailed composition become available, and as knowledge of the association of certain types of particles with specific air pollution effects (e.g., health effects, visibility degradation) improves, it will become possible to write more specific air quality standards for certain classes of particles. In order to devise control strategies to achieve these more specific standards, we will need a better knowledge of both the sources and transport of particles having certain size distributions and composition.

In the past, control strategies have been based largely on source-emissions inventories. At best, source-emissions inventories are barely adequate for estimating the sources of TSP, and they are quite inadequate for the more sophisticated evaluation of aerosol problems, as envisioned above. Source-emissions inventories provide estimates of the total mass of material released by various classes of sources, but even these estimates are subject to large errors because of changes in efficiencies of pollution-control devices as they age or as operating parameters are modified. Furthermore, source-emissions inventories usually give little information about the size distributions of particles released from various sources or about the composition of particles versus size. In addition, inventories provide little information about airborne residence times of particles from various sources. Because of the many limitations of source-emissions inventories, we need much better techniques for associating observed ambient concentrations and size distributions of particles bearing specific chemical species with their ultimate sources.

Over the past few years, there have been enormous improvements in our ability to collect and segregate particles into well-defined size fractions[1,2] and to analyze the separate size groups for 20 or more elements by instrumental, multielement methods, such as x-ray fluorescence[3] (XRF) and instrumental neutron activation analysis[4,5] (INAA). The collection of particles in two separate size fractions with dichotomous samplers and the analysis of collected materials with XRF have become so highly automated that it is now practical to establish large networks of such systems that

*Supported in part by the National Science Foundation RANN Program, under Grant No. ENV 75-02667, and by the Environmental Protection Agency, under Contract No. R806263-01-1.

0077–8923/80/0338–0093 $01.75/1 © 1980, NYAS

routinely collect and analyze samples at several-hour intervals, as was demonstrated in connection with the Regional Air Pollution Study (RAPS) in the St. Louis area.[6]

The operation of networks of such highly sophisticated sampling and analysis systems is providing enormous sets of data on the size distributions and detailed elemental compositions of particles, as a function of time and location, in and around cities. In addition, many of these data sets do or will include simultaneous measurements of pollutant gases, visibility or light scattering, and meteorological data. At the same time that our information base on ambient aerosols has improved, detailed information on the size distributions and compositions of particles released from various types of major air pollution sources has become available. For example, our group at the University of Maryland has obtained detailed information on particles released from coal-fired power plants,[7,8] oil-fired plants,[9] refuse incinerators,[10] motor vehicles,[11] copper smelters,[12] and from a cement plant and a steel mill.[13] Other groups have collected similar data on major sources, especially coal-fired power plants.[14–16]

With the enormous increase in our knowledge of both ambient particles and particles released from various classes of sources, it is important to develop reliable methods for treating these data bases in such a way as to extract the maximum information about the sources of particles of various sizes and compositions, and to determine relationships between emission rates and observed concentrations. From these relationships, we should be able to learn more about the connecting processes, i.e., the dispersion of materials released (both horizontally and vertically), the deposition rates of particles at ground level, and the resultant residence times of particles of different size and composition.

In this paper and those to follow in this session, we focus attention on methods for obtaining maximum information about sources from large data sets on ambient aerosols.

Chemical Element Balances

Two basic approaches have been used for interpretation of large sets of data on ambient aerosols: chemical element balances (CEB) and factor analysis (FA). The basic assumption of CEB is that the concentration of element i in a given particulate sample is given by:[17]

$$C_i = \sum_j m_j X_{ij} \qquad (1)$$

where m_j is the total mass of material in the sample originating from source j, and X_{ij} is the concentration of element i in material released by source j. To calculate the source strengths, m_j, from a set of concentrations, C_i, one must assemble concentration data for particles from each of the sources thought to be important in the area in question. Miller *et al.* used soil dust, marine aerosols, motor vehicle emissions, and particles from fuel oil combustion in an attempt to fit the observed concentration patterns of Pasadena, California, aerosols.[18] Friedlander assumed these components, plus cement dust and particles abraded from tires, in a reevaluation of the Pasadena, California, CEB.[17] Gatz used auto, cement dust, coal, steel, fuel oil, and soil components to resolve the Chicago area atmospheric elemental concentration pattern.[19]

Once the components are chosen, source strengths must be determined by fitting to the concentrations of a number of elements equal to or greater than the number of components. Usually this is done by a least squares method in which there is some kind of weighting of elements, so that elements present in small amounts have a reasonable influence on the fit.

TABLE 1

CHEMICAL ELEMENT BALANCE OF WASHINGTON, D.C., AEROSOLS*

	Contributions from Components (ng/m³)						Total Concentration (ng/m³)		
Element	Soil	Marine	Coal	Oil	Refuse	Motor Vehicle	Predicted	Observed†	Larger‡/Smaller
Elements used in fitting									
Al	1200	< 0.001	720	1	14	—	1940	1680 ± 1100	1.15§
Na	63	280	12	27	82	—	460	470 ± 470	1.0§
Fe	750	< 0.001	510	6.1	6.5	67	1340	1260 ± 940	1.06§
V	1.6	< 0.001	2.2	51	0.031	—	55	54 ± 57	1.0§
Zn	1.7	< 0.001	3.6	3.6	120	20	150	150 ± 90	1.0§
Pb	0.23	< 0.001	2.9	0.86	81	1300	1380	1400 ± 1660	1.01§
Mn	19	< 0.001	2.2	0.22	0.73	—	22	27 ± 16	1.2§
As	0.09	< 0.001	4.4	0.061	0.24	—	4.8	5.7 ± 5.0	1.2§
Remaining elements									
K	230	10	94	1.0	—	—	340	510 ± 350	1.5
Mg	110	37	38	8.6	13	—	210	440 ± 380	2.1
Ca	97	11	66	18	17	67	280	770 ± 500	2.8
Ba	10	0.001	6.6	4.2	0.7	17	39	27 ± 24	1.4
Cl	3.9	510	1.5	27	200	130	870	140 ± 75	6.2§
Br	0.14	1.8	3.0	0.12	1.6	490	500	190 ± 220	2.6§
I	0.08	1.6	3.0	—	—	—	4.7	9.3 ± 7.2	2.0§
Sc	0.22	<0.001	0.25	<0.001	0.001	—	0.47	0.63 ± 0.63	1.3
Ti	77	<0.001	44	0.056	2.0	—	120	120 ± 95	1.0
Cr	1.2	<0.001	1.2	0.13	0.50	—	3.0	11 ± 9	3.7
Co	0.32	<0.001	0.35	0.36	0.007	—	1.0	1.1 ± 1.0	1.1
Ni	0.67	<0.001	1.5	8.6	0.16	—	11	27 ± 28	2.5
Cu	0.34	<0.001	3.0	1.8	1.6	—	6.7	13 ± 12	1.9
Se	0.001	<0.001	1.1	0.076	0.037	—	1.2	3.5 ± 2.7	2.9§
Cd	0.002	<0.001	0.19	0.006	1.5	—	1.7	3.5 ± 2.2	2.1
Sb	0.012	<0.001	0.18	0.015	2.1	—	2.3	9.7 ± 8.5	4.2
La	1.1	<0.001	0.44	0.036	0.004	—	1.6	1.9 ± 1.7	1.2
Ce	1.4	<0.001	0.86	0.030	0.016	—	2.3	3.4 ± 3.1	1.5
Th	0.17	<0.001	0.15	0.004	0.002	—	0.33	0.32 ± 0.24	1.0
								Average	2.0

*Data from Reference 20.
†Uncertainty is standard deviation.
‡Ratio of observed/predicted or predicted/observed, whichever is larger.
§Not included in calculation of average larger/smaller value.

The CEBs described above gave good fits to the major elements, largely because they were used in the fitting routine. However, concentrations of many minor and trace elements were often underpredicted by large factors, especially chalcophile elements such as As, Cd, and Zn. The main problems with these early attempts at CEBs seemed to be that (1) some important sources may not have been included; and (2) the composition of input materials, or of fly ash collected in pollution-control devices, was used for some components because of a lack of information about suspended particles released from the sources.

A much better fit to the Washington, D.C., aerosol composition pattern was obtained by Kowalczyk *et al.*, who use 8 elements to fit six components: soil, marine, motor vehicle, oil, coal, and refuse.[20] The predicted and observed concentrations of the 27 elements of their CEB are shown in TABLE 1. The first 8 elements were used in the least squares fit to determine the source strengths.

The sea salt, oil, refuse, and motor vehicle source strengths are determined largely by the concentrations of Na, V, Zn, and Pb, respectively, but it is difficult to resolve the soil and coal components, which are chemically quite similar for most elements. The marker elements Al and Fe provide a fit to the sum of soil and coal components. Although they are not ideal choices, As and Mn are used to resolve the sum, as Mn is depleted in coal and As has a much higher concentration in coal emissions than in soil.

Of the 19 elements not used in the fit, 4 are considered too volatile to be conservative on the particles (Cl, Br, I, and Se). The quality of the fit is indicated by the larger/smaller ratio (predicted/observed or vice versa, whichever is larger) for the remaining 15 elements. The average value of the larger/smaller ratio (L/S) is 2.0, a considerably better fit than in previous CEB studies. Major reasons for the improvement were (1) addition of the refuse component (which accounts for much of the predicted Zn, Cd, and Sb); (2) the fact that Washington has little industry; (3) the use of true suspended-particle data for most components; and (4) the fact that the 15 test elements include several lithophiles (e.g., Sc, Ti, Th, rare earths) that are usually fitted well if Al, Si, or Fe is used in the fit.

Despite the improvements, the CEB approach still has serious deficiencies. The Washington, D.C., CEB underpredicted Mg and Ca by factors of 2 and 3, respectively, probably indicating the need for a limestone or a cement component. It underpredicted Cr, Cu, Ni, Cd, and Sb by factors of 2 to 4. This may indicate another source not included, but possibly is a result of a more fundamental deficiency of the model—the assumption that elements from a given source remain coherent after release. The most obvious difficulty with that assumption is that the various elements from a particular source may not reside on particles having the same size distribution. For example, Al and many other lithophile elements from coal-fired power plants are borne preferentially by large particles, whereas some of the more volatile elements, such as As and Sb, are predominantly attached to small particles. The latter have longer atmosphere residence times, so at long times after release, they will be enhanced relative to lithophiles in the coal-combustion component.

In regard to size distributions, the observed size distributions of particles bearing the various elements in ambient air can be compared with similar size information on elements on suspended particles from specific sources. If the CEB correctly predicts the major sources of an element, then there should be agreement between the predicted and observed size distributions. For most elements treated by Kowalczyk *et al.*,[20] there is such agreement; e.g., elements such as Pb, Br, As, and Zn are predicted to come from small-particle sources (high-temperature combustion), and elements such as Al, Fe, Sc, rare earths, etc., from large-particle sources (soil and the fly ash particles from coal combustion), and the observed size distributions for these elements are consistent with those predictions (e.g., Reference 5). However, one disturbing result of this comparison is that Mn is predicted to originate from large-particle sources (mainly soil), but size distributions for Mn-bearing particles clearly display both large- and small-particle components.[5] This is a serious problem for the approach of Kowalczyk *et al.*,[20] as they found Mn to be one of the most useful elements for distinguishing between coal and soil components.

Another general weakness of the CEB method is that, in its usual form, it deals only with primary aerosols, i.e., materials released as particles from the sources. It does not conveniently deal with secondary aerosols, i.e., substances such as SO_2, NO_x, and NH_3, that are released as gases but subsequently form particulate matter. Friedlander[17] and Gartrell and Friedlander[21] have attempted to deal with secondary material, but as an "add-on" to the CEB rather than an integral part of it.

Factor Analysis

The other major approach to the interpretation of elemental composition data on ambient aerosols is factor analysis (or, more generally, "multivariate analysis"). In the usual factor analysis (e.g., Reference 22), the concentrations X_{ij} of element i in sample j are transformed to normalized variables:

$$Z_{ij} = \frac{X_{ij} - \overline{X}_i}{\sigma_i} \tag{2}$$

where $\overline{X}_i$ is the average concentration of element i over the entire set of samples, and σ_i is the standard deviation of the X_i distribution. Then one uses computer-fitting methods in order to fit the Z_{ij} values to a set of equations:

$$\begin{aligned} Z_{1j} &= a_{11}F_{1j} + a_{21}F_{2j} + --- + a_{m1}F_{mj} + d_1U_{1j} \\ --- \quad & --- \quad --- \quad --- \quad --- \quad --- \\ Z_{nj} &= a_{1n}F_{1j} + a_{2n}F_{2j} + --- + a_{mn}F_{mj} + d_nU_{nj} \end{aligned} \tag{3}$$

in which the F_{ij}s are the factors common to all of the variables, and the U_{ij}s are factors unique to each element. The *a*s are called the "factor loadings." The objective of the fitting procedure is to determine a number of common factors (m) that will explain most of the observed variation of concentrations of the elements in the set.

Hopke *et al.*[22] applied factor analysis to a set of 18 elements in 90 samples from the Boston area. They found that 77.5% of the total variance in the system could be accounted for by six common factors, with little improvement when more factors were permitted. From the factor loadings, they were able to associate the factors with several sources: soil (mixed with coal), sea salt, oil, auto emissions, and refuse combustion. A sixth factor had large loadings for only Mn and Se and could not be identified with a particular source.

Factor analysis has several advantages over CEB in that no a priori assumptions need to be made about either the number or the composition of the sources. This means, for example, that factor analysis can handle secondary materials that become associated with primary particles between their release and their reception at collection sites. For example, Gaarenstroom *et al.* included NH_4^+, $SO_4^=$, NO_3^-, and

TSP, along with 20 metals, in their factor analysis of data on Tucson, Arizona, aerosol.[23] Ammonium and $SO_4^=$ ions were the strongest members of the second factor extracted from the data at each sampling location. This factor also contained major amounts of Cu and minor amounts of other chalcophile metals. Most of the NO_3^- variation was explained by a different factor that contained little else at most locations.

Another advantage of factor analysis is that no assumptions are made about the composition of the components. Thus, there is no implicit assumption that all elements from a source remain coherent after release. Another important advantage is that variables associated with the samples other than concentrations can be included in the data set, e.g., particle size group, meteorological data, concentrations of gaseous pollutants measured simultaneously with sample collections, b_{scat} or visibility, etc. This potentially useful feature of factor analysis has not been exploited in aerosol studies, but Hopke used many variables in addition to elemental concentration in his interpretation of lake sediment data, e.g., sand, silt, and clay contents, organic carbon, depth, and grain size.[24]

Factor analysis also has weaknesses. Since it works on variations rather than absolute concentrations, it is more effective on data sets in which there are large variations from sample to sample. It is not very effective on data sets in which little variation occurs. Gladney *et al.* applied factor analysis and CEB to data from a network of stations in a rural area of Prince George's County, Maryland.[25] The CEB gave fits to the data with six components that were slightly better than obtained for the Washington, D.C., aerosol composition.[20] However, for most of the data set, factor analyses yielded only two or, at most, three factors. The problem seems to be that, since the data are from a rural network, most of the particles except for soil are transported from some distance and, thus, are sufficiently well mixed that elemental concentrations are rather uniform throughout the network. Even when there is a lot of variation in a system, Henry has noted that interpretative schemes based on the observation of correlations suffer from the fact that concentrations of most species in the air are positively correlated with each other because most are correlated with dispersion parameters imposed by meteorological conditions.[26] Kleinman has attempted to remove this correlation from the data by normalizing actual concentrations to the same average wind speed and mixing height.[27] Henry suggests that the effect be removed by the use of concentrations of elements per unit mass of TSP, rather than absolute concentrations per unit volume of air.[26]

Another problem with factor analysis in its usual form is that one cannot obtain the relative concentrations of elements within the factors. The output indicates only the fraction of the variation of the concentration that is explained by the factor. Attempts to rectify this deficiency are noted below.

The other major weakness of factor analysis is that it will probably be difficult or impossible for it to resolve components that have very similar composition patterns. We noted above the classic problem of resolving soil and coal components with CEB. The factor analyses done to date do not appear to resolve soil and coal. Usually the first factor is soil, probably with an admixture of coal emissions if much coal is consumed in the area. In the factor analysis of Boston aerosols noted above, Hopke *et al.* found one unidentified component with large factor loadings for only Mn and Se.[22] Perhaps this component was compensating the major component (soil and, probably, coal) for the elements that have different relative concentrations for soil and coal.

Note that the Boston data set did not include As. Selenium has a much higher concentration relative to Al in coal emissions than in soil ($\sim 10^{-3}$ vs. 10^{-6}), but was not used in the CEB fitting by Kowalczyk *et al.* because of its volatility.[20]

FUTURE DIRECTIONS

Chemical element balances and factor analysis are powerful and useful approaches for extracting maximum information from multielement data on aerosols. The strengths and weaknesses of the two approaches are rather complementary, so that both are needed for interpretation of large data sets. What is now needed is a common approach that has the advantages of both present approaches. From the factor analysis side, both Henry[26] and Hopke[28] have modified the usual approach to bring the treatment more into agreement with CEB. Formerly, it was common to rotate the eigenvectors of the solution in order to obtain maximum alignment with the axes (i.e., the normalized concentrations), according to the principle of "simple structure" (see Appendix of Reference 22). However, since something is known about the probable compositions of the factors from the CEB approach, more physically understandable results are obtained by rotating the solutions to achieve maximum overlap with test vectors that are similar to the components used in CEB. Results so far obtained by Henry and Hopke suggest that this will be a fruitful approach in bringing these two separate methods to a common ground.[26,28]

On the CEB side, Kowalczyk is attempting to relax the assumption of coherence of elements from a given source by making separate CEB calculations on large and small particles, based on size fractionated data on individual sources and ambient air.[29] Also, our group at Maryland and a group at the Oregon Graduate Center[30] are obtaining data on additional major sources that have not yet been included in CEB calculations. It is clear from Kowalczyk's work, and from similar studies by Dzubay,[31] that some kind of additional component—containing large amounts of Ca and other alkaline earths—is needed to account for those elements. The origin of this component is not clear. Cement plants make a contribution, but agricultural liming, abrasion of highways, and building demolition and construction may provide much of this component. A steel mill component, under investigation in our laboratory,[13] will be needed for many industrial cities.

At the present level of sophistication, we are content to account for most elements to within a factor of two or, in really favorable cases, to $\pm 50\%$. If it becomes necessary or desirable to account for elements more accurately, say to $\pm 20\%$, much more work will have to be done, even on sources for which we now have reasonably good information. Among the sources now treated, oil-fired emissions are subject to great fluctuations caused by variations in the composition of the oil.[32,33] In order to obtain a very accurate accounting of elements contributed in large amounts from that source, one would need analyses of major sources of oil for the area and time of the study. Likewise, Zoller *et al.* have found large variations of composition from one copper smelter to another.[12] Thus, to treat an area affected by copper (or other nonferrous metal) smelters, one would need good data for each individual plant. Although this is a disadvantage for determining source-type contributions, it may be an advantage for identification of contributions from individual plants within a source category.

Another problem requiring further study is that of the composition of material from some sources that is actually present in the plume for a reasonable time. In-stack measurements, for example, have been done on coal-fired power plants,[7,8,14–16] but too little has been done on plumes. A lot of in-stack material is on particles so large that they probably fall out close to the plant. Furthermore, at stack temperatures, large fractions of some volatile elements (Hg, Se, halogens) are surely in the gas phase,[7,34,35] but it isn't known how much may attach to particles upon cooling. Airplane collections of size-segregated particulate material should be performed at various distances downwind in plumes from large sources, although this is difficult because of small amounts collected per pass through a plume, isokinetic sampling requirements, etc. Some major components should be followed for long-term variations of composition, e.g., the motor vehicle component which is changing as lead is phased out, diesel-powered automobiles come into wider use, etc.

The above discussion has focused on the use of elemental patterns to determine source contributions. Little work has been done on the use of molecular species, but some of these species are quite promising as "signatures" of sources, e.g., the series of polynuclear aromatic hydrocarbon (PAH) compounds. About 10 or more of the PAH compounds on particles can be measured by high-pressure liquid chromatography,[36] and with gas chromatographic mass spectrometry, various alkyl homologues can be measured as well.[37] Thus, series such as these, whose patterns differ from one source to another, could be combined with elemental data to provide more definitive source identifications. A preliminary attempt to apply this approach, using PAH compounds, has been reported.[38]

The methods described above offer the possibility for obtaining high quality information on the contributions of various sources to aerosols collected in a particular area. Some very high quality, extensive data sets are needed to refine and test the methods. These sets must contain data on a large number of elements or other chemical species. Obviously, the number of species must be greater than the number of sources treated, and, if one is to test the fit, there must be many additional species. Furthermore, to provide sensitivity of the fits and tests, these cannot be just any group of elements or compounds that are easy to measure. Many lithophile elements, for example, have about the same relative concentrations regardless of the source, so they don't provide too sensitive a test of fitting procedures. It is more challenging, but also more informative, if the data set includes a large number of chalcophile elements, which usually exhibit greater enrichments and variations from one source to another.

Summary

Neutron activation analysis and x-ray fluorescence produce large, reliable data sets on elemental concentrations on particles in urban areas. Variations of concentrations with particle size, time, and location have high information content, but we are just learning to extract maximum information on sources of particles and specific elements from these data. Two methods used for treatment of data are chemical element balances (CEB) and factor analysis. Both are useful, but have important limitations. In its present form, CEB requires an accurate knowledge of the compositions of particles from all major sources, does not handle secondary aerosols conveniently, and assumes that all elements from a source have the same residence time. The CEB method yields the primary aerosol contributions from various types of sources to the total suspended particulate matter concentration (TSP) and to the concentrations of individual elements attached to suspended particles. Factor analysis requires no prior knowledge on sources, handles secondary particles well, and can distinguish elements of different residence times from a source. However, in its usual form, factor analysis does not yield compositions of the components and yields little information from data sets in which variations are small. Recent improvements in both techniques have overcome some disadvantages, but the methods remain complementary. Both approaches are needed to treat large data sets.

References

1. Stevens, R. K., T. G. Dzubay, G. Russwurm & D. Rickel. 1978. Atmos. Environ. **12:** 55.
2. Loo, B. W., W. R. French, R. C. Gatti, F. S. Goulding, J. M. Jaklevic, J. Llacer & A. C. Thompson. 1978. Atmos. Environ. **12:** 759.
3. Jaklevic, J. M. & R. L. Walter. 1977. Comparison of minimum detectable limits among x-ray spectrometers. *In* X-Ray Fluorescence Analysis of Environmental Samples. T. G. Dzubay, Ed.: 63–75. Ann Arbor Science Publishers, Inc. Ann Arbor, Mich.
4. Zoller, W. H. & G. E. Gordon. 1970. Anal Chem. **42:** 257.
5. Gladney, E. S., W. H. Zoller, A. G. Jones & G. E. Gordon. 1974. Environ. Sci. Technol. **8:** 551.
6. Jaklevic, J. M., R. C. Gatti, F. S. Goulding & A. Thompson. 1978. Automated elemental analysis using energy dispersive x-ray fluorescence analysis. Paper No. 186. *In* Proceedings of the 4th Joint Conference on Sensing of Environmental Pollutants, New Orleans, La., 1977. American Chemical Society. Washington, D.C.
7. Gladney, E. S., J. A. Small, G. E. Gordon & W. H. Zoller. 1976. Atmos. Environ. **10:** 1071.
8. Small, J. A. 1976. An Elemental and Morphological Characterization of the Emissions from the Dickerson and Chalk Point Coal-Fired Power Plants. Ph.D. Thesis. Department of Chemistry, University of Maryland. College Park, Md.
9. Mroz, E. J. 1976. The Study of the Elemental Composition of Particulate Emissions from an Oil-Fired Power Plant. Ph.D. Thesis. Department of Chemistry, University of Maryland. College Park, Md.
10. a) Greenberg, R. R., W. H. Zoller & G. E. Gordon. 1978. Environ. Sci. Technol. **12:** 566; b) Greenberg, R. R., G. E. Gordon, W. H. Zoller, R. B. Jacko, D. W. Neuendorf & K. J. Yost. 1978. Environ. Sci. Technol. **12:** 1329.
11. Ondov, J. M. 1974. A Study of Trace Elements on Particulates from Motor Vehicles. Ph.D. Thesis. Department of Chemistry, University of Maryland. College Park, Md.
12. Zoller, W. H., M. Small, M. Germani & J. L. Moyers. 1978. Atmospheric trace element emissions from copper smelters. Presented at the 174th National Meeting of the American Chemical Society, Miami Beach, Fla., September 14, 1978.
13. Small, M. 1978. Unpublished data.
14. Klein, D. H., A. W. Andren, J. A. Carter, J. F. Emery, C. Feldman, W. Fulkerson, W. S. Lyon, J. C. Ogle, Y. Talmi, R. I. Van Hook & N. E. Bolton. 1975. Environ. Sci. Technol. **9:** 973.
15. Kaakinen, J. W., R. M. Jorden, J. H. Lawasani & R. West. 1975. Environ. Sci. Technol. **9:** 862.
16. Natusch, D. F. S., J. R. Wallace & C. A. Evans. 1974. Science **183:** 202.
17. Friedlander, S. K. 1973. Environ. Sci. Technol. **7:** 235.
18. Miller, M. S., S. K. Friedlander & G. M. Hidy. 1972. J. Colloid Interface Sci. **39:** 165.
19. Gatz, D. F. 1975. Atmos. Environ. **9:** 1.
20. Kowalczyk, G. S., C. E. Choquette & G. E. Gordon. 1978. Atmos. Environ. **12:** 1143.
21. Gartrell, G., Jr. & S. K. Friedlander. 1975. Atmos. Environ. **9:** 279.

22. Hopke, P. K., E. S. Gladney, G. E. Gordon, W. H. Zoller & A. G. Jones. 1976. Atmos. Environ. **10:** 1015.
23. Gaarenstroom, P. D., S. P. Perone & J. L. Moyers. 1977. Environ. Sci. Technol. **11:** 795.
24. Hopke, P. K. 1976. J. Environ. Sci. Health **A11:** 367.
25. Gladney, E. S., G. S. Kowalczyk, G. E. Gordon, W. H. Zoller & P. K. Hopke. 1978. Unpublished data.
26. Henry, R. C. 1977. The application of factor analysis to urban aerosol source identification. Presented at 5th Conference on Probability and Statistics, Las Vegas, Nev., November 1977.
27. Kleinman, M. T. 1977. The Apportionment of Sources of Airborne Particulate Matter. Ph.D. Thesis. Environmental Health Science Program, New York University. New York, N.Y.
28. Hopke, P. K. 1977. The use of multivariate statistical techniques to interpret multielement environmental analysis. *In* Proceedings of the 3rd International Conference on Nuclear Methods in Environmental and Energy Research, TID Document No. Conf. 771072: 49–57. University of Missouri. Columbia, Mo.
29. Kowalczyk, G. S. 1979. Concentrations and Sources of Trace Elements on Washington, D.C. Area Atmospheric Particles. Ph.D. Thesis. Department of Chemistry, University of Maryland. College Park, Md.
30. Cooper, J. A. 1978. Private communication.
31. Dzubay, T. G. 1980. Chemical element balance applied to dichotomous sampler data. Ann. N.Y. Acad. Sci. (This volume.)
32. Cahill, R. A. 1974. A Study of the Trace Element Distribution in Petroleum. M.S. Thesis. Department of Chemistry, University of Maryland. College Park, Md.
33. Shah, K. R., R. H. Filby & W. A. Haller. 1970. J. Radioanal. Chem. **6:** 185 and 413.
34. Andren, A. W., D. H. Klein & Y. Talmi. 1975. Environ. Sci. Technol. **9:** 856.
35. Billings, C. E. & W. R. Matson. 1972. Science **176:** 1232.
36. Fox, M. A. & S. W. Staley. 1976. Anal. Chem. **48:** 992.
37. Hites, R. A., R. E. LaFlamme & J. W. Farrington. 1977. Science **198:** 829.
38. Gether, J. & H. M. Seip. 1979. Atmos. Environ. **13:** 87.

SOURCE IDENTIFICATION AND RESOLUTION THROUGH APPLICATION OF FACTOR AND CLUSTER ANALYSIS

Philip K. Hopke

Environmental Research Laboratory
Institute for Environmental Studies
University of Illinois
Urbana, Illinois 61801

INTRODUCTION

The determination of the origins of the elements observed in urban aerosols has been the subject of a number of recent studies. The majority of these investigations employ the chemical balance approach first introduced by S. K. Friedlander and his coworkers.[1,2] A number of the initial efforts[3–5] in source resolution produced poor fits to the data because of the lack of good data on the chemical composition of the particulates emitted by various sources. Improved results were obtained by a more sophisticated model[6] that required much more detailed information regarding gas-to-particle conversion rates and other reaction constants. The extensive efforts of the group at the University of Maryland have led to much better determinations of the elemental profiles of a number of sources and have allowed the group to develop a set of sources that gives a very good fit to a set of Washington, D.C., air particulate elemental concentrations.[7] These approaches all initially assume that the number and chemical nature of the source emissions are known and that the contributions of each source to each sample are the only quantities to be determined.

Although through careful examination of source emissions, it is possible to perform a resolution, an enormous amount of effort is required to make determinations of the composition of each type of source. It would, therefore, be of considerable value if a technique could be developed to permit the identification of the number and nature of the sources contributing to a given aerosol, as well as the ability to estimate the particulate loading contribution of each source. The development and application of multivariate statistical techniques, such as factor analysis, may permit such determinations. The mathematical bases of factor and cluster analysis will be presented, and the application of these techniques to the resolution of a set of aerosol compositions from the Boston metropolitan area will be explored. The details of sample collection and analysis have been reported previously.[8] TABLE 1 summarizes the average elemental compositions and standard deviations over the 90 samples that were taken and analyzed.

FACTOR ANALYSIS

In determining the quantity of a particular element or compound in a specific sample at a definite time, the investigator has randomly removed a sample from a distribution of materials present in the environment. Then, by taking enough samples,

0077–8923/80/0338–0103 $01.75/1 © 1980, NYAS

the distribution of that particular variable in that kind of sample can be described by several parameters commonly used for that purpose, including the mean value of the ith variable:

$$\bar{x}_i = \frac{1}{N} \cdot \sum^{N} x_{ij} \qquad (1)$$

and the second moment of the distribution or the *variance:*

$$s_i^2 = \frac{1}{N-1} \cdot \sum^{N} (x_{ij} - \bar{x}_i)^2 \qquad (2)$$

where N is the number of samples examined. The standard deviation is simply the square root of the sample variance.

TABLE 1

MEAN VALUES AND STANDARD DEVIATIONS OF ELEMENTAL CONCENTRATIONS OVER ALL SAMPLES

Element	Concentration Unit	Mean Value	Standard Deviation
Na	($\mu g\ m^{-3}$)	0.51	0.77
Cl	($\mu g\ m^{-3}$)	1.50	1.1
Br	($ng\ m^{-3}$)	190	120
Al	($\mu g\ m^{-3}$)	1.27	1.03
Sc	($ng\ m^{-3}$)	0.26	0.24
V	($\mu g\ m^{-3}$)	0.725	0.50
Cr	($ng\ m^{-3}$)	3.4	5.5
Mn	($ng\ m^{-3}$)	27	19
Fe	($\mu g\ m^{-3}$)	1.09	1.0
Co	($ng\ m^{-3}$)	1.00	0.7
Zn	($ng\ m^{-3}$)	190	220
Se	($ng\ m^{-3}$)	1.23	1.23
Sb	($ng\ m^{-3}$)	9.1	11
La	($ng\ m^{-3}$)	1.45	1.26
Ce	($ng\ m^{-3}$)	2.2	1.8
Sm	($ng\ m^{-3}$)	0.20	0.17
Eu	($pg\ m^{-3}$)	35	30
Th	($pg\ m^{-3}$)	200	190

For some statistical procedures, it is necessary to remove the effects of using different metrics in describing the various variables, and so the variables are put in standard form. First, the deviation is calculated by subtracting the mean value from each sample value.

$$d_{ij} = x_{ij} - \bar{x}_j \qquad (3)$$

The standardized variable, z_{ij}, can be calculated by dividing the deviation by the standard deviation.

$$z_{ij} = \frac{d_{ij}}{s_j} = \frac{x_{ij} - \bar{x}_j}{s_j} \qquad (4)$$

The standardized value then has a mean value of zero and a standard deviation of unity, and thus all standardized variables have the same metric.

The initial step in the analysis of the data generally requires the calculation of a function that can indicate the degree of interrelationship that exists within the data. Functions exist that can provide this measure between the two variables, when calculated over all of the samples, or between the samples, when calculated over all of the variables. The most well known of these functions is the product-moment correlation coefficient. To be more precise, this function should be referred to as the correlation about the mean. The "correlation coefficient" between two variables x_j and x_k over all N samples is given by:

$$r_{jk} = \frac{\sum^{N} (x_{ji} - \bar{x}_j)(x_{ji} - \bar{x}_k)}{\left[\sum^{N} (x_{ji} - \bar{x}_j)^2 \cdot \sum^{N} (x_{ki} - \bar{x}_k)^2\right]^{1/2}} \tag{5}$$

Utilizing the standardized variables, Equation 5 simplifies to:

$$r_{jk} = \frac{1}{N} \sum^{N} z_{ji} \cdot z_{ki} \tag{6}$$

There are several other measures of interrelationship that can also be utilized. These measures include covariance about the origin:

$$c_{jk} = \sum^{N} x_{ji} \cdot x_{ki} \tag{7}$$

the covariance about the mean:

$$s_{jk} = \sum^{N} d_{ji} \cdot d_{ki} \tag{8}$$

and the correlation about the origin:

$$r'_{jk} = \frac{\sum^{N} x_{ji} \cdot x_{ki}}{\left(\sum^{N} x_{ji}^2 \sum^{N} x_{ki}^2\right)^{1/2}} \tag{9}$$

The matrix of all of the values of one of these functions for a given set of data is called a dispersion matrix and can be obtained from the data matrix, either in its original form or after appropriate transformation. The data matrices contain the data for the n variables measured over the N samples. The correlation about the mean is given by:

$$(R_m) = (Z)(Z)^t \tag{10}$$

where $(Z)^t$ is the transpose of the standardized data matrix (Z). The correlation about the origin:

$$(R_0) = (Z^0)(Z^0)^t = \frac{(X)(X)^t}{(V)(V)^t} \tag{11}$$

where $z_{ij}^0 = x_{ij}/(\Sigma^N x_{ij}^2)^{1/2}$, which is a normalized variable still referenced to the original variable origin, and (V) is a diagonal matrix of the square root of the sums of the squares of the variables. The covariance about the mean is given as:

$$(C_m) = (D)(D)^t \tag{12}$$

where $(D)^t$ is the transpose of the matrix of deviations from the mean, calculated using Equation 3. The covariance about the origin is:

$$(C_0) = (X)(X)^t \tag{13}$$

the simple product of the data matrix by its transpose. As written, these product matrices would be of dimension n by n and would represent the pairwise interrelationships between variables. If the order of the multiplication is reversed, then the resulting N by N dispersion matrices contain the relationships between samples.

There has been some discussion in the literature[9,10] regarding the relative merits of these functions to reflect the total information content of the data. Rozett and Peterson[9] argue that since many types of physical and chemical variables have a real zero, the correlation and covariance about the mean remove the information regarding the location of the original origin by only including differences from the variable mean. The normalization made in calculating the correlation from the covariances causes each variable to have an identical weight in the subsequent analysis. In mass spectrometry, where the variables consist of the various m/e values observed for the fragments of a molecule, the normalization represents a loss of information because the metric is the same for all of the m/e values. In environmental elemental analyses, where concentrations range from trace species at the sub–part-per-million level to major constituents at the percent level, the use of covariance may weight the major constituents too heavily in the subsequent analyses. The choice of function depends heavily on the underlying assumptions made concerning the nature of the parameters being measured.

The goal of factor analysis is to simplify the quantitative description of a system by determining the minimum number of new variables necessary to reproduce various attributes of the data. These procedures reduce the original data matrix from one having n variables necessary to describe the N samples to a matrix with m components or factors ($m < n$) for each of the N samples. In addition to permitting a parsimonious representation of data, this also permits the calculation of the nature of each of the factors.

In factor analysis, the value of the variable observed in the sample is assumed to be a linearly additive function of the contribution from each of the m causalities that actually govern the system's composition. For example, for air particulates, the amount of particulate lead in the air could be considered to be a sum of contributions from several sources, including automobiles, incinerators, and coal-fired power plants, etc.:

$$x_t(\text{Pb}) = c_1(\text{Pb}) + c_2(\text{Pb}) + c_3(\text{Pb}) + \ldots \tag{14}$$

where x_t is the observed total particulate lead per unit volume, $c_1(\text{Pb})$ is the amount of lead per unit volume from source 1, $c_2(\text{Pb})$ is the amount per unit volume from source 2, and so forth for all of the m possible sources.

The individual source contributions can generally be thought of as a product of two cofactors; one that gives the amount of lead in particles emitted by that particular source, and another that gives the amount of particulate matter per unit volume present that can be attributed to that source. Thus, the c_i(Pb)'s can be rewritten as:

$$c_i(\mathrm{Pb}) = a_i(\mathrm{Pb})F_i \tag{15}$$

Expanding this approach to all of the n variables that have been measured for the N samples gives:

$$x_{ij} = a_{i1}F_{1j} + a_{i2}F_{2j} + \ldots + a_{im}F_{mj} \tag{16}$$

where x_{ij} is the value of the ith variable for the jth sample. Equation 16 is the model for the analysis and can be expressed in matrix form as:

$$(X) = (a)(F) \tag{17}$$

where (X) is the $n \times N$ data matrix, (a) is the $n \times m$ factor loading matrix, and (F) is the $m \times N$ factor score matrix. Factor analysis must permit determination of the number of terms required, m, and of one of the two matrices.

The resolution of the data matrix into the product of two cofactor matrices is accomplished by the diagonalization of one of the four dispersion matrices given in Equations 10 to 13, which represent the variable correlations, or from the dispersion matrices that give the sample relationships. In the previously published uses of factor analysis,[8,11] the analysis was made on the matrix of correlations about the mean between the variables. In the terminology of Rozett and Peterson,[9] these are R analyses. If the dispersion matrix is calculated between samples, then it is called a Q analysis. Both analyses aim for the same result; the determination of the (a) and (F) matrices. The R analysis calculates the (a) matrix first, while the Q analysis determines the (F) matrix. The remaining matrix is calculated from the data, using Equation 17. The new results presented below for the Boston aerosol data will be calculated using a Q analysis, while the original report[8] used an R analysis. A more important difference between these two results is the choice of dispersion matrix. The original report used correlation about the mean, while this report will employ correlation about the origin. The implications of this change will be presented below.

The next step in the analysis is to find a matrix that will diagonalize the dispersion matrix (C) such that:

$$(B)^{-1}(C)(B) = (\Lambda) \tag{18}$$

where (Λ) is a diagonal matrix of eigenvalues arranged in order of descending value. If the eigenvalues are not degenerate, the eigenvectors in the (B) matrix are orthogonal, so that $(B)^{-1} = (B)^{t}$. Then, specifically for the correlation about the origin,

$$(B)^{-1}(R_0)(B) = (B)^{-1}(Z^0)^{t}(Z^0)(B) = (\Lambda) = (B)^{t}(Z^0)^{t}(Z^0)(B) \tag{19}$$

Thus, if the (a) matrix in Equation 17 is defined to be $(Z^0)(B)$, then:

$$(B)^{t}(Z^0)^{t}(Z^0)(B) = (a)^{t}(A) = (A) \tag{20}$$

The data can then be expressed as:

$$(Z^0) = (a)(B)^t \tag{21}$$

Using $(Z^0) = (X)/(V)$, from analogy to Equation 17,

$$(F) = (B)^t(V) \tag{22}$$

so that if the matrix that diagonalized (C) can be found, its transpose times the (V) matrix will represent the values for (F) in Equation 17:

$$(X) = (a)(B)^t(V) = (a)(F) \tag{23}$$

which relates the original data to the matrix obtained from diagonalization. If correlation about the mean has been employed, then the diagonalization would yield:

$$(Z) = (a)(B)^t \tag{24}$$

giving a model of the form:

$$\frac{x_{ij} - \bar{x}_i}{s_i} = \sum_{k=1}^{m} a_{ik} F_{kj} \tag{25}$$

Only the deviations from the mean value of an element for a sample can be related to the source factors. Thus, the use of the correlation about the mean precludes a source resolution in the same sense as that of the chemical balance approach and as given in Equation 16. One factor analysis resolves the sources of variance in the system, while the other permits the actual resolution of data.

So far, however, the dimensionality of the system has not yet been reduced. Since the elements are interrelated by being emitted by a few common sources, the data can be reproduced by m factors rather than n variables. In theory, the eigenvalues corresponding to sources should have nonzero values, and the rest of them should be zero. In practice it is often difficult to choose the proper number of factors to retain. In the previous report on these Boston data, six factors were retained, but there were several elements whose variance was not well reproduced, including bromine, selenium, and antimony, indicating the likelihood of additional factors. For the present study, the results of several methods to help choose the proper number of factors are given in TABLE 2. From the apparent large changes in the eigenvalues, the root-mean-square percent error, and the exner function,[12] it appears that seven factors should be retained.

It is then necessary to obtain an interpretation of the nature of these factors for the system being studied. Mathematically, any matrix that performs the diagonalization is as good as any other. However, it is desired to be able to have elemental distributions and source contributions as the physical entities being modeled. In order to obtain this understanding, it is necessary to rotate the factor axes in the reduced dimensional space.

In the previous applications of factor analysis to elemental source identification studies,[8,11] the matrix of factors was rotated in such a way as to maximize the number of values that are zero or unity. This rotation criterion is called "simple structure" (see Reference 8, appendix), and a Varimax rotation was used to achieve it. Simple structure may not be a very useful criterion for true source resolution, since an element may be present in a sample because of several different causal forces. Therefore, factor loadings should not necessarily have values of either 0 or 1, but some intermediate value. This can help clarify the partitioning of variance in the system as seen in the previously published results.

A different approach to the rotation problem has been developed by Malinowski,[13–16] in which the investigator uses his understanding of the system to suggest the nature of the possible factors, utilizing a rotation matrix (R):

$$(X) = (a)(R)(R)^{-1}(F) = (\bar{a})(\bar{F}) \tag{26}$$

where $(\bar{a})$ and $(\bar{F})$ are the rotated matrices. The rotation matrix is determined by trying to maximize the overlap between one of the factors and a suggested test vector by the method of least squares. A test vector has up to n values in it, but the rotation matrix is m by m. Thus, since $n > m$, the problem is overdetermined, and a least squares procedure can be applied. In the case of an elemental source resolution, possible test vectors would be the profiles of elements emitted by specific sources, such as automobiles, coal-fired or oil-fired power plants, soil dust, sea-salt aerosol, municipal incinerators, etc. The function to be minimized is a difference between a column of the (a) matrix and the test vector, $(\bar{\bar{a}})$.

TABLE 2

RESULTS OF THE EIGENVALUE ANALYSIS

Factor Number	Eigenvalue	RMS* % Error	Exner
1	35.8	153	0.5216
2	4.19	112	0.2569
3	1.60	103	0.1774
4	0.87	96	0.0972
5	0.26	86	0.0751
6	0.18	83	0.0422
7	0.12	48	0.0043
8	0.00080	35	0.0029
9	0.00043	26	0.0003
10	0.0000052	19	0.0003
11	0.0000028	14	0.0001
12	0.0000004	13	0.0001

*Root-mean-square.

$$\Delta a_{ik} = \bar{a}_{ik} - \bar{\bar{a}}_{ik} = a_{i1}R_{1k} + a_{i2}R_{2k} + \ldots + a_{ij}R_{jk} - \bar{\bar{a}}_{ik} \tag{27}$$

When this difference is minimized, it can be shown:

$$(R_k) = \begin{vmatrix} R_{1k} \\ R_{2k} \\ \vdots \\ R_{jk} \end{vmatrix} = (1/\lambda_j)\delta_{jk}\,(a)^t\,(\bar{\bar{a}}) \tag{28}$$

TABLE 3
POSSIBLE TEST VECTORS

Element	Soil-1[18]	Soil-2[19]	Soil-3[20]	Oil[21]	Coal[22]	Refuse[23]	Marine[24]	Motor Vehicle[25]
Na	0.045	0.12	0.053	0.53	0.016	0.68	1.00	—
Cl	—	—	0.0033	0.53	0.002	1.67	1.81	0.26
Br	1.0×10^{-4}	—	1.2×10^{-4}	0.0023	0.004	0.013	0.0062	1.0
Al	1.0	1.0	1.0	0.018	1.00	0.12	9.5×10^{-7}	—
Sc	1.5×10^{-4}	3×10^{-4}	1.8×10^{-4}	8×10^{-6}	3.5×10^{-4}	1.1×10^{-5}	3.8×10^{-9}	—
V	—	1.4×10^{-3}	1.3×10^{-3}	1.0	0.030	2.6×10^{-4}	1.9×10^{-7}	—
Mn	0.013	0.006	0.016	0.0043	0.003	0.0061	1.9×10^{-7}	—
Fe	0.53	0.6	0.63	0.12	0.70	0.054	9.5×10^{-7}	0.13
Co	—	4×10^{-4}	2.7×10^{-4}	0.007	4.8×10^{-4}	5.5×10^{-5}	4.8×10^{-8}	—
Zn	0.0012	0.002	0.0014	0.07	0.0050	1.0	9.5×10^{-8}	0.039
Se	—	—	1.1×10^{-6}	0.0015	0.0015	3.1×10^{-4}	3.8×10^{-7}	—
Sb	—	—	1.0×10^{-5}	3×10^{-4}	2.5×10^{-4}	0.0175	4.8×10^{-8}	—
La	—	8×10^{-4}	9.3×10^{-4}	7×10^{-4}	6.0×10^{-4}	3.1×10^{-5}	2.8×10^{-8}	—
Ce	0.001	—	0.0012	6×10^{-4}	0.0012	1.3×10^{-4}	3.8×10^{-8}	—
Sm	—	—	—	—	—	—	—	—
Eu	—	—	—	—	—	—	—	—
Th	—	—	1.4×10^{-4}	7×10^{-5}	2.0×10^{-4}	1.5×10^{-5}	4.8×10^{-9}	—

The quality of alignment can then be determined, and it can be ascertained whether the suggested test vector is or is not a good representation of one of the factors. A series of such vectors can be tested. Also, since the problem is overdetermined, this procedure can be completed without necessarily specifying all of the values in the test vector. The resultant calculated prediction can then be compared with the withheld information to help verify that the test vector is indeed a good representation of a factor. An example of this procedure can be seen in the report of Malinowski and McCue.[16]

After the test vectors have been determined to be probable factors, they can be refined by substituting the predicted value for the original test value. Thus, since we expect differences between the initially proposed test vectors and the actual causal factors, we have a mechanism for determining what the nature is of these actual profiles, by using the data to tell us how to derive the best fit. A combination of these refined test vectors can then be used to reproduce the original data and to calculate the $(\overline{F})$ matrix. The program to perform such an analysis has been written by Malinowski and coworkers[17] and is available from the Quantum Chemistry Program Exchange at Indiana University.

The choice of target vectors is given in TABLE 3. For each vector, the maximum values of the concentrations are normalized to a maximum value of 1.0 for the most abundant element. In addition to these vectors, a uniqueness vector—consisting of unity for one element and zero for the others—was tested for each element. A problem that arises in these studies is that of missing data. In general, it has proved a useful rule of thumb to take one-half of the elemental detection limit as the value for that point. In the case of chromium in these data, there is a sufficiently large number of missing data points that it was excluded from the analysis. The vectors that were found to align were soil-1, oil, auto, marine, incinerator, iron, and aluminum. A refining of the vectors was performed by altering the values of the input vectors until further changes did not improve the average point-by-point percent error or lower the maximum single point error. The average error is 38%. The maximum uncertainty in any single point is approximately 100%.

The particulate loading contributions can thus be calculated, but only after the test vectors are scaled to their actual concentration values. If the total suspended particulate (TSP) values were known for each sample, then the scaling factors could be calculated from a multiple regression of:

$$(\mathrm{TSP})_j = \sum_{k=1}^{m} \beta_k F_{kj} \qquad j = 1, N$$

For these Boston samples, the total particulate weights were not determined, and so estimates for the scaling factors must be made, estimated from other measurements of source emissions. For the soil, oil, marine, motor vehicle, and incinerator factors, the scaling factors are the same as Kowalczyk, Choquette, and Gordon.[7] For the iron and aluminum factors, the scaling factors were arbitrarily set to 70% (Fe_2O_3) and 53% (Al_2O_3), respectively. The final set of scaled vectors and the average percent error for each element are given in TABLE 4. The expected TSP values can be calculated by summing the particulate loading contributions of each source for each sample.

This new factor analysis procedure does permit an analysis similar in nature to the chemical balance method. It also provides a mechanism to identify the number of independent sources and to iteratively obtain the best representation of their elemental profiles. The procedure still has some disadvantages. It cannot provide information regarding sources whose major elemental contributions are not among the elements measured. It might be expected that some road cement might be entrained by motor vehicle traffic and represent a contribution to the system. However, the major elements contributed by cement are Ca and Mg, which were not determined. Therefore, although we can obtain a reasonable idea of the sources of the elements observed, it does not mean that a complete description of that aerosol is obtained. Only a complete analysis, inorganic and organic, would permit a total characterization of the system sources.

TABLE 4

REFINED ELEMENTAL CONCENTRATION VECTORS (%)

Element	Soil-1	Oil	Motor Vehicles	Marine	Incinerator	Iron	Aluminum	Average Error
Na	4.8	0.	0.	31.	0.00065	0.00013	0.	0.0072
Cl	0.114	0.	0.000064	49.	0.00018	0.0019	0.	0.021
Br	0.108	0.	5.7	0.19	0.0023	0.0018	0.	0.087
Al	5.7	0.	0.	0.	0.0043	0.0069	53.	0.16
Sc	0.00034	0.	0.00011	0.	0.	2.8×10^{-5}	7×10^{-6}	62.
V	0.0057	7.0	0.017	0.	0.0014	0.0018	0.	0.011
Mn	0.036	0.039	0.076	0.00031	0.	0.	0.00029	42
Fe	3.0	1×10^{-6}	0.00015	0.00031	0.	70.	0.0013	0.009
Co	0.00083	0.00061	0.0035	3×10^{-6}	0.0013	0.0028	3.7×10^{-5}	59
Zn	0.13	7×10^{-6}	0.0015	0.00026	9.0	0.	0.0084	0.45
Se	0.00031	0.000658	0.0043	0.0017	0.00038	3.6×10^{-5}	0.	64.
Sb	0.0020	0.	0.00019	0.	0.055	9.1×10^{-5}	0.	78.
La	0.00062	5.6×10^{-5}	0.0018	0.	0.00075	3.4×10^{-5}	5.3×10^{-5}	84.
Ce	0.0017	0.0020	0.	0.	0.	5.2×10^{-5}	1×10^{-6}	66.
Sm	0.00026	9×10^{-6}	0.00018	0.	0.	0.00018	7×10^{-6}	75.
Eu	6×10^{-6}	1.9×10^{-5}	3.5×10^{-5}	0.	0.	6×10^{-6}	2×10^{-6}	73.
Th	5.7×10^{-5}	0.	0.00022	0.	0.00038	1×10^{-6}	1×10^{-6}	81.

Cluster Analysis

The cluster analysis has been discussed in detail by Hopke *et al.*[8] in their original analysis of this data set. The cluster analysis is an attempt to systematically group like samples. In this way, some of the source relationships may become clearer because of the pattern of sample associations. The variable axes define a multidimensional space,

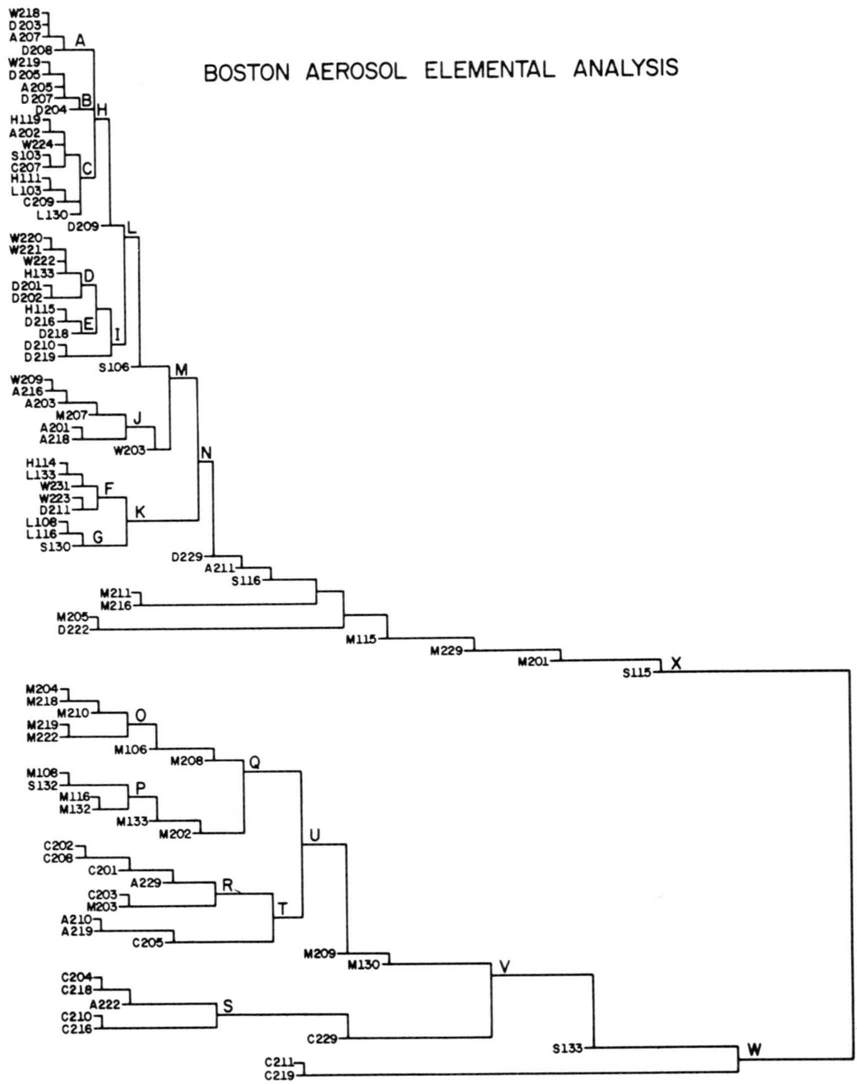

FIGURE 1. Dendrogram for the cluster analysis of the Boston aerosol sample analyses. The horizontal displacement is proportional to the degree of dissimilarity between clusters. (This figure is taken from Reference 8, with permission of the publisher.)

TABLE 5

AVERAGE ELEMENTAL CONCENTRATIONS FOR THE CLUSTERS

Element	H	I	J	K	L	M	N	Q	R	S	T	U	V	W	X
Na ($\mu g/m^3$)	0.802	1.66	0.888	1.46	1.16	1.07	1.14	1.32	1.32	1.68	1.89	1.55	1.60	1.59	1.46
Cl ($\mu g/m^3$)	0.375	0.395	0.225	0.270	0.429	0.393	0.372	0.362	0.348	0.300	0.326	0.347	0.381	0.381	0.580
Br (ng/m^3)	105.	116.	268.	115.	110.	145.	140.	234.	288.	154.	232.	233.	239.	232.	160.
Al ($\mu g/m^3$)	0.325	0.715	0.700	0.918	0.467	0.498	0.571	1.37	1.52	2.72	1.70	1.51	1.77	2.01	0.839
Sc (ng/m^3)	0.060	0.135	0.103	0.183	0.087	0.090	0.106	0.291	0.291	0.586	0.329	0.307	0.376	0.433	0.162
V ($\mu g/m^3$)	0.391	0.209	0.647	0.582	0.316	0.412	0.442	0.887	1.30	1.05	1.21	1.02	1.02	1.03	0.547
Cr (ng/m^3)	0.5	0.5	0.2	1.7	0.5	0.5	0.7	2.0	8.6	12.7	7.6	4.3	5.6	6.8	1.5
Mn (ng/m^3)	10.6	16.5	14.1	36.6	12.8	13.1	17.2	28.5	31.0	38.0	31.3	29.6	31.8	33.4	23.2
Fe ($\mu g/m^3$)	0.302	0.587	0.518	0.757	0.409	0.427	0.484	0.977	1.33	2.68	1.43	1.16	1.47	1.73	0.714
Co (ng/m^3)	0.384	0.452	0.685	0.789	0.404	0.481	0.534	1.03	1.64	1.81	1.55	1.25	1.36	1.48	0.721
Zn (ng/m^3)	0.072	0.075	0.208	0.115	0.073	0.095	0.098	0.110	0.376	12.9	0.319	0.196	0.203	0.249	0.148
Se (ng/m^3)	0.488	1.01	0.791	1.95	0.670	0.690	0.910	0.854	1.52	2.60	1.33	1.05	1.13	1.17	1.26
Sb (ng/m^3)	2.49	2.43	7.39	3.25	2.41	3.17	3.19	12.5	11.6	4.40	12.2	12.4	13.4	14.1	6.15
La (ng/m^3)	0.302	0.695	0.701	0.930	0.442	0.489	0.566	2.08	1.37	2.60	1.46	1.82	2.18	2.43	0.876
Ce (ng/m^3)	0.572	1.24	0.945	1.74	0.806	0.836	0.992	2.93	2.41	4.40	2.47	2.74	3.38	3.53	1.46
Sm (ng/m^3)	0.043	0.095	0.095	0.131	0.062	0.068	0.079	0.261	0.196	0.382	0.214	0.242	0.296	0.333	0.122
Eu (pg/m^3)	9.21	18.5	18.8	24.8	12.8	13.5	15.5	39.6	30.1	75.2	38.2	39.0	48.3	54.9	22.7
Th (pg/m^3)	44.4	94.9	117.	62.0	62.0	69.7	80.8	210.	239.	491.	264.	232.	307.	121.	345.

and each sample can be represented by a single point in that space. Various measures can be defined to describe the relationships between the sample points. It is then possible to look for groups or clusters of points that are close to one another. The procedure has been applied using the squared Euclidean distance as the measure of sample point separation, since, for a given cluster, the average value of this measure gives the mean within-cluster variance. Initially, each point is considered as a separate group. Groups are then merged to give new clusters. The choice of which groups to combine is based on minimizing the new cluster variance. These clusters are sequentially combined until there is only a single cluster containing all of the points. The cluster pattern or dendogram for the Boston area data is given in FIGURE 1. TABLE 5 gives the average concentration data for selected labeled clusters.

The samples fall into two major divisions, with most of the samples taken at locations around the periphery of the city in the upper cluster, and the downtown samples in the lower group. The average concentrations for elements strongly associated with anthropogenic sources, such as V, Br, Cr, Zn, and Sb, are much higher for cluster **W** than for cluster **X**, as would be anticipated from the higher density of activity in the central city area. The interpretations for the subclusters have been previously reported.[8] The cluster analysis permits some assistance in interpreting the factors by organizing the samples into groups that are affected by common sources.

CONCLUSION

In this report, a different approach to factor analysis has been presented. This new technique has the advantage of being able to make a source resolution analogous to that done using the chemical balance procedure. It permits a direct determination of the number of independent sources, the elemental distribution for each source, and the contribution of each source to the total particulate loadings. These initial results encourage the further exploration of the method to refine and calibrate it in such a way as to assure its reliability and accuracy. Such efforts are currently under way and will, hopefully, permit more widespread applications of these multivariate statistical procedures for isolating and understanding the sources contributing to complex urban aerosols.

REFERENCES

1. MILLER, M. S., S. K. FRIEDLANDER & G. M. HIDY. 1972. A chemical element balance for the Pasadena aerosol. J. Colloid Interface Sci. **39**(1): 165.
2. FRIEDLANDER, S. K. 1973. Chemical element balances and identification of air pollution sources. Environ. Sci. Technol. **7**(3): 235.
3. HEINDRYCKX, R. & R. DAMS. 1974. Continental, marine, and anthropogenic contributions to the inorganic composition of the aerosol of an individual zone. J. Radioanal. Chem. **19**: 339.
4. BOGEN, J. 1973. Trace elements in atmospheric aerosol in the Heidelberg area, measured by instrumental neutron activation analysis. Atmos. Environ. **7**: 1117.
5. GATZ, D. 1975. Relative contributions of different sources of urban aerosols: application of a new estimation method to multiple sites in Chicago. Atmos. Environ. **9**: 1.
6. GARTRELL, G. & S. K. FRIEDLANDER. 1975. Relating particulate pollution to sources: the 1972 California aerosol characterization study. Atmos. Environ. **9**: 279.

7. KOWALCZYK, G. S., C. E. CHOQUETTE & G. E. GORDON. 1978. Chemical element balances and identification of air pollution sources in Washington, D.C. Atmos. Environ. **12:** 1143.
8. HOPKE, P. K., E. S. GLADNEY, G. E. GORDON, W. H. ZOLLER & A. G. JONES. 1976. The use of multivariate analysis to identify sources of selected elements in the Boston urban aerosol. Atmos. Environ. **10:** 1015.
9. ROZETT, R. W. & E. M. PETERSEN. 1975. Methods of factor analysis of mass spectra. Anal. Chem. **47:** 1301.
10. DUEWER, D. L., B. R. KOWALSKI & J. L. FASCHING. 1976. Improving the reliability of factor analysis of chemical data by utilizing the measured analytical uncertainty. Anal. Chem. **48:** 2002.
11. GAARENSTROOM, P. D., S. P. PERONE & J. L. MOYERS. 1977. Application of pattern recognition and factor analysis for characterization of atmospheric particulate composition in southwest desert atmosphere. Environ. Sci. Technol. **11**(8): 795.
12. EXNER, O. 1966. Additive physical properties. I. General relationships and problems of statistical nature. Collect. Czech. Chem. Commun. **31:** 3222.
13. WEINER, P. H., E. R. MALINOWSKI & A. R. LEVINSTONE. 1970. Factor analysis of solvent shifts in proton magnetic resonance. J. Phys. Chem. **74**(26): 4537.
14. WEINER, P. H. & E. R. MALINOWSKI. 1971. Anisotropic solvent shifts determined by factor analysis. J. Phys. Chem. **75**(9): 1207.
15. WEINER, P. H. & E. R. MALINOWSKI. 1971. An investigation of the van der Waals effect in nuclear magnetic resonance spectroscopy by factor analysis and the prediction of diamagnetic and paramagnetic susceptibilities. J. Phys. Chem. **75**(20): 3160.
16. MALINOWSKI, E. R. & M. MCCUE. 1977. Qualitative and quantitative determination of suspected components in mixtures by target transformation factor analysis of their mass spectra. Anal. Chem. **49**(2): 284.
17. MALINOWSKI, E. R., D. G. HOWERY, P. H. WEINER, J. M. SOROKA, P. T. FUNKE, R. B. SWIZER & A. LEVINSTONE. 1976. FACTANL—Target Transformation Factor Analysis. Program 320, Quantum Chemistry Program Exchange, Indiana University. Bloomington, Ind.
18. WEDEPOHL, K. H. 1971. Geochemistry. Holt, Rinehart and Winston, Inc. New York, N.Y.
19. SHACKLETTE, H. T., J. C. HAMILTON, J. G. BOERNGEN & J. M. BOWLES. 1971. Geological Survey Professional Paper 574-D. U.S. Government Printing Office. Washington, D.C.
20. THOMAS, S. C. 1977. Size and Composition of Atmospheric Particles in Rural Areas Near Washington, D.C. Ph.D. Thesis. University of Maryland. College Park, Md.
21. MROZ, E. J. 1976. The Study of the Elemental Composition of Particulate Emissions from an Oil-Fired Power Plant. Ph.D. Thesis. University of Maryland. College Park, Md.
22. GLADNEY, E. S., J. A. SMALL, G. E. GORDON & W. H. ZOLLER. 1971. Atmos. Environ. **10:** 1071.
23. GREENBURG, R. R., W. H. ZOLLER & G. E. GORDON. 1978. Environ. Sci. Technol. **12:** 566.
24. GOLDBERG, E. D. 1963. The oceans as a chemical system. *In* The Sea. M. N. Hill, Ed. **2:** Chapter 1. John Wiley & Sons, Inc. New York, N.Y.
25. ONDOV, J. M. 1974. A Study of Trace Elements on Particulates from Motor Vehicles. Ph.D. Thesis. University of Maryland. College Park, Md.

THE USE OF LONG-TERM AND SEASONAL TRENDS OF ELEMENTAL COMPOUNDS AS AN AID TO THE IDENTIFICATION OF SOURCES OF AIRBORNE POLLUTANTS*

Michael T. Kleinman

Environmental Health Laboratory
Rancho Los Amigos Hospital Campus
University of Southern California School of Medicine
Downey, California 90242

Air pollutant concentration trends are often compared with the time of imposition of an air pollution control strategy to provide insights into the relationships between ambient pollutant concentrations and emissions of those pollutants by their sources. Physical parameters, such as temporal variation of meteorological conditions, spatial variation in surface roughness, and micrometeorological changes, however, can obscure the effects on ambient air quality attributable to changes in source emission characteristics.

The dispersion factor, which is the product of the height of the mixing layer and the average wind speed through the mixing depth, measured during a sampling interval, divided by a long-term average value for this product, has been used as a means for normalizing data with respect to atmospheric dispersion changes.[1] This technique has been applied in cities such as New York City,[2] Rio de Janeiro, Brazil,[3] and Denver, Colorado.[4] Dispersion-normalized pollutant concentrations should be proportional to changes in source emissions[1] and can be used to make valid year-to-year or season-to-season comparisons of pollutant concentrations.

Morning mixing depth and wind speed data are available in most major cities, but there are many places where the data needed for dispersion normalization are not available. In such circumstances, average daily inversion heights and surface wind speeds may make suitable alternative variables.

New York City is an ideal place to examine the usefulness of studying trends in dispersion-normalized data. This is because of the large available base of meteorological and air quality data, and because New York City has been extremely active in promulgating and enforcing air quality legislation. In 1967, New York City began a program to improve the quality of air by imposing strict limitations on major pollution sources. Legislation (known as Local Law 14, passed in March 1968) specified a timetable for upgrading or elimination of incinerators, reducing the percentage of sulfur in fuels, and restricting the use of coal as a fuel.

With regard to incineration, the following steps were taken:

1. Burning of wastes on open barges in off-shore waters was prohibited.

2. Incinerators in existing residential, commercial, and institutional buildings were to be upgraded. This included the installation of an auxiliary gas burner,

*Supported by the Electric Power Research Institute, Project No. RP 1225.

0077-8923/80/0338-0116 $01.75/1 © 1980, NYAS

over-fire air, an automatic clock, and a water scrubber. Particulate emissions were to be reduced by 90%.

3. Those incinerators that were not upgraded were to be discontinued, that is, replaced by a compactor or by collection of raw refuse by the Department of Sanitation.

4. Municipal incinerators were required to meet even more stringent conditions. The allowable emission rate was set at 1.25 lb/1000 lb of refuse charged.

5. No new incinerators were to be installed other than municipal incinerators meeting the standards.

A schedule for reducing the lead content of gasoline was implemented and enforced.

A summary of the various deadlines that have been met in part is given in TABLE 1. The reduction of the sulfur content in fuel oils from 2.2% to 1% S was expected to

TABLE 1

NEW YORK CITY POLLUTION SOURCE LIMITS*

	Incineration†	Coal Burning	Wt. % S in Residual Fuel Oil	Pb in Gasoline (grams/gallon)
1967		Prohibition on burning bituminous coal		
1968	>100 dwelling units		2.2%	
1969	>61 dwelling units and all commercial, industrial, and institutional			
1970	>20 dwelling units	All coal burning prohibited‡	1.0	
1971				2
1972			0.5	1
1973			0.3	0.5
1974			>1.0	0

*From Kleinman.[8]

†Maximum number of units per incinerator or incinerator types required to be upgraded.

‡Except in some schools, hospitals, and municipal buildings.

reduce the sulfur dioxide (SO_2) content of the air by a factor of two by 1971. However, the date for compliance to a 1% sulfur limit on all fuels burned in the metropolitan area was moved up to 1970, following a New York State ruling.

It was expected that the particulate emissions would also be reduced, since the ash content of the fuel was reduced coincidentally along with the sulfur. This change was not required by the law. The New York City Division of Air Resources[5] stated that they expected the new 1% fuels to have 50% lower ash content. Consequently, the New York City Division of Air Resources predicted that by 1971, the contribution of fuel-oil-burning sources to the total suspended particulate matter (TSP) would be halved. As shown in FIGURE 1, as the average concentration of sulfur in residual fuel oils burned for generation of heat and power decreased, the average ambient TSP concentrations[5,6] also decreased.

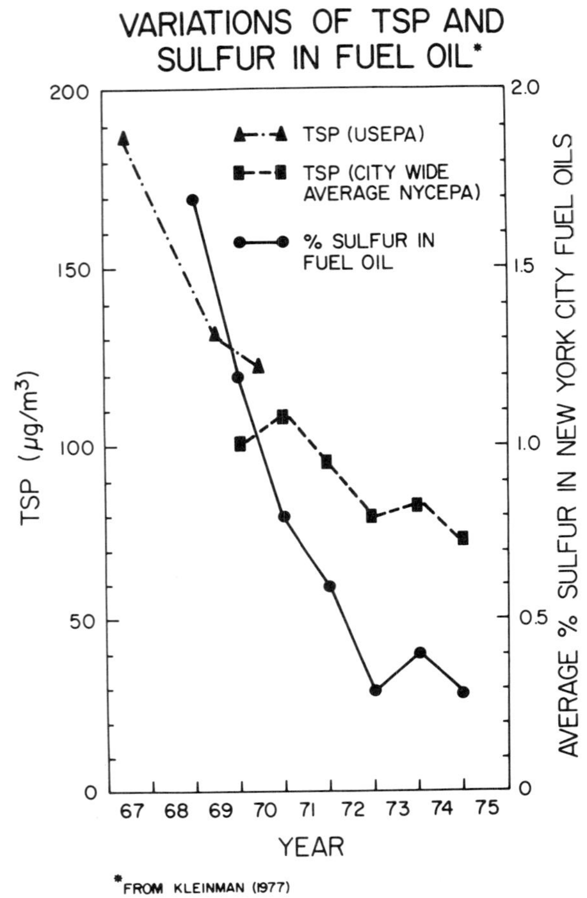

FIGURE 1

The effects of legislative controls on air pollution sources can also be seen in the general decrease of concentrations of trace metals associated with specific air pollution sources.

As shown in TABLE 2, dispersion-normalized trace metal concentrations at four sampling stations in New York City were substantially reduced from 1969 to 1972. Concentrations of vanadium and copper, elements that are believed to be associated with sources affected by the city's air pollution emission codes (fuel-oil burning and incineration, respectively),[7,8] were reduced more than tenfold.

From 1967 through 1971, the sulfur content of fuels burned for space heating was reduced from 2.2% to 1%. The sulfur content of fuels burned for electric-power generation was reduced to 1% in 1966–67 by Consolidated Edison's elimination of coal and use of low-sulfur oils, and remained approximately constant through 1972. From 1972 on, with minor exceptions, all fuels burned in the city for both heat and power contained 0.3% S. Thus, one would expect that the contribution of particulate emissions for oil-burning sources in 1972 and later years should be significantly lower than that in earlier years. For example, in 1968, about 60% of the oil burned was for space heating, and these fuels contained about 2.2% S. The remaining 40% of the fuel was burned for power generation and contained about 1% S. The weighted average sulfur content for fuel oil in 1968 was, therefore, $2.2 \times 0.6 + 1.0 \times 0.4 = 1.7$. In 1972, the fuel oil that was burned contained 0.3% S. In 1968, fuel-oil burning for heat and power combined produced about 46% of the total particles emitted, according to an emissions inventory. In 1972, one would expect that this contribution would have decreased by a factor of 0.3/1.7, or 0.18, and that fuel-oil burning in 1972 would therefore have produced about 8% of the TSP.

A more generalized approach to extracting information from long-term pollution trends involves the use of pattern recognition techniques. These techniques can examine correlations among variables in order to discover underlying associations that can be related to emissions from specific sources. Meteorological variations of a daily, weekly, or seasonal nature will, however, impose artifact correlations among pollutants, by causing pollutant concentrations all to change in the same direction at the same time.

A numerical model for determining the relative importance of various air pollution sources as contributors to TSP was formulated, using weekly NYU Medical Center data.[7] The model apportioned TSP by the following equation:

$$\mathrm{TSP} = b_1\,[\mathrm{Pb}] + b_2\,[\mathrm{V}] + b_3\,[\mathrm{Cu}] + b_4\,[\mathrm{Mn}] + b_5\,[\mathrm{SO_4}^{=}] + b_0 \qquad \mathbf{(1)}$$

where $b_0 \ldots b_5$ were coefficients estimated by a regression analysis technique, and the bracketed substances served as atmospheric tracers for emissions from automobiles (Pb), fuel-oil burning (V), incineration (Cu), soil-like materials and/or coal fly ash (Mn), and secondary aerosol formation ($SO_4^{=}$).

The weekly TSP and trace metal data were negatively correlated to wind speed and morning mixing depth, and to the product of these terms—the dispersion factor.[1] Generally, both parameters and their product exhibit distinct seasonal variations, being lower in summer and higher in winter. Thus, even with seven-day samples, seasonal variations in meteorological parameters cause some artifact correlations among the measured pollutants, which can decrease the accuracy with which we could estimate the regression coefficients in the above equation.

TABLE 2

TRACE METAL CONCENTRATIONS IN NEW YORK CITY AIR*

Element	1968	1969	1972	1973	1974	1975
TSP	150000	190000	67000	74000	78000	61000
Cd	31	14	5	6	7	5
Cr	74	46	9	7	11	9
Cu	301	750	52	53	48	47
Fe	—	—	1300	1500	1600	1300
Mn	86	130	23	26	25	23
Pb	3900	2900	1000	1200	1500	1200
V	1100	1500	55	82	87	44
Zn	3900	1000	320	290	340	320
SO_4^{2-}	—	—	7600	9700	9800	9200

*Normalized for dispersion—ng/m^3.

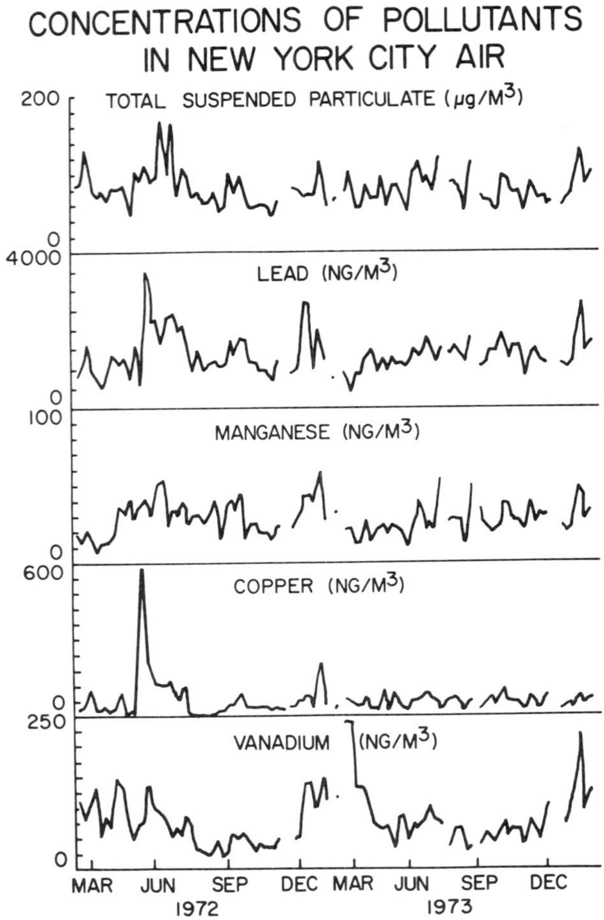

FIGURE 2

It was desirable to search for a method to minimize the effect on pollutant concentrations of extrinsically imposed seasonal meteorological variations. Trijonis *et al.*[9] had suggested that the effect of meteorological fluctuations on trend comparisons could be reduced by studying net concentration changes rather than absolute concentrations. Monthly average New York City suspended particulate data were used to test whether this would also hold true for seasonal variations imposed by meteorological parameters. Net monthly changes (differentials) in TSP, measured at the New York University Medical Center during the years 1972 and 1973, were calculated, and the differential TSP values were apparently uncorrelated to the dispersion factor or to the net changes in the dispersion factor. It therefore appears that by studying changes in concentrations of pollutants rather than the absolute concentrations, one can reduce the impact of seasonal, as well as annual, variations in meteorological parameters related to atmospheric dispersion.

In terms of the previously stated model (Equation 1), using week-to-week changes in pollutant concentrations instead of the absolute values is analogous to differentiating the model with respect to time. Equation 1, when differentiated, can be written as:

$$\frac{d[TSP]}{d(t)} = b_1 \frac{d[Pb]}{d(t)} + b_2 \frac{d[V]}{d(t)} + b_3 \frac{d[Cu]}{d(t)} + b_4 \frac{d[Mn]}{d(t)} + b_5 \frac{d[SO_4^{=}]}{d(t)} \qquad \textbf{(2)}$$

Note that the constant term, b_0, disappears from the differential equation, and that the coefficients b_1 through b_5 are the same as in Equation 1. Since the use of differentials may reduce the associations imposed by seasonal meteorological variations, the estimates of the regression coefficients should be more accurate than those made using absolute values.

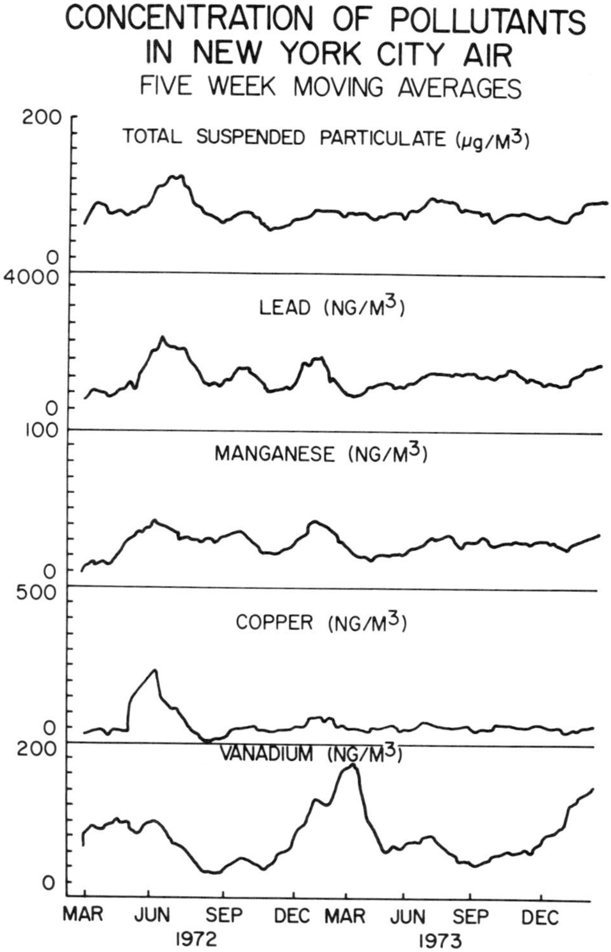

FIGURE 3

RESULTS

Weekly TSP and trace metal data from the New York University Medical Center were used to test the hypothesis that the effects of seasonal meteorological variations on estimated regression coefficients could be minimized by using differential data as opposed to absolute concentrations. The trace metal data used are shown, graphically, in FIGURE 2. Gaps in the data, caused by sampler malfunctions or other losses of samples, were eliminated as follows. Polynomial regression curves were fitted to the available data, and the resulting equations were used to estimate the missing data points. To make seasonal variations more pronounced, five-week moving averages were plotted (FIGURE 3).

These data had previously given satisfactory results with both factor analysis and subsequent multiple regression fitting to develop a linear equation describing the relationship of TSP and five tracers.[8] The data were therefore felt to be appropriate for testing other methods of data analysis. The use of the polynomial fits to fill in missing data caused small changes in the trace metal data distribution. In the case of sulfate, however, several blocks of data were missing, and insertion of computed values imposed an artificial structure to the correlation matrix.

TABLE 3

MATRIX OF CORRELATION COEFFICIENTS (ABSOLUTE WEEKLY VALUES)

	Pb	Mn	Cu	V	SO_4^{2-}	TSP
Pb	1.0	0.496	0.249	0.094	0.090	0.481
Mn		1.0	0.356	0.072	0.181	0.513
Cu			1.0	0.216	0.012	0.380
V				1.0	0.074	0.246
SO_4^{2-}					1.0	0.285

Differentials were calculated as the concentration on week 2 minus the concentration on week 1. Correlation coefficients were calculated for both the differential and the absolute data sets. These coefficients are shown in TABLES 3 and 4. There does not seem to be any obvious pattern in the changes in interelement and element-to-TSP correlations between the two sets of data. Regression analyses were performed with both sets of data, and the regression coefficients estimated are shown in TABLE 5. The coefficients for Pb and V, which are strongly identified with specific sources, automobiles and fuel-oil burning, are in close agreement. The coefficients for Mn and Cu, which are less specifically source related, appear to be different when estimated by the two methods.

The coefficients for sulfate are in agreement by both methods, but incorrect. These coefficients should not be given any physical significance because of the apparent biases introduced in the polynomial fitting process. The coefficient for $SO_4^{=}$ determined for the same data set prior to insertion of the computed missing values is about 1.6, which is appropriate for a sulfate aerosol that may be ammonium sulfate with some added associated material. Filling in missing data by the above techniques is not advocated here, but was used to provide a large, continuous data set to test the

TABLE 4

DIFFERENTIAL VALUES (WEEK 2 − WEEK 1)

	Pb	Mn	Cu	V	SO_4^{2-}	TSP
Pb	1.0	0.221	−0.157	0.377	0.118	0.368
Mn		1.0	0.281	0.291	0.312	0.610
Cu			1.0	0.107	0.074	0.184
V				1.0	0.232	0.372
SO_4^{2-}					1.0	0.359

efficacy of the differentiating methods. The test is satisfactory in that the three most important tracers give coefficients that agree by both methods.

These results suggest that the use of differentials will reduce the impact of extrinsic temporal variations on a regression analysis. The regression coefficients so obtained should be more accurately estimated than those estimated using absolute values. But, was this technique powerful enough to allow us to perform regression analyses on data that we previously were unable to analyze?

Daily TSP and trace metal data were not available on a long-term basis. It was, however, of interest to try to apply this technique to daily data, so it was decided to try to model the formation of ozone (O_3) from its precursors, nitrogen oxides (NO_x) and hydrocarbons, which have been affected by regulatory actions, especially in Los Angeles. The formation and destruction of ozone in the atmosphere depend upon the fairly complex interactions of nitric oxide (NO), nitrogen dioxide (NO_2), reactive nonmethane hydrocarbons (RHC), and several physical parameters, such as solar radiation intensity, temperature, and humidity. Trijonis *et al.*[9] explored some of the relationships between ambient oxidant levels in the Los Angeles area and trends in the emission and atmospheric concentrations of oxidant precursors. They concluded that oxidant concentrations near source areas were sensitive to changes in the RHC/NO_x ratio, while transported oxidants were probably more dependent on the absolute levels of RHC and NO_x. In an investigation of relatively short-term variations of photochemical pollutant concentrations in the St. Louis area, Karl[10] concluded that increased levels of NO in the city on workdays, as opposed to Sundays, resulted in decreased levels of ozone. The NO_2/NO ratio is also an important factor.[11]

TABLE 5

RESULTS OF REGRESSION ANALYSES PERFORMED USING ABSOLUTE AND DIFFERENTIAL SETS OF DATA

Element	Regression Coefficient	
	Absolute	Differential
Pb (ng/m^3)	0.012 (0.0034)*	0.010 (0.0036)
Mn (ng/m^3)	0.578 (0.1726)	1.12 (0.1860)
Cu (ng/m^3)	0.060 (0.0260)	0.021 (0.0257)
V (ng/m^3)	0.080 (0.0434)	0.097 (0.0660)
SO_4^{2-} (μg/m^3)	0.561 (0.2261)	0.504 (0.2460)
Residual (μg/m^3)	32.374	−0.499

*Standard error of coefficient shown in parentheses.

A model for ozone can now be written as:

$$Z_t = b_1 + b_2\,[NO]_t + b_3\,[NO_2]_t + b_4\,[RHC]_t + b_5\,[RHC/NO]_t + b_6\,[RHC/NO_2]_t + b_7\,[NO_2/NO]_t \quad (3)$$

where $b_1 \ldots b_7$ are parameters to be fitted by least squares, Z_t is the net change in ozone concentration from day t to day $(t + 1)$, and the bracketed symbols represent the net changes in pollutant concentrations from day t to day $(t + 1)$.

Pollutant concentration data for the years 1976 through 1978 were obtained from the South Coast Air Quality Management District in Los Angeles. Two months of data from 1976 were used to preliminarily test the model. Net concentration changes were calculated and used to determine estimates of the parameters $b_1 \ldots b_7$, which were statistically significant, by the method of least squares. Regression analysis of the absolute pollutant concentrations failed to yield any statistically significant regression coefficients, nor was the multiple correlation coefficient significant.

The differential model appears to qualitatively support the previously mentioned conclusions.

$$\Delta\,O_3 = -0.045 - 0.033\,[NO] + 0.080\,[NO_2] + 0.030\,[RHC] - 0.043\,[RHC/NO] + 0.043\,[RHC/NO_2] + 0.066\,[NO_2/NO] \quad (4)$$

An increased amount of NO will be associated with a decreased amount of ozone, as shown by the negative coefficient for NO in the model. This supports the findings of Karl.[10] It is of interest to note that the (NO_2/NO) coefficient 0.066 is within a factor of three of the coefficient computed from chemical kinetic data (0.023).[11] This agreement is encouraging, considering the small number of data tested thus far.

Conclusions

Seasonal and annual variations in dispersion-normalized air pollutant concentration data appear to accurately reflect changes in source emission patterns. The normalized concentrations permit one to observe the impact of regulatory changes on ambient air quality without these impacts being obscured by meteorological fluctuations.

Numerical modeling techniques and regression analysis can be powerful tools for extracting information from large sets of air quality data. The use of differential, as opposed to absolute, pollutant concentration values will reduce artifact correlations caused by seasonal, weekly, or daily meteorological fluctuations and will permit more accurate estimation of the regression coefficients. The technique was successfully applied to a set of daily pollution measurements whose absolute concentrations were found not to yield a statistically significant fit by multiple regression. Additional research into the applicability of these techniques is needed. However, these initial findings are extremely encouraging.

Acknowledgments

I would like to thank Theo J. Kneip, Michael P. Jones, and Jack D. Hackney for their suggestions and comments.

References

1. Kleinman, M. T., T. J. Kneip & M. Eisenbud. 1975. Seasonal patterns of airborne particulate concentrations in New York City. Atmos. Environ. **10:** 9.
2. Kleinman, M. T., D. M. Bernstein & T. J. Kneip. 1977. An apparent effect of the oil embargo on total suspended particulate matter in New York City air. J. Air Pollut. Control Assoc. **27:** 65.
3. Trindade, H. A., A. E. Oliveira, W. C. Pfeiffer, H. Londres & C. C. Ribeiro. 1979. Meteorological parameters and concentration of total suspended particulates in the urban area of Rio de Janeiro. (Submitted for publication.)
4. Edwards, H. W. 1978. Seasonal trends in Denver atmospheric lead concentrations. Environ. Sci. Technol. **12:** 687.
5. Division of Air Resources. 1969. Annual Report. New York City Department of Environmental Protection. New York, N.Y.
6. United States Environmental Protection Agency. 1974. Personal correspondence.
7. Kneip, T. J., M. T. Kleinman & M. Eisenbud. 1974. Relative contribution of emission sources to the total airborne particulate in New York City. *In* Proceedings of the Third International Clean Air Congress, Dusseldorf, Federal Republic of Germany, October, 1973: 696–697.
8. Kleinman, M. T. 1977. The Apportionment of Sources of Airborne Particulate Matter. Ph.D. Dissertation. New York University. New York, N.Y.
9. Trijonis, J., T. Peng, G. McRae & L. Lees. 1978. Oxidant and precursor trends in the metropolitan Los Angeles region. Atmos. Environ. **12:** 1413.
10. Karl, T. R. 1979. Day of the week variations of photochemical pollutants in the St. Louis area. Atmos. Environ. **12:** 1657.
11. National Research Council Committee on Medical and Biological Effects of Environmental Pollutants. 1977. Ozone and Other Photochemical Oxidants. National Academy of Sciences. Washington, D. C.

CHEMICAL ELEMENT BALANCE METHOD APPLIED TO DICHOTOMOUS SAMPLER DATA

Thomas G. Dzubay

Environmental Sciences Research Laboratory
U.S. Environmental Protection Agency
Research Triangle Park, North Carolina 27711

Introduction

Between 1970 and 1976, substantial progress was made in reducing the amount of suspended particulate matter in many parts of the United States. During 1970, for example, 60% of the population in the New York Air Quality Control Region was exposed to total suspended particulate concentrations that exceeded the annual National Ambient Air Quality Standard of 75 $\mu g\ m^{-3}$. By 1976, particulate concentrations were sufficiently reduced so that the entire region met the standard.[1] Nonetheless, many regions of the United States are still not meeting the National Air Quality Standard.[1] To make further progress in the control of particulate matter, it would be valuable to know its composition and major sources.

Miller *et al.* have developed a chemical element balance method for resolving the chemical composition of airborne particles into components from several types of emission sources.[2] Data on the concentrations of several elements were used to determine the abundance of sea salt, soil, automobile emissions, and fuel-oil fly ash in total suspended particulate matter at Pasadena, California.[2] Later, Friedlander applied the method for Pasadena, including additional components for Portland cement and rubber tire dust.[3] Winchester and Nifong[4] and Gatz[5] used element balance methods to estimate source components for the Chicago area. Recently, Kowalczyk *et al.* resolved aerosol in the Washington, D.C. area into the following six components: soil, marine, coal, oil, refuse incineration, and motor vehicle emissions.[6] By adding sulfate and nitrate to their predicted six components, Kowalczyk *et al.* were able to account for about 80% of the total suspended particulate mass.[6]

In the present work, the chemical element balance method is applied to aerosols collected in ten dichotomous samplers in the vicinity of St. Louis, Missouri, as part of the Regional Air Pollution Study (RAPS),[7,8] and in the Great Smoky Mountain National Park at Elkmont, Tennessee. The dichotomous sampler collected particles in two size ranges[9] corresponding to the two major modes of the bimodal distribution of aerosol mass reported by Whitby *et al.*[10] The fine fraction collected by the dichotomous sampler consists mostly of particles from combustion sources and secondary aerosol, and the coarse fraction consists mostly of mechanically generated aerosols. Dichotomous samplers[8] in RAPS were operated in the St. Louis vicinity between May 1975 and March 1977 at locations shown in FIGURE 1. The present work is based upon RAPS data for July and August of 1976. For the study in the Smoky Mountains, samples were collected from September 20 to 26 of 1978.

0077–8923/80/0338–0126 $01.75/1

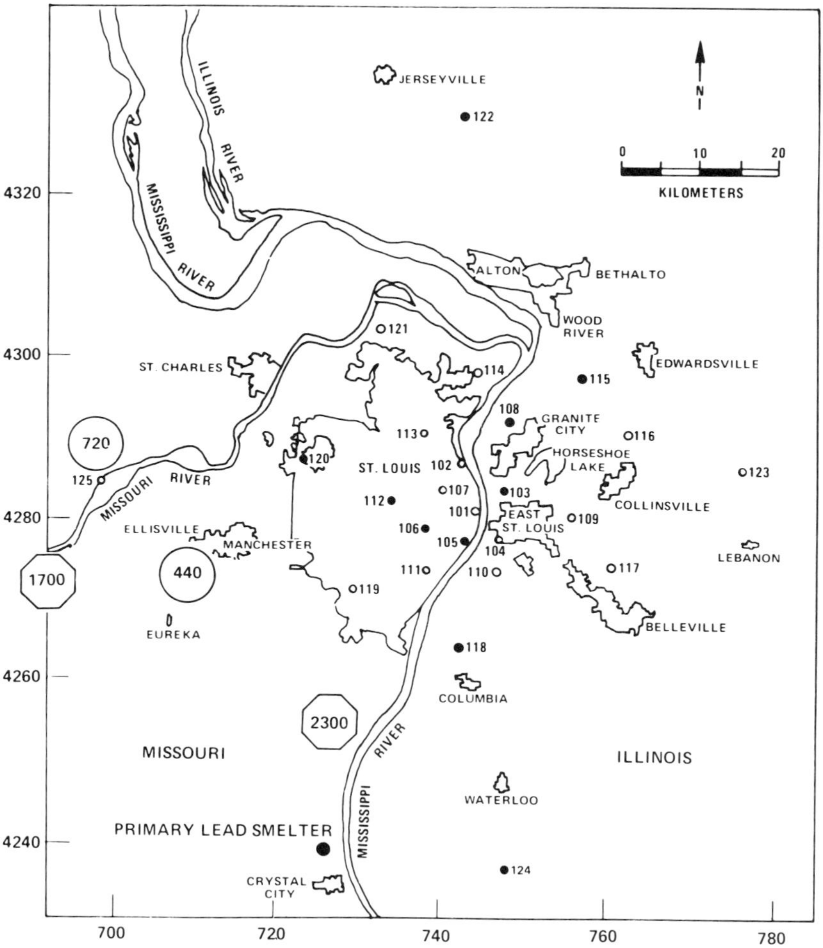

FIGURE 1. A map showing the 25 RAPS stations locations (○ and •). The 10 RAMS stations having dichotomous samplers are represented as solid dots (•). The octagons represent coal-fired power plants, the circles represent quarries, and numbers within indicate estimated particulate emission rates in tons per year obtained from a data base developed by Littman.[14] The axes are expressed as Universal Transverse Mercator (UTM) coordinates for Zone 15.

METHOD

Measurements

All of the results reported here are based on aerosol samples collected in an automated dichotomous sampler developed by Loo *et al.*[11] Each sampler was operated at a flow rate of 50 ℓ min^{-1} and separated particles in the fine (< 2.4 μm) and coarse (2.4 to 20 μm) size ranges. At all ten RAPS sites, aerosol particles were continuously

TABLE 1

SAMPLES COLLECTED AT RAPS SITES DURING 1976 REPRESENTATIVE OF AN AVERAGE AND EVENTS WHEN CONCENTRATIONS OF CERTAIN ELEMENTS WERE UNUSUALLY HIGH*

Site	Two-Month Average 120		Paint Pigment 106		Steel Processing 108		Nonautomotive Lead 105		High Zinc 105		Fireworks 112	
Period	—		0000–1200		1200–1400		0000–0600		1800–2400		1200–2400	
Day	183–238		234		183		233		244		186	
Element	F	C	F	C	F	C	F	C	F	C	F	C
Mass	32,000	22,000	60,000	50,000	32,000	48,000	100,000	58,000	—	82,000	65,000	61,000
Al	130	900	200	1,600	140	970	450	2,400	200	2,700	2,000	3,300
Si	290	3,300	430	4,500	260	5,100	1,100	6,400	420	8,700	670	11,000
P	< 10	< 30	32	270	< 12	120	13	180	60	120	37	140
S	4,000	340	7,500	930	3,600	410	11,000	1,300	5,300	1,800	5,800	2,100
Cl	50	140	180	390	11	350	2,000	1,700	6,200	1,100	560	770
K	110	290	180	400	270	450	840	610	200	680	13,000	4,700
Ca	75	1,650	230	4,500	200	4,600	180	5,000	260	5,700	280	2,600
Ti	25	110	1,300	8,040	12	120	80	140	39	280	160	270
V	< 3	< 3	< 8	40	7	17	50	13	31	50	31	9
Mn	10	30	41	90	24	180	80	85	34	130	26	70
Fe	140	690	870	4,680	370	8,900	590	3,500	260	2,160	140	2,600
Cu	9	4	9	9	16	18	930	80	410	480	70	13
Zn	50	28	100	39	80	130	590	220	10,700	19,300	90	60
Se	2	< 1	8	3	2	< 1	8	2	13	5	2	1
Br	103	41	150	60	160	40	140	50	90	40	440	110
Sr	< 1	4	< 1	10	< 1	9	< 3	11	< 3	35	380	50
Cd	2	< 2	7	< 3	< 3	3	60	8	460	510	< 3	< 3
Sn	4	< 2	7	< 4	6	51	370	80	31	< 9	8	< 4
Sb	< 3	< 3	5	< 4	< 4	5	60	16	< 8	< 9	3,000	110
Ba	< 10	14	< 16	100	< 15	23	< 32	< 34	< 32	< 34	310	130
Pb	560	140	1,110	210	570	280	12,800	2,400	1,800	1,640	1,300	360

*The column headings indicate the nature or most likely source for these events. The results are expressed for the fine (F) and coarse (C) fractions in units of ng m^{-3}.

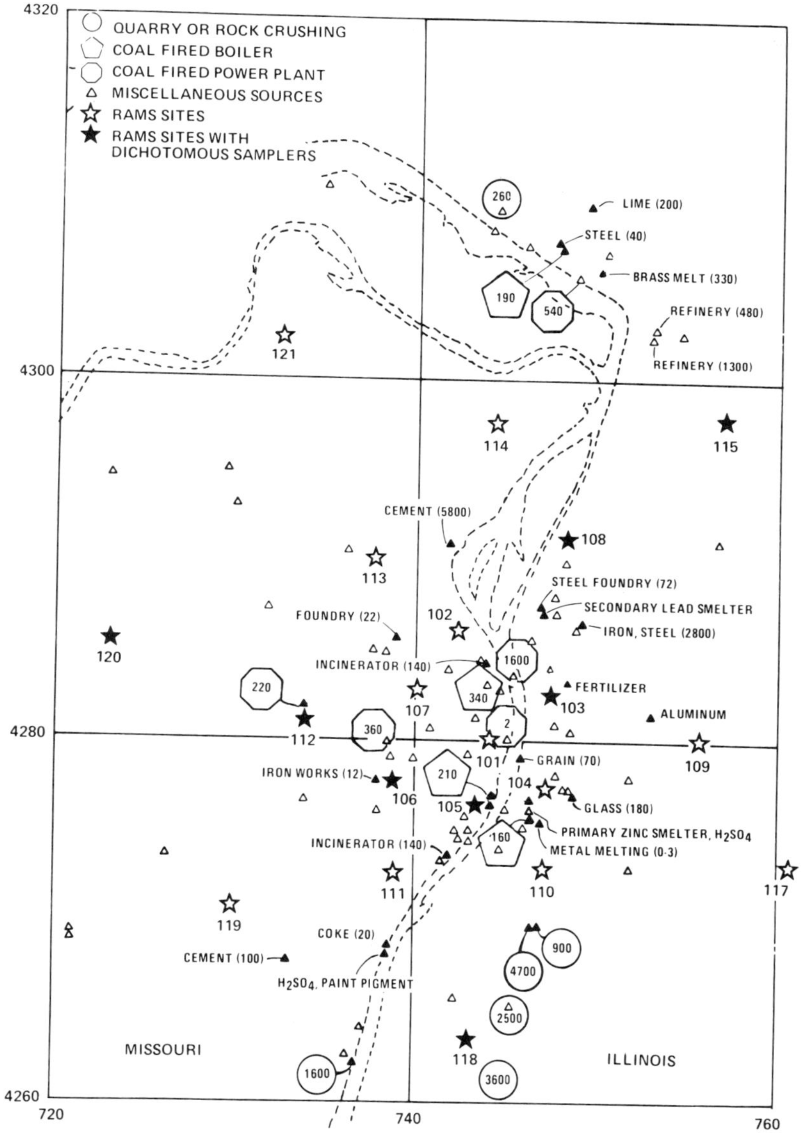

FIGURE 2. Emission sources in the vicinity of St. Louis, Missouri. For selected sources, the emission rates in tons per year are shown in parenthesis or within the symbols and were obtained from a data base developed by Littman.[14] The axes represent UTM coordinates.

collected on cellulose esters membrane filters for 6- or 12-hour sampling periods. In the Smoky Mountains, the filter medium was polytetrafluoroethylene (Teflon), and the sampling periods were 12 hours long.

Samples collected during RAPS were analyzed at the Lawrence Berkeley Laboratory, Berkeley, California, for mass and elemental content, using beta ray attenuation and x-ray fluorescence (XRF) methods, respectively.[8,12,13] The XRF analyses included elements with atomic numbers between 13 and 38, as well as cadmium (Cd), tin (Sn), antimony (Sb), barium (Ba), and lead (Pb). Samples collected in the Smoky Mountains were analyzed in a similar manner, using instruments fabricated at the Lawrence Berkeley Laboratory.[12,13] TABLE 1 illustrates the results for mass and elemental composition of selected samples. In addition to the elements shown, concentration values were also obtained for chromium (Cr), nickel (Ni), and rubidium (Rb).

Components in the Chemical Balance

In carrying out a chemical element balance analysis, one must select components into which the aerosol concentrations can be resolved. For each component, the abundances of all measured elements must be specified. Moreover, the specified abundances should accurately represent the composition of the particulate matter as it exists on the collected samples. Obtaining the required information for each component is the most difficult part of using the chemical element balance method.

FIGURES 1 and 2 indicate many of the particulate emission sources and their mass emission rates in the vicinity of St. Louis, Missouri, and East St. Louis, Illinois. The major sources of particulate mass include numerous quarries and bituminous coal-fired boilers and power plants. The composition of the dust and fly ash emitted by such sources is very similar to that of soil. To resolve the various sources of soil, road dust, quarry dust, and fly ash by the chemical element balance method would require very accurate compositional data, which are not currently available. Therefore, crustal contributions from such sources were lumped together into two components that are hereafter referred to as crustal shale and crustal limestone. The composition of the crustal components is given in TABLE 2, and the various sources that contribute to them are summarized in TABLE 3.

In addition to the two crustal components, four components were included in the calculations: ammonium sulfate, motor vehicles, paint pigment, and steel. Their compositions are shown in TABLE 2. The set of six components sufficiently describes most of the total suspended particulate mass in the St. Louis vicinity, but additional components are needed to describe the relatively small sources of trace elements. Such components were not included because required data on compositions were lacking. The rationale used in selecting the six components is further outlined below.

Ammonium Sulfate Component

Stevens *et al.* analyzed samples collected at several sites across the United States and found that the particulate sulfur in the fine fraction usually consisted of

TABLE 2
COMPOSITION IN mg PER g OF THE SIX COMPONENTS USED IN CHEMICAL ELEMENT BALANCE ANALYSIS APPLIED TO DICHOTOMOUS SAMPLER DATA

Element	Ammonium Sulfate	Motor Vehicles	Crustal, Shale	Crustal, Limestone	Paint Pigment	Steel Industries
C		563	15	114		
Na			9.6	0.4		
Mg		6	15	47		3
Al*			80	4.2		11
Si*			273	24		7
P		0.9	0.7	0.4		1
S*	242	6	2.4	1.2		
Cl		19	0.18	0.15		
K*			26.6	2.7		
Ca*			22.1	302		12
Ti*			4.6	0.4	422	
V			0.13	0.02		
Cr			0.09	0.01		
Mn		0.6	0.85	1.1	4	31
Fe*		1	47.2	3.8	204	570
Ni			0.07	0.02		
Cu			0.05			
Zn		3	0.1	0.02		
Se						
Br*		37		0.01		
Rb			0.14			
Sr			0.3	0.61		
Cd						
Sn			0.01			
Sb						
Ba		0.9	0.58	0.01		
Pb*		148	0.02	0.01		
Other†	758	215	501	498	370	365

*The asterisk indicates the elements used in least squares fit.
†This includes H, N, and O.

ammonium sulfate.[15] However, when precautions were taken to prevent neutralization of any acidity in the samples by ambient ammonia, Stevens *et al.* discovered that the sulfate consisted of a mixture of ammonium sulfate and sulfuric acid.[15] Because precautions were not taken to preserve acidity in stored RAPS samples prior to analysis, it is assumed that any sulfuric acid would have been neutralized to ammonium sulfate. Therefore, an ammonium sulfate component is appropriate for representing the measured particulate sulfur in the fine fraction. However, this does not imply a lack of acidity in the aerosol as it originally existed in the atmosphere.

Although the chemical form of particulate sulfur in the coarse fraction is not well known, the ammonium sulfate component was also used to represent the sulfur in that fraction. If coarse particle sulfur were actually in another chemical form, the resulting error would be negligible due to the low abundance of sulfur in that fraction. For example, TABLE 1 illustrates the results of concentrations averaged over a two-month period at Site 120. Only 8% of the total sulfur occurs in the coarse fraction, and sulfur constitutes less than 2% of the total mass in the coarse fraction. A similar situation exists for the average coarse particle sulfur concentrations at the nine other RAPS sites and at the Smoky Mountain site.

Motor Vehicle Component

The composition of this component is based upon particulate emission rates for cars and trucks measured in highway tunnels by Pierson.[16,17] The emission rates are weighted, assuming that cars and diesel-powered trucks account for 96% and 4% of the traffic, respectively.[18] An emission rate of 10 mg km^{-1} was assumed for lead. Because of the instability of bromine (Br) on auto exhaust particles, a bromine to lead ratio of 0.25 is used, rather than the value 0.39 for fresh auto exhaust.[19] The average of the bromine to lead ratio for all 10 RAPS sites was 0.17. In chemical element balance

TABLE 3

POSSIBLE ORIGINS OF THE COMPONENTS USED IN THE CHEMICAL ELEMENT BALANCE ANALYSIS

Component Name	Possible Origin
Ammonium sulfate	Primary sulfate emissions Atmospheric transformation of SO_2 to sulfate
Crustal—Shale	Wind erosion of soil Anthropogenic suspension of soil Vehicle-generated dust Fly ash Quarrying Manufacturing processes
Crustal—Limestone	Wind erosion of calcium-rich soil Anthropogenic suspension of calcium-rich soil Vehicle-generated cement dust Cement manufacturing Other manufacturing
Steel	Iron and steel production Suspension of iron-rich soil Manufacturing processes

analysis, the magnitude of this component is determined by bromine and lead concentrations in the collected aerosol. Because of possible variations in the relative amounts of car and diesel truck traffic, and because of possible differences in the exhaust particle composition from that assumed for the component, the accuracy of the deduced component mass concentration is estimated to be about ±40%.

Crustal Components

Detailed soil analyses for the St. Louis area were not available to serve as a basis for the crustal components. Even if such data were available, their applicability is unclear due to the multitude of possible sources of crustal matter shown in TABLE 3. Moreover, Gatz has studied aerosol data for St. Louis for 1973 to 1975 using factor analysis and found that calcium (Ca) is not entirely associated with soil.[20] The data in

TABLE 1 indicate that the calcium to aluminum (Al) ratio is substantially higher than that found in the earth's crust. In Mason's average earth crust,[21] the calcium to silicon (Si) ratio is 0.45; in the average U.S. soil composition tabulated by Shacklette *et al.*,[22] that ratio is only 0.36. For the coarse particle data in TABLE 1, the calcium to aluminum ratio is typically about 2 or larger. Thus, at least two components are needed to fit the crustal matter. Here it is represented in terms of two types of sedimentary rock: shale and limestone. The composition shown in TABLE 2 is taken from Tables 6.1 and 6.5 of the book by Mason.[21]

Paint Pigment Component

St. Louis is unique among cities because of occasional high extremes in the titanium concentration of its aerosol. TABLE 1 indicates an event when the titanium (Ti) concentration at RAPS Site 106 was 8040 ng m^{-3}. Rheingrover has identified the source of the Ti as a paint pigment factory[23] (see FIGURE 2). Data in TABLE 2 were deduced by subtracting the shale component from the average of several samples in which the titanium concentrations exceeded 2000 ng m^{-3}. Iron appears to be associated with titanium in the paint pigment component.

Steel Component

According to FIGURE 2, there are steel mills located in the vicinity of RAPS Sites 103 and 108. TABLE 1 illustrates an event at Site 108, in which the iron (Fe) concentration was enriched relative to that of Si. The elemental abundances for the steel component shown in TABLE 2 are representative of emissions from basic oxygen furnaces.[4]

Other Sources

The above six components are unable to totally account for concentrations of trace elements emitted from many of the industrial sources in the St. Louis area. For example, TABLE 1 illustrates a case at RAPS Site 105 on Day 244, in which the total zinc concentration was 30 μg m^{-3}; concentrations of chlorine (Cl), copper (Cu), Cd, and Pb were also substantially elevated. Although a municipal incinerator near Site 105 was a possible source for the high zinc concentration, one may rule out the incinerator as being the dominant source on the basis of a very low Sn to Zn concentration ratio. The Sn to Zn ratio was only 0.001 in this instance, and is much lower than the average value of 0.1 observed by Greenberg *et al.* in studies of incinerator emissions in Washington, D.C. and Chicago.[24] FIGURE 3 illustrates the Ni, Cu, zinc (Zn), and amonium sulfate concentrations for a 9-day period in which the Zn concentrations were rather high. Although there are similarities in the time dependence of the Ni, Cu, and Zn concentrations, those elements may be emitted from a variety of separate sources, such as metal plating and melting operations, as well as the zinc smelters and incinerators noted in FIGURE 2. Although the composition of incinerator emissions is well known,[24] the elemental compositions of the other sources of trace elements are not adequately known for the St. Louis area. Thus, no attempt was made to include such sources in the chemical element balance analysis.

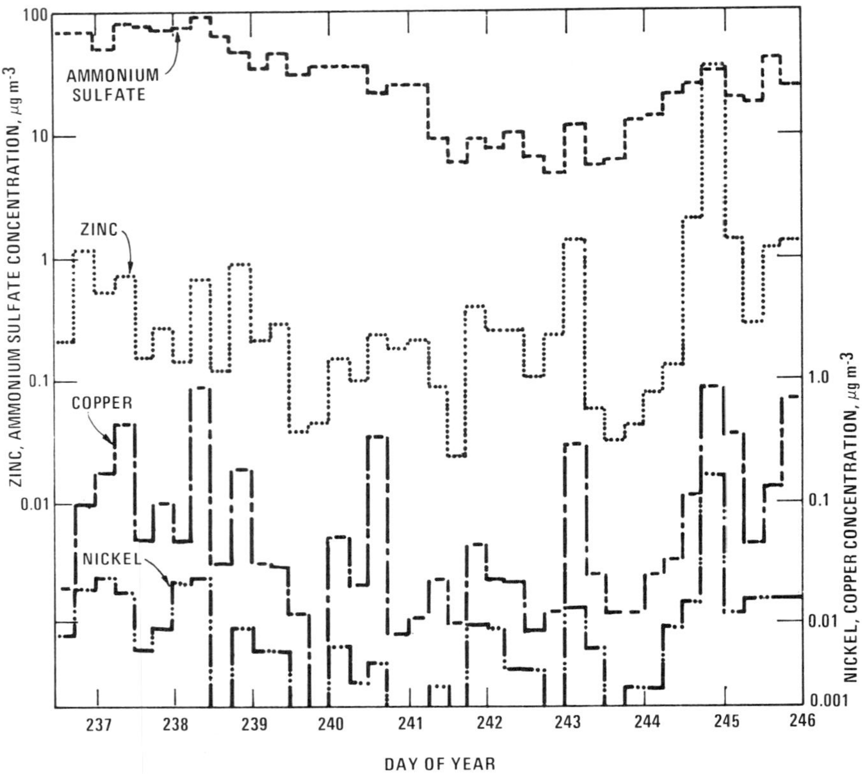

FIGURE 3. Ammonium sulfate, zinc, copper, and nickel concentrations at RAPS Site 105 during Days 236 through 245 of 1976. Plotted are six-hour averages summed over fine and coarse fractions. Each tic mark indicates the beginning of a day.

TABLE 1 documents an event at RAPS Site 105 on Day 233, when a high lead concentration (15 $\mu g\ m^{-3}$) and a low Br to Pb ratio (0.01) indicated the possible influence of the lead smelter near Site 105 (see FIGURE 2). However, this event appeared to be unimportant on an average basis since, for a two-month period, the average Br to Pb ratio at Site 105 was 0.17, in agreement with values at sites farther from lead smelters.

On one occasion, the phosphorus content at RAPS Site 103 exceeded 12 $\mu g\ m^{-3}$, presumably due to a fertilizer plant located 300 m to the northeast. High chlorine concentrations were often observed during the winter, due to salt application on roads to melt snow. On one occasion, the chlorine concentration at Site 112 reached 15 $\mu g\ m^{-3}$. TABLE 1 also illustrates an unusual event at Site 112 on Day 186 (July 4, 1976). Exceptionally high concentrations of Al, K, Sr, Sb, and Ba were observed when fireworks were displayed at an American Bicentennial celebration near the sampling site. For the two-month period considered in this paper, neither of the high phosphorus, road salt, or fireworks events contributed significantly to the average concentrations; hence, components to represent them were excluded from the chemical element balance analysis.

Procedure for Determining the Components

It is assumed that the aerosol composition can be represented as:

$$P_k = \sum_j C_{jk} M_j \tag{1}$$

where P_k is the predicted concentration of element k, C_{jk} is the fractional abundance of element k in component j, M_j is the total concentration of component j, and the sum is carried out for all sources j.

The fractional abundances must be normalized so that:

$$\sum_k C_{jk} = 1$$

when the sum over k is carried out for all elements capable of contributing mass to the aerosol. The component arrays in TABLE 2 satisfy this condition.

The component concentrations M_j are obtained by the technique of linear least squares[25] in which χ^2 is minimized in the expression:

$$\chi^2 = \sum (O_k - P_k)^2/\sigma_k^2 \tag{2}$$

where O_k is the observed concentration of element k, and σ_k is a weighting factor given by:

$$\sigma_k^2 = a_k^2 + (b_k O_k)^2 \tag{3}$$

The sum in Equation 2 is carried out only for those elements k^* for which a fit to the data is demanded, which in the present work only includes Al, Si, S, K, Ca, Ti, Fe, Br, and Pb. The quantities a_k are standard deviations at very low concentrations, and b_k are fractional uncertainties in the concentrations. TABLE 4 lists values for a_k and b_k. The large values of b_k for Al and Si account for large uncertainties in the attenuation corrections needed in the x-ray fluorescence analysis.

The total suspended particulate mass concentration in each size fraction is

TABLE 4

PARAMETERS USED IN THE WEIGHTING FOR THE LEAST SQUARES FIT

Element	a_k (ng m^{-3})	b_k Fine	Coarse
Al	100	0.30	0.35
Si	50	0.25	0.30
S	20	0.10	0.35
K	15	0.08	0.20
Ca	10	0.07	0.16
Ti	7	0.07	0.12
Fe	3	0.05	0.06
Br	1	0.05	0.05
Pb	1	0.05	0.05

TABLE 5

A COMPARISON OF THE AVERAGE MEASURED AND PREDICTED ELEMENTAL AND MASS CONCENTRATIONS AT RAPS SITE 106 FOR AUGUST AND SEPTEMBER 1976†

	Fine Fraction‡		Coarse Fraction‡							
Element	Total Measured	Total Predicted	Total Measured	Total Predicted	Ammonium Sulfate	Motor Vehicles	Shale	Limestone	Paint	Steel
C	§	2,300	§	1,868	0	714	219	934	0	0
Na	§	30	§	142	0	0	138	3	0	0
Mg	§	81	§	613	0	8	216	387	0	2
Al*	203	250	1,209	1,194	0	0	1,153	35	0	6
Si*	458	852	4,470	4,137	0	0	3,935	198	0	4
P	1	6	98	15	0	1	10	3	0	1
S*	4,655	4,655	533	533	481	8	35	10	0	0
Cl	87	74	257	27	0	24	3	1	0	0
K*	261	83	392	406	0	0	383	22	0	0
Ca*	132	132	2,817	2,816	0	0	319	2491	0	7
Ti*	112	112	475	475	0	0	66	3	405	0
V	2	0	4	2	0	0	2	0	0	0
Cr	6	0	9	1	0	0	1	0	0	0
Mn	36	10	35	42	0	1	12	9	3	17
Fe*	275	275	1,218	1,218	0	1	680	31	196	309
Ni	4	0	5	1	0	0	1	0	0	0
Cu	29	0	14	1	0	0	1	0	0	0
Zn	101	11	74	5	0	3	1	0	0	0
As	2	0	1	0	0	0	0	0	0	0
Se	4	0	1	0	0	0	0	0	0	0
Br*	132	147	47	47	0	47	0	0	0	0
Rb	0	0	2	2	0	0	2	0	0	0
Sr	2	1	7	9	0	0	4	5	0	0
Cd	4	0	1	0	0	0	0	0	0	0
Sn	9	0	1	0	0	0	0	0	0	0
Sb	6	0	1	0	0	0	0	0	0	0
Ba	20	5	34	10	0	1	8	0	0	0
Pb*	688	588	189	189	0	188	0	0	0	0
Other¶	§	17,100	§	14,400	1500	270	7,220	4110	360	200
Mass	34,000	27,000	28,000	28,000‖	1980	1270	14,410	8240	960	540

*Elements included in the least squares fit.
†For the coarse fraction, a complete listing of the predicted components is given.
‡In units of ng m^{-3}.
§Not measured.
¶Mainly includes H, N, and O plus 18 other elements.
‖Includes a trace element sum of 430 ng m^{-3} in addition to the sum of the component masses (see Equation 4).

estimated from the equation:

$$M = \sum_j M_j + \sum_k (O_k - P_k) \tag{4}$$

where the first sum is carried out for all components, and the second is a sum for all measured elements not included in the least squares fit. The second term is usually quite small, but for events such as the one having a high zinc concentration (see TABLE 1), the term is needed to account for measured elements that are not predicted in any of the components.

Results and Discussion

Tables 5–6 and Figures 4–8 illustrate the results obtained in the chemical element balance analysis. For the RAPS data, the measured concentrations are averages for July and August of 1976. For the samples collected in the Smoky Mountains, the concentrations are averages for September 20 to 26, 1978.

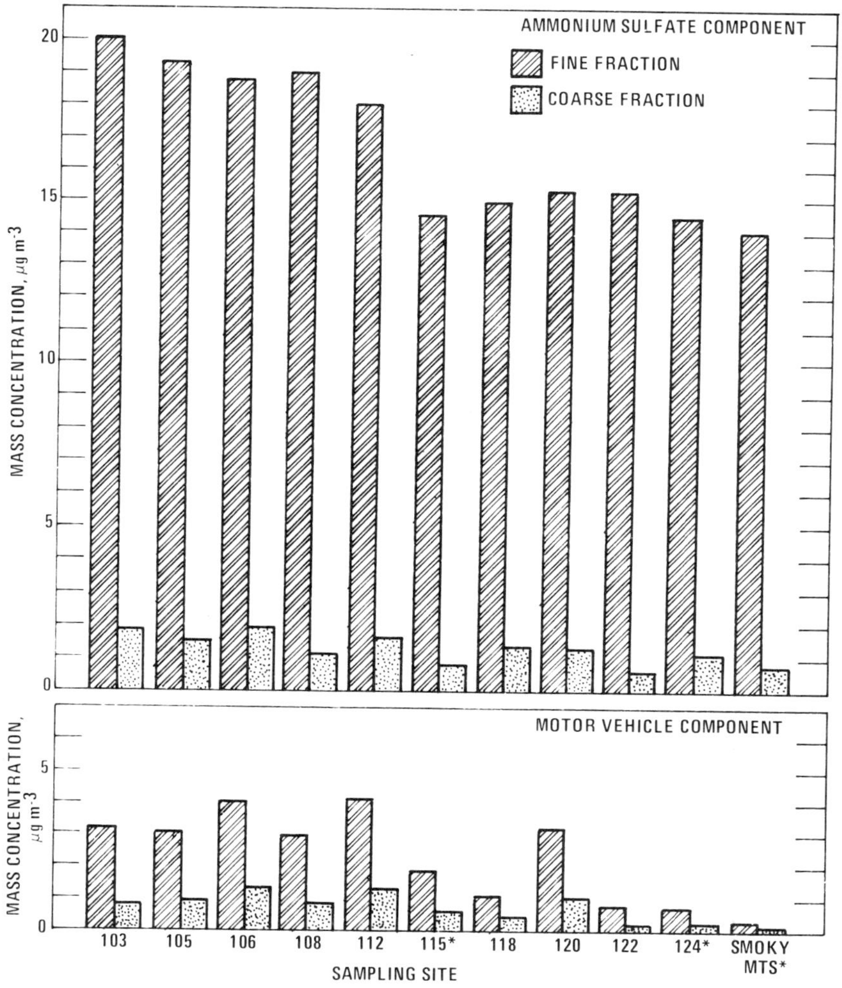

Figure 4. Average ammonium sulfate and motor vehicle components deduced in the chemical element balance analysis. Except where indicated by asterisks, the analysis is based on averages over at least 37 days between July 1 and August 31, 1976. Only 24 days were included at RAPS Site 115. At RAPS Site 124, the fine and coarse fractions covered 41- and 16-day periods, respectively. At the Great Smoky Mountain site, the period covered was September 20 to 26, 1978.

TABLE 5 shows the measured and predicted concentrations for both the fine and coarse fractions at RAPS Site 106. In addition, the composition of each component is shown for the coarse fraction. There is good agreement between the measured and predicted concentrations for the nine elements included in the least squares fit. Exact agreement for sulfur results from the fact that the ammonium sulfate component is determined by sulfur alone. Although the elements Rb and Sr are not included in the least squares fit, their average concentrations are predicted within 20% for the coarse fraction and are due to the shale and limestone components. The observed concentrations of vanadium (V), Cr, Ni, Cu, Zn, selenium (Se), Cd, Sn, Sb, and Ba are much larger than the values predicted from the six components used. Additional components representing the many industrial sources indicated in FIGURE 2 would be needed to account for those elements.

Individual Components

FIGURE 4 shows the average ammonium sulfate component and motor vehicles component at each site. Both components primarily occur in the fine fraction. The average motor vehicles component varies substantially from site to site and is highest at urban and lowest at rural sites. In contrast, the ammonium sulfate component is relatively constant from site to site and has values that range from 15 to 20 $\mu g\ m^{-3}$.

FIGURE 5 shows the average shale and limestone components at each site. Although both components occur primarily in the coarse fraction, there is a substantial difference in the percentages that occur in the fine fraction. Only about 3% of the limestone component occurs in the fine fraction, whereas about 15% of the shale component occurs in that fraction. Any substantial amount of particulate matter in the fine fraction would be expected to originate from combustion sources rather than from mechanically generated aerosol. A consideration of the possible origins of the total shale component indicated in TABLE 3 suggests that fly ash is a possible origin for shale in the fine fraction.

FIGURE 6 shows the steel and paint pigment components. The average steel component ranges from 0.5 to 2.5 $\mu g\ m^{-3}$ at RAPS Sites 103, 105, 106, and 108, which are located in the vicinity of the steel mills shown in FIGURE 2. The average steel component is below 0.2 $\mu g\ m^{-3}$ at RAPS Sites 115, 118, 120, 122, and 124, which are farther from the steel mills. The paint pigment component is largest at RAPS Sites 105, 106, and 112, due to the prevailing southerly wind direction and proximity to the source as shown in FIGURE 2.

Total Mass

FIGURES 7 and 8 illustrate how individual components and their sums compare with the measured mass concentrations in the fine and coarse fractions at each site. TABLE 6 shows the percentage abundances of components in the fine, coarse, and total (fine + coarse) size fractions, averaged for all 10 RAPS stations. The components, in order of decreasing overall abundance, are ammonium sulfate, crustal-shale, crustal-

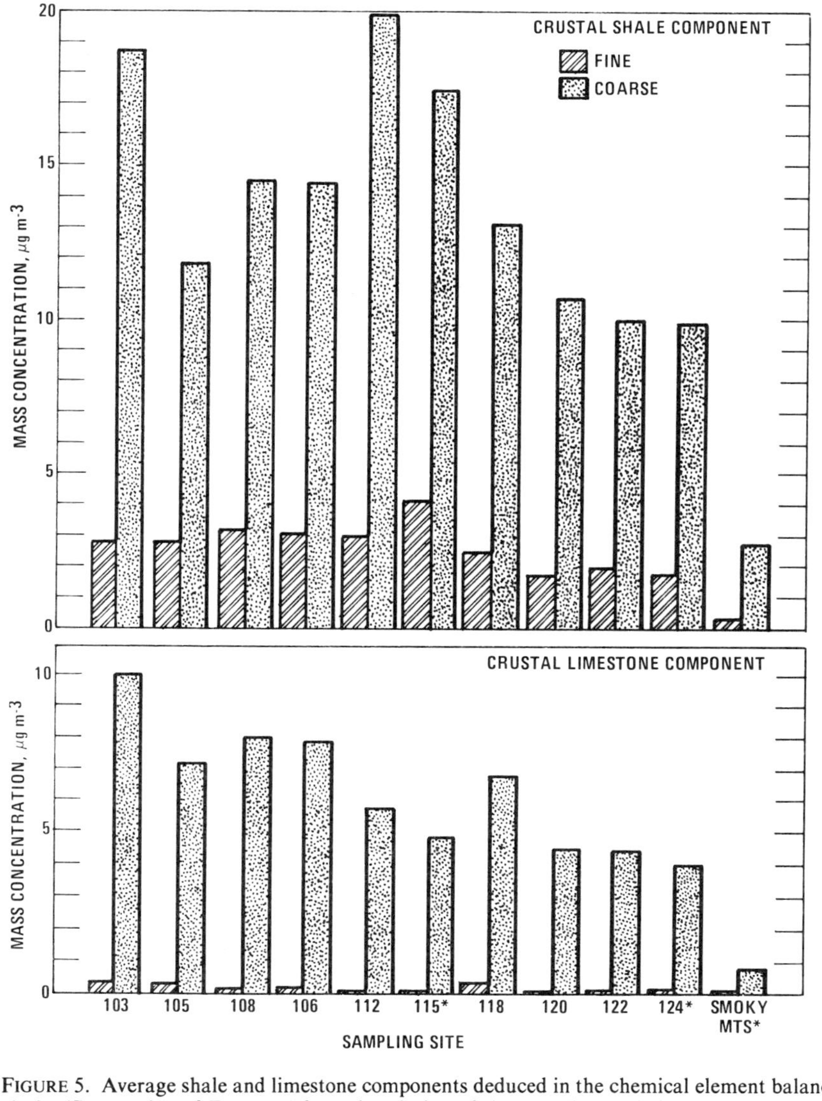

FIGURE 5. Average shale and limestone components deduced in the chemical element balance analysis. (See caption of FIGURE 4 for a description of the sampling periods covered.)

limestone, motor vehicles, steel processing, paint pigment, and trace elements. The trace element concentration is based upon the sum for all elements not included in the least squares fit.

The chemical element balance analysis can account for a substantial part of the total measured mass concentrations. According to TABLE 6, the portions unaccounted

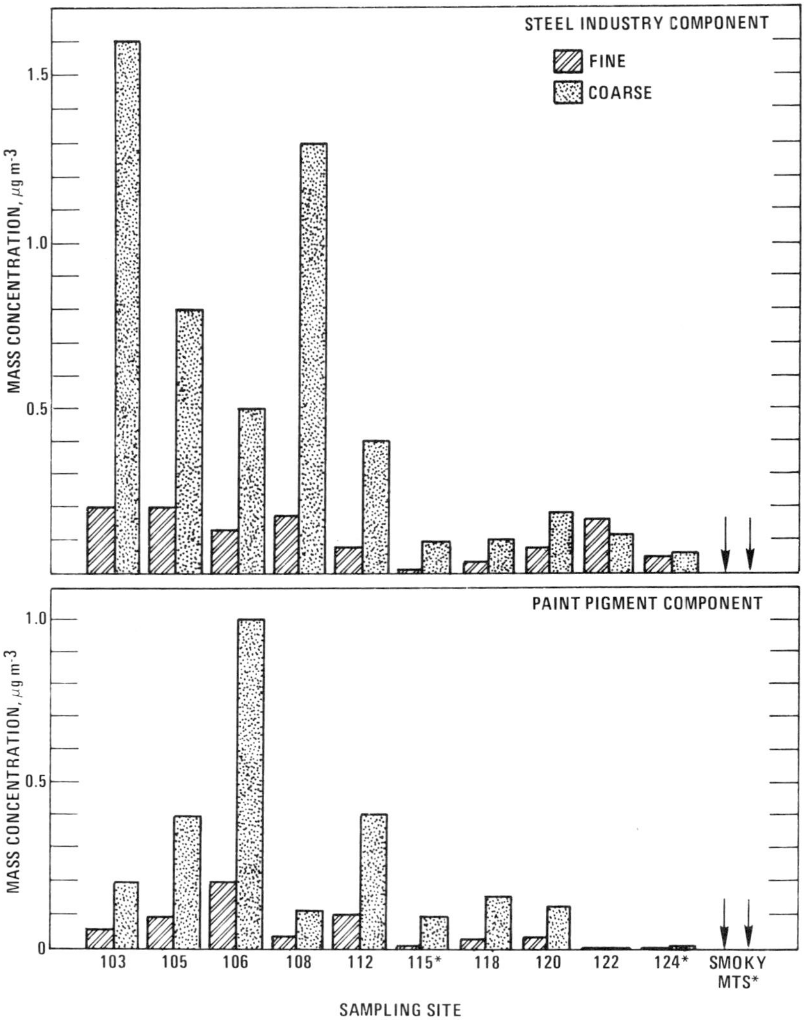

FIGURE 6. Average steel and paint pigment components deduced in the chemical element balance analysis. The arrows indicate values too small to be resolved on the chart. (See caption of FIGURE 4 for a description of the sampling periods.)

for are 22%, 4%, and 13% in the average fine, coarse, and total fractions, respectively. The missing portion may consist of carbon and water, since these are undetectable in the XRF analysis on which the chemical element balance analysis is based. Although carbon is included in some of the components, the amount included may be too small. TABLE 5 shows that the predicted carbon at Site 106 constitutes about 7% of the measured mass in both the fine and coarse fractions. At the other sites, the predicted carbon ranged from 2% to 8% of the mass. In contrast, Lewis and Macias have measured carbon collected in Charleston, West Virginia, using dichotomous samplers and found that carbon constitutes 16% and 12% of the fine and coarse particle fractions, respectively.[26]

Interpretation of the Crustal Components

Although it was demonstrated that the crustal components account for a major portion of the total particulate mass, the origins of the crustal matter are not established. Further measurements might be valuable in distinguishing among the various possible sources suggested in TABLE 3. For example, electron microscopy could be used to distinguish fly ash from soil particles. Both x-ray diffraction and electron microprobe analyses could distinguish among soil types. Also, carbonate analysis could be used to distinguish between calcium-rich soil and cement, since the former contains carbonate, and the latter does not.

SUMMARY AND CONCLUSIONS

Elemental concentrations of size-fractionated aerosol were analyzed by the chemical element balance method using the following six components: ammonium

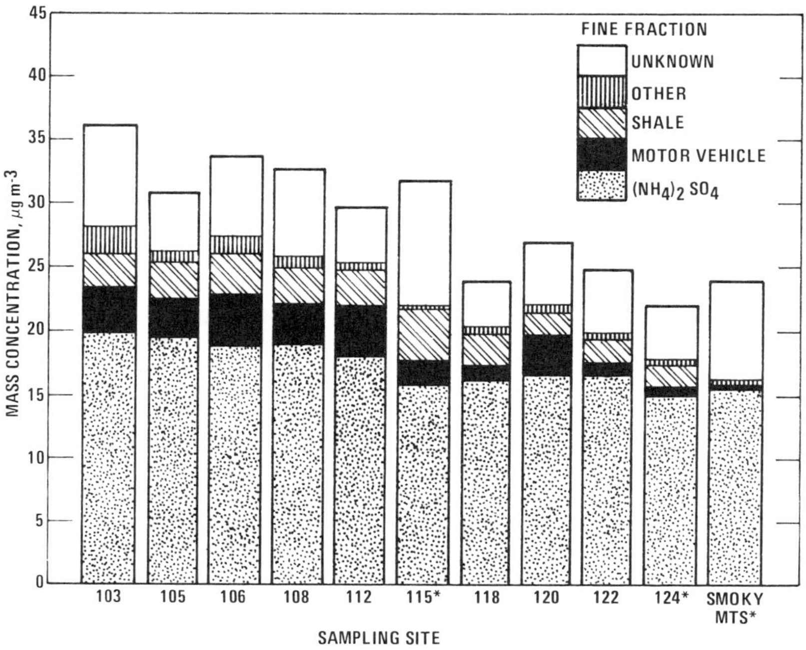

FIGURE 7. A comparison of the average mass concentrations observed in the fine fraction and the component concentrations deduced in the chemical element balance analysis. The component labeled "other" includes the limestone, steel, and paint pigment components and the trace elements. (See caption of FIGURE 4 for a description of the time periods covered.)

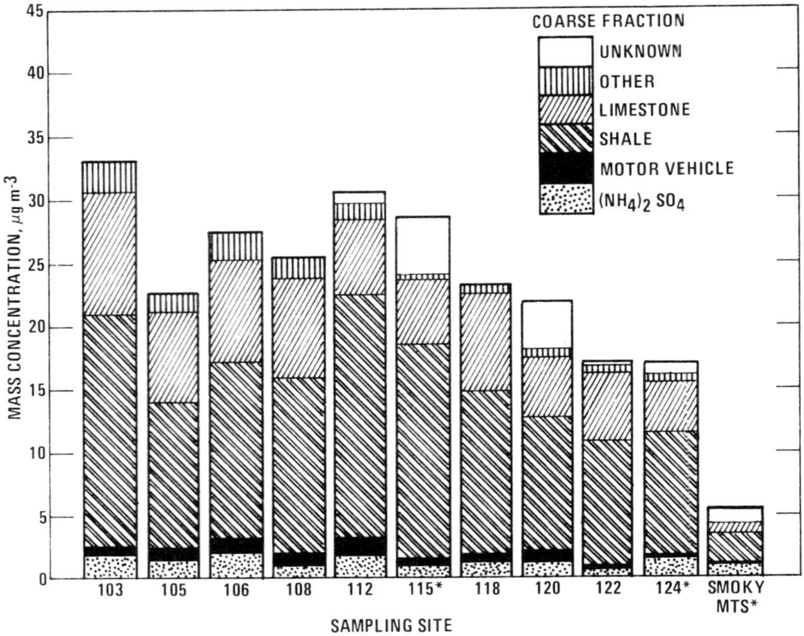

FIGURE 8. A comparison of the average mass concentrations observed in the coarse fraction and the component concentrations deduced in the chemical element balance analysis. The component labeled "other" includes the steel and paint pigment components and the trace elements. (See caption of FIGURE 4 for a description of the time periods covered.)

sulfate, motor vehicles, crustal-shale, crustal-limestone, steel processing, and paint pigment. On the average, during July and August of 1976 in the St. Louis area, ammonium sulfate accounted for nearly 60% of the mass in the fine fraction, and the two crustal components accounted for 83% in the coarse fraction. In the summed fine and coarse fractions, the two crustal components accounted for 43%, and ammonium sulfate accounted for 35%. Motor vehicles, steel, and paint pigment accounted for 6%, 1.2%, and 0.6%, respectively.

TABLE 6

PERCENT ABUNDANCES OF THE COMPONENTS IN THE FINE, COARSE, AND TOTAL (FINE + COARSE) SIZE FRACTIONS, AVERAGED FOR ALL TEN RAPS SITES

Component	Fine Percent	Coarse Percent	Total Percent
Ammonium sulfate	59	6	35
Motor vehicles	8	3	6
Crustal-shale	9	56	30
Crustal-limestone	0.8	27	13
Paint pigment	0.2	1	0.6
Steel processing	0.4	2	1.2
Trace elements	0.6	1	0.9
Unknown	22	4	13

It would be desirable to further resolve the broadly defined crustal components into specific categories, such as fly ash, soil and dust from roads, cement, etc., but their compositional similarities make it difficult to resolve such specific components by the chemical balance method. A more promising approach might be to complement the elemental analysis needed for chemical element balance analysis with additional measurements that characterize particles by their morphology and crystalline structure.

ACKNOWLEDGMENTS

The author is grateful to J. M. Jaklevic, B. W. Loo, R. Gatti, and A. Thompson of the Lawrence Berkeley Laboratory for providing the RAPS data, W. Courtney of Northrop Services, Inc. for providing the Smoky Mountains data, and Adrian Busse of the Environmental Protection Agency for assistance in reading the RAPS data tapes. I also thank Robert Jurgens and Joan Novac for the map and source inventory data shown in FIGURE 2, C. Keadle and C. Winstead for preparing the illustrations, and Ruth Barbour for typing the manuscript.

REFERENCES

1. Monitoring and Data Analysis Division. 1977. National Air Quality and Emissions Trends Report, 1976. Office of Air Quality Planning and Standards Report EPA-450/1-77-002. U.S. Environmental Protection Agency. Research Triangle Park, N.C.
2. MILLER, M. S., S. K. FRIEDLANDER & G. M. HIDY. 1972. A chemical element balance for the Pasadena aerosol. J. Colloid Interface Sci. **39**(1): 165.
3. FRIEDLANDER, S. K. 1973. Chemical element balances and identification of air pollution sources. Environ. Sci. Technol. **7**: 235.
4. WINCHESTER, J. W. & G. D. NIFONG. 1971. Water pollution in Lake Michigan by trace elements from pollution aerosol fallout. Water Air Soil Pollut. **1**: 50.
5. GATZ, D. F. 1975. Relative contributions of different sources of urban aerosols. Atmos. Environ. **9**: 1.
6. KOWALCZYK, G. S., C. E. CHOQUETTE & G. E. GORDON. 1978. Chemical element balances and identification of air pollution sources in Washington, D.C. Atmos. Environ. **12**: 1143.
7. MYERS, R. L. & J. A. REAGAN. 1976. The regional air monitoring systems—St. Louis Missouri USA. *In* Proceedings of the International Conference on Environmental Sensing and Assessment, September 1975.
8. LOO, B. W., W. R. FRENCH, R. C. GATTI, F. S. GOULDING, J. M. JAKLEVIC, J. LLACER & A. C. THOMPSON. 1978. Large scale measurement of airborne particulate sulfur. Atmos. Environ. **12**: 759.
9. DZUBAY, T. G. & R. K. STEVENS. 1975. Ambient air analysis with dichotomous sampler and x-ray fluorescence spectrometer. Environ. Sci. Technol. **9**: 663.
10. WHITBY, K. T., R. B. HUSAR & B. Y. H. LIU. 1972. The aerosol size distribution of Los Angeles smog. J. Colloid Interface Sci. **39**: 177.
11. LOO, B. W., J. M. JAKLEVIC & F. S. GOULDING. 1976. Dichotomous virtual impactors for large scale monitoring of airborne particulate matter. *In* Fine Particles. B. Y. H. Liu, Ed.: 311–350. Academic Press, Inc. New York, N.Y.
12. GOULDING, F. S., J. M. JAKLEVIC & B. W. LOO. 1978. Aerosol Analysis for the Regional Air Pollution Study—Interim Report. Office of Research and Development Report EPA-600/4-78-034. U.S. Environmental Protection Agency. Research Triangle Park, N.C.

13. JAKLEVIC, J. M., D. A. LANDIS & F. S. GOULDING. 1976. Energy dispersive x-ray florescence spectrometry using pulsed x-ray excitation. Adv. X-Ray Anal. **19:** 253.
14. LITTMAN, F. E. 1978. Regional Air Pollution Study—Point Source Methodology and Emission Inventory. Office of Air Quality Planning and Standards Report EPA-600/4-78-042. U.S. Environmental Protection Agency. Research Triangle Park, N.C.
15. STEVENS, R. K., T. G. DZUBAY, G. RUSSWURM & D. RICKEL. 1978. Sampling and analysis of atmospheric sulfates and related species. Atmos. Environ. **12:** 55.
16. PIERSON, W. R. 1978. Private communication.
17. PIERSON W. R. 1979. Particulate organic matter and total carbon from vehicles on the road. *In* Proceedings, Conference on Carbonaceous Particles in the Atmosphere, Lawrence Berkeley Laboratory, Berkeley, Calif.
18. HAEFNER, L. E. 1976. Line and Area Source Emissions from Motor Vehicles in RAPS Program. Office of Air Quality Planning and Standards Report EPA-450/3-77-019. U.S. Environmental Protection Agency. Research Triangle Park, N.C.
19. DZUBAY, T. G., R. K. STEVENS & L. W. RICHARDS. 1979. Composition of aerosol over Los Angeles freeways. Atmos. Environ. **13:** 653.
20. GATZ, D. F. 1978. Identification of aerosol sources in the St. Louis area using factor analysis. J. Appl. Meteorol. **17:** 600.
21. MASON, B. 1966. Principles of Geochemistry. 3rd edit. John Wiley and Sons, Inc. New York, N.Y.
22. SHACKLETTE, H. T., J. C. HAMILTON, J. G. BOERNGEN & J. M. BOWLES. 1971. Elemental Composition of Surficial Materials in the Conterminous United States. Geological Survey Professional Paper 574-D. U.S. Government Printing Office. Washington, D.C.
23. RHEINGROVER, S. W. 1977. A Statistical Model for Titanium Pollution Transport and Dispersion in the Atmophere of St. Louis. M.S. Thesis. Department of Meteorology. Florida State University. Tallahassee, Fla.
24. GREENBERG, R. E., G. E. GORDON & W. H. ZOLLER. 1978. Composition of particles emitted from the Nicosia municipal incinerator. Environ. Sci. Technol. **12:** 1329.
25. BEVINGTON, P. R. 1969. Data Reduction and Error Analysis for the Physical Sciences. McGraw-Hill Book Co. New York, N.Y.
26. LEWIS, C. W. & E. S. MACIAS. 1979. Composition of size fractionated aerosol in Charleston, West Virginia. Atmos. Environ. (To be published.)

AMBIENT SULFATE MEASUREMENTS ON ALLEGHENY MOUNTAIN AND THE QUESTION OF ATMOSPHERIC SULFATE IN THE NORTHEASTERN UNITED STATES

William R. Pierson, Wanda W. Brachaczek, Timothy J. Truex, James W. Butler, and Thomas J. Korniski

Scientific Laboratory
Ford Motor Company
Dearborn, Michigan 48121

Atmospheric sulfate is of continuing concern relating to visibility degradation and acid rainfall[1] and also to health.[2,3] Sulfate concentrations are generally higher in the Northeast than in the rest of the United States.[4-9] This $SO_4^=$ is commonly attributed to atmospheric oxidation of pollutant SO_2. It follows that measures to suppress SO_2 emissions should result in lower SO_2 and $SO_4^=$ atmospheric concentrations. Although SO_2 indeed has been declining, $SO_4^=$ has not.[4,7-12] In fact, rural $SO_4^=$ concentrations generally have increased.[8,9] There are many other anomalies[7-10,12]—for example, the seasonal variation of rural $SO_4^=$ levels (highest in summer) is opposite to that of SO_2 emissions.[10,12] Many explanations have been suggested,[4,9,11-15] but none have yet been proven.

The chief aim of the present study was to contribute to the understanding of the problem just stated. Collateral objectives dealt with $SO_4^=$ in rain and dry deposition, and the relationship of $SO_4^=$ to visibility degradation and atmospheric light scattering. The measurements were carried out in July and August 1977.

Experimental

The strategy in this study was to search for interrelationships of $SO_4^=$, % $SO_2 \rightarrow SO_4^=$ conversion, meteorological properties, rainfall, visibility, b_{scat}, SO_2, aerosol mass, H^+, NO_3^-, a number of elements $1 \leqq Z \leqq 82$, and rainfall H^+ and $SO_4^=$.

Site

The site was an abandoned radio tower, 16 meters high, on the summit of Allegheny Mountain (elevation 838 meters) in southwestern Pennsylvania. The location (latitude 39°57′33″ N, longitude 78°51′9″ W, FIGURE 1) lies within the area of maximum $SO_4^=$ concentration in the Northeast.[8,10] The immediate area is entirely woodland, except that the Pennsylvania Turnpike passes through the mountain 131 meters directly below the base of the tower. The sampling inlets were 15 meters up on the tower, well above the plane of the treetops at 9.2 meters above ground.

0077–8923/80/0338–0145 $01.75/1 © 1980, NYAS

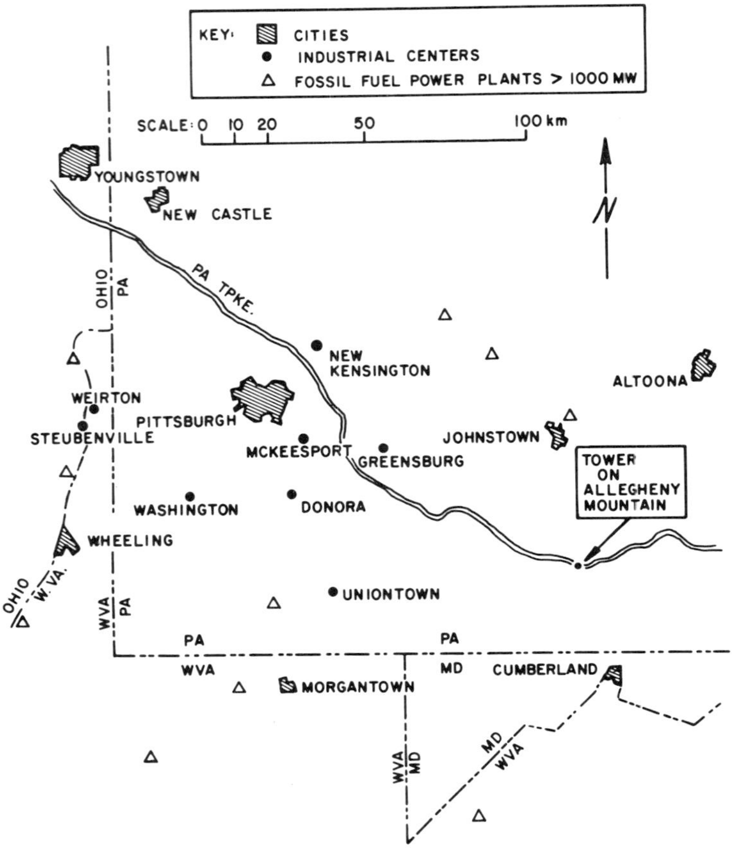

FIGURE 1. Map of the southwestern Pennsylvania vicinity of the Allegheny experiment, showing the location of the tower, major urban areas, and coal-burning power plants.

Equipment

The units installed on the tower were as follows:

A standard high-volume sampler with 8 in. × 10 in. quartz-fiber filters.

A sampler employing 142-mm diameter fluorocarbon membrane filters (Fluoropore PTFE Teflon) of 0.2 μm nominal pore diameter.

A sampler employing 47-mm diameter fluorocarbon membrane filters, 0.2-μm pore size, backed with oxalic acid impregnated glass fiber filters for NH_3 collection.

An Andersen 8-stage cascade impactor and quartz-fiber backup filter.

An impinger containing 3% H_2O_2 for SO_2 collection, with a 25-mm fluorocarbon membrane filter of 0.2-μm pore size upstream to stop $SO_4^{=}$.

A second standard high-volume sampler, with various tandem filter combinations, to evaluate artifact formation and, relocated from the balcony to a landing 4.4 meters above ground, to look for a vertical $SO_4^=$ concentration gradient.

An integrating nephelometer for continuous recording of b_{scat}. The airstream leaving the instrument was 5 to 16°C hotter than the ambient air; thus, the relative humidity of the exit stream could never have exceeded 70%.

A wind vane 3.6 meters above the top of the tower. Wind speed and direction were continuously recorded.

A microcoulometric real-time SO_2 monitor (developed at our laboratory).

Six landmarks 3.2 to 32.2 km away were chosen for estimation of visual range. At the foot of the tower, in a standard weather-station shelter, were instruments for continuous recording of atmospheric pressure, temperature, and relative humidity. In a 20 × 25-meter clearing 57 meters from the tower was a rain gauge for collection and measurement of each rainfall (covered between rains) and a dustfall jar (open except during rain).

Sampling and On-Line Measurement Procedures

For the filter and impinger samples, the experiment was divided into 35 consecutive 12-hour sample-collection runs beginning at 0800 and 2000 EDT each day. Run 1 commenced 2000 Sunday 24 July 1977, and Run 35 ended 0800 Thursday 11 August 1977. Twelve-hour averages of the continuously recorded variables (b_{scat}, meteorological parameters) were calculated to facilitate comparisons of all experimental variables on the same time scale. Adequate sample size required longer sampling periods for the cascade impactor—7.5 and 10 days. Each rainfall was sampled separately and from beginning to end. One overnight dewfall sample was collected in the rain gauge. All samples (filters, etc.) were sealed a few minutes after collection and kept sealed at room temperature until analyzed.

Sample Analysis

The 47- and 142-mm fluorocarbon filters were weighed at controlled temperature and humidity (39%). Analytical methods included ion chromatography, $Ba(ClO_4)_2$ titrimetry, pH measurement, combustion, x-ray fluorescence, neutron activation analysis, inductively coupled argon plasma atomic emission spectroscopy, and two methods for NH_3.

Storage times to extraction and analysis for H^+, NH_4^+, NO_3^-, and/or $SO_4^=$ were 8–9 months for the 142-mm fluorocarbon filters, 7 months for the Hi-vol filters, and 4 months for the impinger samples. Loss of NO_3^- and $SO_4^=$ from the 142-mm filters during storage was zero, as determined by checking the containers for NO_3^- and $SO_4^=$. Losses of $SO_4^=$ to the inner surfaces of the cascade impactors and the 142-mm filter-holder inlets were all very small.

Data Analysis

The main method of data analysis in this experiment was to perform linear least squares regressions of the various parameters against one another to find which ones were interrelated, the criterion for relatedness being a high correlation coefficient (r). Simple regressions of the form:

$$y_i = ax_i + b \quad 1 \leqq i \leqq 35 \qquad \textbf{(1)}$$

were determined, both on the assumption that all the error is in y and on the assumption that the error is shared between y and x in proportion to their magnitudes. The computer program gives a, b, and r and the standard errors (σ) in a and b. A second criterion of relatedness in some cases (in addition to r) is that, within experimental error, the regression line should pass through the origin (i.e., b should be $<\sigma_b$). From r^2, one learns how much of the variation in y is "explained" in terms of x if y does not depend on any other variables.

In some cases (e.g., regression of total N vs. NH_4^+ and NO_3^-), multiple least squares linear regression of the form:

$$x_i = ay_i + bz_i + c \quad 1 \leqq i \leqq 35 \qquad \textbf{(2)}$$

was appropriate. The computer program gives the constants a, b, c, their standard errors, and the simple, partial, and multiple correlation coefficients. One can readily determine how much of x is "explained" by y, by z, or by the two in combination. In still other cases (for example, in assessing the effects of rain or wind direction on $SO_4^=$), other means had to suffice—e.g., simply plotting the data.

Results

The most important results from this experiment are as follows:

1. Aerosol H^+ and $SO_4^=$ were uniquely correlated to each other and to b_{scat}. The $SO_4^=$ was predominantly H_2SO_4, except at the lowest $SO_4^=$ concentrations. At all $SO_4^=$ levels, the $SO_4^=$, H^+, and NH_4^+ can be satisfactorily represented almost wholly as $(NH_4)_xH_ySO_4$ ($x + y = 2$). On the average, $SO_4^=$ was about 35% of the aerosol mass, and material associated with it was another ~25%. While the $SO_4^=$ 12-hour levels fluctuated greatly (1 to 32 $\mu g/m^3$), the aerosol unassociated with $SO_4^=$ was relatively constant at ~15 $\mu g/m^3$.

2. Light scattering was entirely attributable to $SO_4^=$.

3. The filter samples were strongly acidic (pH as low as 3.2 in 50-ml extracts of a

Table 1

Average Values of Certain Meteorological Parameters at Tower on Allegheny Mountain 24 July to 11 August 1977

Wind Speed	7.4 meters/sec	(0 to 22.9)
Wind Direction (*from*)	258° True	
Temperature	18.1°C	(9 to 25)
Relative Humidity	79.6%	(43 to 99)
Pressure	700.6 mm Hg	(695 to 707)
Visual Range	15 km	(<3.2 to >32)
b_{scat}*	2.0×10^{-4} m^{-1}	(10^{-5} to 6.5×10^{-4})

*This is the aerosol scattering only (does not include molecular scattering).

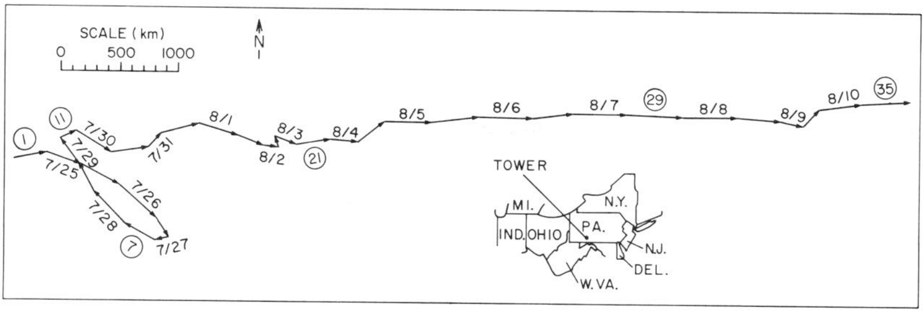

FIGURE 2. Wind trace 2000 EDT 24 July to 0800 EDT 11 August 1977, constructed from the continuous record of wind speed and direction at the tower on Allegheny. Each vector corresponds in direction and length to the net wind run in one 12-hour period. The arrows are placed head to tail to give the total wind run during the experiment. Dates are noted on daytime (0800–2000) runs; run numbers are indicated in a few instances (circled).

142-mm diameter fluorocarbon filter or 1/8 of a quartz-fiber Hi-vol filter) and nearly colorless.

4. There was no evident relation of SO_2 to any other parameter, including $SO_4^=$. The $SO_4^=/SO_2$ mole ratio (average, 0.51) was uninfluenced by humidity or any other meteorological parameter except—possibly to a slight extent—wind direction. There was no day/night alternation in $SO_4^=$, SO_2, or $SO_4^=/SO_2$ ratio.

5. A purge of $SO_4^=$ by an intrusion of clear air from the north had little effect on the concentration of the non-$SO_4^=$ part of the aerosol mass (including most of the individual constituents measured therein), only a modest effect on SO_2, but a pronounced effect on b_{scat} and visual range.

6. There was no evidence for a role of carbon, or the rest of the aerosol mass, in $SO_4^=$ formation. There was no evidence for SO_2 oxidation in solution as a means of $SO_4^=$ formation, with or without transition-metal ion catalysis, with or without NH_3 neutralization as a driving force.

These and other results will be amplified in the following sections.

Weather during the Experiment

Average values of some of the meteorological parameters are listed in TABLE 1. FIGURE 2 shows approximately what the wind run would be like if the tower readings reflected the regional air movement. The experiment was dominated by consistent moderate westerly winds, with one important exception: During a 4-day period, 25–28 July, there was an "episode" of unusually clear air brought down from the north. The "episode" was characterized by wind from the north (finally reversing after stalling at the episode midpoint), minimum temperature, low humidity, no rain, maximum visual range (>32 km), minimum b_{scat} ($1.3 \times 10^{-5}m^{-1}$, FIGURE 3), minimum aerosol H^+ (FIGURE 3), minimum $SO_4^=$ (<1 $\mu g/m^3$, FIGURE 3), and low (but not minimum) aerosol mass loading (FIGURE 3)—mostly reflecting low $SO_4^=$. As illustrated in FIGURE 4, the non-$SO_4^=$ aerosol mass (M − $SO_4^=$) and most of its components

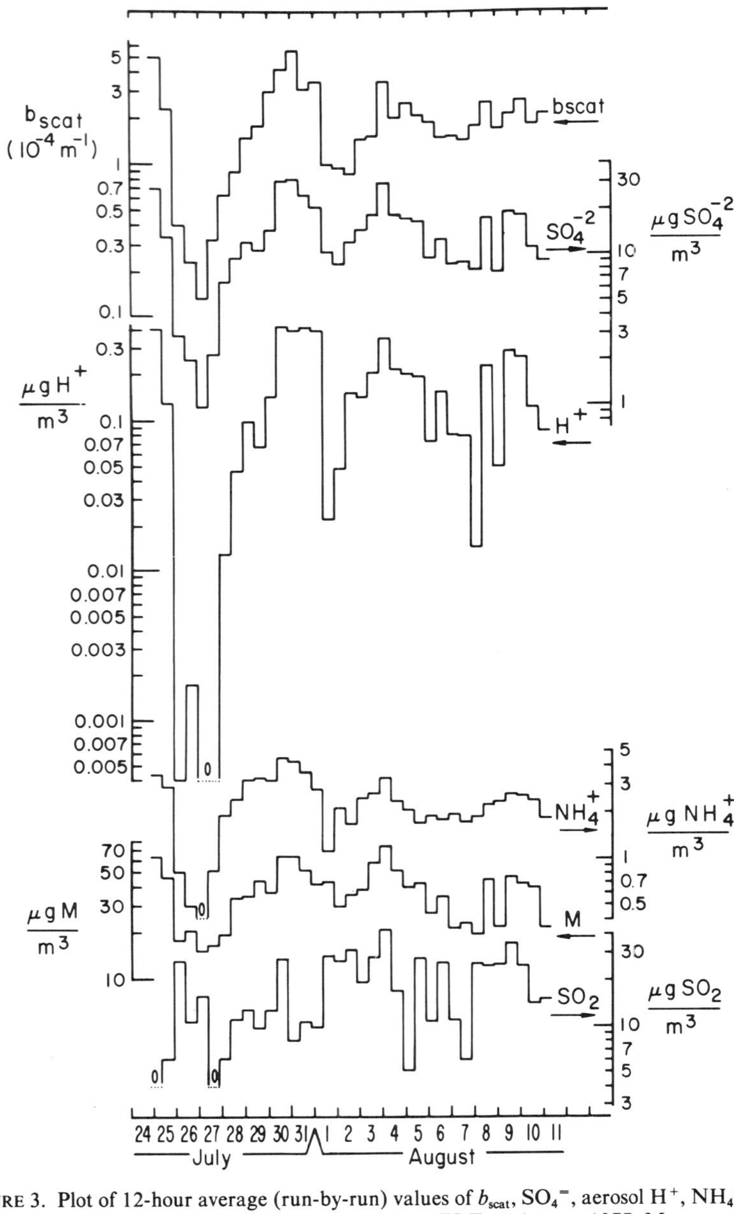

FIGURE 3. Plot of 12-hour average (run-by-run) values of b_{scat}, $SO_4^=$, aerosol H^+, NH_4^+, total aerosol mass (M), and SO_2, 2000 EDT 24 July to 0800 EDT 11 August 1977. M was evaluated at constant 39% humidity.

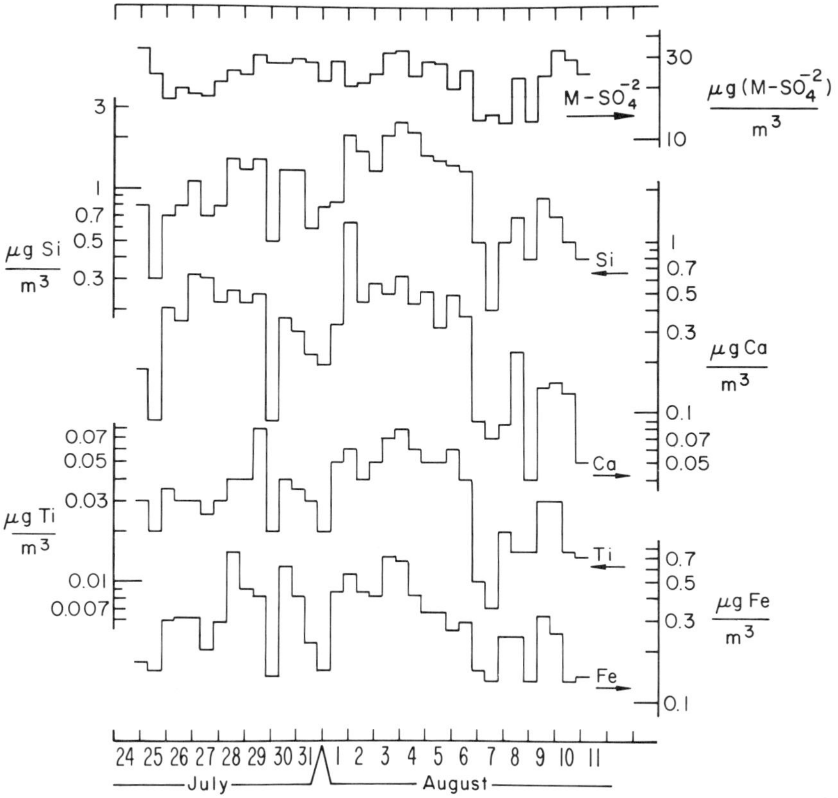

FIGURE 4. Plot of 12-hour average concentrations of total nonsulfate aerosol mass (M − $SO_4^=$) and x-ray fluorescence Si, Ca, Ti, and Fe, 2000 EDT 24 July to 0800 EDT 11 August 1977. M was evaluated at 39% humidity.

remained about the same during this period, and Mg, Ca, and NO_3^- even went through local maxima.

The onset was followed in ~12 hours by a lapse of SO_2 (27 July, FIGURE 3) as the wind origin veered through north around to the east—at which time the $SO_4^=$ seemed to be climbing. Recovery of all parameters to "normal" values accompanied the completion of the wind shift full-circle to its original westerly direction. The $SO_4^=$ recovered from this clear-air purge with a time constant of ~12 hours. The back-trajectory (FIGURE 2) suggests that the "episode" represented a southeastward diversion of air that, on an easterly route, would have passed some 900 km to the north.

As is typical of this locality and its elevation, the average overnight maximum humidity was 91%, and there was always heavy dew on a clear night. Fog was frequent. These are conditions expected to be conducive to $SO_4^=$ formation through SO_2 oxidation in solution.

TABLE 2

AVERAGE LEVELS OF ATMOSPHERIC CONSTITUENTS AT TOWER ON ALLEGHENY MOUNTAIN 24 JULY TO 11 AUGUST 1977

	$\mu g/m^3$	ppb	nanomoles/m^3
SO_2 (g)	18.3	6.4	286
NH_3 (g)*	<0.7	<1.1	<43
Aerosol total mass†	38.0		—
H^+	0.157		155
NH_4^+	2.31		128
NO_3^-‡	(0.53)		(8.5)
SO_4^{2-}	14.06		146
H other than H^+, NH_4^+	1.05		1040
C	0.78		65
N other than NH_4^+, NO_3^-§	(0)		(0)
Na	0.2		9
Mg	0.07		3
Al	0.7		26
Si	1.0		37
P	0.02		0.7
S other than SO_4^{2-}¶	(1.3)		(40)
K	0.07		2
Ca	0.33		8
Ti	0.04		0.7
V	0.003		0.06
Mn	0.009		0.2
Fe	0.32		6
Zn	0.02		0.3
Se	0.003		0.03
Br	~0.02		~0.3
Cd	0.003		0.03
Ba	0.004		0.03
Pb	0.09		0.4

*Maximum observed 12-hour average = 0.86 $\mu g/m^3$ (1.3 ppb or 50 nanomoles/m^3).
†Determined by weighing at 39% relative humidity.
‡On 142-mm fluorocarbon filter; questionable (see text).
§By difference from total N; actually = −0.1 $\mu g/m^3$.
¶Difference between SO_4^{2-} on 142-mm fluorocarbon filter and S by x-ray fluorescence on 47-mm fluorocarbon filter; may not be real.

Average Concentrations

Grand-average concentrations of some of the atmospheric species measured are listed in TABLE 2. It is obvious that, on the average:

The sum of NH_3 and NH_4^+ on an equivalence basis was well short of the amount of $SO_4^=$;

There was less than half enough NH_4^+ for the $SO_4^=$ to have been $(NH_4)_2SO_4$ or for NH_3 neutralization to have been important in $SO_4^=$ formation;

The amount of H^+ was substantial—on a mole basis, as great as (1.06 x) the $SO_4^=$;

The sum of H^+ and NH_4^+ was essentially stoichiometric with the $SO_4^=$;

The amounts of the other cations were so low that H^+ and NH_4^+ must have been the predominant cations for the $SO_4^=$;

The main anion for the H^+ and NH_4^+ must have been $SO_4^=$;

Most of the aerosol S was $SO_4^=$;

NH_4^+ and NO_3^- accounted for almost all of the aerosol N;

The amount of carbon was insignificant as a factor in $SO_4^=$ formation;

Contamination by motor vehicle emissions (e.g., the Turnpike) was low; the Pb level was about the same as the nationwide nonurban average;[5]

The $SO_4^=$ probably was not from urban sources, as the $Pb/SO_4^=$ ratio was lower by at least an order of magnitude than ratios in cities;[5]

The SO_2 probably was not from urban sources, on grounds of Pb/SO_2 ratios.[5]

Within a factor of 2, the relative abundances of Na, Mg, Al, Si, K, Ca, Ti, Mn, and Fe corresponded to their relative abundances in the earth's crust (or in coal fly ash[16]). However, their intercorrelations (TABLE 3) do not indicate a source in common. The low Na/Al or Na/Si ratios argue against a maritime influence.

Given the humidity at which the samples were weighed (39%) and the probable nature of the $SO_4^=$ species, the aerosol mass not recorded in TABLE 2 probably was mostly water. If the $SO_4^=$ in TABLE 2 were H_2SO_4, then its associated water at 39% humidity would be $\sim 16\ \mu g/m^3$, containing oxygen nearly identical to the aerosol mass not identified in TABLE 2 ($15.6\ \mu g/m^3$). The large amount of H present as other than H^+ and NH_4^+ in TABLE 2 is compatible with this; if all of the H other than H^+ and NH_4^+ were attributed to water, then the latter would be $\sim 9\ \mu g/m^3$.

The amount of NO_3^- (TABLE 2) was small. The NO_3^- was correlated to Mg and Ca but not to H^+ or $SO_4^=$ (TABLE 3). The Mg and Ca anticorrelations to H^+ suggest that the two are present as alkaline species, in turn explaining their correlation to NO_3^- (see below).

Concurrent collection of filters at different m^3 per cm^2 of filter surface—either by longer sampling or by different m^3/cm^2/minute flow rates, TABLE 4—indicates that the amount of aerosol NO_3^- found was very dependent on the amount of air to which a unit area of filter surface was exposed, in such a way as to suggest that some other component in the air was displacing NO_3^- (either real or artifact NO_3^-) from the filter. H_2SO_4 obviously answers this description, metathesizing collected NO_3^- to HNO_3, which evaporates, e.g.,

$$NO_3^- + H_2SO_4 \rightarrow HNO_3 \uparrow + HSO_4^- \qquad (3)$$

and/or preventing formation of artifact NO_3^-. Evidence of this has been seen in other work.[17–19]

Strong Relationships between b_{scat}, $SO_4^=$, H^+, and NH_4^+

In TABLE 3, the correlations among the four variables b_{scat}, $SO_4^=$, H^+, and NH_4^+ immediately stand out. (The same applies also for total aerosol N and S, which consist almost entirely of NH_4^+ and $SO_4^=$, respectively.) These strong relationships are evident in the nearly identical time patterns of b_{scat}, $SO_4^=$, H^+, and NH_4^+ in FIGURE 3.

TABLE 3

CORRELATION COEFFICIENTS, TOWER ON ALLEGHENY MOUNTAIN 24 JULY TO 11 AUGUST 1977*†

Temperature (°C)	Humidity (%)	mm Hg	b_{scat}	SO_4^{2-}	H^+	SO_2	Aerosol M	M − SO_4^{2-}	$SO_2 \rightarrow SO_4^{2-}$ (%)	Aerosol H	Aerosol C	NH_4^+	NO_3^-	Mg	Al	Si	Ca	Ti	V	Mn	Fe	Zn	Ba	Pb	$SO_2 + SO_4^{2-}$	
.16	.41	−.11	.04	−.11	−.16	.03	−.18	−.23	−.28	.22	−.43	−.11	−.46	−.38	−.21	−.32	−.38	−.30	−.08	−.29	−.25	−.07	−.03	−.27	−.02	Wind speed (m/s)
	.01	−.35	.38	.48	.50	.03	.49	.44	.31	.35	−.02	.35	−.21	−.25	−.18	.12	−.24	.12	−.06	.07	.25	.21	−.03	−.23	.22	T(°C)
		−.44	.40	.14	.08	.14	−.01	−.16	−.12	.16	−.27	.30	−.50	−.60	−.50	−.33	−.65	−.32	−.16	−.39	−.34	.00	−.13	−.13	.17	Humidity (%)
			−.52	−.45	−.47	−.21	−.43	−.34	−.08	−.24	.20	−.42	.59	.62	.45	.31	.61	.23	.30	.00	.07	.19	.22	.13	−.36	mm Hg
				.90	.88	.08	.76	.52	.45	.17	.02	.83	−.41	−.35	−.20	.03	−.35	−.01	.01	.21	.25	.47	−.07	.21	.44	b_{scat}
					.96	.30	.93	.72	.36	.26	.21	.84	−.18	−.12	−.03	.32	−.10	.24	.01	.50	.56	.61	.16	.32	.65	SO_4^{2-}
						.21	.84	.61	.40	.16	.12	.75	−.25	−.17	−.15	.21	−.12	.10	−.09	.42	.49	.57	.04	.19	.56	H^+
							.34	.34	−.64	.30	.26	.09	−.03	.10	.11	.42	.10	.33	−.23	.43	.48	.24	.51	.09	.92	SO_2
								.93	.30	.34	.33	.77	−.03	.00	.21	.45	.03	.44	.12	.55	.67	.59	.30	.37	.65	M
									.19	.38	.41	.60	.11	.11	.41	.52	.15	.57	.21	.52	.70	.49	.39	.37	.56	M − SO_4^{2-}
										.08	−.06	.39	.04	−.02	−.17	−.13	−.08	−.10	.14	−.06	−.04	.11	−.27	.08	−.37	$SO_2 \rightarrow SO_4^{2-}$ (%)
											.15	.25	−.02	−.06	.09	.18	−.12	.21	.01	.05	.26	.23	.34	.12	.35	Aerosol H
												.34	.55	−.17	.66	.61	.40	.67	.58	.47	.52	.32	.44	.64	.29	Aerosol C
													−.10	−.24	.05	.16	−.23	.16	.35	.37	.28	.57	.07	.52	.42	NH_4^+
														.69	.49	.58	.69	.49	.33	.35	.46	−.04	.36	.37	−.10	NO_3^-
															.52	.63	.95	.61	.15	.53	.56	.06	.44	.33	.03	Mg
																.65	.60	.80	.57	.42	.56	.18	.54	.62	.08	Al
																	.72	.88	.18	.67	.87	.36	.66	.52	.46	Si
																		.67	.12	.53	.61	.07	.47	.32	.06	Ca
																			.30	.60	.81	.25	.67	.56	.36	Ti
																				.06	.10	.19	.20	.56	−.18	V
																					.85	.58	.34	.64	.55	Mn
																						.53	.61	.61	.61	Fe
																							.12	.55	.44	Zn
																								.31	.47	Ba
																									.20	Pb

*There are 35 data points for each parameter tabulated. Critical correlation coefficients are $|r| = 0.334$ at 95% probability of being real, and $|r| = 0.429$ at 99% probability of being real.

†Regressions were performed also with the following aerosol components not included in this table: N, Na, P, S, Cl, K, Sc, Se, Br, Cd, Sb.

TABLE 4

APPARENT OVERALL AVERAGE NO_3^- CONCENTRATIONS AS A FUNCTION OF INTEGRATED AIR FLOW PER UNIT SURFACE AREA OF FILTER

	m^3/cm^2	Apparent $\mu g\ NO_3^-/m^3$
142-mm fluorocarbon filter	0.7	0.53
quartz-fiber Hi-vol filter	2.5	~0.3*
quartz-fiber Hi-vol backup filter	2.5	0.5*
25-mm fluorocarbon filter	5	<0.23†

*Not determined for the whole experiment. The figure here is that projected from the ratio between the Hi-vol and 142-mm filters over that part of the experiment where comparison could be made.

†Detection limit. No NO_3^- detected.

The regression of H^+ onto $SO_4^=$ (concentrations in $\mu g/m^3$):

$$H^+ = (0.0158 \pm 0.0008) \cdot SO_4^= - 0.066 \pm 0.012 \qquad r = 0.956 \qquad (4)$$

clearly indicates that the $SO_4^=$ was mainly H_2SO_4. Equation 4 implies an $H^+/SO_4^=$ mole ratio approaching 1.5 at sufficiently high $SO_4^=$. FIGURE 5 indeed shows that high $H^+/SO_4^=$ ratios were favored at the higher ($\gtrsim 12\ \mu g/m^3$) $SO_4^=$ concentrations. Conversely, high $NH_4^+/SO_4^=$ ratios were favored at the lowest $SO_4^=$ concentrations (possibly connoting increasing neutralization of progressively smaller amounts of H_2SO_4 during collection). Finally, FIGURE 5 shows that the sum of the $H^+/SO_4^=$ and $NH_4^+/SO_4^=$ mole ratios was about 2.0 (actually, 1.94), independent of $SO_4^=$ concentration. Multiple regression of $SO_4^=$ onto H^+ and NH_4^+ (concentrations again in $\mu g/m^3$):

$$SO_4^= = (45 \pm 4) \cdot H^+ + (2.2 \pm 0.5) \cdot NH_4^+ + 1.8 \pm 1.2 \qquad r = 0.972 \qquad (5)$$

shows that 94% of the $SO_4^=$ variance is accounted for by H^+ and NH_4^+: 51 ± 4% by H^+, 37 ± 9% by NH_4^+. Multiple regression of $SO_4^=$ onto H^+ and total aerosol N gives even higher correlation ($r = 0.99$), or effectively all (97%) of the $SO_4^=$ accounted for by H^+ and N; since NH_4^+ constitutes 93 ± 6% of the N, this regression is tantamount to the preceding one. Of the strongly related constituents $SO_4^=$, H^+, and NH_4^+, the H^+ and NH_4^+ were more correlated to $SO_4^=$ than to each other. The three were not highly correlated to any other aerosol constituents (except mass, S, and N).

The $SO_4^=$ in this experiment is thus seen to have existed in the form:

$$(NH_4)_{(2-x)}H_xSO_4 \qquad (6)$$

Only during the clear-air "episode" did x reach zero. Maximum x was 1.76. Half of the time, x was greater than 1.0 (i.e., more acidic than NH_4HSO_4). Especially at $SO_4^=$ levels above ~12 $\mu g/m^3$, the $SO_4^=$ was mostly H_2SO_4.

Sulfate was somewhat related to the non-$SO_4^=$ part of the aerosol mass determined by weighing at 39% humidity:

$$M - SO_4^= = (0.74 \pm 0.12) \cdot SO_4^= + 13.5 \pm 1.7 \qquad r = 0.719 \qquad (7)$$

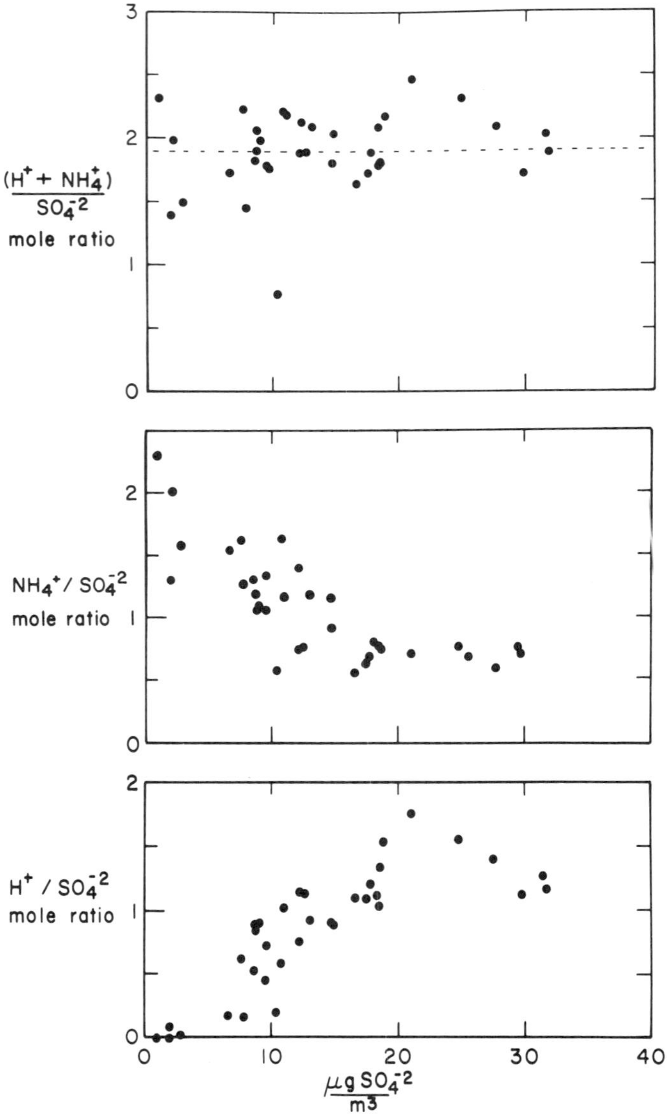

FIGURE 5. Plot of aerosol $H^+/SO_4^=$ and $NH_4^+/SO_4^=$ mole ratios, and their sum $(H^+ + NH_4^+)/SO_4^=$, as a function of $SO_4^=$ concentration.

The $SO_4^=$-associated material—evidently a substantial portion of M − $SO_4^=$—presumably includes water, NH_4^+, and H^+.

Sulfate as the Origin of the Light Scattering

TABLE 3 shows that b_{scat} was related to $SO_4^=$, and to its kindred species H^+, NH_4^+ (and total aerosol N and S), far more than to any other parameter (including humidity or aerosol mass). Multiple regression analyses (b_{scat} in 10^{-4} m^{-1}, concentrations in $\mu g/m^3$) show that b_{scat} was remarkably insensitive to any aerosol constituent but $SO_4^=$ or its associated cations:

$$b_{scat} = (0.1325 \pm 0.0098) \cdot SO_4^= - (0.74 \pm 0.18) \cdot NO_3^- + 0.50 \pm 0.12 \qquad r = 0.935 \tag{8}$$

(94 ± 7% of the b_{scat} is "explained" by $SO_4^=$ alone, on the average);

$$b_{scat} = (5.4 \pm 1.0) \cdot H^+ + (0.49 \pm 0.14) \cdot NH_4^+ - 0.005 \pm 0.334 \qquad r = 0.910 \tag{9}$$

(83% of the b_{scat} variance is "explained" by H^+ together with NH_4^+: 43 ± 8% by H^+, 57 ± 17% by NH_4^+);

$$b_{scat} = (0.169 \pm 0.015) \cdot SO_4^= - (0.040 \pm 0.015) \cdot (M - SO_4^=) + 0.57 \pm 0.37 \qquad r = 0.920 \tag{10}$$

(119 ± 11% of the b_{scat} is "explained" by $SO_4^=$, on the average).

The point is also made by single correlations of b_{scat} with $SO_4^=$ and with (M − $SO_4^=$) (TABLE 3). As shown in FIGURE 6,

$$b_{scat} = (0.139 \pm 0.012) \cdot SO_4^= + 0.03 \pm 0.16 \qquad r = 0.901 \tag{11}$$

Equation 11 implies that b_{scat} was dominated by $SO_4^=$, with a mass scattering coefficient of $0.139 \times 10^{-4}\ m^{-1}/(\mu g/m^3) = 13.9 \pm 1.2\ m^2/g$.

Water surely must have been involved in order that there could have been such a large mass scattering coefficient. This is only to be expected, since H_2SO_4 is hygroscopic—as is NH_4HSO_4 above ~39% humidity.[20] And indeed, b_{scat} is correlated better to the product of $SO_4^=$ and humidity (i.e., $r = 0.957$) than to $SO_4^=$ alone.

Equation 9 shows that the b_{scat} not associated with H^+ and NH_4^+ was insignificant and (when molecular weights are considered) that the per-molecule scattering power of the H^+ salt (presumably H_2SO_4 and its associated water) was comparable to that of the NH_4^+ salt.

Total aerosol mass vs. b_{scat} gave a relatively poor correlation,

$$M = (9.3 \pm 1.4) \cdot b_{scat} + (19.5 \pm 2.7) \qquad r = 0.763 \tag{12}$$

with a large residual term (19.5 $\mu g/m^3$).

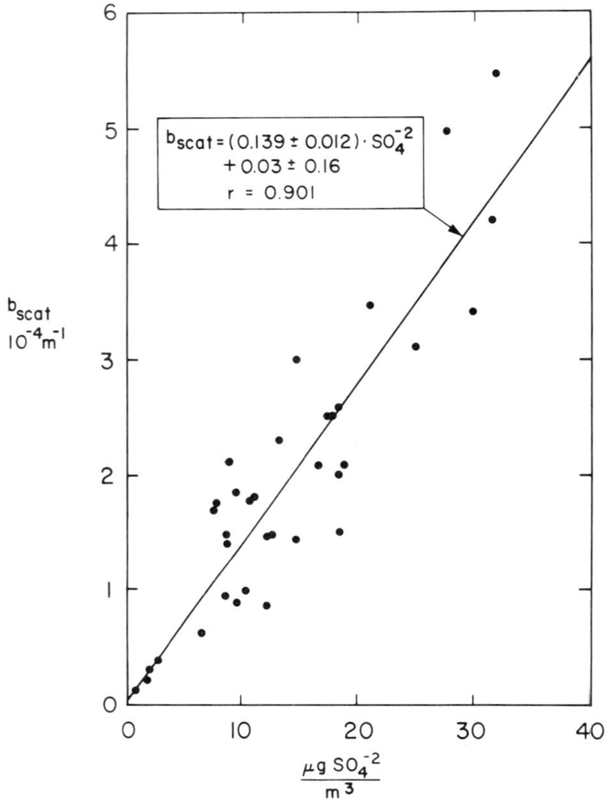

FIGURE 6. Plot of b_{scat} vs. aerosol $SO_4^=$ (12-hour averages).

Relationship between $SO_4^=$ and Visibility

FIGURE 7 reveals a similarity in patterns between visual range L_v and the reciprocal of the $SO_4^=$ concentration, with:

$$L_v \cdot SO_4^= \sim 0.1 \text{ g/m}^2 \quad \textbf{(13)}$$

Undoubtedly, $L_v \cdot SO_4^=$ is a function of humidity (since we find it to be true of $L_v \cdot b_{scat}$). We have shown that b_{scat} tracks $SO_4^=$. Therefore, $SO_4^=$ may be a good index of visibility (and vice versa) *if* humidity is taken into account.

Species unassociated with $SO_4^=$ did not exhibit patterns resembling L_v (compare FIGURES 3, 4, and 7). The point is unequivocal in the 26–27 July clear-air "episode," where the extraordinarily good visibility, low b_{scat}, and low $SO_4^=$ clearly had nothing to do with the non-$SO_4^=$ part of the aerosol mass or with any of its identified components (V being an exception, explained later).

Particle Size Distributions

Further demonstrating the connection between H^+ and $SO_4^=$, their size distributions were similar (FIGURES 8 and 9), with mass median equivalent diameter (MMED) of $\langle \rho^{1/2} d \rangle = 0.84 \pm 0.02$ μm. FIGURE 8 shows a log-normal portion with MMED ~ 0.8 μm and $\sigma_g = 1.7 \pm 0.1$ for both H^+ and $SO_4^=$. The size distribution for the first half of the experiment (first impactor collection) was indistinguishable from that for the second.

Although the $H^+/SO_4^=$ ratio was not a strong function of particle size, it did tend to be highest in the 0.7–1.1 μm range (where $SO_4^=$ was also highest), with $H^+/SO_4^=$ mole ratio 1.45 (FIGURE 9). The lower $H^+/SO_4^=$ at ~6 μm indicates that the larger-particle $SO_4^=$, by contrast, was less acidic than H_2SO_4 or NH_4HSO_4.

The observed MMED is near the optimum size for light scattering per unit mass, in harmony with the close coupling between $SO_4^=$ (and H^+) and b_{scat}.

Diurnal Variations

TABLE 5 reveals no appreciable day/night alternation in $SO_4^=$ or in apparent % $SO_2 \rightarrow SO_4^=$ conversion.

Variations with Elevation

The two runs in which simultaneous measurements were made 4.4 meters and 14 meters above the ground showed 14 m/4.4 m = 1.08 ± 0.14 for $SO_4^=$ and 1.13 ± 0.15 for H^+, i.e., no significant gradient.

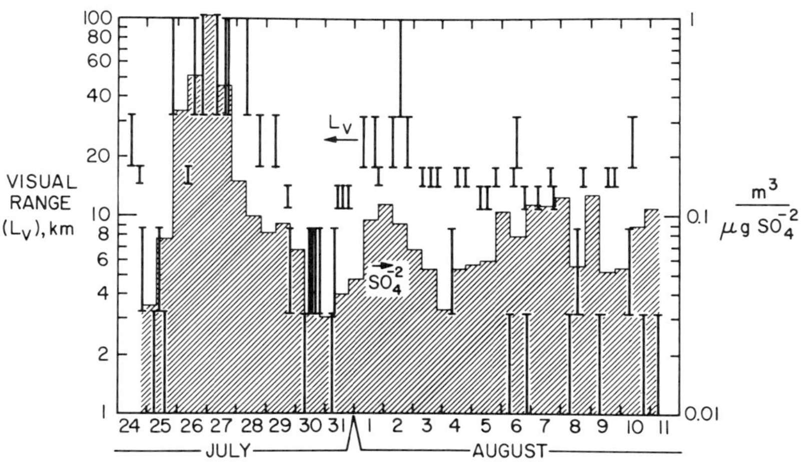

FIGURE 7. Visual ranges L_v (bars) from the tower on Allegheny. Reciprocal of 12-hour $SO_4^=$ is plotted also (histogram), for comparison. In a few of the lowest L_v cases, fog was present.

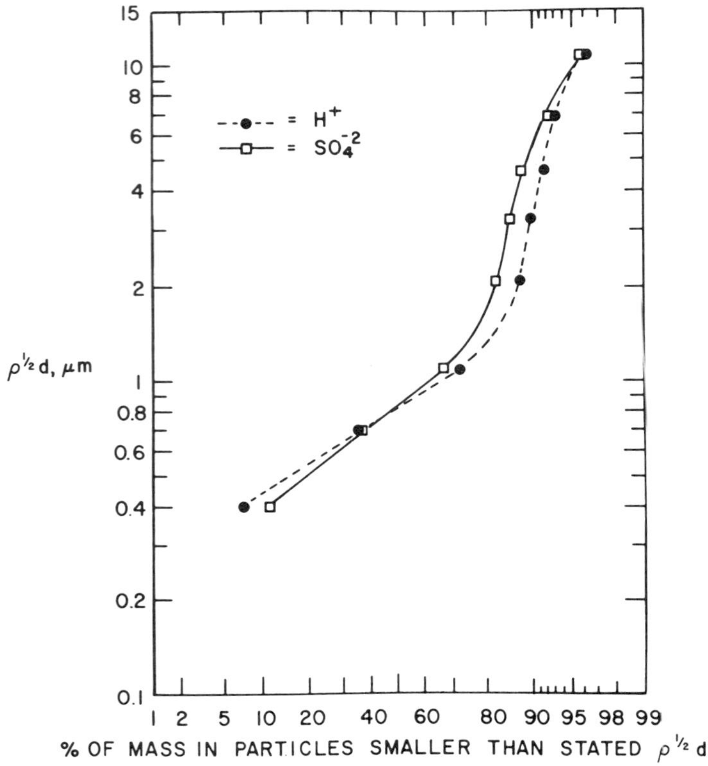

FIGURE 8. Particle size distributions of H^+ and $SO_4^=$ at the tower on Allegheny Mountain, 2000 EDT 24 July to 0800 EDT 11 August 1977.

SO_2 and Its Lack of Correlation to Other Parameters

SO_2 was essentially uncorrelated to $SO_4^=$ and not strongly correlated to any other variable measured (except of course the $SO_2 + SO_4^=$ moles/m^3 total, TABLE 3).

Even though SO_2 was not correlated to aerosol $SO_4^=$ as a rule, there was one important exception, namely, the second half of the clear-air "episode" (Runs 6 through 10, FIGURE 3), when the $SO_4^=$ was in the process of recovering to "normal" levels. For this special case, we find $r = 0.977$. There was no perceptible delay between the regrowth of the SO_2 and the regrowth of the $SO_4^=$. (For example, the $SO_4^=/SO_2$ was nearly constant, FIGURE 10.) Since the recovery time constant was of the order of 12 hours, this means that (1) SO_2 was not the precursor of the $SO_4^=$; or (2) the oxidation of SO_2 to $SO_4^=$ was very fast (much faster than 12 hours); or (3) the clear-air mass was gradually being mixed with dirty air in which most of the presumed SO_2 oxidation had already occurred. The first and third possibilities are supported by the similarity of the Pb regrowth pattern to that of $SO_4^=$.

The poor correlation between NH_4^+ and SO_2 ($r = 0.09$) or between NH_4^+ and humidity ($r = 0.30$) argues that NH_3-promoted heterogeneous SO_2 oxidation[21-23] was not important.

SO_2 Microcoulometer Record

The output of the microcoulometer showed the SO_2 as being generally intermittent, arriving at irregular intervals in puffs with half-widths of ~1 ± 0.5 hours. This behavior suggests that the SO_2 is not fully dispersed and therefore is not of remote origin. But comparison of the microcoulometer and wind-direction traces fails to confirm that the spikes are coming from the known sources in FIGURE 1.

Variation of $SO_4^=$, SO_2, and $SO_4^=/SO_2$ with Wind Speed

There was no significant correlation of $SO_4^=$ or SO_2 with wind speed (TABLE 3). This rules out local stagnation and therefore suggests transport as the likely source of high $SO_4^=$. The SO_2 peak during the 9 August wind stall may be an exception, i.e., a case where local stagnation governed.

If the $SO_4^=$ was produced through oxidation of industrial SO_2 in the air of the region, then the apparent % $SO_2 \rightarrow SO_4^=$ conversion should have been anticorrelated with wind speed, since the latter would be inversely proportional to transport time. In fact, the anticorrelation was very weak (TABLE 3). If winds only from the prevailing direction are considered, namely, the sector 260°–280°, the anticorrelation was still very poor ($r = -0.37$).

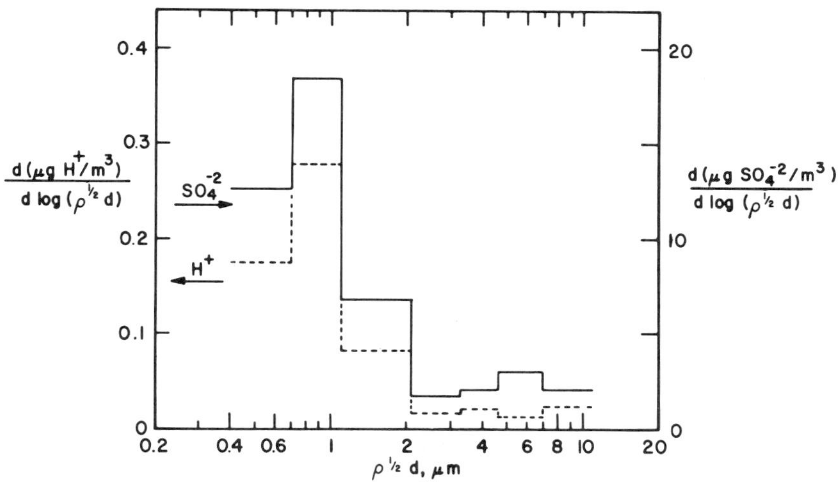

FIGURE 9. Differential size distributions of aerosol H^+ and $SO_4^=$ at the tower, 2000 EDT 24 July to 0800 EDT 11 August 1977.

TABLE 5

DIURNAL VARIATIONS AT TOWER ON ALLEGHENY

	Night/Day*	Night − Day†
Wind speed	1.0	
Temperature		−2.8°C
Humidity		14%
Barometer		0.5 mm Hg
b_{scat}	1.25	
SO_4^{2-}	0.997	
H^+	0.97	
NH_4^+	1.04	
SO_2	1.00	
NO_3^-‡	0.89	
M	0.86	
M − SO_4^{2-}	0.8	
% $SO_2 \rightarrow SO_4^{2-}$	1.00	
H	0.8	
C	0.9	
N	1.04	
Na	0.8	
Mg	0.9	
Al	0.9	
Si	0.95	
S	1.0	
K	0.9	
Ca	1.0	
Ti	0.9	
V	0.9	
Mn	0.8	
Fe	0.8	
Zn	0.7	
Br	1.0	
Ba	1.1	
Pb	1.0	

*(Weighted average of 2000 to 0800 data)/(Weighted average of 0800 to 2000 data).
†(Weighted average of 2000 to 0800 data) − (Weighted average of 0800 to 2000 data).
‡All NO_3^- data are questionable (see text).

Variation of $SO_4^=$, SO_2, and $SO_4^=/SO_2$ with Wind Direction

The directional distributions of $SO_4^=$ and SO_2 were anisotropic and similar (FIGURE 10). But if the clear-air "episode" (Runs 2 through 10) is discounted, the remaining distributions indicate little directionality. The large nearby power plants at 330° to 355° True (FIGURE 1) seemingly contributed little or no SO_2, judging from its directional distribution (FIGURE 10).

If the $SO_4^=$ was produced from atmospheric oxidation of regional industrial SO_2, then $SO_4^=$ should have been more isotropically distributed than SO_2, and the apparent % $SO_2 \rightarrow SO_4^=$ conversion should have been lowest when the wind was blowing from the sources (i.e., when SO_2 was highest). It is not clear that either prediction is correct. Though $SO_4^=/SO_2$ was somewhat anticorrelated with SO_2, the same would be true even if the $SO_4^=$ and SO_2 were independent.

$SO_4^=/SO_2$ Ratios and Implied % $SO_2 \rightarrow SO_4^=$ Conversions

On the assumption that $SO_4^=$ arises from SO_2 oxidation, apparent $SO_2 \rightarrow SO_4^=$ conversion percentages were calculated from $SO_4^=/SO_2$ ratios for correlation to the other parameters. Except for trivial relationships (such as those to $SO_4^=$ itself), no significant correlations with % $SO_2 \rightarrow SO_4^=$ emerged; in particular:

1. No correlation with humidity. Local heterogeneous SO_2 oxidation (in solution or on wet particle surfaces) should have yielded a positive correlation. There should have been a sharp enhancement as the humidity rose above 70%,[24] whereas experimental average conversions below and above 70% humidity were about equal (32% and 37.5%, respectively).
2. Weak anticorrelation with wind speed.
3. Lack of the expected wind-direction dependence.

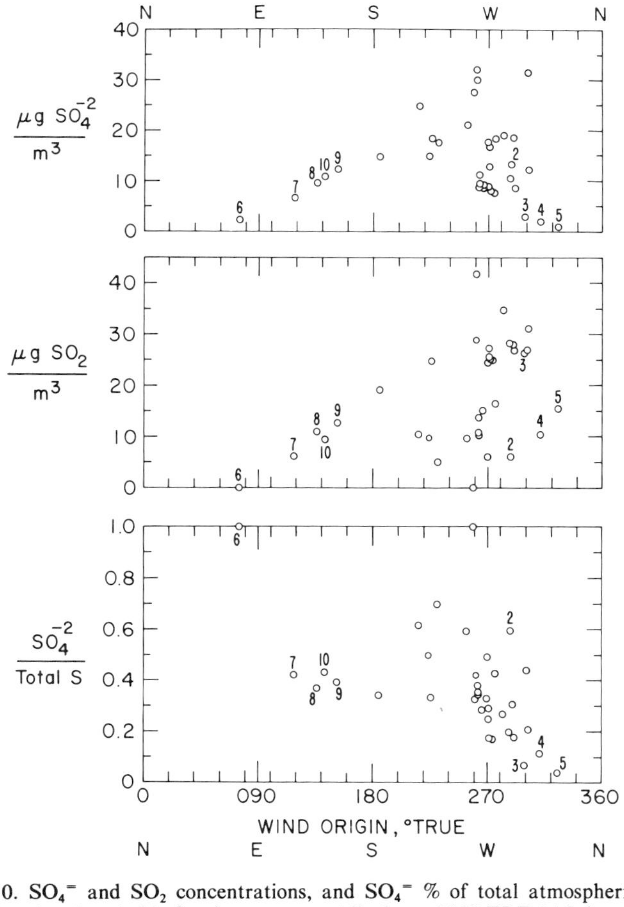

FIGURE 10. $SO_4^=$ and SO_2 concentrations, and $SO_4^=$ % of total atmospheric S, vs. wind direction (direction the wind is *from*) at tower on Allegheny, 2000 EDT 24 July to 0800 EDT 11 August 1977. Numbered points are run numbers during the intrusion of clear northerly air.

4. No correlation with aerosol carbon or non-$SO_4^=$ aerosol mass (i.e., no evidence of increase with the particle surface available for SO_2 adsorption and reaction).

5. No correlation to aerosol transition metals (e.g., Fe, $r = -0.04$) or to any measured cationic species that could have been expected to catalyze the heterogeneous oxidation of SO_2.

6. No correlation to Ca or other species that might have been expected to promote SO_2 oxidation by neutralizing the product. Low abundances also rule out such species as a significant factor.

7. No correlation with aerosol $NH_4^+/SO_4^=$ mass ratio ($r = 0.158$) or with NH_4^+/H^+ ($r = -0.34$). This means that the extent of $SO_2 \rightarrow SO_4^=$ conversion was independent of the extent of neutralization of product H_2SO_4 by NH_3 and, hence, that NH_3-aided heterogeneous SO_2 oxidation[21-23] was unimportant. Since both $NH_4^+/SO_4^=$ (i.e., the extent of neutralization) and % $SO_2 \rightarrow SO_4^=$ (i.e., the extent of oxidation) should increase with air-parcel age as the SO_2 oxidizes in the atmosphere and the product reacts with NH_3, a correlation would have been expected.

Without NH_3 as a factor, then another test of heterogeneous SO_2 oxidation is the ratio H^+/SO_2 between aerosol H^+ and atmospheric SO_2. One would expect a correlation between this ratio and the relative humidity if the aerosol H^+ is mainly H_2SO_4 (as is apparently the case) and if this H_2SO_4 were produced by heterogeneous SO_2 oxidation. H^+/SO_2 would be a better test than $SO_4^=/SO_2$ because H^+/SO_2 would pick out the recent reactions, whereas $SO_4^=/SO_2$ would include aged $SO_4^=$. The observed correlation between H^+/SO_2 and relative humidity was insignificant ($r = -0.09$).

These problems will be addressed further in the discussion section.

Effects of Rain on Aerosol

It rained 12 times during the experiment. The 8 occasions when there was more than a trace accumulation are listed in TABLE 6. The average aerosol $SO_4^=$ on runs during which rain occurred was 94% of the grand average; for SO_2, the figure was 83%. The onset of rain did not cause a decrease in b_{scat}. In other words, rain did not rapidly remove the light-scattering particles (presumably $SO_4^=$), consistent with reported[25] aerosol washout coefficients of 2×10^{-5} to 1×10^{-4} sec^{-1}.

Deposited $SO_4^=$

The rain data are summarized in TABLE 6. The rainfall $SO_4^=$ concentration (micromoles/liter) was correlated to total atmospheric sulfur (SO_2 plus aerosol $SO_4^=$, $\mu g/m^3$):

$$[SO_4^=]_{rain} = (5.5 \pm 2.1) \cdot SO_4^= + (0.86 \pm 0.88) \cdot SO_2 - 18.1 \pm 14.4$$

$$r = 0.896 \qquad \textbf{(14)}$$

But the correlation is accounted for mostly (~65 to 100%) by aerosol $SO_4^=$. The SO_2 contribution is essentially insignificant (partial $r = 0.40$), even though simple

TABLE 6

RAINFALL SAMPLES*†

Rain No.	Start (EDT)	Duration (min)	mm	SO_4^{2-} mg/m²	SO_4^{2-} μmoles/liter
1	1340 7/25	345	4.1	23.0	58.8
2	1739 7/30	471	1.9	9.3	52.0
3	1848 8/6	187	2.7	15.8	60.7
4	0300 8/7	197	9.2	31.6	35.6
5	1700 8/7	175	7.1	37.2	54.1
6	1622 8/8	98	4.5	44.7	104.7
7	0435 8/10	130	0.55	5.6	118.8
8	1500 8/10	268	2.0	10.5	55.3
Dew	1928 8/10	717	—	4.7‡	?

*The four not listed had accumulations insufficient for a sample.

†Minimum rainfall pH observed in Rain Nos. 6 and 7, each pH = 3.67. pH data were not utilized because of organic acid contamination from the butyrate plastic rain gauge.

‡Total SO_4^{2-} = 38 μg in overnight dewfall on an 81 cm² area.

regression indicates otherwise (TABLE 7). Rainfall $SO_4^=$ was correlated to aerosol H^+ and $SO_4^=$ (Equation 14 and TABLE 7).

The $SO_4^=$ concentrations in the rain samples were remarkably similar ($\sigma = \pm 40\%$), declining little if at all with increasing amount of rain. The weighted average rainfall $SO_4^=$ (total $SO_4^=$/total rain) was 58×10^{-6} M. In terms of $SO_4^=$ deposition per unit area of earth's surface, the rainfall $SO_4^=$ was 178 mg/m² in 31 hours of rain, or 1.6 μg $SO_4^=/m^2$/second.

Dew also showed considerable $SO_4^=$ (TABLE 6). The dustfall summed over all dry periods showed only 2.3 mg NO_3^-/m^2 and 9.3 mg $SO_4^=/m^2$ in a dry collection time of 306 hours.

DISCUSSION

Probably the most significant results from the Allegheny experiment are the importance of H_2SO_4 in the aerosol, the dominant role of $SO_4^=$ in atmospheric

TABLE 7

CORRELATION COEFFICIENTS OF RAINFALL SO_4^{2-} CONCENTRATIONS (μMOLES/LITER) WITH OTHER ATMOSPHERIC CONSTITUENTS

	r*
Aerosol H^+ μg/m³	0.95
Aerosol SO_4^{2-} μg/m³	0.87
Atmospheric SO_2 μg/m³	0.72†

*There are eight data points for each parameter tabulated. Critical correlation coefficients are $|r| = 0.707$ at 95% probability of being real and $|r| = 0.834$ at 99% probability of being real.

†Not significant. See explanation in text.

turbidity (b_{scat}), and the failure to document a genetic relationship between atmospheric SO_2 and aerosol $SO_4^=$.

Aerosol H_2SO_4

Much of the filter-sample $SO_4^=$ (even most of it, at high $SO_4^=$) seems to have been H_2SO_4. This result is not surprising in view of pH and $SO_4^=$ values reported in rain (e.g., Reference 26) and in cloud water.[27–29] Since SO_2 wet removal is pH-limited in both solubility[20,28,30] and S(IV) oxidation rate,[21,23,24,28,30–34] some of the lowest rainfall pH values and their connection with rainfall $SO_4^=$ imply existence of atmospheric H_2SO_4. Cloud water is known to contain H_2SO_4.[27–29] Direct evidence of H_2SO_4 in ambient aerosol is also on record.[18,20,35–43]

Measurements of ambient aerosol H^+, NH_4^+, $SO_4^=$, and sometimes NO_3^-, at a variety of sites, generally have shown that NO_3^- is minor,[18,44–46] that NH_4^+ and $SO_4^=$ are correlated,[47] and that there is a semblance of stoichiometry between (H^+ + NH_4^+) and $SO_4^=$ ions[41,42,45–47]—all more or less like the Allegheny results. However, most studies find[18,43,45–47] that H^+ seldom exceeds NH_4^+ on a mole basis, and speciation by benzaldehyde extraction[46] reveals little H_2SO_4. But there is also evidence[46,47] that coarse alkaline collected particles reacting with the acidic particles, during analysis if not also during collection, are suppressing the acidity of the aerosol samples. [At Allegheny, indeed, the $H^+/SO_4^=$ ratio was lowest for the large particles (FIGURE 9).] Generally the NH_3 concentration, low at Allegheny, was not reported in the other studies.

Brosset *et al.*[18,43] have described two types of acid aerosol incursions on the Swedish west coast: Type 1 or "black," and Type 2 or "white." The "white" aerosols sound very much like those at Allegheny: high $SO_4^=$ content, high $H^+/SO_4^=$ ratios, low NO_3^-, and almost colorless. Their white aerosol total mass and $SO_4^=$ concentrations are similar to those at Allegheny. There seems to be a trend of increasing $H^+/SO_4^=$ and decreasing $NH_4^+/SO_4^=$ with higher $SO_4^=$ levels, qualitatively like the pattern at Allegheny (FIGURE 5). Their results demonstrate that acidic $SO_4^=$ aerosol can persist for a long transport time and distance. However, they conclude that no more than ~20% of the $SO_4^=$ was H_2SO_4.

The H_2SO_4 share of the aerosol $SO_4^=$ at Allegheny is likely to have been even greater than the results indicate. There are two reasons:

1. H_2SO_4 collected early in a run will react with NH_3 arriving later. In a 12-hour run this surely must happen, suppressing the measured H_2SO_4 below its correct value.
2. The measured H^+ is only net H^+ after interactions with the rest of the aerosol on the filter and during chemical analysis, and should be less than the H^+ that existed in the uncollected aerosol.

The complementarity of the $NH_4^+/SO_4^=$ and $H^+/SO_4^=$ patterns in FIGURE 5, including the apparent increase of $H^+/SO_4^=$ with increasing $SO_4^=$ concentration, is consistent with initial existence of the atmospheric $SO_4^=$ as H_2SO_4, followed by partial NH_3 neutralization in the air before sampling and/or on the filter during sampling. Stevens *et al.*[45] found that the installation of an NH_3 stripper upstream of their filter resulted in immediately successful analysis for strong acid in filter samples; their result obviously matches the foregoing picture.

Neutralization by atmospheric NH_3 during storage and analysis probably could have had little effect. For example, the amounts of $SO_4^=$ in the filter samples were at least 6 orders of magnitude greater than the amount of atmospheric NH_3 that could have been sealed in with them at the tower. As for the converse process, NH_4^+ loss as NH_3 during storage,[48] one would not expect to create acidic aerosol by NH_3 outgassing to polystyrene and/or polyethylene surfaces.

We doubt that the evident presence of aerosol H_2SO_4 at Allegheny was an anomaly. It persisted too long. And earlier measurements, which began at this site in 1975, have shown 12-hour $SO_4^=$ averages ranging up to 47 $\mu g/m^3$, presumably of the H_2SO_4 type unless the (not measured) atmospheric NH_3 was quite high—enough to supply NH_4^+ concentrations up to, say, 10 $\mu g/m^3$.

Our opinion is that the $H^+/NH_4^+/SO_4^=$ pattern is a concomitant of a dearth of atmospheric NH_3 and is anomalous only insofar as the NH_3 deficit may be. On the assumption of equilibrium with gas-phase NH_3, the filter-sample H^+/NH_4^+ ratios imply atmospheric NH_3 concentrations less than 10 ng/m^3. Accepted estimates of atmospheric NH_3 are generally ~10^2 times higher. However, (1) on the basis of precipitation NH_4^+ and H^+ data, equilibrium NH_3 partial pressures in high-humidity situations are[49,50] only 1–10 ng/m^3 (1–5 ng/m^3 in the vicinity of Allegheny)[49] or even less;[28] and (2) levels below an ~75 ng/m^3 detection limit have been reported in Ohio.[51]

Effect of $SO_4^=$ on b_{scat}

The close relationship observed between b_{scat} and aerosol $SO_4^=$ has been seen by many investigators.[4,10,36,37,52–61] The light scattering cross section per unit mass of $SO_4^=$, 13.9 ± 1.2 m^2/g (Equation 11), is consistent with other data (see TABLE 8). The large $SO_4^=$ mass fraction in the aerosol (up to 50%), tightly distributed in particle size at just about the optimum size for light scattering per unit mass, does much to explain the $SO_4^=$ dominance of b_{scat}.

Deposited $SO_4^=$

The b_{scat} traces and data on rainfall $SO_4^=$ concentration vs. mm rain show that below-cloud aerosol $SO_4^=$ removal by rain was slow at best; and yet the aerosol $SO_4^=$ (and H^+) and rainfall $SO_4^=$ were correlated. Presumably, the below-cloud and in-cloud $SO_4^=$ should be correlated. The SO_2 was correlated weakly, if at all, with the rainfall $SO_4^=$. The simplest inference (if scavenging is a local process) is that the rainfall $SO_4^=$ was mostly the result of (slow) $SO_4^=$ scavenging.

The weighted average rainfall $SO_4^=$ per unit area was 1.6 $\mu g\ SO_4^=/m^2/sec$. Given the average SO_2 and $SO_4^=$ concentrations, the corresponding sulfur deposition velocity ($SO_2 + SO_4^=$) was ~4 cm/sec during rain. This velocity, with a 1000-meter mixing height (an estimated summertime average for the site during rain), corresponds to a total sulfur wet removal coefficient ~4×10^{-5} sec^{-1}. This means that a rain of average duration at Allegheny should remove a substantial fraction (say, 40%) of the atmospheric sulfur. Total sulfur wet removal coefficients in the literature are even larger (1.5×10^{-4} to 3×10^{-4} sec^{-1}).[62]

The observed dewfall $SO_4^=$, 4.7 mg/m^2 overnight (TABLE 6), is readily accounted for by SO_2 deposition. Given the average SO_2 concentration during the night, the deposition velocity was 0.6 cm/sec, which is reasonable for SO_2 dry deposition.

The 9.3 mg/m^2 dustfall $SO_4^=$ collected during the experiment is 19 times less than the rainfall $SO_4^=$ (178 mg/m^2). Given the average atmospheric $SO_4^=$ existing during the experiment, the $SO_4^=$ dry settling velocity was only 0.06 cm/sec. The implied unimportance of gravitational removal of $SO_4^=$ is consistent with the small particle size and the lack of an $SO_4^=$ vertical concentration gradient on the tower.

TABLE 8

VALUES OF a, b, AND CORRELATION COEFFICIENT r IN THE REGRESSION
b_{scat} (10^{-4} m^{-1}) = $a \cdot SO_4^{2-}$ ($\mu g/m^3$) + b
FROM VARIOUS INVESTIGATIONS

Reference	a	b	r
California Air Resources Board[52]	0.07	*	†
Waggoner *et al.*[53]	0.02 to 0.17	†	up to 0.99
Cass[4,54]	0.11 to 0.22	*	0.76 to 0.86
White and Roberts[55,61]	~0.08	*	0.96
Leaderer *et al.*[56]	0.13 ± 0.02	~0.3	0.89
Cobourn *et al.*[38]	~0.3	—	(poor)
Cowherd *et al.*[67]	—	—	0.72 ± 0.12
Husar *et al.*[60]	0.05 to 0.11	—	0.87
Husar *et al.* (cited in Ref. 68)	0.11	3.24‡	0.7
Allegheny 1976 8/10–8/15	0.10 ± 0.04	0.3 ± 0.7	0.65
Tuscarora 1976 8/18–8/24§	0.17 ± 0.01	0.25 ± 0.2	0.97
Present experiment	0.139 ± 0.012	0.03 ± 0.16	0.90

*Only multiple regressions were presented. Additive term generally was not physically reasonable (e.g., <0).

†Not given.

‡St. Louis, summer 1975. Appears to be higher than the other determinations.

§Tuscarora Mountain, 90 km east of Allegheny.

Relationship between SO_2 and Aerosol $SO_4^=$

There was no sign of a genetic relationship between SO_2 and aerosol $SO_4^=$:

1. SO_2 was not correlated with $SO_4^=$;
2. % $SO_2 \rightarrow SO_4^=$—that is, the mole ratio $SO_4^=/(SO_2 + SO_4^=)$—was not correlated with relative humidity;
3. % $SO_2 \rightarrow SO_4^=$ was only weakly anticorrelated with wind speed, even at fixed wind direction;
4. % $SO_2 \rightarrow SO_4^=$ was effectively not correlated to non-$SO_4^=$ aerosol mass, transition metals, alkali or alkaline-earth metals, carbon, etc.;

5. % $SO_2 \rightarrow SO_4^=$ was effectively not correlated with aerosol NH_4^+/H^+ or $NH_4^+/SO_4^=$, and only weakly correlated with NH_4^+;
6. % $SO_2 \rightarrow SO_4^=$ showed no diurnal variation.

The most obvious conclusion is that the $SO_4^=$ does not originate from the SO_2. There is a very strong presumption against this view. Accordingly, we assume that the aerosol $SO_4^=$ does arise from atmospheric SO_2 oxidation and ask how it happened. The answer must satisfy three constraints:

1. With no sources within 40 km (FIGURE 1), the presence of SO_2 in an amount comparable to that of $SO_4^=$ (e.g., FIGURE 10) suggests that the oxidation required at least several hours.
2. The lack of time delay between the SO_2 and $SO_4^=$ buildups in the trailing half of the clear-air "episode" (if anything, the $SO_4^=$ buildup preceded the SO_2 buildup, FIGURE 3) would imply rapid SO_2 oxidation—faster than the ~ 2-day maximum advection time for SO_2 generated, as is supposed, within the Northeast. Further, the lack of correlation between SO_2 and $SO_4^=$ would imply that the SO_2 decay (oxidation plus deposition) was fast compared to the $SO_4^=$ deposition rate. If that is the case, then the $SO_4^=$ and SO_2 indeed can be said to be from different sources—SO_2 from nearer sources, $SO_4^=$ from more distant ones. The pulsed behavior of the SO_2 microcoulometer trace is consistent with that inference.
3. The $SO_4^=$ can hardly have arisen from heterogeneous oxidation of SO_2—wet or dry, catalytic or not, NH_3-promoted or not. The evidence has already been presented: low $NH_4^+/SO_4^=$ ratio; poor NH_4^+ correlation with SO_2 or humidity; poor $SO_4^=$ correlation with SO_2 or humidity; poor correlation between humidity and H^+/SO_2; and poor % $SO_2 \rightarrow SO_4^=$ correlation with $NH_4^+/SO_4^=$, NH_4^+/H^+, NH_4^+, humidity, fog incidence, or nonaerosol mass or components therein (C, V, Mn, etc.). It is obvious why this process was unimportant: the aerosol was too acidic. The pH values of the filter extracts signify aerosol pH values commonly far below 2. At such pH values, SO_2 solubility and S(IV) oxidizability (even with catalysts) are minimal.[20,21,23,24,28,30–34] (The effect of strong oxidizing agents such as O_3 or H_2O_2 at low pH is an issue,[24,30,63] but apparently[63] the O_3 reaction would be too slow except in fogs or clouds.)

The observed lack of correlation between % $SO_2 \rightarrow SO_4^=$ and relative humidity is now understood: if the $SO_4^=$ is too acidic to have formed in solution, then no correlation ought to have been expected.

While heterogeneous SO_2 oxidation could be significant where NH_3 is more abundant, the Allegheny results show that large amounts of $SO_4^=$ can be generated without it. This is consistent with other results.[42] (If NH_3 neutralization—either homogeneous or heterogeneous—were required for $SO_4^=$ formation, then the failure of $SO_4^=$ to respond to SO_2 controls in the Northeast might have been explained as NH_3 producing a stoichiometric amount of $SO_4^=$ regardless of the amount of SO_2 present. Clearly this explanation is not valid, at least at Allegheny.)

For the presumed SO_2 oxidation, therefore, a gas-phase process seems to be the best possibility. Brosset *et al.*[18,43] have arrived at the same conclusion, namely, that the colorless acidic particles of the "white" episodes arose from gas-phase SO_2 oxidation.

Origins of the SO_2 and $SO_4^=$

The next question is the origin of the SO_2, or of the total measured atmospheric sulfur (SO_2 plus $SO_4^=$), at Allegheny. It is already clear from the SO_2/Pb and $SO_4^=$/Pb ratios that cities were the origin of < 10–20% of the total atmospheric sulfur at Allegheny and, at most, ~50% of the $SO_4^=$.

Power plants may not be the major source, either. Sheih[64] has calculated the ambient SO_2 and $SO_4^=$ concentrations attributable to power plants. The maxima are in the vicinity of Allegheny but are still a factor of 4 short of the observed concentrations. The correlations of SO_2 and of artifact $SO_4^=$ with artifact NO_3^- might suggest the power plants as the SO_2 sources, but then the lack of evident SO_2 advection from the large coal-burning installations nearby would be troublesome. Sheih has deduced[64] that the power-plant contribution to nonurban $SO_4^=$ is at most 30%.

Vanadium was not associated with $SO_4^=$, SO_2, or their sum at Allegheny (TABLE 3). The highest V concentrations, an order of magnitude above the average, were associated with wind from approximately 160° True. At Allegheny, the overall $SO_4^=$/V mass ratio, ~4300, was 15–20 times the East-Coast[9] nonurban ratio. The V, like $SO_4^=$, did show a pronounced subsidence during the first half of the clear-air "episode," but only because the source was downwind then. The foregoing does not support the idea that the $SO_4^=/SO_2$ situation in the Northeast has to do with $SO_4^=$ directly emitted in combustion of high-V oil, one of the possible explanations listed in the review in Reference 11.

As to the possibility of other nonurban industrial sources, the problem is that neither the SO_2 nor the $SO_4^=$—nor the molar sum of the two—was correlated to much else in the aerosol (TABLE 3). For example, the total atmospheric sulfur was only indifferently correlated to the nonaerosol mass (r = 0.56, TABLE 3) or to any of its measured constituents.

High $SO_4^=$ concentrations observed in New York State have been traced to prior stagnation in the upper Ohio Valley area.[65,66] The $SO_4^=$ concentration, while unrelated to wind speed at the sites, was correlated to the reciprocal of the speed that the arriving air had had one to three days earlier.[66] This may explain the $SO_4^=$ peak on the night of 3 August at Allegheny (cf., FIGURES 2 and 3).

[**Note added in proof:** Recent trajectory calculations by Perry J. Samson (1979. Ensemble trajectory analysis of summertime sulfate concentrations in the eastern United States. Presented at the EPA Seminar/Workshop on Persistent Elevated Pollution Episodes, Durham, N.C., March; 1979. Ph.D. Thesis. University of Wisconsin.) show that the $SO_4^=$ in the Allegheny experiment can be traced to earlier stagnation (say, 40 hours earlier) over the lower Ohio Valley and eastern Midwest. For example, the $SO_4^=$ peak on the night of 3 August was preceded by a stagnation at the conjunction of Ohio, West Virginia, and Kentucky some 24 hours earlier. The Allegheny $SO_4^=$ (and, by implication, H^+ and b_{scat}), like $SO_4^=$ at other sites in the Northeast, showed a significant correlation with the inverse of wind speed (stagnation) occurring more than 24 hours upstream of the sampling point.]

CONCLUSION

The Allegheny experiment raises some questions as to how the aerosol $SO_4^=$, which seems to have been largely H_2SO_4, could have formed by SO_2 oxidation.

Evidently, it did not form by heterogeneous oxidation of SO_2, catalytically or otherwise (except, conceivably, initially or in the lowest-$SO_4^=$ instances). Indeed, there was no evidence for any genetic connection between SO_2 and aerosol $SO_4^=$.

Given the premise that ambient aerosol $SO_4^=$ does form by SO_2 oxidation, then the proposition most nearly fitting the data is that the oxidation occurred in the gas phase—photochemically or otherwise.

In general, $SO_4^=$ seemed to be associated with transport from the west and not with local stagnation. But the transport does not seem to have been from urban areas. There are problems also in attributing the $SO_4^=$ to power plants or other nonurban industrial sources. Accordingly, we adopt simply the proposition that the $SO_4^=$ at Allegheny stemmed from transport, with or without SO_2 oxidation, from westward nonurban sources of some kind. The sources seem to have been sources of very little else in the aerosol, except insofar as the H^+ and NH_4^+ cannot be excluded as possible coemissions with the $SO_4^=$.

The outstanding chemical feature of the aerosol, besides the magnitude of its $SO_4^=$ component, was the consistent balance between H^+, NH_4^+, and $SO_4^=$, with H^+ being the more abundant cation at $SO_4^=$ levels exceeding $\sim 12\ \mu g/m^3$ and, in some cases, approaching the $H^+/SO_4^=$ ratio of H_2SO_4. (The $H^+/SO_4^=$ ratio could be even higher than observed, owing to possible NH_3 neutralization during sampling.) Whether this behavior is general for the northeastern U.S. comes down to the question of whether the atmospheric NH_3 concentrations were typical, as the aerosol properties observed are consistent with, and follow from, the low atmospheric NH_3 concentration. Accordingly, the situation existing in the present experiment may have been the typical one minus the obfuscating effect of NH_3. The genesis of the atmospheric $SO_4^=$ is thus seen as involving the formation (or emission?) of H_2SO_4.

Aerosol $SO_4^=$ clearly was the key index of atmospheric light scattering (b_{scat})—a conclusion reached by many other investigators—though the $SO_4^=$ relationship to visibility was not as good. While it is only partly clear why $SO_4^=$ should so completely dominate b_{scat}, empirically it seems to be a general phenomenon.

The rainfall $SO_4^=$ appeared to be attributable less to SO_2 scavenging than to $SO_4^=$ scavenging. The rainfall, dew, and dustfall results seem consistent with other published data.

Finally, the acidic aerosol made the NO_3^- problem not so much one of artifact, but the opposite one of loss of aerosol NO_3^- by metathesis to HNO_3 when nonalkaline filters were used.

Acknowledgments

We wish to thank our Ford colleagues Alex D. Colvin, Fred E. McKelvy, Dale Pacynski, and Robert Stordeur for their assistance in the field as well as in the laboratory; Elaine C. Beckwith and Y. T. Lu for the ammonia methodology development and measurements; and Robert A. Gorse, Jr. and Joseph M. Norbeck for setting up the computer programs. Of our many friends at the Pennsylvania Turnpike who have generously cooperated with us so often in the past, we are especially indebted in the present experiment to Charles E. Barron, Robert J. Hauger and his crew at the Allegheny Tunnel, Warren E. Kipp, Guy Oakman, and William J. Raves. We appreciate the many discussions with L. Willard Richards and Robert K. Stevens.

REFERENCES

1. WOLFF, G. T. 1979. J. Air Pollut. Control Assoc. **29:** 26.
2. ROWE, M. D., S. C. MORRIS & L. D. HAMILTON. 1978. J. Air Pollut. Control Assoc. **28:** 772.
3. PERHAC, R. M. 1978. Atmos. Environ. **12:** 641.
4. CASS, G. R. 1977. Ph.D. Thesis. California Institute of Technology. Pasadena, Calif.
5. U.S. Environmental Protection Agency. 1972. Air Quality Data for 1968. U.S. EPA No. APTD-0978. Research Triangle Park, N.C.
6. ALTSHULLER, A. P. 1973. Environ. Sci. Technol. **7:** 709.
7. FRANK, N. H. 1974. Paper #74-245. Presented at the Air Pollution Control Association 67th Annual Meeting, Denver, Co., June 9–13.
8. FRANK, N. H. & N. C. POSSIEL, JR. 1976. Paper #12. Prepr. Pap. Nat. Meet. Div. Environ. Chem., Am. Chem. Soc. **16**(2): 46.
9. ALTSHULLER, A. P. 1976. J. Air Pollut. Control Assoc. **26:** 318.
10. WOLFF, G. T. & P. J. LIOY. 1978. *In* Proceedings of the Air Pollution Control Association Mid-Atlantic States Second Semi-Annual Technical Conference on the Questions of Sulfates, Philadelphia, Penn., April 13–14.
11. WILSON, W. E. 1978. Atmos. Environ. **12:** 537.
12. HIDY, G. M., P. K. MUELLER & E. Y. TONG. 1978. Atmos. Environ. **12:** 735.
13. National Research Council. 1975. Senate Document No 94-4. U.S. Government Printing Office. Washington, D.C.
14. HITCHCOCK, D. R. 1976. J. Air Pollut. Control Assoc. **26:** 210.
15. MACCRACKEN, M. C. 1978. *In* Proceedings of the Air Pollution Control Association Mid-Atlantic States Second Semi-Annual Technical Conference on the Questions of Sulfates, Philadelphia, Penn., April 13–14.
16. KOWALCZYK, G. S., C. E. CHOQUETTE & G. E. GORDON. 1978. Atmos. Environ. **12:** 1143.
17. HARKER, A. B., L. W. RICHARDS & W. E. CLARK. 1977. Atmos. Environ. **11:** 87.
18. BROSSET, C. 1978. Atmos. Environ. **12:** 25.
19. STEVENS, R. K., T. G. DZUBAY, D. T. MAGE, R. BURTON, G. RUSSWURM & E. TEW. 1978. Paper #98. Prepr. Pap. Natl. Meeting Div. Environ. Chem., Am. Chem. Soc. **18**(2): supplementary pages.
20. CHARLSON, R. J., D. S. COVERT, T. V. LARSON & A. P. WAGGONER. 1978. Atmos. Environ. **12:** 39.
21. JUNGE, C. E. & T. G. RYAN. 1958. Q. J. R. Meteorol. Soc. **84:** 46.
22. MILLER, J. M. & R. G. DE PENA. 1971. Paper CP7C. *In* Proceedings of the Second International Clean Air Congress, Washington, D.C., December 6–11, 1970. H. M. Englund & W. T. Beery, Eds.: 375–378. Academic Press, Inc. New York, N.Y.
23. JUDEIKIS, H. S., T. B. STEWART & A. G. WREN. 1978. Atmos. Environ. **12:** 1633.
24. HEGG, D. A. & P. V. HOBBS. 1978. Atmos. Environ. **12:** 241.
25. DAVENPORT, H. M. & L. K. PETERS. 1978. Atmos. Environ. **12:** 997.
26. PACK, D. W. 1978. Geophys. Res. Lett. **5:** 673.
27. HOUGHTON, H. G. 1955. J. Meteorol. **12:** 355.
28. GRAVENHORST, G., TH. JANSSEN-SCHMIDT & D. H. EHHALT. 1978. Atmos. Environ. **12:** 691.
29. ONO, A. 1978. Atmos. Environ. **12:** 753.
30. BEILKE, S. & G. GRAVENHORST. 1978. Atmos. Environ. **12:** 231.
31. MARSH, A. R. W. 1978. Atmos. Environ. **12:** 401.
32. FREIBERG, J. 1975. Atmos. Environ. **9:** 661.
33. OREL, A. E. & J. H. SEINFELD. 1977. Environ. Sci. Technol. **11:** 1000.
34. FUZZI, S. 1978. Atmos. Environ. **12:** 1439.
35. GEORGII, H.-W. 1978. Atmos. Environ. **12:** 681.
36. WEISS, R. E., A. P. WAGGONER, R. J. CHARLSON & N. C. AHLQUIST. 1977. Science **195:** 979. (Also previous work cited therein.)
37. CHARLSON, R. J., A. H. VANDERPOL, D. S. COVERT, A. P. WAGGONER & N. C. AHLQUIST. 1974. Atmos. Environ. **8:** 1257; 1973. Science **184:** 156.

38. COBOURN, W. G., R. B. HUSAR & J. D. HUSAR. 1978. Atmos. Environ. **12:** 89.
39. STEPHENS, E. R. & M. A. PRICE. 1972. J. Colloid Interface Sci. **39:** 272.
40. a) CUNNINGHAM, P. T. & S. A. JOHNSON. 1976. Science **191:** 77; b) CUNNINGHAM, P. T. 1979. Private communication.
41. ATKINS, D. H. F., R. A. COX & A. E. J. EGGLETON. 1972. Nature **235:** 372.
42. GRAVENHORST, G. 1978. Atmos. Environ. **12:** 707.
43. BROSSET, C., K. ANDREASSON & M. FERM. 1975. Atmos. Environ. **9:** 631.
44. MUELLER, P. K. & G. M. HIDY. 1978. *In* Proceedings of the Air Pollution Control Association Mid-Atlantic States Second Semi-Annual Technical Conference on the Questions of Sulfates, Philadelphia, Penn., April 13–14.
45. STEVENS, R. K., T. G. DZUBAY, G. RUSSWURM & D. RICKEL. 1978. Atmos. Environ. **12:** 55.
46. TANNER, R. L., R. CEDERWALL, R. GARBER, D. LEAHY, W. MARLOW, R. MEYERS, M. PHILLIPS & L. NEWMAN. 1977. Atmos. Environ. **11:** 955.
47. TANNER, R. L., R. GARBER, W. MARLOW, B. LEADERER & M. A. LEYKO. 1979. Ann. N.Y. Acad. Sci. **322:** 99.
48. SMITH, J. P., D. GROSJEAN & J. N. PITTS, JR. 1978. J. Air Pollut. Control Assoc. **28:** 930.
49. LAU, N-C. & R. J. CHARLSON. 1977. Atmos. Environ. **11:** 475.
50. DAWSON, G. A. 1978. Atmos. Environ. **12:** 1991.
51. DAVID, D. J., M. C. WILLSON & D. S. RUFFIN. 1977. Paper #15. Prepr. Pap. Nat. Meet. Div. Environ. Chem., Am. Chem. Soc. **17**(1): 41.
52. California Air Resources Board. 1975. An Assessment of the Aerosol-Visibility Problem in the South Coast Air Basin. Staff Report No. 75-20-3, October 28. Sacramento, Calif.
53. WAGGONER, A. P., A. J. VANDERPOL, R. J. CHARLSON, S. LARSEN, L. GRANAT & C. TRÄGÅRDH. 1976. Nature **261:** 120.
54. CASS, G. R. 1976. The Relationship between Sulfate Air Quality and Visibility at Los Angeles. Environmental Quality Memorandum #18. California Institute of Technology. Pasadena, Calif.
55. WHITE, W. H. & P. T. ROBERTS. 1977. Atmos. Environ. **11:** 803.
56. LEADERER, B. P., D. M. BERNSTEIN, J. M. DAISEY, M. T. KLEINMAN, T. J. KNEIP, E. O. KNUTSON, M. LIPPMANN, P. J. LIOY, K. A. RAHN, D. SINCLAIR, R. L. TANNER & G. T. WOLFF. 1978. J. Air Pollut. Control Assoc. **28:** 321.
57. VANDERPOL, A. H., R. J. CHARLSON, A. P. WAGGONER, D. S. COVERT & N. C. AHLQUIST. 1975. Atmos. Environ. **9:** 765.
58. WOLFF, G. T., P. R. MONSON & M. A. FERMAN. 1978. Paper #66. Prepr. Pap. Nat. Meet. Div. Environ. Chem., Am. Chem. Soc. **18**(2): 186.
59. LIPPMANN, M., M. T. KLEINMAN, D. M. BERNSTEIN, G. T. WOLFF & B. P. LEADERER. 1979. Ann. N.Y. Acad. Sci. **322:** 29.
60. HUSAR, R. B., D. E. PATTERSON, J. D. HUSAR, N. V. GILLANI & W. E. WILSON, JR. 1978. Atmos. Environ. **12:** 549.
61. WHITE, W. H. 1976. Nature **264:** 735.
62. HÖGSTRÖM, U. 1978. Atmos. Environ. **12:** 1505.
63. LARSON, T. V., N. R. HORIKE & H. HARRISON. 1978. Atmos. Environ. **12:** 1597.
64. SHEIH, C. M. 1977. Atmos. Environ. **11:** 173.
65. GALVIN, P. J., P. J. SAMSON, P. E. COFFEY & D. ROMANO. 1978. Environ. Sci. Technol. **12:** 580.
66. SAMSON, P. J. 1978. Atmos. Environ. **12:** 1889.
67. COWHERD, H., F. W. LIPFERT & S. C. KAUFMAN. 1978. *In* Proceedings of the Air Pollution Control Association Mid-Atlantic States Second Semi-Annual Technical Conference on the Questions of Sulfates, Philadelphia, Penn., April 13–14.
68. LYONS, W. A., J. C. DOOLEY, JR. & K. T. WHITBY. 1978. Atmos. Environ. **12:** 621.

MODELING OF AEROSOL DYNAMICS: AEROSOL SIZE AND COMPOSITION*

Richard A. Beyak and Thomas W. Peterson†

Department of Chemical Engineering
University of Arizona
Tucson, Arizona 85721

1. Introduction

Particulate pollutants are classed as primary or secondary, depending on their origin from direct particle emission or from chemical transformation of gaseous pollutants. Possible mechanisms of conversion from primary gaseous pollutants to secondary particulate species include homogeneous nucleation (formation of new particles) and condensation (growth of existing particles).

This work examines the change in particle size and chemical composition of a collection of thermodynamically stable particles exposed to various pollutant concentrations in the gas phase. First, a description of the processes by which gaseous pollutants convert to the particulate phase will be given. The specific processes that affect the aerosol particle size and chemical composition are described for a system of constant gas phase pollutant composition. The system is then generalized to include changes in gas phase composition due to transport, dispersion, and chemical reaction. Specific case studies will then be presented of growth and chemical change of aerosol particles with varying initial size and composition and exposed to various specific pollutants.

2. Model Components

The description of growth and chemical change of secondary aerosols in the presence of gaseous pollutants requires a mathematical description of (1) all fundamental processes governing the physical and chemical fate of secondary particulate pollutants, *given* the simultaneous gas phase composition; and (2) the variation of gas phase pollutant concentrations subject to transport, dispersion, and chemical reaction. The former we define as "aerosol processes." The latter we define as "gaseous transport, dispersion, and conversion processes," since these processes describe the change in gas phase concentration of primary pollutants undergoing transport, dispersion, and chemical reaction.

The major emphasis of this work will be on the fundamental descriptions of aerosol processes, since a great deal of study has already been devoted to the area of gaseous pollutant dispersion models. A modification of the simple analytic model developed by Peterson and Seinfeld[1] will be used to describe the change in concentration for gaseous

*This work was supported by National Science Foundation Grant ENG 78-05414.
†To whom correspondence should be addressed.

0077–8923/80/0338–0174 $01.75/1 © 1980, NYAS

pollutants undergoing transport, dispersion, and chemical reaction. The basic elements of this model will be discussed.

A. *Aerosol Processes*

The model developed describes the three processes by which vapor molecules transfer from the gas to the liquid phase. These processes, described in detail by Peterson and Seinfeld,[2] include:

1. Transport from the gas phase to the particle surface;
2. Transport across the gas–liquid interface;
3. Transport and chemical reaction within the liquid phase.

Transport from the Gas Phase to the Particle Surface

It is necessary to include in the formulation of the description of transport from the gas phase to the particle surface a correction for noncontinuum transport based on the Knudsen number:

$$\mathrm{Kn} = \lambda / r \tag{1}$$

This correction is necessary because particle sizes in an atmospheric aerosol distribution can vary over many orders of magnitude (for example, particle diameters between 0.001 and 10 μm). A number of transition regime models have been proposed (Fuchs,[3] Brock,[4] Sampson and Springer,[5] Shankar[6]) to provide a continuous description of transport to particles of all sizes. The formula proposed by Sahni[7] was chosen for this work.

Particle growth by condensation can hence be described by:

$$\frac{\mathrm{d}M}{\mathrm{d}t} = \frac{1}{1 + \ell \mathrm{Kn}} \frac{4\pi r D}{R T_\infty} (p_\infty - p_c) \qquad (2)‡$$

$$\ell = \frac{4/3 + 0.71\ \mathrm{Kn}^{-1}}{1 + \mathrm{Kn}^{-1}} \tag{3}$$

Transport across the Gas–Liquid Interface

Both the Kelvin effect and solute effect influence the equilibrium surface pressure above an aerosol droplet. The Kelvin effect describes the increase in surface vapor pressure in equilibrium with a curved surface over the surface vapor pressure in equilibrium with a flat surface. The solute effect describes the decrease in equilibrium vapor pressure due to the presence of other components in solution. Together, the vapor pressure above the aerosol is described by:

$$p_{c_i} = \gamma_i x_i p_{\mathrm{sat}_i} \exp\left(\frac{2\sigma \bar{v}_i}{r R T_\infty}\right) \tag{4}$$

‡Nomenclature is given in an Appendix.

If the component of interest is water, the water activity is defined as:

$$a_w = \gamma_w x_w \tag{5}$$

In order to relate water activity to composition of a multicomponent aerosol, the correlation of Zdanovskii[8] and Stokes and Robinson[9] for electrolytic solutions is utilized.

$$1 = \sum_i \frac{m_i}{m_{bi}(a_w)} \tag{6}$$

This relation permits calculation of particle size from water activity and particle composition. Binary water activity data over the entire range of interest are necessary in order to use this relationship.

Transport and Chemical Reaction within the Liquid Phase

Once absorbed within the aerosol droplet, vapor species may undergo chemical reaction. Considering that all aerosol particles of interest are less than 10 μm in diameter, it is safe to assume that diffusion within the particle can be ignored.

Chemical reactions within the aerosol can be classified as equilibrium or nonequilibrium. Equilibrium reactions include dissociation reactions, and hence the model includes electroneutrality constraints and ionic strength calculations in order to correctly predict dissociation product concentrations. The nonequilibrium reactions include oxidation mechanisms for dissolved gases and their dissociation products.

The mechanisms necessary to implement the model describing aerosol growth and chemical reaction are now complete, once the information on initial aerosol size and chemical composition are given.

B. *Gaseous Transport, Dispersion, and Conversion Processes*

It is not the purpose of this work to develop a sophisticated dispersion model for the description of temporal and spatial variation of gaseous pollutants. In order to accurately apply the model components describing aerosol growth and chemical change to many realistic atmospheric problems, however, it is necessary to account in some way for changes in gas phase pollutant concentration by dispersion, transport, and chemical reaction. This will be accomplished by obtaining an analytic solution to the steady state dispersion equation, including in the equation dispersion, advection, removal, and reaction terms.

$$\bar{u}\frac{\partial c}{\partial x} - w_s \frac{\partial c}{\partial z} = \frac{\partial}{\partial y}\left(K_H \frac{\partial c}{\partial y}\right) + \frac{\partial}{\partial z}\left(K_v \frac{\partial c}{\partial z}\right) - k(x)c \tag{7}$$

This equation assumes a constant wind velocity, constant vertical and horizontal eddy diffusivities, a vertical settling term, and a conversion rate that can vary with distance. For primary particulate pollutants, $k(x) = 0$, and for primary gaseous pollutants, $w_s = 0$. We solve Equation 7 for a point source of primary gaseous pollutants, with following initial and boundary conditions:

$$c(0, y, z) = \frac{Q}{\bar{u}} \delta(y)\delta(z - z_s) \tag{8}$$

where $\delta(\cdot)$ is the dirac delta function.

$$c(x, y, z) = 0 \quad \text{as } y \rightarrow \pm \infty \tag{9}$$

$$\frac{\partial c}{\partial z} = 0 \quad \text{at } z = H_a \tag{10}$$

$$K_v \frac{\partial c}{\partial z} = v_d c \quad \text{at } z = z_a \tag{11}$$

The boundary condition in Equation 10 implies an elevated temperature inversion at a height H_a, and the boundary condition in Equation 11 implies that the flux of pollutant to the ground by vertical dispersion is matched by surface absorption with deposition velocity v_d.

The solution is obtained in dimensionless form. The following dimensionless variables are defined:

$$X = \frac{K_v x}{\bar{u} H^2}; \; Y = \frac{y}{H}; \; Z = \frac{z - Z_a}{H} \tag{12–14}$$

where $H = H_a - Z_a$.

The dimensionless concentration is given by:

$$C = \frac{\bar{u} H^2 c}{Q} \tag{15}$$

The following dimensionless groups arise:

$$\beta = K_H / K_v \,; N = v_d H / K_v \,; \alpha(x) = k(x) H^2 / K_v \tag{16–18}$$

Note that the dimensionless rate constant $\alpha(x)$ is a function of downwind distance. Equations 7–11 become:

$$\frac{\partial C}{\partial X} = \beta \frac{\partial^2 C}{\partial Y^2} + \frac{\partial^2 C}{\partial Z^2} - \alpha(X) C \tag{19}$$

$$C(0, Y, Z) = \delta(Y)\delta(Z - Z_s) \tag{20}$$

$$C(X, Y, Z) = 0 \quad \text{as } Y \rightarrow \pm \infty \tag{21}$$

$$\frac{\partial C}{\partial Z} = 0 \quad \text{at } Z = 1 \tag{22}$$

$$\frac{\partial C}{\partial Z} = NC \quad \text{at } Z = 0 \tag{23}$$

Solution to (**19**)–(**23**) can be obtained by separation of variables to give:

$$C(X, Y, Z) = \frac{1}{\sqrt{4\beta X}} \exp\left(-\int_0^X \alpha(\xi)\mathrm{d}\xi\right) \exp\left(\frac{-Y^2}{4\beta X}\right) \cdot \sum_{n=1}^{\infty} A_n\phi_n(Z) \exp\left(-\gamma_n^2 X\right) \qquad \textbf{(24)}$$

$$\tan \gamma_n = \frac{N}{\gamma_n} \qquad \textbf{(25)}$$

$$\phi_n(Z) = \cos(\gamma_n Z) + \frac{N}{\gamma_n} \sin(\gamma_n Z) \qquad \textbf{(26)}$$

$$A_n = \frac{\cos(\gamma_n Z_s) + \dfrac{N}{\gamma_n}\sin(\gamma_n Z_s)}{\dfrac{1}{2}\left(1 + \dfrac{N^2}{\gamma_n^2}\right) + \dfrac{\left(1 - \dfrac{N^2}{\gamma_n^2}\right)\sin(2\gamma_n)}{4\gamma_n} + \dfrac{N}{\gamma_n^2}\sin^2\gamma_n} \qquad \textbf{(27)}$$

Since Equation 7 assumes a constant wind velocity in the x-direction, transport time is a unique function of distance downwind, and hence a reaction rate variable in time can be written as a reaction rate variable in x. Given the rate of conversion of gaseous to particulate pollutants, we wish to describe both the change in gaseous pollutant concentration due to reaction and the change in pollutant conversion rate as a function of distance downwind.

We know:

$$\left(\frac{\partial c}{\partial t}\right)_{rxn} = -k(x)c \qquad \textbf{(28)}$$

Since transport time $t = x/\bar{u}$, we can write Equation 28 as:

$$\bar{u}\left(\frac{\partial c}{\partial x}\right)_{rxn} = -k(x)c \qquad \textbf{(29)}$$

Or, in dimensionless form:

$$\left(\frac{\partial C}{\partial X}\right)_{rxn} = -\alpha(X)C \qquad \textbf{(30)}$$

Similarly, let:

$$F(X) = \int_0^X \alpha(\xi)\,\mathrm{d}\xi \qquad \textbf{(31)}$$

Then:

$$\frac{\mathrm{d}F}{\mathrm{d}X} = \alpha(x); \qquad F(0) = 0 \qquad \textbf{(32)}$$

By solving the differential equations (**30**) and (**32**) numerically for successive values of X, the solution $C(X, Y, Z)$ of Equation 24 can be obtained for arbitrary chemical mechanisms of conversion from gaseous to particulate pollutants.

3. Case Studies

The model developed was utilized to study a number of possible cases of aerosol growth and chemical change in polluted atmospheres. The cases considered to date can be classed in two categories. The first category neglects spatial or temporal variation of gaseous pollutants. That is, the calculations include only the aerosol processes as outlined briefly in Section 2A. While realistic atmospheric conditions obviously involve variations in gas phase pollutant concentrations, such simplified calculations provide valuable information regarding variation in particle growth rates and chemical reaction with gas phase pollutant concentrations. Section 3A describes a number of case studies of this form.

The second category includes the variation of gas phase pollutant concentration due to dispersion, advection, removal, and chemical conversion. In so doing, we now consider a complex chemical system, where particle growth and chemical change depend on ambient gas phase conditions, which in turn are affected by the rate of conversion of gases to particles. Section 3B presents a case study where these two coupled processes are considered.

A. *Constant Concentrations of Gas Phase Pollutants*

The model developed was first used to study growth of a marine aerosol exposed to various gaseous pollutants and water vapor. The aerosol is assumed to begin as a thermodynamically stable solution of NaCl and $MgCl_2$ in water, at salt ratios found in seawater. The first calculations assumed the particle was exposed to gaseous pollutants SO_2 and NH_3 and to water vapor (Peterson and Seinfeld[2]). The levels of pollutants assumed were typical of those found in polluted urban atmospheres.

Two possible mechanisms of sulfate formation were considered. First, SO_2 oxidation was assumed to occur only via O_2/O_3 dissolution within the aerosol particle (Larson *et al.*[10]). Second, SO_2 oxidation was assumed to occur via homogeneous gas phase kinetics, possibly photochemical, followed by condensation of H_2SO_4 on existing particles. Both situations can be viewed as idealizations of the process of growth of a marine aerosol as it passes over a polluted urban region of relatively uniform SO_2 and NH_3 concentration.

Subsequent calculations included the possibility of heterogeneous oxidation of nitrogen oxides to nitrates (Peterson and Seinfeld[11]). Although later evidence (Stelson *et al.*[12]) indicates homogeneous nitrate formation may occur in the gas phase, this mechanism was not considered here. The model was used to study the temporal variation of particle size and chemical composition for a range of values of the parameters of the model.

TABLE 1 summarizes the results of these studies. The most acidic aerosol particles are those affected by condensation of sulfuric acid and heterogeneous nitrate formation (Cases B and C). For these particles, the pH is in the range 2–4. Both sulfate and ammonium concentrations are in the 1 M range, and for the case in which

TABLE 1

AEROSOL GROWTH AND CHEMICAL REACTIONS: CONSTANT GAS PHASE CONCENTRATIONS

Case	Initial Aerosol	Exposed to Gases	Relative Humidity	Chemistry	Base Case Values	Results
A	Marine Aerosol	SO_2, NH_3, H_2O	85–90%	SO_2 oxidation via dissolved O_2/O_3	RH = 90% $r_0 = 0.1\ \mu m$ $p_{NH_3} = 0.01$ ppm $p_{SO_2} = 0.01$ ppm	$5 \leq pH \leq 6$ $10^{-4} \leq [SO_4^{2-}] \leq 2 \times 10^{-3}$ M $10^{-4} \leq [NH_4^+] \leq 3 \times 10^{-3}$ M $r \approx 0.1\ \mu m$
B	Marine Aerosol	Same as A plus H_2SO_4	85–99%	Homogeneous gas phase SO_2 oxidation	Same as A $p_{H_2SO_4} = 5 \times 10^{-6}$ ppm	$1.75 \leq pH \leq 3.75$ $0.1 \leq [SO_4^{2-}] \leq 1.4$ M $0.1 \leq [NH_4^+] \leq 2.5$ M $0.2 \leq r \leq 0.8\ \mu m$
C	Marine Aerosol	Same as B plus NO, NO_2	90–91%	Same as B, Liquid phase NO_x oxidation	Same as B $p_{NO} = 0.1$ ppm $p_{NO_2} = 0.05$ ppm	$1.8 \leq pH \leq 2.9$ $0.1 \leq [SO_4^{2-}] \leq 0.8$ M $1.2 \leq [NH_4^+] \leq 1.7$ M $0.1 \leq [NO_3^-] \leq 1.2$ M $0.1 \leq r \leq 1.4\ \mu m$
D	Plume Aerosol	SO_2, NH_3, H_2O	95–97%	Fe^{3+}-catalyzed SO_2 oxidation	RH = 95% $r_0 = 0.1\ \mu m$ $p_{NH_3} = 0.01$ ppm $p_{SO_2} = 0.01$ ppm	$3 \leq pH \leq 4$ $0.01 \leq [SO_4^{2-}] \leq 0.2$ M $0.05 \leq [NH_4^+] \leq 0.4$ M $0.1 \leq r \leq 0.14\ \mu m$

NO_x was considered, nitrate concentrations also fell within this range. Of interest in the case that considered simultaneous sulfate and nitrate formation (Case C) are the observations that (1) near-stoichiometric concentrations of ammonium, nitrate, and sulfate were found only for conditions that included simultaneous gas phase sulfate formation and heterogeneous nitrate formation; (2) particle pH increases with particle radius; and (3) those parameters that tend to increase nitrate levels also tend to decrease sulfate levels, and vice versa. Specifically, sulfate concentration within the particle increased as the gas phase concentrations of NO and H_2SO_4 increased, but decreased as the gas phase concentrations of NH_3, NO_2, and relative humidity increased. On the other hand, nitrate levels rose with relative humidity and gas phase concentrations of NO_2 and NH_3, and decreased with gas phase levels of NO and H_2SO_4.

Also undertaken were growth calculations under conditions that might be experienced downwind from a power plant plume emitting SO_2 (Case D). The metal-ion catalytic oxidation of SO_2 to sulfate is thought to play an important role under these conditions, and therefore the iron-catalyzed mechanism proposed by Freiberg[13] was assumed. The rate of conversion to sulfate and the subsequent rate of particle growth under these conditions are much slower than those experienced for homogeneous gas phase SO_2 oxidation. Calculations for the plume case indicate that growth rates are strongly pH dependent and, hence, significantly influenced by ambient ammonia levels. Particle growth over periods of three to four hours is not negligible, but neither is it as significant as that experienced under H_2SO_4 condensation. Final values for particle pH are in the range 3 to 4.

More detailed description of the results of these simulations, including the variation in particle radius and composition with the parameters of the system (relative humidity, initial particle size, SO_2, NH_3, NO, NO_2, and H_2SO_4), can be found in Peterson and Seinfeld.[2,11]

B. *Variable Gas Phase Pollutants*

The studies summarized in Section 3A indicate the utility of the model developed for aerosol growth and reaction calculations, and provide useful means for interpreting data on size and chemical composition of atmospheric particulate pollutants. A more realistic indication of possible composition and size changes experienced by secondary aerosols, however, must take into account the dispersion, advection, and gas-to-particle conversion processes that affect the concentration of gaseous pollutants, which in turn affect the rate of formation of secondary aerosols. In this section, the trajectory model described in Section 2B, incorporating dispersion, advection, deposition, and reaction of gaseous pollutants, is coupled with the model for physical and chemical behavior of aerosols to predict the behavior of secondary aerosols in a dispersing plume. This model is then used to predict the size and chemical composition of secondary sulfate aerosols within a smelter plume in a region of low humidity. We describe the physical problem to be examined, the procedure for coupling the model components of Sections 2A and 2B, and the application of the coupled models to the problem of interest.

Secondary Aerosol Formation: Smelter Plumes in Arid Regions

We wish to examine a possible mechanism for sulfate formation within the plume emitted from a copper smelter in southern Arizona. This problem is of interest because of the possible detrimental effects on health and visibility of sulfur emissions from copper smelters. While these emissions are predominantly in the form of gaseous SO_2, the more serious impact of particulate sulfate on visibility and health requires that the possible conversion from SO_2 to sulfate be quantified.

Chosen for study is a smelter complex in the central Arizona area, emitting approximately 300 tons/day SO_2, on average. In addition to SO_2, the plume is its own source of water vapor, assumed saturated at the point of emissions. As the plume disperses, the relative humidity approaches that found in the desert background, here assumed to be 30% at 25°C (approximately 9400 ppm). The initial aerosol is assumed to be a solution of sulfuric acid and water, the concentration of which is determined by the equilibrium concentration for a particle of given size. This is calculated by taking into account both the solute and Kelvin effects, as described in Section 2A.

It is further assumed that the mechanism of conversion from SO_2 to sulfate is that given by Freiberg:[13]

$$\frac{d}{dt}[SO_4^{=}] = K_T K_S^2 \frac{[Fe^{3+}][SO_{2(\ell)}]^2}{[H^+]^3} \tag{33}$$

where:

$$K_T = K_0 + K_F[Fe^{3+}]$$

$$K_0 = 9.1 \times 10^3 \text{ mole}^{-1} \, \ell \, \text{min}^{-1}$$

$$K_F = 1.33 \times 10^6 \text{ mole}^{-2} \, \ell^2 \, \text{min}^{-1}$$

$$K_S = 0.0127 \text{ mol } \ell^{-1}$$

The concentrations of SO_2 and water vapor vary with plume travel distance, according to the solution of Equations 24–27. It is further assumed that background NH_3 is also present at a constant concentration. These gases all participate in standard

TABLE 2

EQUILIBRIUM CHEMISTRY

Reaction	Equilibrium Constant	Value at 25°C
$H_2O_{(g)} \rightleftharpoons H_2O_{(\ell)}$	$1/p_{sat_w} = a_w/p_{H_2O}$	31.99 atm^{-1}
$H_2O_{(\ell)} \rightleftharpoons H^+ + OH^-$	$K_w = [H^+][OH^-]\,\gamma_+\gamma_-$	10^{-14} $moles^2\,\ell^{-2}$
$SO_{2(g)} \rightleftharpoons SO_{2(\ell)}$	$K_{hs} = [SO_{2(\ell)}]/p_{SO_2}$	1.24 moles ℓ^{-1} atm^{-1}
$SO_{2(\ell)} \rightleftharpoons H^+ + HSO_3^-$	$K_{1s} = [H^+][HSO_3^-]\,\gamma_+\gamma_-/[SO_{2(\ell)}]$	0.0127 moles ℓ^{-1}
$HSO_3^- \rightleftharpoons H^+ + SO_3^{2-}$	$K_{2s} = [H^+][SO_3^{2-}]\,\gamma_+\gamma_{2-}/[HSO_3^-]\,\gamma_-$	6.24×10^{-8} moles ℓ^{-1}
$NH_{3(g)} \rightleftharpoons NH_{3(\ell)}$	$K_{hA} = [NH_{3(\ell)}]/p_{NH_3}$	57. moles ℓ^{-1} atm^{-1}
$NH_{3(\ell)} \rightleftharpoons NH_4^+ + OH^-$	$K_{1A} = [NH_4^+][OH^-]\,\gamma_+\gamma_-/[NH_{3(\ell)}]$	1.774×10^{-5} moles ℓ^{-1}

TABLE 3
AEROSOL GROWTH IN ARID REGIONS: SIMULATIONS PERFORMED

Case	p_{NH_3} (ppm)	SO_2 Source Strength (tons/day)	Sulfate Formation Rate (moles ℓ^{-1} min^{-1})
A (Base Case)	0.01	290	Equation 33
B	0.01	290	.01 × Equation 33
C	0.01	290	100 × Equation 33
D	0.01	150	Equation 33
E	0.01	500	Equation 33
F	0.001	290	Equation 33
G	0.1	290	Equation 33

equilibrium chemistry (TABLE 2), as well as the liquid phase oxidation to $SO_4^=$ described by Equation 33. The model is first applied to the base case (Case A, TABLE 3) of parameters selected and then is used to study the effect of changes in these parameter values. In Cases B and C, the effect of increasing or decreasing the sulfate formation rate is examined. In Cases D and E, the SO_2 emissions rate is studied; and in Cases F and G, the effect of background ammonia concentration is studied. In each case, only one parameter value is altered, allowing the effect of that single parameter to be determined. Also examined is the change in particle conditions with changing initial particle radius.

Results

The results of the simulations described in TABLE 3 are presented in TABLE 4. Presented are variations of particle radius, sulfate and ammonium ion concentrations, and total sulfate for each case.

It is significant to note that appreciable initial particle growth occurs only for Cases C and G. Recall, Case C represents large SO_2 oxidation rates, and Case G represents high background ammonia concentration. Both conditions lead to increased sulfate in the particle, which, by lowering the water activity, causes an increase in particle growth. As the water content in the plume drops (relative humidity approaches ambient), all particles shrink.

The sulfate and ammonium concentrations within the particle also show significant increases only for increases in the SO_2 oxidation rate and background ammonia concentrations. Increasing the SO_2 source strength does not affect the sulfate concentration as significantly as one might expect. Interestingly, significant decreases in SO_2 reaction rate (B) and background ammonia levels (F) below those values used in the base case do not significantly change the sulfate and ammonium levels. The secondary sulfate formation rate in the base case (according to Equation 33) is already relatively insignificant, and therefore further decreases in this rate, whether by alteration of PH or by changing the proposed mechanism, will not affect the sulfate concentration.

Similar results are seen in the variation of particle pH, at least initially, although only the changes in background ammonia concentration significantly affect final particle pH. For these cases (F and G), final particle pH is approximately 1.2 and 3.0,

TABLE 4

AEROSOL GROWTH IN ARID REGIONS: SUMMARY OF RESULTS

Case	Radius (μm)			$[SO_4^{2-}]$ (moles ℓ^{-1})		$[NH_4^+]$ (moles ℓ^{-1})		pH		Total Sulfate (moles $\times 10^{-20}$)		
	Initial	Peak	Final	Initial	Final	Initial	Final	Initial	Final	Initial	Final	Ratio $\left(\frac{\text{final}}{\text{initial}}\right)$
A	0.100	0.100	0.0416	0.225	3.256	0.452	6.504	3.337	2.166	90.6	98.2	1.084
B	0.100	0.100	0.0409	0.219	3.168	0.439	6.329	3.351	2.177	90.6	90.8	1.002
C	0.100	0.1283	0.0570	0.355	4.325	0.710	8.641	3.142	2.049	90.6	335.5	3.703
D	0.100	0.100	0.0410	0.222	3.193	0.446	6.378	3.345	2.174	90.6	92.7	1.023
E	0.100	0.100	0.0427	0.234	3.393	0.472	6.780	3.319	2.150	90.6	110.7	1.222
F	0.100	0.100	0.0408	0.223	3.186	0.442	6.305	2.347	1.181	90.6	90.6	1.000
G	0.100	0.1190	0.0599	0.635	4.800	1.280	9.600	3.888	3.005	90.6	431.0	4.757

respectively. For all other cases, particles are emitted from the stack at approximately 3.3 pH and reach a final value in the range 2.0 to 2.2. This is a much more acidic particle than predicted under higher humidity conditions.

Finally, examination of total sulfate in the particle indicates that unless (1) the mechanism of sulfate formation is significantly faster than the mechanism of Equation 33 (Case C), or (2) the background ammonia level is of the order $p_{NH_3} = 0.1$ ppm or greater (Case G), then at least 80% of the final sulfate, for particles of final radius approximately 0.04 μm, is primary sulfate. For high SO_2 oxidation rates (Case C) and high background ammonia (Case G), the final sulfates are approximately 70% and 80% *secondary* sulfate, respectively.

It is also of interest to study the change of the particle size distribution and sulfate

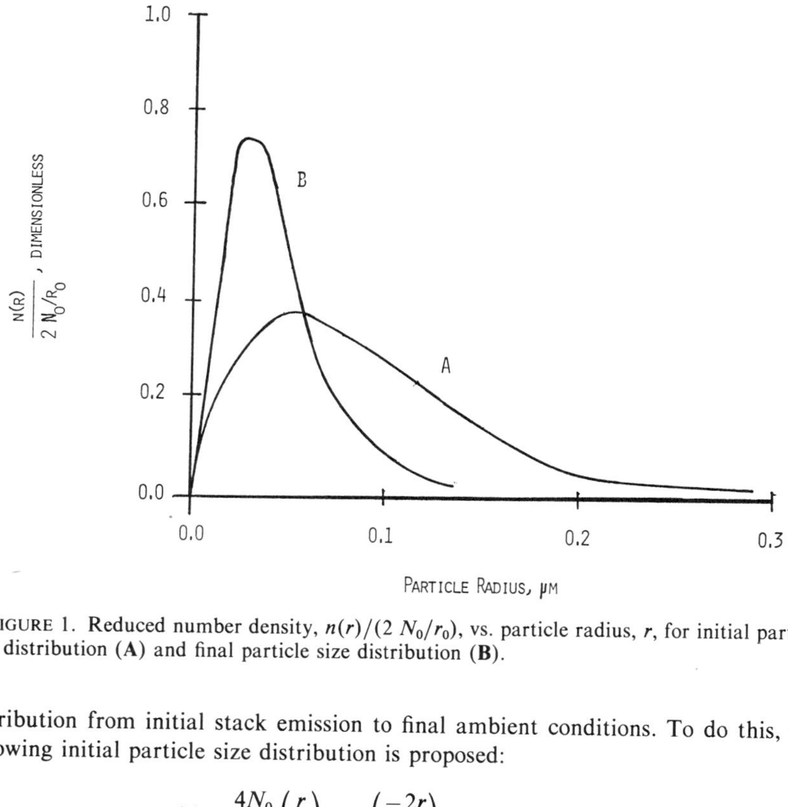

FIGURE 1. Reduced number density, $n(r)/(2\ N_0/r_0)$, vs. particle radius, r, for initial particle size distribution (**A**) and final particle size distribution (**B**).

distribution from initial stack emission to final ambient conditions. To do this, the following initial particle size distribution is proposed:

$$n(r) = \frac{4N_0}{r_0}\left(\frac{r}{r_0}\right)\exp\left(\frac{-2r}{r_0}\right)\ [=]\ \#\ \text{cm}^{-3}\ \mu\text{m}^{-1} \qquad \textbf{(34)}$$

If one assumes particle growth to be the only mechanism affecting particle size, then it is possible to study the particle size distribution. FIGURE 1 shows $n(r)/(2\ N_0/r_0)$ vs. r, initially and for long times, assuming $r_0 = 0.1$ μm. If we define the sulfate density function as:

$$S(r) = \frac{4}{3}\pi r^3[SO_4^{=}] \times 10^{-15}\, n(r)\ [=]\ \text{moles cm}^{-3}\ \mu\text{m}^{-1} \qquad \textbf{(35)}$$

it is possible to study the change in size-dependent chemical composition. FIGURE 2 shows $S(r)/(2\ N_0/r_0)$ vs. r, initially and for long times.

Both the size distribution and sulfate distribution shift significantly to the smaller particles, in equilibrium with the low background relative humidity. Of interest is the ratio of initial to final particle sulfate as a function of particle size, as given in TABLE 5. Here we see that for particles with final radius less than approximately 0.03 μm, nearly all sulfate is primary sulfate. For particles greater than about 0.05 μm, however, secondary sulfate becomes significantly more important. Particles greater than 0.1 μm are at least 75% secondary sulfate, and the largest particle simulated (initial radius 0.4 μm, final radius 0.132 μm) is 88% secondary sulfate.

The reason for a larger fraction of secondary sulfate in the larger particles is initial particle pH. Most of the secondary sulfate is formed in the near-plume region and, hence, is strongly affected by initial pH. Because of the Kelvin effect, smaller particles of mixtures of H_2SO_4 and H_2O are more acidic than larger particles. Since the SO_2 oxidation mechanism of Equation 33 is pH limited, the smaller particles form proportionally less secondary sulfate.

Conclusions

By coupling the model for aerosol growth and chemical reaction, as outlined in Section 2A, with a simple trajectory model for gaseous transport, dispersion, and

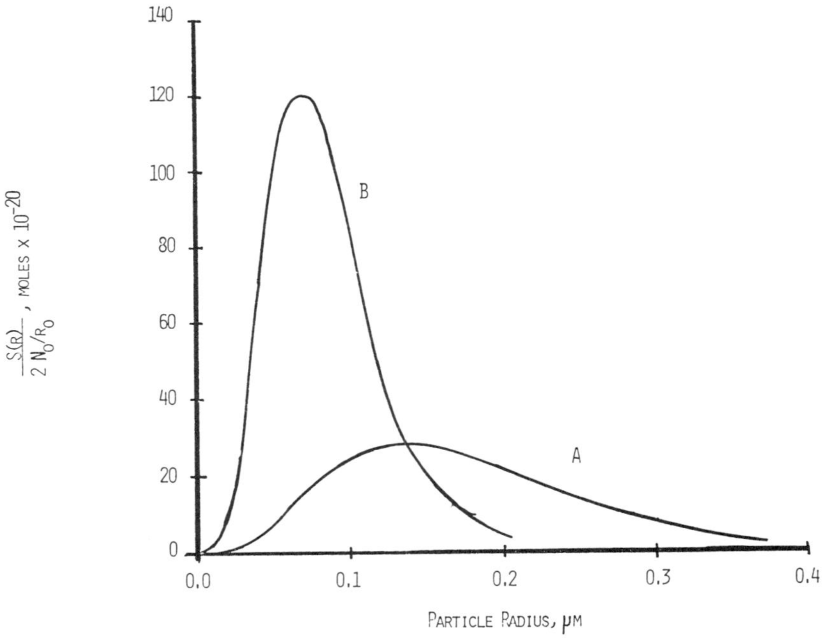

FIGURE 2. Reduced sulfate density, $S(r)/(2\ N_0/r_0)$, vs. particle radius, r, for initial sulfate distribution (**A**) and final sulfate distribution (**B**).

TABLE 5
AEROSOL GROWTH IN ARID REGIONS: SIZE AND COMPOSITION DISTRIBUTIONS

Radius (μm)			Total Sulfate (moles × 10^{-20})		Sulfate Ratio
Initial	Peak	Final	Initial	Final	($\frac{final}{initial}$)
0.0064	0.0064	0.0053	0.200	0.200	1.000
0.010	0.010	0.0080	0.684	0.684	1.000
0.044	0.044	0.0236	17.5	17.5	1.003
0.060	0.060	0.0292	33.0	33.3	1.010
0.100	0.100	0.0416	90.6	98.2	1.084
0.133	0.134	0.0517	150.0	202.1	1.347
0.155	0.156	0.0575	200.0	283.0	1.415
0.200	0.212	0.0730	320.0	629.5	1.967
0.310	0.323	0.1046	500.0	2107.4	4.215
0.400	0.428	0.1321	560.0	4511.5	8.056

conversion, as outlined in Section 2B, one is capable of studying growth and chemical change of aerosol particles emitted from various sources under a wide variety of physical and chemical conditions. This model is specifically applied to the problem of aerosol growth/shrinkage and chemical change in an arid environment. The results of this study promote the following observations:

1. All particles shrink as the water content of the plume approaches ambient relative humidity due to dispersion.
2. For a wide range of physical conditions, particle pH in the range 2.0–2.2 results for an ambient relative humidity of 30%. Significant deviations from that range result only from large changes in background ammonia concentrations.
3. Most secondary sulfate formation occurs in the near-plume regions, where SO_2 concentrations are highest.
4. A larger fraction of secondary sulfate is found in larger particles, due to the strongly acidic nature of smaller particles and the effect of particle pH on SO_2 oxidation rate.
5. For a given particle size, the major influences on secondary sulfate are the SO_2 reaction rate and the background ammonia concentration. The magnitude of SO_2 emissions has a much less significant effect on the magnitude of final secondary sulfate.

SUMMARY

This work describes a mechanism of evolution of size and composition of an aerosol particle. The model developed includes transfer of gaseous pollutants to the surface of existing aerosol particles, transfer of these pollutants across the gas–particle interface, and reaction of dissolved species within the aerosol particle. Reaction mechanisms considered include homogeneous gas phase chemistry and heterogeneous solution chemistry. Although not considered here, the model is also capable of representing surface-catalyzed chemistry at the gas–particle interface.

Studies are made of aerosol particles evolving within an urban plume (area sources) and those evolving within power plant and smelter plumes (point sources). Evolution of particle size and chemical composition is also studied for regions of varying relative humidities, with particular emphasis on aqueous chemistry in the low-humidity environments of the southwestern U.S.

A simulation of a copper smelting complex in central Arizona indicates that all particles shrink from their original size as plume humidity approaches background relative humidity. Most secondary sulfate formation occurs in the near-plume region, and the final particle pH is in the range 2.0–2.2 for a wide range of physical conditions.

References

1. Peterson, T. W. & J. H. Seinfeld. 1977. Mathematical model for transport, interconversion and removal of gaseous and particulate air pollutants—application to the urban plume. Atmos. Environ. **11:** 1171.
2. Peterson, T. W. & J. H. Seinfeld. 1979. Aerosol dynamics in an urban atmosphere. Adv. Environ. Sci. and Technol. J. N. Pitts, Ed.: 10.
3. Fuchs, N. A. 1959. Evaporation and Droplet Growth in Gaseous Media. Pergamon Press, Inc. Oxford, England.
4. Brock, J. R. 1966. Diffusion to particles in the near free molecule region. J. Colloid Interface Sci. **22:** 513.
5. Sampson, R. E. & G. S. Springer. 1969. Condensation on and evaporation from droplets by a moment method. J. Fluid Mech. **36**(3): 577.
6. Shankar, P. N. 1970. A kinetic theory of steady condensation. J. Fluid Mech. **40**(2): 385.
7. Sahni, D. C. 1966. The effect of a black sphere on the flux distribution in an infinite moderator. J. Nucl. Energy Parts A/B **20:** 915.
8. Zdanovskii, A. B. 1936. Tr. Solyanoi Lab. Vses. Inst. Galurgii Akad. Nauk SSSR **6.**
9. Stokes, R. H. & R. A. Robinson. 1966. Interactions in aqueous nonelectrolyte solutions. I. Solute solvent equilibria. J. Phys. Chem. **70**(7): 2126.
10. Larson, T. V., N. R. Horike & H. Harrison. 1978. Oxidation of sulfur dioxide by oxygen and ozone in aqueous solution: A kinetic study with significance to atmospheric rate processes. Atmos. Environ. **12:** 1597.
11. Peterson, T. W. & J. H. Seinfeld. 1979. Calculation of sulfate and nitrate levels in a growing, reacting aerosol. Am. Inst. Chem. Eng. J. **25**(5): 831.
12. Stelson, A. W., S. K. Friedlander & J. H. Seinfeld. 1979. A note on the equilibrium relationship between ammonia and nitric acid and particulate ammonium nitrate. Atmos. Environ. **13:** 369.
13. Freiberg, J. 1975. The mechanism of iron catalyzed oxidation of SO_2 in oxygenated solutions. Atmos. Environ. **9:** 661.

Appendix

Nomenclature

a_w	= water activity, dimensionless.
c	= gas phase pollutant concentration, gr cm^{-3}.
D	= binary molecular diffusivity, cm^2 sec^{-1}.
H_a	= inversion height, cm.
k	= pseudo-first-order reaction rate constant for gas-to-particle conversion, sec^{-1}.
Kn	= Knudsen number, defined by Equation 1, dimensionless.
K_H	= horizontal eddy diffusivity, cm^2 sec^{-1}.
K_v	= vertical eddy diffusivity, cm^2 sec^{-1}.
M_i	= moles of species i in aerosol particle, gmoles.
m_i	= molality of species i in aerosol particle, gmoles ℓ^{-1}.
m_{bi}	= binary molality of species i at a given water activity, gmole ℓ^{-1}.
$n(r)$	= number density of aerosol particles, # cm^{-3} μm^{-1}.
p_{c_i}	= partial pressure of species i above aerosol particle, atm.
p_{sat_i}	= saturation vapor pressure of species i, atm.
p_{∞_i}	= ambient partial pressure of species i, atm.
Q	= source strength of primary pollutant, gr sec^{-1}.
r	= particle radius, cm.
r_0	= mean initial particle radius, cm.
R	= universal gas constant, cm^3 atm gmole^{-1} °K^{-1}.
$S(r)$	= sulfate density of aerosol particles, gmoles cm^{-3} μm^{-1}.
T_∞	= ambient temperature, °K.
$\bar{u}$	= average wind speed, cm sec^{-1}.
$\bar{v}_i$	= molar volume of species i, cm^3 gmole^{-1}.
v_d	= deposition velocity, cm sec^{-1}.
w_s	= settling velocity, cm sec^{-1}.
x_i	= mole fraction species i in aerosol particle, dimensionless.
Z_a	= height at which deposition velocity is measured, cm.
z_s	= emissions source height, cm.
γ_i	= activity of species i, dimensionless.
λ	= mean free path of background gas, cm.
σ	= surface tension of aerosol particles, dyne/cm.
$[\cdot]$	= liquid concentration of a given species, gmoles ℓ^{-1}.

SOURCE APPORTIONMENT FOR URBAN AEROSOLS: GENERAL DISCUSSION*

Glen E. Gordon, *Moderator*

Department of Chemistry
University of Maryland
College Park, Maryland 20742

T. J. KNEIP (*New York University Medical Center, New York, N.Y.*): Dr. Gordon, the old national air sampling network indicated considerably higher manganese in cities that were burning coal in the past, and I'm surprised that your data now show essentially no manganese contribution from coal.

G. E. GORDON: It is conceivable that small-particle manganese is a small component coming from certain kinds of coal-fired plants. I've not ruled that out entirely; but in the plant we looked at, Chalk Point, there didn't appear to be a lot of small-particle manganese. However, a plant that has a very effective electrostatic precipitator that takes out most of the particulates tends to fractionate those that do get out. So it is possible that coal-fired plants with very effective pollution controls would produce a fine-particle manganese component.

T. J. KNEIP: Well, I'm thinking of data that go back to the days when coal-fired plants didn't have controls, and they do indicate that cities widely spread over the United States had much higher manganese; iron and manganese both were elevated in coal-burning cities.

G. E. GORDON: The absolute quantity will be elevated, but I don't see much manganese coming from the coal-fired plant. It's terribly depleted in coal.

T. J. KNEIP: Also, when we factor the data for New York City, as you are aware, we come up with copper as an independent tracer. We think it is an incinerator tracer, and you don't think that can be possible. But we also find that zinc is not independent; it's tied into several sources in some way in New York City. I think this points out the problems of the difference in locations and of the differences in sources in different locations. I wonder, what is the altitude of your sampler, since you are getting elevated manganese in the city?

G. E. GORDON: We have the samplers at various altitudes. We haven't gotten to the stage yet of looking separately at the high-altitude and low-altitude samplers, but we are planning to do that.

J. W. WINCHESTER (*Florida State University, Tallahassee, Fla.*): Dr. Gordon, the chemical element balance results you showed probably are the best of any in achieving a balance. But I've always been bothered by the chemical element balance as you've done it, because you prejudge the outcome. You select certain elements that seem, intuitively, more likely to be associated with some source or other, and then see how the other elements fit.

Could you reverse the procedure, picking out some of the other elements that seem to fit the best, and redo the calculation to see whether the major elements that you have chosen in advance as marker elements in fact turn out to be? Could you then do a regression analysis of the chosen element against the linear combination of other

*This is a compilation of the discussions held after the presentation of each of the papers in this section.

0077–8923/80/0338–0190 $01.75/1 © 1980, NYAS

elements to see whether the coefficients turn out to be as large as we saw this morning?

G. E. GORDON: Yes, there are a lot of things to try. You know, we've just got this data set in pretty good shape and are doing different things with it. That is certainly one thing that we could try. Sheldon Friedlander, I think, suggested that we try using virtually all elements in the fit, although I've tended to say we need something left over to test the fit with. I don't know what would happen if we threw all the elements into the original fitting program.

J. W. WINCHESTER: In your opening remarks, you mentioned that a complete statistical analysis based on large data sets would be the nicest to use. Do you have any idea of how large a data set you would like to see? One hundred or so samples is pretty big.

G. E. GORDON: Earlier, as you know, we did a similar kind of job on just 10 samples in Washington, and the final results really haven't changed very much, going from 10 samples to 130 samples. In some ways, that is encouraging; it says that we don't have to take this many data in the future, if we really understand the source materials. On the other hand, it's sort of dull not to have something new come out when you do more samples.

Now, there are a lot of other things, like the urban–rural ratio, that we couldn't do with 10 samples. But the basic picture hasn't changed dramatically between 10 samples and 130 samples.

T. G. DZUBAY (*U.S. Environmental Protection Agency, Research Triangle* Park, *N.C.*): Dr. Gordon, can I infer from what you are saying, then, that you are not finding large variations in the relative contributions of these various source categories from sample to sample?

G. E. GORDON: Well, we are confining all of our effort, to date, on our work in Washington, D.C.

T. G. DZUBAY: I'm speaking about within a city or at a specific site.

G. E. GORDON: There is one source that I am very skeptical about, and that's the oil source, because the composition of oil varies quite a bit from one batch to another and from one area to another. If you really want to get down to extremely high accuracy, you have to sample the major supplies of oil being burned in a given area. I think the oil component is the one that worries me the most.

T. G. DZUBAY: If I might carry this a bit further. In your slide for an average—or it may have been a day, I'm not sure—you showed soil at about 20 micrograms per cubic meter and some other source categories at about 5–10–20. Do the relative amounts of those change much from day to day, when you run balances on individual days?

G. E. GORDON: Yes, they will change somewhat. We've got all the individual ones, and I just gave you the summary here. We've done a chemical balance on each sample, so we've done this thing 130 times, and I'm just giving you the average. Certainly, there are day-to-day variations—considerable ones—in the strengths of various sources. You've got things like the dispersion factor, and so forth.

T. G. DZUBAY: Are those variations significant enough? You know how people tend to develop control strategies based on annual analyses, because a 24-hour analysis typically is very difficult to do. Have you seen anything in your analyses that would make you leery of that approach?

G. E. GORDON: I don't know that our approach is really answering the question you've asked. I'm not sure.

UNIDENTIFIED SPEAKER: Dr. Hopke, did you try to plot the values of your various factors on a map for each of the sites, just to see if there were any obvious relationships between what you thought those factors were and, perhaps, point sources in the area?

P. K. HOPKE (*University of Illinois, Urbana, Ill.*): I haven't had a chance, here, to discuss our cluster analysis, where we did try to group samples on the basis of composition. We do get nice grouping of the perimeter samples and the in-town samples. One of the other things I'm hoping to do is to look separately at the factor analysis of those in-town samples that fell into one group and the peripheral samples that fell into the other.

UNIDENTIFIED SPEAKER: I think this question might have been a little more applicable to Dr. Gordon's talk, but maybe you can contribute something to it, Dr. Hopke. In situations where chemical element balance gives you ambiguous results, and even factor analysis doesn't pull you out, would there be any profit in looking at stable chemical compounds? For example, might it be true that coal manganese is in the form of the sulfide and soil manganese is in the form of the oxide? Has anybody examined this sort of a possibility?

P. K. HOPKE: Chemical speciation studies of air particulates are still very much in their infancy, and there is not much that I've seen with regard to people doing that sort of analysis.

G.E. GORDON: Dr. Hopke, I'd also like to address that. An important thing that I had to leave out of my talk here was the fact that there is no reason to confine this kind of identification procedure to elements *per se*. I think that some series of organic compounds, such as the PAH [polynuclear aromatic hydrocarbon] compounds that Dr. Daisey described, might be tremendous as a handle on sources.

Another thing is that there is a whole new data set from John Cooper in Portland. One interesting identification procedure he's been looking at is the carbon-14/total carbon ratio, which helps distinguish, for example, between wood-smoke and coal-smoke particles. So there are a lot of other things besides just elements.

S. DATTNER (*Texas Air Control Board, Austin, Tex.*): Dr. Hopke, I have done a lot of factor analysis, and one thing you have to be very careful about is exact calibration of your instrument. In my factor analysis, about half of my factors turn out to be extremely low level interferences. Have you checked into your studies about interferences causing those ambiguities?

P. K. HOPKE: Well, these are neutron activation results and are not going to be quite the same kinds of interelement effects that your x-ray fluorescence results are giving. We haven't really looked at the nature of the noise eigen-covectors. Obviously, there are a large number of values with vectors with small eigenvalues, which are your noise. We haven't really looked in detail at those noise vectors.

G. E. GORDON: I might say, in regard to your question, that Jarvis Moyers, in his resolution of the Tucson data by factor analysis, found one factor that was a group of elements that he took out together in a group separation scheme. He thought that might be the explanation.

P. K. HOPKE: Dr. Kleinman, another possible approach you could use would be to normalize your data by dividing by the TSP [total suspended particulate matter]. In other words, if you are worried about things like dilutant factors, that is an alternative method—if you have TSPs available. We didn't have TSPs for Boston, so we didn't use it. I like your idea, and I think we will give it a try and see whether that helps sharpen up some of the things we've done.

M. T. KLEINMAN (*University of Southern California School of Medicine, Downey, Calif.*): Thank you. One of the problems with dividing through by a TSP value is that you divide everything through by a constant that in turn adds some more correlation to the overall correlation structure—because now you are multiplying or dividing everything by the same number. I've tried that on just a test data set, and it does indeed raise the correlation several points.

P. K. HOPKE: I saw this first proposed by Ron Henry. He had done it on a partial data set for Portland and had definitely improved his results. It did seem to be of some value. So it may depend on the data set.

M. T. KLEINMAN: That was what I was going to say, Dr. Hopke. If the TSP already has several components, and you are trying to get the source components of the TSP—and one of the factors causing the variation is meteorology—then if you use the TSP as the normalizing variable, you can't go back and apportion the sources of the TSP very well; you've already built it in.

J. W. WINCHESTER: Dr. Kleinman, your use of the differentiated data set is extremely interesting. Two people in our program have done something that may be very similar to what you've done. They have tried to look at the residuals, or departures, from the smooth-curve trend through the data set. They plot a smooth curve by a time series analysis approach, not by a preconceived function. Then they look at the residuals from this and do a principal component analysis. Their concept is that everything goes up and down together, as you pointed out, because of meteorology. But, over and above that, you have source effects where some elements are coming from their sources with certain associations. Their approach—and perhaps your differentiated data set, if I understand it correctly—is looking at associations that are there before meteorology affects the concentrations. If you get correlations that are stronger or weaker than you have in the measured data set, they are perhaps more representative of what happens at the source. This is something we all may ponder a bit more when we are looking at a measured data set that is not necessarily what we are interested in. We may want to know more about what comes out of sources, and those associations are blurred by the meteorology. However, your approach, and perhaps this treatment (they call it a "shock treatment," from some use of the word in the statistical literature), looks at these residuals as the real data set.

M. T. KLEINMAN: I initially thought of trying this using a smooth curve and subtracting out the smooth curve. The only reason I didn't try it for this presentation was that the differentiated data set gave exactly the same beta coefficient. I thought it would be a way to determine whether I was doing any drastic damage to my original concept of apportioning the sources using the various tracers. I wasn't quite sure what subtracting out a smooth curve would do.

J. W. WINCHESTER: One other thing that has impressed me in thinking about this, is that the principal components from this analysis or from the raw data set don't necessarily have to have a physical significance in themselves; but the associations of the elements—the patterns of association with the principal components—may tell you about the clustering of the elements. For example, maybe you think all lead should be from automobiles; yet the principal components analysis shows a little bit of lead in principal component number 1, and number 2, and number 3, and number 4, and so forth. But if you want to know if lead and bromine are associated, then you'd like to know how the elements are correlated with the principal components, and whether those correlations are associated with each other on component 1 vs. component 2, or component 2 vs. component 3, and so forth.

So the physical significance lies in the associations of elements with respect to the principal components, and not necessarily on one's insisting that they are all within one principal component as a physical reality. The statistical procedures may enable you to get those insights.

M. LIPPMANN (*New York University Medical Center, New York, N.Y.*): Dr.

Dzubay, can you estimate from any data, or just from your own speculation, how much of that ammonium sulfate is due to SO_2 emissions farther west, and how much is due to biological decay of hydrogen sulfide or some such natural cause?

T. G. DZUBAY: One thing that you do notice is that there isn't a very big increment for St. Louis. There is a 5 microgram per cubic meter increment above a background of 15. That is all that I can say for sure, the rest is speculation. Use your imagination. My guess is that a good share of this is coming from upwind of the city.

C. BROSSET (*Swedish Water and Air Pollution Research Laboratory, Gothenburg, Sweden*): I will help you a little, because this is very similar to the situation in Sweden. In Sweden, 90%, or even 95%, of sulfate is imported from outside.

W. R. PIERSON (*Ford Motor Company, Dearborn, Mich.*): Dr. Dzubay, one thing I may not have mentioned to you in the past is that, in our tunnel studies, we did find a family of elements that always changed together and that consisted of such things as sodium, magnesium, calcium, potassium, scandium, titanium, and iron. I don't know anything about the sizes, because we weren't doing any size fractionation. I presume that they were coarse. If I assume that it's soil material, and then throw in the rest of it intuitively, it amounts to about 10% by mass of the vehicle aerosol. Now that is just in highway driving; I don't know how generalizable it is.

T. G. DZUBAY: If that's a road inside of a tunnel—not exposed to weathering—what happens on a road exposed to rainfall and everything else?

W. R. PIERSON: Presumably, it will be higher on the outside.

UNIDENTIFIED SPEAKER: Dr. Dzubay, it is very gratifying to find that remarkable agreement between the total mass and the components, and it is also gratifying to see that ideas that began as sort of academic exercises are actually turning out to be useful in EPA regulatory practice, or at least getting to that point.

I think that the carbon aspect—the organic component—is still a major unknown. This cropped up at the meeting that Dr. Gordon and I attended in Norfolk on the toxic substances, where we were speculating on the possibility of the use of this type of approach for identifying sources of organic compounds. Also, Dr. Macias and I have been talking about the need for doing an improved carbon balance. We tried carbon balances in the early stages of our measurements, without very much success. But now there is a lot more information, and I think that is where the main emphasis ought to be placed, for many reasons, some of which you just established.

G. E. GORDON: Dr. Dzubay, one thing I noticed was that you had very low vanadium, which seems to be a very sharp marker of oil. You must use very little oil in St. Louis.

T. G. DZUBAY: When they use residual oil there, it has to be barged in; it won't flow through pipes very well. If they don't bring it in on barges, they don't have it.

M. LIPPMANN: Dr. Pierson, perhaps I missed it, but did you say you did, or did not, have ozone concentration data at this site?

W. R. PIERSON: We had two or three big problems in this experiment, and the biggest one was that we didn't measure ozone. I hadn't realized, when we started, that it was important.

K. A. RAHN (*University of Rhode Island, Kingston, R.I.*): Dr. Pierson, contrary to previous discussion, I believe that your explanation is completely consistent with transport. The last few sentences of your paper could have been taken from the talk that I'm going to give here for the Arctic aerosol, where transport is the key to explaining sulfate. We find tenfold enrichment of the sulfate to vanadium ratio in the Arctic, compared to the sources. As far as I can see, that is completely consistent with oxidation of sulfur dioxide along the way. So it looks exactly like an aged aerosol.

P. T. CUNNINGHAM (*Argonne National Laboratory, Argonne, Ill.*): Dr. Pierson, as part of the MAPS program, we've been sampling at Pennsylvania State University since January of 1977. For the two months that you were looking at that site, the monthly average of the hydrogen to sulfate ratio that we observed was 0.6. That is slightly lower than what you were reporting, but we have the same problems. We cannot find more acidity than is present, though it is certainly possible that what we are observing is less acidity than is present.

I might point out that, for the two years of data that we have, there are two periods when the acidity is maximized. If you happen to be in the maximum for 1977, it maximizes again in the winter months; we observed that in early 1978, and then it recurred with a maximum in the summer. So there are actually two maxima in the degree of acidity that we observed at Penn State. The University of Virginia data correlate with these data very well; so it appears to be very much a regional phenomenon and not a local one.

W. R. PIERSON: You are talking about the aerosol sulfate?

P. T. CUNNINGHAM: We're talking about the aerosol sulfate.

Now, the other factor that you mentioned was the uncoupling between SO_2 and sulfate, and the apparent recoupling of this and rainfall. I should point out that our studies of the oxygen-18 composition of aerosol sulfate compared to that found in rainfall would completely support that contention.

W. R. PIERSON: Thank you.

C. BROSSET: Dr. Pierson, I would only confirm that everything you said is absolutely equal to what we have found in Sweden. I'm very pleased to see that these phenomena are not occurring only in Sweden. Now, you mentioned a neutralization with ammonia that you had to be a little careful with, because, of course, if you have sampled acid particles and you get an excess of ammonia, it is neutralized. However, if you get clean, ammonia-free air, ammonia will be removed. So what you really get on your filter probably reflects the concentration of ammonia during the last sampling hours, because the ammonia in that determines the ratio of hydrogen to ammonium in your sulfate.

W. R. PIERSON: Exactly. The ammonium that comes in near the end of the run is critical.

P. J. LIOY (*New York University Medical Center, New York, N.Y.*): Dr. Pierson, I'm very pleased with the correlations that you had for sulfate and SO_2. In the 1976 aerosol study, we found, basically, that there was very little relationship between the sulfate and SO_2. I think this substantiates the idea of transport, especially when you put it into the perspective of a trajectory analysis and work back about five or six days. I think something close to what Perry Samson has done on ensemble trajectory analysis might be helpful for you to figure out why the decoupling occurred.

We also sampled in the 1977 period, and I think you saw the data from Roger Tanner for High Point in a similar period of time. The period where you had that major episode was quite consistent with what we saw at High Point, and the analyses have indicated that the trajectories put the source region back toward the Ohio Valley.

W. R. PIERSON: I would just like to make a couple of comments. One is that I was aware of your work, and there is an awful lot of stuff on this subject. I didn't try to tie in how well we agreed with the literature, because there is just so much of it. The other point is about using back trajectory. I'm going to try to give my sulfate data to Perry Samson, have him play this trajectory game, and see what he can get out of it.

K. T. WHITBY (*University of Minnesota, Minneapolis, Minn.*): Just two comments, Dr. Pierson. The first one concerns the hysteresis of ammonia. We have an instrument, which we now call the "tandem differential mobility analyzer," that does this similarly to what has been done by Charlson. The only conditions under which you get no change in size as you increase humidity are if the particle is fairly pure ammonium sulfate and has been brought down to below 5 or 6% relative humidity. In a real atmosphere, this doesn't happen, particularly in the eastern U.S. What this means is that ammonium sulfate in the real mixtures and sulfuric acid cannot be distinguished in their behavior by running up and down from 30–40% and 100% relative humidity, which would be the range of conditions. You couldn't really distinguish it.

W. R. PIERSON: The ammonium sulfate deliquescent point is suppressed somehow.

K. T. WHITBY: Yes. The second point is that I wholeheartedly agree with the fact that there is no relationship between SO_2 and sulfate at these rural sites. Last summer, we had the continuous sulfate-aerosol sulfur and continuous gas sulfur instrument running, and the data were a real revelation. A couple of times, we were in the power-plant plume at the end of the day when the plume lifted, and the SO_2 concentration disappeared an hour or so before the aerosol sulfur from the same plume. We concluded that, when the atmosphere stratifies at the end of the day, it takes a rather short time for the SO_2 to be cleaned out of the lower 100 meters or so of the atmosphere. So, if you are running samplers that are averaging over a whole day, I would be exceedingly surprised if you were to find a correlation between SO_2 and the aerosol sulfur.

W. R. PIERSON: As a matter of fact, we were running a microcoulometer for SO_2 measurement at the same time, and we saw the SO_2 coming in, in spikes.

G. E. GORDON: Dr. Pierson, this is the first time that I've ever heard of a clean air episode. That is a new one for me.

C. BROSSET: Dr. Peterson, I saw your table, but I didn't quite understand the balance that you get in your final composition. How was the balance between hydrogen, ammonium, and sulfate?

T. W. PETERSON (*University of Arizona, Tucson, Ariz.*): Actually, we saw that you could almost completely neutralize the sulfate with ammonia.

C. BROSSET: Yes, that is correct because as I saw that you have a pH of 2, I just looked for the ammonium concentration, but it was neutralized. It has to be that, because you had a very high ammonia concentration. How was ammonia measured?

T. W. PETERSON: I'm really not sure of the procedure that was used. As I said, I took the number from some work that was done by Lawson in the southern Arizona area.

C. BROSSET: Oh, I see. It seems a little high. Thank you.

L. A. BARRIE (*Atmospheric Environment Service, Downsview, Ont., Canada*): Dr. Peterson, I also have a question concerning the ammonia. It's clear from your results that ammonia affects them, and they are fairly sensitive to ammonia. Did you assume ammonia was constant?

T. W. PETERSON: Yes. As I said, we had no indication that the plume was a source of ammonia, and we did take into account that there was ammonia depletion based on absorption by the particle. But again, that really governs only the diffusion of the ammonia to the particle surface, because, for simplicity, we assumed that there was essentially an infinite background source of ammonia.

L. A. BARRIE: Do you think that the entrainment of ammonia by the plume could have some effect on your results?

T. W. PETERSON: Certainly. I think it's obviously going to significantly decrease the magnitude that is going to reach the center line, although, again, we did take that into consideration. We assumed a ratio of aerosol volume to air volume, in order to recalculate the total ammonia takeup by the secondary aerosol.

L. A. BARRIE: And reduce the air concentrations accordingly.

T. W. PETERSON: Right.

UNIDENTIFIED SPEAKER: Dr. Peterson, I just want to know where that water came from. Did you have a scrubber?

T. W. PETERSON: There are a number of sources of water; the most obvious would be the sulfuric acid conversion plants at the smelters.

A REVIEW OF URBAN PLUME STUDIES

Francis A. Schiermeier*

Environmental Sciences Research Laboratory
U.S. Environmental Protection Agency
Research Triangle Park, North Carolina 27711

Previous speakers at this conference have already discussed the characteristics and source apportionment of urban aerosols. The theme of this session, "the urban plume," encompasses the physical and chemical transformations occurring as the plume is transported downwind while subjected to varying meteorological regimes of dispersion, radiation, and humidity. It should be noted that although the emphasis of this session is the urban plume, it is frequently difficult to distinguish the urban from surrounding power plant plumes, when performing field measurements. Some of the results presented in the next few papers will therefore deal with plumes from both source types. In order to provide a setting for the presentations, I will offer an overview of recent field studies sponsored by the Environmental Protection Agency (EPA) and aimed at determining pollutant transport, transformation, and removal rates.

During recent years, air pollution scientists have become increasingly aware of the relatively high pollutant loadings existing not only near, but also far downwind of, urban areas. The rising sulfate concentrations in the northeastern United States are a case in point. Of additional concern, however, are recent formulations of national energy goals that advocate a possible shift from oil and natural gas to more abundant coal as a fuel source for electrical generating stations and industrial plants, a move likely to produce increased levels of sulfur oxides, nitrogen oxides, and other pollutants in the atmosphere. Thus, the EPA has initiated a series of research studies based on the need to determine quantitatively the impact on local air quality of emitted pollutants and their transformation products from distant sources.

Perhaps the most comprehensive EPA field study performed to date was the Regional Air Pollution Study (RAPS). This program was conducted in the St. Louis, Missouri/Illinois metropolitan area during the years of 1974–1977. The main objective of RAPS was to develop, evaluate, and validate air quality simulation models on a regional scale, covering urban and rural, stationary and mobile sources. In addition to comprehensive monitoring of all criteria and selected noncriteria pollutants, a series of experiments was performed directed toward understanding and subsequently describing the relationships between atmospheric dynamic, kinematic, and energetic processes that occur in the urban boundary layer and their resultant impacts on the transport and dispersion of pollutants.[1]

Allied temporally with RAPS was the EPA's Midwest Interstate Sulfur Transport and Transformation (MISTT) project, which examined the long range behavior of the St. Louis urban and surrounding power plant plumes. Although the complete mix of pollutants from the entire urban complex was measured, the focus was on oxides of

*On assignment from the National Oceanic and Atmospheric Administration.

0077-8923/80/0338-0198 $01.75/1 © 1980, NYAS

sulfur, with much of the experimental work devoted to the dynamics of aerosol formation. The principal objective of MISTT was the quantitative estimation of the kinetics of SO_2 oxidation and ground removal in large plumes, and hence the determination of the sulfur mass balance during plume transport.[2]

These projects were succeeded by the current EPA research study, Sulfur Transport and Transformation in the Environment (STATE). Combining the disciplines of chemistry and meteorology, the STATE program will develop the relationships between source emissions of sulfur/nitrogen oxides and their attendent concentration levels in the northeastern United States. This primary goal of STATE encompasses two objectives, each delineated by consideration of the scale at which pollutant plumes from discrete sources or source complexes can be distinguished from one another.

STATE plume studies, encompassing travel distances up to several hundred kilometers, have the objective of quantifying spatial distributions and physicochemical forms of pollutants from identifiable single sources by means of studies of dispersion, transformation, and removal processes acting on emitted pollutants. The first STATE experiment on this scale was the Tennessee Plume Study (TPS), performed this past August on the plume from the Cumberland Power Plant located in northwestern Tennessee.[3]

One speaker this morning will present results from TPS, but basically STATE relates to a transport scale more applicably described in tomorrow's synoptic scale sessions, as evidenced by the second objective. STATE regional studies, encompassing areas up to subcontinental scale, have the objective of quantifying the cumulative impact on local air quality of the many discrete sources contributing to the pollutant loading of an air mass as it moves over source regions. The first STATE research study on this scale is Persistent Elevated Pollution Episodes (PEPE), planned for the summer of 1980 in the northeastern United States.

Another current EPA research study is Visibility Impairment due to Sulfur Transport and Transformation in the Atmosphere (VISTTA), presently being conducted in Arizona, New Mexico, and Utah. In this area, where plume scale quickly becomes regional scale, measurements are being taken of power plant and copper smelter plumes in an attempt to relate the physical and chemical forms of primary and secondary aerosols to visibility impairment. Hence, the emphasis of VISTTA is directed toward the optical effects of aerosols, rather than toward characterization *per se*.

This list of EPA-sponsored field research studies is not intended to be exhaustive but was compiled to provide a synopsis of the programs in which the ensuing speakers have participated. Also to be discussed are some results from the Multi-State Atmospheric Power Production Pollution Study (MAP3S), sponsored by the Department of Energy (DOE), and the Sulfate Regional Experiment (SURE), funded by the Electric Power Research Institute (EPRI).

The goal of the MAP3S program is to develop and demonstrate an improved, verified capability to simulate the present and potential future changes in pollutant concentration, atmospheric behavior, and precipitation chemistry as a result of pollutant releases to the atmosphere from large-scale power production processes, primarily coal combustion. The study domain is the high population, energy intensive northeastern quadrant of the United States.[4] The SURE program is attempting to define the relation between emitted primary pollutants and regional ambient concentrations of secondary products. Emphasis is on identifying the contribution of the electric power industry in the northeastern United States.[5]

The EPA, DOE, and EPRI programs are coordinated through frequent meetings of the program directors. While the programs have some aspects in common, each has areas of special concern and responsibility. For example, DOE is primarily responsible for deposition processes, EPA for transformation measurements, and EPRI for a monitoring network. Thus, the program directors insure that gaps are filled and unnecessary duplications prevented. The three groups together provide a comprehensive national program to study the transport, transformation, and removal of energy-related pollution.

A common goal of the field research studies described here is the determination of various aspects of transport and transformation that pollutants undergo in the atmosphere, at dilutions and time scales that are difficult to simulate in the laboratory. Perhaps this resolve is most clearly expressed by Calvert *et al.*: "It is now generally recognized that the development of scientifically sound and lasting strategy for 'sulfate' aerosol control requires that we understand the nature of the various atmospheric reactants which oxidize SO_2, the chemical nature of the products formed in its reactions, the rates at which these reactions will occur in a given atmosphere, as well as the physical processes which transport these gases and aerosols about the atmosphere and those which remove them from it."[6]

Similar arguments in recent years have resulted in numerous field research studies, each attempting to augment existing knowledge of some physical, chemical, or meteorological aspect of air pollution. But an inevitable result of this increased activity is that each completed field study tends to generate additional questions. The rapidly expanding data banks do indeed provide insight into pollution processes, but they also manage to elucidate those areas where supplementary information is required for increased comprehension. Justifiably, these requirements for additional field measurements generate successive research studies.

One must never lose sight, however, of the ultimate goal, i.e., the development and verification of quantitative relationships between sources of pollution and ambient air quality measurements, so as to allow control activities to become more sophisticated and selective. General control actions need to be tested through these relationships in order to develop strategies that provide the desired level of control for the lowest cost. Ideally, such quantitative relationships would even allow impact on air quality to become a factor in community and industrial planning for future growth.

References

1. Schiermeier, F. A. 1978. Air monitoring milestones: RAPS' field measurements are in. Environ. Sci. Technol. **12:** 644.
2. Wilson, W. E. 1978. Sulfates in the atmosphere: a progress report on Project MISTT. Atmos. Environ. **12:** 537.
3. Schiermeier, F. A., W. E. Wilson, F. Pooler, J. K. S. Ching & J. F. Clarke. 1979. Sulfur transport and transformation in the environment (STATE): a major EPA research program. Bull. Am. Meteorol. Soc. **60:** 1302.

4. MacCracken, M. C. 1978. MAP3S: an investigation of atmospheric energy related pollutants in the northeastern United States. Atmos. Environ. **12:** 649.
5. Perhac, R. M. 1978. Sulfate regional experiment in northeastern United States: the 'SURE' program. Atmos. Environ. **12:** 641.
6. Calvert, J. G., Fu Su, J. W. Bottenheim & O. P. Strausz. 1978. Mechanism of the homogeneous oxidation of sulfur dioxide in the troposphere. Atmos. Environ. **12:** 197.

AEROSOL COMPOSITION OF URBAN PLUMES PASSING OVER A RURAL MONITORING SITE

Thomas G. Ellestad

Environmental Sciences Research Laboratory
U.S. Environmental Protection Agency
Research Triangle Park, North Carolina 27711

INTRODUCTION

It has been demonstrated that an urban plume—the aggregate emissions of a metropolitan area—can be readily identified at distances on the order of 100 kilometers downwind of a city.[1] In 1975, as part of the U.S. Environmental Protection Agency's Project MISTT (Midwest Interstate Sulfur Transformation and Transport), an aircraft was used to map the major constituents of the St. Louis urban plume as it traveled several hundred kilometers downwind. The objective was to study the mass flow rates and chemical transformations of such parameters as sulfur dioxide, sulfates, and ozone at various downwind distances.

As vital as an aircraft is for plume mapping, however, there are severe size, weight, and power restrictions placed on all instruments aboard. Many sensitive and useful analytical methods cannot meet the requirements. Therefore, the Aerosol Research Branch placed an extensively equipped ground station, and invited two universities to make specialized measurements, at a site about 100 kilometers north of St. Louis, taking a chance that winds would bring the urban plume to the site. Objectives of the ground station included identifying which parameters could serve as tracers of an urban plume and measuring the aerosol composition and gaseous concentrations of the plume.

SITE DESCRIPTION

In choosing a site, consideration was given to the likelihood of intercepting the St. Louis urban plume, to downwind distance, and to the absence of major sources between St. Louis and the site. The field laboratory was placed about 2 kilometers west of Glasgow, Illinois (population: 100), which corresponds to 104 kilometers north-northwest of the St. Louis Gateway Arch. The site's latitude was 39°33′ N, its longitude 90°30′ W, and its altitude about 165 meters. The laboratory was atop one of the local hills that overlook the Illinois River basin to the west. The surrounding farmland produced mainly corn and soybeans, with some cattle grazing and hog production. Roads within 10 kilometers included numerous unpaved farm access roads with extremely light traffic, paved county-state roads with only local traffic, and U.S. Highway 36-54 nine kilometers to the north. A limestone-quarry asphalt plant was located about 3 kilometers southwest of the site, but never had any obvious influence during the sampling period because of wind directions.

0077-8923/80/0338-0202 $01.75/1 © 1980, NYAS

Laboratory Description

The EPA field laboratory was a 40-foot long semitrailer, equipped with 10-meter-high sample stacks and having sufficient room inside to house the gas analyzers, filter samplers, and a computerized data acquisition system. *In-situ* aerosol analyzers located in the trailer were operated by personnel from the University of Minnesota. The parameters measured, along with instrument manufacturer, model number, and calibration reference, are listed in Table 1. Mass concentration was obtained from 47-mm glass fiber filters changed every two hours. The filters were equilibrated at 35% relative humidity for 24 hours before the initial and final weighings on a Mettler

Table 1
Measurements

Parameter	Manufacturer	Model	Calibration Reference
SO_2	Meloy	SA-160	NBS permeation tube
O_3	Bendix	8002	Titration with NO
NO–NO_x	Teco	series 14	NO reference cylinder
CO	Beckman	6800	Reference cylinder
b_{scat}	Meteorology Research, Inc.	1550	Fluorocarbon-12
Nuclei concentration	EG&G	RICH-100	See Liu[2]
Fine aerosol size distribution	Thermo-Systems	3030	See Liu[3]
Coarse aerosol size distribution	Royco	220 & 245	Latex spheres
Optical radiation	Eppley	2	Eppley
Winds	Meteorology Research, Inc.	Vectorvane	Wind tunnel
Mass concentration	—	—	See text
Fractionated elemental concentrations (Al, Si, P, S, Cl, K, Ca, Ti, V, Cr, Mn, Fe, Co, Ni, Cu, Zn, Ga, Ge, As, Se, Br, Rb, Sr, Cd, Sn, Sb, Ba, W, Hg, Pb)	—	—	See Stevens[4]

ME-22 microbalance. Fractionated elemental concentrations were measured by x-ray fluorescence from samples produced by an automatic dichotomous sampler (ADS). The ADS separates fine ($< 2.5\ \mu m$) particles from coarse ($> 2.5\ \mu m$) particles by virtual impaction and collects each fraction on its own filter. Pairs of cellulose filters were changed every two hours. Because of the two-hour integration period imposed by the mass and elemental concentration data, all other parameters were averaged over the same intervals before analysis.

After mass determination from the 47-mm glass fiber filters, the filters were analyzed by ion chromatography for sulfate ion. As a rough correction for SO_2

adsorption by the basic glass fiber filters, identical secondary filters had been run behind the primary filters. Sulfate values from the secondaries were then subtracted from their corresponding primary sulfate values.

Also at the Glasgow site was a field gas chromatography lab, operated by Washington State University (WSU) personnel. Frequent analyses of $CFCl_3$ (fluorocarbon-11), carbon tetrachloride, and a myriad of heavy and light hydrocarbons were performed. A data summary and interpretive report has been published.[5] The analysis appearing in this paper uses the fluorocarbon-11 and total nonmethane hydrocarbon data from the WSU lab.

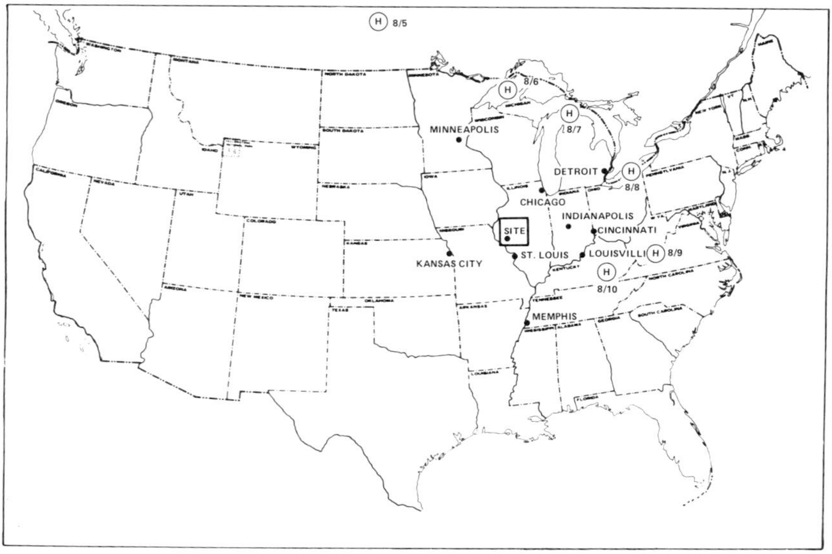

FIGURE 1. Movement of high pressure system responsible for transporting air to Glasgow, Illinois on August 5–10, 1975.

SAMPLING OVERVIEW

The field program was to be conducted between July 15 and August 15, 1975. Because of the labor involved in operating the field site, it was not operated continuously during the study. Only when favorable meteorological conditions were expected was complete data collection attempted. There were four such intensive periods during the study, accounting for about 13 days of sampling.

The most interesting intensive sampling period occurred August 5–10, 1975. During these days, a high pressure center moved from 1600 km north-northwest of the site to 700 km east-southeast of the site, creating wind trajectories on successive days of northeasterly, easterly, southeasterly, and southerly. We thus had the opportunity to sample air from distant cities, a few relatively close point sources, and of course the St. Louis urban plume.

FIGURE 2. Wind trajectories for arrivals at 0100 and 0700 CDT, August 6, 1975. Bumps denote three-hour increments of transport.

ANALYSIS

FIGURE 1 shows the movement of the high pressure system that transported air to Glasgow on August 5–10, 1975. Winds on August 5 were quite variable. However, by August 6, the high pressure center was over Lake Superior and had established

FIGURE 3. Wind trajectories for arrivals at 0100 and 0700 CDT, August 7, 1975.

FIGURE 4. Wind trajectories for arrivals at 0700 and 1300 CDT, August 7, 1975.

well-defined winds throughout the Midwest, transporting air from the Lake Michigan area to Glasgow. On the following days, the high pressure center moved to northern Lake Michigan (August 7), to Lake Erie (August 8), to West Virginia (August 9), and finally to eastern Kentucky (August 10). The path of the high pressure's slow migration caused its anticyclonic winds to sweep slowly from northeasterly to southerly at Glasgow over the intensive sampling period.

FIGURE 5. Wind trajectories for arrivals at 0700 and 1300 CDT, August 8, 1975.

FIGURE 6. Wind trajectories for arrivals at 1300 and 1900 CDT, August 8, 1975.

To aid interpretation of the data, wind trajectories were calculated using the program of J. Heffter of NOAA.[6] Trajectories for Glasgow arrival times of 0100, 0700, 1300, and 1900 CDT were computed backwards in time for 72 hours for a transport layer of 1500 meters height. Selected trajectories are presented in FIGURES 2–8. Each bump on the trajectories indicates three hours backwards in time from the arrival time at Glasgow. The trajectories are believed to be sufficiently accurate to

FIGURE 7. Wind trajectories for arrivals at 0100 and 0700 CDT, August 9, 1975.

FIGURE 8. Wind trajectories for arrivals at 1300 and 1900 CDT, August 9, 1975, and 0100 CDT, August 10, 1975.

show the transport of the sampled air parcels from the urban source areas to Glasgow.

Time plots of key parameters are presented in FIGURES 9, 10, and 11. In FIGURE 9, the fine sulfur data are from x-ray fluorescence of dichotomous filters, while the sulfate data are from ion chromatography of glass fiber filters. Neglecting the obvious peaks, which usually represent urban plumes, one can see a general decline in many variables from August 5th until the 6th. Once the winds established a pattern and sufficient time had passed for transport, the base lines of most variables showed a steady increase from August 7th through the 10th. The increase was most likely due to the urban-industrial emissions of Detroit, Cleveland, Pittsburgh, Cincinnati, Louisville, and Indianapolis, which provided the upwind air for viewing the St. Louis urban plume on August 8–10. An interesting observation is that the base line of condensation nuclei showed exactly the opposite behavior. Its decrease from the 7th to the 10th was probably the result of coagulation and reduced photochemical activity as the regional aerosol aged.

Sulfur dioxide showed no base line above the minimum detectable concentration (5 ppb). Only when the St. Louis urban plume appeared, or when we sampled downwind of the large power plant 80 km east of Glasgow, were SO_2 concentrations measurable. Those periods correspond to transport times of 5–10 hours from source to Glasgow.

The peaks of fine aerosol lead concentration do not correspond to the expected diurnal emission pattern of lead (plus transport time), because of the distance from the source to Glasgow. Over such distances, winds are much more important than temporal emission patterns in determining concentrations at the sampling site.

Urban plumes are identified by simultaneous rises of at least several of the following: fluorocarbon-11 (F-11), carbon monoxide, total nonmethane hydrocarbons

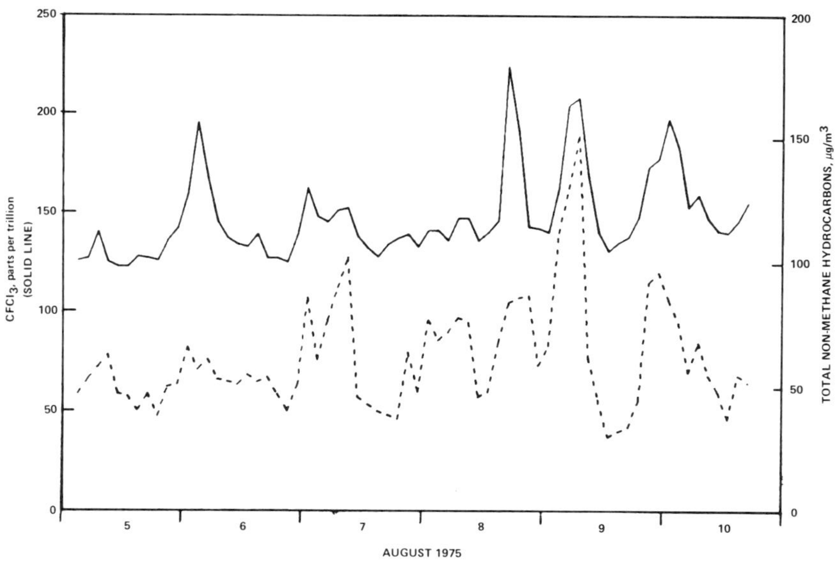

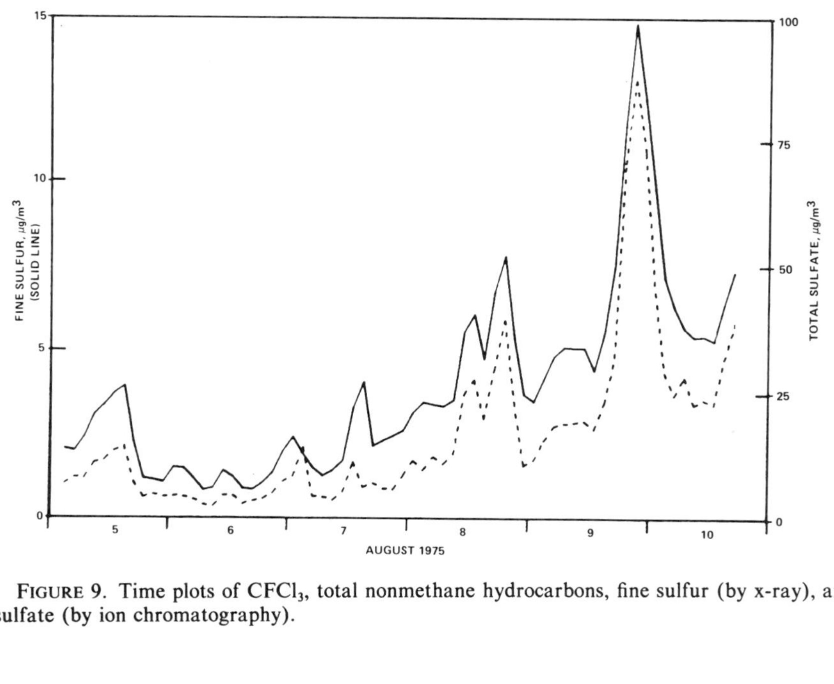

FIGURE 9. Time plots of $CFCl_3$, total nonmethane hydrocarbons, fine sulfur (by x-ray), and sulfate (by ion chromatography).

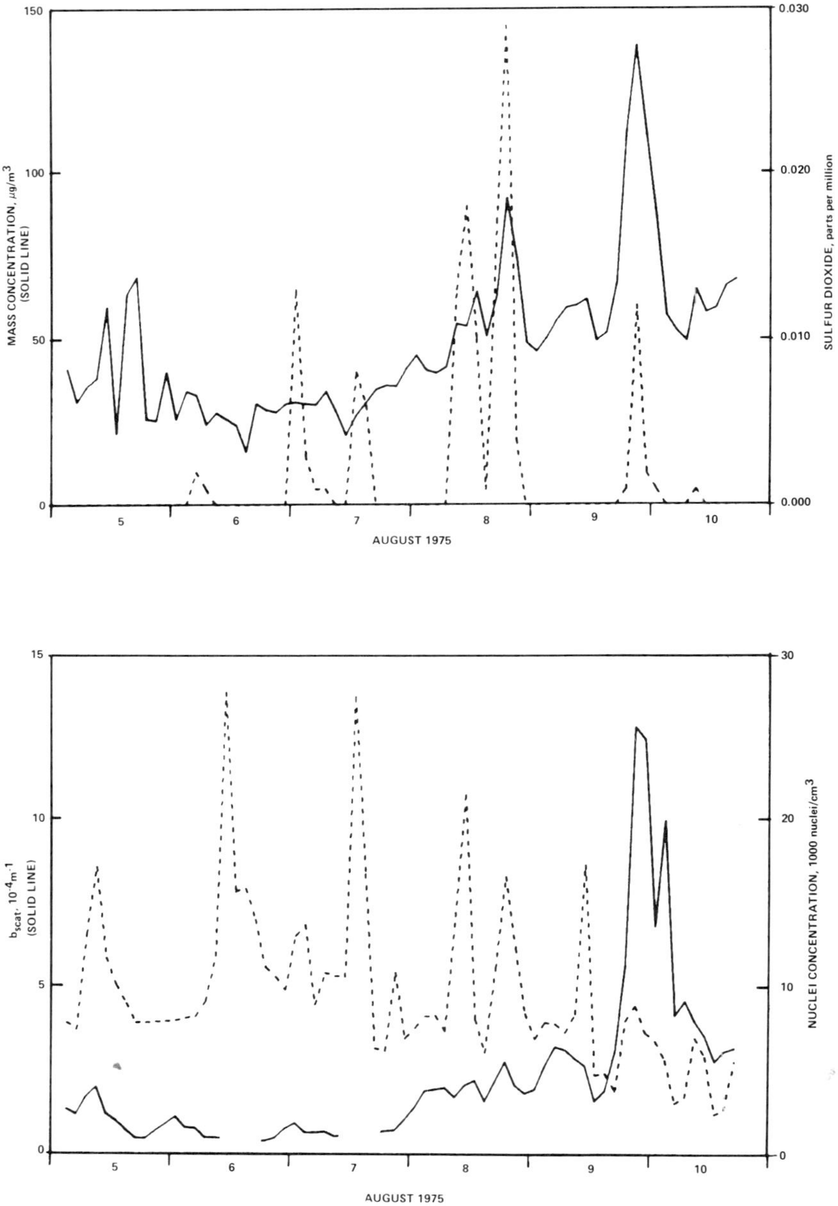

FIGURE 10. Time plots of mass, sulfur dioxide, b_{scat}, and condensation nuclei.

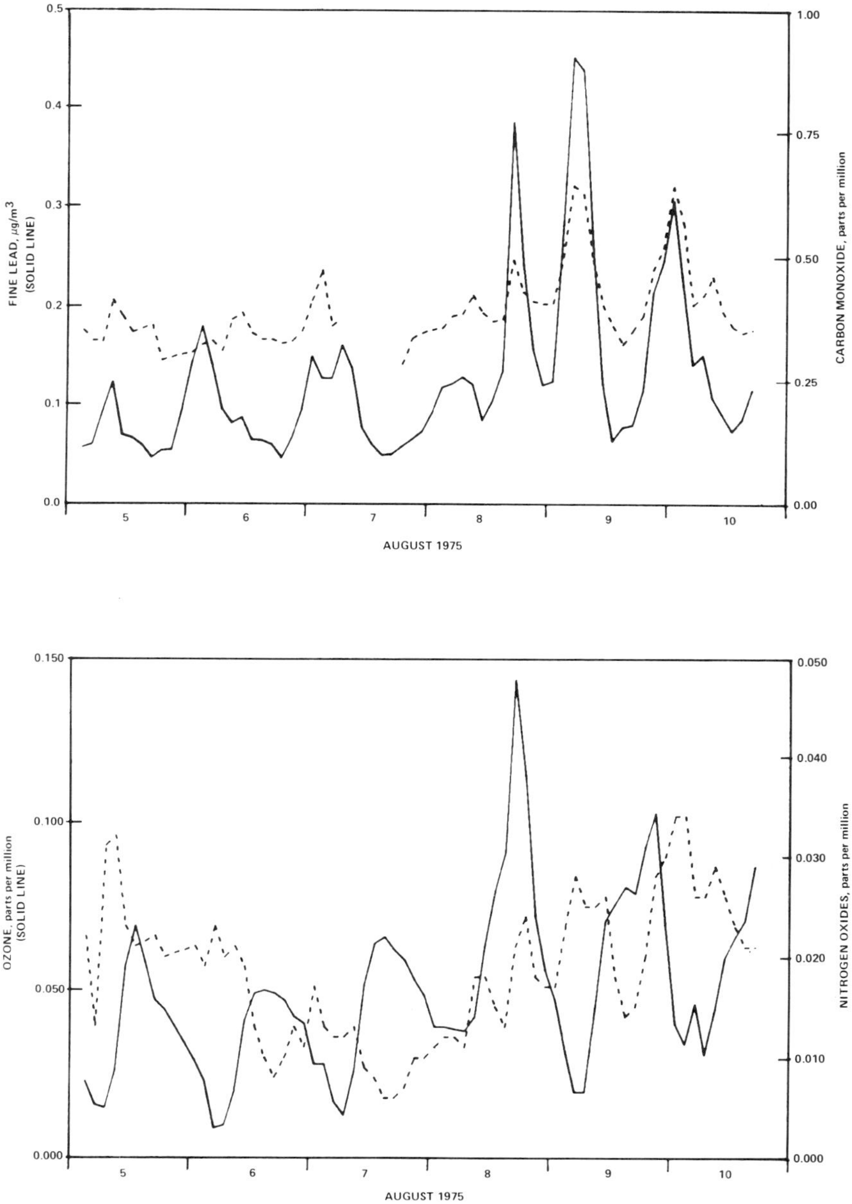

FIGURE 11. Time plots of fine lead, carbon monoxide, ozone, and nitrogen oxides.

(TNMHC), fine aerosol lead, fine aerosol sulfur, and, to a lesser extent, total mass concentration and b_{scat}. The importance of b_{scat} as an urban plume indicator is somewhat less than may be inferred from its time plot, since a portion of its rise is often due to relative humidity rising above 80%. Ozone concentration is not a clear indicator of urban plumes, although several times it has shown undeniable influence on them. Ozone's problem as an indicator results from its diurnal production in the photochemical smog cycle. This point applies only to remote ground stations and not to aircraft sampling, in which ozone is often an excellent indicator of an urban plume, since a plane can make a pass through a plume before the background ozone production has changed significantly.

The following paragraphs give an analysis of seven plume episodes at Glasgow. The plume episodes of August 5 are not analyzed because of the poorly defined wind fields on that date.

On August 6, 1975, from 0200–0400 CDT, definite rises were seen in F-11, TNMHC, fine lead, and b_{scat}, with a minor rise in fine sulfur. The wind trajectory indicates transport from the Chicago-Milwaukee area (350 km distant), with a transport time of about 12 hours (see FIGURE 2). The 3300 MW power plants 130 km northeast of Glasgow apparently did not influence the plume we saw, given no rise of SO_2, NO, and NO_x. Assuming the particulate sulfur occurred as sulfates, one can estimate the mass of sulfate and its cation (hereafter referred to as XSO_4) by multiplying *sulfur* values by 3.6 [the average of 3.1 for H_2SO_4, 3.6 for $(NH_4)HSO_4$, and 4.1 for $(NH_4)_2SO_4$]. One then finds that XSO_4 mass accounted for about 16% of the total mass loading during this plume, which is a lower percentage than for most other plumes we saw. The rise of b_{scat} was largely due to the rise of relative humidity, which had risen to 92% during the plume. This is corroborated by size distribution measurements, which show a submicrometer surface area increase from ~100 to ~300 $\mu m^2/cm^3$ as the plume arrived, followed by a decrease to ~200 $\mu m^2/cm^3$ as the plume passed. The change of b_{scat} is roughly proportional to the change in aerosol surface area in the 0.1 to 1.0 μm range, since both vary as the square of particle radius. Elemental peaks in the plume also include fine Zn, K, and Br. Delayed elemental peaks (probably delayed because of spatial separation at the source area) included coarse Al, P, K, and Fe and fine Fe, Ti, and Mn. Trace elements detectable only in the plume were fine Co and Sb, and coarse Ge and Sn.

On August 7, 1975, from 0000–0200 CDT, distinct peaks of F-11, TNMHC, SO_2, NO_2, fine sulfur, fine lead, and b_{scat} were detected. The wind trajectories indicate transport from the Indianapolis area (370 km distant), with a transport time of about 24 hours (see FIGURE 3). It is likely that the 1800 MW power plants 80 km east of Glasgow contributed to the plume, based on SO_2 readings and the high contribution of XSO_4 to the increase in mass loading. This plume exhibited additional elemental peaks of fine K, Ca, Fe, Cu, Zn, and Br and a peak of coarse Fe. Delayed by a two-hour interval were rises in fine and coarse manganese and carbon monoxide. The rise in b_{scat} was caused by the increased sulfur loading and the humidity (84% during the plume episode). Aerosol surface area from 0.1 to 1.0 μm fell from ~300 to ~200 $\mu m^2/cm^3$ after the plume passed, despite the humidity's remaining above 80% until midmorning. Trace elements included fine Sn, Sb, and Se and coarse Ge.

The next urban plume (August 7, 0800–1000) was also from the Indianapolis area, but this time the winds had developed sufficiently southerly travel to avoid the power plants 80 km east of Glasgow (see FIGURE 4). Transport time from the Indianapolis area had decreased to about 20 hours. Distinct peaks were seen in F-11, TNMHC, and fine Pb and Br. Unfortunately, carbon monoxide and b_{scat} were off-line for calibration. Small peaks existed for fine Cu and coarse Ca, and for trace elements of fine Sn and coarse Ge, Sn, Rb, and Cd. Unlike the previous plume, there was no peak of fine S and no change in aerosol surface area.

The first of the major St. Louis urban plume episodes occurred from 1000–1200, August 8, 1975. Transport time from St. Louis was 6–9 hours (see FIGURE 5). Fine sulfur showed a 60% rise, and SO_2 rose sharply. The rise of these parameters preceded those of F-11, TNMHC, and CO. Peaks that accompanied the earlier arrival of sulfur included fine Al, Fe, and Mn and coarse Al, Si, K, and Fe. b_{scat}, which had been elevated by the 80–85% humidity of the early morning hours, did not drop as the humidity fell to 55%, probably because of the large rise in fine S. Note from TABLE 2 that about 52% of the increase of mass loading is estimated to be due to XSO_4. Trace

TABLE 2

ESTIMATED CONTRIBUTIONS OF SULFATE AND ITS CATIONS TO MASS LOADINGS USING X-RAY SULFUR DATA

Date	Time, CDT	Contribution to Total Loading	Contribution to Increase of Loading	Probable Source Area	Transport Time (hr)
8/6/75	0200–0400	16%	—	Chicago	12
8/7/75	0000–0200	28%	60%	nearby power plants/ Indianapolis	5/24
8/7/75	0800–1000	19%	—	Indianapolis	20
8/8/75	1000–1200	36%	52%	St. Louis	6–9
8/8/75	1600–1800	35%	55%	St. Louis	6–8
8/9/75	0400–0600	31%	38%	St. Louis	6–9
8/9/75	2000–2200	39%	40%	St. Louis	6

elements of this plume included fine Co, Se, As, and Sb and coarse Sn, Cd, and Sr. The drop of fine Pb and Br during the plume is unexplained.

A second passage of the St. Louis plume occurred on August 8 from 1600–1800. Transport time was 6–8 hours (see FIGURE 6). The first parameters to rise were F-11 TNMHC, CO, fine K, Fe, Cu, Zn, Ti, Pb, and Br, and coarse Fe. These parameters were followed two hours later by SO_2, fine S and Mn, coarse Ca, mass, NO_2, and b_{scat}. Note that SO_2, fine S, and fine Mn arrived as the delayed peaks this time, instead of as initial peaks. This was probably due to spatial separation of the sources and a changing wind field, although the accuracy of the wind trajectories is not sufficient to permit verification. It also implies that the manganese emissions occur near the sulfur emissions. This second St. Louis plume also demonstrates that b_{scat} can be dramatically affected, even when the humidity is well below the 80% deliquescent point of ammonium sulfate aerosol. Humidity during the b_{scat} peak was 49%, and yet surface area of 0.1–1.0 μm aerosol rose by about 85%. Clearly, water was not responsible for the rise of b_{scat}. TABLE 2 estimates that XSO_4 accounted for 55% of the increase in mass loading. Trace elements of this plume included fine Co, Ga, Se, Sn, and Sb and coarse Cr, Sr, Cd, and Sn.

The third St. Louis plume episode occurred on August 9, 1975, from 0400–0600, with transport times of 6–9 hours (see FIGURE 7). Initial peaks were seen in CO, b_{scat}, NO_2, fine S, P, K, Fe, Cu, Zn, Ti, Mn, Br, and Pb, and coarse Ti. Delayed peaks included fine Al, mass, F-11, and TNMHC. This plume exhibited very strong peaks of CO, fine Pb, F-11, and TNMHC yet, unlike other St. Louis plumes, showed no SO_2 and, by microscopy, a somewhat decreased loading of fly ash. Trace elements included fine Se, As, and Sn and coarse Sn and Sr.

The last of the St. Louis urban plume episodes was by far the most dramatic. It occurred on Saturday, August 9, 1975, from 2000–2200. The wind trajectory indicates possible strong influence of the 2400 MW Labadie power plant (see FIGURE 8). Transport time was about six hours. Initial peaks included fine Si, P, S, K, Fe, Cu, Zn, and Mn, coarse Al, Si, P, K, Ca, Fe, and Ti, mass concentration, O_3, TNMHC, SO_2, and b_{scat}. Delayed peaks were seen in fine Al, Pb, and Br, CO, F-11, NO, and NO_2. The most notable rises were those of fine sulfur (170%), b_{scat} (570%), and 0.1–1.0 μm surface area (280%), compared to their values six hours earlier. Humidity during this peak averaged 75%, near, but somewhat below, the deliquescent point of ammonium sulfate. Microscopy showed that sulfate particles had increased greatly in size and abundance. Trace elements of the plume were fine As, Se, Sn, and Sb and coarse As and Sn.

Correlation coefficients were computed among all measured parameters for the six days of data, including some from associate investigators. Key parameter and selected coefficients are shown in TABLES 3 and 4. Among the obvious highlights are the strong correlations between fine sulfur and both mass concentration and b_{scat}. Fine Pb was well correlated with CO, F-11, and TNMHC. Sulfur dioxide exhibited poor correlations because of its undetectable base-line values and because it was not present at all during one of the major St. Louis plume episodes. Nevertheless, SO_2 did increase to measurable levels during many of the major plume episodes. None of the nonmeteorological parameters correlated well with either wind direction or wind speed. Common mineral elements (Si, Ca, Al, Fe, K) showed moderate coefficients with one another. The high correlation between fine P and fine S was unexpected. Electron microscopy was employed to attempt to locate phosphorus-containing particles but met with little success because of the surface roughness of the cellulose filter.

Size distributions during plume episodes showed, as expected, well-aged aerosol. Particles below 0.1 μm diameter rarely accounted for more than 10–20% of the total

TABLE 3

KEY PARAMETER CORRELATION COEFFICIENTS

	SO_2	CO	F-11	TNMHC	Mass	b_{scat}	Fine Pb	Fine S
SO_2	1.00	0.17	0.34	0.18	0.33	0.14	0.24	0.32
CO		1.00	0.73	0.73	0.51	0.60	0.86	0.54
F-11			1.00	0.71	0.42	0.47	0.91	0.45
TNMHC				1.00	0.26	0.35	0.84	0.24
Mass					1.00	0.79	0.41	0.93
b_{scat}						1.00	0.45	0.87
Fine Pb							1.00	0.43
Fine S								1.00

TABLE 4
SELECTED CORRELATION COEFFICIENTS

O_3 & SO_2: 0.48	Mass & fine Si: 0.62
O_3 & fine S: 0.61	Mass & coarse Si: 0.72
O_3 & F-11: 0.11	Mass & fine P: 0.75
	Mass & total Al: 0.60
b_{scat} & NO_x: 0.60	Mass & total Si: 0.73
b_{scat} & fine Si: 0.66	
b_{scat} & fine P: 0.72	Coarse Si & coarse Al: 0.62
	Coarse Si & coarse K: 0.74
Fine P & fine S: 0.82	Coarse Si & coarse Ca: 0.68
Fine P & mass: 0.75	Coarse Si & coarse Fe: 0.56
Fine P & b_{scat}: 0.72	
Fine P & F-11: 0.48	Fine S & coarse Al: 0.49
	Fine S & coarse Si: 0.65
Fine Zn & F-11: 0.56	Fine S & coarse K: 0.23
Fine Zn & fine Pb: 0.63	Fine S & coarse Ca: 0.36
Fine Zn & fine S: 0.31	Fine S & coarse Fe: 0.37
Fine Mn & F-11: 0.51	
Fine Mn & fine Fe: 0.69	
Fine Mn & fine Pb: 0.56	
Fine Mn & fine S: 0.50	

aerosol surface area, and rarely more than 5% of total aerosol volume. Particles between 0.1 and 1.0 μm accounted for most of the remainder of total surface area and typically 60–80% of total aerosol volume. The modes of 0.1 to 1.0 μm aerosol were at about 0.2 μm for surface and about 0.3 μm for volume. Aerosol larger than 1.0 μm formed a volume mode at about 5–7 μm, and microscopy showed the particles to be primarily minerals and fly ash.

TABLE 5 presents concentrations and the ratios of fine to total for all detectable elements during a 42-hour period that included all four St. Louis urban plume episodes. It presents averages during the plume episodes and at other times. Bromine values are omitted because of volatilization loss during filter storage.[7] Since a certain amount of subjectivity was introduced in defining when plume episodes began and ended, TABLE 5 provides only estimates of elemental loadings and of the ratios of fine to total.

Sulfate values from glass fiber filters have not been used much in this analysis because of the probable errors in such values during SO_2 peaks. This likelihood leads to a question of whether or not mass concentration values (derived from the same filters) are anomalously high at times. The possibility exists; however, it would only enhance the contribution of sulfur and sulfate to mass loadings (as presented in TABLE 2), since increases in mass loadings during plumes would then be smaller.

Sulfur appears to have been the most prevalent element by mass for elements above atomic number 12. This is true both when urban plumes were present and when they were not. Typically, 95% of the sulfur occurs in particles less than 2.5 μm diameter. During the four major St. Louis plume episodes, sulfur accounted for about 10% of the total mass loading. Assuming the sulfur occurred as sulfate, 30–40% of the total mass loading during a plume was due to sulfate and its cations. Looking at the increase in mass loading due to the plume, sulfate and its cations accounted for

40–55% of the mass increase. The plume from Chicago and the second one from Indianapolis showed about half the contribution of sulfur and sulfates to mass loadings compared with the plumes from St. Louis, probably due to dispersion and removal during the longer transport. Since the sulfur compounds tend to reside in the size range of < 2.5 μm, they contribute heavily to the light scattering properties of the aerosol.

The elements Si, Ca, K, Al, and Fe occurred predominantly as coarse particles. Concentrations of Si, Al, and Ca showed some tendency to rise with the arrival of an urban plume, but they were not reliable indicators. Fe and K were more reliable, partially because both had a larger fraction of their mass as fine particles. Microscopy

TABLE 5

UNFRACTIONATED ELEMENTAL CONCENTRATIONS FROM 0600, AUGUST 8 to 2400, AUGUST 9, 1975

Element	Average during St. Louis Plumes (μg/m^3)	Percent <2.5 μm	Average at Other Times (μg/m^3)	Percent <2.5 μm
S	8.5	98	4.7	98
Si	1.03	7	0.92	7
Ca	1.21	5	1.21	11
K	0.27	36	0.23	35
Al	0.53	15	0.35	4
Fe	0.53	23	0.36	21
Pb	0.28	90	0.16	90
P	0.116	96	0.073	83
Mn	0.023	100	0.011	100
Zn	0.074	90	0.064	81
Cu	0.018	100	0.014	100
Ti	0.041	34	0.020	0
Sn	0.012	50	*	—
Se	0.005	100	*	—
Sb	0.014	100	*	—
Co	0.008	100	*	—
As	0.015	80	*	—
Sr	0.008	0	*	—
Cr	0.017	30	*	—
Cd	0.008	0	*	—
Ba	no peak	—	0.141	50

*Not detectable.

of two St. Louis plume episodes showed the coarse particles to be largely mineral dust, with a mode about 5–7 μm. Fly ash was also seen by microscopy and was estimated to be a significant mass contributor. Its size ranged from 0.2–30 μm, with a mode at about 1 μm. Even accounting for its density greater than unity, much of the fly ash was collected as fine (< 2.5 μm) aerosol by the dichotomous sampler.

Minor components of the aerosol included Pb and Br, which are believed due to automobile emissions and which occurred primarily in the fine fraction. Fine P, Mn, and Zn were other elements that were usually detectable and peaked during plume passages. Coarse and fine Ti and fine Cu displayed peaks during a few of the urban plumes. Barium was often detectable in coarse and fine fractions but showed no correlations with urban plumes.

Trace elements that were usually detectable only during urban plumes included fine Sn, Se, Sb, Co, and As and coarse Sn, Sr, Ge, and Cd.

CONCLUSIONS

1. Gaseous and elemental aerosol concentrations at a ground site 100 km from a major urban area were clearly influenced by the latter's urban plume. There is strong evidence that urban plumes are detectable at 350 km from the urban area.

2. Sulfur compounds accounted for a significant portion (30–40%) of the mass loading within the St. Louis urban plume. These compounds, presumably sulfate, reside almost entirely in the size range below 2.5 μm.

3. For our ground station, the most reliable tracers of an urban plume were fine Pb, fluorocarbon-11, total nonmethane hydrocarbons, and carbon monoxide. Fine S and SO_2 were ambiguous indicators because of the rural locations of many major SO_2 sources. Total mass concentration showed some reliability for 100 km transport but not for 350 km. Ozone and b_{scat} are ambiguous indicators for a ground station because of natural ozone production and humidity effects, respectively.

4. Evidence was seen of a several-day-long regional buildup of fine sulfur, b_{scat}, mass, ozone, and NO_x and, to a lesser extent, carbon monoxide and fluorocarbon-11.

SUMMARY

A field study was conducted 100 km north of St. Louis, Mo. in August 1975 to measure aerosol composition and gaseous concentrations of urban plumes passing the site. Coarse and fine aerosol elemental concentrations, b_{scat}, meteorological data, and concentrations of SO_2, CO, O_3, and NO–NO_x were measured and analyzed with fluorocarbon-11, hydrocarbon, and size distribution data from associate investigators. Urban plumes were identified by simultaneous rises of at least several of the following: fluorocarbon-11, CO, total nonmethane hydrocarbons, fine Pb, fine S, and, to a lesser extent, total mass concentration and b_{scat}. The urban plume of St. Louis was clearly seen several times, and those of Chicago and Indianapolis are believed to have been detected. Fine aerosol sulfur accounts for a significant portion of the St. Louis urban plume. Assuming the sulfur occurs as sulfates, about 30–40% of the plume mass is sulfate and its cations. These sulfur compounds reside almost entirely in the size range below 2.5 μm. Other significant mass components that increased were Si, Al, K, and Fe, which occurred mainly as coarse particles (>2.5 μm), and Pb and P, which appeared mostly as fine particles. Minor components that peaked included Mn, Zn, Cu, and Ti. Trace elements usually detectable only during a plume included Se, Sb, Sn, As, and Co in the fine sizes and Sn, Sr, Cd, and Cr in the coarse.

ACKNOWLEDGMENTS

The author acknowledges the following contributions to this project: W. E. Wilson (EPA) for project planning and financial and administrative support; T. G. Dzubay (EPA) for the x-ray fluorescence sampler and analysis; the Particle Technology Lab of the University of Minnesota for *in-situ* aerosol analysis; D. Coventry (EPA) for the wind trajectory calculations; and the staff of EPA's Aerosol Research Branch for operating and supporting the field laboratory.

References

1. Wilson, W. E., R. J. Charlson, R. B. Husar, K. T. Whitby & D. Blumenthal. 1977. Sulfates in the Atmosphere. Report No. EPA-600/7-77-021. U.S. Environmental Protection Agency. Research Triangle Park, N.C.
2. Liu, B. Y. H. & D. Y. H. Pui. 1974. J. Colloid Interface Sci. **47:** 155.
3. Liu, B. Y. H. & D. Y. H. Pui. 1975. J. Aerosol Sci. **6:** 249.
4. Stevens, R. K., T. G. Dzubay, G. Russworm & D. Rickel. 1978. Atmos. Environ. **12:** 55.
5. Rasmussen, R. A., R. Chatfield & M. Holdren. 1977. Hydrocarbon and Oxidant Chemistry Observed at a Site near St. Louis. Report No. EPA-600/7-77-056. U.S. Environmental Protection Agency. Research Triangle Park, N.C.
6. Heffter, J. L., A. D. Taylor & G. J. Ferber. 1975. A Regional–Continental Scale Transport, Diffusion, and Deposition Model. NOAA Technical Memorandum ERL-ARL-50. National Oceanic and Atmospheric Administration. Silver Springs, Md.
7. Dzubay, T. G., R. K. Stevens & L. W. Richards. 1979. Composition of aerosols over Los Angeles freeways. Atmos. Environ. **13:** 653.

AEROSOL FORMATION IN URBAN PLUMES OVER LAKE MICHIGAN*

David F. Miller†

Department of Chemistry
Battelle's Columbus Laboratories
Columbus, Ohio 43201

Abdul J. Alkezweeny

Department of Atmospheric Sciences
Battelle's Pacific Northwest Laboratories
Richland, Washington 99352

INTRODUCTION

Several major U.S. studies are in progress to obtain more definitive information on the transformation of SO_2 to sulfate aerosols. To accommodate modeling efforts aimed at predicting sulfate levels over large geographical areas, relatively simple empirical rate expressions have been idealized. Achievement of that goal is presently remote, for the fact is that very little sound information exists with which to formulate any useful expression for describing SO_2 oxidation as a dependent variable.

Smog chamber experiments conducted in the absence of meteorological factors suggest that the rates for the homogeneous routes of SO_2 oxidation depend most strongly on the nonmethane hydrocarbon (NMHC)–nitrogen oxide (NO_x) ratios.[1] Kinetic modeling of gas phase atmospheric chemistry indicates that SO_2 oxidation rates are proportional to the concentrations of three key radicals, OH, HO_2, and RO_2,[2,3] which in turn are coupled in a complex fashion to carbon-hydrogen-nitrogen precursors, namely hydrocarbons, CO, and nitrogen oxides.

Studies of SO_2 oxidation under actual atmospheric conditions are quite limited in details, and in many cases, data have not been obtained for precursor conditions. Thus, it is understandable that a consensus has not been formed about the factors affecting SO_2 oxidation rates from the set of reports, TABLE 1, that place the oxidation rates from near zero to upwards of 32%/hr.

As might be expected, the highest oxidation rates have been reported for the more highly polluted conditions that occur in urban areas. For example, Roberts and Friedlander[5] reported maximum oxidation rates of 12 to 13%/hr for the Los Angeles basin. Alkezweeny and Powell[7] reported maximum oxidation rates of 10 to 12%/hr in the urban plume of St. Louis. Breeding *et al.*[9] reported SO_2 half-lives corresponding to fractional oxidation rates of 5–30%/hr. In their study, the highest rate was measured near noon, and the NMHC concentration that day was a maximum (750 ppb) for their study period.

*Sponsored by the U.S. Department of Energy under the MAP3S program. Contract EY-76-C-06-1830.

†Present affiliation: Atmospheric Science Center, Desert Research Institute, P.O. Box 60220, Reno, Nevada 89506.

0077–8923/80/0338–0219 $01.75/1 © 1980, NYAS

TABLE 1

CONVERSION RATES OF SO_2 TO SULFATE (% hr^{-1})

Range	Average	Location	Reference
6–25	16.6	Rouen, France	Benarie *et al.*[4]
1.2–13.0	7.1	Los Angeles, California	Roberts and Friedlander[5]
31	31	Budapest, Hungary	Meszaros *et al.*[6]*
10–14	12	St. Louis, Missouri	Alkezweeny and Powell[7]
8–11.5	9.8	St. Louis, Missouri	Alkezweeny[8]
5.3–32	16.4	St. Louis, Missouri	Breeding *et al.*[9]†
1–9	4	Milwaukee, Wisconsin	This study
0.6–4.4	1.68	Arnhem-Armsterdam, Netherlands	Elshout *et al.*[10]
0.28–1.73	0.72	Western Europe	Eliassen *et al.*[11]
0.73	0.73	Ohio Valley	Lavery *et al.*[12]

*Calculated from their half-life data.
†Calculated from their data by Alkezweeny.[7]

The work of White *et al.*[13] led to somewhat lower estimates of SO_2 oxidation rates for the St. Louis plume. Isaksen's[14] kinetic analyses of White's data indicated rates of about 5%/hr and a diurnally integrated conversion of about 25%. Data collected in conjunction with the SURE program have led to even lower estimates of the oxidation rate (~ 0.7%/hr) for the Ohio Valley.[12]

It is clear that part of the reason for the apparently divergent rates results from comparing reports where, in one case, SO_2 oxidation rates are maxima derived over 1-hr periods, while in other cases, the rates represent maxima of averages taken over periods of a day or more. Nonetheless, a disconcerting amount of discrepancy remains among the rate data, particularly for polluted urban air.

In most of the studies reviewed, SO_2 oxidation rates were adjusted to account for expanding mixing volumes and for sulfur losses due to deposition. In some cases, entrainment of fresh pollutants from local sources also appeared to have influenced the data.

In this paper, we report results of field measurements designed to minimize those complications. As part of the MAP3S program, this study has focused on measurements of sulfate aerosol formation in plumes originating from Milwaukee, Wisconsin, and transported over Lake Michigan.

EXPERIMENTAL

During the summer months, the temperature of Lake Michigan is usually well below the daytime air temperature, and this cooling effect, together with a smooth lake surface, generally results in stable offshore air flows. That feature, as well as isolation from downwind pollutant sources, creates an advantageous situation for studying atmospheric reactions. Under such conditions, dry deposition and diffusion become rather insignificant parameters and can generally be ignored in the interpretative processes.

Aerometric data were collected on board a DC-3 aircraft specially equipped for air surveillance. Real-time measurements were made for temperature, dew point, wind speed, wind direction, standard aircraft parameters (altitude, air speed, position), O_3, SO_2, NO, NO_x, total light scattering, particles > 0.01 μm, and total condensation nuclei concentration (CNC). Time-integrated air samples were also captured in Tedlar bags and sent to the U.S. Environmental Protection Agency for hydrocarbon analyses. Three high-volume samplers were used for aerosol collections. Collections for $SO_4^=$ were taken on IPC filters and analyzed by x-ray fluorescence. Selected aerosol samples were also analyzed for trace metals, nitrate, and ammonium. A twin-engine Cessna sometimes flew simultaneous missions and was used to collect high-volume aerosol samples and SO_2 data, upwind and outside the plumes.

Results and Discussion

Aerometric Data and SO_2 Oxidation Rates

The experiments to be described involved measurements of the Milwaukee plumes (August 27, 1976, August 28, 1976, and July 20, 1977) as they were transported by westerly winds. The location of each plume cross section from these flights is identified in FIGURE 1. On July 20, 1977, data were also obtained for a power plant plume within the urban plume.

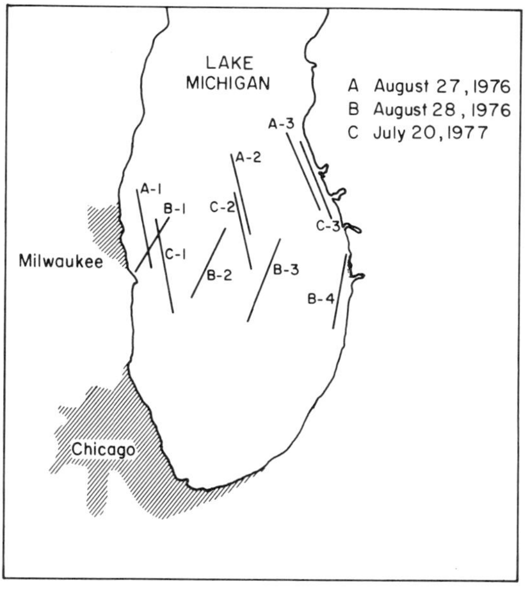

FIGURE 1. Location of urban plume cross sections as monitored by aircraft.

Detailed hydrocarbon data were obtained for the flights on August 27 and 28, and some exemplary data for these cases are shown in TABLE 2. Hydrocarbon data were not obtained for July 20, and in order to provide interpretations for that particular event, it was necessary to use data obtained on other days having similar conditions.

TABLE 2

REPRESENTATIVE HYDROCARBON DISTRIBUTIONS OF SAMPLES COLLECTED ON AUGUST 27 AND 28, 1976

Hydrocarbon	Concentration, ppb (vol/vol)	
	August 27*	August 28†
ethane	7.5	2.3
ethylene	9.5	1.4
propane	9.7	1.3
acetylene	4.0	1.9
isobutane	1.3	0.4
n-butane	5.9	1.5
propylene	1.3	0.7
isobutylene	1.0	0
t-butene-2	1.7	0
c-butene-2 plus butadiene	1.0	0
isopentane	5.8	1.3
n-pentane	2.3	0.5
pentene-1	0.8	0.1
cyclopentane	2.4	1.2
2 methylpentane	1.3	0.5
3 methylpentane	0.5	0.2
n-hexane	0.5	0.4
2,4 dimethylpentane	3.8	1.9
2,3 dimethylpentane	6.5	4.3
3 methylhexane	0.4	0.2
2,2,4 trimethylpentane	10.0	6.8
n-heptane	0.3	0.2
m-methylcyclohexane	1.0	0.2
toluene	5.6	7.7
nonane	0.7	0.5
ethylbenzene	0.4	0.4
m,p-xylene	1.6	2.9
o-xylene	0.5	0.8
isopropylbenzene	0.1	0.1
m,p-ethyltoluene	0.3	0.3
1,3,5 trimethylbenzene	0.2	0.1
1,2,4 trimethylbenzene	0.5	0.9
Total Concentration, ppb (v/v)	88.4	41.0
Total Concentration, ppbC	420	245
Average Molar reactivity	15.6	14.7

*Average concentrations for cross section No. 1 for samples collected at three altitudes: 2000, 2500, and 3000 ft (MSL).

†Average concentrations for cross section No. 2 for samples collected at two altitudes: 1000 and 1500 ft (MSL).

TABLE 3

SUMMARY OF DATA COLLECTED ON AUGUST 27 AND 28, 1976

Transport Time (hr)	Downwind Section	NMHC (ppbC)	NMHC Reactivity*	NO_x (ppb)	SO_2 (ppb)	O_3 (ppb)	SO_4^{2-} ($\mu g/m^3$)	b_{scat} ($10^{-4}/m$)
		Milwaukee flight 8-27-76, noon:						
0	1	420	15.6	5.4	17.5	70	14.7	1.53
1.5	2	274	—	5.6	14.0	91	15.9	1.78
2.5	3	219	—	4.8	13.6	105	24.0	2.16
		Milwaukee flight 8-28-76, noon:						
0	1	270	14.7	9.3	22.4	13	1.7	0.21
1	2	245	—	7.2	19.7	24	2.2	0.21
2	3	227	—	6.2	18.3	48	1.9	0.21
3	4	220	—	5.2	14.9	49	0.9	0.20

*Reactivity values are averages for the hydrocarbon mixtures, and they correspond to the average reaction rate of hydrocarbons with OH radicals.

Pertinent data from the experiments of August 27 and 28, 1976, are summarized in TABLE 3. Gross rates of SO_2 oxidation to sulfate aerosol computed from these data are summarized in TABLE 4.

TABLE 4

HOURLY AVERAGED RATES OF SO_2 OXIDATION FOR URBAN PLUMES TRANSPORTED ACROSS LAKE MICHIGAN

Reaction Time (hr)	Cross Section Interval	R_{SO_2}—S* (%/hr)	R_{SO_2}—b_{scat}† (%/hr)
	Milwaukee flight 8-27-76, noon:		
1.5	1–2	4.0	8.6
1.0	2–3	11.7	9.3
2.5	1–3	7.1	8.8
	Milwaukee flight 8-28-76, noon:		
1.0	1–2	0.89	0.70
1.0	2–3	−0.19	0.39
1.0	3–4	−1.09	0.93
3.0	1–4	−0.13	0.67

*Rate based on time-integrated sulfate aerosol collections and time-averaged SO_2 concentrations.

†Rate based on sulfate levels inferred from light scattering data and time-averaged SO_2 concentrations.

The transformation rates are expressed in terms of a first-order mechanism:

$$\frac{d}{dt}(SO_4^{=}) = -kSO_2$$

$$k = -\left[\frac{\alpha_2 - \alpha_1}{\beta_2 - \beta_1} \ln \frac{\beta_1}{\beta_2}\right] / (t_2 - t_1)$$

where:

$$\alpha = SO_4^=/S_t, \quad \beta = SO_2/S_t, \quad S_t = SO_4^= + SO_2$$

and t is time in a Lagrangian sense.

As noted earlier, sulfate concentrations were determined by x-ray fluorescence analyses of time-integrated aerosol collections on filters. Because in some instances the sulfate filter data were incomplete or suspect, oxidation rates were also computed from sulfate concentrations inferred from time-averaged nephelometry data. White and Roberts[15] showed that for Los Angeles, the light scattering response to sulfate aerosol amounted to about 0.22×10^{-4} m^{-1} per ppb sulfate. From plume studies in St. Louis, Husar *et al.*[16] found the relationship to be about 0.19×10^{-4} m^{-1} per ppb sulfate. We arbitrarily adopted a relationship of 0.20×10^{-4} m^{-1} per ppb S particulate. Rates of

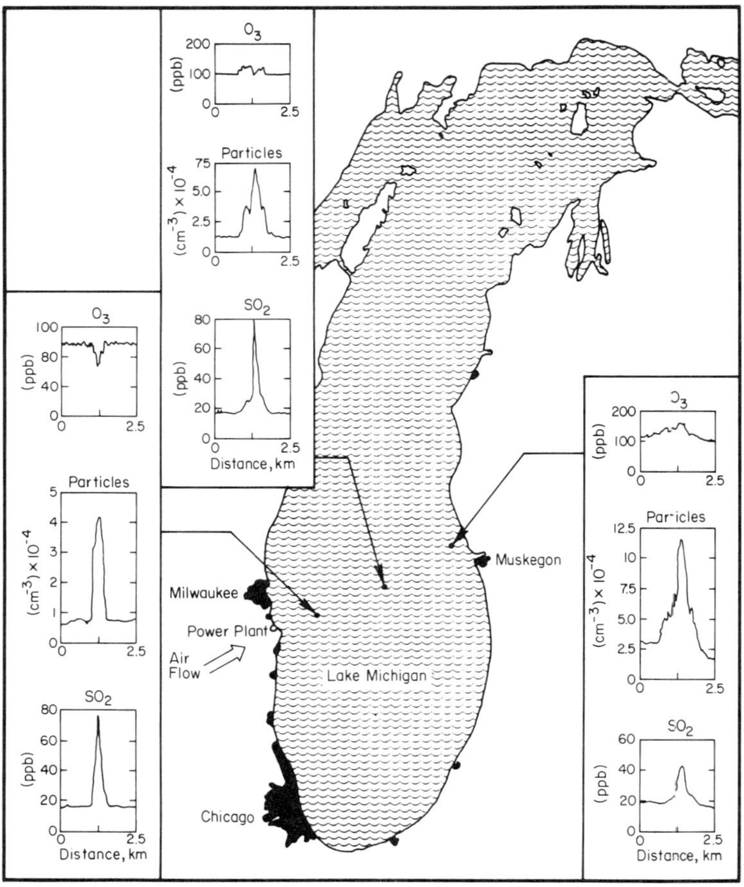

FIGURE 2. Recordings of O_3, particles (> 0.01 μm), and SO_2 concentrations in the downwind cross sections of a power plant plume transported across Lake Michigan.

TABLE 5

AVERAGE SPECIE CONCENTRATIONS IN A DISPERSED POWER PLANT PLUME AND AMBIENT AIR

Transport Time of Power Plant Plume (hr)	Cross Section Interval	NO_x (ppb)	O_3 (ppb)	SO_2 (ppb)	SO_4^{2-} (ppb)	b_{scat} $(10^{-4}/m)$
1	Section 1					
	Total Plume	20.4	89	42	—	4.17
	Ambient	10	95	15	34.4	4.11
	Net Plume	10.4	−6	27	—	0.06
2.5	Section 2					
	Total Plume	15.4	112	28	—	3.90
	Ambient	10	98	14	37.4	3.50
	Net Plume	5.4	14	14	—	0.40
4	Section 3					
	Total Plume	14.4	130	25	—	3.60
	Ambient	10	102	14	—	3.12
	Net Plume	4.4	28	11	—	0.48

SO_2 oxidation calculated by this relationship appear in the right-hand column of TABLE 4.

As indicated in TABLE 4, the SO_2 oxidation rates based on sulfate and light scattering data are comparable but not identical. It seems safe to conclude that on the afternoon of the 27th, the average oxidation rate was in the range of 6–8%/hr, while for the same period on the following day, the average rate was < 1%/hr. That, indeed, is a dramatic difference in SO_2 transformation rates, which, as discussed further on in the paper, we cannot adequately account for on the basis of differences in meteorological effects and/or differences in the measured quantities of precursors. Clear skies prevailed on both days; the average temperature and relative humidity were 28°C/58% and 25°C/56%, respectively, for the 27th and 28th. Gas chromatographic data suggest that fresh combustion-related hydrocarbons were present in both downwind air masses at the time of our initial sampling (approximately noon). The most significant difference for the two days appeared to be in the quality of the upwind air masses. The very low concentrations of upwind O_3 and sulfate on the 28th suggest that the air mass passing through Milwaukee that day was very clean; a condition we have normally found with northwesterly transport winds in that area. In contrast, the southwesterly flow of air moving through Milwaukee on the 27th seemed typical of well-aged polluted air, as evidenced by the high levels of O_3, particulate sulfate, and light scattering. It is conceivable that important differences in unmeasured pollutants (e.g., HONO, aldehydes, H_2O_2) existed—between the relatively clean Canadian air and the contaminated midwestern air—that may have had profound effects on the rate and intensity of photochemical smog formation, including SO_2 oxidation.

On July 20, 1977, aerometric data were obtained for a coherent power plant plume in transport within the Milwaukee plume; again, the airflow was from the southwest. Recordings of SO_2, O_3, and particle concentrations for three downwind cross sections of the plumes are reproduced in FIGURE 2. Average values for light scattering (b_{scat}) and for SO_2, NO_x, and O_3 levels are presented in TABLE 5. Rates of SO_2 oxidation (calculated as described earlier) for the ambient air and for the power plant plume (net plume) are given in TABLE 6.

As indicated in TABLE 5, only a limited amount of sulfate data was available from this experiment, and it was obtained for the first two cross sections of ambient air. Rates of SO_2 oxidation for the power plant plume were calculated, using average cross-sectional values for light scattering.

Note that during the course of transport across the lake, nephelometer readings for the ambient air diminished with time, while the average readings for the total plume (ambient plus power plant plume) increased with age (see TABLE 5). Assuming a relatively constant mixing height for this period, the nephelometry data might normally be interpreted to say that the SO_2 oxidation rate was insignificantly low in the ambient air. In contrast, the limited sulfate data indicate a substantial SO_2 oxidation rate of about 4%/hr for the ambient condition. Based on other indicators of photochemical smog activity that day, we expect that the latter rate is more likely correct. The gradual decline in ambient light scattering could possibly be related to subtle changes in the size distribution of sulfate aerosols or to changes in aerosol volume due to reduced relative humidities in the afternoon. It is clear from the nephelometry data that during the sampling period, and particularly after the power plant plume had aged for > 1 hr, the overall rate of production of sulfate in the power plant plume was greater than in the ambient air. Possible reasons for that situation are discussed further in the section on modeling.

TABLE 6

RATES OF SO_2 OXIDATION IN A POWER PLANT PLUME AND AMBIENT AIR

Reaction Time (hr)	Cross Section Interval	R_{SO_2}—Ambient* (%/hr)	R_{SO_2}—Net Plume† (%/hr)
1.5	1–2	4.0	8.2
1.5	2–3	—	4.3
3.0	1–3	—	6.2

*Rate based on sulfate and SO_2 data.
†Rate based on nephelometer and SO_2 data.

Kinetic Modeling

In an attempt to give cause for the various oxidation rates observed in these experiments, a chemical kinetics model was used to simulate the pollutant conditions. The chemical mechanism, listed in TABLE 7, is similar to the type employed for smog chamber studies, and although the code successfully replicates the general features of photochemical smog, it consists of many reactions and rate constants that are uncertain.

The code is based upon the assumption of a well-mixed system, and it can emulate plume conditions only by diluting the plume parcel as time progresses. It is well

Table 7

Description of Kinetic Model

Reaction	Rate Constant*	Units†
NO_2 = NO + O	.270E + 00	min^{-1}
NO + NO_2 + H_2O = 2HONO	.220E − 08	$ppm^{-2}\ min^{-1}$
HONO + HONO = NO + NO_2	.140E − 02	
HONO + = NO + HO	.540E − 01	min^{-1}
NO + O_3 = NO_2	.230E + 02	
NO + HO = HONO	.890E + 04	
NO_2 + HO = $HONO_2$	.190E + 05	
CO + HO = HO_2	.440E + 03‡	
NO + HO_2 = NO_2 + HO	.300E + 04	
NO_2 + HO_2 = HO_2NO_2	.120E + 04	
HO_2NO_2 = NO_2 + HO_2	.670E + 01§	min^{-1}
NO_2 + O_3 = NO_3	.460E − 01	
NO_3 + NO = $2NO_2$	.130E + 05	
NO_3 + NO_2 = N_2O_5	.560E + 04	
N_2O_5 = NO_3 + NO_2	.170E + 02§	min^{-1}
NO + NO = $2NO_2$	.150E − 03	
NMHC + HO = RO_2	.140E + 05‡	
CH_4 + HO = RO_2	.120E + 02‡	
ALD = 0.75 HO_2 + 0.25 RO_2 + CO	.300E − 02‡	min^{-1}
ALD + HO = 0.50 HO_2 + 0.5 RCO_3 + 0.5 CO	.100E + 05‡	
RO_2 + NO = NO_2 + RO	.400E + 04	
RCO_3 + NO = NO_2 + RO_2	.400E + 04	
RCO_3 + NO_2 = PAN	.150E + 04	
PAN = NO_2 + RCO_3	.270E − 01§	min^{-1}
RO + NO = RONO	.250E + 03	
RO_2 + NO_2 = RO_2NO_2	.160E + 04	
RO_2NO_2 = RO_2 + NO_2	.670E + 01§	min^{-1}
RO_2NO_2 = NITRATE	.400E − 01¶	min^{-1}
RONO = RO + NO	.810E − 01	min^{-1}
HO_2NO_2 = NITRATE	.400E − 01¶	min^{-1}
HO_2 + HO_2 = H_2O_2	.830E + 04	
HO_2 + RO_2 = RO_2H + 0.5 RO + 0.5 HO	.680E + 03‡	
RO_2 + RO_2 = RO + 0.5 ALC + 0.5 ALD	.680E + 03‡	
RO = HO + ALD	.500E + 04‡	
H_2O_2 = 2HO	.150E − 02	min^{-1}
HO + HO_2 = H_2O	.120E + 05	
HO + HO = H_2O_2	.650E + 05	
NO + O = NO_2	.300E + 04	
NO_2 + O = NO_3	.300E + 04	
NO_2 + O = NO	.150E + 05	
ALD + O = HO + HO_2 + 0.5 RCO_3	.740E + 03‡	
O = O3	.330E + 07	
N_2O_5 + H_2O = $2HONO_2$	.100E − 04	
N_2O_5 = NITRATE	.400E − 01¶	min^{-1}
O_3 = O1D	.227E − 02	min^{-1}
O1D = O	.735E + 11	
O1D + H_2O = 2HO	.510E + 06	
O_3 + HO_2 = HO	.150E + 01	
H_2O_2 + HO = HO_2	.120E + 04	
HO + SO_2 = SULFATE + HO_2	.160E + 04	
HO_2 + SO_2 = SULFATE + HO	.120E + 01	
RO_2 + SO_2 = SULFATE + RO	.200E + 01	
O + SO_2 = SULFATE	.840E + 02	

*$h\nu$, M, and O_2 assumed.

†Units are $ppm^{-1}\ min^{-1}$ unless noted.

‡Surrogate reaction; rate constant represents rate limiting step.

§Thermal decomposition.

¶Heterogeneous reaction on aerosol surface—hypothetical.

recognized that the applicability of such one-dimensional models is severely limited when complex meteorological conditions exist.

To simulate the events on August 27 and 28, the detailed hydrocarbon data were lumped into a single term and assigned a rate constant with OH based on the average rate constant (reactivity)[17] computed for the entire distribution. Simulations were performed in two steps. In order to simulate P.M. conditions for which input data were available, it was necessary to first simulate the A.M. period; this provides, in theory, the intermediate species that must carry into the P.M. simulation to avoid creating an unrealistic induction period at the start of a P.M. simulation. Thus, for each day, the code was executed twice; the second time with the specific chemical data obtained for the first downwind cross section and the theoretical data tailored to support those conditions. For example, to simulate the afternoon of August 27, 1976, the code was initiated with the following concentrations of input species.

The emphasized species were observed; the remainder are theoretical. Concentration units are ppm (vol/vol), and the notation 0.66 (−3) represents 6.6×10^{-2}. **NMHC** 0.88(−1), **CO** = 0.70, **NO** = 0.10(−2), $\mathbf{NO_2}$ = 0.42(−2), $\mathbf{O_3}$ = 0.70(−1), $\mathbf{SO_2}$ = 0.17(−1), $\mathbf{H_2O}$ 0.20(+5), $\mathbf{CH_4}$ = 0.17(+1), ALD = 0.30(−1), HO_2NO_2 = 0.80(−4), HONO = 0.17(−4), N_2O_5 = 0.23(−5), RO_2NO_2 = 0.84(−4), HO = 0.15(−6), HO_2 = 0.96(−4), RO_2 = 0.79(−4), and H_2O_2 = 0.47(−2).

Results for the August 27 simulation are shown in the top panel of FIGURE 3, together with the corresponding data observed by aircraft sampling. The observed rates of SO_2 oxidation (FIGURE 3) are averages derived from the nephelometry data.

The modeling results for O_3 (denoted O_3-Theory, FIGURE 3) compare reasonably well with the atmospheric data (O_3-Observed). However, the model underestimates the rate of SO_2 oxidation by nearly a factor of two. In this particular case, the simulation results indicate that peroxy radicals accounted for about 70% of the total SO_2 oxidation rate. We incorporated conservative values for the rate constants for those reactions, based on some of our recent smog chamber findings. The work of Whitbeck *et al.*[18] indicates that the rate of SO_2 oxidation by RO_2 may be more than twice the value we used. Had the higher value been used, the theoretical oxidation rate for SO_2 would have been around 6%/hr.

A more difficult problem of interpreting the urban rate data for SO_2 appeared when attempts were made to simulate the data for August 28, 1976. Some of those results are illustrated in the lower panel of FIGURE 3. In this case, the model slightly overestimates the O_3 levels, but it grossly overestimates the rate of SO_2 oxidation. Under the conditions specified for the 28th, OH radicals played the dominant role in SO_2 oxidation (85% vs. 15% for the peroxy species), according to the simulation results. Thus, it appears that OH concentrations in the simulation became much higher than they might actually have been on the day in question. Based on our present experience in applying this type model to atmospheric data, we must confess that the lone differences observed in the NMHC and NO_x data for the 27th and 28th are not nearly large enough to account for the differences observed in the SO_2 oxidation rates for these two days.

The final simulation to be considered is for July 20, 1977. A 2-step modeling procedure similar to that previously described was also used here. In the first step, the ambient pollution conditions were simulated. In the second step, the power plant plume conditions representing the first downwind cross section were made to integrate with the conditions simulated for the ambient air. The simulation results are shown in FIGURE 4, together with data points for the ozone concentrations measured that afternoon in the power plant plume and ambient air. Simulation conditions were adjusted so that the theoretical ozone profile for ambient air matched the actual ozone

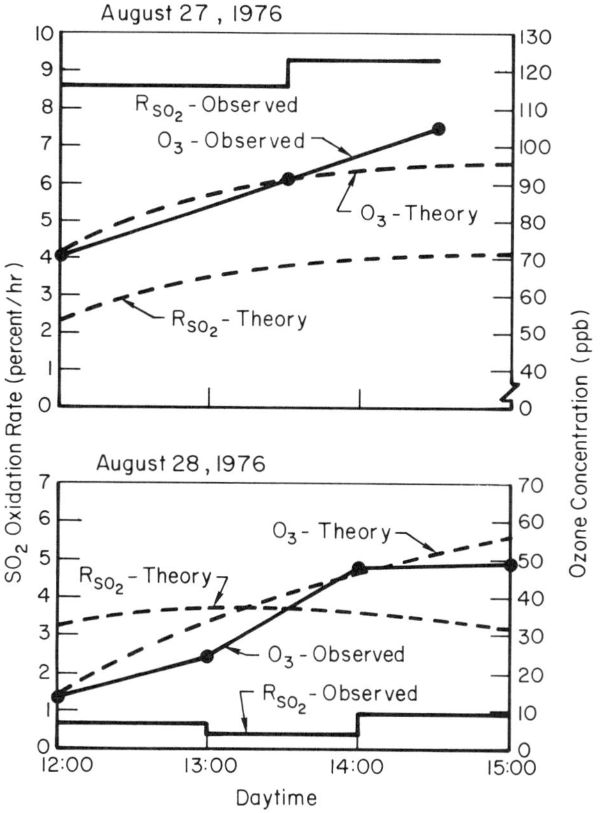

FIGURE 3. Comparisons between observed and theoretical (computer simulation) profiles of the rates of SO_2 oxidation and O_3 concentrations for two days of different photochemical smog intensities.

data for the first cross section. With those conditions set, the predicted rate of SO_2 oxidation reached 4.5%/hr for ambient air (recall that there were insufficient sulfate data with which to check these estimates).

When power plant plume emissions were integrated into the simulation (at a plume age of 1 hr, FIGURE 4), the model predicted an eventual increase in ozone above ambient levels and an increase in the rate of SO_2 oxidation. The rise in O_3 has been accounted for previously on the basis of NMHC and NO_x effects.[19] According to the kinetic theory applied here, the increase in the SO_2 oxidation rate for this particular plume condition is attributed to the role of NO_x in the following way. During the afternoon, when the ambient air mass approaches its peak in O_3, most of the SO_2 oxidation proceeds via the peroxy radicals (HO_2 and RO_2). If, however, there is an influx of precursor pollutants, such as NO, the potential exists for more rapid regeneration of OH and for increased rates of hydrocarbon and SO_2 oxidation via that

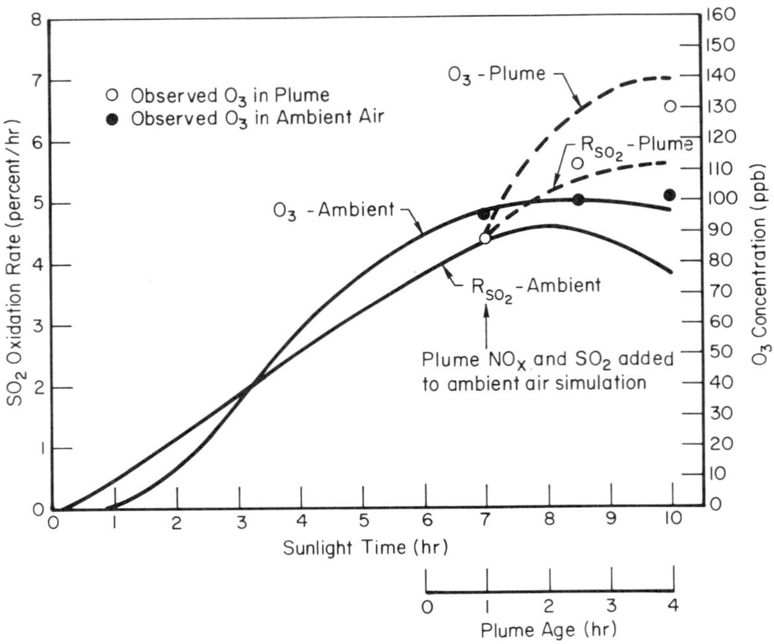

FIGURE 4. Computer simulations of the downwind effects of power plant emissions on the average rates of SO_2 oxidation and O_3 formation.

radical. In the specific simulation conducted here for the plume mixture, OH was responsible for 70% of the SO_2 oxidation rate in the region of the dispersed power plant plume, as opposed to only 30% for the ambient condition.

CONCLUSION

Rates of SO_2 oxidation in isolated urban plumes appear to be quite variable. Kinetic analyses of the data have not yet been successful in accounting for the rate dependence. We feel that much of the problem with the kinetic simulation approach relates to an inadequate understanding of the free radical chemistry that is coupled strongly to the nitrogen oxides cycles. In addition, a more extensive body of data is needed for study episodes, since it appears that the quality of an upwind air mass affects the subsequent rates of SO_2 oxidation when fresh urban-type pollutants become entrained. A preferred analytical approach to this problem would be to measure the free radical (HO_2, RO_2, OH) concentrations in real time, and thereby derive instantaneous rates for SO_2 oxidation.

ACKNOWLEDGMENTS

The authors express their sincere appreciation to Dr. William Lonneman and his staff at the U.S. Environmental Protection Agency for performing the detailed hydrocarbon analyses.

REFERENCES

1. MILLER, D. F. 1978. Precursor effects on SO_2 oxidation. Atmos. Environ. **12:** 273.
2. CALVERT, J. G., F. SU, J. W. BOTTENHEIM & O. P. STRAUSZ. 1978. Mechanism of the homogeneous oxidation of sulfur dioxide in the troposphere. Atmos. Environ. **12:** 197.
3. EGGLETON, A. E. J. & R. A. COX. 1978. Homogeneous oxidation of sulfur compounds in the atmosphere. Atmos. Environ. **12:** 227.
4. BENARIE, M., A. NONAT & T. MENARD. 1972. The transformation of sulphur dioxide into sulphuric acid in relation to the climatology of an urban/industrial area (Rouen, France). Paper presented at the Clean Air Conference, Melbourne, Australia, May 15–18.
5. ROBERTS, P. T. & S. K. FRIEDLANDER. 1975. Conversion of SO_2 to sulfur particulate in the Los Angeles atmosphere. Environ. Health Prospective **10:** 103.
6. MESZAROS, E., D. J. MOORE & J. P. LODGE, JR. 1977. Sulfur dioxide–sulfate relationships in Budapest. Atmos. Environ. **11:** 345.
7. ALKEZWEENY, A. J. & D. C. POWELL. 1977. Estimation of transformation rate of SO_2 to SO_4 from atmospheric concentration data. Atmos. Environ. **11:** 179.
8. ALKEZWEENY, A. J. 1978. Measurement of aerosol particles and trace gases in METROMEX. J. Appl. Meteorol. **17:** 609.
9. BREEDING, R. J., H. B. KLONIS, J. P. LODGE, JR., J. B. PATE, D. C. SHEESLEY, T. R. ENGLERT & D. R. SEARS. 1976. Measurements of atmospheric pollutants in the St. Louis area. Atmos. Environ. **10:** 181.
10. ELSHOUT, A. J., J. W. VILJEER & H. VAN DUUREN. 1978. Sulphates and sulphuric acid in the years 1971–1976 in the Netherlands. Atmos. Environ. **12:** 785.
11. ELIASSEN, A. & J. SALTBONES. 1975. Decay and transformation rates of SO_2, as estimated from emission data, trajectories and measured air concentrations. Atmos Environ. **9:** 425.
12. LAVERY, T. F., G. M. HIDY, R. L. BASKETT & J. THRASHER. 1979. Occurrence of long range transport of sulfur oxides in the northeastern United States. Paper presented at 4th Symposium on Turbulence, Diffusion, and Air Pollution, American Meteorological Society.
13. WHITE, W. H., J. A. ANDERSON, D. L. BLUMENTHAL, R. B. HUSAR, N. V. GILLANI, J. D. HUSAR & W. E. WILSON, JR. 1976. Formation and transport of secondary air pollutants: ozone and aerosols in the St. Louis urban plume. Science **194:** 187.
14. ISAKSEN, I. S. A., E. HESSTVEDT & O. HOV. 1978. A chemical model for urban plumes: test for ozone and particulate sulfur formation in St. Louis urban plume. Atmos. Environ. **12:** 599.
15. WHITE, W. H. & P. T. ROBERTS. 1975. The nature and origins of visibility-reducing aerosols in Los Angeles. Paper presented at the 68th Annual Meeting of the Air Pollution Control Association, Boston, Mass.
16. HUSAR, R. B., D. E. PATTERSON, J. D. HUSAR, N. V. GILLANI & W. E. WILSON, JR. 1978. Sulfur budget of a power plant plume. Atmos. Environ. **12:** 549.
17. DARNELL, K. R., A. C. LLOYD, A. M. WINER & J. N. PITTS. 1976. Reactivity scale for

atmospheric hydrocarbons based on reaction with the hydroxyl radical. Environ. Sci. Technol. **10:** 692.

18. Whitbeck, M. R., J. W. Bottenheim, S. Z. Levine & J. G. Calvert. 1976. A kinetic study of the CH_3O_3 and $(CH_3)_3CO_2$ radical reactions by kinetic flash spectroscopy. In Abstracts of 12th Informal Conference of Photochemistry, National Bureau of Standards. Gaithersberg, Md.
19. Miller, D. F., A. J. Alkezweeny, J. M. Hales & R. N. Lee. 1978. Ozone formation related to power plant emissions. Science **202:** 1186.

SIZE AND COMPOSITION OF VISIBILITY-REDUCING AEROSOLS IN SOUTHWESTERN PLUMES*

Edward S. Macias†‡

Department of Chemistry
Washington University
St. Louis, Missouri 63130

Donald L. Blumenthal and Jerry A. Anderson

Meteorology Research, Inc.
Santa Rosa, California 95401

Bruce K. Cantrell

SRI International
Menlo Park, California 94025

INTRODUCTION

The southwestern United States desert and mountain areas are generally characterized by very good visibility (>100 km). Until recent years, scenic vistas of natural landmarks and mountains with a visual range of over 100 miles were common. These vistas have been considered a major resource of the Southwest, and numerous national parks, forests, and monuments have been created to preserve them. The tourism resulting from the Southwest's scenery has played a major role in the economy of the region.

As the population of the region has grown, there has been an apparent reduction in the average visibility.[1] Residents of the Southwest are concerned about this loss and point to various anthropogenic activities as the cause. This concern has led to the August 1977 amendments to the Clean Air Act, which contain a section stating "as a national goal the prevention of any future, and the remedying of any existing, impairment of visibility in mandatory Class I Federal areas which impairment results from man-made air pollution."§ Previously, however, comprehensive studies had not been performed to quantify and determine the cause of the current problem or to provide the basis for a control strategy to keep the problem from getting worse.

Visibility impairment in the atmosphere is due primarily to the presence of small particles, which scatter and absorb light, and, secondarily, to NO_2, which absorbs blue light.[2] In the Southwest, a regional decrease in visibility is probably due to the long range transport of aerosol or gaseous precursors into unpopulated pristine areas. A second possible cause of regional visibility impairment may be windblown dust. Both types of particles may be of anthropogenic origin. Some of the varied activities that

*Supported by the Federal Interagency Energy/Environment Research and Development program, through a contract with the EPA-ESRL Office of Regional Studies.

†Consultant to Meteorology Research, Inc.

‡Work performed while on leave at California Institute of Technology, Pasadena, Calif.

§Section 169A of the Clean Air Act.

0077–8923/80/0338–0233 $01.75/1 © 1980, NYAS

may contribute to the decline in visibility are: increased population and the resulting increase in automotive emissions; emissions from power plants; emissions from smelters; disturbance of the topsoil and desert pavement due to agriculture, grazing, and off-road vehicles; and mining activities.

Project VISTTA,¶ sponsored by the EPA Environmental Science Research Laboratory in Research Triangle Park, North Carolina, has been initiated to determine the sources of visibility reduction in the Southwest and to help resolve the potential conflict between the congressional mandate for improved visibility and the need for increased energy production.

In this paper, we describe the results of the fall 1977 aircraft measurements of Project VISTTA. The Southwest visibility-reducing aerosol is characterized with regard to spatial distribution, particle size distribution, and chemical composition as a function of size. The emissions from a copper smelter and a coal-fired power plant in the region, as measured in elevated plumes, are characterized with regard to chemical composition as a function of size. The contribution of various species to light scattering is also quantitatively assessed.

Experimental Description

The initial exploratory field program of Project VISTTA was carried out October 1–10, 1977. Measurements were taken within the mixing layer (<500 m above ground level), using the instrumented MRI Beechcraft Queen Air aircraft. This twin-engine aircraft has minimum and maximum sampling speeds of 180 and 290 km/hr, respectively. With a full instrument complement, the maximum sampling time is over four hours. The aircraft was equipped to continuously measure and record on magnetic tape light scattering coefficient, Aitken nuclei count, ozone, sulfur dioxide, nitrogen oxides, temperature, dew point, turbulence, pressure (altitude), and navigational parameters. Aerosol size distribution measurements were made over a range of 0.01 to 30 μm, using three size-selective aerosol sensors. An electrical aerosol analyzer and optical particle counter were used in conjunction with an automated "grab bag" sampling system designed to collect and hold a discrete static air sample for analysis.[3] An axial scattering probe was operated in the free stream near the nose of the aircraft for particle size measurements in the range of 2 to 30 μm. Multistage impactor and two-stage filter sampler samples were also collected, in order to determine the elemental composition as a function of size. The details of the aircraft instrumentation are given in Table 1. Visual range measurements were carried out by viewing distant landmarks and verified by optical photography.

The multistage impactor, a Lundgren-type impactor specially modified for aircraft sampling by investigators at the University of California at Davis (UCD), was mounted in the nose assembly of the aircraft. The impactor stages had 50% collection efficiency for particle diameters of 4, 2, 1, and 0.5 μm. Impactor samples were collected on greased Mylar film. A Nuclepore final filter was used to collect all particles not collected on the four impaction stages. The impactor samples were analyzed by particle-induced x-ray emission (PIXE)[4] at the Crocker Nuclear Lab

¶VISTTA is an acronym for Visibility Impairment due to Sulfur Transport and Transformation in the Atmosphere.

cyclotron at UCD. This analysis gives elemental concentration of elements with atomic numbers of 13 (Al) or greater.

The fine particle samples from the TWOMASS two-stage sampler[5] were analyzed by flash vaporization–flame photometric detection (FV–FPD)[6] by J. D. Husar and Associates (St. Louis, Missouri).

The field program consisted of seven sampling days. The experimental plan included flights of the region on three full days and parts of one other, flights in the plume of a copper smelter (San Manuel smelter) on three days, and flights in the plume of a coal-fired power plant (Mohave power plant) on one day.

TABLE 1

BEECHCRAFT QUEEN AIR INSTRUMENTATION

Measured Parameter	Instrument
Aerosols—Integral Size:	
Light Scattering	MRI 1562 Nephelometer (modified by UW for high sensitivity)
Aitken Nuclei	Environment One Condensation Nuclei Monitor
Aerosols—Differential Size:	
Particle Diameter;	
0.01–1.0 μm	TSI 3030 Electrical Aerosol Size Analyzer (EAA)
0.5–4.0 μm	Royco 218 Optical Particle Counter (OPC)
2.0–30.0 μm	Particle Measurement Systems Axial Scattering Probe (ASP)
Aerosol Samples:	
Two Stage Sampler	MRI TWOMASS Sampler
Multistage Impactor	UCD Impactor System
Gases:	
Ozone	REM 612 O_3
Sulfur Dioxide	Meloy 285 Total Sulfur Monitor
Nitrogen Oxides	Monitor Labs 8440 NO–NO_x
Other:	
Dew Point	Cambridge Systems 137
Temperature	MRI Airborne Instrument Package
Turbulence	MRI Airborne Instrument Package
Altitude	MRI Airborne Instrument Package
Indicated Air Speed	MRI Airborne Instrument Package
Position: VOR DME	Aircraft Navigation System
Visual Range	Optical Photography

The regional flights were designed to characterize the visibility-reducing aerosol on flight paths of over 300 km. On two days, the total flight paths were over 1000 km. All flights were within the mixing layer. However, occasional spirals were made to define the mixing layer structure. The visibility along these paths was documented with a high-sensitivity nephelometer, a camera, and with visual range measurements from the aircraft to prominent landmarks.

Smelter plume flights were carried out at several downwind distances up to 127 km from the San Manuel smelter. Flights in the 1580 MW Mohave power plant plume were performed at distances as far as 62 km downwind of the plant.

The plume flight patterns were designed to provide a detailed characterization of the plume at discrete distances downwind from the source. At each distance, horizontal traverses were made through the plume normal to the plume axis at several elevations. Traverses were long enough to include background air on either side of the plume. Vertical spirals were also performed to help characterize the plume structure. In addition to cross-sectional plume traverses, orbital flights within the plume of up to one hour were carried out at each distance, to gather heavily loaded impactor and filter samples. Orbits upwind of the source were flown to gather background samples.

Single theodolite pilot balloon (pibal) releases were carried out at various times on every sampling day, to determine wind direction and speed as a function of altitude in the sampling area.

Results and Discussion

Southwest Background Measurements

Spatial Distribution

The character of visibility-reducing aerosols in the Southwest was determined in long regional flights, in order to measure a regional background. These flights were made between or upwind of large point sources of pollutants, such as urban areas and power plants. All flights were at a constant altitude to within ±25 m. On October 5, 1977, flights were made between Flagstaff (Arizona)–Tuba City (Arizona)–Island in the Sky (Utah)–Farmington (New Mexico)–Grants (N. Mex.)–Flagstaff, as shown in Figure 1. The flights on October 9 followed the same 1080-km path in the reverse direction. Shorter flights in the region (~260 km) were flown on October 8 and 10. On all days, the visibility and air quality were quite good (average visual range = 140 km). This resulted in very low values for all measured aerosol and gas parameters except ozone, which was present in concentrations representative of clean air [36 ppb (O_3)]. Therefore, only data from the longer flights on October 5 and 9, which have more meaningful averages, are discussed in this paper. The averages for each day were determined by averaging over each segment of the flight and then averaging the flight segment averages. The NO, NO_x, and SO_2 concentrations were below the sensitivity of the monitors throughout the region (<5 ppb NO and NO_x, <1 ppb SO_2). The b_{scat} values, although quite low, were detectable with the aircraft instrumentation.

The spatial distribution of the visibility-reducing aerosol is most easily determined from the frequency distribution of the light scattering coefficient due to particles, b_{sp}, as shown in Figure 2. These values are determined from the 10-sec average nephelometer readings, with the Rayleigh scattering contribution to the scattering coefficient at the flight altitude subtracted.

The average value of b_{sp} on both sampling days was rather low, ranging from $(14 \pm 1) \times 10^{-6}$ m^{-1} on October 9 to $(17 \pm 3) \times 10^{-6}$ m^{-1} on October 5. Although these values are low, they are above the minimum sensitivity of the instrument [$b_{sp}(\min) \simeq 5 \times 10^{-6}$ m^{-1}]. The Rayleigh scattering, or clean air, contribution at the sampling altitude ($\sim 11 \times 10^{-6}$ m^{-1}) was, on the average, 42% of the total light scattering coefficient during the very clear days on which these regional flights were made. The average measured visual range in Mesa Verde and other national parks in the region in 1976[7] was about equal to the average visual range calculated from the scattering coefficient on the VISTTA sampling days (~150 km). Therefore, the VISTTA data can be considered typical for the region.

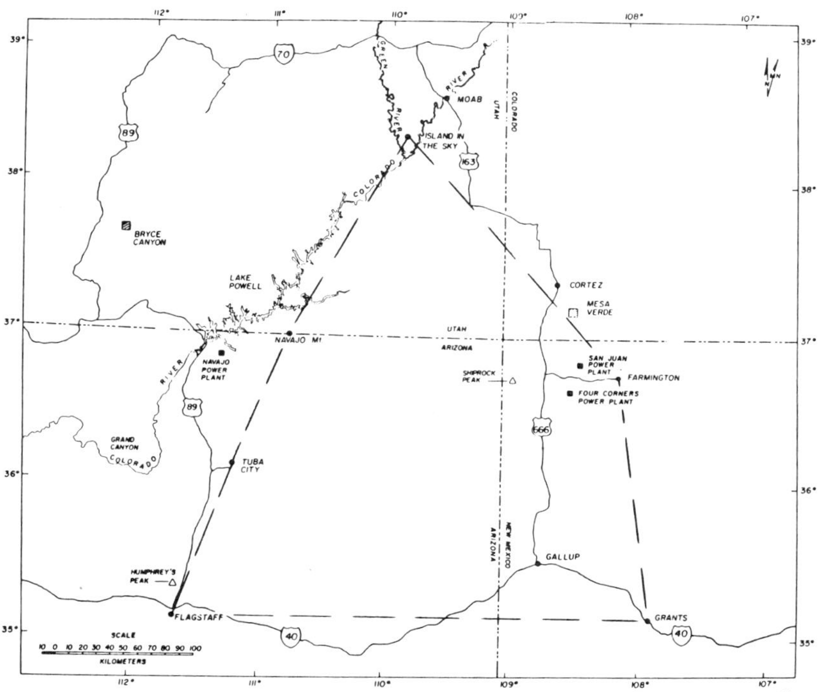

FIGURE 1. Flight path of VISTTA regional flights on October 5 and 9, 1977. The entire flight path was ~1080 km.

The coefficient of variation of the b_{sp} frequency distribution was 7% on October 9 and 18% on October 5 over the entire 1080-km flight path, which indicates that the aerosol was homogeneous on both days.

Size Distribution

The average size distribution of background Southwest aerosol measured on the aircraft during the October 9 regional flight is shown in FIGURE 3. The distribution is

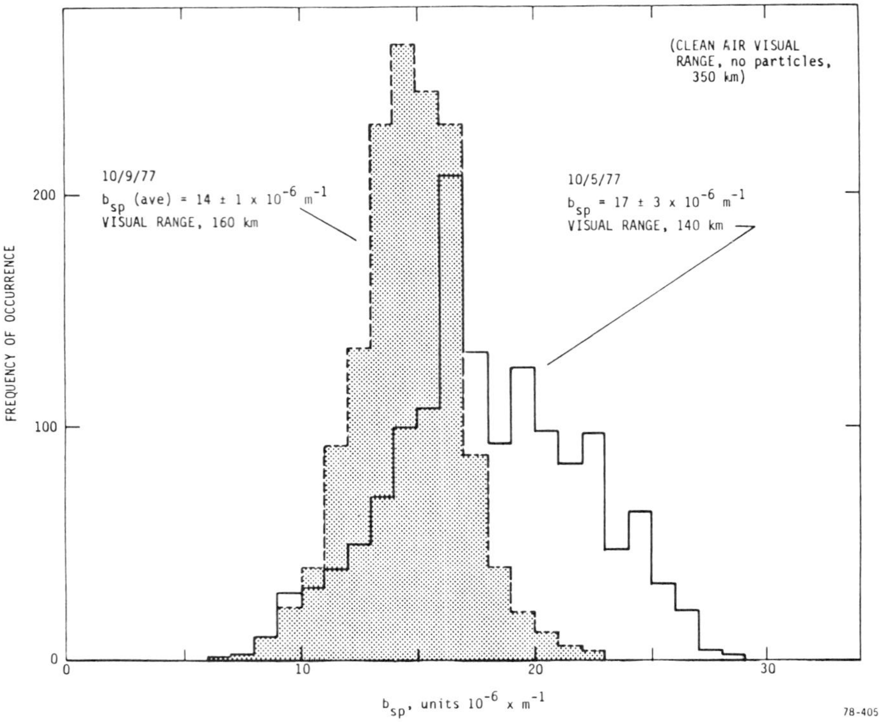

FIGURE 2. Frequency distribution of the light scattering coefficient due to particles, b_{sp}, measured in regional flights on October 5 and 9, 1977. Measurements were averaged over 10 sec (1 km) for the 1080-km flight path for each day. The b_{sp} averages on October 5 and 9 were 17 and 14×10^{-6} m^{-1}, with standard deviations of 3 and 1×10^{-6} m^{-1}, respectively, indicating that the aerosol was homogeneous on these days.

bimodal. The aerosol size distribution data were reduced, using procedures described elsewhere.[8] The size distribution data were collected every five minutes during the entire flight. All measured size distributions were averaged before extracting integral parameters using a log-normal approximation technique. There is some question about altitude effects on these instruments, which may lead to errors in the size distribution measurements. However, we do not think these errors are large enough to alter the conclusions of this paper. We are presently looking into the magnitude of this flow rate effect. The fine particle, or accumulation, mode had a geometric volume mean size, D_p, of 0.25 μm and a geometric standard deviation, σ, of 2.0. The coarse mode geometric volume mean size is 5.5 μm, with a σ of 2.2. The volume concentration of particles is 2.6 $\mu m^3/cm^3$ and 2.5 $\mu m^3/cm^3$ for the fine and coarse modes, respectively. The accumulation mode mean size is slightly smaller than the continental average of 0.3 μm.[9] However, the coarse particle mean size is the same, within experimental error, as the continental average of 6 μm expected for particles from similar sources, such as windblown dust.

Elemental Composition

The elemental composition of the visibility-reducing aerosol in the Southwest was determined from impactor and filter samples collected on regional flights October 5 and 9, 1977. As discussed previously, the aerosol was quite homogeneous throughout the region during this sampling, making a regional average meaningful. The elemental concentrations, divided into fine particles ($D_p < 1 \mu m$) and coarse particles ($D_p > 1 \mu m$), averaged for all samples collected on October 5 and 9 are given in TABLE 2. The rationale for the separation at 1 μm is based on the high scattering efficiency for particles with diameters less than 1 μm. Furthermore, particles in each of these size ranges generally have different chemical compositions. All other elements not listed were below the minimum detectable limits. The elemental concentration of fine particle sulfur was measured by FV–FPD analysis of TWOMASS samples. All other data are from PIXE analysis of impactor samples. There was some discrepancy between the values from the two techniques, but the FV–FPD values appeared more reliable. The coarse particle fraction is dominated by crustal elements, such as silicon, aluminum, calcium, and iron. Sulfur, silicon, and potassium are predominant elemental species in the fine particles.

The size distribution of individual elemental species, determined from regional impactor samples collected on the October 5 and 9 regional flights, is shown in FIGURE 4. Aluminum, calcium, and iron all show very similar size distributions, with the highest concentration of each element in particles with diameters greater than 4 μm. The concentration on the final filter ($D_p < 0.5 \mu m$) was below the minimum detectable limit for all these elements. Silicon has a size distribution similar to aluminum down to

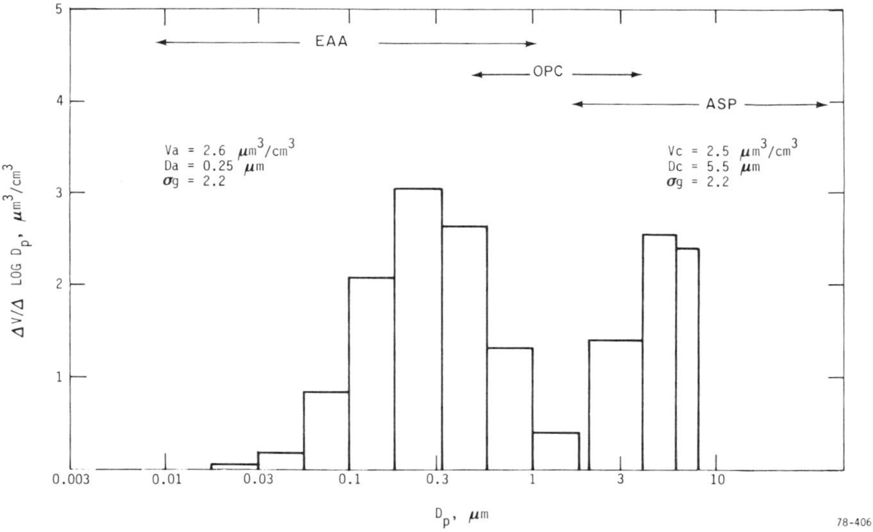

FIGURE 3. Average size distribution of background Southwest aerosol, measured on the October 9, 1977 regional flight.

TABLE 2

AVERAGE ELEMENTAL CONCENTRATION OF AEROSOL IN THE SOUTHWEST REGION, OCTOBER 5 AND 9, 1977

Element*	Fine Particles† ($\mu g/m^3$)	Coarse Particles‡ ($\mu g/m^3$)
Al	0.01	0.28
Si	0.72	1.12
S	0.68	0.06
Cl	0.11	0.01
K	0.21	0.09
Ca	0.03	0.24
Ti	0.08	0.01
V	0.04	0.01
Cr	—	0.01
Fe	0.004	0.16
Zn	0.03	0.003
Zr	—	0.02
Pd	0.02	0.01
Ba	0.10	0.01
Pb	0.09	0.004

*The following elements were detected in only a few samples, and, therefore, it is not possible to determine a meaningful average concentration: Fine Particles—Cu, Mn, Ni, Rb, Co, Hg, Mo, Pt. Coarse Particles—Cu, Mn, Se, Ni, Rb, Co, Mo, Br, Pt.

†Fine particle ($D_p < 1$ μm) concentrations were determined from the sum of impactor stage 4 and the final filter. Sulfur values were determined from TWOMASS samples.

‡Coarse particle ($D_p > 1$ μm) concentrations were determined from the sum of impactor stages 1, 2, and 3.

1 μm; below that diameter, the concentration of silicon increases. The concentration of small particle silicon ($D_p < 0.5$ μm) is nearly equal to the large particle value ($D_p > 4$ μm). These data indicate that silicon is a major contributor to the fine particle mass and, as discussed below, to visibility reduction. However, subsequent measurements have shown lower silicon concentrations. From the source characterization determined in this work, fine particle silicon results from the condensation of volatile silicon emitted from high-temperature combustion sources.

Chemical Composition

The total aerosol mass was not measured, but it can be estimated from the measured elemental concentrations, and the oxygen concentration estimated with assumptions of the chemical form of the measured elements. The sulfur can be expressed as sulfate because the FV–FPD method is specific for water-soluble sulfate.[8] It is also assumed that ammonium is the only ion associated with the sulfate. Other elements are assumed to exist as oxides. This leads to the mass balance listed in TABLE 3 and FIGURE 5. The mass estimates may be low due to unmeasured elements, particularly those with $Z < 13$. The total mass in each size range can also be estimated from the aerosol volume determined by integrating each mode in the differential size distribution and by estimating the aerosol density from the chemical composition. A

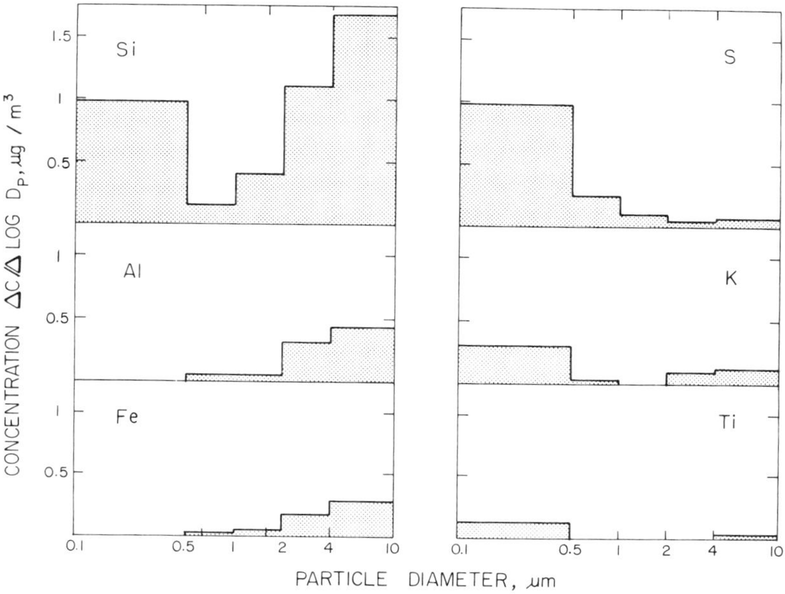

FIGURE 4. Average size distribution of individual elemental species, determined from impactor samples collected on October 5 and 9 regional flights. Si has a size distribution similar to Al, Ca, and Fe down to 1 μm; below that diameter, the concentration of Si increases.

TABLE 3

CHEMICAL SPECIES BALANCE FOR THE SOUTHWEST REGION—OCTOBER 5 AND 9, 1977

	Fine Particles		Coarse Particles	
Compound	Concentration ($\mu g/m^3$)	Contribution* (%)	Concentration ($\mu g/m^3$)	Contribution (%)
Al_2O_3	0.02	0.4	0.53	13.4
SiO_2	1.54	29.2	2.40	60.6
$(NH_4)_2SO_4$	2.81	53.3	0.25	6.3
K	0.21	4.0	0.09	2.3
CaO	0.04	0.8	0.34	8.6
TiO_2	0.13	2.5	0.02	0.5
Fe_2O_3	0.006	0.1	0.23	5.8
ZnO	0.04	0.8	0.004	0.1
PbO	0.10	1.9	0.004	0.1
Other (in normal oxide form)	0.37	7.0	0.09	2.3
Total	5.27	100.0	3.96	100.0

*The percent contribution is based on the mass estimated from the sum of the chemical concentrations.

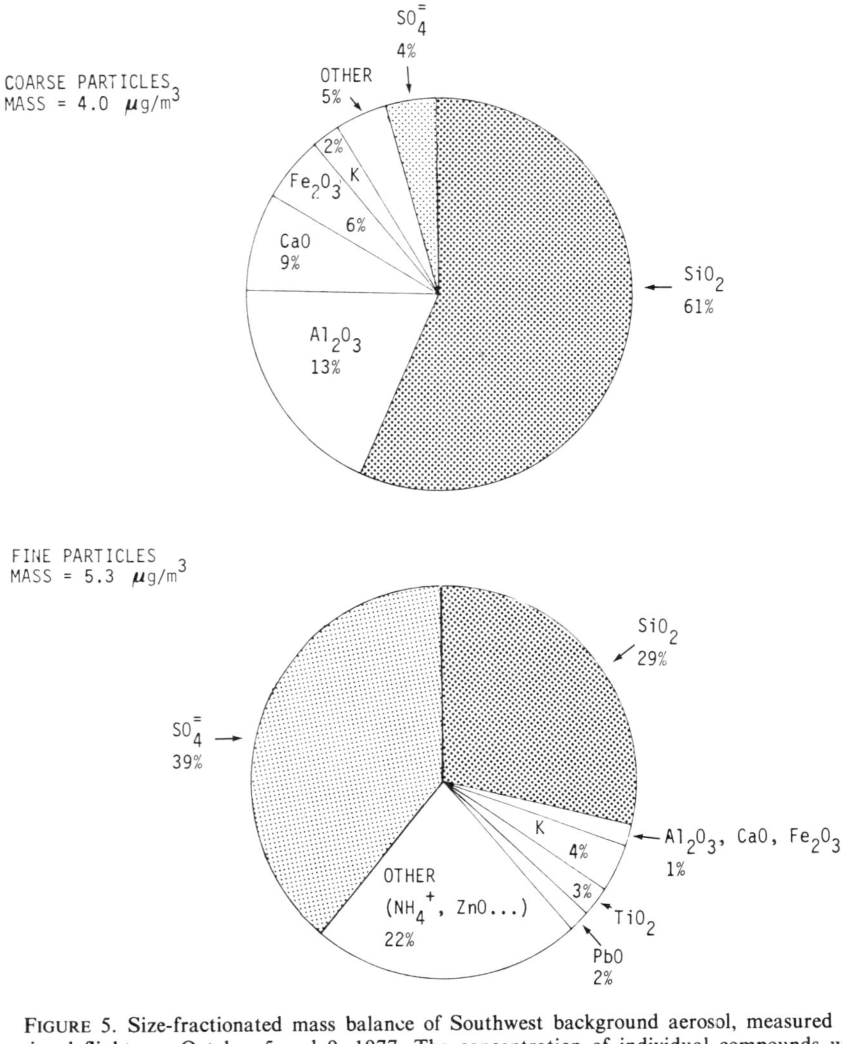

FIGURE 5. Size-fractionated mass balance of Southwest background aerosol, measured on regional flights on October 5 and 9, 1977. The concentration of individual compounds was determined from the measured elemental concentration and assumed chemical form. The total aerosol mass in each size fraction was determined from the sum of the individual chemical components. Comparison with aerosol volume measurements indicates that the mass estimated from composition measurements represents 93% and 64% of the fine and coarse particle mass, respectively.

comparison of these two approaches indicates that the elemental mass balance accounts for ~93% and ~64% of the fine and coarse particle mass derived from volume, respectively. The assumptions of chemical form, although somewhat arbitrary, lead to a conservative estimate of the total mass. For example, iron has been assumed to be present exclusively as Fe_2O_3, when in reality it is likely that FeO is present as well. This is the case for crustal material.[10] Comparison of fine particle mass and b_{sp} allows a further check on the mass estimate. Extensive measurements in the western United States indicate that the fine particle mass/b_{sp} ratio is normally 0.32 to 0.34 g/m.[11] The ratio for the VISTTA regional data is 0.34 g/m, which is a further indication that the fine particle mass estimate is reasonable.

The total measured aerosol mass in both size ranges (9.23 $\mu g/m^3$), determined from the sum of the concentration of each compound, is lower than the average total suspended particulate value ($\sim$20 $\mu g/m^3$) measured in several national parks in the region.[1] It appears from volume and measurements that this discrepancy is not primarily due to fine particle mass estimates. It is more likely that the coarse particle mass estimate is low due to unmeasured constituents. Furthermore, coarse particles may be diminished due to settling at 300 to 500 m above the ground, where the samples were collected.

Probably the most important species unaccounted for are nitrates, carbon compounds, and water. The average annual concentration of nitrate in the nonurban Southwest region was about 20% of the sulfate value in the early 1970s, as measured in Grand Canyon and other remote southwestern national parks by the National Air Surveillance Network.[7] These values may be subject to error due to artifact sulfate and nitrate effects on filters. Very few data exist on the concentration of carbon in Southwest ambient aerosols, but it is a significant component of eastern urban aerosols.[12]

Enrichment Factors

The predominant species in the coarse particles are elements found in high concentration in soil and crustal material. Elemental enrichment factors, given in TABLE 4, help to characterize the source of these particles. These factors are calculated as follows:

$$\text{Enrichment Factor (EF)} = \frac{(X/\text{Al})\ \text{Aerosol}}{(X/\text{Al})\ \text{Crust}}$$

Aluminum was chosen as the reference element because its major source is likely to be soil. However, aluminum is also present in fly ash from coal combustion.[13] Enrichment factors >1 indicate an enrichment of that element relative to crustal abundances; values <1 indicate a depletion relative to crust. Because of the uncertainty in the data, EF values between 0.5 and 2 indicate no significant difference from crustal abundance. It can be seen from TABLE 4 that Si, K, Ca, Ti, V, and Fe are all near the crustal abundance in the coarse particles relative to Al. These elements represent 93% of measured coarse particle mass. The other coarse particle species are enriched, indicating an anthropogenic source. This analysis is complicated by the fact that many of these elements are present in coal fly ash, with abundances similar to crustal material.

All fine particle species except iron are enriched relative to the crust (normalized to aluminum), with enrichment factors ranging from 7 to 3×10^7. The enriched fine particle species represent 99.5% of the measured fine particle mass, which indicates that the vast majority of fine particles are not of crustal origin.

TABLE 4

ENRICHMENT FACTORS

Element	Enrichment Factor: Fine Particles ($D_p \leq 1$ μm)		Enrichment Factor: Coarse Particles ($D_p \geq 1$ μm)	
Southwest Background				
Al	1.0	Crustal Abundance	1.0	Crustal Abundance
Fe	0.7		0.9	
Si	21	Enriched Relative to the Crust	1.2	
K	66		1.0	
Ca	7		1.9	
Ti	1×10^2		0.7	
S	2×10^4		67	Enriched Relative to the Crust
Cl	7×10^3		23	
V	2×10^3		29	
Cr	—		13	
Zn	4×10^3		35	
Zr	—		3×10^5	
Pd	2×10^7		7	
Ba	2×10^3		3×10^5	
Pt	3×10^7		88	
Pb	6×10^4			
San Manuel Smelter 60 km Downwind (10/4/77)				
Al	1.0	Enriched	1.0	Crustal Abundance
Si	32		0.9	
K	1×10^2		1.0	
Ca	—		2	
Ti	37		0.3	
Mn	—		0.9	
Fe	3		0.9	
S	7×10^4		—	
Cr	2×10^4		26	Enriched
Cu	—		46	
Se	4×10^7		—	
Mohave Power Plant 32 km Downwind (10/8/77)				
Al	1.0	Crustal Abundance	1.0	Crustal Abundance
Si	2		0.7	
K	—		0.5	
Ca	—		2	
Ti	—		0.5	
Fe	—		0.6	
S	2×10^3	Enriched	—	
Cr	—		28	Enriched
Ni	4×10^3		7	
Pb	—		128	

Light Scattering Budget

The detailed contributions to light scattering as a function of particle size can be determined from Mie calculations[14] of the light scattering efficiency and the measured particle size distribution. The sum of the calculated b over all particle sizes can be compared to the measured light scattering coefficient from the integrating nephelometer. Mie scattering calculations, combined with particle size and composition measurements, permit the determination of the individual contribution to visibility reduction for particles of a given composition and size (light scattering budget). This approach can also serve as a test of the consistency of the various measurements.[15] The following expression was evaluated in the scattering coefficient calculation:

$$\bar{b}_{sp} = \int G(D_p) \frac{dV}{d \log D_p} d \log D_p$$

where G is the light scattering per unit volume of aerosol. The quantity G is calculated using Mie scattering functions.

The contributions to the light scattering coefficient as a function of particle size for the average Southwest regional aerosol (October 9, 1977), calculated from the average regional aerosol size distribution, are given in FIGURE 6. A refractive index, $m = 1.54$, was used in this calculation, which is slightly lower than the mass-weighted values of 1.57 and 1.65 for fine and coarse particles, respectively, calculated from aerosol composition measurements. The calculation was done for light of 525 nm wavelength. The sum over all calculated contributions yields:

$$\sum b_{sp} = (16 \pm 2) \times 10^{-6}\ m^{-1}$$

which agrees well with the average value of light scattering due to particles, measured with the integrating nephelometer (with the contribution of Rayleigh scattering due to gases removed), of:

$$b_{sp}(\text{measured}) = 14 \pm 1 \times 10^{-6}\ m^{-1}$$

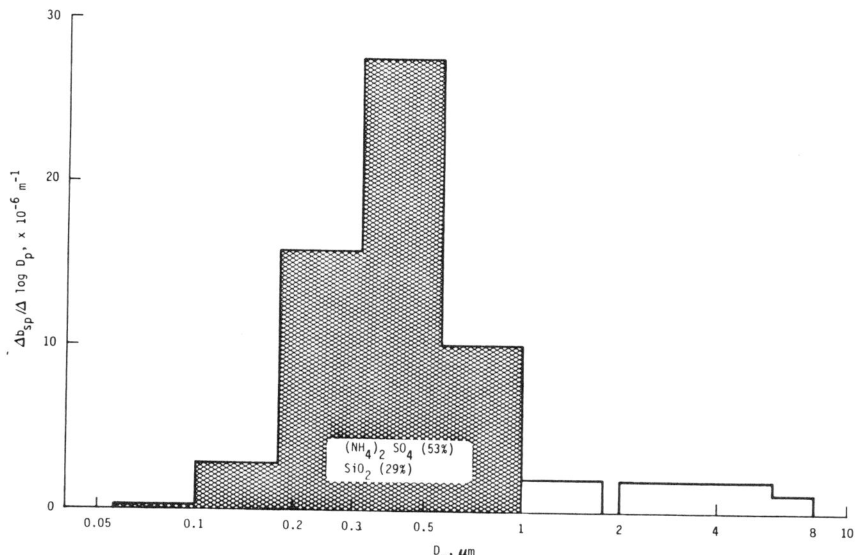

FIGURE 6. Light scattering contribution as a function of size for the Southwest region, October 9, 1977. The fine particle contribution, which represents 93% of the total b_{sp}, is shaded in the figure. The major components of the fine particles ($D_p \leq 1$ μm) are $(NH_4)_2SO_4$ (53%) and SiO_2 (29%).

TABLE 5
LIGHT SCATTERING BUDGET

Component	D_p (μm)	b_{scat} ($\times 10^{-6}$ m^{-1})	Contribution to Total b_{scat} (%)
Southwest Region 10/9/77 (Visual Range ~160 km)			
Air Molecules	—	11	44
$(NH_4)_2SO_4$*	0.1 to 2.0	7	28
SiO_2†	0.1 to 1.0	4	16
Other Compounds‡	0.1 to 1.0	2	8
Coarse Particles	1.0 to 20.0	1	4
		25×10^{-6} m^{-1}	100
San Manuel Smelter (62 km downwind) 10/4/77			
$(NH_4)_2SO_4$	0.1 to 1.0	41	43
SiO_2	0.1 to 1.0	9.5	10
Other Compounds	0.1 to 1.0	5.5	6
Coarse Particles	1.0 to 20.0	39	41
		95×10^{-6} m^{-1}	100
Mohave Power Plant (32 km downwind) 10/8/77			
$(NH_4)_2SO_4$	0.1 to 1.0	3	11
SiO_2	0.1 to 1.0	2	7
Other Compounds	0.1 to 1.0	1	4
Coarse Particles§	1.0 to 20.0	21	78
		27×10^{-6} m^{-1}	100

*Assumes that all fine particle sulfate exists as ammonium sulfate.
†Assumes that all fine particle silicon exists as SiO_2.
‡Given in TABLE 3.
§Determined from b_{scat} (total) − b_{scat} (fine particles).

The visibility budget for the Southwest background aerosol on October 9, 1977, determined from the Mie calculations, indicates that 52% of the light scattering was due to fine particles ($D_p \leqq 1$ μm), 44% was due to Rayleigh scattering from gases, and 4% was due to coarse particles. Considering only the light scattering due to particles, 93% is due to fine particles. The detailed light scattering budget, summarized in TABLE 5, was constructed using the results of the fine particle mass balance and assumes that sulfur and silicon compounds have the same size distribution as the total fine particle aerosol. This is reasonable, since these compounds represent 82% of the measured fine particle mass.

The ratio of fine particle sulfate mass to the total b_{sp} value for the Southwest region, measured on October 9, 1977, is 0.15 g ($SO_4^=$)/m. A similar calculation for the fine particle SiO_2 mass/b_{sp} ratio yields 0.11 g (SiO_2)/m. Perhaps a more interesting calculation is the ratio of fine particle sulfate or SiO_2 mass to the contribution of that species to b_{sp}, as determined in the light scattering budget. This ratio is 0.4 g/m for both fine particle sulfate and SiO_2.‖ This calculation assumes that sulfur and silicon have the same size distribution.

‖The inverse of this ratio, i.e., the contribution of $SO_4^=$ or SiO_2 to b_{sp} divided by the fine particle $SO_4^=$ or SiO_2 mass, is (2.5×10^{-6} m^{-1})/(μg/m) for $SO_4^=$ and for SiO_2.

Plume Measurements

In order to characterize the emissions from specific sources in the region, measurements were made in plumes from the San Manuel copper smelter on October 1, 2, and 4, 1977, and the Mohave power plant October 8, 1977. Separate orbital flights in the plume and in the background, as well as plume traverses, were carried out at each distance downwind of the source.

To help evaluate the data presented in this section on source characterization, it is instructive to compare the emission rates of the San Manuel smelter and Mohave power plant. In 1975, the Mohave power plant emitted an average of 66 tons/day of SO_x and 77 tons/day of NO_x.[16] In 1977, the San Manuel smelter emitted 557 tons/day of sulfur expressed as SO_2.[17]

Plume Impact

The impact of a plume on visibility can be estimated by comparing the visual range with the plume present to the visual range through the background without the plume. The parameters used in this semiquantitative calculation are shown in FIGURE 7. The subscripts p and b indicate plume and background parameters, respectively. The light scattering coefficient in the plume, b, includes the background contribution.

The visual range, V, is defined as "the distance, under daylight conditions, at which the apparent contrast between a specified type of target and its background becomes just equal to the threshold contrast of an observer."[18] For the following analysis, the target and background are a black object viewed against the horizon sky, and the threshold contrast of a typical observer, 0.02, is used.[19] In this case, the visual range is given by:

$$V = \frac{-\ln 0.02}{\bar{b}_{\text{ext}}} = \frac{3.9}{\bar{b}_{\text{ext}}}$$

where $\bar{b}_{\text{ext}}$ is the average extinction coefficient along the sight path. This equation assumes uniform illumination along the sight path and neglects multiple scattering.

The following calculations assume that the contribution of absorption to extinction is small, so that the equation for visual range can be rewritten as $Vb_{\text{scat}} = 3.9$. For the case depicted above of a plume imbedded in background air, the corresponding

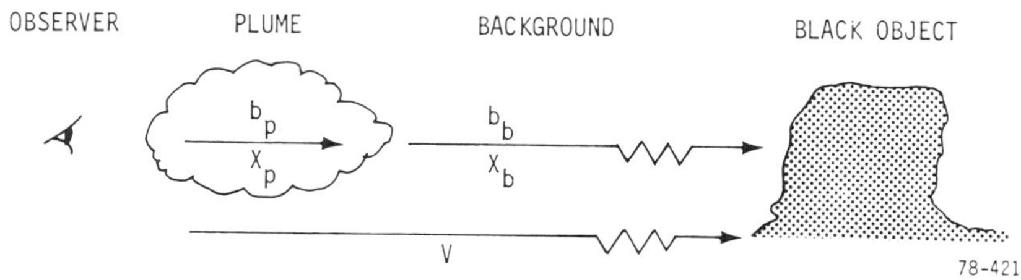

FIGURE 7. Schematic diagram of parameters used in semiquantitative calculation of plume impact. The subscripts p and b indicate plume and background parameters, respectively.

equation is:

$$X_p b_p + X_b b_b = 3.9$$

where b_p and b_b are the average scattering coefficients in the plume and background, respectively, and X_p is the measured plume width. The quantity $X_p b_p$, termed the optical depth of the plume (τ_p), is the integral of scattering across the plume, i.e., $\tau_p = \int b_p(x)dx$. The visual range looking through the plume was calculated using the equation:

$$V = X_p + X_b = X_p + \frac{3.9 - \tau_p}{b_b}$$

The values for X_p, b_p, and b_b were determined from measurements at the lowest altitude at which plume traverses were made. The plume opacity is defined in terms of the plume optical depth as $1 - \exp(-\tau)$.

The visual range and plume optics calculations are summarized in TABLE 6 and FIGURE 8. The visual range through the copper smelter plume is reduced at 8 km downwind of the stack from 135 km in the background to 13 km with the plume present (90% reduction). The visibility reduction decreases further downwind of the plant. At 60 km downwind, the visual range is 82 km with the plume present (39% visibility reduction relative to the background). The visibility reduction with the plume present is nearly equal at 60 and 127 km downwind of the smelter. The power plant plume exhibits quite different visual behavior. The reduction of the background visual range is less at 32 km than at 60 km downwind (11% vs. 25%, respectively). However, the power plant plume was measured under meteorological conditions that varied during the day, resulting in erratic plume behavior.

The difference in the reduction of the background visual range between these plumes may be the result of greater emissions released from the smelter stack and of greater dispersion of the Mohave plume on October 8.

TABLE 6
PLUME IMPACT DATA

	Downwind Distance (km)	Plume Width (km)	Plume Optical Depth	Plume Opacity	Visual Range With Plume (km)	Visual Range Background	Visual Range Reduction due to Plume (%)
San Manuel Smelter	8	16	4.9	0.99	12	135	90
	32*	20	2.8	0.94	54	135	60
	60	25	2.2	0.89	82	135	39
	127	54	3.1	0.95	78	135	42
Mohave Power Plant	32	8	0.66	0.48	98	110	11
	60	27	1.92	0.85	82	110	25

*Plume width at 32 km downwind, interpolated from measurements at 8 and 60 km downwind.

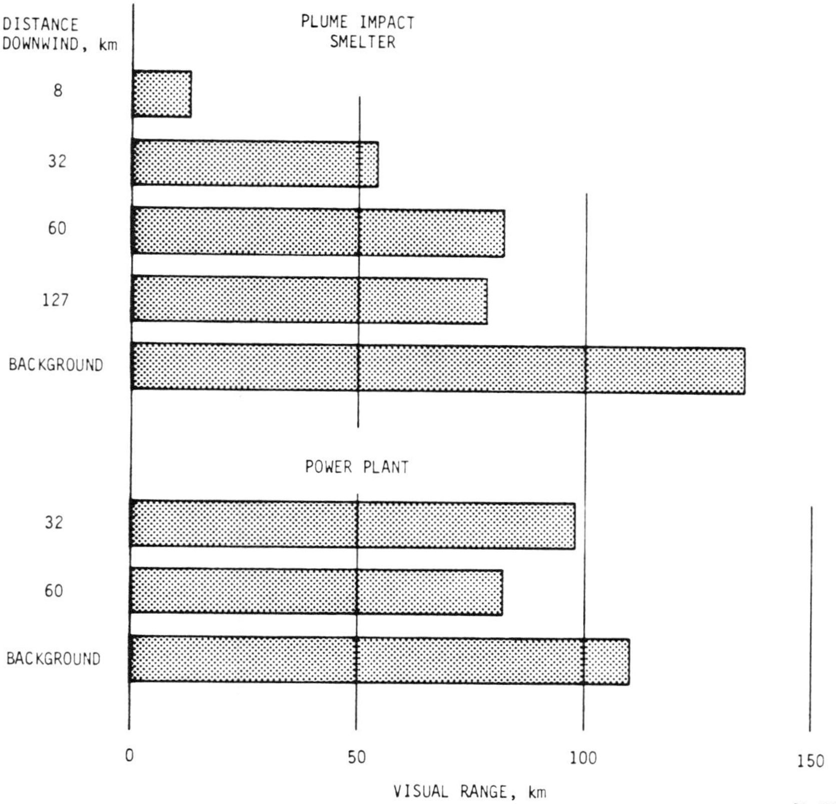

FIGURE 8. Visual range calculations for the smelter and power plant, at several distances downwind, compared to the background visual range.

Elemental Composition in Plumes

The elemental composition of the plume excess aerosol, i.e., the plume aerosol with the background subtracted, was determined from the analysis of samples collected in orbital plume flights. The background concentrations were determined in orbital flights upwind of the plume. The aerosol elemental compositions for the San Manuel smelter plume at 60 km downwind and the Mohave power plant at 32 km downwind are given in TABLE 7. The largest plume excess elemental constituents of the fine particle aerosol in both plumes are sulfur and silicon. Silicon, aluminum, calcium, and iron are major constituents of the plume excess coarse particles. These data indicate that coal-fired power plants and copper smelters in the Southwest may be major sources of the fine particle silicon as well as sulfur, the two major constitutents of the Southwest visibility-reducing aerosol.

Plume excess aerosol enrichment factors for the San Manuel smelter and Mohave power plant are given in TABLE 4. Coarse particle Si, K, Ca, Ti, and Fe in both plumes

TABLE 7

PLUME EXCESS* AEROSOL ELEMENTAL CONCENTRATION

	San Manuel Smelter 60 km downwind (10/4/77)		Mohave Power Plant 32 km downwind (10/8/77)	
Element†	Fine Particles‡ ($\mu g/m^3$)	Coarse Particles§ ($\mu g/m^3$)	Fine Particles ($\mu g/m^3$)	Coarse Particles ($\mu g/m^3$)
Al	<0.01	1.91	0.05	1.46
Si	1.08	6.12	0.34	3.67
S	2.29	—	0.30	—
K	0.43	0.60	—	0.25
Ca	—	1.58	—	1.16
Ti	0.02	0.03	—	0.04
Mn	—	0.02	—	—
Cr	0.20	0.06	—	0.05
Fe	0.02	1.02	—	0.55
Cu	—	0.06	—	—
Ni	—	—	0.20	0.01
Se	0.26	—	—	—
Pb	—	—	—	0.03

*Plume excess values were determined by subtracting background concentrations from plume concentrations of each element.

†Molybdenum was detected in concentrations above background but at the minimum detectable limit in the Mohave power plant plume.

‡Fine particle ($D_p \leq 1$ μm) concentrations were determined from the sum of impactor stage 4 and the final filter. Sulfur values were determined from TWOMASS samples.

§Coarse particle ($D_p \geq 1$ μm) concentrations were determined from the sum of impactor stages 1, 2, and 3.

and coarse particle Mn in the smelter plume are present in crustal abundances relative to Al. Thus, the crustal abundances of these elements in Southwest background coarse particles may be an indication that the source of these particles is crustal weathering or fly ash. Plume excess coarse particle Cr, Ni, Cu, and Pb are all enriched relative to crustal abundances. All plume excess fine particle species are enriched relative to the crust, except for fine particle Si in the Mohave plume.

The size distribution of plume excess elemental constituents shows many similarities with the regional data. Aluminum, calcium, and iron all exhibit the highest concentrations for particles with diameters >4 μm. Silicon has a size distribution similar to aluminum down to 1 μm but increases below that diameter. Sulfur is mainly concentrated in fine particles.

The total aerosol mass was not measured but can be estimated from the elemental concentrations and mean aerosol volume, as described previously. A comparison of these two approaches shows that in the San Manuel smelter on October 4, 1977, at 62 km downwind, 100% of the plume excess fine particle mass, determined from plume excess aerosol volume (assuming a mass-weighted density of 2 g/cm^3), can be accounted for from the measured constituents. The coarse particle volume was not measured in this plume. Approximately 73% of the plume excess fine particle mass is sulfate (expressed as ammonium sulfate), and 17% is silicon (expressed as SiO_2). In the Mohave power plant plume on October 8, 1977, only about 50% of the fine particle mass determined from aerosol volume can be accounted for by the measured constituents. Of this 50% at 32 km downwind, plume excess sulfate [expressed as $(NH_4)_2SO_4$] accounts for 52% of the mass inferred from measured aerosol composition, and plume excess fine particle silicon (expressed as SiO_2) accounts for 31%.

Sulfur Transformation

The conversion of SO_2 to particulate sulfate in plumes can be estimated from the measurements of gaseous SO_2, particulate sulfate, light scattering coefficient, and mean accumulation mode volume. A number of methods to determine SO_2 conversion rates in power plant plumes have been used in the past.[20] The approach used here compares the fraction of particulate sulfur of the total sulfur, S_p/S_T, to plume age, estimated from analysis of winds aloft. Plume excess sulfur values with background

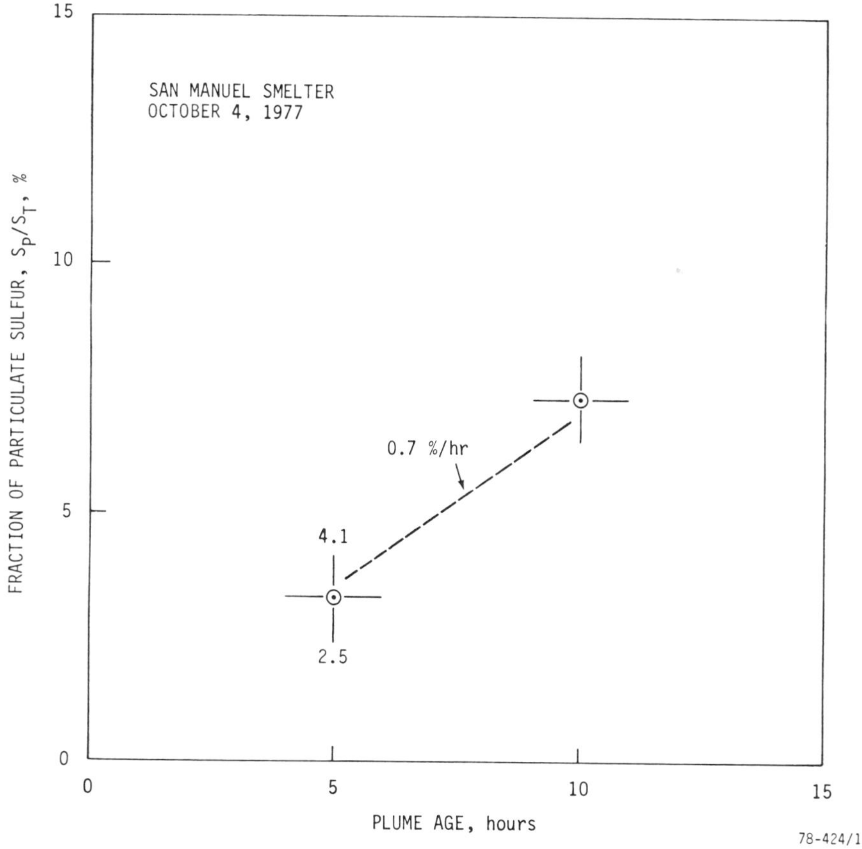

FIGURE 9. Determination of SO_2 conversion rate for the San Manuel smelter on October 4, 1977, from the fraction of particulate sulfur vs. plume age. Measurements were made from 0900 to 1230 MST, between 60 and 127 km from the plant.

subtracted were used in this calculation. The SO_2 conversion rate, measured in the San Manuel smelter plume on October 4, 1977, from 0900 to 1230 MST, was measured to be 0.7 ± 0.2%/hour between 60 and 127 km downwind from the plant (FIGURE 9). This calculation assumes no SO_2 deposition between measurements. However, even as much as 20% SO_2 deposition would not lower the conversion rate below the stated uncertainty. The SO_2 conversion rate was not calculated for the Mohave power plant plume because of poor plume resolution and erratic wind behavior encountered on October 8, 1977.

This single measurement of the $SO_2 \rightarrow SO_4^{=}$ conversion rate should not be considered typical without many additional measurements under a variety of conditions. It is interesting to note, however, that this value is within the range of conversion rates measured in plants in the Midwest.[21]

A number of other parameters can be used as qualitative indicators of SO_2 conversion, such as plume excess b_{sp}/S_g, S_p/S_g, and mean accumulation mode volume $(V_a)/S_g$ ratios, and accumulation mode aerosol size, D_{Va}. These indicators are shown graphically in FIGURE 10, as a function of distance downwind from the plant, for the San Manuel smelter and Mohave power plant plumes on October 4, 1977, and October 8, 1977, respectively. The three plume excess parameter ratios all increase downwind in both plumes, indicating that sulfur aerosol is being formed. The mean accumulation mode size decreases downwind from both plants, which also indicates fresh aerosol is being formed.

One method of assessing the effect of sulfur transformation on visibility is to determine the improvement of visual range if no SO_2 were converted to sulfate. This can be estimated for the San Manuel smelter at 20 and 32 km downwind on October 2, 1977, assuming that the b_{sp}/SO_2 ratio remains constant beyond 8 km downwind. The visual range with the plume present is then calculated, using b_{scat} and assuming no conversion, as described previously. This approach will slightly overestimate the effect of aerosol dilution (and therefore visual range) due to loss of SO_2 from the plume because of deposition and transformation. The results of this calculation are shown in FIGURE 11. At 32 km downwind from the smelter on October 2, the visual range would increase by nearly a factor of two if no SO_2 were converted to sulfate. At 60 km downwind, the visual range would be increased by 36% if there were no conversion. Thus, sulfur transformation has a large effect on visibility impairment within at least 60 km of the smelter.

Light Scattering Budget

The detailed light scattering contributions to visibility impairment as a function of particle size due to a smelter and power plant plume can be determined, using Mie calculations and measured plume excess aerosol size distributions, in an analogous manner to the treatment of regional data. However, there are several problems with the plume data. During the San Manuel smelter plume flights, only the fine particle size distribution was measured ($D_p \leqq 1\ \mu m$). During the Mohave power plant flights on October 8, 1977, the entire size distribution was measured, but windblown dust was occasionally mixing into the plume. Therefore, the plume visibility budgets must be considered as much more tentative than the regional visibility budget.

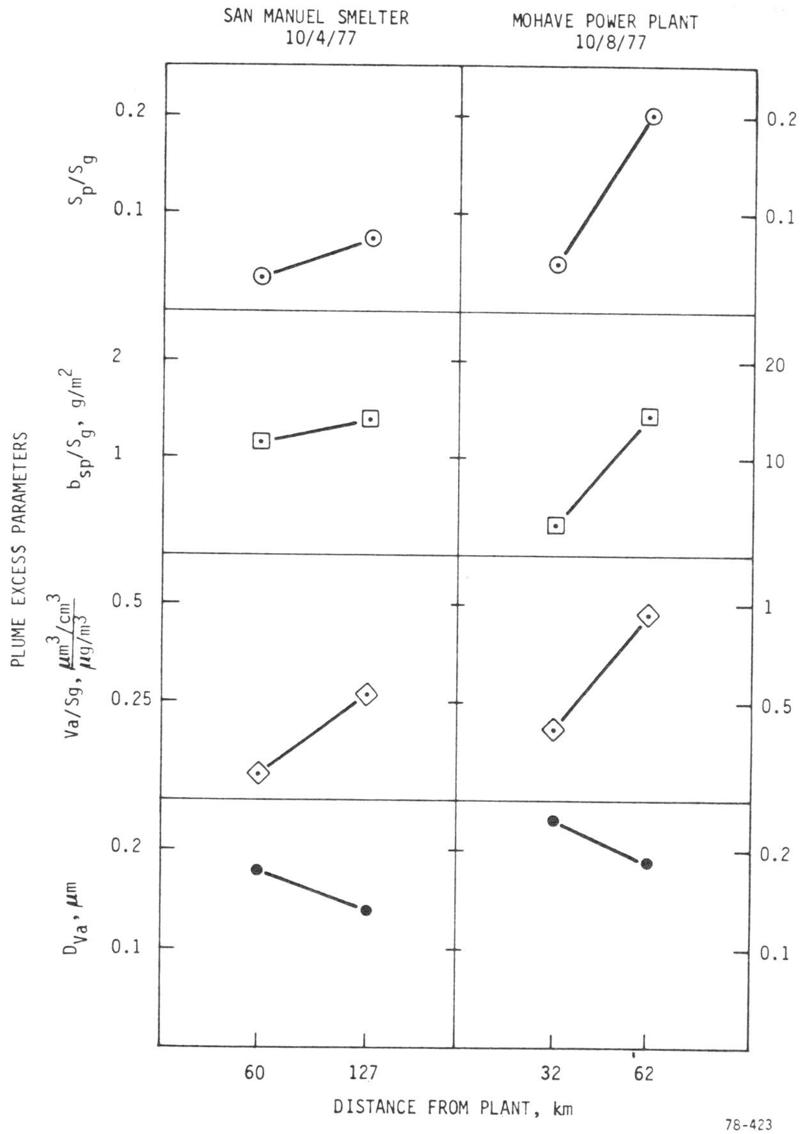

FIGURE 10. Qualitative indicators of SO_2 conversion in plumes, plotted vs. distance from plant.

In the San Manuel smelter plume at 62 km downwind on October 4, 1977, the plume excess b_{sp} value synthesized from only the measured fine particle size distribution ($D_p \leqq 1\ \mu m$), using Mie scattering functions, was $\Sigma b_{sp} = 56 \times 10^{-6}\ m^{-1}$. This is 59% of measured value: b_{sp}(measured) $= 95 \times 10^{-6}\ m^{-1}$. The detailed fine particle visibility budget given in TABLE 5 was determined from chemical composition

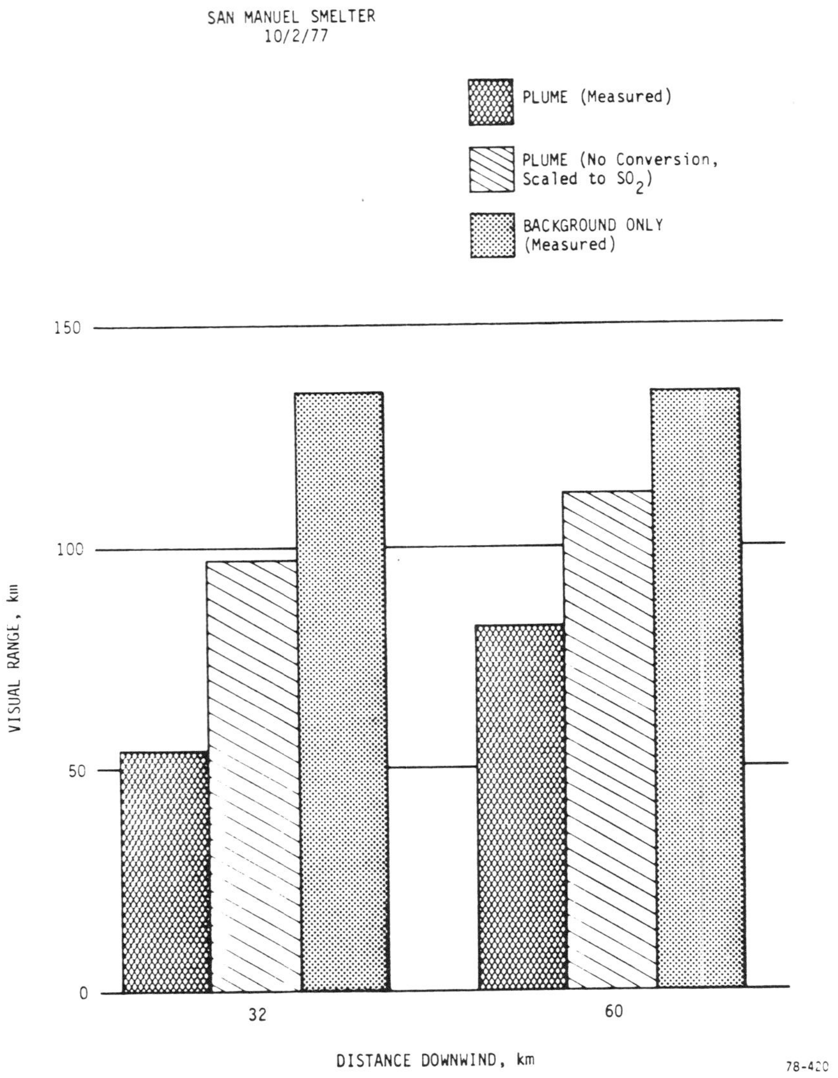

FIGURE 11. Visual range through the San Manuel smelter plume, with and without SO_2 conversion and through background only, for two downwind distances.

and size distribution, using Mie scattering calculations. Sulfate [assumed to be in the chemical form of $(NH_4)_2SO_4$] accounted for 43% of the total b_{sp} in the smelter plume of October 4, 1977; SiO_2 accounted for 10%.

The Mohave power plant plume excess b_{sp} calculation, on October 8, 1977, at 32 km downwind, was $\Sigma b_{sp}(\text{calc}) = 27 \times 10^{-6}\ m^{-1}$, while the measured value was $18 \times 10^{-6}\ m^{-1}$. This difference may be due to the loss of large particles in the nephelometer sampling line.

Calculations of the light scattering budget for the Mohave power plant are complicated by the fact that aerosol composition calculations account for only about 50% of the fine particle aerosol mass as estimated from aerosol volume measurements. Therefore, the estimated contributions of individual chemical species to b_{sp} may be somewhat high. The visibility budget for the Mohave power plant plume, at 32 km downwind on October 8, 1977, is quite different from the smelter results. Coarse particles account for 78% of the calculated b_{sp}. This high value is at least partially due to windblown dust mixing into the power plant plume at the time of the measurement. Fine particle ammonium sulfate accounted for 11% of b_{sp} $\{b_{sp}[(NH_4)_2SO_4] = 3 \times 10^{-6}\ m^{-1}\}$. Fine particle silicon accounted for 7% of b_{sp} in the Mohave plant.

SUMMARY

The atmospheric visibility-reducing aerosol in the Southwest has been experimentally characterized with respect to particle size, composition, and contribution to light scattering. Measurements were made on seven days within the mixing layer in the Southwest region, in the plume of a copper smelter (San Manuel smelter) and in the plume of a coal-fired power plant (Mohave power plant), using the MRI instrumented Beechcraft Queen Air aircraft. The aircraft was equipped to measure, and record on magnetic tape, the light scattering coefficient, Aitken nuclei count, size distribution, ozone, sulfur dioxide, nitrogen oxides, temperature, dew point, turbulence, pressure (altitude), and navigational parameters. Multistage impactor and size-fractionated filter samples were also collected, in order to determine aerosol elemental composition as a function of size. Visual range estimates were obtained by viewing distant landmarks and verified by optical photography.

Results of the study are summarized below:

1. On two regional flights over large parts of the Southwest, the visibility-reducing aerosol was quite homogeneous throughout the entire region, indicating that the visibility impairment was of regional extent.
2. The aerosol size distribution throughout the region on October 5 and 9, 1977, was bimodal, with a geometric volume mean size for the fine particles of ~0.25 microns, which is slightly smaller than the continental background aerosol average mean size (0.3 microns). The measured coarse particle mode volumetric mean size was ~5.5 microns, which is equal, within error, to the continental background.
3. The elemental size distribution, from both regional and plume impactor samples, indicated that aluminum, calcium, and iron are present predominantly in coarse particles; sulfur and titanium are present predominantly in fine partcles; and silicon and potassium have substantial concentrations in both modes.
4. Sulfur and silicon were found in nearly equal concentrations in the Southwest background data and were the elements present in the highest concentrations in the fine particles ($D_p < 1\ \mu m$). The detailed aerosol mass balance determined for the Southwest region showed that sulfur (expressed as ammonium sulfate) and silicon (expressed as SiO_2) amounted to ~53% and ~29% of the fine particle mass, respectively.
5. A total of ~93% of the measured coarse particle ($D_p > 1\ \mu m$) mass in the Southwest region was composed of elements that were present in the same abundances

relative to aluminum as in the earth's crust. This indicates that the source of these particles is either windblown dust or material with an elemental composition nearly that of crustal material, such as fly ash.

6. The light scattering budget for Southwest background aerosol on October 9, 1977, indicated that ~52% of the light scattering was due to fine particles, ~44% was due to Rayleigh scattering from gases, and ~4% was due to coarse particles. Considering only the light scattering due to particles, ~93% is due to fine particles that are composed mainly of sulfates [~53% as $(NH_4)_2SO_4$] and silicon compounds (~29% as SiO_2).

7. Mie scattering calculations of the light scattering coefficient due to particles, using the measured averaged regional size distribution, were in good agreement with the values measured with a nephelometer. This is a quantitative indication of the internal consistency of the size distribution and b_{scat} measurements, which adds credibility to the calculated light scattering budget.

8. Plume excess fine particle aerosol, i.e., the point source emission plume aerosol with the background subtracted, was composed largely of sulfur and silicon compounds for both the smelter and power plant plumes. The major elemental species in the plume excess coarse particle aerosol were aluminum, silicon, potassium, calcium, and iron, in approximately crustal abundances in both plumes. These results indicate that fine particle Si may be a good tracer for primary combustion aerosol in smelter and power plant plumes.

9. A simple semiquantitative calculation of visual plume impact is described that compares the visual range with and without the plume present. The smelter plume caused a ~90% reduction of visual range relative to the background visual range (135 km) at 8 km downwind from the plant. Increased b_{scat} and sulfate levels relative to background concentrations and a ~42% reduction of visual range were observed as far as 127 km downwind of the smelter plume on October 4, 1977. Measurements at 60 km downwind of the Mohave power plant plume on October 8, 1977, showed a ~25% reduction of visual range.

10. Sulfate aerosol was formed in Southwest power plant and smelter plumes. The measured SO_2 conversion rate from the San Manuel copper smelter on one morning between 0900 and 1230 (MST) was 0.7 ± 0.2%/hour between 60 and 127 km downwind.

11. The plume excess visibility budget indicated that fine particle sulfur and silicon species contributed ~53% to the excess b_{scat} in the San Manuel smelter plume. In the Mohave power plant plume, coarse particles were the major contributors to the excess b_{scat}, which may have been at least partially due to windblown dust on the sampling day.

Acknowledgments

We appreciate the guidance provided by William Wilson, Director of the Office of Regional Studies, and the field support provided by Tom Ellestad of EPA. The analysis herein was greatly enhanced by consultation with, and guidance from, S. K. Friedlander and R. Flagan of Caltech. We also appreciate the field support provided by J. A. McDonald, the data processing support of J. Roebuck, and the analysis support of J. Ogren and G. Markowski. Part of the data analysis has been supported by the U.S. EPA under Grant No. R 802160 to Caltech.

References

1. Trijonis, J. & K. Yuan. 1978. Visibility in the Southwest: An Exploration of the Historical Data Base. Report No. EPA-600/3-78-039. U.S. Environmental Protection Agency. Research Triangle Park, N.C.
2. Charlson, R., A. Waggoner & J. Thielke. Visibility Protection for Class I Areas: The Theoretical Basis. (To be published.)
3. Blumenthal, D. L., J. A. Ogren & J. A. Anderson. 1978. Airborne sampling system for plume monitoring. Atmos. Environ. **12:** 613.
4. Cahill, T. A. 1975. *In* New Uses for Ion Accelerators. J. Ziegler, Ed.: 1–72.
5. Macias, E. S. & R. B. Husar. 1976. Atmospheric particulate mass measurement with beta attenuation mass monitor. Environ Sci. Technol. **10:** 904.
6. Husar, J. D., R. B. Husar & P. K. Stubits. 1975. Determination of submicrogram amounts of atmospheric particulate sulfur. Anal. Chem. **47:** 2062.
7. Latimer, D. A., R. W. Bergstron, S. R. Hayes, M.-K. Liu, J. H. Seinfeld, G. Z. Whitten, M. A. Wojcik & M. J. Hillyer. 1978. The Development of Mathematical Models for the Prediction of Anthropogenic Visibility Impairment. Systems Applications, Inc. Report EF 78-63.
8. Cantrell, B. K. & K. T. Whitby. 1978. Aerosol size distributions and aerosol volume formation for a coal-fired power plant plume. Atmos. Environ. **12:** 323.
9. Whitby, K. T. 1978. The physical characterization of sulfur aerosols. Atmos. Environ. **12:** 135.
10. Mason, B. 1966. Principles of Geochemistry. J. Wiley and Sons, Inc. New York, N.Y.
11. Waggoner, A. P. 1978. Personal communications to E. S. Macias.
12. Lewis, C. W. & E. S. Macias. Composition of size-fractionated aerosol in Charleston, West Virginia. Atmos. Environ. (In press.)
13. Lyon, W. S. 1977. Trace Element Measurements at the Coal-Fired Steam Plant. CRC Press. Cleveland, Ohio.
14. Mie, G. 1908. Beiträge zur Optik trüber Medien, speziell Kolloidaler Metallösungen. Ann. Physik **25:** 377.
15. Ensor, D. S., R. J. Charlson, N. C. Ahlquist, K. T. Whitby, R. B. Husar & B. Y. H. Liu. 1972. Multiwavelength nephelometer measurements in Los Angeles smog aerosol. I. Comparison of calculated and measured light scattering. J. Colloid Interface Sci. **39:** 242.
16. Marians, M. & J. Trijonis. 1978. Empirical Studies of the Relationship between Emissions and Visibility in the Southwest. Preliminary draft of report to the U.S. Environmental Protection Agency.
17. Larson, N. I. & P. E. Billings. 1978. Second Annual Report on Arizona Copper Smelter Air Pollution Control Technology. Arizona Department of Health Services. Phoenix, Ariz.
18. Huschke, R. E., Ed. 1959. Glossary of Meteorology. American Meteorological Society. Boston, Mass.
19. Middleton, W. E. K. 1952. Vision Through the Atmosphere. University of Toronto Press. Toronto, Canada.
20. Wilson, W. E. 1978. Sulfates in the atmosphere: a progress report on project MISTT. Atmos. Environ. **12:** 537.
21. Husar, R. B., D. E. Patterson, J. D. Husar, N. V. Gillani & W. E. Wilson, Jr. 1978. Sulfur budget of a power plant plume. Atmos. Environ. **12:** 549.

AEROSOL FORMATION IN URBAN PLUMES*

Kenneth T. Whitby

Mechanical Engineering Department
University of Minnesota
Minneapolis, Minnesota 55455

Introduction

We, along with an increasing army of researchers, have been measuring aerosols and other pollutants in plumes, on roadways, in cities, in the country, and, in fact, almost everywhere that airplanes can fly and surface vehicles can travel. As the data have accumulated, we, along with everyone else, have become increasingly concerned with what it all means, not just in a qualitative sense, but in a quantitative sense that will permit our patient sponsors to develop better air pollution control measures.

One of the most important questions of the day is exactly how to measure the contribution of identifiable sources to the increasing haze that now covers a good part of the eastern United States in summer. This concern has resulted in a number of large studies of point sources, such as power plants. However, the multitude of small sources in every urban area also combine to form an urban plume that not only contributes to the pollution in the downwind areas immediately under it, but also to the general haze after the plume has lost its identity. Furthermore, we have only a sketchy idea of what such plumes contribute to the general haze on the days after they have lost their identity. Although it is well known that secondary aerosol mass and O_3 are formed in the urban plume, quantitative relationships between primary precursors and secondary products are scarce.

The 1977 Clean Air Act amendments pertaining to visibility, and the increasing pressures on states to reduce O_3 concentrations by implementing control measures, make it imperative that we quantitate the contribution of the urban plume to area-wide pollution.

This paper will review some of the things the Minnesota group and its collaborators have learned from field studies about aerosol growth in urban plumes and will present new material from the analysis of aircraft data obtained in the St. Louis urban plume in 1975. Since others will be discussing the chemistry, I will concentrate on the physical size distribution and concentration changes during the early stages, with particular emphasis on the role of nuclei formation in governing the size distribution and the rates of aerosol volume formation.

Since the nomenclature, the methods for calculating aerosol growth rates from the plume data, and the methods for calculating nucleation rates from aerosol size distribution data have been published,[1,2] methodology will not be discussed in much detail here.

Before going into the analysis of the 1975 St. Louis urban plume, important conclusions from other encounters with urban plumes will be reviewed.

*Supported by Environmental Protection Agency Grant No. R-803851-03.

0077–8923/80/0338–0258 $01.75/1

TABLE 1
CHARACTERISTIC VOLUME CONCENTRATIONS IN THE THREE MODES OF THE AEROSOL SIZE DISTRIBUTIONS MEASURED IN DIFFERENT LOCATION CATEGORIES

Category	Modal Volume Concentration ($\mu m^3\ cm^{-3}$)		
	Nuclei, V_n	Accumulation, V_a	Coarse Particle, V_c
Marine background	0.0005	0.1	12
Clean continental background	0.006	1.5	5
Average background	0.04	4.5	26
Background + aged urban plume	0.03	44	27
Background + local auto	0.6	3	39
Urban average	0.6	38	31
Urban + freeway	9	38	43

URBAN AEROSOL AND PLUME REVIEW

From operating ground-based mobile laboratories in California, Minnesota, Colorado, Missouri, Michigan, and, during the summer of 1978, in Tennessee and Alabama, we have learned a number of important things about urban aerosols, their concentration, and their size distribution.

TABLE 1 summarizes the average concentrations broken down into seven categories of location, and TABLE 2 gives the grand average volume mean size, DGV, and σ_g for each mode.† Further discussion of these data may be obtained in the California Aerosol Characterization Experiment (ACHEX) book.[3]

Following are the most important conclusions relevant to urban plume behavior from our stratification and categorization of a large amount of surface field data:

1. The urban plume only contributes significantly to the mass of the accumulation mode. From TABLE 1, note that the background plus urban plume average of 44 $\mu m^3\ cm^{-3}$ is essentially equal to the urban average of 38. However, the nuclei mode average of 0.03 $\mu m^3\ cm^{-3}$ is essentially the same as for average background.

2. The volume in the nuclei mode is governed primarily by the proximity of combustion sources. Note that the background plus local auto of 0.6 $\mu m^3\ cm^{-3}$ is the same as the 0.6 for the urban average. The nuclei mode decays by coagulation after a few hours, so that concentrations above 0.05 $\mu m^3\ cm^{-3}$ are always an indication of local sources of primary nuclei mode particles.

TABLE 2
AVERAGE GEOMETRIC VOLUME MEAN SIZE, DGV, AND σ_g FOR THE THREE MODES

Mode	DGV (μm)	σ_g
Nuclei	0.03 ± 0.007	1.6 ± 0.1
Accumulation	0.29 ± 0.06	2.0 ± 0.1
Coarse particle	6.3 ± 2.3	2.3 ± 0.2

†Nomenclature is given in an Appendix.

3. The concentration in the coarse mode is also governed by local sources for all but exceptional conditions. Except for dust storm conditions, most of the coarse particles apparently come from agriculture, roads, and other surface sources, which are more or less continuous in time and location due to their association with human activities.

Urban Plume Encounters

In addition to purposeful encounters with urban plumes, we have also encountered them accidentally and have learned a little about the general contribution of these plumes to the general aerosol observed in rural areas. Some of the more interesting observations are described below:

1. In 1970, while participating in the Second International Workshop on Cloud and Ice Nuclei in Ft. Collins, Colorado, we obtained an interesting alternation of measurements in and out of the Ft. Collins urban plume.[4] The Aitken nuclei (ANC) trace and size distribution for the night of August 8 is shown in FIGURE 1. When we

NIGHT-AUGUST 13-14, 1970

	WIND		no/cm³	S-μm²/cm³				V-μm³/cm³	
RUNS	DIR.	VEL.	CNC	S2	S3	S4	ST	V3-	VT
138-165	160°	10.6	2272	9 9	49.1	26.4	86	2.5	30
139-156	162°	5.3	11,230	39 9	86 5	41.1	168	4 1	50

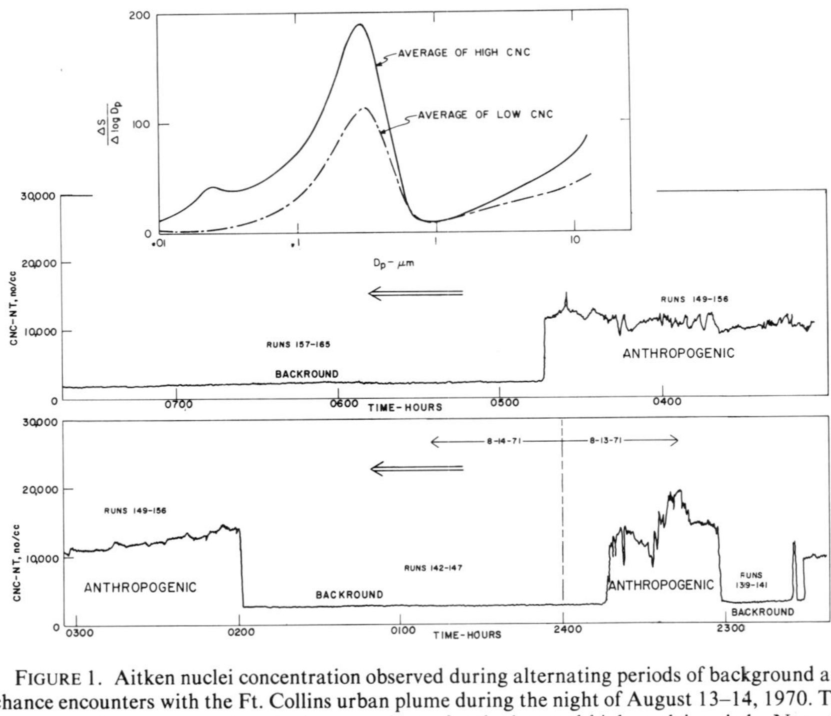

FIGURE 1. Aitken nuclei concentration observed during alternating periods of background and chance encounters with the Ft. Collins urban plume during the night of August 13–14, 1970. The average surface area size distributions are shown for the low and high nuclei periods. Note the much larger surface area in the 0.01 to 0.1 μm range during the urban plume encounters.

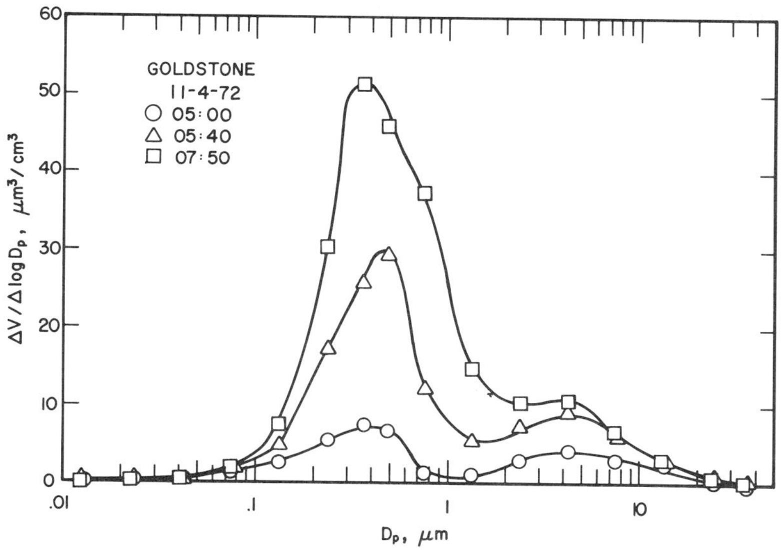

FIGURE 2. Incursion of aged aerosol from the south coast basin of California. The volume concentration increased from 4 to 32 μm^3 cm^{-3} in a span of less than three hours.

were in the Ft. Collins plume, the ANC was about 11×10^3, but when we were sampling continental background, the ANC averaged 2.2×10^3 and was very constant in time. This degree of ANC constancy is only observed when the air mass has traveled at least several hours since the last combustion aerosol contamination.

2. On the 3rd of November, 1972, while sampling at the Goldstone tracking station in California,[5] we observed a pronounced incursion of accumulation mode aerosol from the Los Angeles basin to the southwest. From FIGURE 2, it can be seen that in the space of three hours, V_a increased from about 4 to 32 μm^3 cm^{-3}. $V_a = 32$ is comparable to the average V_a observed in Pasadena during the Pasadena smog experiment. Since there was no increase of the nuclei mode during this incursion, it is clear that the aerosol was well aged.

The fact that the accumulation mode size did not change significantly suggests that the slow coagulation of the accumulation mode was being offset by a slow, but significant, new nuclei production rate.

3. During the 1972 ACHEX experiment,[3] we operated a large mobile laboratory at the sites shown in FIGURE 3. It will be seen that the *average* aerosol concentrations, even at such background sites as Goldstone and Hunter-Liggett, are well above clean continental. Examination of the data suggests that, on the average, all of these sites were influenced by urban areas at least one-fourth of the time, even though it is difficult to say conclusively that we were sampling in an urban plume.

AEROSOL GROWTH IN THE ST. LOUIS URBAN PLUME

During the summer of 1975, our group participated with other investigators in the measurement of the St. Louis urban plume by aircraft and ground mobile laboratory,

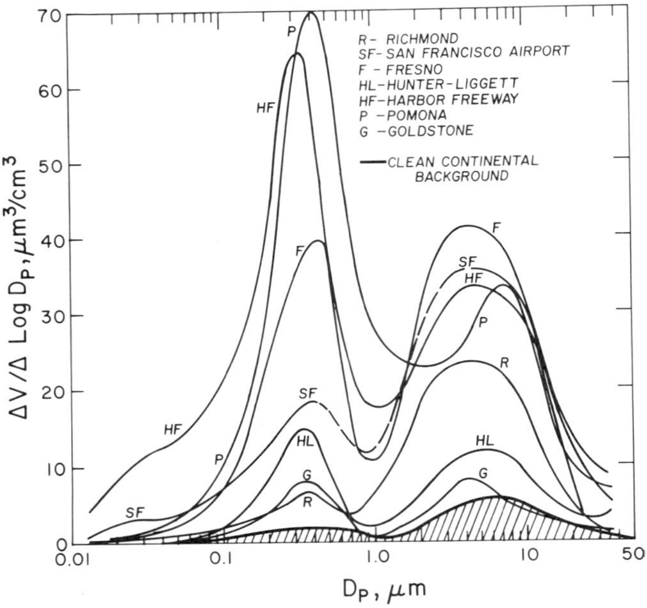

FIGURE 3. Grand average volume size distributions for seven of the sites, characterized by the mobile laboratory during the ACHEX experiment in 1972. A size distribution for clean continental aerosol is shown at the bottom for comparison. Note that even the background sites of Goldstone and Hunter-Liggett are well above clean background.

as part of EPA Project MISTT (Midwest Interstate Sulfur Transformation and Transport). Our part of the project was to furnish and reduce the data from an aerosol sizing system aboard the MRI (Meteorology Research, Inc., Altadena, Calif.) Cessna 206[9] and to help operate the large EPA mobile laboratory located at Glasgow, Missouri, about 100 miles north of St. Louis.[6] The aircraft urban plume b_{scat} and O_3 results for July 18 have already been partially described by White *et al.*,[7] and the ozone and aerosol sulfur have been modeled by Isaksen *et al.*[8] Many of the results from the EPA Mobile Laboratory (EPAML) will be described at this meeting by Ellestad.[6] In the section below, I will present size distribution and aerosol growth results from our part of the aircraft measurements and a few comments on the size distributions from the EPAML during some suspected encounters with the St. Louis urban plume.

Experimental

The aircraft and the instrument system have been described by Blumenthal *et al.*,[9] with further information on the aerosol measuring system by Cantrell and Whitby.[1] The aerosol measuring system uses four sensors: an Environment/One (Schenectady, N. Y.) Aitken nuclei counter, an MRI integrating nephelometer for integral aerosol measurements, a TSI, Inc. (St. Paul, Minn.) Model 3030 electrical aerosol analyzer

(EAA) for the 0.0056 to 1 μm size range, and a modified Royco (Menlo Park, Calif.) 218 optical particle counter for the 0.56 to 10 μm size range. Actually, data from the latter are only usable up to 3 μm, because of losses in the sampling system. Samples for the latter two instruments are grabbed and held for the measurement cycle of the instruments by an automatic bag system. On July 18, the aircraft made four spirals and 19 passes at five different distances from the St. Louis Gateway Arch toward the northeast. The sky was clear but hazy, and the wind was strong, averaging 44 km hr^{-1} during the five cross sections reported. A total of 120 distributions were measured, the first at 0649 and the last at 1855. Thus, data were obtained throughout most of the daily cycle of ultraviolet radiation. Since the wind velocities were so high, and the farthest traverse was made at 160 km, the measurements represent a maximum travel time of about 4.6 hours.

Data Analysis

The data analysis procedure used is the same as that reported by Cantrell and Whitby[1] for the Labadie power plant plume. White *et al.*'s[7] b_{scat} flows, Fb_{scat}, were converted to aerosol volume flows, FV_a, by multiplying Fb_{scat} by the average V_a/b_{scat} for each of the five cross sections. V_g, the gas flow at each cross section, was obtained by dividing Fb_{scat} by the average b_{scat} for each cross section. TABLE 3 presents the important plume transport parameters calculated in this way. When computing the average plume transport times, one hour was somewhat arbitrarily added to the pass distance from the arch for the modeling calculation in FIGURE 11, since there is a large urban area upwind of the arch.

V_a, the volume in the accumulation mode, was selected for characterizing the significant aerosol mass–related parameter, for the reasons discussed recently by Whitby.[2] V_a was determined by the trimodal fitting procedures discussed previously.[2] QN_n, the rate of nuclei formation, was calculated from the ANC and the number in the accumulation mode by the previously described procedures,[10] except that an improved algorithm for characterizing the Fuchs coagulation matrix was used, and where there were no ANC data, *NT* calculated from the EAA size distribution was used. Because the Aitken nuclei counter and the EAA were not calibrated as well in 1975 as they were in 1976, the accuracy of the calculated QN_ns is not as good.

TABLE 3

IMPORTANT PLUME TRANSPORT PARAMETERS

Pass Number	Mean Pass Time	Pass Distance (km)*	Travel Time (hr)†	V_g ($km^3\ hr^{-1}$)	FV_a ($cm^3\ hr^{-1}$)
1	0915	28	1.6	167	1.56×10^6
2	1240	66	2.5	227	3.49×10^6
3	1403	92	3.1	369	4.52×10^6
4	1647	127	3.9	738	7.59×10^6
5	1805	160	4.6	762	6.39×10^6

*From St. Louis Gateway Arch.

†44 km, corresponding to 1 hour travel time, was added to the distance from the arch before computing these times.

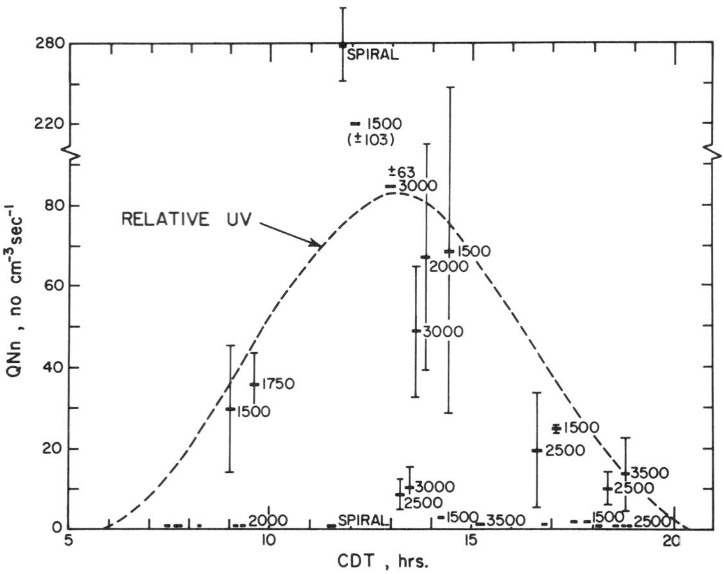

FIGURE 4. The nuclei formation rate as a function of time of sampling for the St. Louis urban plume on July 18, 1975. The numbers by each bar are the altitudes at which the sampling occurred in feet above mean sea level. Most of the numbers are averages of three to five individual measurements. The relative ultraviolet curve was calculated from measurements made by the EPAML at Glasgow, Mo.

Absolute accuracy is probably no better than a factor of two and may be as low as a factor of three. However, since the values vary from less than 0.3 in the background to over 200 in the plume, the lack of absolute accuracy does not affect the conclusions discussed later.

After QN_n had been calculated, a diurnal pattern was evident. Therefore, the ultraviolet intensity (UV) at ground level was obtained from the EPAML, which was operating at Glasgow that same day.

Plume Model for Calculating V_a *and* dV_a/dt

Early in the data analysis, it was observed that QN_n correlated approximately with time of day (FIGURE 4). In order to investigate the possibility that dV_a/dt might be related to UV, it was necessary to develop a simple plume model for calculating the aerosol volume concentration in the plume, V_a. Because the nuclei mode decays rather rapidly, QN_n is approximately proportional to the instantaneous rate of nuclei formation. On the other hand, accumulation mode aerosol is removed so slowly from the atmosphere that a plume model that takes into account the aerosol formed in the entire plume as well as plume dilution is necessary to determine whether dV/dt is a function of UV. Now, FV_a is given by:

$$FV_a = V_a\, V_g \tag{1}$$

and

$$\frac{dFV_a}{dt} = \frac{dV_a}{dt} V_g + \frac{dV_g}{dt} V_a \tag{2}$$

The hypothesis to be tested is that:

$$\frac{dV_a}{dt} = K\,UV \tag{3}$$

UV is an experimental variable that can be characterized adequately on July 18 by:

$$UV = 3.24 \exp - \left[\frac{t - 13.1}{3.93}\right]^2 \tag{4}$$

where V_a = accumulation mode volume concentration; dV_a/dt = rate of volume formation of V_a; FV_a = flow of V_a at a given plume cross section; t = central daylight time, hours; V_g = plume gas flow at travel time t_t; UV = ultraviolet intensity as a function of time; and K = constant.

For convenience, V_g was also characterized by three linear segments, as shown in FIGURE 5. In the morning, before breakup, vertical mixing is weak, and V_g can only increase slowly by horizontal mixing. When breakup occurs, V_g increases rapidly. As the day progresses, the mixing height increases slowly, and the plume volume again increases slowly. The intercepts a_1, a_2, and a_3 and the slopes b_1, b_2, and b_3 were estimated from a consideration of both the observed V_g and the mixing height data. Isaksen *et al.*,[8] however, assumed that the plume was instantaneously mixed to the top of the mixing height early in the morning. After FV_a was calculated by numerical integration from $t = 0$ to t_t, V_a was calculated from $V_a = FV_a/V_g$ for each cross section time. The value of K was set equal to 2.2, since this gave the right average values of V_a.

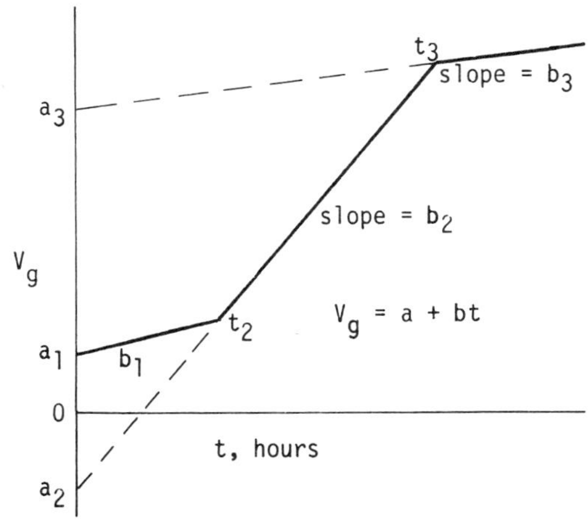

FIGURE 5. Three linear segment plume gas flow models used in the plume model calculations.

TABLE 4

AEROSOL ACCUMULATION MODE AND AEROSOL GROWTH PARAMETERS

								From Model	
Pass	DGV* (μm)	σ_{ga}*	V_a* ($\mu m^3\ cm^{-3}$)	V_{ae}† ($\mu m^3\ cm^{-3}$)	V_{ae} + BK‡ ($\mu m^3\ cm^{-3}$)	$\overline{QN_n}$ (cm^{-3})	V_a/b_{scat} ($\mu m^3\ cm^{-3}\ 10^4\ m^{-4}$)	dV/dt ($\mu m^3\ cm^{-3}\ hr^{-1}$)	V_{ae} + BK ($\mu m^3\ cm^{-3}$)
BK	0.38	1.87	5.62			<0.5			
1	0.27	2.00	9.18	9.35	12.8	22	4.04	2.5	8.7
2	0.26	2.17	18.7	15.39	21.0	188	5.56	6.4	20.7
3	0.27	2.06	14.4	12.25	17.9	40	4.08	6.2	16.3
4	0.28	1.9	19.5	10.29	20.6	22	3.34	2.8	16.9
5	0.30	2.07	11.3	8.38	14.0	12	2.90	1.5	16.8

*Three highest in plume.

†$V_{ae} = FV_a/V_g$.

‡$V_a = V_{ae} + V_{ab}$. BK = background.

Thus, only the diurnal variation of dV_a/dt that is related to *UV* can be tested, and not the ability to predict the correct absolute values of V_a.

Results and Discussion

Size Distributions at the Five Cross Sections

The size distributions in the background outside the plume and in the plume were averaged, and the variation of the parameters of the accumulation mode with distance was investigated.

DGV_a, σ_{ga}, and V_a are presented in the first four columns of TABLE 4. Also, DGV_a vs. x is plotted in FIGURE 6. From FIGURE 6, it is noted that the size is not only

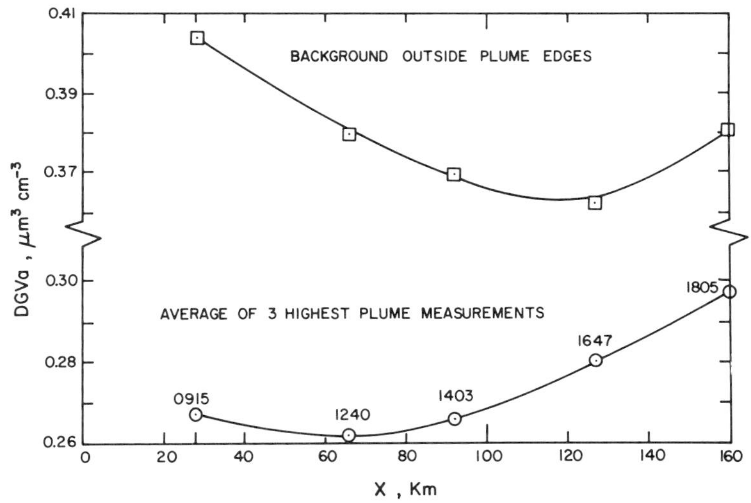

FIGURE 6. Accumulation mode geometric mean size as a function of distance from the St. Louis Gateway Arch for the background outside the plume and for the three highest plume measurements. The minimum size in the plume is observed at about 1300 at the time of maximum *UV* and greatest nucleation rate.

significantly smaller in the plume, but that the size in the plume is smallest at about noon, just when *UV* is the greatest. DGV_a was also found to correlate with V_a in the plume, according to $DGV_a = 0.44 - 0.01\ V_a$, $r^2 = 0.6$. Thus, just as was found in the Labadie power plant plume,[1] the accumulation mode size is the smallest when the aerosol formation rate is the highest. The size change outside the plume was smaller and appeared to reach a minimum somewhat later than in the plume. This latter observation is not too significant, because it assumes that the aircraft was accurately sampling a true background. There is no guarantee that this was true.

A typical background and average plume volume size distribution are shown in FIGURE 7. The smaller mean size of the plume aerosol is obvious. This is due to

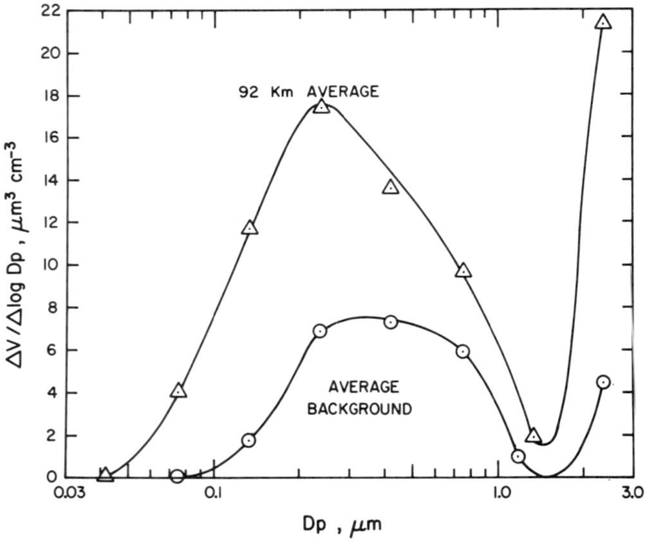

FIGURE 7. Average background and 92 km plume average volume size distributions. The plume size distributions at the other distances are similar, except for the different volumes. Because of the limitations of the sampling system, only about 1/5 of the volume in the coarse mode was sampled.

condensation on new nuclei entering the size distribution from sizes on the order of 0.01 μm.

From FIGURE 7, it is clear that only a portion of the coarse particle mode was measured due to the sampling system cut-off at about 3.5 μm. Using partial modal fitting procedures, in which the total volume in the coarse mode is estimated from fitting a log-normal distribution to the two available points, and assuming that $DGV_c = 6.3$ μm and $\sigma_{gc} = 2.3$, the average values previously determined by Whitby,[2,3] V_c was estimated to be 9 for the background, 12 for 66 km, and 16 μm^3 cm^{-3} for 92 km. These were the only distances where optical counter data were available. The accuracy of these estimates is about ±30%. Thus, there appears to be no great increase in V_c in the plume. It is likely that the small increase observed is due to dust transport from vertical mixing in the middle of the day, when the 66- and 92-km measurements were made. The background measurement was made at 0649, before mixing could have transported dust to the measurement altitude.

The average number distributions for background and the different distances are presented in TABLE 5.

Nucleation Rates

Nucleation rates, QN_n, were calculated for both the background and the plume, for groups of data that appeared to be relatively homogeneous with respect to concentration and altitude. Results are presented in FIGURE 4, along with a relative *UV* curve scaled so that it forms an envelope for all data except the very high QN_ns around 1200.

For some unknown reason, for both the spiral and the pass at 1500 ft made at about 1200, QN_n values of over 200 cm^{-3} sec^{-1} were observed. All of the rest of the calculations yielded averages below 90. QN_n values for the background outside the plume edges are also shown. These show that the nucleation rate does indeed approach low values below 1 or 2 cm^{-3} sec^{-1} in the background throughout the entire day. Thus, QN_n values larger than about 2 cm^{-3} sec^{-1} are clearly associated with the urban plume. Average QN_ns for the background and for each cross section are shown in TABLE 4.

FIGURE 8 shows the correlation between QN_n and the experimental dV_a/dt at each cross section. Except for the very high values mentioned earlier at 1240, the correlation is reasonably good. If the slope through the first three points only is used, the slope of 2.97×10^4 μm^{-3}—representing the number of nuclei formed per μm^3 of new aerosol volume formed—can be compared with the value of 3.1×10^4 for the Labadie power plant plume.[10] The similarity of this ratio for the power plant plume and the urban plume suggests that the ratio is physically governed. The similarity may also be due to the fact that the secondary aerosol in both the Labadie and St. Louis urban plumes is mostly sulfate.

As a further test of the hypothesis that the formation of secondary pollutants in the

TABLE 5

AVERAGE NUMBER SIZE DISTRIBUTIONS FOR THE BACKGROUND AND FOR THE 3 HIGHEST READINGS AT EACH OF THE 5 CROSS SECTIONS

	Number in Size Range (no. cm^{-3})					
D_p (μm)	Average Background	28 km	66 km	92 km	127 km	160 km
.0056						
	20372	47633	1.25E5*	1.76E5	1.76E5	1.10E5
.01						
	2695	7343	1.097E4	1.41E4	7.77E4	8453
.0178						
	392	3279	4917	2830	1857	1757
.0316						
	333	2400	9867	4353	4023	2443
.056						
	297	2897	7230	4607	2537	2543
.1						
	359	1477	2813	2367	3143	1637
.178						
	245	408	794	626	995	484
.316						
	47	53.9	110	87.7	124	75.9
.56						
	6.7	7.8	15.0	10.9	13.7	10.1
1						
	0.16	—	0.32	0.36	—	—
1.78						
	0.16	—	0.33	0.77	—	—
3.16						
	1.5×10^{-3}	—	0.07	0.006	—	—
5.6						

*1.25E5 = 1.25×10^5.

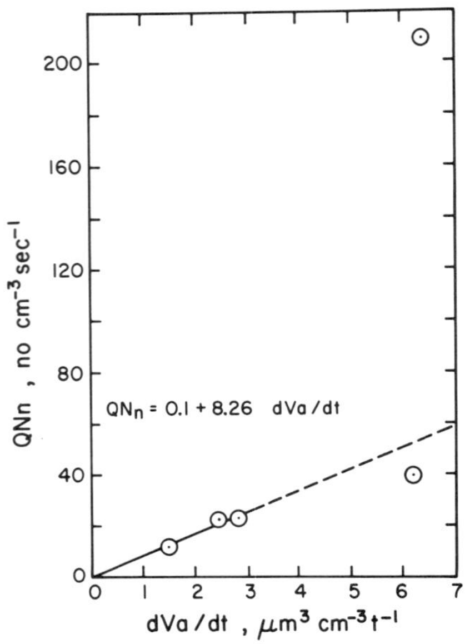

FIGURE 8. Correlation of the nucleation rate, QN_n, with the aerosol volume formation rate, dV_a/dt. The very high value is for the cross section in the late morning. The reason for this very high QN_n value is not known.

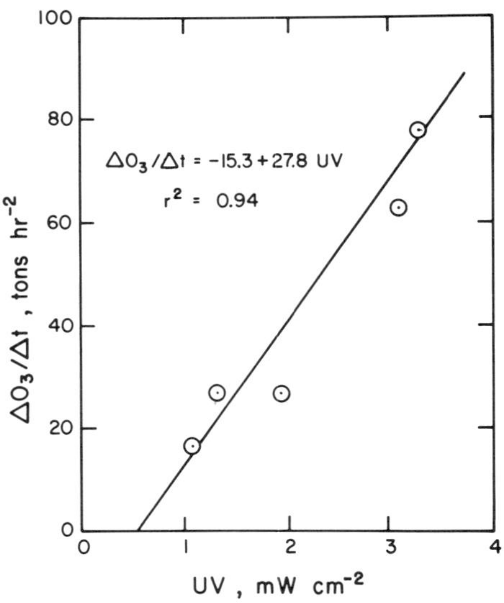

FIGURE 9. Correlation of the excess ozone formation rate with *UV*. This good correlation shows that the excess ozone is formed photochemically.

urban plume was primarily photochemical, the ozone formation rate, as published by White *et al.*,[7] was correlated with *UV*. FIGURE 9 shows that the correlation is good.

Plume Aerosol Volume Growth Rate

One of the prime objectives of our work was to determine the rate at which aerosol volume is formed in the plume. There are several ways to calculate dV_a/dt. One way is to calculate the flow at each cross section measured, and then calculate that formed between each cross section. This method, which was used previously for the Labadie power plant plume,[1] works well if the gas flows are known accurately. However, if there is uncertainty in the plume gas flows, there is considerable scatter in the

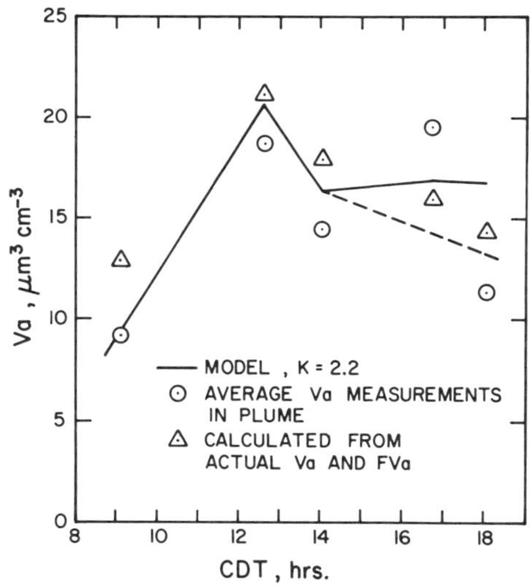

FIGURE 10. Experimental and model accumulation mode aerosol volumes as a function of the time of measurement of the cross sections. The dashed line for the model results from increasing the plume gas flow by 32% above that from the model of FIGURE 5.

calculated dV_a/dt's. Therefore, another method was chosen, which uses the model discussed earlier to determine K in Equation 3 by comparison of aerosol concentrations calculated from the model with the actual concentrations measured in the plume. K was adjusted until the average experimental concentrations agreed with those from the model.

In making these calculations, only the first two segments of the V_g model in FIGURE 5 were used, since the maximum travel time was 4.6 hours. For this calculation: $a_1 = 1.21 \times 10^{17}$; $b_1 = 7.05 \times 10^{16}$; $a_2 = -5.97 \times 10^{17}$; $b_2 = 4.67 \times 10^{17}$; and $t_2 = 2.8$ hr. Results are shown in FIGURE 10. The V_as shown are the actual concentrations, since the average background of 5.6 $\mu m^3\ cm^{-3}$ has been added to the excess V_as calculated from the model.

Experimental V_as calculated by two different procedures are shown. The circles are the average of the actual measured aerosol volumes in the plume. The triangles are the V_as calculated from $V_a = (FV_a/V_g)$ + Average Background.

Agreement between the experimental values and the model results is reasonably good, except that the actual concentrations at 1800 are significantly less than those predicted by the model. Probably this results from an underprediction of the plume gas flow by the model. If it is assumed that the plume has expanded up to the top of the mixing layer by the end of the day, then the model underpredicts the gas flow by about 32%. If the last model concentrations are reduced by multiplication by 0.68, the dashed line in FIGURE 10 is obtained. This is in better agreement with the measured values.

Another adjustment was made to the travel times during the model calculations. The original distance measurements were from the St. Louis Arch, which is on the

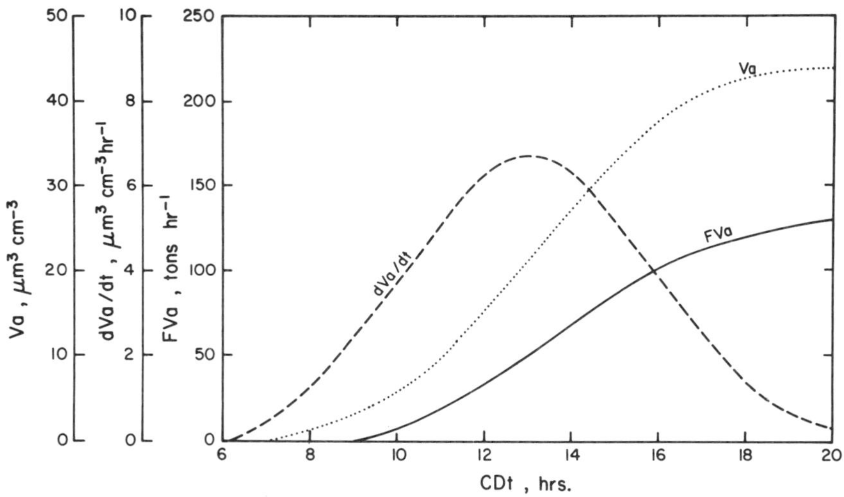

FIGURE 11. Model calculations of the excess V_a, FV_a, and dV_a/dt for a plume air volume leaving St. Louis at 0600. Most of the growth occurs in the afternoon. The final excess aerosol flow is about 130 metric tons hr^{-1}.

downwind side of the St. Louis metropolitan area. Therefore, one hour of travel time was added to the times calculated from the distances from the arch. The actual travel times used are shown in TABLE 3.

The best fit K was 2.2 $\mu m^3\ cm^{-1}\ hr^{-1}\ mw^{-1}$. Using this K and the three-segment model for V_g with: $a_1 = 1.21 \times 10^{17}$; $b_1 = 7.05 \times 10^{16}$; $a_2 = -5.97 \times 10^{17}$; $b_2 = 4.67 \times 10^{17}$; $a_3 = 2 \times 10^{18}$; $b_3 = 7.05 \times 10^{16}$; $t_2 = 1.9$ hr; and $t_3 = 6.6$ hr; the results shown in FIGURE 11 were computed by integrating from 0600 to the time shown. The peak dV_a/dt predicted is 6.7 $\mu m^3\ cm^{-3}\ hr^{-1}$ at about 1300. The maximum aerosol flow, FV_a, is 130 metric tons hr^{-1} for the plume volume that left the city at 0600. The predicted aerosol volume concentration is 43 $\mu m^3\ cm^{-3}$. As pointed out above, this is probably 20% to 30% high, because the model underpredicts the plume gas flow late in the day when the mixing height stops increasing.

The total secondary aerosol formed in the urban plume from 0600 to 2000 CDT, corresponding to one irradiation day, was also computed by integrating the aerosol produced in plume volumes starting out at different times during the day. For an aerosol density of 1.0 gm cm^{-3}, the total mass produced was 1000 metric tons per day, or about 1 pound/day per capita, since the population of the St. Louis metropolitan area is about two million.

The average dV_a/dt from 0600 to 2000 was 3.1 μm^3 cm^{-3} hr^{-1}. Recently, McMurry *et al.*[11] estimated dV_a/dt's of from 1 to 2, from aerosol growth measurements made during August 1978 in the Great Smoky Mountains in Tennessee. These values are only about one-tenth of the values estimated by Husar *et al.*[12] for Los Angeles on a smoggy day. For the moderately hazy conditions existing in the Smokies and in St. Louis on July 18, 1975, dV_a/dt's in the 1 to 3 μm^3 cm^{-3} hr^{-1} range appear to be reasonable, since it seems to take 4 to 7 days for a hazy air mass to form. The average V_a/b_{scat} is equal to 4 μm^3 cm^{-3} 10^{-4} m^{-1} (TABLE 3). Husar *et al.*[13] have pointed out that it takes about 5 to 7 days for the b_{scat} to increase to 6 or so in the summer in the east central United States. Since 7 days $\times$ 3 μm^3 cm^{-3} = 21 μm^3 cm^{-3}, a b_{scat} of 6 would represent an aerosol volume concentration in the 18 to 24 μm^3 cm^{-3} range.

Comparisons of V_a with Particulate Sulfur, S_p

White *et al.*[7] stated that they found S_p values as high as 7.5 μg m^{-3} for the 127-km traverse. From TABLE 4, it is seen that the average V_a is 20.6 μm^3 cm^{-3} during this traverse. At the 70% to 85% relative humidities existing during this cross section, this V_a would correspond to about 7.5 μg m^{-3} of H_2SO_4. Also, the average background V_a of 5.6 μm^3 cm^{-3} corresponds to about 2 μg m^{-3} of H_2SO_4, as compared with White *et al.*'s estimates[7] of below 3.6 μg m^{-3}. It must be concluded that the S_p values in White *et al.*'s paper[7] must really be sulfate and not sulfur. If their values were sulfate, then the 7.5 μg m^{-3} of H_2SO_4 in equilibrium with water at 75% relative humidity (RH) would correspond to V_a = 21 μm^3 cm^{-3}, in satisfactory agreement with the measured value of 20.6.

Certainly, within the accuracy of these plume measurements, our results support the conclusion that all of the aerosol can be accounted for by sulfur compounds.

CONCLUSIONS

1. There are few places in the United States where aerosol from urban areas in general does not increase the average accumulation mode concentration.
2. Nucleation rates exceeding 200 cm^{-3} sec^{-1} were found in the St. Louis urban plume on July 18, 1975. The nucleation rate was found to correlate with the ultraviolet intensity.
3. Using a simple model in which the accumulation mode aerosol volume formation rate was assumed to be proportional to the ultraviolet intensity, aerosol concentrations similar to those observed were predicted.
4. Using the above model, it was estimated that St. Louis produces about 1000 metric tons of secondary aerosol in one irradiation day.
5. Our results agree with those of White *et al.*[7] and Isaksen *et al.*[8] that the entire secondary aerosol can be accounted for by assuming it is a sulfate such as H_2SO_4.

SUMMARY

This paper summarizes some of the results from encounters with urban plumes during field studies in California, Missouri, and Tennessee. Evidence indicates that there are no areas within a few hundred km of urban areas that are not influenced by aerosols from large urban areas part of the time.

Analysis of aerosol measurements in the St. Louis urban plume on July 18, 1975 showed nuclei formation rates of up to several hundred cm^{-3} hr^{-3}. Nuclei formation rates and ozone formation rates correlated well with the ground-level ultraviolet intensity. Aerosol volume formation rates of up to 6.7 μm^3 cm^{-3} hr^{-1} during the middle of the day were observed. These also appeared to correlate well with the ultraviolet radiation. Using a simple aerosol growth model based on the experimental results, the total weight of secondary aerosol formed in the plume during one irradiation cycle was found to be 1000 metric tons day^{-1}.

ACKNOWLEDGMENTS

I should like to acknowledge Dr. Bruce Cantrell's work during the experimental work in St. Louis, and the work of Loren Tarvestad and Rajagopal Vijayakumar in doing the data reduction.

REFERENCES

1. CANTRELL, B. K. & K. T. WHITBY. 1978. Aerosol size distributions and aerosol volume formation rates for a coal-fired power plant plume. Atmos. Environ. **12:** 323.
2. WHITBY, K. T. 1978. The physical characteristics of sulfur aerosols. Atmos. Environ. **12:** 135.
3. WHITBY, K. T. & G. M. SVERDRUP. 1979. California aerosols: their physical and chemical characteristics. *In* the ACHEX Hutchinson Memorial Volume. (To be published.)
4. WHITBY, K. T. & R. B. HUSAR. 1970. The particle size spectra of atmospheric aerosols. Presented at the 2nd International Workshop on Condensation and Ice Nuclei, Ft. Collins, Colo., August 5–19.
5. SVERDRUP, G. M., K. T. WHITBY & W. E. CLARK. 1975. Characterization of California aerosols. II. Aerosol size distribution measurements in the Mojave Desert. Atmos. Environ. **9:** 483.
6. ELLESTAD, T. G. 1980. Aerosol composition of urban plumes passing over a rural monitoring site. Ann. N.Y. Acad. Sci. (This volume.)
7. WHITE, W. H., J. A. ANDERSON, D. L. BLUMENTHAL, R. B. HUSAR, J. D. HUSAR & W. E. WILSON. 1976. Formation and transport of secondary air pollutants: ozone and aerosols in the St. Louis urban plume. Science **194:** 187.
8. ISAKSEN, I. S. A., E. HESSTVEDT & O. HOV. 1978. A chemical model for urban plumes: test for ozone and particulate sulfur formation in St. Louis urban plume. Atmos. Environ. **12:** 599.
9. BLUMENTHAL, D. L., J. A. OGREN & J. A. ANDERSON. 1978. Airborne sampling system for plume monitoring. Atmos. Environ. **12:** 613.
10. WHITBY, K. T., B. K. CANTRELL & D. B. KITTELSON. 1978. Nuclei formation rates in a coal-fired power plant plume. Atmos. Environ. **12:** 313.
11. MCMURRY, P. H., B. Y. H. LIU & K. T. WHITBY. 1978. Recent progress in fine particle research at the University of Minnesota. Presented at the Gesellschaft für Aerosolforschung (GAF) Conference, Vienna, Austria, Sept. 26–28.

12. HUSAR, R. B., K. T. WHITBY & B. Y. H. LIU. 1972. The physical mechanisms governing the dynamics of Los Angeles smog aerosol. J. Colloid Interface Sci. **39:** 211.
13. HUSAR, R. B., N. V. GILLANI, J. D. HUSAR, C. C. PALEY & P. N. TURCU. 1976. Long-range transport of pollutants observed through visibility contour maps, weather maps and trajectory analysis. Presented at the 3rd Symposium on Atmospheric Turbulence, Diffusion, and Air Quality, Raleigh, N.C.

APPENDIX

NOMENCLATURE

a_1, a_2, a_3	= constants in plume gas flow model (see FIGURE 5), $cm^3\ hr^{-1}$.
ANC	= number concentration measured by Aitken nuclei counter, no. cm^{-3}.
b_1, b_2, b_3	= slopes in plume gas flow model (see FIGURE 5), $cm^3\ hr^{-2}$.
b_{scat}	= light scattering coefficient, m^{-4}.
DGV	= volume geometric mean diameter, μm.
Fb_{scat}	= excess b_{scat} flow, $10^{-4}\ m^{-1}\ km^{-3}$ hr.
FV_a	= excess V_a flow in plume, $\mu m^3\ cm^{-3}\ hr^{-1}$.
K	= constant, $\mu m^3\ cm^{-1}\ hr^{-1}\ mw^{-1}$.
NT	= total number concentration, no. cm^{-3}.
QN_n	= nucleation rate, no. cm^{-3} sec, or no. cm^{-3} hr.
S_p	= aerosol sulfur concentration, $\mu g\ m^{-3}$.
t	= central daylight time, hr.
t_2 & t_3	= See FIGURE 5, hr.
t_t	= plume travel time, hr.
UV	= ultraviolet intensity at ground level, $mV\ cm^{-2}$.
V	= volume concentration, $\mu m^3\ cm^{-3}$.
V_a	= accumulation mode volume concentration, $\mu m^3\ cm^{-3}$.
V_{ae}	= excess V_a above background, $\mu m^3\ cm^{-3}$.
V_g	= plume gas flow volume, $km^3\ hr^{-1}$ or $cm^3\ hr^{-1}$.
x	= distance, km.
σ_g	= geometric standard deviation.

Subscripts

a, c, n	= accumulation, coarse particle, and nuclei modes, respectively.
b	= background.
e	= excess over background.
g	= gas.
p	= plume.

FORMATION AND TRANSPORT OF OZONE AND AEROSOLS IN POWER PLANT PLUMES*

Noor V. Gillani

Mechanical Engineering Department
Washington University
St. Louis, Missouri 63130

William E. Wilson

Regional Field Studies Office
Environmental Protection Agency
Research Triangle Park, North Carolina 27711

INTRODUCTION

In industrial regions, it is common for ozone and light scattering aerosols to occur concurrently. Seasonally, both ozone and aerosol sulfate concentrations exhibit peaks in the summer months. Diurnally, the formation of both is maximum on warm sunny afternoons. In the eastern United States, the spatial distribution patterns of ozone and sulfates are also believed to have a similar regional character. The cooccurrence of elevated levels of ozone and light scattering aerosols, particularly sulfates, has been shown in regional scale episodes,[1] in large urban plumes,[2] in aging power plant plumes,[3–5] and in smog chamber experiments.[6,7]

The regional distribution of ozone and sulfates is now generally accepted as being due largely to the long range transport and transformations of the precursors (SO_x, NO_x, hydrocarbons, etc.) emitted from large urban-industrial and power plant sources.[8,9] Owing to the interaction of many plumes over large areas, and the complex nature of atmospheric transport and chemistry, the task of designing and implementing a rational control strategy for secondary pollution, based on the control of precursor emissions, is a most difficult one. Definitive scientific results about the formation and transport of ozone and fine aerosols (sulfates, nitrates, as well as organics) are needed.

White *et al.*[2] examined the formation of ozone and light scattering aerosols during the daytime transport of a well-defined urban plume of St. Louis out to nearly 200 km under particularly favorable transport conditions. In that analysis of aircraft measurements, it was observed that following an initial period ($\sim$1 hr) of limited activity, ozone and light scattering aerosols, particularly sulfates, were formed in the plume substantially in excess of the background levels. The main impact region was more than 50 km downwind of the source area. For various practical reasons, the plume was not sampled beyond 200 km. This range was exceeded in the flight of the DaVinci II balloon[10] embedded in the St. Louis plume. The secondary pollution that formed in the plume during the afternoon was transported overnight in a layer decoupled from the ground. The elevated pollution of St. Louis was then entrained into the growing mixing layer of the next morning and brought down to ground in western Indiana.

*Supported by U.S. Environmental Protection Agency Grant No. R805918.

0077–8923/80/0338–0276 $01.75/1 © 1980, NYAS

Gillani *et al.*[4] and Gillani[11] described in quantitative detail the formation and transport of sulfates in the Labadie power plant plume over distances exceeding 300 km on two separate occasions (July 9 and 18, 1976; see FIGURE 1). Both times, plume transport was tracked through day and night, providing unique data that led to direct evidence and documentation of the dominant role of sunlight in the oxidation of SO_2 to sulfates in the atmosphere (FIGURE 2; upper). The maximum gas-to-particle conversion rate of sulfur was observed in the afternoon hours, when ozone formation also peaked to levels well in excess of the background. The dependence of sulfate formation

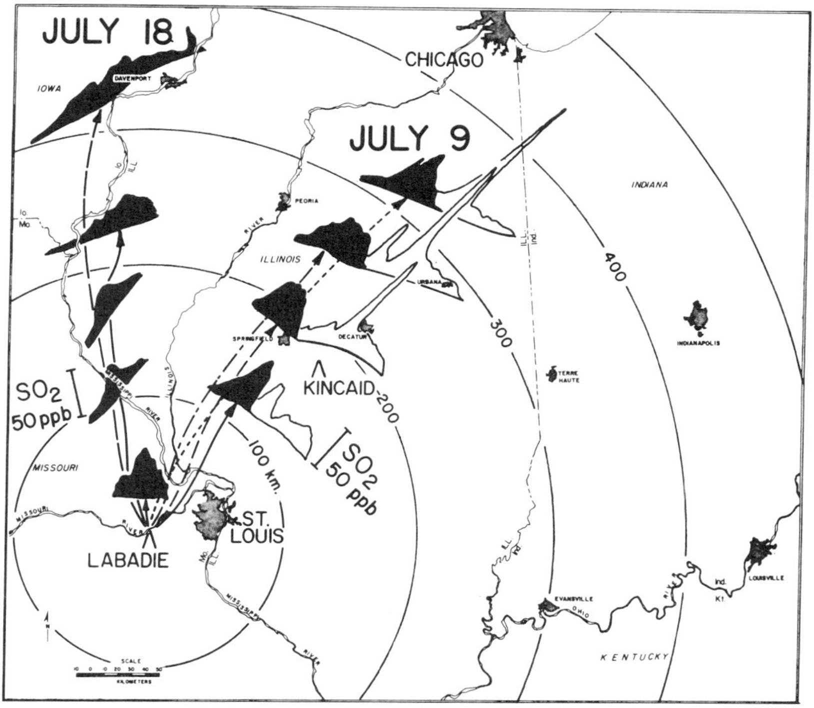

FIGURE 1. Crosswind plume profiles of SO_2 during selected constant-altitude aircraft traverses on July 9 and July 18, 1976. The Labadie power plant plume sections are shaded. Also shown are backward trajectories for the Labadie plume (from Reference 4).

rate on sunlight, while crucial, is not unique. On July 9 and 18, different amounts of particulate sulfur formed in the plume for the same sunlight dose. A number of factors other than sunlight, but driven by sunlight, could also influence sulfate formation rate. Chief among these factors are enhanced ambient and dew point temperatures, atmospheric mixing, convective cloud formation, and photochemistry. Consequently, the roles played by all these factors except clouds (no data available) were examined. The difference in sulfate formation rates for the two days could be explained by differences in mixing and photochemical activity (ozone formation). Mixing and ozone formation were significantly more intense on July 18, the day with the higher

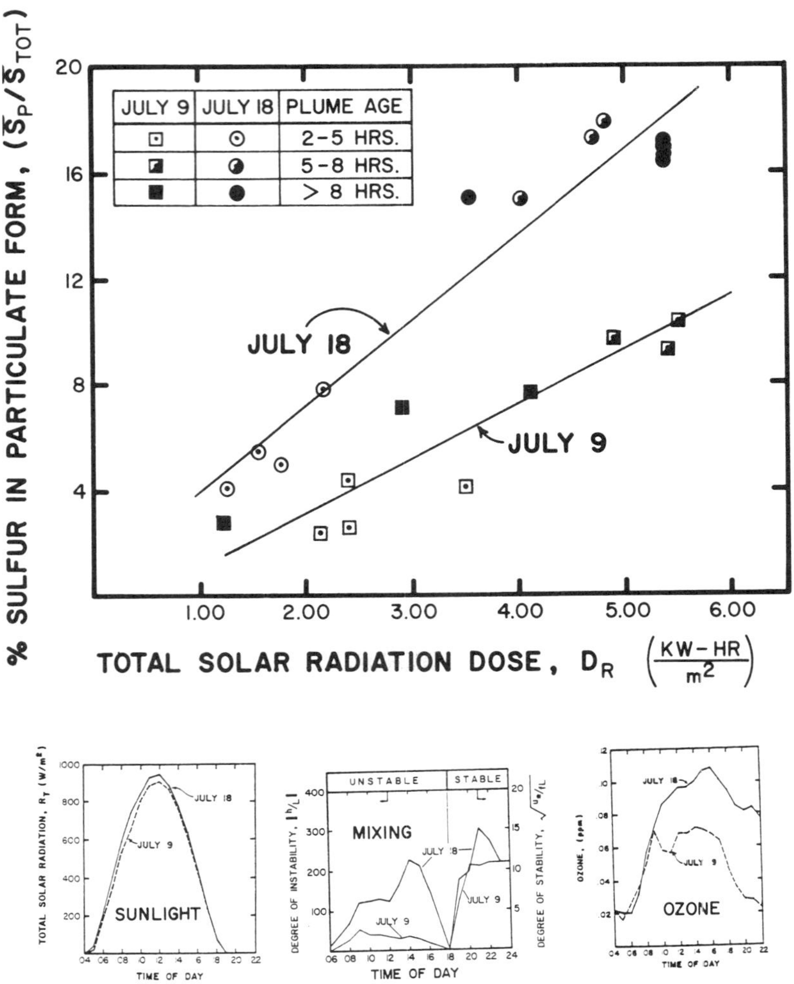

FIGURE 2. **Upper:** Ratio of particulate-to-total sulfur concentrations (average for traverse) as a function of total plume solar radiation dose, D_R (time integral of incoming solar radiation flux, R_T, between plume release time and plume sampling time). **Lower:** Comparisons of the temporal variations of incoming solar radiation flux, R_T, atmospheric stability parameter, and ozone concentrations for July 9 and 18, as measured at RAMS Station 118 (from Reference 4).

sulfate formation rate (FIGURE 2; lower). Based on the motivation provided by the above results, Husar *et al.*[12] examined the Labadie plume data of five more days and verified the diurnal variability of the sulfate formation rate. Then, using this observed diurnal pattern, they extrapolated the results and suggested that as much as 30–45% of the sulfur emissions from the power plant may ultimately be converted to sulfates. This process would take several days of plume transport ranging over a regional scale.

The urban and power plant plume data referred to in the above studies[2,4,9,11,12] were all collected as part of Project MISTT, sponsored by the U.S. Environmental Protection Agency (EPA). In this paper, the MISTT data of the Labadie power plant plume are reexamined to investigate the ozone-aerosol link more closely, and also to identify the ambient conditions most conducive to the generation of the smog-haze pollution. Hopefully, such information will aid in designing control measures that properly reflect the role of interactions between anthropogenic emissions of the precursors and the ambient atmosphere. Preliminary observations in the data collected as part of EPA's Project STATE/Tennessee Plume Study (August 1978) appear to support the conclusions drawn in this paper.

EXPERIMENTAL

The results presented in this paper are based primarily on the MISTT data of Labadie plume transport on 9, 14, 18, and 30 July, 1976. The data were collected aboard the following two aircraft: Washington University (WU) AeroCommander (July 9, 18); Meteorology Research, Inc. (MRI) Cessna 206 (July 14, 18, 30). The aircraft were instrumented for continuous measurements of a variety of air quality (gaseous as well as aerosol), meteorological, and navigational parameters. Particulate sulfur measurements, however, were based on filter sample collections with subsequent laboratory analysis. Grab samples of the ambient air were also collected in the MRI aircraft and subsequently analyzed for hydrocarbon species. More detailed descriptions of the airborne platforms, instrumentation, and measurements are given in References 12 and 13. In the MISTT plume study, the standard aircraft sampling pattern included crosswind plume traverses and vertical spirals at different distances downwind of the source, supplemented by sampling in the background air.

For the analysis of the MISTT data, use is also made of the pertinent data of RAMS Station 118 (a rural station to the south of the metropolitan St. Louis area) of the Regional Air Pollution Study (RAPS), also sponsored concurrently by EPA.

The Labadie power plant of the Union Electric Company is located 60 km west of the Gateway Arch which is at the Mississippi riverfront in downtown St. Louis. It is a coal-fired, base-load plant with a rated capacity of 2400 MW. The plant has four boiler units feeding into three 214-m stacks. The monthly average sulfur content of the fuel in July 1976 was 3.17%. The sulfur emission rates for the four days of interest here ranged between 10 and 14 kg/s (as SO_2).

Labadie plume transport for July 9 and 18 is shown in FIGURE 1, and for July 14 and 30 is shown in FIGURE 3. The July 9 plume mixed into a warm, moist, and somewhat stable gulf air mass. Wind speeds ranged from 6–8 m/s in the morning, increased gradually in the afternoon, and reached 20 m/s by midnight. The July 18 air mass belonged to a slow circulation (afternoon wind speed, 3–4 m/s) around a continental high pressure cell to the east. Skies in the sampling area remained very clear, and visibility remained unrestricted throughout the day. On July 14 and 30, air mass trajectories were generally from Texas and Oklahoma. On July 14, the sampling area was in a wedge formed by a warm front to the east and a cold front to the west. Skies were generally clear, except for patchy fog in the morning. A semistationary front continued to oscillate in the sampling area through most of July 30. Skies were partly cloudy to cloudy during the sampling period.

Approach of Data Analysis

A central hypothesis and conclusion of this paper is that the condition of the background air mass into which the plume is emitted and plume-background interactions are of paramount significance in the formation of both ozone and aerosols during the long range transport of power plant plumes. During the photochemically active daylight hours, chemical reactions between two species that are both of primary origin will become rapidly quenched[14] as the plume dilutes. Thereafter, and for the entire subsequent plume transport, the primary plume species (e.g., SO_2, NO_x) will

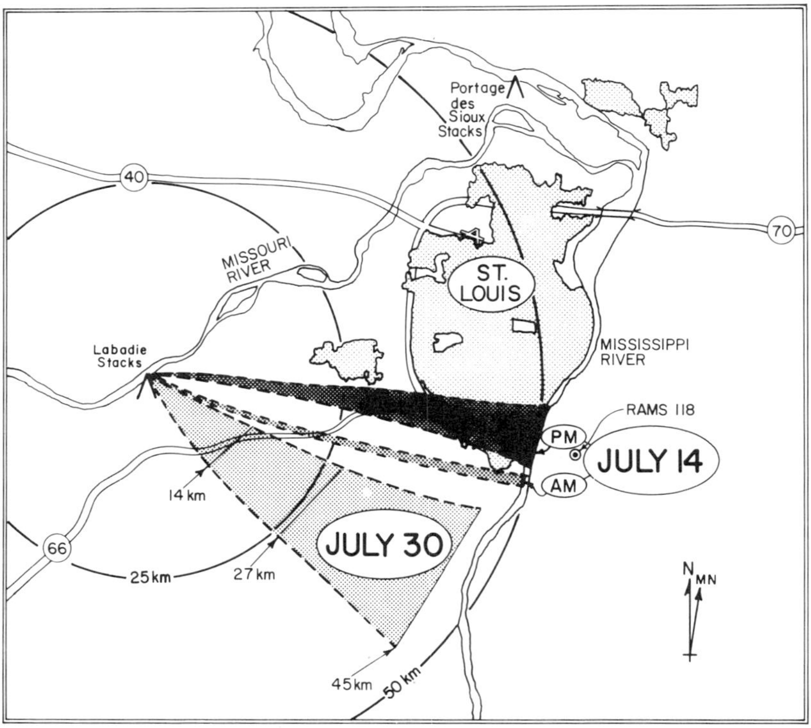

FIGURE 3. Labadie plume transport and aircraft sampling sections for July 14 and July 30, 1976.

react ever increasingly with reactants of background or secondary origin. The local ratio of background to plume air is a crucial parameter, governing the formation rate of ozone as well as aerosols. This ratio varies along the plume centerline for increasing distance of plume transport, as well as along crosswind plume traverses at a given plume cross section. In past studies, in general, integrals of crosswind plume data or integrals of cross-sectional plume data at increasing downwind distances have been compared to infer plume kinetics.[2,4,11] This approach is hampered by two practical shortcomings with respect to investigation of plume-background interactions. Firstly, the maximum downwind range covered has been limited to 300 km under the most favorable transport conditions. Hence, only a fraction of the full range between 100% plume air (at the source) and 100% background air (very far downwind) is covered. Secondly, the measurements are at discrete points of the spectrum because the aircraft measurements are not truly Lagrangian. In contrast, each crosswind plume traverse by the aircraft during the early hours of plume transport yields a continuous profile of almost the entire spectrum between nearly 100% plume air (at the core) and 100% background air (at plume edges).

An additional attractive feature of the crosswind plume data at early transport times is that they can be obtained in a single plume traverse by the aircraft within a few minutes. The data thus approximate a steady state "snapshot" of a plume section with respect to pollutant concentrations, while the effect of variable mixing is highlighted. The plume core approximates conditions of plume-to-background ratio that would exist in a stable, coherent plume, while the plume edges approximate conditions representative of a rapidly dispersing plume. In other words, the role of mixing in plume chemistry is isolated for study, while other factors remain in a steady state. Needless to say, the role of mixing in plume-background interactions is crucial. Plume-background interactions are probably especially important in power plant secondary pollution formation, because an essential ingredient (reactive hydrocarbons) for significant photochemical ozone-aerosol formation is missing in power plant primary emissions and must be entrained by mixing from the background air.

For the reasons stated above, the emphasis in this paper is on examination of crosswind plume profiles of primary, as well as secondary, species.

Our approach in data examination and presentation is motivated by one other hypothesis, which concerns the practical significance and utility of ozone in assessing aerosol formation rate. The chemistry that leads to aerosol formation in plumes is very complex. A very large number of chemical species have been identified as relevant reactants, and an even larger number of reactions have been postulated. Many of these species have never been measured in the atmosphere, and many of the reaction rates are either not known at all, or are known with little confidence. For practical simulations at the present time, therefore, it is necessary to resort to lumped parametrizations and empirical relations, based on a very limited number of species acting as surrogates for the many. The development of such parametrizations or empirical relations, to be used as submodels representing the chemical transformation terms in long range transport models, is one of the principal, and most pressing, goals of many current research programs. In this paper, data are presented that suggest a basis for the possible use of ozone as a surrogate for the radical species believed to be the principal agents for the oxidation of SO_2 and NO_2 in the gas phase. The theoretical and empirical bases for such use of ozone as a surrogate quantity are briefly examined. Finally, the conditions believed to be conducive to ozone generation in power plant plumes are identified and are shown to be also favorable to aerosol formation.

Ozone-Aerosol Link: Theoretical Basis

Aerosol Formation

Sulfate, nitrate, and organic fine aerosols are formed from the precursor gases (SO_x, NO_x, HC, CO) in the atmosphere. The oxidation of SO_2 in the gas phase is believed to occur principally by three key free radicals: $OH\cdot$, $HO_2\cdot$, and $RO_2\cdot$.[15,16] If the atmosphere is "clean," i.e., hydrocarbon-poor, the role of $OH\cdot$ is predominant. $HO_2\cdot$ and $RO_2\cdot$ will become increasingly important as more reactive hydrocarbons become available.† In the liquid phase, O_3 and H_2O_2 may be the dominant oxidizing agents during long range plume transport.[17,18] All five oxidizing species named above, viz., $OH\cdot$, $HO_2\cdot$, $RO_2\cdot$, O_3, and H_2O_2, are photochemically generated. Of the five, only O_3 has been measured extensively in the atmosphere. The difficulty in measuring the other species in the atmosphere is a significant hindrance to a full understanding of the mechanisms responsible for the conversion of SO_2 to sulfates. It is also for this reason that the validity of ozone as a reasonable surrogate for the radical species would be a substantial advantage.

The oxidation of NO_2 by $OH\cdot$ or O_3 and of NO by $HO_2\cdot$ can lead to the formation of HNO_3.[19] HNO_3 has a high vapor pressure at ambient temperatures and cannot exist as aerosol except in the presence of water or sulfuric acid droplets.[19] The formation of particulate nitrate may result from gas phase reactions of HNO_3 with NH_3 or from liquid phase oxidation of dissolved NO_x.

Ozone Formation

The only proven chemical source of O_3 in the troposphere is the photolysis of NO_2:

$$\begin{aligned} NO_2 + h\nu &\longrightarrow NO + O(^3P) \\ O(^3P) + O_2 &\xrightarrow{N_2} O_3 \end{aligned} \qquad (1)$$

Its principal sink in the plume is the reaction with NO:

$$O_3 + NO \longrightarrow NO_2 + O_2 \qquad (2)$$

In a fresh plume, ambient ozone is rapidly depleted by reaction with NO, leading to the well-known ozone deficit. During the daytime, a gradual recovery of ozone will occur as NO_2 is produced, and a photostationary state is rapidly reached in which the concentration of ozone is directly proportional to the NO_2/NO concentration ratio. No net production of ozone can take place in this photolytic cycle in the absence of competing reactions.

In several recent studies, a net generation of ozone in excess of the background level has been observed in power plant plumes.[3-5,20,21] This can occur under certain conditions when the ozone source reaction (**1**) continues while ozone depletion (Reaction 2) is slowed down by the following competing reactions involving the peroxy radicals $RO_2\cdot$ and $HO_2\cdot$:

$$HO_2\cdot + NO \longrightarrow NO_2 + OH\cdot \qquad (3)$$

$$RO_2\cdot + NO \longrightarrow NO_2 + RO\cdot \qquad (4)$$

†Very recent laboratory studies[26,27] have suggested that the role of $HO_2\cdot$ in the direct oxidation of SO_2, as well as in the breakdown of hydrocarbons and the subsequent formation of ozone, is of relatively negligible significance compared to the role of $OH\cdot$.

In other words, these peroxy radicals participate in the oxidation of SO_2 as well as in the generation of excess plume ozone.† This is the first important link between ozone and sulfate formation in plumes. A second link concerns the source of the peroxy radicals. Both $HO_2\cdot$ and $RO_2\cdot$ are formed during the consumption of reactive hydrocarbons entrained into the plume from the background air by mixing. During the early hours of plume irradiation, $OH\cdot$ is by far the most important agent that attacks the hydrocarbons to generate $HO_2\cdot$ and $RO_2\cdot$.[19] Thus, the hydroxyl radical ($OH\cdot$), which plays an important role in the formation of sulfates, is also a key participant in the generation of ozone in the plume.

The free radical link between ozone and aerosol generation in the power plant plume thus depends on plume NO_x, hydrocarbons in the background air, and sufficient mixing to entrain the hydrocarbons into the plume. Thus, the ratio of background-to-plume air is related to the ratio $NMHC/NO_x$,‡ which has been shown in smog chamber studies to be a strong factor in the homogeneous oxidation of SO_2.[22]

If the plume transport is in a "clean" atmosphere, the ozone-aerosol link is more direct. In this case, $OH\cdot$ is the predominant agent for oxidizing SO_2, and O_3 is its principal source:

$$\begin{aligned} O_3 + h\nu &\longrightarrow O(^1D) + O_2 \\ O(^1D) + H_2O &\longrightarrow 2OH\cdot \end{aligned} \tag{5}$$

In a clean background, however, ozone generation and recovery in the plume will be limited, and the overall rate of sulfate formation will also be expected to be slower than for the case of plume transport in a polluted air mass.

In summary, the radicals $OH\cdot$ and $RO\cdot$ are the common link between the generation of ozone and of aerosols in the power plant plume when the following conditions are met: precursors SO_2, NO_x in the plume; precursor hydrocarbons in the background; mixing, to entrain the hydrocarbons into the plume; and sunlight, to drive the photochemical processes.

OZONE-AEROSOL LINK—PLUME OBSERVATIONS

During the summer, ozone and aerosol formation in plumes is most intense after the first hour or so of plume transport during the afternoon. In the July 1976 field program of Project MISTT, aircraft sampling was performed in a clearly identifiable Labadie plume for two or more hours of afternoon transport on six separate days—July 5, 9, 12, 14, 18, and 30. On all six afternoons, ozone generation in excess of the background level was realized. In this paper, the data of July 9, 14, 18, and 30 are selected for presentation because, between them, they include the extreme conditions and, together, they provide data that are representative of the summer conditions in the Midwest.

†Very recent laboratory studies[26,27] have suggested that the role of $HO_2\cdot$ in the direct oxidation of SO_2, as well as in the breakdown of hydrocarbons and the subsequent formation of ozone, is of relatively negligible significance compared to the role of $OH\cdot$.

‡NMHC = nonmethane hydrocarbon.

TABLE 1

SUMMARY OF AIRCRAFT SAMPLING, AMBIENT AIR QUALITY, AND SULFUR GAS-TO-PARTICLE CONVERSION RATES ($R_{S_g-S_p}$) IN THE LABADIE POWER PLANT PLUME (JULY 1976)

Date	Downwind Distance (km)	Approximate Sampling Time	Plume Age (hr)	Background Levels b_{scat} (10^{-4} m^{-1})	Background Levels TNMHC† (μg/m^3)	Background Levels Ozone (ppb)	$R_{S_g-S_p}$* (% hr^{-1})
July 9	(70)‡	1045–1200	3.0	2.1	NA	60	0.8
(WU)	105	1230–1245	4.5	2.0	NA	60	0.9
	135	1530–1630	6.5	1.7	NA	60	1.6
	190	2000–2100	8.5	1.5	NA	60	0.9
	250	2130–2200	9.8	1.5	NA	55	0.7
	310	2345–0015	9.2	1.6	NA	55	0.3
July 14	47	0720–0745	1.3	1.0	147	65	0.1
(MRI)	50	0800–0830	1.4	1.0	NA	65	0.3
	(50)	1200–1300	2.8	1.0	229	70	0.8
July 18	(30)	1420–1530§	2.2	0.6	148¶	100	3.2
(WU)	110	1540–1620	6.0	0.5	NA	100	3.0
	160	1815–1930	8.5	0.5	NA	85	2.0
	220	1945–2015	10.0	0.5	NA	80	1.6
	320	2345–0015	12.0	0.5	NA	70	1.2
July 30	14	1150–1215	1.0	1.3	164	75	1.0
(MRI)	(27)	1230–1310	2.5	1.3	159	78	1.3
	45	1405–1420	4.2	1.3	NA	78	3.0

*Average conversion rate between plume release time and plume sampling time.

†Total nonmethane hydrocarbons (light + heavy hydrocarbon species). Samples collected in MRI aircraft.

‡(x) = first observation of ozone bulge in Labadie plume core at x km from source.

§WU and MRI.

¶Not including light hydrocarbons.

On July 9 and 18, sampling was done over a long range (FIGURE 1), while on July 14 and 30, sampling was over a short range (FIGURE 3). TABLE 1 summarizes the sampling times and locations, the ambient air quality, and the sulfur gas-to-particle conversion rates for all four days. Gas-to-particle conversion of sulfur and the ozone bulge formation are taken as indices of secondary pollutant formation in the plume, while the ambient ozone level is interpreted as an index of the potential for photochemical smog formation in the background. The sulfur conversion rates on all four days peaked in the afternoon, the nighttime and early morning rates being the lowest. Of the four days, July 18 and 30 were the higher conversion days (peak $R_{S_g-S_p} \geqq 3\%$ hr^{-1}), while July 9 and 14 were relatively lower conversion days (peak $R_{S_g-S_p} \leqq 1.6\%$ hr^{-1}). Ambient ozone was significantly higher on the days of higher plume sulfur conversion. Since the peak conversion rates are averaged over transport times varying from 2.2 hr (July 18) to 6.5 hr (July 9), such a comparison of the rates can be misleading. Consequently, plume conditions are compared at the times when an ozone bulge is first observed over the entire plume. The downwind distance where such a condition is first observed on each day is circled in TABLE 1.

Plume and ambient data at the first observation of the ozone bulge are isolated in TABLE 2. On all four days, plume ozone exceeded background ozone over an entire crosswind traverse for the first time within 2 to 3 hours of transport. The earliest time of observation of an ozone bulge was about noon. On July 18 and 30, the bulge appeared within 30 km of the source, while on July 9 and 14, 50 km or more of transport was required. In terms of sulfur conversion rate, July 18 stands out with a value of 3.2% hr^{-1}, July 30 comes in a distant second at 1.3% hr^{-1}, and July 9 and 14 trail last with 0.8% hr^{-1}. The background ozone values show a similar ranking. Data of the transport of the Milwaukee urban plume over Lake Michigan for two summer days also showed a similar behavior, with $R_{S_g\text{-}S_p}$ being 4–11% hr^{-1} and <1% hr^{-1}, respectively, corresponding to background ozone levels of 95 ppb and <50 ppb.[23] Evidently, plume sulfate formation relates well to the photochemical smog formation potential of the background.

Conditions Conducive to Ozone and Sulfate Formation

The principal measured parameters likely to influence ozone and aerosol formation during long range transport of the Labadie plume under midwestern summer conditions are believed to be the following:

Primary emissions: SO_2, NO_x;

Background: nonmethane hydrocarbons, ozone, b_{scat};

Meteorological: sunlight, low-level clouds (no quantitative data), temperature, relative humidity (RH), and mixing.

Of these quantities, the emissions of SO_2 and NO_x were only slightly variable for the four days. However, it has previously been reported that ozone bulges are more likely to form in the plumes of intermediate-size power plants, compared to plumes of large power plants with higher NO_x emissions.[20] The significance of the meteorological and background parameters is assessed below.

FIGURE 4 shows the hourly data of incoming solar radiation, temperature, and relative humidity, measured at the rural, ground-level Station 118 of the RAMS network of the St. Louis Regional Air Pollution Study. Superimposed on each daily trace of each parameter is a heavy solid line, covering the transport period from plume release to the time of first observation of the ozone bulge (see TABLE 2).

The *solar radiation* pattern is nearly the same for all days except July 30, when

TABLE 2

AMBIENT CONDITIONS AND $R_{S_g\text{-}S_p}$ AT FIRST OBSERVATION OF OZONE BULGE IN THE LABADIE PLUME CORE (JULY 1976)

Date	Downwind Distance (km)	Approximate Transport Period	Plume Age (hr)	Background Levels: b_{scat} (10^{-4} m^{-1})	Background Levels: TNMHC ($\mu g/m^3$)	Background Levels: Ozone (ppb)	$R_{S_g\text{-}S_{p_1}}$ (% hr^{-1})
July 9	70	0850–1150	3.0	2.1	NA	60	0.8
July 14	50	0900–1150	2.8	1.0	229	70	0.8
July 18	30	1350–1600	2.2	0.6	148*	100	3.2
July 30	27	1000–1230	2.5	1.3	159	78	1.3

*Not including light hydrocarbons.

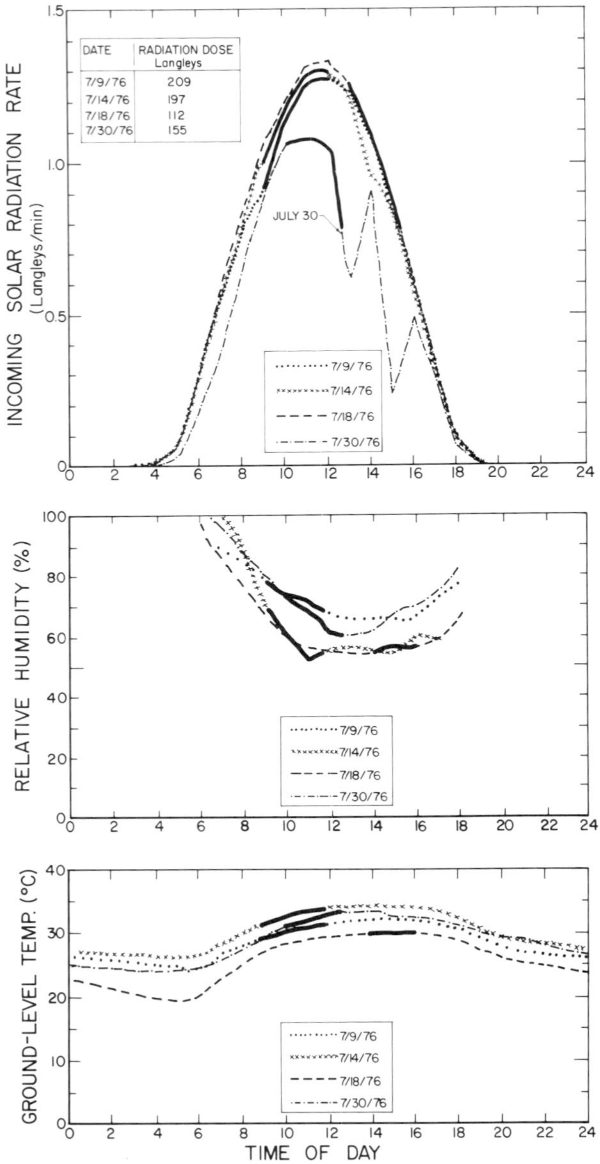

FIGURE 4. Comparison of solar radiation flux, ambient temperature, and relative humidity for July 9, 14, 18, and 30, 1976, as measured at RAMS Station 118. The heavy solid line on each trace emphasizes the conditions between plume release and the first observation of the ozone bulge in the Labadie plume.

the skies become progressively more cloudy during the day. The other days had more or less clear skies (FIGURE 4). There was, nevertheless, sufficient total plume exposure to sunlight (155 Ly) on July 30 to result in the formation of an ozone bulge in at least one of five plume traverses at 27 km. Approximately the same Lagrangian plume parcel was sampled at 27 km and again at 45 km. The sulfur conversion rate between these two sampled distances exceeded 3% hr^{-1} between about 1230 and 1400. The participation of clouds in sulfate formation during these hours on July 30 is likely. As will be shown later, however, substantial photochemical activity was quite evidently also taking place during these hours. The data of July 18 show that a sunlight dose as low as 112 Ly is sufficient for an ozone bulge to form, if other conditions are particularly favorable.

Temperatures were high (28–34°C) on all four days during the afternoon hours of sampling (FIGURE 4). While no clear positive correlation between sulfate formation rate and temperature is evident (the lowest afternoon temperature of 28°C was recorded at Station 118 on the day with the highest measured sulfate formation rate, July 18), it is also evident that ozone bulge formation is favored on hot days with ground-level temperatures exceeding 28°C. The same observation was also made by Keifer.[20]

Relative humidity (RH) ranged between 57% and 70% at Station 118 during the early plume transport periods leading to the ozone bulge formation (FIGURE 4). On July 5, for which plume data are not presented here, ground-level RH was as low as 49% in the afternoon while plume ozone bulge was evidenced in the aircraft data. As with temperature, no clear positive correlation is observed between ozone-aerosol formation rate and RH.

Of the meteorological parameters, *atmospheric mixing* appears to show the most direct and strong positive correlation with ozone and sulfate formation rates in the plume. In FIGURE 5, the intensity of mixing in the horizontal is compared for the four days in terms of horizontal plume spread, and mixing height is taken as a measure of vertical mixing (see inset). The crosswind plume profiles shown are for SO_2 distribution at the altitude where the peak SO_2 concentration was observed in the plume section where the ozone bulge was first observed. Horizontal as well as vertical mixing was strongest by a wide margin on July 18 (the day of most rapid ozone and sulfate formation in the plume), followed by July 30, and finally July 9 and 14.

Mixing is presumably so important because it dilutes plume NO_X, increases plume *NMHC* by entrainment from the background, and thereby causes a gradual increase in the ratio $NMHC/NO_x$, which controls ozone and aerosol formation under warm, sunny conditions. A wide variety of hydrocarbon species of varying reactivity are present in the atmosphere. An attempt was made in the hydrocarbon analysis (performed at Washington State University) to identify as many individual species as possible. Unfortunately, the bulk of the NMHC concentration was due to species not individually identified. The total NMHC concentrations for the four days (see Table 2) show no clear positive correlation with the rates of ozone and aerosol generation in the plume. Both in Project MISTT, 1976, and in the Tennessee Plume Study of Project STATE, 1978, ozone bulges appeared on nearly every afternoon of plume sampling. Presumably, both in the Missouri-Illinois area and in the Kentucky-Tennessee area, the lower atmosphere contains sufficient reactive hydrocarbons on

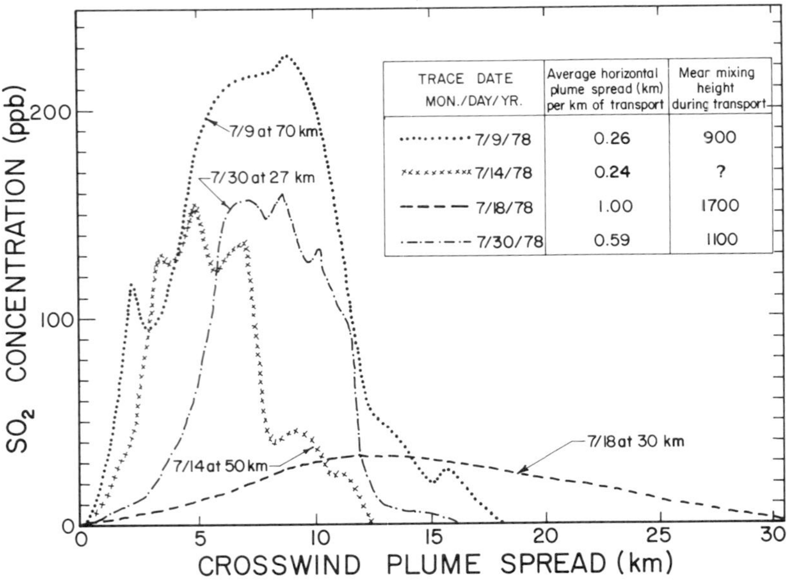

FIGURE 5. Comparison of mixing intensity for Labadie plume transport on July 9, 14, 18, 30. Crosswind profiles of SO_2 through the traverse closest to plume centerline in the cross section where ozone bulge was first observed. **Inset:** Comparison of horizontal and vertical mixing parameters.

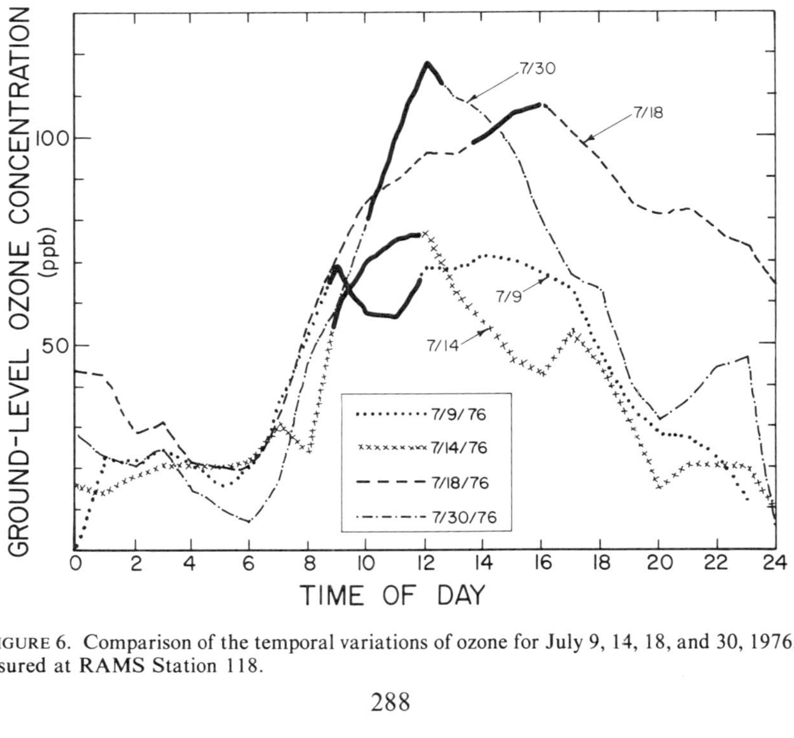

FIGURE 6. Comparison of the temporal variations of ozone for July 9, 14, 18, and 30, 1976, as measured at RAMS Station 118.

summer afternoons to permit ozone bulges to occur with far greater regularity than was previously believed.

The *background ozone* levels listed in TABLES 1 and 2 are based on aircraft data. The hourly ground-level ozone data at the rural Station 118 are shown for the four days in FIGURE 6. For the well-mixed conditions of the afternoon hours, ground-level data represent the conditions aloft reasonably well. Both the ground and the air data show clearly that high background ozone levels correspond to rapid secondary pollutant generation (July 18, 30) in the plume.

The *background* b_{scat} values (TABLE 2) for the four days do not show a strong

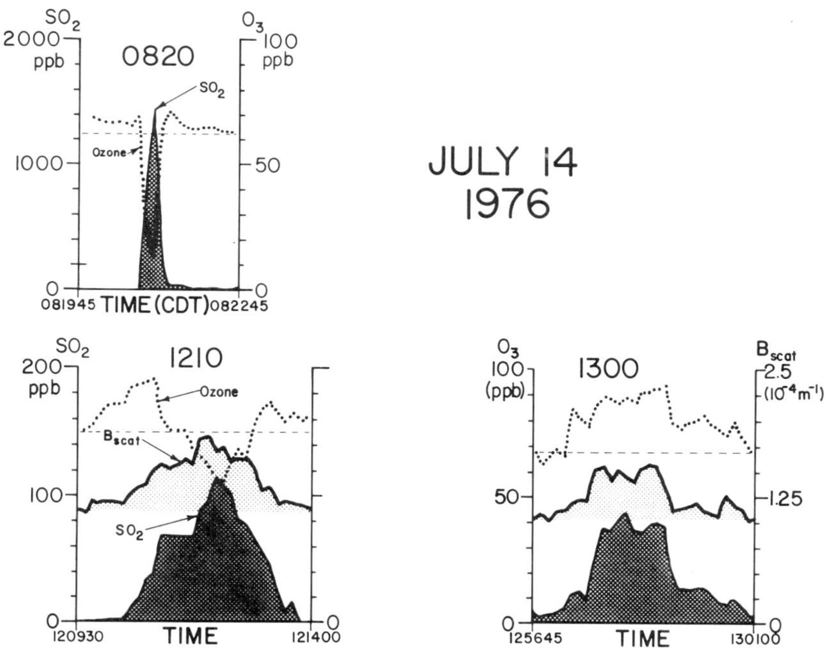

FIGURE 7. Crosswind plume data for 610 m MSL traverses at 50 km radial distance from Labadie at three different times on July 14, 1976. The three stages of secondary pollution formation are illustrated.

positive correlation with sulfate formation rate in the plume. None of the days, however, was particularly hazy. The air mass of July 18 (highest sulfur conversion rate) was the least hazy ($b_{scat} = 0.6 \times 10^{-4}\ m^{-1}$) but also had the highest background ozone. It was also the only one associated with a continental high pressure cell.

Stages in the Formation of Ozone and Aerosols

Three distinct chronological stages may be identified in the formation of ozone and aerosols during power plant plume transport in daylight. These three stages are illustrated in FIGURE 7, which shows crosswind plume profiles of SO_2, O_3, and b_{scat}

obtained during aircraft traverses made in the Labadie plume on July 14. All three traverses were made at about the same location, 50 km from the source (see FIGURE 3), at the same altitude (610 m MSL), but at different times of the day. The first traverse at 0820 represents the "early" stage (plume age is 1.4 hr), in which the primary emissions are concentrated in a narrow, coherent plume within which ozone is largely depleted as a result of its reaction with NO to form NO_2. There is little other interaction between the plume and the background, and $R_{S_g\text{-}S_p}$ is low (0.3% hr^{-1}). The traverse at 1210 belongs to an "intermediate" stage, characterized by ozone bulges at the fringes of the plume resulting from strong interactions there between plume NO_x and the reactive hydrocarbons from the background. The ratio $NMHC/NO_x$ is logically highest at the fringes. In the transition from the early to the intermediate stage, ozone deficit in the core of the plume is gradually recovered, more or less in keeping with the photostationary state cycle involving NO, NO_2, and O_3. The traverse at 1300 is the "fully developed" stage (plume age is 2.8 hr), in which background species have diffused through to the plume core and the ozone bulge has formed over the entire profile. The transitory early and intermediate stages have given way to a fully developed plume profile, which would persist as long as photochemistry could proceed. The distributions of the radical species linking ozone and aerosol formation have presumably reached steady state, evidenced by the high degree of correlation now existing between the ozone and b_{scat} profiles. The average value of $R_{S_g\text{-}S_p}$ is now 0.8% hr^{-1}.

The three stages are once again illustrated in the data of July 30 (FIGURE 8). Here,

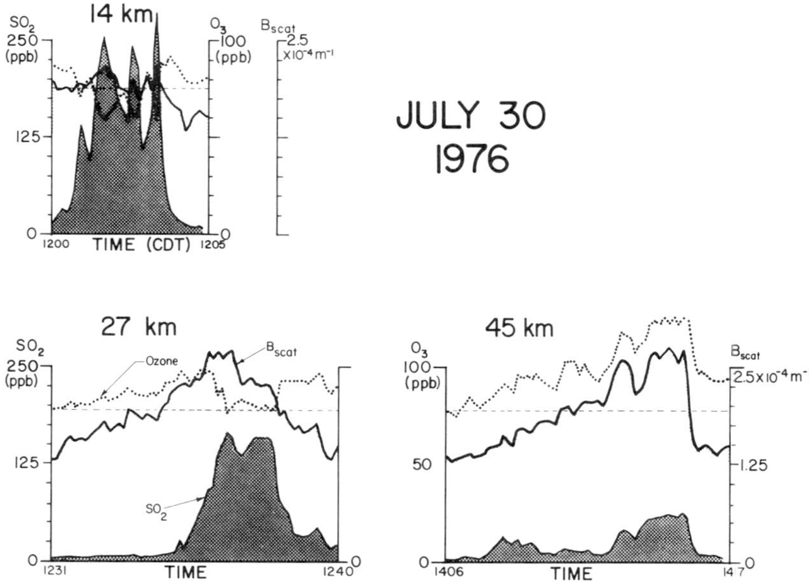

FIGURE 8. Crosswind plume data for traverses at 14, 27, and 45 km radial distances for Labadie on July 30, 1976. The "early," "intermediate," and "fully developed" stages of secondary pollution formation are illustrated. Observe, in particular, the fringe enhancement of secondary pollution parameters relative to SO_2 in the data at 27 km.

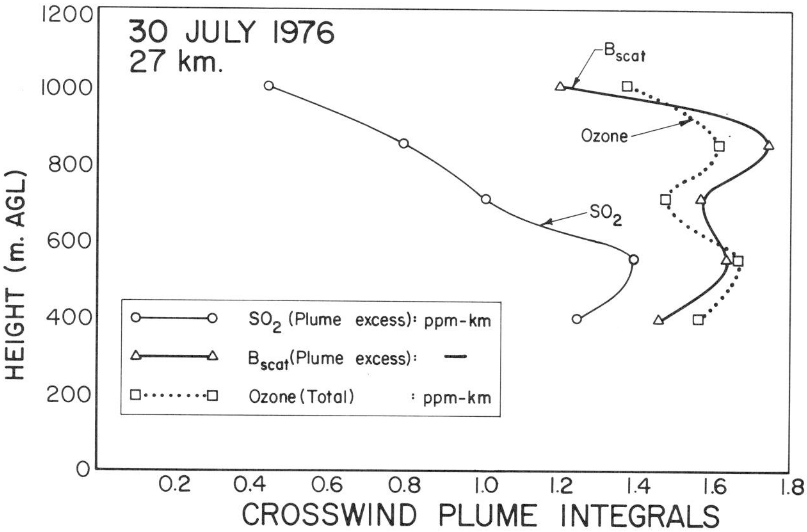

FIGURE 9. Vertical profiles of the Labadie plume of July 30, 1976 at 27 km. Each data point represents the crosswind plume integral of the parameter over a given traverse trajectory. Observe the enhancement of the secondary pollution parameters at the upper edge of the SO_2 plume.

quasi-Lagrangian plume profiles are shown at 14 km, 27 km, and 45 km radial distances from the source and correspond to plume ages of about 1, 2.5, and 4.2 hours, respectively. An important consequence of plume-background interactions is illustrated in the plot of the intermediate stage at 27 km. The ozone bulges at the fringes indicate the generation of the peroxy radicals there, which causes a higher *rate* of aerosol formation there than at plume core. The b_{scat}/SO_2 ratio is thus higher at the fringes, and the aerosol plume has the appearance of being substantially broader than the SO_2 plume. The same fringe effect is obtained also in the vertical plume profiles at the intermediate stage, shown in FIGURE 9. Observe that the highest value of the ratio b_{scat}/SO_2 is near plume top, where SO_2 is low. The conversion rate of sulfur increases from 1.0 to 1.3 to 3.0% hr^{-1} as the plume develops through the three stages. Similar stages in the development of the power plant plume have also been observed by Davis *et al.*[24]

Two striking illustrations of fringe-enhanced aerosol formation by homogeneous gas phase mechanisms are shown in FIGURE 10. Observe the "spikes" of the Aitken Nuclei Count (ANC) near plume edges. Fringe enhancement of aerosol production also has been investigated recently by White.[25]

The time it takes for the plume to develop to the mature third stage depends largely on the availability of the reactive hydrocarbons in the background and the intensity of mixing. Under summer daytime conditions, this process would take a few hours. Thereafter, the mature conditions would persist until the sunlight level became insufficient to drive the photochemistry. A question of obvious importance for long range plume transport is: What happens to an aged plume after a hard day's journey into the night?

The Labadie plume transport was sampled during the day and night on July 9 (see FIGURE 1 and TABLE 1). Crosswind plume profiles at 70 km (ca. 1100), 135 km (ca. 1600), 190 km (ca. 2000), and 310 km (ca. midnight) are shown in FIGURE 11. The profile at 70 km is in the intermediate stage. The developed profiles at 135 km and 190

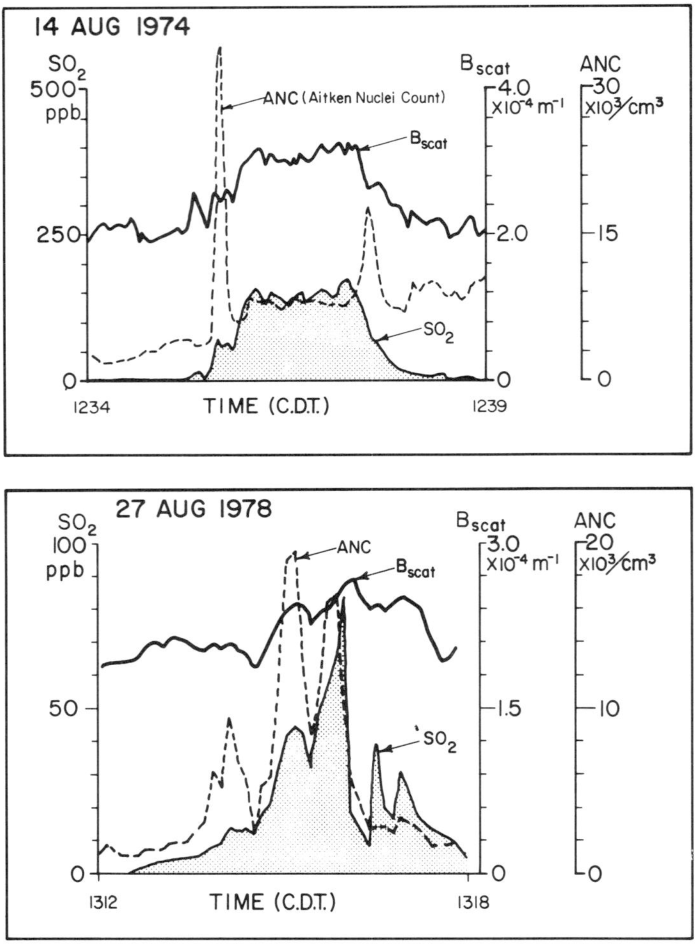

FIGURE 10. Examples of enhanced homogeneous nucleation (Aitken Nuclei Count) at plume fringes for the Labadie plume of 8/14/1974 (upper) and the Cumberland plume of 8/27/1978 (lower).

km are at different ages. The 190 km and 310 km profiles, though not belonging to the same plume parcel, illustrate what happens to an aged daytime plume during the night. There is no new generation of ozone or of the photochemical radical species at night. Plume NO level is also largely depleted following many hours of daytime

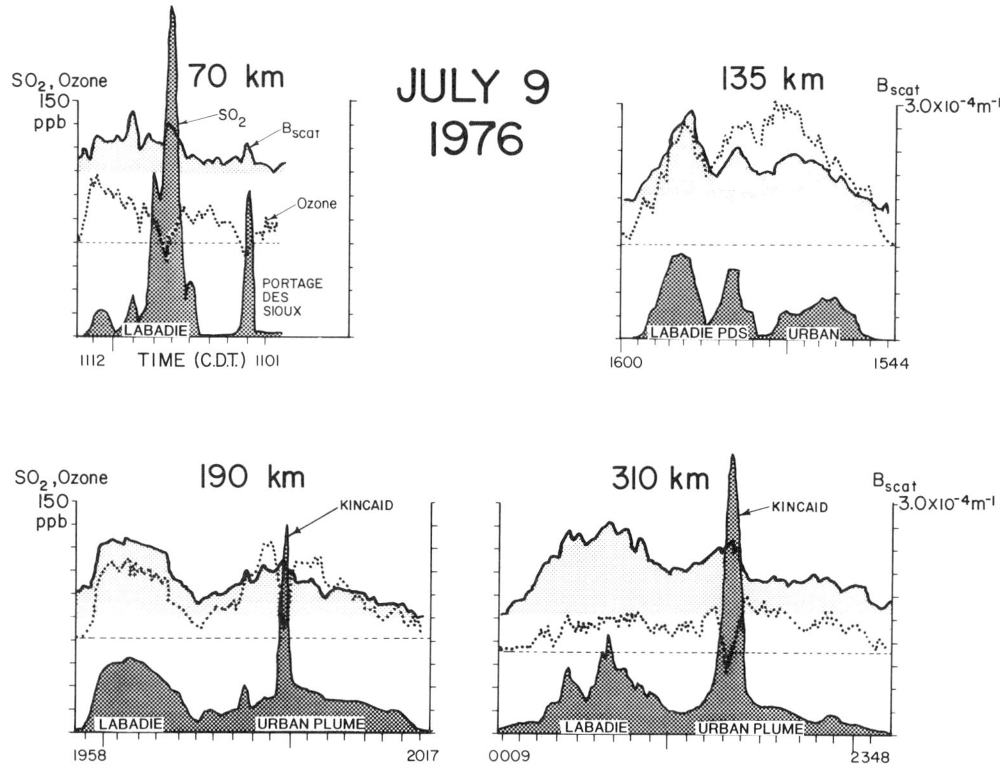

FIGURE 11. Crosswind plume data of the St. Louis plumes of July 9, 1976 at 70, 135, 190, and 310 km radial distances from the Labadie stacks. Daytime generation of secondary pollution, and its dispersion by shear at night.

transport. Consequently, ozone depletion is also slow, if any. One would expect logically, therefore, that the plume ozone bulge would be maintained and transported overnight. By midnight on July 9, however, the ozone bulge of the afternoon had diminished and broadened substantially. The most likely explanation, we suggest, is the wind shear effect. During the night, a considerable directional shear in the wind caused the plume to spread out over a large volume, and led to the termination of the mission. In the absence of strong wind shear, the secondary formations could be expected to carry over into the next day largely intact.

Oxidation of NO_2

No data were collected of HNO_3 or particulate nitrate concentrations in Project MISTT. However, based on the NO_2 data and the reaction kinetics of NO_2 oxidation, a substantial amount of HNO_3 formation may be expected to occur. NO_2 oxidation occurs primarily by OH· and O_3. The oxidation of NO_2 by OH· is more than an order of magnitude faster than the oxidation of SO_2 by OH·.[16] Consequently, whenever the ratio NO_2/SO_2 exceeds 0.1 in the plume, nitric acid formation may exceed sulfate formation. In the Labadie plume of July 30, the NO_2/SO_2 ratio increased from an average 0.18 at 14 km, to 0.22 at 27 km, and 0.33 at 45 km. In the first four hours of daytime plume transport, nitric acid formation not only may have exceeded sulfate formation, but may have done so increasingly with plume transport. This dominance could have persisted for much longer times. If sufficient NH_3 is available from the background, significant formation of particulate nitrates may also result. These data serve to point out the need for more measurements and investigation of this subject.

Summary and Conclusions

A case is made, based on theoretical and empirical grounds, that a close link exists between the formation of ozone and aerosols in power plant plumes. The condition of the background air mass (NMHC loading), the extent of plume-background interactions, and photochemical processes are suggested as factors of paramount importance in the cogeneration of ozone and aerosols in plumes on warm summer days in the U.S. Midwest. The ambient ozone level is a good indicator of background reactivity. Ambient ozone and mixing are the major easily measurable factors that relate to the $NMHC/NO_x$ ratio in the plume and, hence, to the rates of generation of secondary pollutants. A moderate amount of sunlight exposure (~ 110 Langleys), ground-level temperature above 28°C, and relative humidity above 50% are sufficient conditions for rapid formation of excess plume ozone and aerosols. Ozone bulges appear in power plant plumes on warm summer afternoons quite regularly in the Midwest.

The radical species OH· and RO_2· are possibly the principal intermediates in the ozone-aerosol link. The role of OH· is dominant in the early stages of plume transport and during transport in a relatively clean background. RO_2· gains in importance with increasing plume transport and with increasing amount of reactive hydrocarbons in the background. So does ozone. The practical significance of ozone—not only in its own right, but also in the formation and long range transport of aerosols—is potentially great. Until the important radical species are easily measured in the atmosphere, ozone may serve as a good surrogate for them in the development of empirical parametrization of the submodule for chemical transformations in long range transport models for aerosols.

Based on the plume data of Project MISTT, three distinct stages of daytime plume development are identified. An early stage is dominated by the primary emissions and an ozone deficit. An intermediate stage is dominated by enhanced formation of ozone and aerosols in plume fringes as the early manifestation of entrainment of background species into the plume. Ultimately, a quasi-steady, fully developed stage is attained, in which the entire plume has matching levels of the background species. As the stages develop, secondary pollution formation rate in the plume also increases rapidly. In the mature stage, secondary pollution dominates in the plume as it gradually advances to contribute to the regional scale background. Except under strong wind shear, the secondary pollutants may jet through the night, unabated in concentration.

Total nitrate formation may well exceed aerosol sulfate formation during the first several hours of plume transport. This subject deserves much more attention in the future.

Acknowledgments

We thank participants of Project MISTT. The data from the Regional Air Pollution Study were provided to us by U.S. EPA, for which we are grateful. The drawings were made by David Chisholm.

References

1. Husar, R. B., D. E. Patterson, C. C. Paley & N. V. Gillani. 1977. Ozone in hazy air masses. Paper No. 6–5. *In* Proceedings of the International Conference on Photochemical Pollution. U.S. EPA Pub. No. EPA–600/3-7-00lb. Research Triangle Park, N.C.
2. White, W. H., J. A. Anderson, D. L. Blumenthal, R. B. Husar, N. V. Gillani, J. D. Husar & W. E. Wilson. 1976. Formation and transport of secondary air pollutants: ozone and aerosols in the St. Louis urban plume. Science **194:** 187.
3. Davis, D. D., G. Smith & G. Klauber. 1974. Trace gas analysis of power plant plumes via aircraft measurements: O_3, NO_x, SO_2 chemistry. Science **186:** 733.
4. Gillani, N. V., R. B. Husar, J. D. Husar, D. E. Patterson & W. E. Wilson. 1978. Project MISTT: Kinetics of particulates sulfur formation in a power plant plume out to 300 km. Atmos. Environ. **12:** 589.
5. Lusis, M. A., K. G. Anlauf, L. A. Barrie & H. A. Wiebe. 1978. Plume chemistry studies at a northern Alberta power plant. Atmos. Environ. **12:** 2429.
6. Wilson, W. E., D. F. Miller, A. Levy & R. K. Stone. 1973. The effect of fuel composition on atmospheric aerosol due to auto exhaust. J. Air Pollut. Control Assoc. **23:** 949.
7. Kochmond, W. C., J. Y. Yang, D. B. Kittelson, K. T. Whitby & K. L. Demerjian. 1977. Aerosol formation in simple photochemical systems. Adv. Environ. Sci. Technol. **8:** 101.
8. Altshuller, A. P. 1977. Formation and removal of SO_2 and oxidants from the atmosphere. Adv. Environ. Sci. Technol. **8:** 9.
9. Wilson, W. E. 1978. Sulfates in the atmosphere: a progress report for Project MISTT. Atmos. Environ. **12:** 537.

10. ZAK, B. D. *et al.* 1979. Project Da Vinci: A Study of Long Range Air Pollution Using a Balloon-Borne Lagrangian Measurement Platform. Report No. SAND 78-0403. Sandia Laboratories. Albuquerque, N. Mex.
11. GILLANI, N. V. 1978. Project MISTT: Mesoscale plume modeling of the dispersion, transformation, and ground removal of SO_2. Atmos. Environ. **12:** 569.
12. HUSAR, R. B., D. E. PATTERSON, J. D. HUSAR, N. V. GILLANI & W. E. WILSON. 1978. Sulfur budget in large plumes. Atmos. Environ. **12:** 549.
13. BLUMENTHAL, D. L., J. A. OGREN & J. A. ANDERSON. 1978. Airborne sampling system for Project MISTT. Atmos Environ. **12:** 613.
14. SCHWARTZ, S. E. & L. NEWMAN. 1978. Processes limiting oxidation of sulfur dioxide in stack plumes. Environ. Sci. Technol. **12:** 67.
15. CALVERT, J. G., F. SU, J. W. BOTTENHEIM & O. P. STRAUSZ. 1978. Mechanism of the homogeneous oxidation of sulfur dioxide in the troposphere. Atmos. Environ. **12:** 197.
16. EGGLETON, A. E. & R. A. COX. 1978. Homogeneous oxidation of sulphur compounds in the atmosphere. Atmos. Environ. **12:** 227.
17. PENKETT, S. A., B. M. JONES, K. A. BRICE & A. E. EGGLETON. 1979. The importance of atmospheric ozone and hydrogen peroxide in oxidizing sulphur dioxide in cloud and rain water. Atmos. Environ. **13:** 123.
18. ERICKSON, R. E., L. M. YATES, R. L. CLARK & D. MCEWEN. 1977. The reaction of sulfur dioxide with ozone in water and its possible atmospheric significance. Atmos. Environ. **11:** 813.
19. FINLAYSON-PITTS, B. J. & J. N. PITTS. 1977. The chemical basis of air quality: kinetics and mechanisms of photochemical air pollution and application to control strategies. Adv. Environ. Sci. Technol. **8:** 75.
20. KEIFER, W. S. 1977. The Generation of Ozone in Plumes from Large Point Sources. Ph.D. Dissertation. University of Maryland. College Park, Md.
21. MILLER, D. F., A. J. ALKEZWEENY, J. M. HALES & R. N. LEE. 1978. Ozone formation related to power plant emissions. Science **202:** 1186.
22. MILLER, D. F. 1978. Precursor effects on SO_2 oxidation. Atmos. Environ. **12:** 273.
23. MILLER, D. F. & A. J. ALKEZWEENY. 1980. Aerosol formation in urban plumes over Lake Michigan. Ann. N.Y. Acad. Sci. (This volume.)
24. DAVIS, D. D., W. HEAPS, D. PHILEN & T. MCGEE. 1979. Boundary layer measurements of the OH radical in the vicinity of an isolated power plant plume: SO_2 and NO_2 chemical conversion times. Atmos. Environ. **13:** 1197.
25. WHITE, W. H. 1979. On the crosswind distribution of secondary aerosol in a power plant plume. (Unpublished manuscript.)
26. GRAHAM, R., A. M. WINER, R. ATKINSON & J. N. PITTS. 1979. Rate constants for the reaction of HO_2 with HO_2, SO_2, CO, N_2O, trans-2-butene and 2,3 dimethyl-2-butene at 300°K. J. Phys. Chem. **83**(12): 1563.
27. BURROWS, J. P., D. I. CLIFF, G. W. HARRIS, B. A. THRUSH & J. P. T. WILKINSON. 1979. Atmospheric reactions of the HO_2 radical studied by laser magnetic resonance spectroscopy. Proc. R. Soc. London Ser. A **368:** 463.

SULFATE FORMATION IN URBAN PLUMES*

John W. Winchester

Department of Oceanography
Florida State University
Tallahassee, Florida 32306

INTRODUCTION

In the concept of an "urban plume," the combined air pollutants from several major emission sources in an urban area are considered to be carried, with air masses moving across the city, out over nonurban land or water. A plume may be distinguished from the unpolluted nonurban atmosphere, which is not directly downwind of a city, by its chemical composition and high concentrations of certain pollutants. However, in the eastern U.S., where the most abundant particulate air pollutants in the plume may be sulfate aerosol formed by the oxidation of gaseous sulfur oxides, the urban source areas are spaced closely enough so that the intervening nonurban regions are often under the combined influence of several urban plumes. The resulting generally high nonurban average pollutant concentrations, especially for sulfate, greatly complicate the task of tracing the transport of individual urban plumes and, of particular importance, of determining the extent of sulfate formation within them. Consequently, larger scale field studies may be desirable, such as by networks of nonurban sampling stations throughout the eastern U.S. region, measurements over the ocean downwind of the entire urbanized northeastern U.S., or a comparison of sulfate concentrations in nonurban northern and southern hemisphere locations, in order to determine the magnitude of sulfate formation and the presence of other air pollutants in composite urban plumes. In this paper, the results of such studies carried out at Florida State University will be briefly summarized. A more complete bibliography of pertinent literature is given in the references cited.

THE EASTERN UNITED STATES NONURBAN SULFATE AEROSOL

A starting point for tracing the formation of sulfate in urban plumes and its transport over nonurban regions is to document the concentration of nonurban sulfate in the eastern U.S., its variability in space and time, the relation between sulfate concentration and other observable quantities such as visibility, and evidence for the dispersion of the combined nonurban sulfate-laden air mass outward in its fringe areas, such as into Florida. Two examples of such investigations are summarized here.

1. Cohn[1] has reported the measurement of aerosol sulfur concentrations from a network of sampling stations ranging from Kansas to New Hampshire, with most

*Portions of the research reviewed here have been supported by the U.S. Environmental Protection Agency, the Florida Sulfur Oxides Study, the National Institutes of Health, the National Science Foundation, the North Atlantic Treaty Organization, the U.S. Office of Naval Research, and the U.S. Department of Energy.

0077–8923/80/0338–0297 $01.75/1 © 1980, NYAS

stations located between St. Louis, Missouri, and Pennsylvania. The concentrations, which were measured in samples collected by the time sequence filter "streaker" sampler using proton-induced x-ray emission (PIXE), showed evidence of large-scale coherence between sites. Such coherence is expected if sulfur oxides and sulfate particulate matter are transported efficiently on a regional scale. The concentrations at the different sampling stations may be plotted on maps at two-hour intervals, the time resolution offered by the streaker samplers, and, considering the air flow regimes inferred from published data of the National Weather Service, the concentrations exhibit patterns that can be contoured. The contoured patterns are seen to vary smoothly with time, suggesting that large-scale transport of aerosol and gaseous sulfur oxides is taking place between urban centers in response to air flow.

Such patterns provide evidence that significant reductions in atmospheric visibility occur when aerosol sulfur concentrations rise above about 2 $\mu g/m^3$ (equivalent to 6 $\mu g/m^3$ sulfate, $SO_4^=$). In order to examine this possibility quantitatively, Cohn selected measurements made on two successive days, July 19 and 20, 1976, at the seven locations listed in TABLE 1. On these days, a strong west to east component of air flow was present without rainfall. Early on July 19, a frontal system was situated to the south of the sampling network and was moving toward the southeast, and late on July 20, another frontal system was approaching from the northwest. However, no frontal activity occurred within the sampling network during these days. Relative humidity was between 70% and 90% at all seven stations on the mornings of both days.

From TABLE 1, it is seen that visibility on July 19 ranged from 6 to 15 miles but on July 20 ranged between 2 and 6 miles, except for the westernmost site in Kansas. A more complete study of all visibility data throughout this region shows that on the second day, visibility was generally poor throughout the region, with frequent reports of haze, in contrast to the first day, when only at selected locations were poor visibility and haze reported. No systematic difference between relative humidity on the two days is apparent. The aerosol sulfur concentrations summarized in TABLE 1 indicate that at Manhattan, Kansas, the concentrations were much lower than at other sites, but from St. Louis eastward, the concentrations were generally a few thousand nanograms per cubic meter. Especially in the easternmost parts of the network, the concentrations were substantially higher on the second day. A more thorough analysis

TABLE 1

VISIBILITY, RELATIVE HUMIDITY, AND AEROSOL SULFUR CONCENTRATIONS AT NETWORK SAMPLING SITES, 0600 HOURS, JULY 19 AND 20, 1976

Station Number	Location	Visibility (miles)		% Relative Humidity		Sulfur Concentration (ng/m³)	
		7/19	7/20	7/19	7/20	7/19	7/20
1	Manhattan, Kansas	15	15	90	90	410	600
2	St. Louis, Missouri	6	6	71	76	3900	2100
4	Argonne, Illinois	9	6	73	74	2500	5700
5	Remington, Indiana	10	4	76	71	—	5700
7	Angola, Indiana	8	6	90	84	—	3300
9	Delaware, Ohio	15	2	81	90	700	2300
10	Meadville, Pennsylvania	15	3	70	84	1300	2700

of the complete time record, with 2-hour resolution, of sulfur concentrations on both days at all sites and the corresponding variations in visibility, together with the results presented in TABLE 1, indicates that aerosol sulfur concentration is a significant factor in reducing atmospheric visibility. The analysis permits a provisional estimate to be made that when sulfur exceeds about 2000 ng/m^3, and the relative humidity is sufficiently high, a significant reduction in visibility occurs. High relative humidity with low sulfur concentration is not associated as strongly with visibility reduction, nor is high sulfur concentration in very dry air. The facts that, on July 20, poor visibility and haze were reported over the very large area of the network and that the correspondences in sulfur concentrations between sites suggested long range transport indicate that the sulfate formed in urban plumes from different urban areas may combine and contribute to a general visibility reduction over the region under appropriate humidity conditions.

2. A sampling network in the state of Florida offers an excellent opportunity to watch for signs that the composite eastern U.S. sulfate aerosol mass can migrate into Florida from the north. In a sampling network[2] illustrated in FIGURE 1, measurements of the aerosol sulfur concentration as a function of particle size were made over three selected two-week intervals in 1976 and 1977.[3] The average concentrations found for the first period, July–August 1976, for each of the particle size fractions are shown in

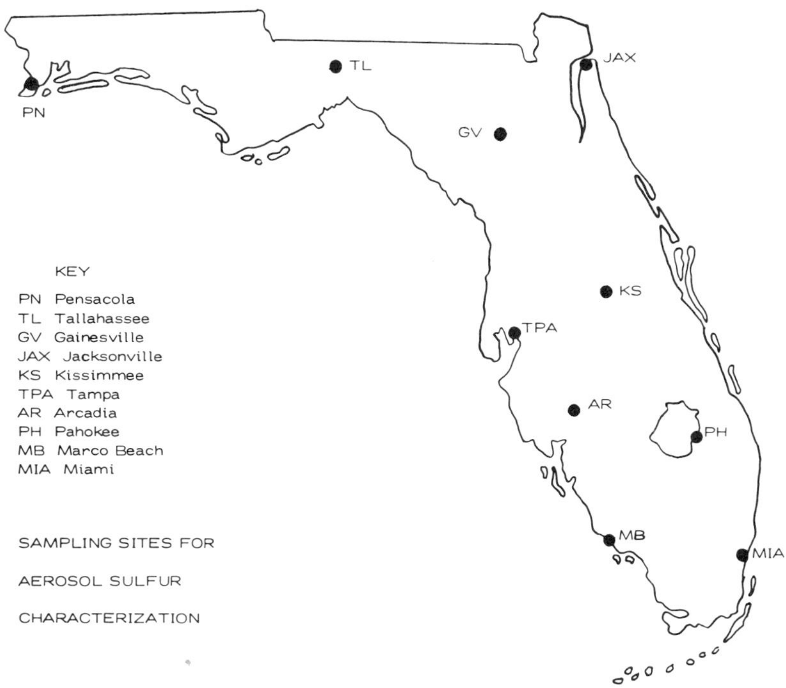

FIGURE 1. Index map of Florida showing locations of ground-level sampling sites. Samplers at PN, GV, JAX, KS, AR, PH, and MB were located at small airports, and at TL, TPA, and MIA within the cities.

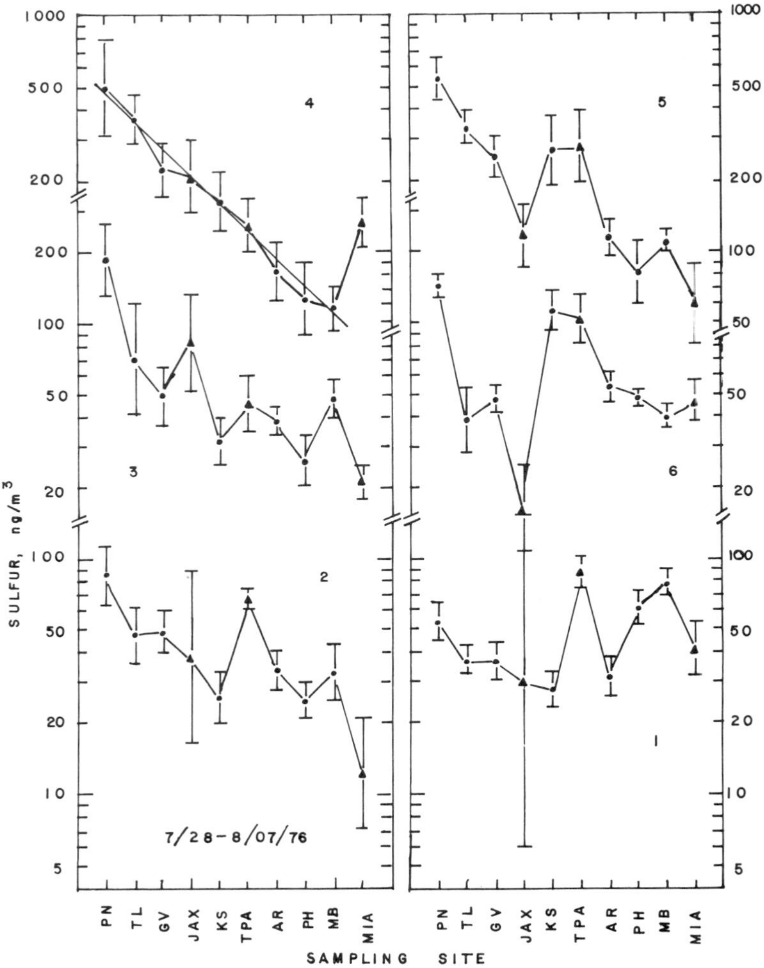

FIGURE 2. Sulfur during the summer, July–August 1976, period in each particle size range: **6**, < 0.25 μm; **5**, 0.25–0.5 μm; **4**, 0.5–1 μm; **3**, 1–2 μm; **2**, 2–4 μm; **1**, > 4 μm aerodynamic diameter. Points represent geometric means, with the standard deviations of the means, for groups of approximately six samples taken. Points are plotted against sampling site index. Absolute calibration of concentrations is significant to a factor of 2.

FIGURE 2. In this plot, the average sulfur concentration is found to exhibit a trend with sampling location, from Pensacola to Miami along the length of Florida. If a sampling site index is used as the horizontal coordinate, the concentration in size class 4, 0.5–1 μm aerodynamic diameter, is found to exhibit a monotonic decrease from northwest to south Florida. No marked difference between urban and nonurban sites is seen in the trend for this particle size range. Less smooth trends are seen for some of the other size classes as well, although for the largest- and smallest-particle size classes, no monotonic decrease from northwest to south is observed. Sea salt sulfate, which may contribute much of the largest particle sulfate, comprises a minor amount of the total aerosol sulfur concentration observed at any of the sites. The fine-particle size range, which comprises the bulk of the sulfur concentration observed at any location, may be formed by gaseous sulfur oxide conversion to sulfate in the atmosphere, ultimately traceable to terrestrial pollution sources.

A careful study of meteorological conditions at each location indicates that air masses from states lying to the north of Florida pass over the north Florida sites more frequently than over those sites in the southern part of the state. Thus, in the averaging of approximately six samples collected at each site over the two-week period, a greater percentage of samples of northern air is represented in northern Florida than in the south. Thus, an apparent decreasing average influence of north air flow regime is seen as one moves southward in the state. The decreasing sulfur concentration suggests that Florida is bathed chronically in a sulfate aerosol concentration gradient, due to advection of sulfate formed in combined urban plumes of cities in the eastern U.S. Superimposed on this regional scale gradient may be local urban effects within Florida. However, apparently in the 0.5–1 μm diameter range, these influences are relatively minor, although in some of the other size classes they may be dominant.

Measurements in December 1976 and April 1977 tend to reaffirm the general interpretation presented here. Cascade impactor samples from December show a less pronounced, though still significant, sulfur concentration decrease from north to south of a magnitude that is expected during this season of deeply penetrating cold fronts. Streaker samples collected during the December and April periods allow a detailed examination of the correspondence of sulfur concentration at specific times with air flow patterns inferred from synoptic weather charts.[4] The times when highest sulfur concentrations are observed are generally those when northerly air flow trajectories are likely. In addition, some occurrences of high sulfur concentrations in Tampa are also apparently due to local air pollution effects, although such local effects have not yet been observed at the other Florida locations. The general finding that Florida's nonurban sulfate background is due primarily to transport of sulfate formed in urban plumes from outside the state to the north is of considerable importance in designing strategy for sulfate air pollution control. The existence of a gradient from north to south in sulfate concentration also suggests that epidemiological studies may be carried out in Florida in the search for a correlation between air quality and respiratory disease in the population that resides in this gradient.

The Distance Scale of Sulfate Transport

The preceding examples suggest a large distance scale for the formation and transport of sulfate in urban plumes. They also suggest that, on a continental scale, North America may be more or less contaminated with sulfate formed from sulfur oxide emissions in our urban areas. Thus, in order to find examples of "clean air," one might have to travel away from the continent. The study of Florida indicates that the southernmost locations, which are heavily bathed in marine air, have fine particle sulfur concentrations that may be closer to those of the natural atmosphere than can be found elsewhere in the eastern U.S. However, the full extent of urban sulfate transport may not be fully appreciated unless measurements in more remote maritime and continental atmospheres are made. Two examples may illustrate this point.

1. On two occasions, we have had the opportunity to collect aerosol samples using cascade impactors from towers located on oceanic islands.[5,6] Some results from these

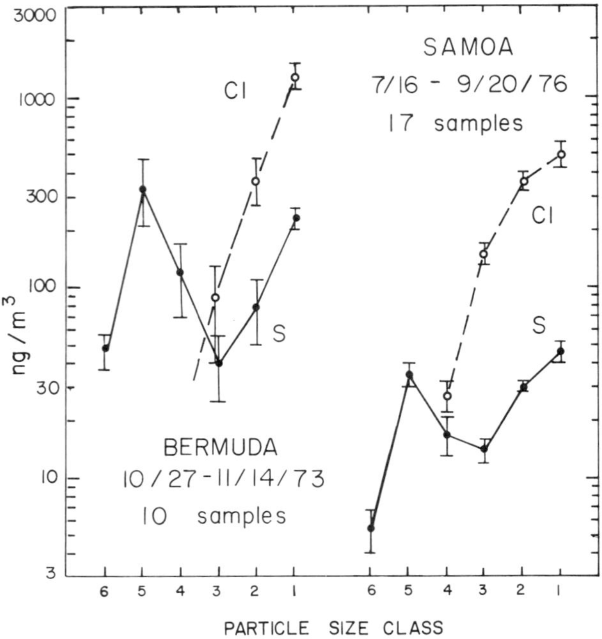

FIGURE 3. Comparison of sulfur and chlorine concentrations, as geometric means ($\pm\sigma_{\bar{x}}$), in marine aerosols as a function of particle size, from towers on Bermuda and Samoa. Classes correspond to aerodynamic diameter ranges as in FIGURE 2. (On Samoa, the measured size class 1 concentrations are less than the true values, because of inefficient particle sampling by cascade impactors at high wind speeds.)

studies are shown in FIGURE 3. Bermuda is located in the western North Atlantic Ocean and Samoa is situated in the central South Pacific. In both locations, it may be presumed that the chloride concentration is predominantly due to sea salt injected into the atmosphere by the action of bubbles bursting at the sea surface. At the two locations, the Cl concentrations are quite similar (except for somewhat lower apparent largest-particle Cl concentrations in Samoa, an artifact due to reduced sampling efficiency at high wind speed). The sulfur concentration may be presumed to result from two sources; one, the sea surface, and the other, the conversion from gaseous to particulate sulfate aerosol in the atmosphere. At Bermuda, the concentrations of sulfur are considerably higher than those observed in Samoa, for small particles by about a factor of ten. These results suggest that a composite urban plume of continental dimensions, possibly largely from North America, has transported gaseous sulfur oxides and sulfate aerosol through the atmosphere over the ocean more than one thousand kilometers from the nearest sources. Sulfate formed in the urban plume may be manifested as fine aerosol particles and also possibly as larger particles, if gaseous absorption by marine aerosol droplets is followed by oxidation to sulfate.

Additional measurements made in remote continental atmospheres from South America[7] and southern Africa[8] indicate similar, possibly even lower, aerosol sulfur concentrations than in marine air. These concentrations, 50–100 ng/m^3, are one hundred times lower than concentrations of particulate sulfur often encountered in the nonurban eastern U.S. atmosphere. These remote atmospheric studies, then, enable us to appreciate how large is the extent of our modification of the composition of the atmosphere in urbanized regions—sulfate concentrations may be one hundred times above the natural level.

2. On a ship midway between Bermuda and New York in the western North Atlantic, Darzi[9] obtained a series of 27 cascade impactor aerosol samples, of 9- to 68-hour sampling times, over a three-week period when air flow varied in its direction, including five samples that were collected approximately downwind of northeast U.S. urban centers. Thus, samples were obtained of a composite urban plume as well as of the atmosphere more typical of the western North Atlantic outside of the urban plume. As with the results in Bermuda, most of the sulfur in the samples was found to be in the fine particle mode, not the coarse particles typical of sea spray, and apparently mostly of pollution origin. However, the five "downwind" samples, collected when a significant fraction of the air had passed over eastern U.S. cities 1–2 days previously, exhibited an anomalously high percentage of the smaller-particle size sulfur contained in the ultrafine < 0.25 μm aerodynamic diameter size range. As shown in FIGURE 4, 50–100% additional ultrafine sulfur, over and above that in the 0.25–2 μm diameter fine range, was observed, whereas the other samples typically had

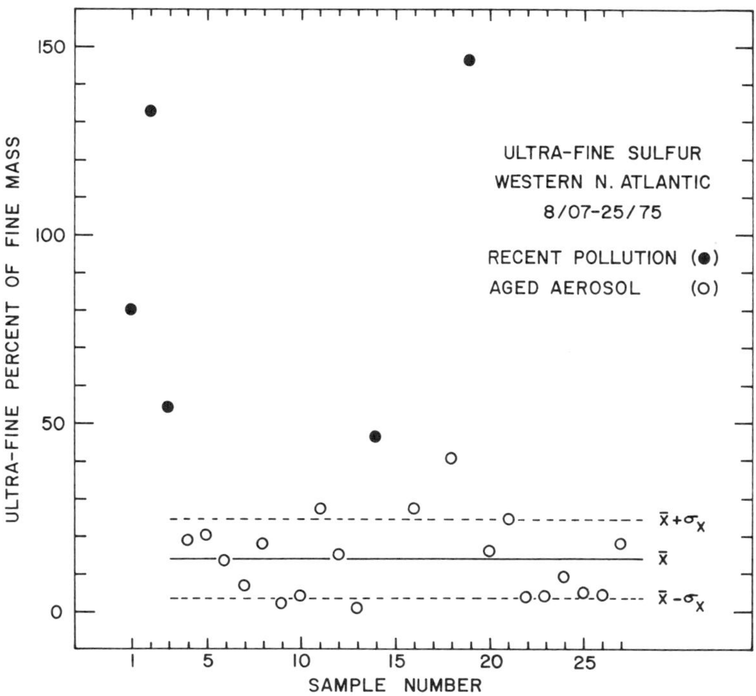

FIGURE 4. Proportions of sulfur in the ultrafine-particle size fraction (class 6, < 0.25 μm) relative to the fine range (classes 5, 4, & 3; 0.25–2 μm) from shipboard on the North Atlantic midway between New York and Bermuda. "Recent pollution" refers to samples collected during times when air trajectories had passed over the New York area 1–2 days previously. "Aged aerosol" refers to samples collected from air that was located over water at least 2 days.

less than 25% additional ultrafine sulfur. These results suggest that either sulfate was being formed directly in the urban plume over the ocean by gas-to-particle conversion, initially producing very small particles, or, if the smallest particles were formed in the urban areas themselves, they had not had sufficient time for coagulation to larger particles during the 1–2 days transport time over the ocean surface. This result of the urban plume could not be detected by comparing total sulfate aerosol concentrations, or even those in the fine-particle size range, since the concentrations were not greater in the "downwind" samples than in the others. Instead, the effect of the plume could be detected by examining the relative percentages of ultrafine-particle size sulfur. Apparently, significant kinetic processes were occurring on a time scale comparable to the transport time over a 1000 km or greater distance out to sea.

The Time Scale of Sulfate Formation in Plumes

The previous examples suggest that the time scale of sulfate formation in plumes may be sufficiently long for transport to occur over hundreds and perhaps thousands of kilometers during the formation process. It would be desirable to design field experiments that would more precisely indicate the time scale and how it varies with geographic and meteorological conditions. Two examples of such field experiments are summarized here.

1. In northwest New Mexico, Ferek[10] directly sampled the plume from a large coal-fired power plant, using ground-based, time sequence, streaker filter samplers. Although the exact direction of air flow was variable, there were times that could later be identified when air flow was consistently along a specified direction. Streaker samplers, which have a two-hour time resolution, were set up in a small array and allowed to operate automatically for several consecutive one-week periods. Then, after examining meteorological records for those time periods when air flow was found to occur in a desired trajectory across the array of samplers, the samples corresponding to a single plume as it passed from one sampler to the next were selected for analysis. In his test of this idea, Ferek examined trends in aerosol sulfur and associated fly ash element concentrations in the plume for successive two-hour periods on August 20, 1976. The results are summarized in FIGURE 5.

Using an upwind station as a regional background reference, the concentrations of all elements are found to vary with distance downwind at the three sampling sites situated at 5, 12, and 23 km from the power plant stack. Al and Si are representative of several elements found within fine fly ash particles. S represents sulfate formed in the plume from gaseous SO_2, which comprised the bulk of the plume sulfur. By assuming that all constituents are diluted equally as the plume expands during its transport, the relative concentration of aerosol sulfur, aluminum, and silicon may represent sulfate formation in the plume. With knowledge of the speed of the plume and the initial stack gas composition of the power plant, the rate of conversion of SO_2 to sulfate can be estimated; for this day, a rate of 0.7% per hour is obtained. Although we may not assume that this value from New Mexico can be considered typical of all conversion rates in all plumes, it is consistent with measurements reported elsewhere in the literature. Moreover, the technique of using ground-based samplers is simple and, in principle, could be applied to other locations. Its further application to plume

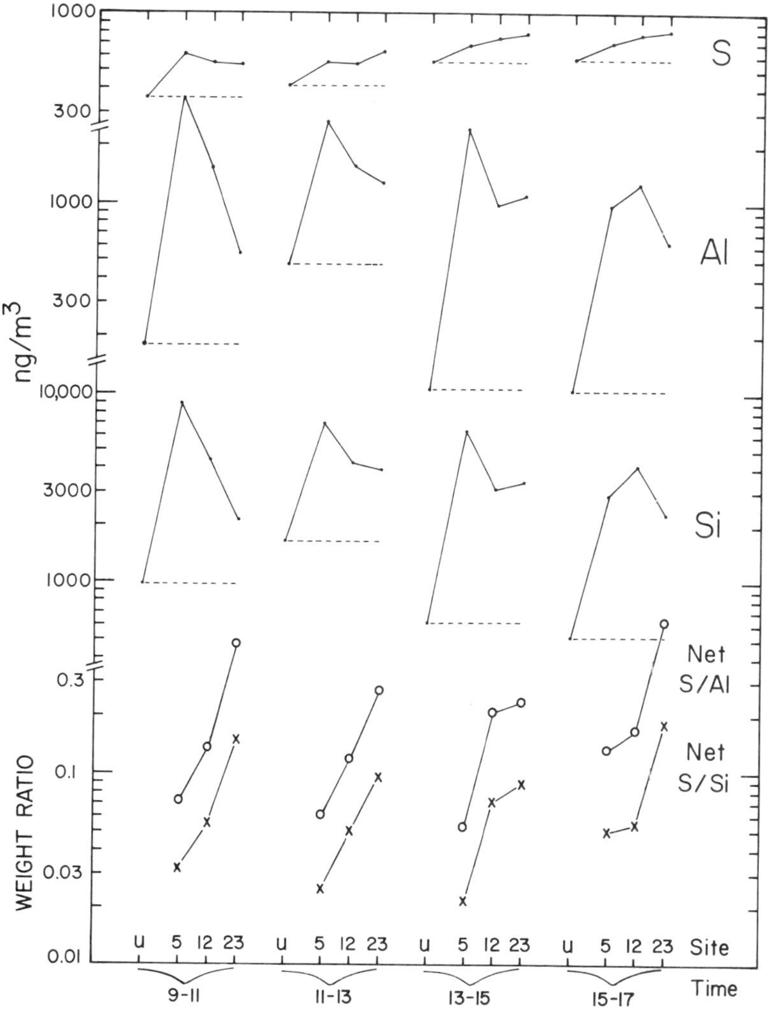

FIGURE 5. Sulfur, aluminum, and silicon concentrations in aerosols from NW New Mexico at ground-level sites located upwind (u) and downwind (at 5, 12, and 23 km) of a large coal-fired electric power plant, during four successive 2-hour intervals on August 20, 1976. Net Al, Si, and S components above upwind concentrations represent fly ash and conversion of gaseous SO_2 to sulfate in the plume. From their ratios, a conversion rate of 0.7% per hour is estimated.

studies may indicate more fully how the rate of sulfate formation in such plumes varies with environmental conditions.

2. The time variability of sulfur concentration in aerosols near the ground in the eastern U.S. is apparently a complicated function of atmospheric environmental factors. In order to examine the atmospheric chemical relationships in detail, we are currently carrying out routine sampling of aerosols at the nine Class I sites of the

EPRI-SURE network ranging from Tennessee to Massachusetts. The samples, collected with streaker samplers, permit PIXE analysis for sulfur and associated trace elements, with 2-hour time resolution.

In FIGURE 6, data from the first two weeks of October 1977 at SURE Site 6, Giles County, Tennessee, are presented.[11] In this example, in many ways typical of nonurban sulfur concentration measurements we have made at other locations in the network, sulfur varies in a seemingly complicated way over a factor of ten in concentration and rarely shows correlation with any of the other trace elements also measured. However, by close examination of the data, several correlations with meteorological data may be discovered. In the first place, the highest concentrations

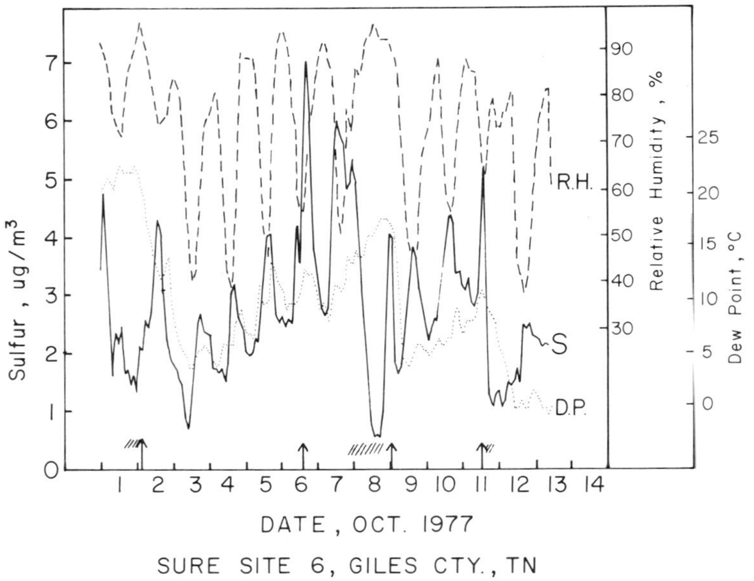

FIGURE 6. Sulfur concentrations measured at 2-hour intervals at SURE Site 6, Giles County, Tennessee, October 1–13, 1977. Also shown are trends of relative humidity, dew point, and times of four frontal passages (arrows) and rainfall (hatched). Sulfur exhibits daytime maxima, when RH exhibits minima and when dew point rises. Sulfur maxima also occur during three of the frontal passages.

often are found near the times of frontal passage. The occurrence of rainfall may tend to decrease concentrations. In this example, additional maxima in sulfur concentrations are seen to occur during the daytime, when relative humidity values exhibit minima. Moreover, the dew point, which is a measure of the absolute humidity of the atmosphere, exhibits maxima at times corresponding to decreasing aerosol sulfur concentration.

At the present time, we cannot present a detailed interpretation of the chemical and dynamic factors that influence sulfate aerosol concentrations in the atmosphere. Undoubtedly, the factors are more than a few in number and may interact in a complex and subtle fashion. In seeking to identify them, it is important to make aerosol concentration measurements on a time scale comparable to significant variations in humidity, trace gas concentrations, and other atmospheric parameters, which may be such controlling factors. Although at this time we cannot make a decision as to the kinetic mechanisms controlling sulfate concentration, the evidence appears to indicate that mechanisms on both a relatively short time scale, e.g., hourly, and a much longer time scale, e.g., a day or more, may be operative.

Summary

The detection of sulfate formed in urban plumes of cities in the eastern U.S. is complicated by the presence of a high nonurban sulfate background caused by the combined effect of all anthropogenic sulfur oxide sources in the region. This pollution background tends to obscure the contributions of individual urban plumes.

The time period for appreciable SO_2 conversion to sulfates may be comparable to that required for plume transport over tens to hundreds of kilometers. Nonurban submicrometer sulfur concentrations, which are a significant factor in reducing atmospheric visibility, vary significantly over a few hours of time at any site and, between sites in a network, show correspondences that suggest long range transport.

In Florida, a north-to-south average concentration decrease is observed, which may result from the averaging or mixing of polluted continental air with cleaner marine air. However, the Bermuda marine atmosphere contains substantially greater fine particle sulfate concentrations than found in more remote locations, e.g., Samoa and continental South America, indicating significant urban pollution transport over the North Atlantic, 1,000 km or more from land.

From shipboard, between New York and Bermuda, anomalously high relative concentrations of aerosol sulfur in $< 0.25\ \mu m$ aerodynamic diameter particles are observed in parcels of air that had passed over polluted northeastern U.S. urban areas 1–2 days previously, suggesting conversion of SO_2 to sulfate over the ocean.

Within the plume from a large coal-fired power plant in New Mexico, the rate of SO_2 conversion to sulfate can be estimated from observed increases in aerosol sulfur to fly ash element ratios with time and distance up to 23 kilometers from the stack. The rate, $\sim 0.7\%$ per hour, which is calculated assuming average plant emissions during the sampling period, is consistent with the large-scale formation and transport of sulfate in urban plumes over the eastern U.S. and out over the North Atlantic Ocean.

At nonurban eastern U.S. sites, aerosol sulfur concentrations often vary inversely with relative humidity and directly with dew point and, in addition, show diurnal patterns as well as maxima during passage of cold fronts. These variations suggest that sulfate concentrations in the urban plume can vary during transport, as a result of both chemical and dynamic processes.

References

1. Cohn, S. L. 1978. Visibility and Air Pollution Sulfur Transport over an Eastern U.S. Network. M.S. Thesis. Department of Meteorology, Florida State University. Tallahassee, Fla.

2. AHLBERG, M. S., A. C. D. LESLIE & J. W. WINCHESTER. 1978. Characteristics of sulfur aerosol in Florida as determined by PIXE analysis. Atmos. Environ. **12**(1–3): 773.
3. LESLIE, A. C. D., M. S. AHLBERG, J. W. WINCHESTER & J. W. NELSON. 1978. Aerosol characterization for sulfur oxide health effects assessment. Atmos. Environ. **12**(1–3): 729.
4. ANNEGARN, H. J., A. C. D. LESLIE, J. W. WINCHESTER & G. S. YOUNG. Air pollution transport and aerosol sulfur concentration variability in Florida. (To be published.)
5. MEINERT, D. L. & J. W. WINCHESTER. 1977. Chemical relationships in the North Atlantic marine aerosol. J. Geophys. Res. **82:** 1778.
6. MAENHAUT, W., M. DARZI & J. W. WINCHESTER. Sea spray and non sea spray aerosol components in the marine atmosphere of Samoa. (To be published.)
7. LAWSON, D. R. & J. W. WINCHESTER. 1978. Sulfur and trace metal concentrations in aerosols from the South American continent. Geophys. Res. Lett. **5**(3): 195.
8. ANNEGARN, H. J., R. VAN GRIEKEN, P. VAN ESPEN, F. VON BLOTTNITZ, J. P. F. SELLSCHOP, J. W. WINCHESTER & W. MAENHAUT. 1978. Background aerosol composition at Gobabeb, South West Africa. Madoqua **2:** 107.
9. DARZI, M. 1977. Terrestrial Influences on Aerosol Composition over the Western North Atlantic Ocean. M.S. Thesis. Department of Oceanography, Florida State University. Tallahassee, Fla.
10. FEREK, R. J. 1978. Sulfur Conversion in a Power Plant Plume Determined by Aerosol Composition Relationships. M.S. Thesis. Department of Oceanography, Florida State University. Tallahassee, Fla.
11. RHEINGROVER, S. W., S. L. COHN, G. S. YOUNG & A. C. D. LESLIE. The correspondence between aerosol sulfur concentration and weather parameters in a nonurban eastern U.S. network. (To be published.)

THE URBAN PLUME: GENERAL DISCUSSION*

Francis A. Schiermeier, *Moderator*

Environmental Sciences Research Laboratory
U.S. Environmental Protection Agency
Research Triangle Park, North Carolina 27711

P. K. MUELLER (*Environmental Research & Technology, Inc., Westlake Village, Calif.*): Dr. Ellestad, is the diurnal pattern of NO_x—either NO, or NO_2, or the sum of the two—influenced by the urban plume?

T. G. ELLESTAD (*U.S. Environmental Protection Agency, Research Triangle Park, N.C.*): I don't recall seeing much other than a shift to NO_x during the daytime when ozone was present.

P. K. MUELLER: Did you see a peak only on days when there was influence from another plume?

T. G. ELLESTAD: For the most part, yes. It did correspond quite a bit to urban plume passages.

P. K. MUELLER: Do you have any simultaneous nitrate?

T. G. ELLESTAD: No, I'm afraid we don't. These data were in 1975, and we weren't quite up to making accurate nitrate measurements at that time.

R. L. TANNER (*Brookhaven National Laboratories, Upton, N.Y.*): Dr. Ellestad, do you have any ammonium data from that period, and, if so, would you expect to find sulfur with some more analysis?

T. G. ELLESTAD: The only ammonium data that we had were collected on a 24-hour basis. We ran a high volume filter. We also ran a Hi-vol with an Anderson head impactor and, from the backup filters of that impactor, obtained 24-hour ammonium values. However, they were not very useful compared to the 2-hour resolution data.

J. W. WINCHESTER (*Florida State University, Tallahassee, Fla.*): Dr. Ellestad, a very nice paper and a very clear indication of the urban plume.

For the aerosol composition, you mentioned that phosphorus and sulfur in the fine mode were correlated. We have found it difficult to resolve x-ray peaks of phosphorus from those of sulfur in the presence of a fair amount of sulfur. I wonder whether you are sure that phosphorus was detected, or if there is a possibility that it's an artifact of the x-ray spectrum resolution.

T. G. ELLESTAD: I'll have to refer that question to Tom Dzubay, since he is the expert on the x-ray fluorescence analysis. He was the one that did the analysis.

J. W. WINCHESTER: I'll just leave it for the record that I raised it as a question. It might be worth scrutinizing the data again to make sure, because we've had this problem.

My second comment is on lead. You mentioned that lead was being detected and indicated, of course, that it is from automobiles. I would say that it is mostly from automobiles, but St. Louis has considerable nonferrous metallurgical operations in smelting. In our analysis of RAPS data, it appears that a significant fraction of the lead in the city is due to the smelting of lead ores. Therefore, to be strictly correct, it might be best to say "mostly from automobiles" in the text.

T. G. ELLESTAD: That might help explain the bromine values, which I did not

*This is a compilation of the discussions held after the presentation of each of the papers in this section.

0077–8923/80/0338–0309 $01.75/1 © 1980, NYAS

present. One reason was that, when we redid the x-ray analysis after a couple of years had passed (it was desirable to do that because of improvements in the sensitivity and accuracy of the x-ray fluorescence technique), the bromine had decayed quite a bit, I guess through volatilization. We got rather poor correlations between lead and bromine, and I was thinking that the decay of the bromine was the major reason. But certainly, if there were other lead emissions in St. Louis, that would tend to decorrelate them as well.

J. W. WINCHESTER: We published our results in the *Journal of Applied Meteorology* last year on the RAPS data. It is not a large fraction of the lead but is on the order of 10% or so.

G. T. WOLFF (*General Motors Research Laboratories, Warren, Mich.*): Dr. Miller, I'd like to comment about your numbers on the ozone half-life. I disagree with the low number of a few hours.

D. F. MILLER (*Battelle's Columbus Laboratories, Columbus, Ohio*): No, I meant to say a few minutes.

G. T. WOLFF: A few minutes? The upper number now seems to be over 100 hours depending on the conditions. The DaVinci experiment, which was done in the St. Louis plume, an aging plume, came up with over 100 hours for the half-life. From flights that were conducted by the Interstate Sanitation Commission a few years ago in an overnight aged plume from the Philadelphia area, we calculated an ozone half-life on the order of 50 hours. So, under certain conditions, the ozone certainly can persist and be transported long distances. If there is a similarity in the formation mechanism between the ozone and the SO_4, you will see it. We have seen that correlation in New York and New Jersey.

D. F. MILLER: Perhaps we differ in our interpretation of the half-life or the *half time*. One of the governing rates of ozone depletion during the day is the photolysis of ozone. That *half-life* is about 8 hours; there is no way, during the day, that ozone could last longer than that.

What you are saying is that ozone is destroyed and replenished so there is little net loss; but the actual life of a single ozone molecule is limited to about 8 hours.

S. K. FRIEDLANDER (*University of California, Los Angeles, Calif.*): Actually, Dr. Miller, I felt quite encouraged by your results, because you came within an order of magnitude of the correct rate for the conversion and the formation of sulfate. How did the wind field in the aerodynamics, mixing, and so on, enter into the calculations, or didn't they?

D. F. MILLER: They didn't. The calculations—the kinetic simulations—are done assuming perfect mixing. For the plume simulation, all we can do, in effect, is dilute it at some rate that seems reasonable; but we can't do any kind of fancy mixing with the simulation.

S. K. FRIEDLANDER: Were the meteorological conditions significantly different on the two days? I think you were three times too low on the first day, and on the second day, you were three times too high in your predictions.

D. F. MILLER: Yes. In fact, a conclusion we arrived at was that, on the day when the rate of SO_2 oxidation was very low, the air was very clean from the North and the wind speeds from the flow across the lake were much the same. But I think the real difference in that case—what we were really seeing—was a conduction period of photochemical oxidation. Although we had freshly entrained hydrocarbons as the air swept through Milwaukee, they just hadn't had time to react during the two or three hours that we followed them across the lake. Whereas on the other day, we had a well-aged, polluted air mass that came through Milwaukee. Photochemistry was already well under way from the moment the sun came out, and we saw a highly elevated rate of oxidation. But a model just isn't quite able to cope with those differences.

G. M. KLAUBER (*Environmental Measurements, Inc., Arnold, Md.*): Dr. Miller, on July 20, did you have any indication as to whether the power plant plume was in, or separated from, the urban plume; from either hydrocarbon or, perhaps, lead and bromine measurements?

D. F. MILLER: It wasn't separated from the urban plume. The urban plume was fairly constant and widespread on this particular day, but I don't know if there are any lead or bromine data that could discern the two. Most of the aerosol collections were for the entire width of the traverse, and they weren't separated from the power plant area.

G. M. KLAUBER: In the Tennessee plume study, all three of the chemical analysis aircraft found ozone bulges, and this is out away from major hydrocarbon sources. Those hydrocarbon measurements that we have give very low numbers. These ozone bulges in the power plant plumes were regular, everyday occurrences.

T. E. GRAEDEL (*Bell Laboratories, Murray Hill, N.J.*): Dr. Miller, how did you select your rate constant for SO_2 plus RO_2?

D. F. MILLER: We're doing some chamber work right now, in which we are doing a series of competition experiments. The best rate that we get from those results is about 2 reciprocal ppm minutes. Jack Calvert's recommendation is, I think, 5.8 or thereabouts for that particular reaction. So we used a rate that was less than half of what he would recommend.

L. NEWMAN (*Brookhaven National Laboratories, Upton, N.Y.*): I have a comment, Dr. Miller. As I understand it, in meteorology, we have the "good weather effect," namely, that meteorologists only like to go out when the weather is good; so much of what we know about meteorology is during good weather conditions.

I'm beginning to get the feeling that we are having the "high oxidation effect" in urban plume studies. We go out to look at the oxidation in urban plumes during high oxidation conditions, notably when high photochemical conditions can occur. So there is a word of caution I'd like to add here. When we compare oxidation rates from one worker to another, looking for numbers for overall oxidation rates, we had better average this over a whole diurnal cycle and remember that the overall oxidation rate appears to be on the order of 1% or less per hour.

T. G. DZUBAY (*U.S. Environmental Protection Agency, Research Triangle Park, N.C.*): Dr. Macias, could you say anything more about why the fine fraction is relatively so rich in silicon in the area where you were? Did they not have precipitators there, or what was the situation?

E. S. MACIAS (*California Institute of Technology, Pasadena, Calif.*): This is fairly speculative at this point, but I don't think this is the silicon that you normally think of in fly ash, i.e., large-particle silicon, which is presumably due to some sort of a melting of mineral inclusions in coal or other minerals, for example, the copper ore.

In this case, what is happening is that you are getting rather high temperatures, and I think that the silicon is actually volatilizing in the combustion zone of the smelter or power plant and then condensing somewhere just after this combustion area, when it cools off just a bit. Temperatures are certainly high enough in modern coal-fired power plants and in smelters. So I think that you are getting a volatilization and formation of fine-particle silicon that would be totally different than the coarse-particle silicon that you often see.

Now, the plume measurements make it look like there is a lot of silicon coming out of the smelters, and some coming out of the power plant.

T. G. DZUBAY: In the East or in the Midwest, you don't seem to see as much silicon in the fine fraction, so there has to be something in the West that enhances it.

E. S. MACIAS: There are two things I would say. One is that you don't have the large copper smelters in the East. Remember I said, just looking at SO_x emissions, that 2,000 tons per day are coming from smelters, and only a 1,000 tons per day are coming from all the power plants in the region. Many of the power plants are really outside of the air shed that we are studying. They are in the eastern part of Colorado.

Two is that the coal that is burned in the Southwest has about ½% sulfur typically. Presumably with less sulfur, there may be a slightly higher ash and silicon content. It has been mentioned that much of the coal burned in the East has higher sulfur and actually slightly lower ash.

So I don't really think the silicon is all coming from power plants, but in fact, if you look at the sulfur to silicon that comes out—or even that is in the coal, if you like—it begins to look like a trend from eastern coal to western coal.

T. G. DZUBAY: You gave percentages for the amount of silicon at each of those two locations you showed on your viewgraphs. Do you have the actual micrograms-per-cubic-meter values for those two? because the TSP [total suspended particulate matter] is markedly different.

E. S. MACIAS: That is a good point. I don't have the values right here but I can give them to you later. More important, I think that is one of the big differences in the Southwest. Remember that we're talking about light scattering coefficients of 25×10^{-6} (or 0.25×10^{-4}) and fine particles of about 5 to 6 micrograms per cubic meter. In the East you may be looking at 25–30–50–60 micrograms per cubic meter. So, in fact, what we are saying is that there aren't many other sources in the Southwest. There aren't the large population areas.

T. G. DZUBAY: From a rough calculation, it appeared that the amount of silicon was the same in both areas, just the percentage of the TSP was different.

E. S. MACIAS: A good point; that's roughly right. Actually, there's slightly less silicon, but not a whole lot less.

D. GILLETTE (*National Center for Atmospheric Research, Boulder, Colo.*): Dr. Macias, I'd like to suggest that one possibility for the source of small silicon-containing particles is clay and that this clay doesn't necessarily come from smelter or power plant operations but from windblown soil and from automobile traffic on the soil.

E. S. MACIAS: Thanks. I would agree completely that that is certainly a possibility. I hope that we can address that with some different techniques in some of the later experiments.

R. JAENICKE (*Max-Planck-Institut für Chemie, Mainz, Federal Republic of Germany*): Dr. Whitby, as you can imagine, I'm very interested in your background aerosol. Can you give us an idea of how you obtained these values? Are they the lowest values you ever have observed, and what is your estimate based on?

K. T. WHITBY (*University of Minnesota, Minneapolis, Minn.*): We went through the California, Colorado, Minnesota, and Michigan data; we saw a bottom value of size spectrum that was quite consistent between all these different places and with those kinds of numbers. That's the basis for it. What I'm saying is that constancy and low value are an index of some kind of a universal continental background.

R. JAENICKE: The only question that remains is nature, of course, with the large variability in the background condition of whatever it is. What would be your estimate about these variations?

K. T. WHITBY: We've made attempts several times, as part of the ACHEX program and also as part of our own work in Minnesota, to find naturally generated aerosols. There is no problem in finding naturally generated dust—there is lots of that around—but in terms of trying to find naturally generated accumulation mode aerosol, we've never seen it.

S. K. FRIEDLANDER: I agree with you, Dr. Whitby, on the importance of studies of new-particle formation—nuclei formation—although I'm not sure why. We've been working on that for the last four or five years. In fact, Peter McMurry, who is now with the faculty at the University of Minnesota, did his Ph.D. in our group on that problem.

One of the basic questions that you ask is Why should new particles form in the presence of existing particles? When you have centers of condensation, you have the competition and the existence of particles, as in an urban area or in an urban plume, which tend to suppress new particle formation.

I'm going to talk about that later. I think it is the most interesting theoretical problem around, in so far as nucleation is concerned, and I have been trying to sort out why it would have an important atmospheric effect. The one explanation I can come up with is that it will certainly change the cloud condensation nuclei balance. So it might have some significant meteorological effects through nucleation effects. Are there other important effects?

K. T. WHITBY: Well, as you know, Dr. Friedlander, we're working on this, too. In fact, Dr. McMurry and I have Gilbert's program running now. We also have another program in which we have put these nucleation rates, and we are looking at the dynamics. The indications are that the numbers I showed you are entirely consistent with those models. There's no obvious contradiction based on our first attempts at running these numbers into those nucleation aerosol growth models.

S. K. FRIEDLANDER: But from a practical point of view, what are the implications? As you pointed out very nicely, only 5% of the new mass—new aerosol—is coming in through new particles. Most of it is plastering itself onto existing particles, at least in the plume. It might be different in the Southwest-type problem.

K. T. WHITBY: All right, let me review some of the conclusions we have come to on this. I didn't say anything about these, because they are in an unpublished *Atmospheric Environment* paper on nucleation.

First, it appears to us now that nucleation is ubiquitous in the atmosphere during the daylight hours. It looks like you have a certain degree of nucleation over the remotest parts of the ocean as well as in these plumes. This can explain why the accumulation mode size is so stationary. This had puzzled us for a long time.

It appears that whenever you have volume and mass formation coupled with it, almost in this fixed ratio or some ratio close to that, you have nuclei formation; and so the condensational size growth is just nicely balanced by the new nuclei coming in, which tend to go up and down with it.

So we see this as the reason why the accumulation mode size doesn't change very much from the remotest oceans to the plume. Now, the only exceptions are when you get into a 300 microgram per cubic meter accumulation mode episode—in Los Angeles, and we've seen one of those also in St. Louis. Then the sheer magnitude of that causes accumulation mode coagulation to grow up to 0.6–0.8 μ. But, probably 99.9% of the time in the atmosphere, the accumulation mode size sits around 0.3 μ, regulated by a simultaneous condensation that tends to cause it to grow one way and nucleation which tends to pull it the other way. This is our conclusion.

T. J. KNEIP (*New York University Medical Center, New York, N.Y.*): Dr. Whitby, would you speculate about forest fires or volcanoes as natural sources of accumulation mode?

K. T. WHITBY: We have no data on that. I would guess that, in the continental U.S., particularly east of the Mississippi, these are minor. In the West—I don't really want to speculate on that at this time. Certainly they are there. However, the question is not whether the volcano produces them, but whether they get down into the troposphere before they are removed by precipitation and so forth. Accumulation mode is cleaned out of the atmosphere by precipitation, 75% of it probably. So the question is of transport and removal, not whether it's produced.

P. K. MUELLER: Dr. Whitby, you gave us a grand average for K, and I have two related questions. How is it affected by water vapor pressure in the atmosphere? How is it affected by other meteorological variables?

K. T. WHITBY: The K that I used was only what came out of this one analysis. Remember, this is only one plume, one piece of datum. We have a couple of days more of data that we are going to analyze on this, but this is my first attempt to link ultraviolet radiation with volume formation. So it is just one number—one day—and that is all it is.

L. NEWMAN: I would like to thank Dr. Gillani for finally taking the mystery out of ozone bulges and for making us all recognize that power plants do indeed provide fuel for the formation of ozone, in the form of NO_x. It is indeed important to add hydrocarbons, among other things, as reactants necessary to the formation of ozone bulges. If I understand correctly, one can also conclude that the conditions for ozone bulges are somewhat similar to the conditions of high photochemical oxidation rates of SO_2 to sulfate, but that an ozone bulge is not necessarily a consequence of the oxidation of SO_2 to sulfate.

N. V. GILLANI (*Washington University, St. Louis, Mo.*): I agree with you. I think that the ozone bulge probably forms before the aerosol formation rate is accelerated. It is simply an indication of the photochemical activity that is taking place.

D. F. MILLER: Dr. Gillani, I'd like to make a comment about the role of ozone. You seem to implicate ozone as the predominant OH source in the power plant plume, and that's really not the case.

What I want to draw your attention to is the photolysis of ozone to give $O(^1D)$ and reaction with water to give OH. Now, that reaction is important in very clean air as a source of OH, but it is not relevant in the power plant plume situation. Ozone does not act directly as a source for OH under those conditions.

That same reaction is the one I was referring to earlier as the one that limits the lifetime of ozone. I looked up the rate constant for that, and it is 2×10^{-3} reciprocal minutes, which gives a lifetime of 8.3 hours for ozone.

E. S. MACIAS: Dr. Winchester, I'd like to comment on your conversion rate of SO_2 at a power plant in the Four Corners region of the United States. We measured the conversion rate in a smelter in Arizona, and found it to be about 0.7 percent per hour—also in an arid region of the U.S. I also looked in that plume, trying to do a sulfur to silicon ratio somewhat similar to what you did, and found a conversion rate that agreed with the more complete measurement of particulate sulfur to total particulate-plus-gaseous sulfur. But I'm a little concerned about your measurements, if you were looking at coarse-particle species. We looked at fine-particle silicon, and, as you go farther downwind, the coarse-particle aluminum, for example, is likely to be affected somewhat differently than the fine-particle sulfur and, in fact, may not follow the same dispersion pattern. In other words, there may be deposition problems with coarse-particle species. I would suggest, if possible, that you look at a fine-particle species.

J. W. WINCHESTER: This is a good point. In this calculation, which is part of a feasibility study, the assumption was made that all the particles, coarse and fine, were equally conservative within the plume during its transits. Now, there may be departures from uniform behavior; indeed, we have puzzled that over. I think the consistency of the data suggests that distribution differences may not be a major factor. You don't see any big differences between the nearest power stations and the farthest power stations. However, a more refined experiment, making particle-size-selected time records, could be done.

G. T. WOLFF: Dr. Winchester, from that last slide, you said that the highest sulfate values occurred during the frontal passage. Was that before the frontal passage or just after the frontal passage?

J. W. WINCHESTER: I can't be sure. I think our calibration on absolute clock basis is good to a half an hour at the sampling site; maybe better than that. We are very careful, but I don't think we can guarantee better than half an hour.

Then we match meteorologic data—from humidity, for example—from a nearby station that is a couple of kilometers away, depending on the station.

Then the front is obtained from the weather maps that we get from National Weather Service. These are huge weather maps, with the meteorologist's line drawn through where he thinks the front is. We look at that, and we have a six-hourly map, and we notice the progress and interpolate. This is subject to a couple of hours of error, just by interpolation, let alone whether his lines are drawn right.

So it seems to be close to the time of the frontal passage, but precisely before or after—I don't know. I've tried to draw the arrows on there as honestly as I could, and the first arrow showed a front a couple of hours before the maximum was seen, but I don't know whether that is real or not. The others lined up remarkably well, with no prejudging of the situation.

K. T. WHITBY: Dr. Winchester, I would like to reinforce Dr. Macias' concern about the settling. We've done a little modeling of this kind of thing, and there's a very large difference in the behavior of the coarse particles, which presumably carry most of the trace metals leaving the stack, and the sulfur compounds, which are going to be in the accumulation mode. I would think one would have to correct for that before one could use such a ratio.

I also have a question. What is the upper size limit of your ultrafine-particle range?

J. W. WINCHESTER: One-quarter micron aerodynamic diameter.

K. T. WHITBY: That is the middle of the accumulation mode.

J. W. WINCHESTER: It's not the nucleation mode, admittedly, so I don't use those terms; but at least it includes the nucleation mode.

K. T. WHITBY: That is what is on the filter?

J. W. WINCHESTER: Yes.

K. T. WHITBY: After the last impactor stage?

J. W. WINCHESTER: Right.

K. T. WHITBY: Also, Dr. Mueller had asked me a question about the sulfate formation rate, so I did a little thinking about that. First, I would like to state what I think the rates might be and see if you can confirm them.

If you take the urban plume data that we have, plus the Smoky Mountain data, plus some other things that I didn't show, it looks like the formation rate of sulfur in the fresh urban plume comes out to be about 3 micrograms per cubic meter per day. The second day, the rate seems to be half of that—1.5 micrograms per cubic meter per day. We have some shaky evidence to indicate that the third day is probably about like the second day. Now, from all these measurements that you've made, do you have any ideas on these numbers? Are they too high or too low, and do you see a differential from day 1 to day 2 to day 3?

J. W. WINCHESTER: I was going to ask you, after your interesting talk, whether you could convert those numbers into percent of SO_2 conversion to sulfate per day, under the assumption that this is the major contributor to the particles. I guess you are talking about a few percent or less per hour?

K. T. WHITBY: You can do that in the fresh plume, but the difficulty comes in that the instruments are all reading zero on the second and third day for SO_2. On that basis, you have very high conversion rates, and I guess I would like to raise this as a kind of question mark here. I'm intimately familiar with the problems in measuring SO_2, having babysat one of the best methods last summer. I would like to point out that I think that there's at least a factor of 2 or 3 uncertainty in these conversion rates at the low levels just due to inaccuracies in the measurements. So I have shifted my thinking and believe we should look at the rate of appearance of sulfur or sulfate at the low level and forget about conversion rate. We really can't measure the sulfur as SO_2.

UNIDENTIFIED SPEAKER: Dr. Winchester, I'm very impressed with the measure of conversion rates that you got in your ground-level sampling of a power plant plume. I have a feeling that the deposition problems are not going to be so great as to change the interpretation markedly. You do indeed have deposition of SO_2 and of particles that aren't considered. But I don't think that is going to be a major change. However, I would like to draw your attention to a paper appearing about seven years ago by, I believe, Weber in Frankfurt. We did a similar study but came out with very different results. I never understood that, and I guess I never believed it. I could never understand why it was wrong.

What he did was measure CO_2 and SO_2 at the ground-level station. He looked for spikes in the CO_2, and these always corresponded with spikes of the SO_2. From that, he was able to calculate apparent rates of oxidation of SO_2 by virtue of the fact that the ratio of the SO_2 to the CO_2 at his measuring point was different from the SO_2 to CO_2 ratio that was present upon emission from the power plant.

The problem is that he came out with very high numbers, like 10 or 20% per hour, and I've never understood what was wrong. I would suggest that you look at those data and see if you can interpret them.

J. W. WINCHESTER: If I understand correctly, he was looking at the rate of disappearance of SO_2?

UNIDENTIFIED SPEAKER: Essentially correct.

J. W. WINCHESTER: And that can be disappearance by diffusion to the ground or conversion to particles?

UNIDENTIFIED SPEAKER: But that is going to have to be a very large disappearance.

J. W. WINCHESTER: Yes, and his overall disappearance rate was 10 or 20%. Now in a couple of tens of kilometers, the decreases are not enormous. You are talking about tiny differences in the first place, and measurement errors then get to be significant. Thank you, I'll try to look at that.

P. K. MUELLER: Dr. Winchester, I just want to say very quickly that some SURE data were shown here this morning that we haven't had a chance to study yet. There is a lot of new information collected as a function of season—30 days a month for 4 months. We should be able to determine some net conversion rates from that information when we have worked out the matching trajectories and the emissions along the path on each day during the time periods studied. It is going to be a while before we get all the data together. Thank you.

NATURAL AEROSOLS*

Ruprecht Jaenicke†

Max-Planck-Institut für Chemie
D-6500 Mainz, Federal Republic of Germany

INTRODUCTION

In their pioneering work, Schmauss and Wigand[1] stated that the atmosphere only rarely is a molecular-disperse system. Only after extended rainfalls and snowfalls can it come close to that condition. Usually it is a colloid-disperse system, having gas as the continuous phase, with solid and liquid material in finest distributions dissolved or suspended. This is true not only for aerosols influenced by man, but for the natural aerosol in general. The aerosol load of the atmosphere usually is smaller by one or two orders of magnitude than the load of atmospheric trace gases. However, the effects of the aerosol—such as its contribution to the cloud formation and to the budget of the atmospheric radiation—cannot be overlooked and are crucial for life on earth. To know the atmospheric aerosol, a large number of points must be studied:

1. Sources of the aerosol;
2. Aging while airborne;
3. Residence time;
4. Sinks of the aerosol;
5. Global distribution;
6. Properties and composition of the aerosol;
7. Effects of the aerosol and secular variations.

Our knowledge about the natural aerosol is quite limited. Studies of the aerosol began some 50 years ago but have been done in a systematic way only since 20 years. However, measurements always carry the possibility of being influenced by anthropogenic aerosols, and we will see later to what extent it is possible to study the natural aerosol not influenced by man.

SOURCES

There are two major processes by which aerosol particles and aerosol mass are formed. One is disintegration and dispersion of bulk material, such as the weathering of crustal material, the disintegration and dispersion of biomass, and the production and dispersion of sea-salt droplets. The second process is the modification of airborne substances in such a way that new particle mass is formed. Such processes are gas-to-particle conversion by condensable gaseous species and the evaporation of cloud elements and rain droplets to leave behind aerosol particles. These different types of

*Supported by the German Science Foundation (DFG) through its Sonderforschungsbereich 73: Atmospheric Trace Components.

†Present affiliation: Institut für Meteorologie, Universität Mainz, D-6500 Mainz, Federal Republic of Germany.

0077–8923/80/0338–0317 $01.75/1 © 1980, NYAS

sources lead to a certain chemical composition of the aerosol and determine the size distribution; the disintegration of bulk material mainly forms particles larger than 0.1 μm in radius, while the airborne modification usually results in particles smaller than 1 μm in radius.

Depending on their linear dimensions, the sources can be subdivided into several categories. The surfaces of the oceans and of the continents act as *area sources.* Volcanoes and isolated meteorological events, like thunderstorms and low pressure systems, act as *point sources.* The gas-to-particle conversion and the droplet-to-particle conversion must be considered as *volume sources.*

Still the most complete study about the strength of natural aerosol sources is given in the SMIC Report[2] and results in an estimate of some 2000 × 10^6 tons per year released by nature, as compared to some 300 × 10^6 tons per year produced by anthropogenic activities. This estimate, however, is somewhat problematic because it

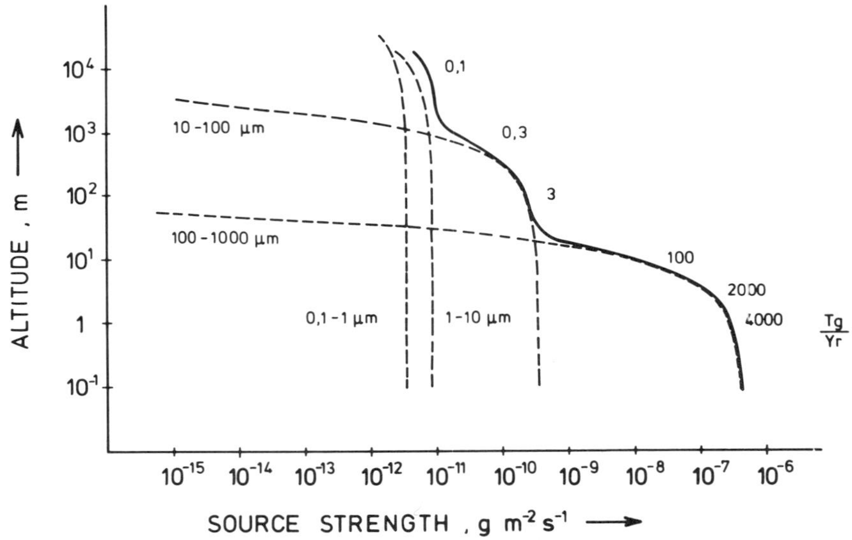

FIGURE 1. The apparent source strength for aerosol particles, as a function of altitude. For various particle size ranges (given in μm) this was calculated on the basis of an average aerosol mass and the limited residence time of particles (FIGURE 2). At various altitudes, the total source strength is given in Tg/yr. The apparent source strength varies rapidly with altitude, and only above some 10 m is the source strength less dependent with altitude.

is limited to particles smaller than 20 μm in radius. This can be demonstrated in the following example. The total particulate load of the atmosphere in natural aerosol is roughly 20 μg m^{-3}.[3] As we have seen above, the particles larger than 0.1 μm mainly come from an area source, the surface of the oceans and the continents. From this source, the particles must be transported horizontally and vertically (a rate of 2 cm s^{-1} is assumed for the purpose of this calculation). As we will see later, the particles have a certain residence time in the atmosphere. All this leads to FIGURE 1, where the apparent strength of the area source is a function of the altitude. It is obvious that the largest particles cannot be transported to higher altitudes, while the smallest particles under consideration are more or less evenly distributed with altitude. The accuracy of the figures given in FIGURE 1 is of minor importance. However, I would like to stress the fact that the apparent strength of the assumed area source at 10 m altitude is roughly two orders of magnitude smaller than at 1 m altitude. This might explain some variations in the source-strength data given in Reference 2.

Following Reference 2, the volume source from gas-to-particle conversion is of equal importance to the area sources mentioned. However, no data are given about the strength of the droplet-to-particle conversion as a particle source. The formation of cloud droplets around an aerosol particle acts as sink on the one side. On the other side, such a droplet takes up gaseous substances from the atmosphere and will evaporate in 9 out of 10 cases.[4] So it acts as a volume source for particles. This source certainly has a maximum in the cloud layer, but no estimates are available.

AGING

After being released as particles from the source, the aerosol is acted upon by a number of different processes, which are summarized as *aging*. The aerosol is transported in the atmosphere through meteorological events, and mixed with other aerosols or clean air; this is called dilution. The particles might collide with each other due to thermal diffusion (coagulation), become center condensation nuclei of cloud droplets (rain-out), be collected by falling rain droplets (wash-out), or sediment out by their own vertical velocity or by impact on ground obstacles (dry removal). The particles might gain mass through condensation of gaseous species or might lose mass due to evaporation.

Only some of these processes are known to a certain extent. Some of them occur only in the lower layers of the troposphere with an abundance of clouds, rain, and ground obstacles. Coagulation is most rapid in highly concentrated aerosols. The most rapid transport and the most effective dilution certainly take place if an aerosol penetrates into the higher layers of the atmosphere, where the air masses are transported rapidly without the friction from the earth's surface. Major effects of this aging are the formation of the aerosol size distribution with its unique features, the changing of the chemical composition of the aerosol—which might be quite different from that of the source, and the formation of rather uniform aerosol bodies, which are summarized as continental, maritime, and background aerosols.

RESIDENCE TIME

Most atmospheric processes described under "aging" affect the time during which an aerosol particle remains airborne. This time is called *residence time*. As for gaseous species, one single residence time was estimated for the aerosol[4,5] in the past. This, however, describes only that fraction of the aerosol most sensitive to the process used for estimation. In addition, it is a figure that averages over the very short residence times of coarse particles (with high deposition velocities) and the much longer residence times of smaller particles (with practically no deposition velocities). For aerosols with a small particle load, the residence time was calculated as a function of particle size.[6] This resulted in the analytical expression:

$$\frac{1}{\tau} = \frac{1}{C_F}\left(\frac{r}{R}\right)^2 + \frac{1}{C_D}\left(\frac{r}{R}\right)^{-2} + \frac{1}{\tau_{wet}} \quad \textbf{(1)}$$

which is shown in FIGURE 2 with: τ = residence time, in s; r = particle radius, in μm; $R = 0.3$ μm, standard radius; C_F, C_D = constants, in s; and τ_{wet} = time limit, in s, mainly controlled by wet removal processes.

Equation 1 is the mathematical description of the compilation of experimental findings and estimates about the residence times of atmospheric aerosols. The residence time in this sense means the time a particle is suspended as an entity. After this time, its mass is removed from the atmosphere or incorporated into another particle. Except for τ_{wet}, which is somewhat coupled to the wet removal processes, the constants in Equation 1 cannot be interpreted in terms of physical meaning. The constant C_F, coupled to that branch of the curve where sedimentation is of importance, seems to be rather independent from the particle concentration. So it applies in background aerosols as well as in continental aerosols. It is, however, coupled to the height of the homogeneous dust layer, which was estimated to be 1.5 km. The constant C_D, in the branch mainly controlled by coagulation, depends on the particle concentration. As can be seen from Reference 7, the residence time—defined as the ratio of the particle concentration over the change of concentration with time due to

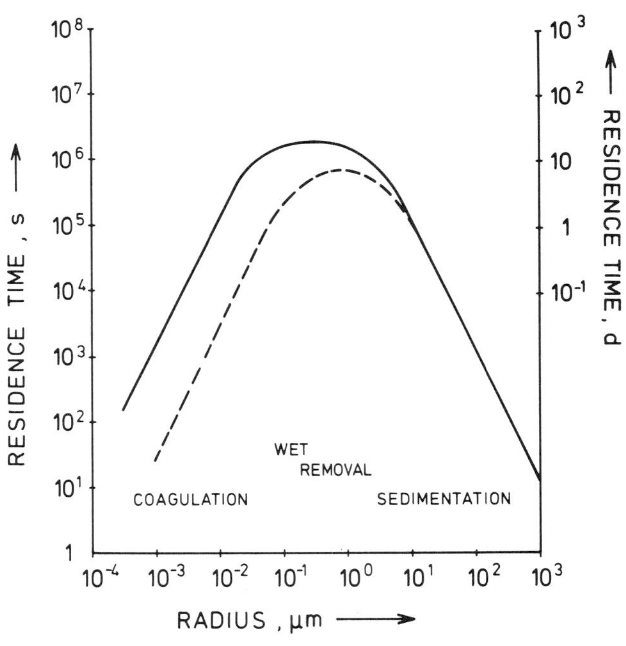

FIGURE 2. The residence time of aerosol particles, based on Equation 1. Two different cases have been calculated: —background aerosol, N = 300 cm^{-3}, τ_{wet} = 21 d; ---continental aerosol, N = 15000 cm^{-3}, τ_{wet} = 8 d. In both cases, the height of the homogeneous aerosol layer was assumed to be 1.5 km. This only has effects on the sedimentation branch of the curve.

TABLE 1

DISPLACEMENT OF AEROSOLS*

Particle Radius (μm)	Horizontal Distance (km)	Vertical Elevation (m)
0.001	8	20
0.01	800	2,000
0.1	8000	20,000
1	8000	20,000
10	800	2,000
100	8	20

*For the residence times given in FIGURE 2, the horizontal and vertical displacement of an aerosol particle is calculated. The actual numbers are of minor importance; only it should be shown that the particles in the range 0.1–1 μm radius can be transported intercontinentally and be distributed in the whole troposphere.

coagulation—is inversely proportional to the particle concentration. Thus, C_D depends on the concentration of the aerosol. FIGURE 2 indicates the ranges of residence times we have to expect in the natural aerosol. The residence time is shortest for the smallest and the largest particles. The particles in the range 0.1–1 μm have the longest residence times. Based on these residence times, FIGURE 1 was calculated.

With these residence times, it is possible to calculate how far particles from a point source—i.e., anthropogenic activities—might be transported. Under the assumption of a horizontal transport velocity of 8 m s^{-1} and a vertical uplift of 2 cm s^{-1}, which seems to be reasonable for atmospheric conditions, TABLE 1 shows (the actual figures are rather unimportant) that particles in the range 0.1–1 μm can be spread over the vertical extension of the troposphere and transported from one continent to the other. This range certainly is most easily influenced by man on a worldwide basis.

A good example of this intercontinental transport is the desert dust. The transport of desert particles from the Sahara can be followed across the Atlantic Ocean to the Lesser Antilles. This transport occurs only for the particles in the size range 0.1–1 μm.[37,24] Smaller and larger particles are transported only in the mesoscale; at selected locations, these particles certainly can be studied free of anthropogenic influences.

It should be mentioned, however, that some gaseous precursors of the smallest particles might have residence times long enough for worldwide (or hemispheric) mixing and distribution. Particles produced from these precursors are, of course, not free of anthropogenic influences if the precursors are released by man. To a certain extent, this is the case for sulfur and might apply to the transport of sulfurous aerosol to Scandinavia from central Europe.[27]

SINKS

The processes responsible for aging and residence times of aerosol particles are sink processes. These sinks—similar to the sources—act as *volume sinks* (so the formation of clouds) and *area sinks* (so the removal by dry deposition). In addition, one could consider the transport of aerosol particles into other volumes of the atmosphere as a sink, too. In principle, such a process could be the transport of aerosol particles into the stratosphere. FIGURE 1 would favor such a sink; however, FIGURE 1 does not take into account the circulation of the atmosphere. And indeed, it is assumed at present[8] that the stratospheric dust layer is more or less produced *in situ* by gas-to-particle conversion of SO_2, rather than by transport of particles. In this sense, the removal of this stratospheric aerosol with a residence time of roughly 2–3 years must be considered as a source of tropospheric aerosol particles. Because up to now only gas-to-particle conversion has been reported in the atmosphere, and never particle-to-gas conversion (an exception is water vapor), more particulate mass must be removed from the atmosphere than was injected. It has been estimated that roughly 50% is removed in the above-mentioned area sink, while 50% is collected by wet removal.[36]

GLOBAL DISTRIBUTION

The interaction of source function, aging, residence time, and sink function in connection with atmospheric transport processes leads to a global distribution of the atmospheric aerosol. It is now rather well established that four different aerosol bodies exist in the atmosphere. One is the *stratospheric aerosol.* As explained earlier, this aerosol is separated from those in the troposphere by the tropopause. The meteorological conditions at the tropopause permit only a small leakage of particles. The intrusion of gaseous precursors of aerosol particles into the stratosphere was discussed earlier.

Based on measurements of the concentration of condensation nuclei (a rather early, but comprehensive, study was published in 1939[9]), *continental aerosols* and those of the free atmosphere (this was later[10] called *background aerosol*) were distinguished. For the condensation nuclei, it means that the background aerosol can be observed at the surface of the ocean, because sea spray particles are hardly present in this size range.[11] From chemical evidence,[12,13] it became obvious that the *maritime aerosol* should be distinguished from the background aerosol. It was shown convincingly[12,14] that sea salt is practically nonexistent above the cloud layer. This differentiation of atmospheric aerosols then leads to FIGURE 3. For the large particles and the condensation nuclei, the model vertical distribution of the particle concentration is given. The aerosols mentioned above, maritime, continental, background, and stratospheric, can be seen. While only three aerosols can be seen for the condensation nuclei, the large particles show four aerosols.

In studies of the atmosphere, this concept of aerosol bodies has proven very helpful. Of course, the borders are not very sharp, and often *transition aerosols*[15] can be observed. In these four aerosols, the chemical composition and the size distributions are quite different and will be discussed next.

PROPERTIES AND COMPOSITION

The properties and the chemical composition of the natural aerosols are produced by the combined effects of sources, sinks, and aging of the aerosols. At present, it is not known how the effects act together to produce the observed properties.

The size distribution of the aerosol most often has attracted researchers—as

Reference 16 shows—to investigate how it is generated. The "self-preserving" size distribution[19] is one of these cases but is only applicable if sufficient time is given to the aerosol for evolution from its source functions, which rarely is the case in the atmosphere. The size distribution of the aerosols is not yet fully understood. The most recent compilation of measurements,[6] in the form of model distributions, shows for all aerosols the typical concentration decrease with increasing particle size for radii larger than 0.1 μm. Smaller particles show a more or less constant concentration, with some

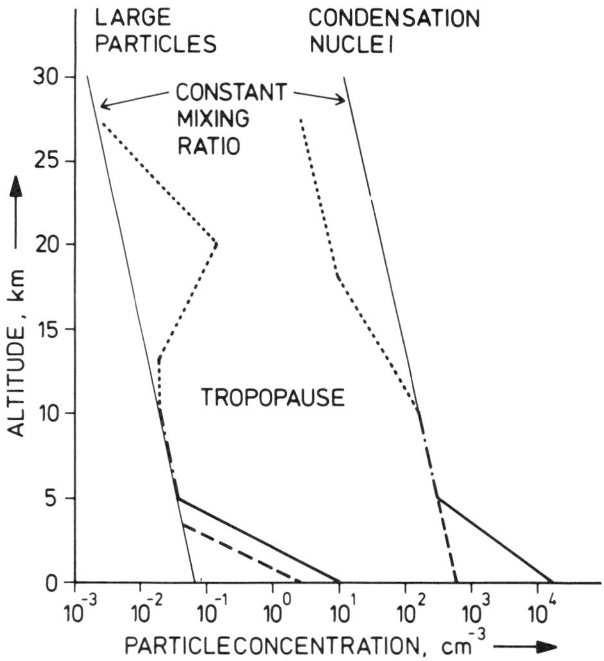

FIGURE 3. Model of the vertical distribution of the natural aerosol. The particle concentration is given for condensation nuclei ($r < 0.1$ μm, determining the number concentration) and for large particles (0.1 μm $< r <$ 1 μm, proportional to the aerosol mass). Four aerosol bodies can be distinguished: — continental; – – maritime; –·– background; and ···· stratospheric aerosols. For the condensation nuclei, the background and maritime aerosols are equal.

distinct maxima and minima. The smallest particles that are observed ($\sim 10^{-3}$ μm) are constantly present in the aerosol. That only can be understood if a continuous aerosol source is effective.[20]

FIGURE 4 shows model size distributions for three aerosols: background, maritime, and (remote) continental aerosols. In contrast to earlier presentations,[4] these distributions are given as a combination of number and volume size distributions. This is done to emphasize those parts of the distributions that are suppressed if only the number or volume size distribution is presented. In a number size distribution of the maritime aerosol, usually the sea spray is suppressed, because it is limited to particles larger

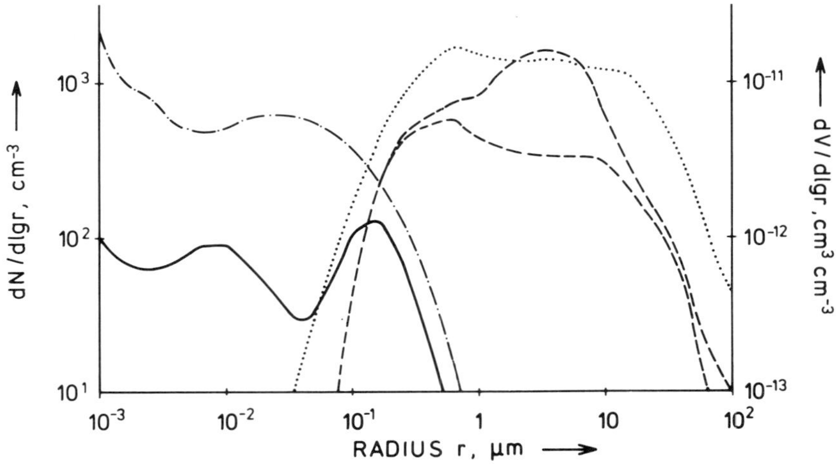

FIGURE 4. Model aerosol size distributions given as number ($dN/d\log r$) or volume ($dV/d\log r$) distribution. Number distribution for: background (maritime) —; and remote continental –·–, Volume distribution for: background, lower branch — — —; maritime, upper branch — — —; and remote continental ········ aerosols.

than 0.1 μm in radius and, thus, to that branch of the distribution with the steep concentration decrease.

The author is convinced that these difficulties in presentation of the size distributions have caused a number of empty controversies. It is certainly because of the work of Whitby[18] that more emphasis is placed, at present, on volume and other distributions.

FIGURE 4 shows clearly that background and maritime aerosols differ only in the volume distribution for particles larger than 0.1 μm. Between 1 and 10 μm, maritime and remote continental aerosols differ only fractionally in volume distribution. The chemical compositions of these particles, however, are different.

FIGURE 4 indicates how problematic it is to describe those distributions with a power function, as it was done in Junge's[17] pioneering work. This kind of mathematical expression was used for a long time and served well for a number of applications. Its limitations, however, were understood rather early.

Currently, the size distributions are often described with normal distributions.[21,22] FIGURE 5 shows how well the model distribution for background aerosol (FIGURE 4) can be described with normal distributions. Shown are number, surface, and volume distribution. The mathematical expression is:

$$n^* = \frac{dN}{d\log r}$$

$$= n_1 f(r, R_1, S_1) + \frac{1}{4\pi r^2} n_2 f(r, R_2, S_2) + \frac{1}{\frac{4}{3}\pi r^3} n_3 f(r, R_3, S_3) \quad \textbf{(2)}$$

$$f(r, R_i, S_i) = \frac{1}{S_i\sqrt{2\pi}} \exp\left\{-\frac{[\log(r/R_i)]^2}{2S_i^2}\right\}$$

This expression was selected because, at present, there is no reason known why the processes responsible for the shape of the aerosol size distribution tend to produce normal distributions in one of the moments: number, surface, or volume only. The above expression produces a normal distribution for each of the moments, with some additional influence from the other moments.

In TABLE 2, the parameters of the above expression are listed for the three aerosols mentioned above: background, maritime, and remote continental. These values have been obtained by minimizing:

$$\sum_i \log \frac{n^{*\prime}(r_i)}{n^*(r_i)} = \min$$

with $n^{*\prime}(r_i)$ given the values shown in FIGURE 4. The purpose of the table is to give a parameterized version of the model distributions, for further calculations. The

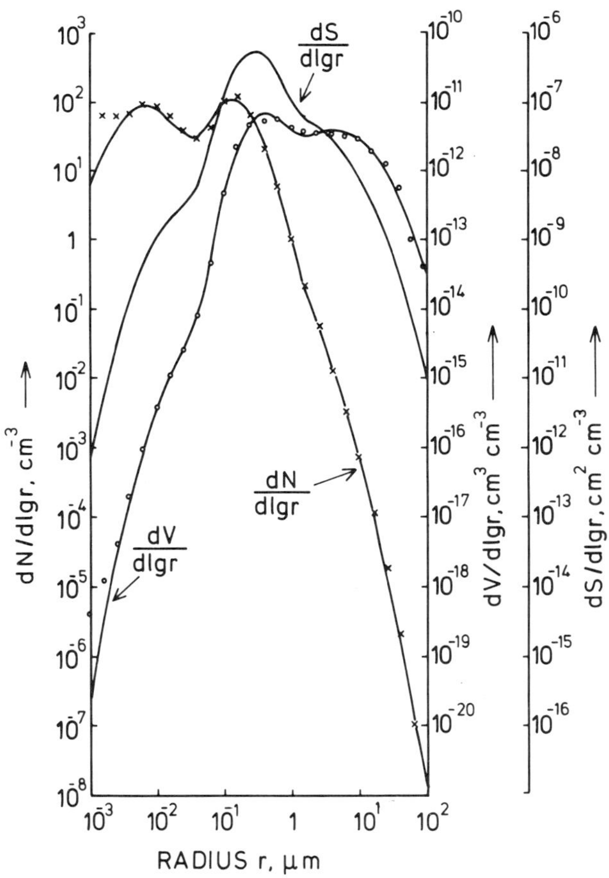

FIGURE 5. For the background aerosol, the model size distribution (FIGURE 4, x – x number, o – o volume) is described by an analytical function (Equation 2). Shown are number ($dN/d\log r$), surface ($dS/d\log r$), and volume ($dV/d\log r$) distribution.

stratospheric aerosol is not included in this table because it is explored only occasionally and undergoes rapid changes due to the nature of its production. The main sources are volcanic eruptions, forming particles mainly by gas-to-particle conversion (see Reference 23). The number size distribution is believed to be normal, with a maximum around 0.3 μm.[16,27]

FIGURES 4 and 5 show clearly that the range 0.1–30 μm is the most important size range for the volume of the aerosol and, consequently, the mass. Any statements about the chemical composition, therefore, must be interpreted in this sense. Continental aerosols consist of roughly 40% water-insoluble substances, mainly minerals[25] and biogenic material.[26] Some 10–20% is ether-soluble organic material.[3] Roughly half of the total organic material is water soluble, and 30% of the total aerosol mass has to be considered as inorganic and water insoluble. The water-soluble fraction mainly consists of inorganic salts with the following ions: $SO_4^=$, NO_3^-, Cl^-, NH_4^+, Ca^{++}, Mg^{++}, Na^+, and K^+. The presence of ammonium sulfate has been postulated, and thus the natural aerosol is coupled to the cycle of natural sulfur.[28] If the background

TABLE 2

THE PARAMETERS FOR THREE AEROSOLS—MARITIME, BACKGROUND, AND REMOTE CONTINENTAL—FOR EQUATION 2

Aerosol	n_1 (cm^{-3})	R_1 (cm)	S_1	n_2 (cm^{-1})	R_2 (cm)	S_2	n_3	R_3 (cm)	S_3
Maritime = Background + Sea Spray	8.20×10^{1}	7.37×10^{-7}	3.51×10^{-1}	2.74×10^{-7}	2.53×10^{-5}	2.30×10^{-1}	1.57×10^{-11}	3.06×10^{-4}	4.65×10^{-1}
Background	8.70×10^{1}	7.63×10^{-7}	3.72×10^{-1}	3.71×10^{-7}	2.94×10^{-5}	2.67×10^{-1}	4.51×10^{-12}	4.56×10^{-4}	4.40×10^{-1}
Remote Continental	1.24×10^{3}	1.42×10^{-6}	5.70×10^{-1}	1.51×10^{-1}	1.87×10^{-6}	8.36×10^{-3}	7.88×10^{-12}	1.96×10^{-4}	2.77×10^{-2}

aerosol is seen as an aged continental aerosol, its chemical composition should be roughly the same.[12] In the maritime aerosol, sea salt is added and, as FIGURE 4 shows, it then makes up the main fraction of the aerosol mass.[29]

Not covered by this survey is the composition of Aitken particles (with radii smaller 0.1 μm). These particles contribute only 5% to the aerosol mass. We now have a good deal of evidence that sulfur and organic compounds are the main constituents of these particles.[30] This is a first indication about the production mechanism for gas-to-particle conversion, which must be effective in the natural aerosols.

An immediate consequence of the chemical composition of the aerosol is its behavior in a humid atmosphere. Water-soluble substances in the aerosol particles adsorb water vapor from the humid atmosphere and form droplets of solution. This process starts at rather low relative humidities[31] and leads to a growth of the aerosol particles. This growth is rather small at low humidities but is important at high humidities. With more water vapor available, these liquid aerosol particles finally form cloud droplets. As an effect of the mixed nature of the aerosol particles with numerous ions, the growth with increasing humidity is rather monotone but smaller than with pure salts.

Effects of the Natural Aerosol and Secular Variations

There are numerous effects of the aerosol, and they have been summarized frequently.[2] It would go beyond the scope of the paper to cover all effects. However, some general remarks seem to be necessary.

Usually the aerosol has no effects through individual particles. Mass and surface of such a particle are much too small to start any macroscopic effect. Any aerosol effect is caused by the particles contained in a certain volume of air. This can be described by:

$$E = E\left(\int_V \int_{r=0}^{r=\infty} e(r) w(r) n^*(r) \, \mathrm{d} \log r \, \mathrm{d}V\right) \tag{3}$$

The effect E is a function of the aerosol size distribution $n^*(r)$ weighted with $w(r)$ and $e(r)$ and integrated over all radii and a certain volume V; $e(r) = 1$, if the number of the particles contributes to the effect; or $e(r) = 4\pi r^2$ or $e(r) = (4/3)\pi r^3$, if surface or volume are involved, respectively; $w(r)$ is a weighting function characterizing the effect. Some examples are: mass, $e(r) = \rho(r)$ with $\rho(r)$ = particle bulk density; extinction of radiation, $e(r) = \sigma$ with σ = volume extinction coefficient; electrical conductivity of the air, $e(r) = \beta$ with β = ion attachment coefficient.

Some effects of the atmospheric aerosol, such as the extinction of sun radiation, are influenced by all particles within a beam of sun radiation. Thus, the effect must be integrated over the whole atmosphere, rather than over a limited volume:

$$E = E\left(\int_{\mathrm{atm}} \int_{r=0}^{r=\infty} e(r) w(r) n^*(r) \, \mathrm{d} \log r \, \mathrm{d(atm)}\right) \tag{4}$$

All the functions $e(r)$ and $w(r)$ are weighting functions, and thus only certain size ranges of the aerosol make contributions essential to the effect. As mentioned above, the mass is influenced only by particles of 0.1–30 μm size. This has consequences upon the question of in how much detail an aerosol must be observed or measured in order to understand a certain effect. It is useless to study the Aitken particle concentration if optical effects of the atmospheric aerosol are to be explained.[30]

This discussion clearly shows how carefully any reports about secular trends in the aerosol must be discussed. In 1970, it was reported[32] that the concentration of the atmospheric aerosol has increased since the beginning of the century. This conclusion was made from monitoring of the electrical conductivity of the atmosphere. Electrical conductivity is a function of the number of small ions in the atmosphere, and this is controlled by the presence of aerosol particles. If one calculates the weighting function of this effect of the aerosol, it turns out that only particles with radii 0.02–0.4 μm contribute to the effect. A decrease in conductivity thus only means that, in this range of particle sizes, the concentration was increased. Such an increase in concentration is very likely, because particles in this size range might be transported far over the oceans, as shown earlier. The conclusion, however, that the aerosol in its entirety has changed cannot be supported.

It could be shown[33] that the atmospheric turbidity decreased from the beginning of the century until the mid-1930s. Since turbidity is influenced by particles in the range 0.1–1 μm, a decrease in the concentration of only these particles can be concluded. Indeed, it could be shown[33–35] that the stratospheric aerosol around 0.3 μm declined in concentration due to low volcanic activity.

CONCLUSIONS

The atmospheric natural aerosol is of very complex structure, and it is impossible to describe it with single parameters. If we generalize the aerosol with a small number of aerosol types, some of its properties can be given as first approximations. We still do not fully understand why the natural aerosol is as we observe it.

SUMMARY

The natural aerosol can be described as four different aerosols: stratospheric, background, maritime, and continental aerosol. Two major processes, the dispersion of mechanically disintegrated material and the transformation of airborne precursors, are the sources of the natural aerosol. The freshly formed aerosol is transported, mixed, and aged, and removed from the atmosphere after a certain residence time. This time depends on the particle size and varies between hours and days. This has consequences as to how far a particle can be transported in the atmosphere and makes it possible to estimate the influence of anthropogenic particulate sources upon the natural aerosol.

REFERENCES

1. SCHMAUSS, A. & A. WIGAND. 1929. Die Atmosphäre als Kolloid. Viehweg-Verlag. Braunschweig, Germany.
2. SMIC. 1971. Inadvertant Climate Modification. Report of the Study of Man's Impact on Climate. MIT Press. Cambridge and London.
3. KETSERIDIS, G., J. HAHN, R. JAENICKE & C. JUNGE. 1976. The organic constituents of atmospheric particulate matter. Atmos. Environ. **10:** 603.
4. JUNGE, C. 1963. Air Chemistry and Radioactivity. Academic Press, Inc. New York and London.
5. MARTELL, E. A. & H. E. MOORE. 1974. Tropospheric aerosol residence times: a critical review. J. Rech. Atmos. **8:** 903.
6. JAENICKE, R. 1978. Über die Dynamik atmosphärischer Aitkenteilchen. Ber. Bunsenges. Phys. Chem. **82:** 1198.
7. FUCHS, N. A. 1964. The Mechanics of Aerosols. Pergamon Press, Inc. New York, N.Y.
8. JUNGE, C. 1975. Stratosphärische Aerosole. Promet **5:** 29.
9. BURCKHARDT, H. & H. FLOHN. 1939. Die atmosphärischen Kondensationskerne. Springer-Verlag. Berlin, Germany.
10. JUNGE, C. 1963. Large scale distribution of condensation nuclei in the troposphere. J. Rech. Atmos. **1:** 185.
11. MESZAROS, A. & K. VISSY. 1974. Concentration, size distribution and chemical nature of atmospheric aerosol particles in remote oceanic areas. J. Aerosol Sci. **5:** 101.
12. GILLETTE, D. A. & I. H. BLIFFORD. 1971. Composition of tropospheric aerosols as a function of altitude. J. Atmos. Sci. **28:** 1199.
13. BLANCHARD, D. C. & A. H. WOODCOCK. 1957. Bubble formation and modification in the sea and its meteorological significance. Tellus **9:** 145.
14. DINGER, J. E., H. B. HOWELL & T. W. WOIJECIECHOWSKI. 1970. On the source and composition of cloud nuclei in a subsident air mass over the North Atlantic. J. Atmos. Sci. **27:** 791.
15. JAENICKE, R. 1977. Continental aerosols over the ocean. *In* Proceedings of the Symposium on Radiation in the Atmosphere, 19–28 August 1976. H. J. Bolle, Ed.: 29–31. Garmisch-Partenkirchen, Science Press.

16. THOMEY, S. 1977. Atmospheric Aerosols. Elsevier Press, Inc. Amsterdam, Oxford, New York.
17. JUNGE, C. 1952. Gesetzmässigkeiten in der Grössenverteilung atmosphärischer Aerosole über dem Kontinent. Ber. Dtsch. Wetterdienstes, U.S. Zone **35:** 261.
18. WILLEKE, K. & K. T. WHITBY. 1975. Atmospheric aerosols: size distribution interpretation. J. Air Pollut. Control Assoc. **25:** 529.
19. FRIEDLANDER, S. K. & C. S. WANG. 1966. J. Colloid Interface Sci. **22:** 126.
20. FRIEDLANDER, S. K. 1980. Future aerosols of the Southwest: implications for future aerosol research. Ann. N.Y. Acad. Sci. (This volume.)
21. DAVIES, C. N. 1974. Size distribution of atmospheric particles. J. Aerosol Sci. **5:** 293.
22. JAENICKE, R. & C. N. DAVIES. 1976. The mathematical expression of the size distribution of atmospheric aerosols. J. Aerosol Sci. **7:** 255.
23. FRIEND, J. P., J. LEIFER & M. TRICHON. 1973. J. Atmos. Sci. **30:** 465.
24. REITER, R., R. SLADKOVIC & K. PÖTZL. 1978. Chemische Komponenten des Reinluftaerosols in Abhängigkeit von Luftmassencharakter und meteorologischen Bedingungen. Ber. Bunsenges. Phys. Chem. **82:** 1188.
25. GILLETTE, D. 1980. Major contributions of natural primary continental aerosols: source mechanisms. Ann. N.Y. Acad. Sci. (This volume.)
26. HAHN, J. 1980. Organic constituents of natural aerosols. Ann. N.Y. Acad. Sci. (This volume.)
27. OTTAR, B. 1980. The long range transport of sulfurous aerosols to Scandinavia. Ann. N.Y. Acad. Sci. (This volume.)
28. FRIEND, J. P. 1979. Sulfur compounds and their distribution. Paper No. 24, presented at the Conference on Aerosols: Anthropogenic and Natural—Sources and Transport, New York Academy of Sciences/Air Pollution Control Association, New York, N.Y., January 9–12.
29. BLANCHARD, D. C. & A. H. WOODCOCK. 1980. The production, concentration, and vertical distribution of the sea-salt aerosol. Ann. N.Y. Acad. Sci. (This volume.)
30. JAENICKE, R. 1978. Physical properties of atmospheric particulate sulfur compounds. Atmos. Environ. **12:** 161.
31. WINKLER, P. & C. JUNGE. 1972. The growth of atmospheric aerosol particles as a function of relative humidity. I. Method and measurements at different locations. J. Rech. Atmos. **6:** 617.
32. COBB, W. E. & H. J. WELLS. 1970. The electrical conductivity of oceanic air and its correlation to global atmospheric pollution. J. Atmos. Sci. **27:** 814.
33. JAENICKE, R. & F. KASTEN. 1978. Estimation of atmospheric turbidity from the burned traces of the Campbell-Stokes sunshine recorder. Appl. Opt. **17:** 2617.
34. LAMB, H. H. 1970. Volcanic dust in the atmosphere; with a chronology and assessment of its meteorological significance. Philos. Trans. R. Soc. London, Ser. A **266:** 425.
35. HAMMER, C. U. 1977. Past volcanism revealed by Greenland ice sheet impurities. Nature **270:** 482.
36. BEILKE, S. 1975. Die Abscheideprozesse der Spurenstoffe aus der Atmosphäre. Promet **5**(2/3): 35.
37. SCHÜTZ, L. 1980. Long range transport of desert dust with special emphasis on the Sahara. Ann. N.Y. Acad. Sci. (This volume.)

THE PRODUCTION, CONCENTRATION, AND VERTICAL DISTRIBUTION OF THE SEA-SALT AEROSOL*

Duncan C. Blanchard

Atmospheric Sciences Research Center
State University of New York at Albany
Albany, New York 12222

Alfred H. Woodcock

Hawaii Institute of Geophysics†
University of Hawaii
Honolulu, Hawaii 96822

Early interest in the sea-salt aerosol centered on its role in the formation of rain,[59,64] but in recent years, interest has spread to the role of sea salt in the evaporation of water from the surface of the sea[36] and in the enrichment of heavy metals,[19] viruses,[2] bacteria,[7] and inert organic material,[6] and to the harmful effects of sea salt on ships' turbines.[49] This paper is both a brief review of what is known about atmospheric sea salt, its origin and concentration, and a medium to present new data on its vertical distribution. No attempt will be made to discuss the influence of the continents on the salt aerosol. What follows assumes a quasi steady state between the ocean and the atmosphere, with a long air trajectory over warmer water. Though space limitations prevent us from even mentioning many fine studies of the salt aerosol, we would be remiss if we did not refer readers to the pioneering work of Eriksson,[21,22] Junge,[32] and Toba.[52–54]

The salt aerosol is thought to originate primarily from the bursting of bubbles, and so we begin with bubble size distributions in the upper few meters of the sea.

Bubble Size Distributions

The first measurements of bubble spectra in the sea were made by Blanchard and Woodcock.[12] They considered four mechanisms of bubble production: (1) raindrops; (2) snowflakes; (3) supersaturations of seawater by temperature change; and (4) whitecaps. Inasmuch as the first two require precipitation, they are important only on an intermittent and local scale. Mechanism 3 requires rapid temperature changes, which may not be commonplace in the sea. The last mechanism, that of whitecaps, is the major one for the production of bubbles. Though not completely understood, the breaking of a wave entrains large amounts of air into the water to produce a bubble size distribution. Blanchard and Woodcock obtained their distributions just beneath the surface of the water a few seconds after small waves had broken. Although the spectrum was from less than 100 μm to several millimeters diameter, it was heavily

*Supported by National Science Foundation Grant No. ATM78-00782, and by the Office of Naval Research.

†Hawaii Institute of Geophysics Contribution No. 954.

0077–8923/80/0338–0330 $01.75/1 © 1980, NYAS

weighted toward the small end, varying inversely with the fourth or fifth power of bubble diameter for bubbles < 0.5 mm. Most of the bubbles were < 200 μm. The concentration of all bubbles was about 10^8 m^{-3}.

Subsequent measurements were made by others,[35,41] though not directly beneath a breaking wave. The most recent are those of Johnson,[30,31] who suspended a camera at various depths from a float and took pictures every 30 seconds of a known volume of water. In measurements obtained during winds of from 10 to about 13 m s^{-1}, he found the bubble concentration to decrease exponentially by a factor of 30, from 4.8×10^5 m^{-3} to 1.6×10^4 m^{-3}, as the camera was lowered from a depth of 0.7 to 4 m. Though the concentration underwent a 30-fold decrease, the diameter of the peak concentration did not change; it remained at about 100 μm. Over this same depth range, the maximum observed bubble diameter decreased by less than a factor of three, from 600 to 220 μm.

Measurements of both maximum and minimum bubble sizes must be regarded with some suspicion. The largest bubbles produced in some breaking waves clearly exceed a centimeter in diameter, as is obvious to anyone who has observed closely a wave breaking at sea. The concentration of these giant bubbles is extremely low; thus, they would not be seen by Johnson's[31] camera. Blanchard and Woodcock avoided these bubbles since their bubble trap could not accommodate the large end of the bubble spectrum. At the other end of the spectrum, detection of the minimum bubble size is made difficult both by the problems of simply seeing such small bubbles, and because surface curvature effects[12,30] cause small bubbles to go into solution in less than a minute, even in water 100% saturated with air. Consequently, neither the size of the smallest bubble produced by whitecaps nor its concentration is known, though bubbles of < 20 or 30 μm have been observed by the above-mentioned investigators.

Johnson's maximum bubble concentration is about 100 times less than that found by Blanchard and Woodcock, but there is no difficulty in reconciling the difference. The former measurements were not made in a breaking wave, where the bubbles are produced, while the latter were. On average, breaking waves cover at most only a few percent of the surface of the sea. Some of the bubbles produced by these waves apparently are carried by turbulent mixing to regions where there are no whitecaps. This mixing is accompanied by a drastic decrease in bubble concentration. The concentrations found by Johnson[31] probably represent a pseudo–steady state that exists in the entire surface layers of the sea when waves are breaking. Superimposed on this steady state concentration are the relatively short-lived, and much higher, concentrations that originate "where the action is," in the chaotic ejection of air into the sea during the formation of whitecaps.

Production of the Sea-Salt Aerosol

Of all the components that make up the particulate material introduced into the global atmosphere each year, that of the sea-salt aerosol is one of the, if not *the*, largest. Estimates[1,3,21,51] of the yearly production range from 10^9 to 10^{10} metric tons.

Before proceeding further, we must make clear what constitutes the sea-salt aerosol. It consists of a variety of particles whose names suggest either how they are formed at the surface of the sea or their physical condition in the atmosphere.

Occasionally there is confusion in the literature with such names as film drops, jet drops, seawater drops, brine drops, hygroscopic salt drops, sea-salt nuclei, and sea-salt particles. Most of the drops of seawater that enter the atmosphere are thought to be jet and film drops from bursting bubbles. Presumably, though not necessarily, these drops have a composition close to that of seawater, i.e., about 3.5% by weight is sea salt, most of which (over 85%) is sodium chloride. Since the equilibrium relative humidity over a seawater surface is 98%, jet and film drops usually begin to evaporate when they enter the atmosphere. But the sea salt in the drops does not change; thus, the salt concentration increases, and the drop vapor pressure decreases until it is that of the environment. Evaporation ceases; the drop is now a brine drop. If the relative humidity of the atmosphere is less than about 70 to 74%, the drop becomes supersaturated with sea salt,[56] and a phase change may occur to produce a sea-salt particle. However, if the drop initially is < 1 μm, the Kelvin effect[40] dictates that it can become supersaturated at humidities $> 75\%$. The equivalent diameter of a salt particle is one-quarter that of the parent seawater drop.

Thus, depending on the relative humidity and particle size, the particulate material produced by the sea can exist in the atmosphere as seawater drops, brine drops, or sea-salt particles. Since salt mass is the only conservative property, they are all called sea-salt particles, with either the size or the mass of the particles given. The mass does not include any water that might be associated with the sea salt.

Bubble Dissolution

Many of the bubbles produced by whitecaps do not return to the surface of the sea. As already mentioned, surface curvature effects cause bubbles, especially small ones, to go quickly into solution. Johnson[30] calculated a critical depth curve that shows the depth beyond which a bubble goes completely into solution before it can reach the surface. For a 100-μm diameter bubble in seawater at 0°C, the critical depth is only 0.3 m. This depth increases slightly with temperature.

As bubbles rise through seawater, dissolved organic material is adsorbed onto their surfaces.[6,23,26] If they go into solution, it is likely that an organic particle will remain behind.[6,30] Indeed, in laboratory experiments, Johnson[30] finds that all bubbles that go into solution produce organic particles. Some stop short of complete dissolution at sizes < 10 μm, apparently stabilized by a rigid organic monolayer on their surfaces.

Bubbles that do not reach or exceed the critical depth carry the adsorbed organics to the surface of the sea, where the bubbles burst to produce a sea-salt aerosol.

Percentage of Sea Surface Producing Sea-Salt Particles

It is a common observation that the higher the wind speed, the more numerous are the whitecaps, but there is much scatter in graphs of whitecap coverage versus wind speed.[3,43,55] Toba and Chaen[55] believe that the wave period is a better indicator of whitecap coverage than is wind speed. The most extensive work on oceanic whitecapping is that of Monahan.[43] He found, as did others, that whitecaps begin to appear at wind speeds of about 3 m s^{-1}, and at wind speeds of 8 m s^{-1}, about 1% of the sea is covered with whitecaps. An earlier estimate[3] had suggested that, on average, the world ocean had a whitecap coverage of about 3.5%, but, in view of Monahan's more extensive work, this estimate may be somewhat high. However, the problem is further complicated when we try to define just what is meant by a whitecap. Does it include only the "white water" that is so apparent immediately after a wave breaks, or does it also include the whitish blue area of bubbles[4,60] that can be seen rising to the surface for a minute or more after a wave breaks? Does it include the foam streaks?[48]

Must we conclude that only about 1–3% of the surface of the sea is producing sea-salt particles? No. Johnson[31] has found a background bubble distribution from which bubbles can rise to the entire surface of the sea. An estimate can be made of the importance of these bubbles. At a depth of 0.7 m, all bubbles $<140\ \mu m$ diameter go into solution before reaching the surface.[30] These bubbles account for 78% of the bubble concentration ($4.8 \times 10^5\ m^{-3}$) found at that depth at wind speeds of 10 to 13 m sec^{-1}. Thus, the 22% that rise to the surface are about 380 times less in concentration than those found at about 0.1 m in a whitecap produced by much smaller winds.[12] However, since background bubbles burst at 100% of the surface of the sea, while the whitecap bubbles cover only about 1%, we calculate that the former may contribute about 10% of that of the latter to salt particle production. The relative contribution of the two should be a function of wind speed. Though not quite correct, we have assumed that the average rise speed of background bubbles that reach the surface is the same as that of the whitecap bubbles.

Ejection of Sea-Salt Particles from the Sea

The flux of whitecap-induced bubbles to the surface of the sea[12] is about $3 \times 10^5\ m^{-2}\ s^{-1}$. Upon bursting, the surface free energy of the bubble is converted into kinetic energy in the form of a jet of water that rises rapidly from the bottom of the collapsing bubble cavity.[3,39] The jet becomes unstable and, depending upon bubble size, breaks up into from 1 to 10 drops. These drops, roughly one-tenth the bubble diameter, are called the jet drops. The maximum drop ejection height increases with bubble size, reaching nearly 20 cm for 2-mm bubbles. For larger bubbles, the ejection height decreases, and for bubbles $>$ 7 or 8 mm, no jet drops are produced.

The above data came from lab experiments where bubbles rose from capillary tips only a centimeter or two before bursting at the surface. Whitecap-induced bubbles are often in the water for many tens of seconds and may rise a meter or more before reaching the surface. During this time, adsorption of dissolved organics from the seawater can lower the surface free energy, resulting in a decrease in ejection height and a change in jet drop size. Recent work[8] has shown that drop ejection height is a function of the amount of dissolved organic carbon in the water, and that the ejection height can decrease by a factor of two after only 10 s of bubble aging. Curiously, these pronounced changes in ejection height with bubble age are restricted mainly to bubbles of about a millimeter and larger. Smaller bubbles, which presumably produce most of the sea-salt particles, are little affected.

The other class of drops produced by bubble bursting are the film drops. These arise from the bursting of the film of water that separates the air in the bubble from the atmosphere. Unlike jet drops, whose numbers generally decrease with bubble size, the number of film drops increases rapidly with bubble size.[3,16] Bubbles < 0.3 mm do not appear to produce any film drops, but a 6-mm bubble can produce a maximum of about 1000. Little is known of the sizes of film drops, though one study[11] found that the spectrum from a 0.7-mm bubble peaked at about 5 μm diameter. To complicate matters, bubble clusters and organic monolayers, both on the surface of the bubble and on the surface of the sea, modify film drop production.[3,24,44]

We can estimate the rate of production of jet and film drops. Assuming that each bubble produces five jet drops, jet drop production in the whitecap regions of the sea is 1.5×10^6 m^{-2} s^{-1}. The film drop production has been estimated[3] to be about 2×10^4 m^{-2} s^{-1}, lower by a factor of 75. However, these estimates of film drop production are on the low side, since experimental difficulties prevented the measurement of bubbles greater than about 0.5 mm. Though less numerous, much larger bubbles exist in the sea. They can be seen by eye, and Cipriano[15] has determined their distribution in a laboratory study of bubble formation by breaking waves.

The limits of the size distribution of particles produced by the sea are not known. The largest drops, produced not by bubbles but by splashing, are about a centimeter in diameter, but obviously they return immediately to the sea and play little direct role in atmospheric processes. The aerosol spectrum in the marine atmosphere extends to particles of at least 100 μm radius,[33] some of which probably are sea salt. From what is known about the sea-salt particle distribution in hurricane winds,[58,60] one should not be surprised to find that such giant particles can be produced by the sea.

Several investigators[15,45,46,62] believe that the smallest particle produced by the sea is < 0.1 μm radius, and one investigator[46] has evidence that the smallest particle is about 0.005 μm radius. Years ago, it was suggested[17,57] that small particles could be produced in the atmosphere when a solution droplet changed phase to produce a salt particle, but later experiments with a seawater aerosol failed to confirm this secondary production.[9,15]

The smaller the sea-salt particle, the higher is the ratio of organic carbon to salt.[25,27] Though there is strong evidence that gas-particle interactions can explain this high ratio on small particles,[34] the possibility cannot yet be ruled out that the smallest sea-salt particles are primarily organic when they leave the sea.[5]

Sea-Salt Concentration and Wind Speed

The first detailed studies of the size distribution and mass concentration of sea-salt particles were made by Woodcock.[58,60,68] He made many of these measurements windward of the Hawaiian Islands, in clear air between the clouds, and at altitudes of about 600 to 800 m, approximately that of cloud base. All his data were obtained from microscope observations of sea-salt particles collected by exposing small glass slides from an aircraft. The isopiestic method (a controlled humidity chamber) was used to get the particle size distribution. Repeated comparisons over the years[58,59,68,69] of salt concentrations obtained by this method to those obtained by direct chemical titration of the chlorides on the glass slides gave ample proof that sea salt, and not other substances, was being collected.

Woodcock's salt mass concentrations as a function of wind speed, redrawn from his original work, are given in Figure 1. Today, over 25 years later, they remain the only extensive data set of their kind. All observations were made upwind of the Hawaiian Islands, with the exception of those for winds of 35 m s^{-1}, which were obtained from the top of a lighthouse during a hurricane in Florida.[58] If we assume that during hurricane winds, the salt concentrations at an altitude of about 40 m (where Woodcock made his measurements) are about the same as those at 600 m, then it appears from FIGURE 1 that over the range of wind speeds from about 5 to 35 m s^{-1}, the salt concentration increases exponentially with wind speed. At lower wind speeds, Woodcock's single observation shows that the salt concentration drops rapidly. This is

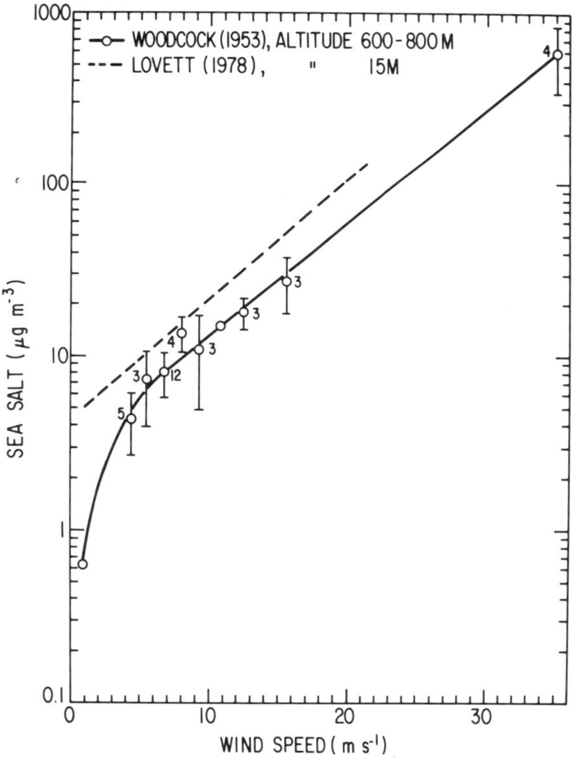

FIGURE 1. Sea-salt concentration at two altitudes as a function of wind speed near the surface of the sea. Redrawn in part from Woodcock.[60] The number of observations averaged for each data point is given (if more than one), as well as standard deviation bars ($\bar{x} \pm \sigma$).

not surprising, for at winds less than about 3 m s^{-1}, no bubbles are produced by the sea.

The standard deviation bars in FIGURE 1 indicate that one should not be surprised to find differences of a factor of two or more in salt concentrations from one day to the next, even though the winds remain the same. In fact, these differences show up in the same air mass. In a series of seven samples obtained one after the other, each sample averaging over 2.7 km, Woodcock[60] found differences of a factor of two. This variability has been observed ever since, in the numerous investigations of salt concentrations in marine air. There are many reasons why this should be so, not the least of which is that, because of the sea-salt residence time of one to two days, the salt concentration is dependent less on the local winds and more on the integrated winds that the air mass experienced in the prior two days. Variations are also produced by subsidence caused by large cloud systems that may be near the sampling region and by differences in the depth of the mixing layer and in the number and height of the clouds that develop in the cloud layer.

The dashed line in FIGURE 1 is the best fit line for salt concentrations obtained by Lovett[37,38] at an altitude of 15 m from three weather ships in the North Atlantic nearly 700 km west of Ireland. Lovett's data are not only the most recent to appear, but are the most numerous and cover the largest range of winds, excluding Woodcock's hurricane observations.[40] In the course of about a year, using filters with isokinetic sampling heads, Lovett determined salt concentrations from 1,821 samples obtained in winds of from 1 to 21 m s^{-1}! His concentrations, like Woodcock's, increase exponentially with wind speed but are greater than Woodcock's. The latter is not surprising, as we might expect to find the largest concentrations nearest the surface of the sea. Unlike Woodcock, Lovett did not find a rapid decrease in salt at winds of < 5 m s^{-1}. Possibly this is due to a higher variability in his winds, as opposed to the trade winds of Hawaii, which are among the most constant in the world. In winds that vary rapidly, a relatively high salt concentration at low winds might have been produced by higher winds a day or two before.

Another interesting aspect of the salt-wind relationship is seen in FIGURE 2, where the same data of FIGURE 1 are shown on a log-log plot. Remembering what was just said about Lovett's data at winds of < 5 m s^{-1}, it appears that the salt concentration increases with about the first power of the wind for speeds < 10 m s^{-1} but more rapidly after that, increasing with about the third power at 20 m s^{-1}, and with more than the fourth power as the winds approach hurricane speed. Qualitatively, this makes sense. The mixing upward of the salt increases rapidly with wind speed, while the source area, the whitecaps, increases with about the square of the wind speed.[3]

The variability of the salt concentration at a given wind speed (FIGURE 1) is also found near the surface of the sea. This is shown in FIGURE 3, which gives salt concentrations obtained from a tower on the windward shore of the island of Oahu, Hawaii.[10] The data were derived from an analysis of salt collected on both stationary and rotating wires, the latter to decrease exposure time and to increase the collection efficiency of the salt particles for the wire. At the highest wind speeds shown here, the decrease in salt concentration with altitude is clear, but the concentration gradient decreases with decreasing wind speed, and at speeds of $<$ about 4 m s^{-1}, the gradient disappears. This is contrary to what is shown in FIGURES 1 and 2, but the apparent discrepancy is, as discussed earlier, explainable in part by the constancy of the Hawaiian trade winds.

Recently, we have wondered whether salt concentrations obtained from island stations are really representative of those over the open sea at the same wind speed. Of course, we have long been aware of the problems associated with salt ejection from a surf zone along the shore,[20] and with drainage winds from large islands, and have taken pains to avoid them. The question here is whether islands modify the wind speed from what it is on the open sea. With an island as small as Oahu, one might think not, since the winds show little change from night to day. However, preliminary results of an analysis presently under way, of the winds reported from ships upwind of Oahu during the 60 days in 1970 when the data of FIGURE 3 were obtained, indicate that winds at sea are about 1 m s^{-1} higher than those crossing the windward shore. Similar results were found during another study of winds near Oahu.[63] This would tend to shift

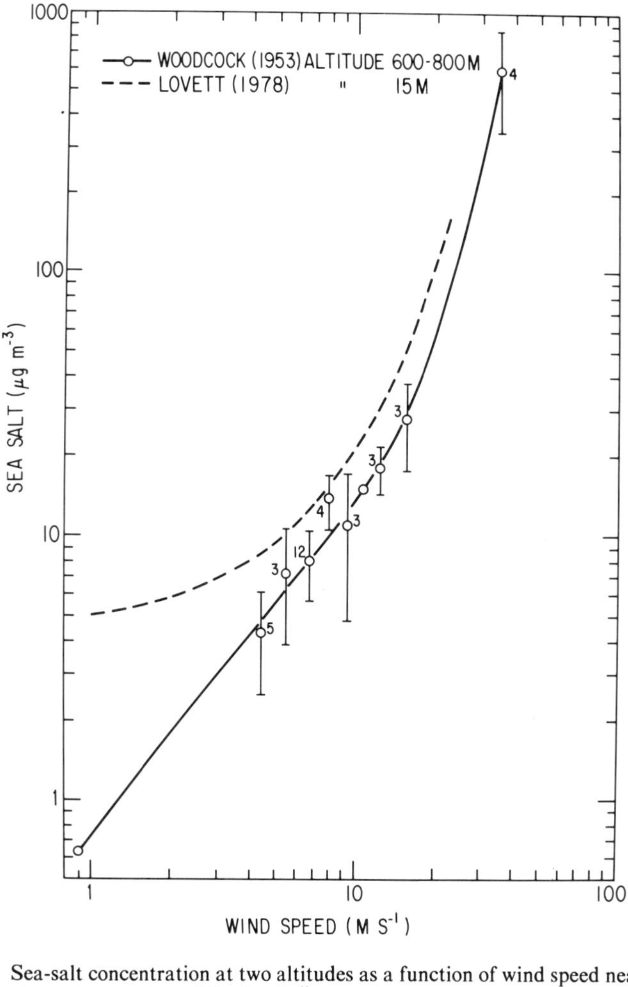

FIGURE 2. Sea-salt concentration at two altitudes as a function of wind speed near the surface of the sea. Redrawn in part from Woodcock.[60]

the data points in FIGURE 3 to the right, to a position about 1 m s^{-1} greater than they are now, suggesting that over the open sea, the salt concentration was somewhat less than that indicated in FIGURE 3. Because of this, caution should be used in the extrapolation to the open ocean of salt-wind data obtained from island stations.

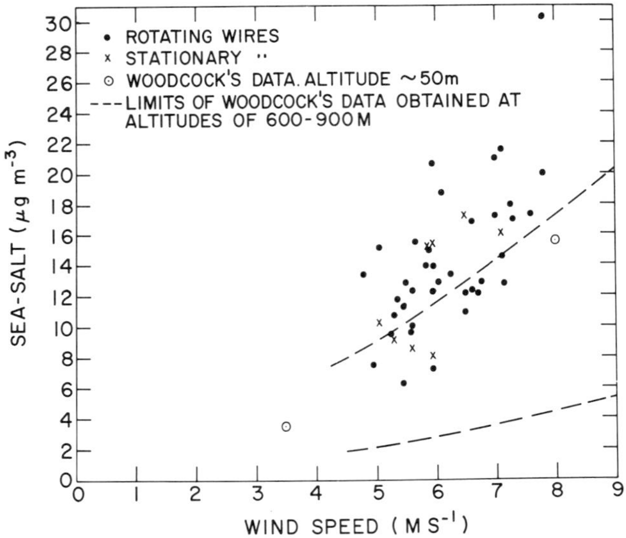

FIGURE 3. Sea-salt concentration as a function of wind speed near the surface of the sea at Oahu, Hawaii.

SIZE DISTRIBUTION OF SEA-SALT PARTICLES

The sea-salt concentrations of FIGURE 1 were obtained from the particle size distributions.[60] These distributions are shown in FIGURE 4 for Beaufort wind forces of 1, 3, 4, 5, 7, and 12, which represent mean winds at the 10-meter level of 1, 4.4, 6.7, 9.3, 15.5, and 35 m s^{-1}, respectively. These cumulative distributions for altitudes of 600–800 m cover a range of particle weight from 1 to 10^5 $\mu\mu$g, representing drop sizes when they leave the sea of about 2 to over 80 μm radius, and which produce dry sea-salt particles of from 0.5 to over 20 μm radius. It is interesting that the number of particles > 2 μm radius (as a seawater drop) increases from about 3×10^5 m^{-3} to 7.5×10^5 m^{-3}, a factor of only 2.5, while the winds increase from 4.4 to 15.5 m s^{-1}. However, at the large end of the spectrum, the increase is more dramatic. The number of particles > 40 μm radius increases from about 0.1 to over 100 m^{-3}, a factor $> 10^3$. This could mean that as the wind increases, relatively more larger particles are produced by the sea; but the most probable interpretation is that particles at the large end of the spectrum, with their shorter residence time, respond more quickly to changes in wind speed. Consequently, the large end of the spectrum should be much nearer a steady state distribution than the small end. A steady state over the entire spectrum may be rare, even in locations like Hawaii, where the winds are relatively constant. For example, the particle distributions for winds of force 1, about 1 m s^{-1}, cannot be steady state, for at such low wind speeds, no bubbles and thus no salt particles are produced by the sea.

Later work by Woodcock[62] and Junge[33] extended salt particle distributions to sizes smaller than shown in FIGURE 4. There appear to be from 20 to 30 $\times$ 10^6 particles per cubic meter larger than about 0.05 μm dry radius. This salt particle count of 20 to 30 cm^{-3} is but a small fraction of the total particle count of 200 to 300 cm^{-3} found in marine air far from the continents,[10,29] and raises the question of whether there are particles of oceanic origin < 0.05 μm that contribute significantly to the total count. Most investigators believe not,[18,33,34] but their arguments are based on the assumption that these small, sea-produced particles are primarily sea salt. As discussed earlier, there is reason to believe they may be mostly organic.[5]

The sea-salt particle distributions in FIGURE 4 are mostly for altitudes of 600 to 800 m. What might they be near the surface of the sea for equivalent wind speeds? The only extensive data set we know of was obtained by Chaen,[14] who made numerous measurements from ships in the Indian Ocean and the East China Sea. FIGURE 5 shows his distributions for an altitude of 6 m; superimposed on them are the

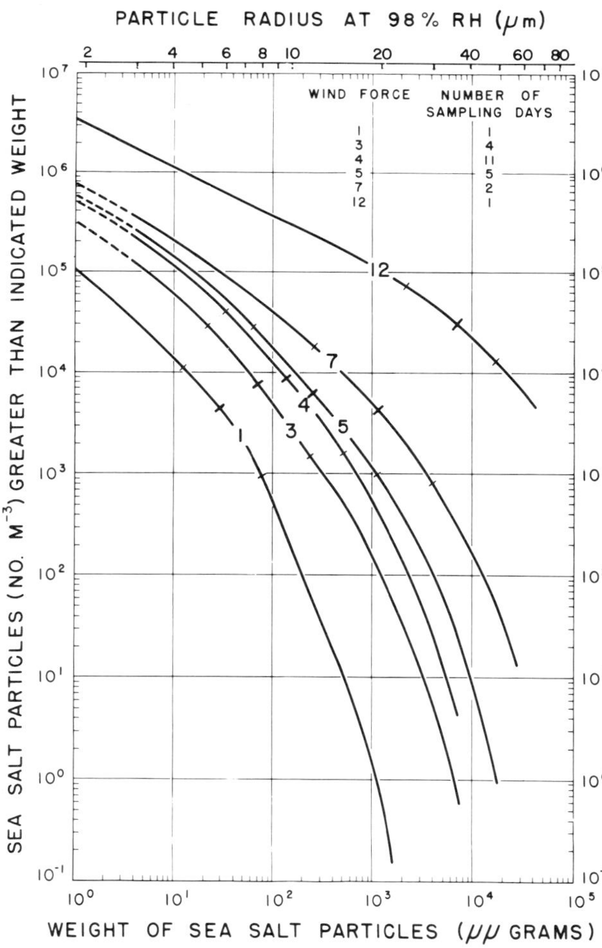

FIGURE 4. Sea-salt particle cumulative distributions at altitudes of 600–800 m above the sea as a function of wind force.

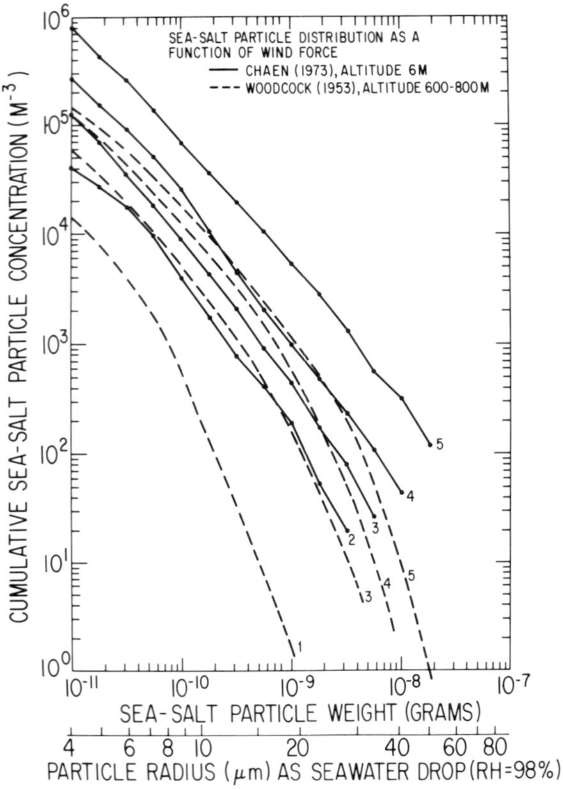

FIGURE 5. Sea-salt particle cumulative distributions both near the sea and aloft as a function of wind force.

distributions of FIGURE 4. Since Chaen did not measure particles $< 10^{-11}$ g, the cumulative curves start there, and not at 10^{-12} g as in FIGURE 4. Though hard to tell at a glance, comparison of the curves at the two altitudes and for a given wind force shows, as might be expected, and indeed was found long ago,[60] that the number of particles generally decreases with height. At the large end of the spectrum, where we have particles of about 10^{-8} g (10 μm dry radius), the decrease is most rapid. Again, this is as expected.

Junge[32] found it convenient to plot aerosol size distributions on a log-log plot of $dN/d\log r$ versus $\log r$, where N is the total number of particles smaller than radius r. He often found a straight line represented by $dN/d\log r = cr^{-\beta}$, where c was a constant and $\beta = 3$. Since many investigators present data in this manner, we have converted the cumulative curves of FIGURES 4 and 5 into a Junge-type $dN/d\log r$ graph; these curves are shown in FIGURE 6, along with a line whose slope $\beta = 3$. It is apparent that the salt particle distributions, both near the sea and aloft, have a $\beta \simeq 3$ for particles of about 1 μm radius, but for larger particles, around 10 μm, β becomes 4 or 5, especially for the distributions at altitudes of over 600 m. Years ago, Eriksson[21] found a similar thing for the distributions aloft and pointed out that the exponent of r must be a function of r.

The Sea-Salt Inversion

In 1962, Woodcock[61] published extensive data on sea-salt concentration as a function of altitude and wind speed. With 268 observations obtained in the North Atlantic Ocean near Massachusetts, Florida, Bermuda, and the West Indies, and in the Pacific Ocean near Hawaii and Australia, he presented three curves, each for a different wind speed, that showed how the sea-salt concentration varied from about 50

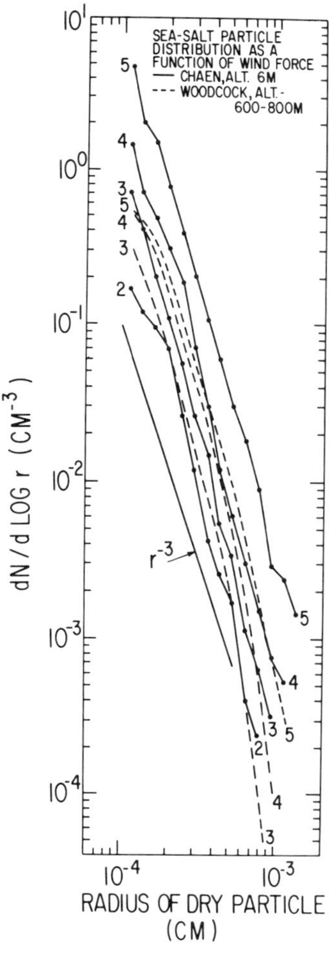

FIGURE 6. Junge distributions for sea-salt particles as a function of wind force and altitude. The number to the left and right of each distribution denotes the wind force.

to over 2500 m altitude. We decided to reanalyze these data and add to them about 75 subsequent observations made by Woodcock, including a new series from the Gulf of Alaska. With more data, we used smaller altitude intervals at lower elevations where the observations were more numerous. The results are shown in FIGURE 7, where the average salt concentration, and the standard deviation ($\bar{x} \pm \sigma$), is given as a function of altitude for average wind speeds of 1, 3.5, 8, and 14 m s^{-1}.

The most striking feature of these curves is the appearance of a salt inversion from 300 to about 600 m for winds of 3.5 and 8 m s^{-1}. Though the spread of the data around the mean is high, as it was in FIGURES 1–3, the inversion appears to be real, with salt concentrations between 500 and 600 m nearly 1.4 times that at 300 m. There is no clear evidence of a salt inversion for winds of 14 m s^{-1}, but with only five observations below 500 m, one cannot draw any conclusions. Above the salt inversion, the decrease of salt concentration with height for all wind speeds is similar to that observed earlier.

The data of FIGURE 7 were separated into those from both the Atlantic and Pacific Oceans and were replotted. The salt inversion appeared in both graphs. There is

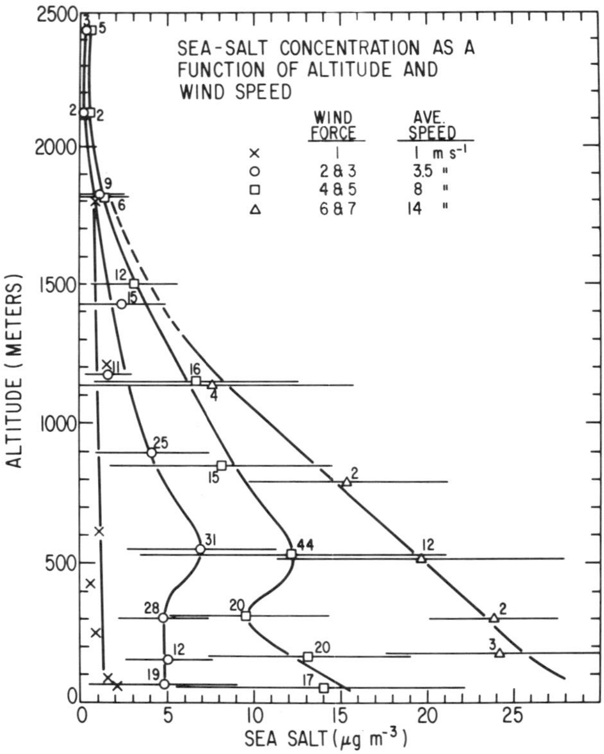

FIGURE 7. Sea-salt concentration as a function of altitude and surface wind speed. The number of observations averaged for each data point is given (if more than one), as well as standard deviation bars ($\bar{x} \pm \sigma$). A salt inversion appears at the two intermediate wind speeds.

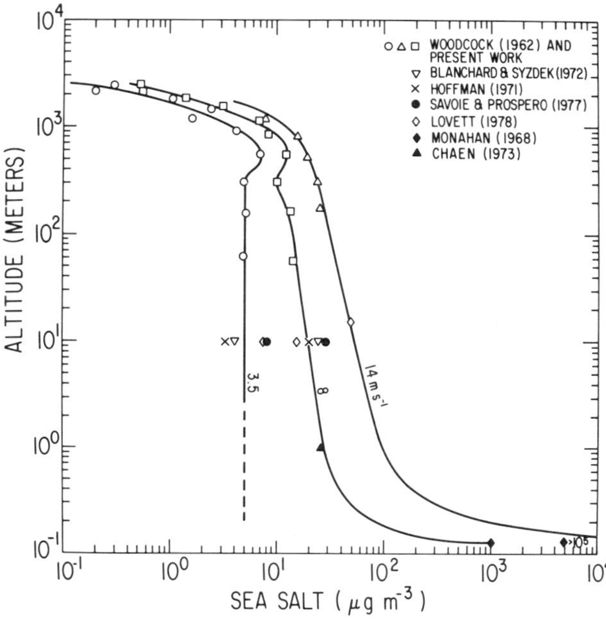

FIGURE 8. Compilation of data on sea-salt concentration as a function of altitude and wind speed. The altitude ranges from < 0.2 m to > 2000 m.

additional work that, in retrospect, helps confirm the existence of a salt inversion. One of us (AHW), several years ago, made salt sampling flights on seven separate days in the Hawaii area. The purpose of the flights was to gather more evidence to confirm the role of small sea-salt particles in the formation of rain.[66] On each flight, the salt concentration was determined both beneath cloud base and about 300 m nearer the sea, the latter location not necessarily directly below the cloud but in the same general area. The two sampling altitudes averaged 625 and 315 m, and the average ratio of the salt concentration of the higher to the lower level was about 1.5. These two levels, by chance, define the salt inversion shown in FIGURE 7. The ratio of 1.5 is close to that found above for the salt inversion.

Before Woodcock published his 1962 paper, he let Eriksson use the data. Eriksson[21] found perturbations in the salt concentration at about 500 m but felt that the concentration decreased continually from there down to the sea. We find no evidence of this direct correlation with altitude below 300 m.

The data of FIGURE 7, minus the standard deviation bars, are given in FIGURE 8, along with data averaged from other studies.[10,14,28,37,38,42,50] This shows how the salt load varies from < 0.2 m above the sea to > 2000 m. At the 10-m level, the differences in the averages of the several investigations show clearly how variable the salt loading can be for a given wind speed. Of course, they may also reflect differences, for reasons not yet clear, due to geographical location. These data, along with Chaen's[14] data at 1 meter, suggest a power law relation between salt and height above the sea, extending from 1 m up to about 300 m. Such a law was predicted by Toba.[53] Over this height range, and for winds of from 3.5 to 14 m s^{-1}, the average sea-salt concentration, s, can be expressed by $s = 5(6.3 \times 10^{-6}H)^{(0.21 - 0.39 \log W)}$, where H is the height in meters and W the wind speed in meters per second.

Several features of the curves in FIGURE 8 relate quite closely with various regimes or layers in the marine subtropical atmosphere. These layers, shown schematically in FIGURE 9 along with the salt curve for winds of 8 m s^{-1}, suggest reasons for the height variation of sea salt. Below 0.2 m, salt concentrations are very high but decrease rapidly within the first meter. This is the layer of direct influence of the large jet drops, most of which rise ≦0.2 m and return directly to the sea. A few of these giant drops may be carried by turbulence to heights of about a meter before falling back to the sea. From there up to cloud base at 600 to 800 m, we have the subcloud layer,[13,47] where the air is well mixed and the lapse rate adiabatic, except perhaps for the last

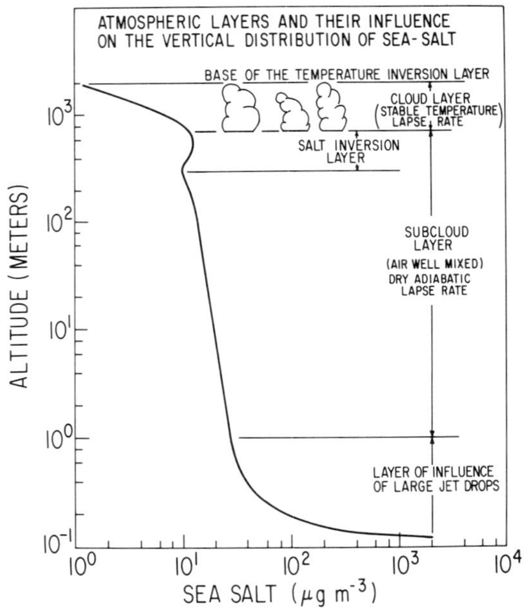

FIGURE 9. The influence of various atmospheric layers on the vertical distribution of sea salt. The curve shows the salt concentration for a wind speed of 8 m s^{-1}

100 m or so, where the air becomes more stable. In this layer, we find the power law relating salt with height. The cloud layer lies atop the subcloud layer and extends from cloud base to the temperature inversion at about 2000 m. Here the temperature lapse rate is stable. It causes a rapid decrease of salt with height, reaching negligible values at the temperature inversion.

The clue to the formation of the salt inversion layer is that its maximum is located just beneath the altitude of cloud base. We see at least two mechanisms by which the salt inversion is maintained. The first is by the fine, mist-like drizzle that is often observed falling from the base of the clouds, even the nonraining clouds.[67] This drizzle contains salt that was carried up from near the sea. The water evaporates, leaving small salt particles in the air. These particles do not remain directly beneath the clouds, since, due to wind shear, the clouds move faster than the subcloud air, and a salt inversion is produced. The second mechanism is that of an accumulation zone, in which high concentrations of salt, carried up in parcels from near the sea, grow rapidly as they rise to more humid regions.[65] Some of the salt particles will attain fall speeds that become significant compared with the updraft speed, and so produce an accumulation or concentration of salt.

There is much more that can be said about the salt inversion, but due to space limitations in this paper, we will pursue this interesting subject elsewhere.

Conclusions

The ocean is the most prolific of all natural sources in the production of atmospheric particulate material.[3,19] The continuous cycling of sea-salt particles between the ocean and the atmosphere involves numerous geochemical processes, from the enrichment of heavy metals in the salt particles to the formation of rain. Understanding of these processes requires more knowledge of how the particles are produced and distributed throughout the atmosphere. Studies of bubble spectra in the sea must be continued. We need to know these spectra as a function of wind speed, and we must determine what fraction of the bubbles formed in whitecaps escapes dissolution to arrive at the surface to produce salt particles. More knowledge is needed of the adsorption to bubbles of dissolved organic material in the sea, for the material carried to the surface by bubbles[6,26] no doubt accounts for the chemical[19] and biological[2,7] fractionation that is found on the sea-produced particles. This adsorption may also cause, by virtue of decreasing the bubble surface free energy, significant changes in the jet and film drop size spectrum produced by bursting bubbles.

Are jet drops more important than film drops in the production of salt particles? How important are spray drops torn off by the wind from the tops of waves? What is the smallest particle produced by the sea? Is secondary particle production at the phase change important? How much atmospheric sea salt does the ocean produce per year? Though attempts have been made to answer all these questions, in some cases they have not been very convincing.

For given wind speeds, are there geographical differences in the production of sea salt? How important is water temperature? We need to know more about the vertical distribution of sea salt. What is the role of clouds in the formation of the salt inversion? Can a salt inversion exist in the absence of clouds? These are just a few of the questions that beg for investigation and that must be answered in our quest for further understanding of the production and distribution of sea-salt particles in the marine atmosphere.

References

1. Ahmed, A. F. M. 1978. Master's Thesis. University of Rhode Island. Kingston, R.I.
2. Baylor, E. R., M. B. Baylor, D. C. Blanchard, L. D. Syzdek & C. Appel. 1977. Science **198:** 575.

3. BLANCHARD, D. C. 1963. Prog. Oceanogr. **1:** 71.
4. BLANCHARD, D. C. 1971. J. Atmos. Sci. **28:** 645.
5. BLANCHARD, D. C. 1971. J. Atmos. Sci. **28:** 811.
6. BLANCHARD, D. C. 1975. Adv. Chem. Ser. **145:** 360.
7. BLANCHARD, D. C. 1978. Pure Appl. Geophys. **116:** 302.
8. BLANCHARD, D. C. & E. J. HOFFMAN. 1978. J. Geophys. Res. **83:** 6187.
9. BLANCHARD, D. C. & A. T. SPENCER. 1964. J. Atmos. Sci. **21:** 182.
10. BLANCHARD, D. C. & L. D. SYZDEK. 1972. J. Phys. Oceanogr. **2:** 255.
11. BLANCHARD, D. C. & L. D. SYZDEK. 1975. Limnol. Oceanogr. **20:** 762.
12. BLANCHARD, D. C. & A. H. WOODCOCK. 1957. Tellus **9:** 145.
13. BUNKER, A. H., B. HAURWITZ, J. S. MALKUS & H. STOMMEL. 1949. Pap. Phys. Oceanogr. Meteorol. **11**(1). Massachusetts Institute of Technology and Woods Hole Oceanographic Institution. Boston, Mass.
14. CHAEN, M. 1973. Mem. Fac. Fish. Kagoshima Univ. **22:** 49.
15. CIPRIANO, R. 1979. Ph.D. Thesis. State University of New York. Albany, N.Y.
16. DAY, J. A. 1964. Q. J. R. Meteorol. Soc. **90:** 72.
17. DESSENS, H. 1946. C. R. Acad. Sci. Paris Ser. D **223:** 915.
18. DINGER, J. E., H. B. HOWELL & T. A. WOJCIECHOWSKI. 1970. J. Atmos. Sci. **27:** 791.
19. DUCE, R. A. & E. J. HOFFMAN. 1976. Ann. Rev. Earth Planet. Sci. **4:** 187.
20. DUCE, R. A. & A. H. WOODCOCK. 1971. Tellus **23:** 427.
21. ERIKSSON, E. 1959. Tellus **11:** 375.
22. ERIKSSON, E. 1960. Tellus **12:** 63.
23. GARRETT, W. D. 1967. Deep-Sea Res. **14:** 661.
24. GARRETT, W. D. 1968. J. Geophys. Res. **73:** 5145.
25. HOFFMAN, E. J. & R. A. DUCE. 1974. J. Geophys. Res. **79:** 4474.
26. HOFFMAN, E. J. & R. A. DUCE. 1976. J. Geophys. Res. **81:** 3667.
27. HOFFMAN, E. J. & R. A. DUCE. 1977. Geophys. Res. Lett. **4:** 449.
28. HOFFMAN, G. L. 1971. Ph.D. Thesis. University of Hawaii. Honolulu, Hawaii.
29. HOGAN, A. W. 1976. J. Appl. Meteorol. **15:** 611.
30. JOHNSON, B. 1979. Ph.D. Thesis. Dalhousie University. Halifax, Nova Scotia, Canada.
31. JOHNSON, B. & R. C. COOKE. 1979. Unpublished manuscript.
32. JUNGE, C. E. 1963. Air Chemistry and Radioactivity. Academic Press, Inc. New York, N.Y.
33. JUNGE, C. E. 1972. J. Geophys. Res. **77:** 5183.
34. KETSERIDIS, G., J. HAHN, R. JAENICKE & C. JUNGE. 1976. Atmos. Environ. **10:** 603.
35. KOLOVAYEV, P. A. 1976. Oceanology **15:** 659.
36. LAI, R. J. & O. H. SHEMDIN. 1974. J. Geophys. Res. **79:** 3055.
37. LOVETT, R. F. 1975. M.S. Thesis. Heriot-Watt University. Edinburgh, Scotland.
38. LOVETT, R. F. 1978. Tellus **30:** 358.
39. MACINTYRE, F. 1972. J. Geophys. Res. **77:** 5211.
40. MASON, B. J. 1971. The Physics of Clouds. Oxford University Press. Fair Lawn, N.J.
41. MEDWIN, H. 1977. J. Geophys. Res. **82:** 971.
42. MONAHAN, E. C. 1968. J. Geophys. Res. **73:** 1127.
43. MONAHAN, E. C. 1971. J. Phys. Oceanogr. **1:** 139.
44. PATERSON, M. P. & K. T. SPILLANE. 1969. Q. J. R. Meteorol. Soc. **95:** 526.
45. PODZIMEK, J. 1977. J. Rech. Atmos. **11:** 227.
46. RADKE, L. F. 1977. *In* Proceedings of the International Conference on Atmospheric Aerosols, Condensation and Ice Nuclei, University College, Galway, Ireland. (In press.)
47. RIEHL, H. 1954. Tropical Meteorology. McGraw-Hill Book Co. New York, N.Y.
48. ROSS, D. B. & V. CARDONE. 1974. J. Geophys. Res. **79:** 444.
49. RUSKIN, R. E., R. K. JECK, F. K. LEPPLE & W. A. VON WALD. 1978. NRL Mem. Report No. 3804. Naval Research Laboratory. Washington, D.C.
50. SAVOIE, D. L. & J. M. PROSPERO. 1977. J. Geophys. Res. **82:** 5954.
51. SELEZNEVA, E. S. 1976. Personal communication.
52. TOBA, Y. 1965. Tellus **17:** 131.
53. TOBA, Y. 1965. Tellus **17:** 365.
54. TOBA, Y. 1966. Tellus **18:** 132.

55. Toba, Y. & M. Chaen. 1973. Rec. Oceanogr. Works Jpn. **12:** 1.
56. Twomey, S. 1953. J. Appl. Phys. **24:** 1099.
57. Twomey, S. & K. N. McMaster. 1955. Tellus **7:** 458.
58. Woodcock, A. H. 1950. J. Meteorol. **7:** 397.
59. Woodcock, A. H. 1952. J. Meteorol. **9:** 200.
60. Woodcock, A. H. 1953. J. Meteorol. **10:** 362.
61. Woodcock, A. H. 1962. *In* The Sea. M.N. Hill, Ed. **1:** 305–312. Wiley-Interscience. New York, N.Y.
62. Woodcock, A. H. 1972. J. Geophys. Res. **77:** 5316.
63. Woodcock, A. H. 1975. Mon. Weather Rev. **103:** 334.
64. Woodcock, A. H. & D. C. Blanchard. 1955. Tellus **7:** 437.
65. Woodcock, A. H., D. C. Blanchard & C. G. H. Rooth. 1963. J. Atmos. Sci. **20:** 159.
66. Woodcock, A. H., R. A. Duce & J. L. Moyers. 1971. J. Atmos. Sci. **28:** 1252.
67. Woodcock, A. H. & R. A. Duce. 1972. J. Rech. Atmos. **6:** 639.
68. Woodcock, A. H. & M. M. Gifford. 1949. J. Mar. Res. **8:** 177.
69. Woodcock, A. H. & A. T. Spencer. 1961. J. Geophys. Res. **66:** 2873.

MAJOR CONTRIBUTIONS OF NATURAL PRIMARY CONTINENTAL AEROSOLS: SOURCE MECHANISMS

Dale Gillette*

National Center for Atmospheric Research†
Boulder, Colorado 80307

INTRODUCTION

Global aerosol budgets have become of increasing interest to researchers investigating problems of climate and air pollution. Natural sources of primary continental aerosols are particularly important, since they supply the baseline concentrations to which man's contribution is added. Although estimates of global aerosol emissions have been made, [1,2] recent research has pointed up the great temporal variability in natural aerosol concentrations. Indeed, the sporadic nature of many of the strong continental emissions, coupled with the enormous mass of material in these emissions, suggests that baseline aerosol concentrations should be considered in two parts: (1) the day-to-day normal concentrations; and (2) the large increases due to natural, but sporadic, emissions.

In this paper, I will review some recent findings on the mechanisms of primary continental aerosols. I will concentrate on particles smaller than 20 μm radius and will therefore not consider biological particles. My emphasis will be on wind-generated soil particles because of their dominant position as primary continental particles.

AEROSOL EMISSIONS FROM THE WIND EROSION OF SOIL

Definitions of Soil Particle Motions

A summary of soil particle motions is given in FIGURE 1. FIGURE 1a shows the creeping motion of a coarse particle close to threshold velocity. In **1b,** the particle is lifted into the air by turbulent fluctuations, and in **1c,** the particle collides with the surface and bounces. In **1d,** a collision is followed by creeping and another lift. Figure 1e shows the breaking off of smaller particles that were encrusted on the colliding particle's surface. In **1f,** a coarse particle collision is followed by splashing of fine particles into the air. Taken together, **1e** and **1f** illustrate sandblasting. In **1g,** many of the above movements are combined, and the fine particles emitted are carried high into the air. FIGURE 1h shows an example of a nonerodible element.

Scaling Considerations

Since wind plays a double role in both generation and transport of aerosol particles, it might be helpful to begin with a dimensional analysis.

*Present affiliation: CIRES (Cooperative Institute for Research in Environmental Sciences), University of Colorado, Boulder, Colo. 80309.

†The National Center for Atmospheric Research is sponsored by the National Science Foundation.

0077–8923/80/0338–0348 $01.75/1 © 1980, NYAS

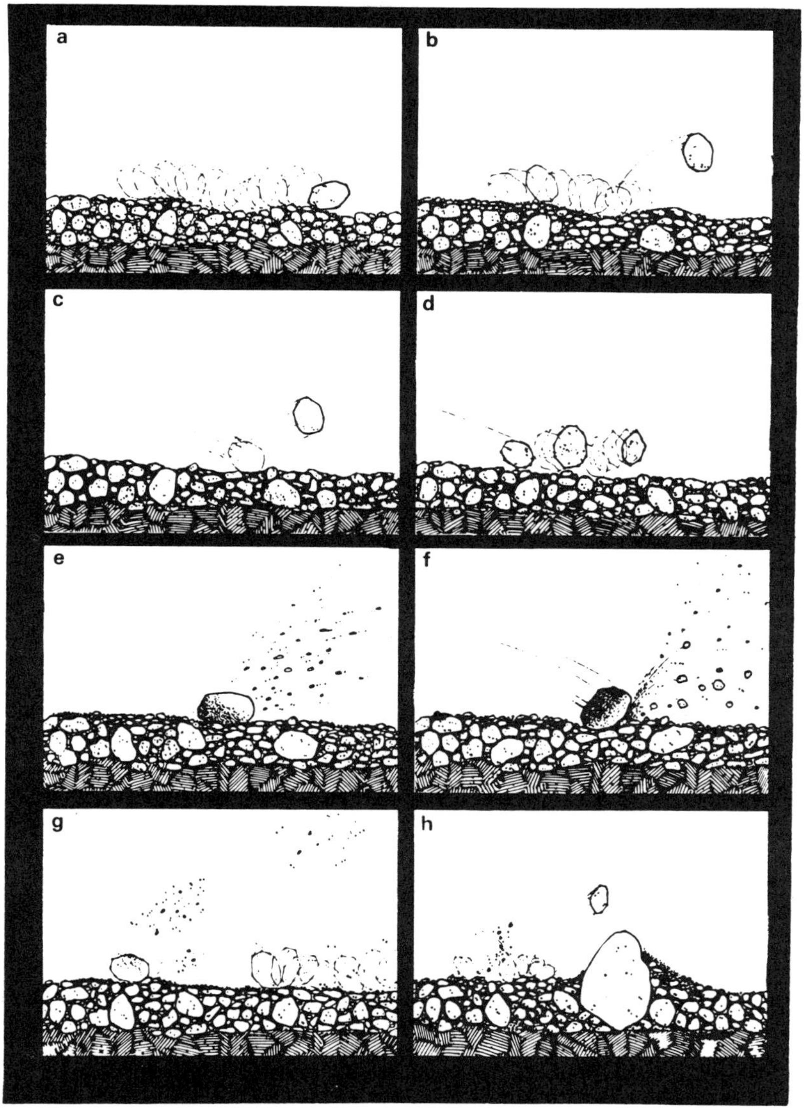

FIGURE 1. Soil particle motions during wind erosion (after Gillette[15]): (a) Creeping motion of a particle moving with wind speed slightly greater than threshold; (b) A coarse particle lifted into the air by turbulent air fluctuations; (c) An airborne particle collides with the surface and bounces (saltation motion); (d) An airborne particle collides with the surface, rolls, and is then relifted by air fluctuations; (e) A particle collision followed by breaking off of smaller particles that were encrusted on the colliding particle's surface; (f) A particle collision followed by "splashing" of the soil; (g) A combination of c, d, e, and f; (h) A nonerodible element giving protection to the soil downwind.

In the turbulent field in which erosion is taking place, a more significant dimension than the roughness height of the particle is the length of particle travel, l. Unfortunately, this length is not normally measured. We can, however, measure flux of momentum, τ, and friction velocity, u_*.

Thus the dimension quantities are:

$$\text{velocity: } u_* = K[\overline{u}(z_2) - \overline{u}(z_1)]/\ln\left(\frac{z_2}{z_1}\right),$$

where K is von Kármán's constant, z_1 and z_2 are two heights in a neutrally stratified atmosphere that are well above any roughness elements in the field, and $\overline{u}(z)$ is a mean wind speed at height z;

$$\text{time: } \frac{u_*}{g},$$

where g is the acceleration of gravity;

$$\text{length: } \frac{{u_*}^2}{g}; \text{ and}$$

mass density: ρ, the mass density of air.

Now, by definition, the momentum flux is $\rho {u_*}^2$.

By dimensional analysis, the horizontal flux of sand, q, having the dimensions of mass/length $\times$ time, may be written:

$$q = \frac{\rho {u_*}^3}{g}$$

An expression for the flux of kinetic energy onto the surface from hits by sand grains is:

$$\text{Flux}_A = \rho {u_*}^3$$

Thus, if the vertical flux of fine particles is proportional to the flux of kinetic energy delivered to the soil surface, then the ratio of the vertical flux of fine material to the horizontal flux of sand should remain constant with wind speed. Physically, such a situation would reflect a soil surface that is uniformly susceptible to destruction by sandblasting. An example of such a surface would be a loose soil well mixed with fine and coarse particles. If the ratio of the vertical flux of fine material to the horizontal flux of sand increases with wind speed, the soil is probably being pulverized or disaggregated by energetic sandblasting, thereby producing a higher proportion of fine to coarse materials. If, on the other hand, the ratio of the vertical flux of fine material to the horizontal sand flux decreases with wind speed (i.e., the proportion of fine material to coarse material in the eroding soil decreases), it may mean that fine particles are being rapidly depleted at high wind speeds. An example of such a soil is one in which an unerodible layer is being reached but sandblasting continues from an upwind source of coarse material.

Threshold Velocities for Wind-Generated Fugitive Dust

The wind speed at which the generation of dust begins is that for which aerodynamic forces are sufficient to overcome those forces holding individual particles in the soil. A number of theoretical and experimental studies have been made for idealized particle systems (e.g., References 3–7). These studies have not considered natural soil surfaces, however, or man's impact upon soil surfaces and aggregate storage. Gillette *et al.*[8] measured threshold velocities on dry soils that were in both undisturbed and disturbed condition. The disturbance was a pulverization of the surface by the wheels of a 3/4 ton pickup truck. A list of threshold velocities versus soil types is given in TABLE 1.

The ordering of threshold velocities suggests that the threshold velocity is determined by the availability of loose, erodible, sand-sized particles at the surface. It also suggests that the threshold velocity is increased (erosion decreased) by the effect

TABLE 1
WIND SPEED FOR INITIATION OF DUST PRODUCTION VERSUS UNDISTURBED OR DISTURBED SOIL TYPE

Approximate Range of u_* (cm s^{-1})	Soil Type
20–40	1. Disturbed soils having less than 50% clay and less than 20% pebble (<1 cm diameter) cover.
25–40	2. Tilled bare sand soils.
40–65	3. Disturbed pebbly soils.
45–70	4. Bare clay soils that have been disaggregated by natural forces.
40–150	5. Disturbed soils having a high salt content or more than 50% clay.
140–200	6. Undisturbed sandy soils having a crust and soils covered with fine gravel.
>150	7. Undisturbed soils having more than 50% clay and surface crusts and salt-crusted soils.
>180	8. Soils covered by coarse (>5 cm) pebbles.

of aerodynamic partitioning of wind stress by nonerodible elements, such as pebbles and larger objects, and by the cementation of the soil by clay and salts.

Marshall[9] studied the effect of pebble-like, nonerodible elements in absorption of wind stress. He found that for a configuration in which the silhouette area of the nonerodible roughness element per floor area per element (L_c) was greater than 0.1, most of the wind momentum was absorbed by the elements, and little stress was available for movement of fine particles lying between nonerodible elements. Lyles and Allison,[14] studying the threshold velocity of sand between nonerodible dowel rods, found a similar effect for $L_c > 0.1$. Thus, it would seem that even a relatively sparse pebble, brush, or boulder cover will give a large amount of protection from fugitive dust production by the wind.

A wind tunnel experiment by Gillette[10] investigated the effects of wind and sandblasting on consolidated and unconsolidated samples of six different dry soil textures. Surface crusts had the effect of raising threshold velocities to values beyond the wind tunnel range. Sandblasting with spherical glass beads was sufficient to break the surface crust of the samples that had surface textures of fine sand and loamy fine sand; erosion was initiated within a few minutes of exposure. The surface crust samples having finer surface textures were not disintegrated by sandblasting in the wind tunnel exposures (10 min at a friction velocity of about 100 cm s^{-1}). Threshold velocities of the loose soil samples varied, with sandy and powdery soils eroding at lower velocities than soils having some remaining aggregation.

Bisal and Ferguson[11] investigated the effect of nonerodible aggregates on threshold wind velocity (in this case, wind velocity at 30.5 cm above the soil surface). The empirical relationship was:

$$\ln V_T = 6.0438 + 0.2332\ C$$

where V_T is the threshold velocity in cm s^{-1} at 30.5 cm, and C is the percentage of soil mass in a sample in aggregates larger than 1 mm. On the basis of analysis, I believe that the effect observed by Bisal and Ferguson is a result not of the partitioning of momentum by nonerodible elements but of increased soil cohesion, which requires larger wind forces for the onset of wind erosion.

Since finer textured soils in general tend to have more stable large soil aggregates, we would expect finer soil textures to correlate with higher threshold velocities. Mineralogical effects, such as the cracking and breakage of montmorillonitic clays, would tend to reduce aggregate size and reduce threshold velocity, however.

The Transport of Dust into the Air

Sand-sized particles are important in wind input of fugitive dust, since their availability as loose individual particles largely determines the threshold velocity of erosion, and since they carry most of the mass of moving soil near the ground, acting as sandblasters of the surface. However, only a few centimeters above the surface, the concentration of sand decreases greatly,[12] and the sand is not transported very far compared to dust smaller than 20 μm.[13]

The production of dust smaller than 20 μm and the total soil movement of a number of soil textures under natural conditions are shown in FIGURE 2. In the figure, F_a' is the vertical flux of particles smaller than 20 μm, q is the horizontal flux of all particles moving with the wind up to height of 76 cm, and F_a'/q' is the ratio of vertical flux of fine particles to measured total movement of soil.‡ The soil textures are shown in the figure. Total particle flux for all soil textures asymptotically approaches a law first proposed by Bagnold[5] (i.e., sand flux varies as the friction velocity cubed, u_*^3). The fine ($d < 20$ μm) particle flux is more variable. In the plot of F_a'/q' it is shown that for sandy loam soils, an increasing part of total soil movement is fine particle movement as the wind speed increases. In general, soils having finer textures produced more fine dust for unit soil movement, except those fine-textured soils whose mineral components formed small aggregates that were hard enough to resist impact breakage at mean wind speeds up to 25 m s^{-1}. Fine particles are generated by breakage of soil aggregates or by the splashing of loose fine material by saltating sand grains.[15]

‡A method of calculating total soil movement for individual soils is given in Reference 13.

Test soils 1, 2, 4, and 5 (FIGURE 2) showed no increase with wind speed in the ratio of vertical flux of fine particles to the horizontal flux of sand. Such a result was expected (see scaling considerations) for a loose soil of well-mixed fine and coarse particles, which would not change proportions with increased sandblasting energy. Indeed, all of these soils were classified as having fine sand textures. For soils 3, 6, and 7 (loamy sand or sandy loam textures), the ratios of vertical flux of fine material to horizontal flux of sand increased with wind speed. This would be expected for

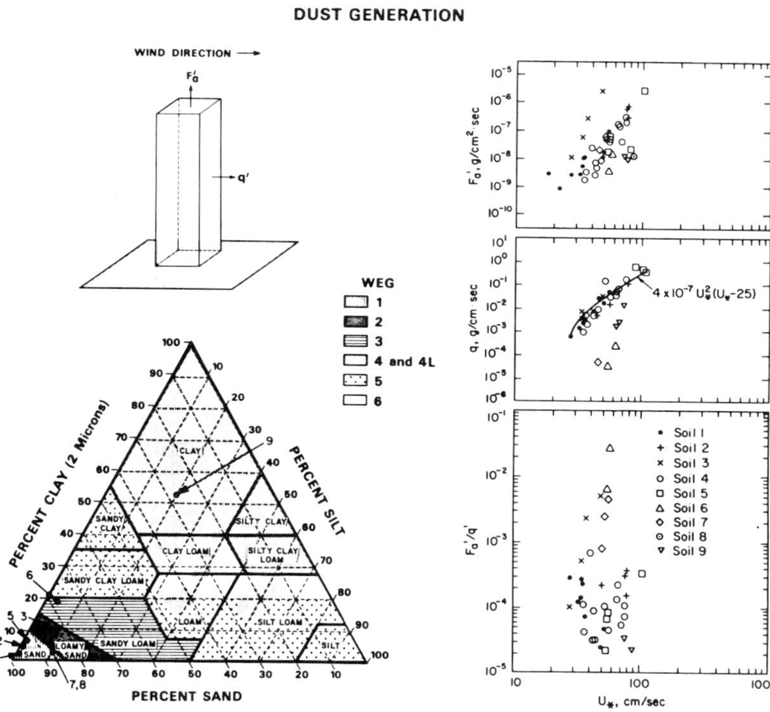

FIGURE 2. Textures of sampled soils; total soil movement vs. wind friction velocity; ratio of vertical flux of particles smaller than 0.02 mm to total soil movement per unit area per time vs. wind friction velocity (after Gillette[13]).

aggregated soil that is easily broken and that changes its proportion of fine material to coarser aggregates as sandblasting energy is increased. The clay soil (9) probably had aggregates of sufficient hardness that increasing wind speed did not disaggregate the soil enough to show an increase of fine to coarse particles with increasing wind speed.

The effect of increased dust particle flux with increasing fetch up to a certain distance was shown by Gillette,[10] as it had been shown earlier for sand flux by Chepil.[16] Briefly, the reason for such an effect is the efficiency of momentum transfer to the soil surface by saltating sand grains, which in turn propels particles into the air. This fine- and coarse-particle emitter supplements direct aerodynamic pickup of soil grains. Since this "avalanche effect" must be related to the saltation length, it is a nonlinear function of wind speed (see section on scaling considerations).

Gillette and Porch[17] used fast-response instrumentation to resolve fluctuations of the vertical wind speed, horizontal wind speed, and aerosol concentration of a portion of the fine airborne particles during a wind erosion episode in a farm field. Each of the above quantities was reduced by the average value and divided by the standard deviation of that quantity. The three-dimensional probability distribution for these fluctuations so altered is schematically shown in FIGURE 3 for emission of dust into the air. Although fluctuations appear in all octants of the probability distribution, the

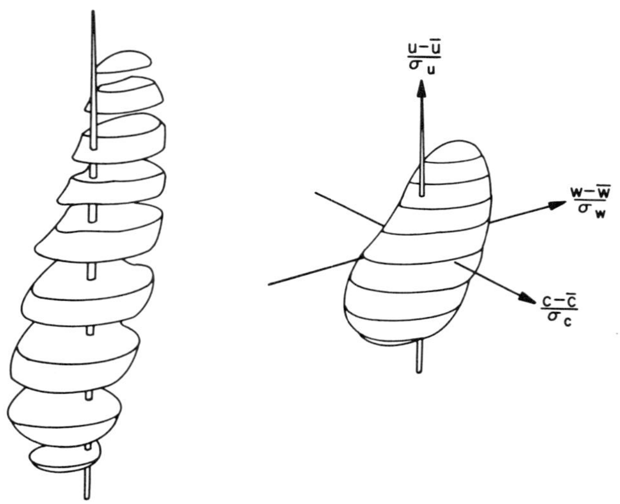

FIGURE 3. A schematic view of a three-dimensional probability distribution of vertical velocity wind fluctuations, particulate concentration, and horizontal wind speed fluctuations for a case in which dust is being emitted to the air. Each quantity has been reduced by its mean, and that reduced quantity divided by its standard deviation. The data from this drawing are constructed from work by Gillette and Porch.[17]

dominating octants for this case when particles are being transported upward are: (1) negative horizontal wind, negative vertical wind, negative concentration fluctuation; and (2) positive horizontal wind, positive vertical wind, positive concentrations. Physically, this corresponds to downward moving clean air during slow horizontal wind periods and to upward moving dirty air during faster horizontal wind periods.

Size Distributions of Soil-Derived Dust

The size distribution of soil-derived aerosol may be roughly characterized as a bimodal log-normal distribution when particle volume is plotted vs. particle size. The larger particles, having a mode around 50 μm, are sand grains that are moving close to the surface and whose sedimentation velocities are of the order of the vertical velocity fluctuations of the air. The finer particles, having a mode between 1 and 10 μm, are particles produced by the sandblasting of sand grains on the soil surface.[18]

Large-Scale Dust Input

Total mass of dust produced in two dust storms in 1972 in the southwestern United States was computed from aircraft-obtained dust concentrations and meteorological analysis of dusty air.[19] Total mass produced by the two storms was estimated to be 0.8×10^{12} g, which is only 0.16% of the yearly production of dust estimated by Goldberg from consideration of pelagic sediment depositions.[1] The average vertical flux, computed from total mass, storm duration, and area of eroding land, was consistent with measured values from other dust storms. Thus, if dust storm frequency were around 10 per year, the southwestern United States would be responsible for only about four million tons per year, or less than one percent of Goldberg's total estimated yearly production of 500 million tons. However, in considering the vast amounts of dust produced in other systems (e.g., the Saharan system reported by Schütz[20]), the magnitude of dust production in the United States is probably not great when compared to the total global input.

Other Sources of Primary Continental Particles

Although soil particle emissions are thought to be the dominant source of primary continental aerosol production, other sources of aerosols, such as fires, volcanoes, and biological emissions, are important. Most of these sources are irregular in time and nonuniform; thus, they may be dominant in a given location and time. I will not cover biological emissions in this paper, since most biological particles fall out of the atmosphere soon enough to have regional rather than global significance.

Both fires and volcanoes emit particles into a buoyant plume, which may be carried far into the atmosphere. Particle emissions in the case of fires depend on wind insofar as the fuel has completely burned. The temperature of the flame front and the nature of the fuel are very important in particle emissions from fires. The distinction made in the literature is between controlled fires, in which the advance of the flaming zone is not rapid and more of the fuel is burned at high temperature (efficient combustion), and wildfires, in which much of the fuel is not burned within the flame zone and much is consumed at lower temperature by smoldering combustion (less efficient combustion).

Emission factors (90% fuel mass as smoke) were given for the two categories of fires.

	Ward *et al.*[21]	Vines *et al.*[22]
Wildfires	7.5	
Controlled fires	2.5	1.8, $\sigma = 0.4$

A global estimate of annual smoke production may be given by assuming that the global production is four times that of U.S. production.[24] Using Ward's[21] estimate of U.S. production, the yearly global production is approximately 16 million tons.

Size distribution of the smoke is shown below.

	% Smaller Than	
Type	1 μm	0.4 μm
Slash pine[23]	95	70
Palmetto-Galberry[21]	91	53

Several investigators have found the mode of the mass-size distribution to be about 0.1 μm.

Volcanic emissions, while irregular in time, are of very great importance globally during the time of their activity. Some recent estimates from aircraft-obtained data of volcanic emissions were given by Stith *et al.*[25] for differing types of volcanic activity. Paroxysmal emissions are the result of magmatic movements and occur on a time scale of minutes. Most of the mass of emitted particles is in particles smaller than 1 μm. Intereruptive (between paroxysmal) events have time scales of days; the size distributions of aerosols from these events are between those of paroxysmal and those of posteruptive events. Posteruptive events follow eruptions and have time scales of days to years. Particle emissions are primarily of submicron particles. An estimate of particle emissions is given in the following table.

Volcanic Particle Emissions[26]

Event	Paroxysmal	Intereruptive	Posteruptive
Order of Emission (tons/s)	400	0.3	0.04

Worldwide emission rates of particle production by volcanoes was estimated by Bach[26] to be 4–150 million tons per year.

CONCLUSIONS

The striking facts about the sources of continental aerosols outlined above are the sporadic nature of the emissions and the enormous magnitudes of the aerosol input during short periods of time. Dust production progresses at least as the cube of wind friction velocity above a threshold value for soil wind erosion. Volcanoes have input of 400 tons s^{-1} for brief paroxysmal eruptions, and wildfires contribute 7.5% of fuel as particulate mass. Clearly, these events lead to extremely high aerosol loading, and any effort to compare total anthropogenic aerosol production to natural continental production may be misleading, since most of the production takes place in a very limited time duration.

I would then suggest the computation of two emission inventories; one for total production, and another for the majority of time when large-scale dust storms, volcanic emissions, and forest fires are not taking place. This natural production would then provide a better "baseline" aerosol production value with which to compare anthropogenic emissions.

SUMMARY

Details of production of natural primary continental aerosols are reviewed. Emission of soil particles into the air by the wind is emphasized, but fire and volcanic emissions are also considered. All of the primary continental sources considered produce very large amounts of aerosols over very short periods of time and thus do not represent a normal background aerosol input. It is suggested that two emission inventories be prepared: one for total emissions and one for the majority of time when no major dust storm, volcano, or forest fire is producing particles.

REFERENCES

1. HIDY, G. & J. BROCK. 1970. An assessment of the global sources of tropospheric aerosols. Presented at the 2nd Clean Air Congress, International Union of Air Pollution Prevention Associations, Washington, D.C., December, 1970.
2. PETERSEN, J. & C. JUNGE. 1971. Sources of particulate matter in the atmosphere. *In* Man's Impact on the Climate. W. H. Matthews, W. W. Kellog & G. D. Robinson, Eds.: 310–320. M.I.T. Press. Cambridge, Mass.
3. PUNJRATH, J. S. & D. R. HELDMAN. 1971. Mechanisms of small particle reentrainment from flat surfaces. J. Aerosol Sci. **3:** 429.
4. IVERSEN, J. D., J. B. POLLACK, R. GREELEY & B. R. WHITE. 1976. Saltation threshold on Mars; the effect of interparticle force, surface roughness and low atmospheric density. Icarus **29:** 381.
5. BAGNOLD, R. A. 1941. The Physics of Blown and Sand and Desert Dunes. Methuen. London, England.
6. CHEPIL, W. S. 1951. Properties of soil which influence wind erosion. IV. State of dry aggregate structure. Soil Sci. **72:** 387.
7. GREELEY, R., J. D. IVERSON, J. B. POLLACK, N. UDOVICH & B. WHITE. 1973. Wind tunnel studies of Martian aerolian processes. NASA Tech. **62:** 297.
8. GILLETTE, D. A., J. ADAMS, A. ENDO & D. SMITH. 1979. Threshold wind velocities for natural desert soil surfaces before and after disturbances by offroad vehicles. *In* Proceedings, International Powder and Bulk Solids Handling and Processing Conference, Philadelphia, Penn., May 15–17, 1979. Industrial and Scientific Conference Management, Inc. Chicago, Ill.
9. MARSHALL, J. 1971. Drag measurements in roughness arrays of varying density and distribution. Agric. Meteorol. **8:** 269.
10. GILLETTE, D. A. 1978. A wind tunnel simulation of the erosion of soil: effect of soil texture, sandblasting, wind speed, and soil consolidation on dust production. Atmos. Environ. **12:** 1735.
11. BISAL, F. & W. FERGUSON. 1970. Effect of nonerodible aggregates and wheat stubble on initiation of soil drifting. Can. J. Soil Sci. **50:** 31.
12. GILLETTE, D. A. & P. A. GOODWIN. 1974. Microscale transport of sand-sized aggregates eroded by wind. J. Geophys. Res. **79:** 4080.
13. GILLETTE, D. A. 1978. Fine particulate emissions due to wind erosion. Trans. ASAE **20:** 890.
14. LYLES, L. & B. ALLISON. 1976. Wind erosion. The protective role of simulated standing stubble. Trans. ASAE **19:** 61.
15. GILLETTE, D. A. & T. R. WALKER. 1977. Characteristics of airborne particles produced by wind erosion of sandy soil, high plains of west Texas. Soil Sci. **123**(2): 97.
16. CHEPIL, W. S. 1957. Sedimentary characteristics of dust storms. III. Composition of suspended dust. Am. J. Sci. **255:** 206.
17. GILLETTE, D. A. & W. PORCH. 1978. The role of vertical and horizontal wind fluctuations in the production and transport of dust by wind erosion. J. Geophys. Res. **83:** 409.

18. Patterson, E. M. & D. A. Gillette. 1977. Commonalities in measured size distributions for aerosols having a soil-derived component. J. Geophys. Res. **82:** 2074.
19. Gillette, D. A., R. N. Clayton, T. K. Mayeda, M. L. Jackson & K. Sridhar. 1978. Tropospheric aerosols from some major dust storms of the southwestern United States. J. Appl. Meteorol. **17:** 832.
20. Schütz, L. 1980. Long range transport of desert dust with special emphasis on the Sahara. Ann. N.Y. Acad. Sci. (This volume.)
21. Ward, D., C. McMahon & R. Johansen. 1976. An update on particulate emissions from forest fires. Presented at the 69th Annual Meeting of the Air Pollution Control Association, Portland, Oregon, July, 1976.
22. Vines, R. G., L. Gibson, A. Hatch, N. King, D. MacArthur, D. Packham & R. Taylor. 1971. On the Nature, Properties, and Behavior of Bush-Fire Smoke. Div. Appl. Chem. Tech. Paper No. 1. CSIRO. Melbourne, Australia.
23. McMahon, C. & P. Ryan. 1976. Some chemical and physical characteristics of emissions from forest fires. Presented at the 69th Annual Meeting of the Air Pollution Control Association, Portland, Oregon, July, 1976.
24. Robinson, E. & R. Robbins. 1971. Emissions, Concentrations, and Fate of Particulate Atmospheric Pollutants. API Publication No. 4076. Stanford Research Institute. Menlo Park, Calif.
25. Stith, J., P. Hobbs & L. Radke. 1978. Airborne particle and gas measurements in the emissions from six volcanoes. J. Geophys. Res. **83:** 4009.
26. Bach, W. 1976. Global air pollution and climatic change. Rev. Geophys. Space Phys. **14:** 429.

ORGANIC CONSTITUENTS OF NATURAL AEROSOLS*

Jürgen Hahn

Max-Planck-Institut für Chemie
D-6500 Mainz, Federal Republic of Germany

INTRODUCTION

The blue haze often observed over forested areas led Went[1] to suggest the potential importance of vegetation as a major source of atmospheric hydrocarbons. Nearly 20 years have passed since, and although interest in the organic chemistry of the troposphere has been growing in recent years, the composition, distribution, and cycle of atmospheric organic matter have remained largely obscure. In an excellent review, Duce[2] recently defined the state of knowledge in this field of air chemistry as primitive at best. Information on vapor phase organic matter in remote areas, or even in nonurban continental areas, is extremely sparse. More data are available on airborne particulate organic matter. But even here, most studies have dealt with the urban situation or have concentrated on specific organic pollutants, such as DDT and other pesticides, polychlorobiphenyls (PCBs), and some polycyclic hydrocarbons. The understanding of the sources, sinks, and residence time and distribution in the troposphere of airborne particulate organic matter as a whole or of its various constituents is still very limited.

This paper summarizes what is known about the organic constituents of natural aerosol particles and the abundance of particulate organic matter in the troposphere, and attempts, despite large uncertainties, to describe a cycle of tropospheric particulate organic matter.

COMPOSITION OF PARTICULATE ORGANIC MATTER IN CLEAN AIR

There is a multitude of organic compounds that may be emitted to the atmosphere from various sources. In addition, a variety of compounds may be formed by chemical reactions in the troposphere itself. Therefore, it is not surprising that airborne particulate organic matter (POM) appears to represent a complex mixture of hundreds of individual compounds. The degree of complexity as far as the number of individual compounds is concerned does not seem to change much as one proceeds from polluted regions into clean air areas.

There are no data on the total composition of POM from clean air. But a few data are available for the ether-extractable organic matter (EEOM) of POM. In the gas chromatographic (GC) analysis of EEOM obtained from aerosol particles collected during a cruise in the southern North Atlantic, Ketseridis *et al.*[3] found 110–145 individual compounds. The total number of organic compounds making up EEOM must be larger than that, since not all constituents of EEOM are volatile enough for GC analysis. The identified compounds are listed in TABLE 1. The abundance of these

*Supported by the German Science Foundation (DFG) through its Sonderforschungsbereich 73: Atmospheric Trace Components.

0077–8923/80/0338–0359 $01.75/1 © 1980, NYAS

compounds is given both in nanograms per standard cubic meter of air and in per mil of EEOM. The average values given in TABLE 1 are based on measurements that were summarized in an earlier publication.[4] For the average values in TABLE 1, only 5 out of 7 samples were considered, disregarding the highest and the lowest values.

In the fraction of the aliphatic hydrocarbons, which accounted for about 28% of EEOM, *n*-paraffins from C_{10} to C_{28} were found. In the aromatic fraction, accounting for about 18% of EEOM, 13 different aromatic hydrocarbons with 2- to 4-ring structures were identified. Polycyclic hydrocarbons with more than 4 rings, such as 1,2-benzopyrene, 3,4-benzopyrene, and perylene, could not be detected. The fraction of the polar compounds made up about 27% of EEOM. From the 20–30 GC peaks, 7 were identified as shown in TABLE 1. The organic acids accounted for about 20% of

TABLE 1

INDIVIDUAL COMPOUNDS IDENTIFIED IN THE ETHER-EXTRACTABLE ORGANIC MATTER (EEOM) FROM AEROSOL PARTICLES COLLECTED IN THE SOUTHERN NORTH ATLANTIC

I. Aliphatic Hydrocarbons

	c_1*	c_2†
n-decane	0.55	0.73
n-undecane	1.9	2.5
n-dodecane	0.92	1.2
n-tridecane	0.85	1.1
n-tetradecane	1.2	1.6
n-pentadecane	0.70	0.93
n-hexadecane	0.58	0.77
n-heptadecane	1.0	1.3
n-octadecane	0.69	0.91
n-nonadecane	2.1	2.8
n-eicosane	0.01	0.01
n-heneicosane	0.61	0.81
n-docosane	1.3	1.7
n-tricosane	0.81	1.1
n-tetracosane	0.24	0.32
n-pentacosane	0.10	0.13
n-hexacosane	0.39	0.52
n-heptacosane	0.01	0.01
n-octacosane	<0.01	<0.01
n-nonacosane	<0.01	<0.01
n-triacontane	<0.01	<0.01
Σ	14.0	18.5

II. Aromatic Hydrocarbons

	c_1	c_2
naphthalene	2.7	3.6
2-methylnaphthalene	1.8	2.4
acenaphthylene	0.38	0.50
acenaphthene	0.25	0.33
fluorene	0.31	0.41
phenanthrene	0.05	0.07
anthracene	0.49	0.65
2-methylanthracene	2.2	2.9
fluoranthene	1.4	1.9
pyrene	0.24	0.32
1.2-benzofluorene	0.78	1.0
chrysene	0.37	0.49
1.2-benzoanthracene	0.28	0.37
3.4-benzopyrene	<0.01	<0.01
1.2-benzopyrene	<0.01	<0.01
perylene	<0.01	<0.01
Σ	11.3	14.9

III. Polar Compounds

	c_1	c_2
coumarine	0.04	0.05
peri-naphthenone	0.11	0.15
xanthone	0.22	0.29
anthrone	0.04	0.05
flavone	0.07	0.09
benzoanthrone	0.24	0.32
carbazole	0.04	0.05
Σ	0.76	1.0

TABLE 1 (continued)

IV. Organic Acids			V. Organic Bases		
	c_1	c_2		c_1	c_2
pelargonic	2.5	3.3	pyridine	n.d.	n.d.
capric	3.5	4.6	quinoline	<0.01	<0.01
hendecanoic	6.1	8.0	iso-quinoline	<0.01	<0.01
lauric	2.0	2.6	aniline	<0.01	<0.01
tridecanoic	1.6	2.1	indole	0.20	0.27
myristic	1.3	1.7	2-methylindole	0.20	0.27
pentadecanoic	3.8	5.0	7-methylindole	0.17	0.23
palmitic	5.3	7.0	2.3-dimethylindole	0.14	0.19
margaric	2.1	2.8	2.5-dimethylindole	0.15	0.20
stearic	2.4	3.2	2.4-dimethylquinoline	0.41	0.54
oleic	1.7	2.2	2.6-dimethylquinoline	0.02	0.03
linoleic	3.1	4.1	2.8-dimethylquinoline	0.78	1.0
linolenic	1.1	1.5	α-naphthylamine	<0.01	<0.01
nonadecanoic	1.1	1.5	benzo(h)quinoline	<0.01	<0.01
arachidic	1.1	1.5	acridine	<0.01	<0.01
heneicosanoic	0.01	0.01	phenanthridine	<0.01	<0.01
behenic	0.82	1.1			
Σ	25.4	33.6	Σ	2.1	2.8
			Σ identified	53.6	71

*c_1 in nanograms per m^3 STP air, averages of five samples.
†c_2 in per mil of EEOM, averages of five samples.

EEOM; straight-chain fatty acids from C_9 to C_{22} were identified. The even-numbered acids were found to be predominant, a result indicating a biological origin of these acids. The fraction of the organic bases was found to be small, making up only about 7% EEOM. Nevertheless, the GC analysis of this fraction gave 15–20 individual peaks, 8 of which could be identified.

As shown in TABLE 1, less than 10% of the total mass of EEOM could be identified. Considering the fact that EEOM is only a fraction of POM (which might, however, account for as much as ½ of POM, as will be shown below), it is evident that we don't know much about the composition of POM in clean air, yet.

As to the origin of the identified compounds, it is not clear which of these compounds are exclusively man-made and which compounds may serve as tracers for the various sources of POM. The data base is still too small for any conclusive statements. It looks as if the 5- and 6-ring polycylic hydrocarbons have no natural sources. The even-to-odd predominance in the fraction of the organic acids suggests a biological origin of a major portion of these compounds. Tentatively, one may define the following features of the different large-scale sources of POM as sufficiently long-lived characteristics.

Land Sources (vegetation)

1. Straight-chain paraffins (*n*-alkanes): C_{23}–C_{35} predominant with a marked preference for the odd carbon numbers.

2. Straight-chain fatty acids (carboxylic acids): C_{22}–C_{32} predominant with a marked preference for the even carbon numbers.

Aquatic Sources (microbiota of marine and lacustrine environments)

1. Straight-chain paraffins (*n*-alkanes): C_{12}–C_{24} predominant, major maximum at C_{17}, no significant carbon number preference.
2. Branched-chain paraffins (*i*-alkanes): mono-methyl-substituted *n*-alkanes C_{13}–C_{23} predominant.
3. Straight-chain fatty acids (carboxylic acids): C_{12}–C_{20} predominant with a marked preference for the even carbon numbers, major maximum at C_{16}.

Geochemical Sources (petroleum seepages)

1. Complex mixture of hydrocarbons C_{10}–C_{30} with no significant carbon number preference, maximum varying from C_{15} to C_{22}.

Man-Made Sources (combustion, oil spills, industrial effluents and wastes, municipal wastes)

1. Complex mixture of petroleum-derived hydrocarbons C_{10}–C_{30} with no significant carbon number preference, maximum varying from C_{15} to C_{22}.
2. Aromatic hydrocarbons with 5- and 6-ring structures; polychlorinated biphenyls (PCBs) and other halogenated organic compounds.

There may be more and better characteristics for the different sources of POM, but this is all one can say at present with some certainty.

If one considers, instead of individual compounds, all neutrals as one group (including aliphatic and aromatic hydrocarbons, alcohols, esters, and other polar compounds), the organic acids as a second group, and the organic bases as a third group, the situation appears much less complex, at least as far as EEOM is concerned. All samples that we have analyzed so far [3,5,6] have shown the neutrals to be the main fraction of EEOM, followed by the organic acids. The group of the organic bases was the smallest in all cases. Considering only samples collected in remote areas (Cape Verde Islands, 1973; Adrigole, Irish southwest coast, 1973; southern North Atlantic, 1973; Jungfrau-Joch, Switzerland, 1973; Mace Head, Irish west coast, 1975; Loop Head, Irish west coast, 1977), EEOM of clean air aerosol particles appears to consist of about 60% neutral compounds, about 30% organic acids, and about 10% organic bases, on the average.

The portion of POM that is insoluble in ether may consist of plant debris, pollen, bituminous material of high molecular weight, and others. In nonurban continental areas, appreciable amounts of soot may be found, which, originating predominantly from anthropogenic sources, travel long distances.

Abundance of Particulate Organic Matter in the Troposphere

Apart from studies in urban areas, relatively few data are available on the abundance of POM in the troposphere. Moreover, POM has never been measured

TABLE 2

AVERAGE CONCENTRATIONS OF AIRBORNE TOTAL PARTICULATE MATTER (TPM), ETHER-EXTRACTABLE ORGANIC MATTER (EEOM), AND OF THE GROUPS OF NEUTRAL COMPOUNDS, ORGANIC ACIDS, AND ORGANIC BASES ON THE BASIS OF MEASUREMENTS AT VARIOUS LOCATIONS[3,5,6]

(All values in $\mu g/m^3$ STP air)

Location	Elevation above Sea Level (m)	Number of Samples	TPM	EEOM	Neutrals	Acids	Bases
Mainz W. Germany Jan. 1972	140	3	150 (120–175)	27.1 (23.5–30.5)	15.0 (12.4–17.1)	10.1 (8.2–12.8)	2.0 (0.98–3.0)
Deuselbach W. Germany June 1972	500	2	21 (18.5–23.5)	2.0 (1.3–2.7)	1.1 (0.75–1.5)	0.69 (0.45–0.93)	0.20 (0.12–0.27)
Adrigole Irish SW Coast Feb. 1973	15	6	16.5 (10.4–26.0)	2.0 (0.90–4.2)	0.94 (0.21–2.3)	0.88 (<0.01–1.1)	0.20 (<0.01–0.86)
Sal Cape Verde Isl. July 1973	20	2	94 (83–106)	1.6 (1.1–2.1)	0.77 (0.54–1.0)	0.54 (0.26–0.81)	0.31 (0.26–0.37)
Meteor Exp., II. Trop; NE-Atlantic Nov.–Dec. 1973	20	4	96 (87–111)	0.91 (0.31–1.9)	0.70 (0.18–1.5)	0.17 (0.06–0.30)	0.05 (0.04–0.10)
Jungfrau-Joch Switzerland Sept. 1973	3,600	2	17.6 (2.7–32.4)	0.90 (0.03–1.8)	0.57 (0.02–1.1)	0.28 (0.01–0.42)	0.05 (<0.01–0.10)
Mace Head Irish W. Coast Aug.–Sept. 1975	30	8	8.6 (3.8–13.5)	0.81 (0.45–1.2)	0.52 (0.45–0.80)	0.23 (0.10–0.35)	0.04 (0.03–0.08)
Meteor Exp., I. Trop. N-Atlantic Oct.–Nov. 1973	20	4	16.7 (10.2–20.8)	0.80 (0.15–1.6)	0.49 (0.06–0.85)	0.25 (0.05–0.59)	0.05 (<0.01–0.08)
Loop Head Irish W. Coast June 1977	70	16	20.5 (9.4–41.3)	0.73 (0.43–1.3)	0.38 (0.20–0.77)	0.25 (0.16–0.45)	0.10 (0.03–0.21)

directly. Due to the analytical techniques frequently used, many of the data available for remote areas are restricted to measurements of particulate organic carbon (POC). There are only a very limited number of cases where, besides POC, the abundance of solvent-extractable organic matter and its fractions has been studied. All measurements in remote areas have been carried out near ground or near the ocean surface. They cover, however, at least for marine air, a reasonably wide geographical distribution.

In TABLE 2, average concentrations of total particulate matter (TPM), of EEOM, and of the groups of neutral compounds, organic acids, and organic bases calculated from data that we collected in various locations[3,5,6] are given. The values are listed in the order of decreasing EEOM concentration. The range of scatter is given in parentheses. For comparison, two values from the urban area of Mainz, West Germany, with westerly winds in January 1972[3] are included. The comparatively high values of TPM, observed on the Cape Verde Islands in July 1973 and during the second part of the Meteor cruise in the eastern tropical North Atlantic in November 1973, were caused by the advection of Sahara dust, which increased TPM substantially but had only a relatively small effect on the EEOM concentrations. For Loop Head, we were able to subdivide the samples collected according to air masses into one

TABLE 3
AVERAGE CONCENTRATIONS OF AIRBORNE PARTICULATE ORGANIC CARBON (POC) IN MARINE AREAS (after Duce[2])

Location	Elevation above Sea Level (m)	Number of Samples	POC ($\mu g/m^3$ STP)	Reference
Loop Head Irish West Coast	70	6*	0.57 (0.28–0.86)	Eichmann *et al.*, 1979[6]
Bermuda	40	8	0.29 (0.15–0.47)	Hoffmann and Duce, 1974[8]
Bermuda	40	8	0.37 (0.15–0.78)	Hoffmann and Duce, 1977[9]
Tropical North Atlantic	20	5	0.59 (0.33–0.93)	Ketseridis *et al.*, 1976[3]
Hawaii	20	7	0.39 (0.36–0.43)	Hoffmann and Duce, 1977[9]
Eastern Tropical North Pacific	10	3	0.49 (0.22–0.74)†	Barger and Garrett, 1976[20]
San Nicolas Isl. off California	20	7	0.90 (0.15–1.8)	Hidy *et al.*, 1974[22]
Samoa	50	9	0.22 (0.13–0.41)	Hoffmann and Duce, 1977[9]
Eastern Tropical South Pacific	10	9	0.32 (0.07–0.53)†	Barger and Garrett, 1976[20]

*Samples taken in marine air.

†Calculated from chloroform-extractable organic matter, using a conversion factor of 0.9 according to Duce.[2]

TABLE 4

AVERAGE CONCENTRATIONS OF AIRBORNE PARTICULATE ORGANIC CARBON (POC) IN NONURBAN CONTINENTAL AREAS (after Duce[2])

Location	Elevation above Sea Level (m)	Number of Samples	POC ($\mu g/m^3$ STP)	Reference
Deuselbach W. Germany	500	2	2.5*	Ketseridis *et al.*, 1976[3]
Adrigole Irish SW Coast	15	5	2.5*	Ketseridis *et al.*, 1976[3]
Jungfrau-Joch Switzerland	3,600	2	1.0*	Ketseridis *et al.*, 1976[3]
Loop Head Irish West Coast	70	3†	4.5 (3.0–6.0)	Eichmann *et al.*, 1979[6]
Nonurban U.S., Total	—	28	2.2 (0.60–5.0)	USPHS, 1968[21]
Western nonurban U.S.	—	14	1.8 (0.60–3.1)	USPHS, 1968[21]
Eastern nonurban U.S.	—	14	2.3 (1.3–5.0)	USPHS, 1968[21]

*Calculated from ether-extractable organic matter, using a conversion factor of 1.2 according to Duce.[2]

†Samples taken in continental air.

group representing marine (North Atlantic) air, a second group representing marine air with some continental influence, and a third group representing continental air with a possible small marine influence. The average EEOM concentration for marine air over Loop Head was found to be as low as 0.51 (0.18–0.84) $\mu g/m^3$ STP air, while marine air with continental influence gave an EEOM average of 0.93 (0.48–1.38) $\mu g/m^3$ STP air; and continental air, 1.20 (0.69–1.71) $\mu g/m^3$ STP air.[6]

Three things may be noted from TABLE 2:

1. As was already mentioned, the average group composition of EEOM remains strikingly uniform as one travels from the urban area of Mainz into remote areas. Current knowledge is not sufficient to determine whether this uniformity results from the injection, and subsequent dilution, of major amounts of EEOM into the atmosphere over densely populated areas, or from the formation of major amounts of EEOM in the troposphere by the oxidation of vapor phase compounds and gas-to-particle conversion.

2. EEOM concentrations decrease by more than a factor of 10 as one proceeds from the urban area of Mainz to Deuselbach, about 100 km west. Usually, Deuselbach has typical central European nonurban air. As one moves farther away from densely populated areas, EEOM concentrations decrease to values around 1 $\mu g/m^3$ STP air.

3. Over the ocean, EEOM concentrations are even lower. Considering a possible slight increase of EEOM concentrations due to dust advection, the average for marine air may be assumed to be close to 0.6 $\mu g/m^3$ STP air. (Very recently, we were able to

collect particulate samples on the southwest coast of Tasmania, in the Southern Hemisphere. The average EEOM concentration found in Tasmania[7] was virtually identical with the average for marine air over Loop Head as given above.

Available data on POC concentrations in marine areas are summarized in TABLE 3. Considering the fact that the data were obtained by using different methods, the scatter is small. Not all of the data listed in TABLE 3 (and in TABLE 4, see below) were measured directly. Some were calculated from solvent-extractable organic matter, using conversion factors.[2] These conversion factors are for noncarbonate carbon and thus include elemental, as well as organic, carbon. Most of the POC data given in this paper include elemental carbon. The data that we very recently obtained from samples collected in Tasmania gave an average POC concentration close to 0.6 $\mu g/m^3$ STP.[7] One may assume, therefore, an average POC near surface concentration of about 0.6 $\mu g/m^3$ STP for marine air.

The few data available for POC concentrations in nonurban continental air are listed in TABLE 4. Average values cover the range from 1 to about 5 $\mu g/m^3$ STP. In remote continental areas that are not contaminated by man-made emissions, POC concentrations are close to the lower end of this range.[2] The comparatively high Loop Head average was apparently due to anthropogenic influence, which appears to have a more pronounced effect on POC than on EEOM concentrations. [While EEOM doubled as continental air replaced the marine air over Loop Head, POC increased almost by a factor of 8 from 0.57 to 4.5 $\mu g/m^3$ STP. This increase was accompanied by a rise in vapor phase xylene concentrations from 10 nanograms per m^3 (STP) to about 130 nanograms per m^3 (STP) air. Xylenes are believed to be exclusively man-made.] In unpolluted continental air, the average POC concentration near ground may be estimated to be 1.0–1.5 $\mu g/m^3$ STP. However, because of the relatively far-reaching effects of air pollution, the average POC concentration for most nonurban continental areas of the Northern Hemisphere is probably higher than the estimated values. The data in TABLE 4 suggest an average value of between 1.5 and 2.5 $\mu g/m^3$ STP air.

The distribution of POC with particle size has recently been investigated by Hoffman and Duce[9] for marine areas. The results obtained from measurements of organic carbon in airborne particulate matter that they collected at Bermuda, Hawaii, and Samoa were quite similar. A distribution that appears to be typical for POC in marine air is shown in FIGURE 1 (taken from Hoffmann and Duce[9]). It turns out that 80–90% of the total POC mass is associated with particles having radii <1 μm. An aerosol generated from seawater (by bursting bubbles) in the laboratory gave a quite different size distribution of POC, whereas the size distribution of the sea-salt mass (determined as sodium) was quite similar to that of the ambient aerosols. In the laboratory study, the major mass of POC was found on particles with radii of 1–3 μm; the major mass of sea salt also occurred on particles of this size. This suggests, as Hoffmann and Duce[9] pointed out, that most of the particulate organic matter in marine aerosols is not associated with sea salt.

The distribution of EEOM with particle size in marine aerosols was examined by Ketseridis and Eichmann[5] at Mace Head on the Irish west coast. Using a high-volume sampler and an impactor, simultaneous samples of both particulate matter covering the entire particle size range and particulate matter of the size range <1 μm radius were collected. During the sampling period in August–September 1975, the sea was calm. The various particulate samples were analyzed for the content of EEOM, neutral compounds (N), organic acids (A), and organic bases (B). The results are shown in Figure 2 (taken from Eichmann and Ketseridis[5]). The data obtained for the entire particle size range are plotted on the left-hand side of the figure. Over the 12-day sampling period (eight samples), the TPM concentration varied by about 50% around a mean of 8.6 $\mu g/m^3$ STP air. EEOM accounted for about 10% of TPM and showed similar variations with time, as did the groups of the neutrals, acids, and bases. By and large, the EEOM content of TPM and the group composition of EEOM remained constant over the sampling period.

At the bottom of FIGURE 2 are plotted the average wind speed and the average particle number concentration during sample collection. The wind direction is indicated by flags. The correlation between TPM concentration and wind speed is

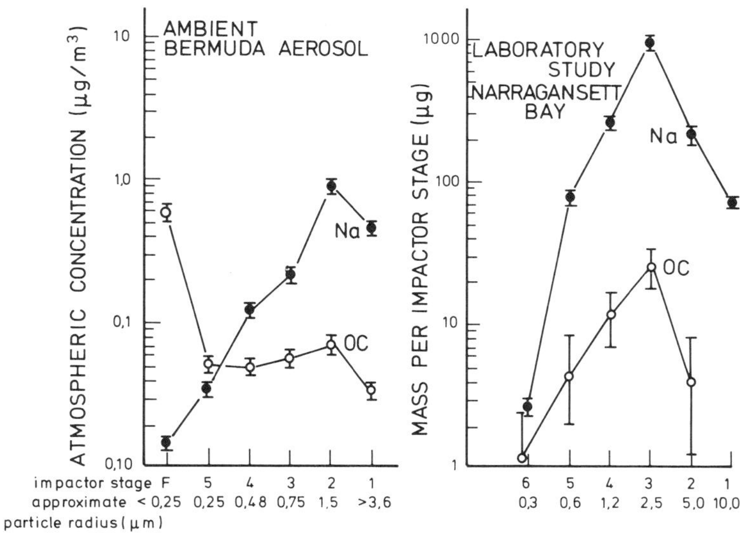

FIGURE 1. Comparison of the distribution of particulate organic carbon (OC) and sodium (Na) with particle size, observed in ambient marine air and in a laboratory-generated aerosol (taken from Hoffmann and Duce[9]).

rather poor. The correlation between TPM concentration and particle number concentration for the first nine days of the sampling period is better than one would expect, considering the different distribution patterns of number concentration and mass with particle size.[10,13]

On the right-hand side of FIGURE 2, the data obtained for the particles with radii <1 μm are plotted. The small particle TPM concentration shows a variation with time similar to that of the "all particle" TPM concentration. But the two TPM time trends are not completely correlated. The same holds for the small particle EEOM and the groups of the neutral compounds, acids, and bases. Considering the accuracy of the measurements, the differences in time trends and group compositions are barely significant, with two exceptions:

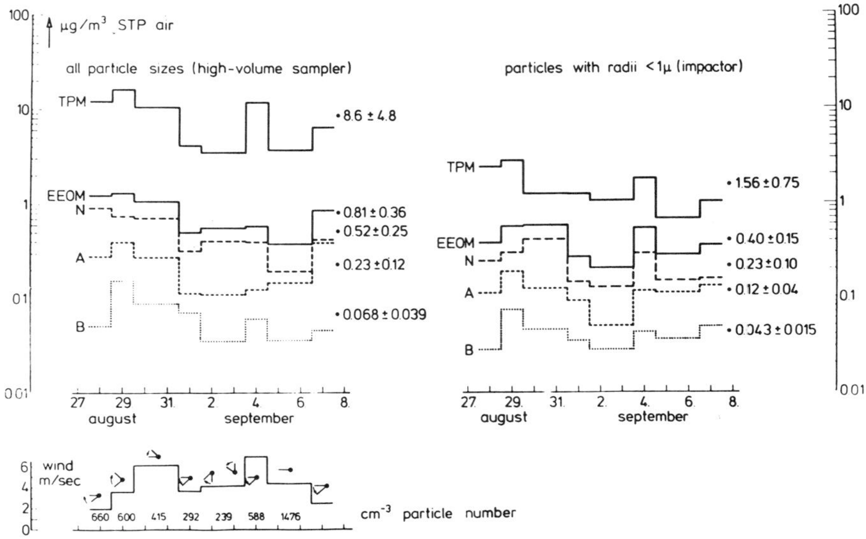

FIGURE 2. Distribution of airborne total particulate matter (TPM), ether-extractable organic matter (EEOM), and the groups of neutral compounds (N), organic acids (A), and organic bases (B) in EEOM for two different size ranges of aerosol particles collected at Mace Head, Irish west coast, in August–September 1975. Also given at the bottom is the variation of the total particle number concentration with the wind speed over the time period of sampling (taken from Ketseridis and Eichmann[5]).

First, close inspection of the various curves reveals that the acids and bases appear to be better correlated with TPM than are the neutral compounds. The neutrals, as the major fraction of EEOM, are thus responsible for major deviations of EEOM from the TPM time trend. This is particularly conspicuous in both particle size ranges with the samples collected from August 29 to August 31 (48-hour samples). It is tempting to conclude that the large-scale source of the neutral compounds that come with aerosol particles in marine air is different, at least to some degree, from the source of the organic acids and bases, which seem to be associated with sea salt. (At a location such as Mace Head, most of the total particulate matter in marine air is sea salt.)

Second, and this appears to be fairly certain, the small particle TPM (radii < 1 μm) contains much more EEOM than the large particle TPM (radii $> 1 \mu$m). Using the data of FIGURE 2, it turns out that the EEOM was more or less equally distributed between the ranges of the large and small particles. However, the small particle TPM, which accounted for about 20% of the total TPM, contained about 25% EEOM, whereas the large particle TPM contained only about 6% EEOM.

The fractions of the neutral compounds obtained from the Mace Head EEOM were further analyzed for the normal paraffins (*n*-alkanes) from C_{10} to C_{28}. As with the EEOM as a whole, it was found that the normal paraffins were more or less equally distributed between the large and small particles. Also, there were no significant differences in the mean distribution patterns of the normal paraffins between the two size ranges of particles. This uniformity suggests that the distribution (over the two particle size ranges) of many of the compounds that make up the EEOM (n-paraffins in particular) is not primarily determined by the characteristics of the (primary) sources of particles on the earth's surface, but by interactions (adsorption/desorption processes) between the particulate and the vapor phase in the troposphere.

According to a prediction by Junge,[11] who considered saturation vapor pressures and the surface area of aerosol particles, the major portion of atmospheric organic compounds with saturation vapor pressures down to about 10^{-7} torr are expected to be in the vapor phase in clean air. This prediction has been confirmed by the results of simultaneous measurements of vapor phase and particulate normal paraffins (n-alkanes) at Loop Head on the Irish west coast in June 1977.[6] Some of the results of these measurements are shown in FIGURE 3. As mentioned earlier, three different air mass categories could be defined during the sampling period at Loop Head. Accordingly, there are three different curves in the upper part of FIGURE 3, representing mean distributions of particulate n-alkanes from C_{10} to C_{28} in marine air, in marine air with some continental (coastal?) influence, and in continental air with little marine influence. It may be noted that the n-alkane concentrations were highest in continental air, with a perceptible predominance of odd carbon numbers for n-alkanes larger than C_{16}. This observation is in accordance with the above-noted characteristics of large-scale sources of atmospheric organic matter. The lowest

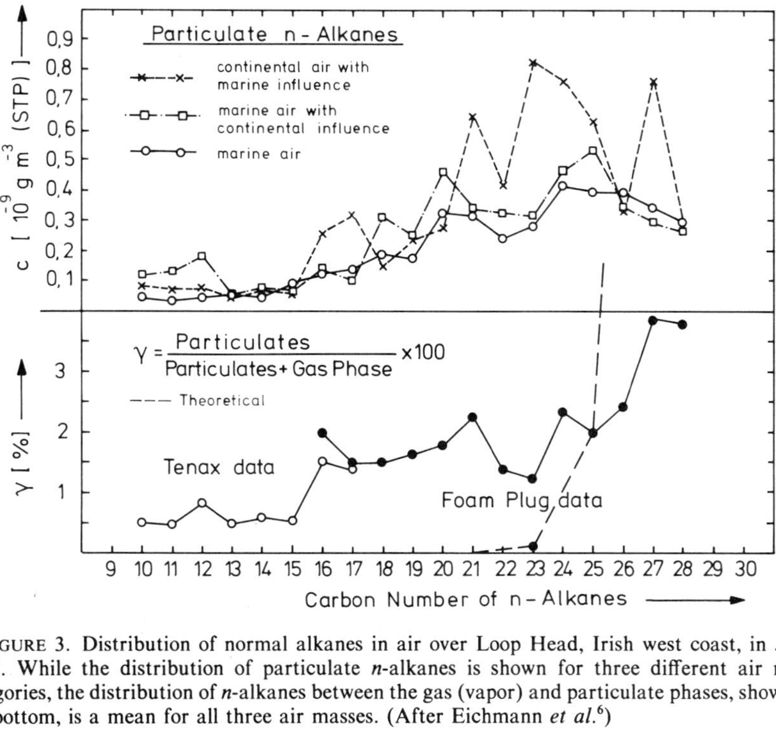

FIGURE 3. Distribution of normal alkanes in air over Loop Head, Irish west coast, in June 1977. While the distribution of particulate n-alkanes is shown for three different air mass categories, the distribution of n-alkanes between the gas (vapor) and particulate phases, shown at the bottom, is a mean for all three air masses. (After Eichmann *et al.*[6])

concentrations of *n*-alkanes were observed in marine air masses. Although the C_{16}, C_{18}, C_{20}, and C_{24} *n*-alkanes appear to be somewhat more abundant than the neighboring odd *n*-alkanes, no significant carbon number predominance exists when the total range of *n*-alkanes from C_{10} to C_{28} is considered. In marine air with some continental (perhaps only coastal) influence (air of this category traveled along the Irish coastline in north–south or south–north directions), the C_{16}, C_{18}, and C_{20} *n*-alkanes are clearly more pronounced than the C_{15}, C_{17}, C_{19}, and C_{21} *n*-alkanes. In all three cases, the concentration of particulate *n*-alkanes increases with increasing carbon number up to about C_{25}.

In the lower part of FIGURE 3, the ratio γ in percent (γ = particulate *n*-alkanes/particulate + vapor phase *n*-alkanes) has been plotted as a function of carbon number. The vapor phase concentrations of *n*-alkanes were measured using two different analytical methods, employing TENAX® adsorption tubes for the collection of *n*-alkanes up to C_{17} and polyurethane foam plugs for the collection of *n*-alkanes larger than C_{15}. For the calculation of γ, means of the concentration of *n*-alkanes in all three air mass categories were used. Although the uncertainty of data is increased by such a calculation, a relationship between the magnitude of γ and the carbon number, i.e., the saturation vapor pressure, is clearly perceptible. The relative proportions of *n*-alkanes in the particulate phase are very small but increase more or less systematically with carbon number, from about 0.5% to about 4%, as the saturation vapor pressure of *n*-alkanes decreases from about 2.5 torr to 4×10^{-8} torr (at 20°C). Although the tendency is the same as Junge's approach[11] predicted, the magnitude of the increase with decreasing saturation vapor pressure is far too small.

For a surface area of aerosol particles of 2×10^{-7} cm^2 per cm^3 (STP) air, which one encounters in very clean air, one would expect the following γ values if only saturation vapor pressure and surface area control the relative proportions of compounds adsorbed from the vapor phase onto particles (dashed curve in FIGURE 3 marked "theoretical"):

n-Alkane	Saturation Vapor Pressure	γ
C_{21}	$\sim 10^{-4}$ torr	0.01%
C_{23}	$\sim 10^{-5}$ torr	0.1%
C_{25}	$\sim 10^{-6}$ torr	2%
C_{27}	$\sim 10^{-7}$ torr	21%
C_{29}	$\sim 10^{-8}$ torr	72%

According to this model, the particle surfaces were supersaturated with *n*-alkanes smaller than C_{25} but were undersaturated with *n*-alkanes larger than C_{25}. Supersaturation is easily explained by irreversible adsorption of the one or two first layers of molecules, by chemisorption, or by a possible underestimation of the actual surface area of particles. Undersaturation, in view of the small surface area assumed in the calculation of γ values, is much harder to understand. One effect that could account for some of the "undersaturation" is the elevation of the vapor pressure of adsorbed compounds due to the curvature of the particles (the small particles in particular). This effect, however, can never account for the factors up to 10 between the observed and the calculated γ values. Thus, we are left with two possibilities; either the model is completely wrong, or chemical reactions restrict the amounts of *n*-alkanes adsorbed onto aerosol particles. The results of a recent study of photolysis rates of chlorofluoromethanes adsorbed onto dust particles, by Ausloos *et al.*,[12] lend support to the second alternative. Unfortunately, there are not enough data, and information on heterogenous reactions in the troposphere is very limited. Nevertheless, this possibility deserves attention because of the important consequences to the cycle of hydrocarbons in the troposphere.

The mean POM concentration in marine and in nonurban continental air may be estimated from POC data by applying a conversion factor. Very few attempts have been made to determine the elemental composition of POM in remote areas. Thus, the magnitude of the conversion factor (f_1) is not well defined. Disregarding organometal complexes, f_1 should be between 1 and 5. A value of 5 would apply if all POM consists of compounds with as many heteroatoms per carbon atom as in, for example, urea (NH_2CONH_2). If sugars or celluloses make up the whole of POM, f_1 would be as high as 2.5. With an elemental composition of POM similar to that of the organic matter in seawater, which was estimated to be $(CH_2O)_{106}(NH_3)_{16}(H_3PO_4)$,[13] f_1 would be close to

TABLE 5
ETHER-EXTRACTABLE PORTION (EEOM) OF TROPOSPHERIC PARTICULATE ORGANIC MATTER

Area	Particle Size Range	Scale Height (m)	Ground Level Concentration ($\mu g/m^3$ STP)	EEOM Mass (10^{12} g $\hat{=}$ Tg)
Marine	$r > 0.2\ \mu m$	1,500	0.30 (0.15–0.60)	0.17 (0.08–0.33)
	$r < 0.2\ \mu m$	8,000	0.30 (0.15–0.60)	0.86 (0.43–1.73)
	all particles	—	0.60 (0.30–1.20)	1.03 (0.51–2.06)
Continental (nonurban)	$r > 0.2\ \mu m$	1,500	0.45 (0.23–0.90)	0.10 (0.05–0.21)
	$r < 0.2\ \mu m$	2,000	0.45 (0.23–0.90)	0.14 (0.07–0.27)
	$r < 0.2\ \mu m$	8,000	0.30 (0.15–0.60)	0.37 (0.18–0.74)
	all particles	—	1.20 (0.60–2.40)	0.61 (0.30–1.22)
Global	—	—	—	1.6 (0.8–3.2)

2.8. Judging from the data on the group composition of the ether-extractable portion of POM, the hydrocarbons appear to constitute a major fraction. Therefore, f_1 should be lower than 2.8. From measurements in atmospheric dust collected on an ice island in the Arctic, Went[1] estimated the elemental composition of the POM to be somewhere between $C_6H_{16}N_2$ and $C_6H_{26}N_2O_5$. Hence, f_1 would be between 1.6 and 2.9. Allowing for the occurrence of some coaly material and soot, particularly in continental areas, one may estimate the conversion factor f_1 to be about 1.5 [e.g., $(CH_2O + CH_2 + C)/3C$] for POM in nonurban continental air, and about 1.8 [e.g., $(CH_2O + CH_2)/2C$] for POM in marine air. With these conversion factors, the mean POM concentration in marine air, 20–100 meters above the sea surface, may be estimated to be about 1 $\mu g/m^3$ STP, and the mean POM concentration in nonurban continental air near ground to be about 3 $\mu g/m^3$ STP.

In summary, one may sketch the following picture of POM in clean air. Clean air POM is a very complex mixture of solvent-extractable and insoluble organic material.

The composition of POM is largely unknown, except for a small fraction of the ether-extractable portion. The major mass of POM is associated with small particles, probably with particles having radii between 0.1 and 1 μm. POM, as reflected by POC data, is relatively equally distributed over the remote areas. POM may be associated to some extent with windblown dust, but does not seem to be associated with sea-salt particles (at least not in appreciable amounts). Regardless of the origin of air masses, the ether-extractable portion of POM exhibits a rather uniform composition with respect to the groups of neutral compounds, organic acids, and organic bases. The group composition of EEOM does not change significantly, even when the air is contaminated. Constituents of POM that are sufficiently volatile interact with the vapor phase. Conversely, appreciable amounts of POM may originate from the vapor phase. This holds particularly for the neutral fraction of the EEOM. The vertical distribution of POM in the troposphere is virtually unknown.

Tropospheric Burden and Budget of Particulate Organic Matter

Finally, we will attempt, despite large uncertainties, to estimate the tropospheric burden of EEOM, POC, and POM and to consider the tropospheric budget of POC.

If one assumes a vertical distribution of particulate organic matter similar to the vertical distribution of aerosol particles as given by Jaenicke,[14] the tropospheric burden of EEOM may be estimated as shown in Table 5. It was estimated above that the mean EEOM concentration in marine air near the sea surface is about 0.6 $\mu g/m^3$ STP; and in nonurban continental air, including man-made injections, about 1.2 $\mu g/m^3$ STP. The estimates may be considered correct within a factor of 2.

According to the vertical distribution of aerosol particles by Jaenicke, one may define somewhat arbitrarily four categories of aerosol particles:

1. Marine aerosol particles with radii >0.2 μm and a scale height of 1,500 meters;

2. Nonurban continental aerosol particles with radii >0.2 μm and a scale height of 1,500 meters;

Table 6

Mass of Tropospheric Particulate Organic Carbon (POC)

Area	Particle Size Range	Scale Height (m)	Ground Level Concentration ($\mu g/m^3$ STP)	POC Mass (10^{12} g ≙ Tg)
Marine	$r > 0.2$ μm	1,500	0.20 (0.10–0.40)	0.11 (0.05–0.22)
	$r < 0.2$ μm	8,000	0.40 (0.20–0.80)	1.15 (0.57–2.30)
	all particles	—	0.60 (0.30–1.20)	1.26 (0.62–2.52)
Continental (nonurban)	$r > 0.2$ μm	1,500	0.80 (0.40–1.60)	0.18 (0.09–0.36)
	$r < 0.2$ μm	2,000	0.80 (0.40–1.60)	0.24 (0.12–0.48)
	$r < 0.2$ μm	8,000	0.40 (0.20–0.80)	0.48 (0.24–0.96)
	all particles	—	2.0 (1.0–4.0)	0.90 (0.45–1.80)
Global	—	—	—	2.2 (1.1–4.3)

TABLE 7

MASSES OF TROPOSPHERIC PARTICULATE ORGANIC CARBON (POC), PARTICULATE ORGANIC MATTER (POM), AND ETHER-EXTRACTABLE ORGANIC MATTER (EEOM)

Area	Particle Size Range	EEOM Mass (10^{12} g ≙ Tg)	POC Mass (10^{12} g ≙ Tg)	f_1 = POM/POC	POM Mass (10^{12} g ≙ Tg)
Marine	$r > 0.2$ μm	0.17 (0.08–0.33)	0.11 (0.05–0.22)	2.5 (CH_2O/C)	0.28 (0.14–0.55)
	$r < 0.2$ μm	0.86 (0.43–1.73)	1.15 (0.57–2.30)	1.2 (CH_2/C)	1.38 (0.68–2.76)
	all particles	1.03 (0.51–2.06)	1.26 (0.62–2.52)	—	1.66 (0.82–3.31)
Continental (nonurban)	$r > 0.2$ μm	0.10 (0.05–0.21)	0.18 (0.09–0.36)	2.0 [$(2CH_2O + C)/3C$]	0.36 (0.18–0.72)
	$r < 0.2$ μm	0.14 (0.07–0.27)	0.24 (0.12–0.48)	2.0 [$(2CH_2O + C)/3C$]	0.48 (0.24–0.96)
	$r < 0.2$ μm	0.37 (0.18–0.74)	0.48 (0.24–0.96)	1.2 (CH_2/C)	0.58 (0.29–1.15)
	all particles	0.61 (0.30–1.22)	0.90 (0.45–1.80)	—	1.42 (0.71–2.83)
Global	—	1.6 (0.8–3.2)	2.2 (1.1–4.3)	(1.4)	3.1 (1.5–6.2)

3. Nonurban continental aerosol particles with radii <0.2 μm and a scale height of 2,000 meters;

4. Background aerosol particles (over both marine and continental areas) with radii <0.2 μm and a scale height of 8,000 meters.

Since there are not sufficient data on the particle size distribution of EEOM, the estimated mean EEOM concentration may be more or less equally distributed over the different categories of particles, as shown in the third column of TABLE 5. Considering the extension of the marine and continental areas, EEOM masses result as given in the fourth column of the table. Summation of the different masses yields a global tropospheric EEOM mass of 1.6 teragrams (Tg) with a range of uncertainty of 0.8–3.2 Tg.

The tropospheric burden of POC may be estimated in a similar way, as shown in TABLE 6. The estimated ground level POC concentrations of 0.6 $\mu g/m^3$ STP air in marine areas and 2.0 $\mu g/m^3$ STP air in nonurban continental areas, including man-made injections, have been split up for the different categories of aerosol particles, considering the data on the particle size distribution of POC by Hoffmann and Duce.[9] For continental areas, it was assumed that the relative proportion of POC on the large particles is somewhat greater than in marine areas. Summation of the different POC masses, in the fourth column of TABLE 6, yields a global tropospheric POC mass of 2.2 Tg with a range of uncertainty of 1.1–4.3 Tg. The value for the global tropospheric POC mass is exactly the same as the estimate by Duce,[2] which was obtained in a somewhat different way. This agreement of better than a factor of 2 is, of course, fortuitous.

In TABLE 7, the estimates of TABLES 5 and 6 are summarized, together with POC-mass-derived estimates for the tropospheric POM burden. The values of the POM mass were obtained using conversion factors (f_1 = POM/POC) from 1.2 to 2.5, which consider a possible variation of the elemental composition of POM with large-scale sources, as indicated in the fourth column of TABLE 7. The value obtained for the global tropospheric POM mass is 3.1 Tg with a range of uncertainty of 1.5–6.2 Tg.

The tropospheric budget of POC may be estimated as shown in FIGURE 4. The various fluxes are given in teragrams POC per year. The oceans, soil and rock, forest and bush fires, and vegetation have been identified as (primary) large-scale sources of POM. The magnitude of the source strengths is not well known. Because of the large uncertainties, no distinction has been made in this budget between large and small particle POC. The numbers given for the various input rates of POC are based on an assessment by Duce.[2] The range of uncertainty has been assumed to be one order of magnitude for the natural sources. This might be somewhat optimistic, in view of the limited information available on these sources. The number for the man-made input rate of POC was estimated by Duce for the 1973–1974 period. Increased emission controls may lower this number. The strength of the total source (or sink) of POC has

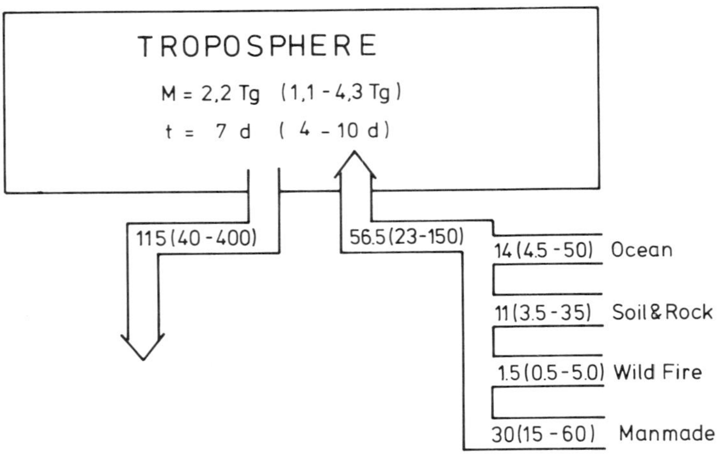

FIGURE 4. Budget of airborne particulate organic carbon (POC) for the global troposphere. (If not marked otherwise, all values are in 10^{12} g (Tg) per year.)

been obtained from the estimates of the tropospheric POC burden and the mean tropospheric residence time of aerosol particles. The generally accepted mean tropospheric aerosol residence time is 4–7 days.[15–19] Since, as we discussed earlier, the major mass of POC is apparently associated with small particles, the estimate has been shifted to somewhat higher values, assuming that the mean residence time of POC in the troposphere is 7 days. With this number, the total source strength is found to be 115 Tg POC/yr (160 Tg POM/yr).

In other words, there is a discrepancy of about 60 Tg/yr to the sum of the POC input rates of the marine, terrestrial, and man-made sources. The discrepancy is very likely due to the fact that a major (secondary) source of POC has not been considered. As was discussed by Duce,[2] this source may be found in gas-to-particle conversion processes in the troposphere. If this is true, gas-to-particle conversion must be considered to be the main source of tropospheric POC and, thus, of tropospheric particulate organic matter. This would explain the observation of (1) a marked concentration of POC in the small particle range; (2) a rather uniform level of near-surface POC over the oceans; (3) a uniform level of EEOM in marine air and in continental air in remote areas; and (4) a rather uniform group composition of EEOM, regardless of the sampling location.

SUMMARY

Particulate organic matter (POM) is found to vary around 1 $\mu g/m^3$ STP in marine air and around 3 $\mu g/m^3$ in nonurban continental air near ground. On the average, the ether-extractable fraction of POM accounts for about 0.6 $\mu g/m^3$ STP in marine air and for about 1 $\mu g/m^3$ STP in nonurban continental air. Although the ether-extractable organic matter (EEOM) is apparently a complex mixture of hydrocarbons, alcohols, acids, esters, organic bases, and other polar organic compounds, the composition of EEOM in terms of neutral compounds (60%), organic acids (30%), and organic bases (10%) remains remarkably constant in ground level air. Large particles appear to contain less organic matter than small particles. The tropospheric POM mass may be estimated to be 1.5–6.2 Tg (Tg = 10^{12} g). The total global POM production may be as high as 550 Tg/yr, with a most likely value of 160 Tg/yr. Gas-to-particle conversion appears to be the main source of POM.

ACKNOWLEDGMENTS

I would like to thank Dr. G. Ketseridis and Dr. R. Jaenicke for valuable discussions and suggestions.

REFERENCES

1. WENT, F. W. 1960. Organic matter in the atmosphere and its possible relation to petroleum formation. Proc. Nat. Acad. Sci. U.S.A. **46:** 212.
2. DUCE, R. A. 1978. Speculations on the budget of particulate and vapor phase non-methane organic carbon in the global troposphere. Pure Appl. Geophys. **116:** 244.
3. KETSERIDIS, G., J. HAHN, R. JAENICKE & C. JUNGE. 1976. The organic constituents of atmospheric particulate matter. Atmos. Environ. **10:** 603.
4. KETSERIDIS, G. & R. JAENICKE. 1978. Organische Beimengungen in atmosphärischer Reinluft: ein Beitrag zur Budget-Abschätzung. *In* Organische Verunreinigungen in der Umwelt. K. Aurand, H. Hässelbarth, E. Lahmann, G. Müller & W. Niemitz, Eds.: 379–390. Erich Schmidt Verlag. Berlin, Federal Republic of Germany.
5. KETSERIDIS, G. & R. EICHMANN. 1978. Organic compounds in aerosol samples. Pure Appl. Geophys. **116:** 274.
6. EICHMANN, R., P. NEULING, G. KETSERIDIS, J. HAHN, R. JAENICKE & C. JUNGE. 1979. *n*-Alkane studies in the troposphere. I. Gas and particulate concentrations in North Atlantic air. Atmos. Environ. **113:** 587.
7. EICHMANN, R., G. KETSERIDIS, G. SCHEBESKE, R. JAENICKE, J. HAHN, P. WARNECK & C. JUNGE. *n*-Alkane studies in the troposphere. II. Gas and particulate concentrations in marine air from the Indian ocean and from Antarctic regions. (In press.)
8. HOFFMAN, E. J. & R. A. DUCE. 1974. The organic carbon content of marine aerosols collected in Bermuda. J. Geophys. Res. **79:** 4474.

9. HOFFMAN, E. J. & R. A. DUCE. 1977. Organic carbon in the marine atmospheric particulate matter: concentration and particle size distribution. Geophys. Res. Lett. **4:** 449.
10. JAENICKE, R. 1978. Physical properties of atmospheric particulate sulfur compounds. Atmos. Environ. **12:** 161.
11. JUNGE. C. 1977. Basic considerations about trace constituents in the atmosphere as related to the fate of global pollutants. Adv. Environ. Sci. Technol. **8**(1): 7.
12. AUSLOOS, P., R. E. REBBERT & L. GLASGOW. 1977. Photodecomposition of chloromethanes adsorbed on silica surfaces. J. Res. Nat. Bur. Stand. **82:** 1.
13. RICHARDS, F. A. 1965. Anoxic basins and fjords. *In* Chemical Oceanography. J. P. Riley & G. Skirrow, Eds. **1:** 623–625. Academic Press, Inc. London and New York.
14. JAENICKE, R. 1980. Natural aerosols. Ann. N.Y. Acad. Sci. (This volume.)
15. MOORE, H. E., E. A. MARTELL & S. E. POET. 1973. ^{222}Ru, ^{210}Pb, ^{210}Bi and ^{210}Po profiles and aerosol residence times versus altitude. J. Geophys. Res. **78:** 7065.
16. MOORE, H. E., E. A. MARTELL & S. E. POET. 1976. Sources of ^{210}Po in the atmosphere. Environ. Sci. Technol. **10:** 586.
17. MARTELL, E. A. & H. E. MOORE. 1974. Tropospheric aerosol residence times: a critical review. J. Rech. Atmos. **8:** 903.
18. TUREKIAN, K. K., J. NOZAKI & L. K. BENNINGER. 1977. Geochemistry of atmospheric radon and radon products. *In* Annual Review of Earth and Planetary Sciences. F. A. Donath, F. G. Stehli & G. W. Wetherill, Eds. **5:** 227–255. Annual Reviews, Inc. Palo Alto, Calif.
19. JAENICKE, R. 1978. Über die Dynamik atmosphärischer Aitkenteilchen. Ber. Bunsenges. Phys. Chem. **82.** (In press.)
20. BARGER, W. R. & W. D. GARRETT. 1976. Surface active organic material in the marine atmosphere. J. Geophys. Res. **81:** 3151.
21. U.S. Public Health Service. 1968. Air quality data from the National Air Surveillance Networks, 1966. National Air Pollution Control Association. Durham, N.C.
22. HIDY, G. M., P. K. MUELLER, H. H. WANG, J. KARNEY, S. TWISS, M. IMADA & A. ALCACER. 1974. Observations of aerosols over southern California coastal waters. J. Appl. Meteorol. **13:** 96.

NATURAL AEROSOLS: GENERAL DISCUSSION*

Ruprecht Jaenicke, *Moderator*

Max-Planck-Institut für Chemie
D-6500 Mainz, Federal Republic of Germany

K. T. WHITBY (*University of Minnesota, Minneapolis, Minn.*): Dr. Jaenicke, you made the observation that the nuclei mode, or the Aitken-nuclei-size particles, could come from either sulfuric or organic species. Do you have any more comments about the evidence for the amount of this in the background aerosol that comes from sulfuric versus organic? I believe that there is evidence that sulfur species contribute to it; but what about organic species?

R. JAENICKE: We have some clues about that. One is that we separated the Aitken mass from the total aerosol and determined the chemistry to a certain extent. We saw sulfur with the x-ray fluorescence method and also with neutron activation. But in the case of the organics, we had only two clues: one was obtained by Duce and Hoffman, who showed that the organic fraction of the aerosol increases as the particles become smaller and smaller; and the second was that we determined the presence of organic carbon in Aitken particles. But it is still a puzzle, because we don't know what kinds of compounds are present. We only can say that there must be something that contains sulfur and organic carbon. But there can't possibly be ammonium sulfate, because that is too heavy. If you convert the measured sulfur into ammonium sulfate, there is much more than we have measured as the total mass of the Aitken particle.

T. E. GRAEDEL (*Bell Laboratories, Murray Hill, N.J.*): Dr. Blanchard, you quoted a figure for the mass of marine aerosol generated by bubble bursting. Can you combine this with estimates of the organics in the surface layer, to estimate how much of that might be organic in nature?

D. C. BLANCHARD (*Atmospheric Sciences Research Center, State University of New York, Albany, N.Y.*): The surface layer? Do you mean the droplet surface or on the sea? I suppose you can, but I think that's the wrong way to go about it. I think it's more important to look at what is on the surface of the bubble while it is scavenging up through the surface of the sea. Eva Hoffman at the University of Rhode Island has done some experiments, which have been published in JGR [*Journal of Geophysical Research*], showing extremely high fractionations on the jet drops and sea-salt particles.

L. BARRIE (*Atmospheric Environment Service, Downsview, Ont., Canada*): Just a question, Dr. Blanchard, concerning that salt inversion below cloud base. Could that also be explained by convergence of the air below an individual convective cloud? Do you observe this as a continuance feature, even between convective clouds?

D. C. BLANCHARD: We just stumbled on the salt inversion about a month or two ago. It came while I was analyzing a lot of these data, and I haven't thought too much about it. I don't know what to say about your suggestion, but I know that the inversion shows up in the analysis of the Atlantic data. On the Pacific data, it seems to be a real feature. It doesn't show up in the data obtained near the continents, where we obviously don't have any steady state between the atmosphere and the ocean.

S. K. FRIEDLANDER (*University of California, Los Angeles, Calif.*): Have you or Dr. Jaenicke done a sodium balance? That is, you have an estimate of the sodium

*This is a compilation of the discussions held after the presentation of each of the papers in this section.

being emitted from the surface. And presumably there are data on sodium that come out of the information you have on natural background aerosols. The two should match up in terms of the total mass for sodium.

D. C. BLANCHARD: No, not really; though I suppose you can. It is difficult right now to try to make estimates of these fluxes between the earth and the atmosphere. As Dr. Jaenicke mentioned in his paper, it depends upon what level you are talking about. But there have been estimates that range over at least an order of magnitude.

P. K. MUELLER (*Environmental Research & Technology, Inc., Westlake Village, Calif.*): Dr. Hahn, do you have any information on graphitic carbon content on your remote carbonaceous material samples?

J. HAHN (*Max-Planck-Institut für Chemie, Mainz, Federal Republic of Germany*): We have just recently started to measure TOC [total organic carbon] and organic matter, so it is a bit early to say something about that. However, from the available data, it looks as if we have some elemental carbon even in remote areas. Over the oceans, if you are getting out of the coastal regions, it seems that there is no appreciable amount of elemental carbon. However, I don't want to be pinned down on these things because it is actually too early to say.

S. N. TSOUKALAS (*U.S. Forest Service, Macon, Ga.*): Dr. Hahn, could you briefly describe the methods used for separation of the extract into the various chemical classes that you mentioned.

J. HAHN: We did it by classical procedures.

S. N. TSOUKALAS: You mean as a base extraction?

J. HAHN: Right.

S. N. TSOUKALAS: Do you find, after base extraction, a large precipitate that, although it is insoluble in the ether, resulted from soluble material in the ether?

J. HAHN: No, we have never seen that.

S. N. TSOUKALAS: How about the neutral compounds; how did you separate them?

J. HAHN: We do it in two ways. We use a neutral fraction and analyze this further by gas chromotography [GC], as we do also for the paraffins. We also use the GC method and thin-layer chromatography method.

S. N. TSOUKALAS: Thin-layer? But this fraction contains hundreds of compounds; I don't know how thin-layer will be effective.

J. HAHN: Well, we separate the neutral fraction into one fraction that is the aliphatic portion, one fraction that is the aromatic portion, and one fraction that is the polar compounds. You can do this by two-dimensional thin-layer chromatography. Then we take those fractions and analyze them further, either again on thin-layer plates for the polar cyclic aromatics, or with the GC columns.

UNIDENTIFIED SPEAKER: Dr. Gillette, you showed, on one of the last slides, that there was a ratio of roughly 1 to 100,000 between the mass fluxes in the vertical and those in the horizontal. Could you comment on that, especially in relation to the picture I showed about the production as a function of altitude? That means if you go to higher altitude, then of course the production rate is smaller, because not that much is penetrating vertically.

D. GILLETTE (*National Center for Atmospheric Research, Boulder, Colo.*): In order to dimensionally make the two quantities similar, I divided the quantity cubed by the height of the collecting slit, which was 30 centimeters. So actually we could multiply that ratio by 30 to get a better number.

MESOSCALE AND SYNOPTIC SCALE TRANSPORT OF AEROSOLS

George T. Wolff

Environmental Science Department
General Motors Research Laboratories
Warren, Michigan 48090

INTRODUCTION

The study of air pollution focuses on three elements: the emission of a contaminant; the transport (and possible transformation en route) of the contaminant to a receptor; and the production of an adverse effect of the contaminant on the receptor. Until the last two decades, our perception of transport was almost exclusively confined to the relatively small area in the proximity of the source. In recent years, however, an increasing amount of evidence suggesting possible ecological effects of anthropogenic effluents far removed from their sources has been uncovered. Consequently, the scope of pollutant transport has been extended from relatively small areas (microscale) to areas hundreds and thousands of kilometers (mesoscale and synoptic scale) away from the effluent source.

The purposes of this paper are to present an overview of mesoscale and synoptic scale transport of aerosols, to identify the characteristics of aerosols that are transported, and to discuss which aerosol species possess these characteristics. Before we begin, however, it is necessary to define the different scales of transport.

SCALES OF POLLUTANT TRANSPORT

The scale of transport is defined by the distance over which the pollutants travel in concentrations that are identifiable above background concentrations. The four scales are listed in TABLE 1. For pollutants that persist long enough to raise the global background, global transport is important. Within certain weather systems (in high or low pressure systems), where particular contaminants accumulate and are transported, synoptic scale transport is important. In cases where urban emissions form a plume that is discernible downwind for a hundred kilometers or so, mesoscale transport is occurring.

Microscale, mesoscale, and synoptic scale transport of pollutants occur within the boundary, or mixing, layer, which is generally less than 3000 m thick.

MESOSCALE TRANSPORT

Meteorologists define a mesoscale weather system as one that is ten to a few hundred kilometers in size.[1] Within this scale, transport is associated with two types of phenomena: the urban plume and mesoscale weather systems. Composite plumes originating from urban areas, or from other large area sources, and individual plumes from extremely large point sources have been observed to travel tens to several

0077-8923/80/0338-0379 $01.75/1

TABLE 1
TRANSPORT REGIMES

Type of Transport	Distance (km)
1. microscale	< 10
2. mesoscale	10–several hundred
3. synoptic or macroscale	several hundred to several thousand
4. global	greater than several thousand

hundred kilometers. Since this phenomenon will be discussed by others,[2] it will not be discussed further here.

Transport of pollutants within mesoscale weather systems, such as sea-land breezes, mountain-valley breezes, and urban-heat-island systems, has been observed to occur. The transport of particulate matter in the Chicago lake breeze has been documented by Lyons and Olsson.[3] A typical lake-sea breeze is depicted in FIGURE 1. Due to the thermal gradient between the land and the body of water under periods of light synoptic pressure gradients, a sea breeze develops during the daytime hours and diminishes around sunset. Overnight, a lighter reverse circulation may develop. Pollutants emitted during the day near the coast travel inland, are lifted, return aloft, descend over the water, and travel back inland.

The most prominent example of particulate transport via a mountain-valley breeze is the Denver situation.[4–7] As illustrated in FIGURE 2, Denver is situated in a valley. The typical diurnal scenario is as follows. Overnight, the heavy colder air from the mountain tops flows down into the valley. Within the valley, there is a gentle decreasing slope to the northeast, which results in a weak flow to the northeast. Because of the net convergence of cooler air in Denver overnight, an intense inversion develops; the typical heights are shown in FIGURE 2. Pollutants, including particulates, accumulate under these conditions, and by morning, a distinct "brown cloud" is

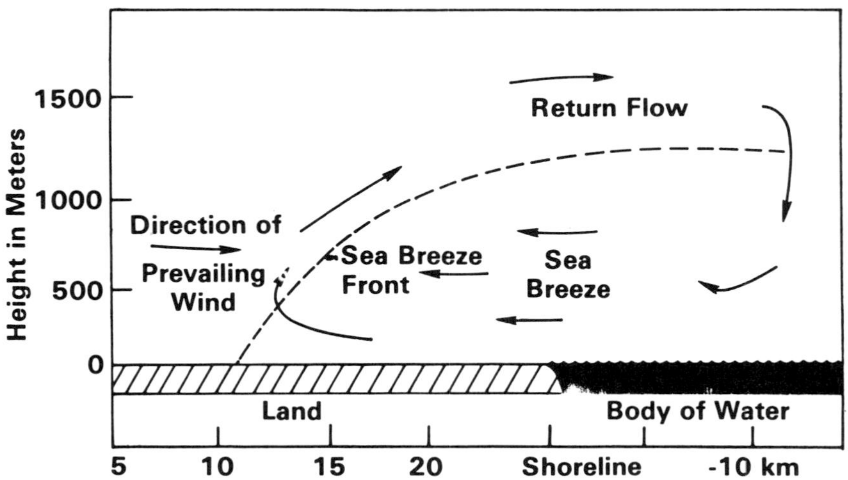

FIGURE 1. Typical sea or lake breeze circulation.

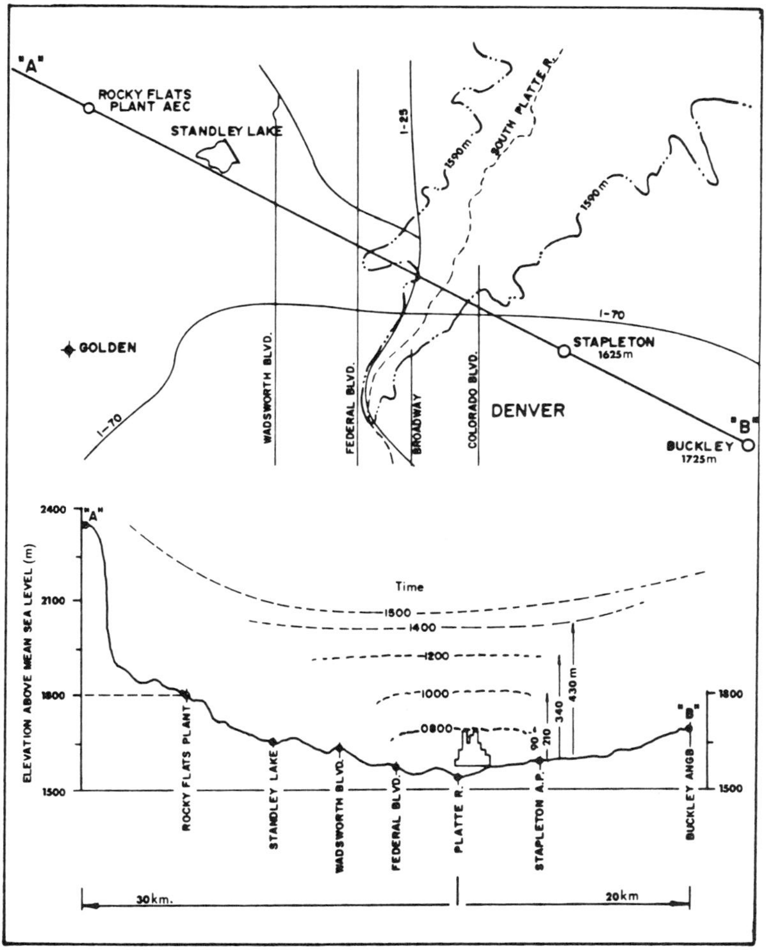

FIGURE 2. The Denver area's topography is conducive to the development of the mountain-valley circulation. Typical mixing depths at different times of the day are also shown (from Crow[4]).

situated over northeast Denver (see FIGURE 3a). At about 11:00 A.M., the mountain-valley flow reverses as the mountain peaks are heated, and the cloud travels southwestward (FIGURE 3b). In midafternoon, the mountain peaks begin to cool, and the downslope flow again occurs around 4:00 P.M. If daytime ventilation is not sufficient to disperse the cloud, it travels back to the northeast and intensifies overnight. Incidentally, the "brown cloud" is thought to be due to particulates and is currently the subject of an intensive investigation conducted by General Motors Research Laboratories and the Motor Vehicle Manufacturers' Association.

Denver also provides an example of the urban-heat-island circulation.[4,5] The convergent flow into Denver, in combination with the urban-heat-island effect, has

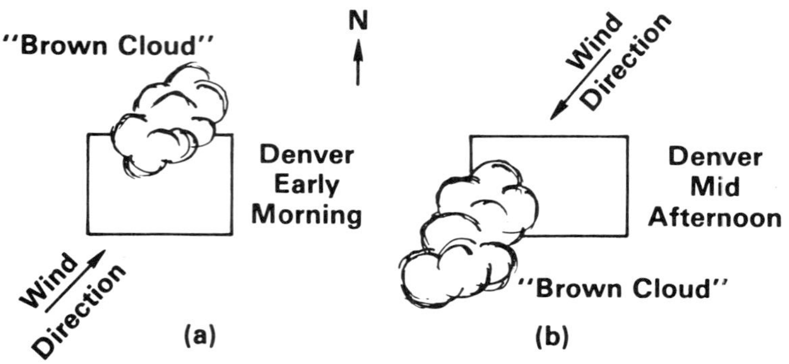

FIGURE 3. Position of the Denver "brown cloud" in the morning (a) and afternoon (b) on a typical day.

resulted in a "stove-pipe effect" that results in the transport of pollutants aloft to the northeast of Denver. This is depicted in FIGURE 4.

SYNOPTIC, OR MACROSCALE, TRANSPORT

The best examples of synoptic, or macroscale, systems are the migrating high and low pressure systems that are on the order of several thousand kilometers in size. Transport of material within such systems has been recognized for some time by scientists studying the fate of radioactive debris, spores, pollen, seeds, and insects.[8] In the 1970s, synoptic scale transport of relatively high concentrations of ozone,[9,10] sulfates,[11–19] and total suspended particulates[9,11,12,15,16,18,19] was identified in North America and in Europe.

The reason for the recent interest in synoptic scale transport is twofold. First, it is only in the 1970s that extensive monitoring of rural areas has been undertaken. The measured levels of some contaminants in rural areas could not be explained by local emissions or by mesoscale transport. Second, due to effective control strategies adopted in many urban areas for sulfur oxides and particulate emissions, concentra-

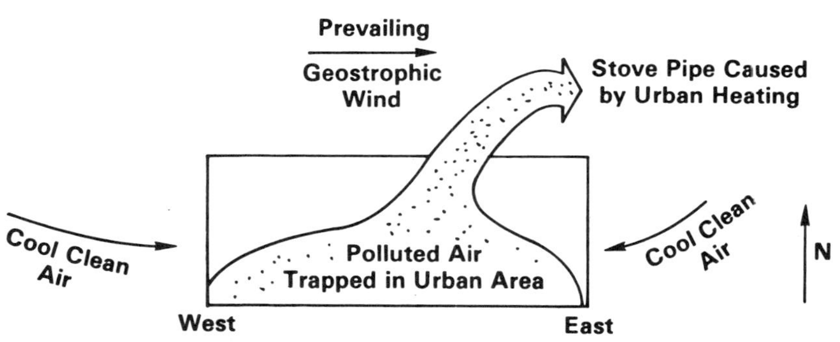

FIGURE 4. An example of the "stove-pipe" effect (from Crow[4]).

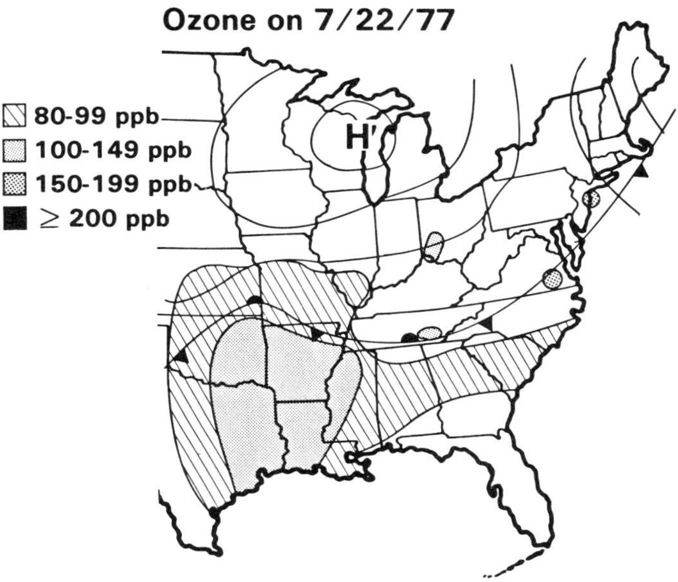

FIGURE 5a. Daily maximum 1-hour ozone concentrations on July 22, 1977.

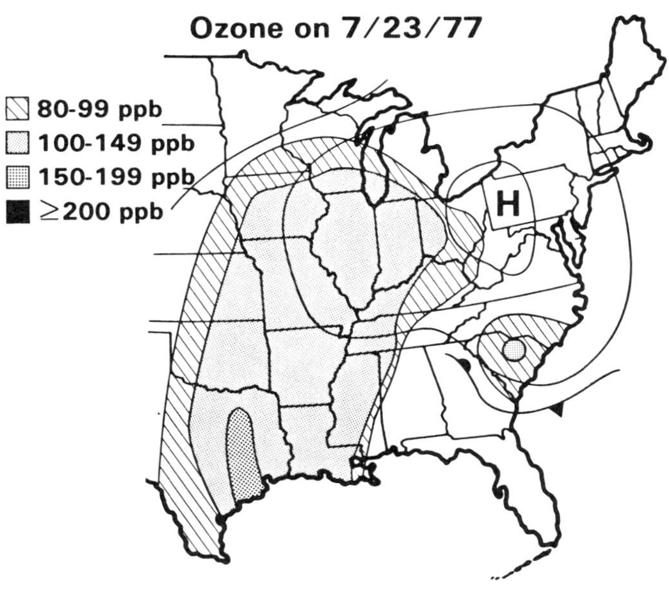

FIGURE 5b. Daily maximum 1-hour ozone concentrations on July 23, 1977.

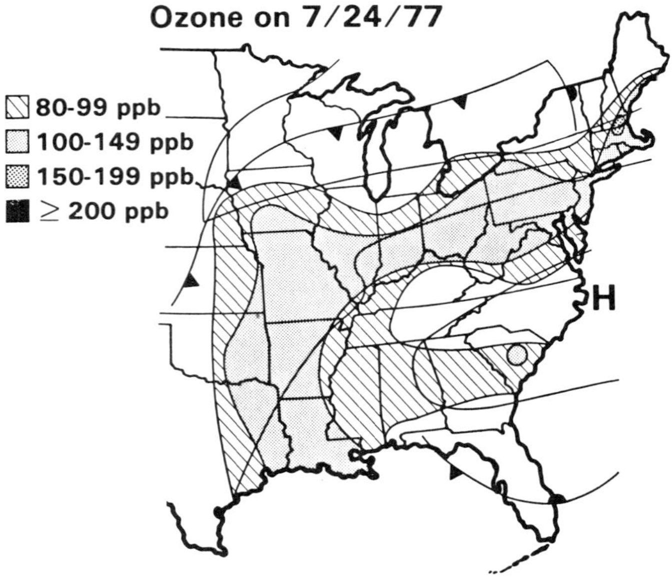

FIGURE 5c. Daily maximum 1-hour ozone concentrations on July 24, 1977.

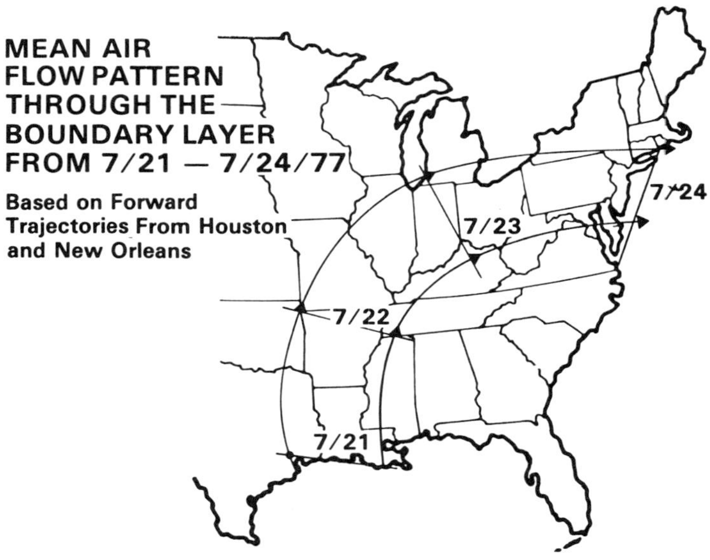

FIGURE 5d. Daily maximum 1-hour ozone concentrations on air parcel trajectories for the period July 21–24, 1977.

tions of these pollutants have been reduced to levels where the background, or transported, contribution has become significant and identifiable. Even in an urban area as large as New York City, 40% of the total suspended particulate and 70% of the sulfate observed at certain times have been attributed to synoptic scale transport.[16]

Synoptic scale transport can be associated with either high pressure or low pressure systems. Pollution episodes are generally associated with transport in high pressure (or fair weather) systems, which are typically several thousand kilometers in diameter. A dramatic example of synoptic scale ozone transport is shown in FIGURES 5a–5d. The circulation around a high pressure system is clockwise. FIGURES 5a–5c demonstrate that the air over Texas and Louisiana, containing high concentrations of ozone, was advected into the New York area in just three days. FIGURE 5d shows the calculated air parcel trajectories, based on Heffter's model.[20]

AEROSOL CHARACTERISTICS

Names	Transient Nuclei	Accumulation Range	Large Particles
Size (μm) .001	.01	0.1 1.0	10 100
		Haze Reduces Visibility Contains Bulk of Aerosol Mass	
SOURCES	Combustion Nucleation	Combustion Condensation Coagulation Chemical Reactions	Mechanical Processes Wind Blown Dust Sea Spray Buffalo Effect
FATE	Rapidly coagulate heterogeneously with accumulation mode particles	Stable for days	Settle Out
	Less than 1 hour	Days	Hours → Minutes

FIGURE 6. Some important aerosol characteristics (adapted, in part, from Willeke & Whitby[23]).

Acid precipitation is an example of synoptic scale transport associated with low pressure systems. The precipitation acidity is dependent upon the trajectory of the air parcels advected into the circulation around the low.[21] The acidity is primarily due to transported sulfuric and nitric acid.[22]

AEROSOL PROPERTIES RELATING TO TRANSPORT

In order for particles to be transported a long distance, they must be stable enough to remain in the atmosphere for many hours or days. Aerosol characteristics important in transport are illustrated in FIGURE 6. In general, the particles that are most important in long range transport within the boundary layer have aerodynamic diameters between 0.1 μm and 3.0 μm. This size range is known as the accumulation range,[23] and particles within this range have lifetimes in the atmosphere on the order of days. Particles in the accumulation range are the principal visibility-reducing particles. This visibility-affecting property has been utilized in tracking the synoptic scale movement of hazy air masses.[9,15,16,19]

TABLE 2 lists some of the particulate species commonly found in the accumulation mode in the ambient air.

TABLE 2
CHEMICAL SPECIES THAT OCCUR PRIMARILY IN AEROSOLS IN THE ACCUMULATION MODE

Species	Reference
sulfate	24
carbon	25
Pb	26
Zn	26
Ti	26
Hg	26
V	26
Mn	26
Ni	26
Cd	26
As	26

SUMMARY

Recent air pollution field studies have suggested the importance of mesoscale and synoptic scale transport of pollutants to both urban and rural air quality. Mesoscale transport occurs within the boundary layer for distances of between ten and a few hundred kilometers. Examples of mesoscale transport systems include: the urban plume, the sea breeze, the mountain-valley wind cycle, and the urban-heat-island circulation. Transport of aerosols in each of these systems has been observed.

Synoptic scale transport is on the order of a few hundred to a few thousand kilometers and is generally associated with high or low pressure systems. There have been many documented cases of aerosol accumulation and transport associated with high pressure systems. Most acid precipitation events are manifestations of synoptic scale transport associated with low pressure systems.

Particles in the 0.1–3 μm diameter size range, which is known as the accumulation mode, have lifetimes in the atmosphere on the order of days. For this reason, these particles are the most important in terms of synoptic and mesoscale transport. They are also the particles primarily responsible for visible haze.

REFERENCES

1. SLADE, D. H., Ed. 1968. Meteorology and Atomic Energy. U.S. Atomic Energy Commission. Washington, D.C.

2. SCHIERMEIER, F. A. 1980. A review of urban plume studies. Ann. N.Y. Acad. Sci. (This volume.)
3. LYONS, W. A. & L. E. OLSSON. 1972. Mesoscale air pollution transport in the Chicago lake breeze. J. Air Pollut. Control Assoc. **22**(11): 876.
4. CROW, L. W. 1976. Airflow study related to EPA field monitoring program. Denver Metropolitan Area, November 1973. *In* Denver Air Pollution Study 1973. Publication No. EPA-600/9-76-007 **1:** 3. U.S. Environmental Protection Agency. Research Triangle Park, N.C.
5. FERMAN, M. A., R. S. EISINGER & P. R. MONSON. 1977. Characterization of Denver air quality. *In* Denver Air Pollution Study 1973. Publication No. EPA-600/9-77-001 **2:** 57. U.S. Environmental Protection Agency. Research Triangle Park, N.C.
6. KLIMISCH, R. L., M. A. FERMAN, P. R. MONSON & G. T. WOLFF. 1978. Air Pollution in Denver. Publication No. GMR-2668. General Motors Research Laboratories. Warren, Mich.
7. RIEHL, H. & D. HERKHOF. 1972. Some aspects of Denver meteorology. J. Appl. Meteorol. **11:** 1040.
8. PASQUILL, F. 1974. Atmospheric Diffusion. 2nd edit. Ellis Horwood Ltd. Chichester, England.
9. WOLFF, G. T., P. J. LIOY, G. D. WIGHT, R. E. MEYERS & R. T. CEDERWALL. 1977. An investigation of long-range transport of O_3 across the midwestern and eastern U.S. Atmos. Environ. **11:** 797.
10. CHUNG, Y-S. 1977. Ground-level ozone and regional transport of air pollutants. J. Appl. Meteorol. **16:** 1127.
11. BROSSET, C., K. ANDREASSON & M. FERM. 1975. The nature and possible origin of acid particles observed at the Swedish West Coast. Atmos. Environ. **9:** 631.
12. BROSSET, C. 1976. Airborne particles: black and white episodes. Ambio **5:** 157.
13. OECD. 1977. Long range transport of air pollutants. Organization for Economic Cooperation and Development. Paris, France.
14. LAFLEUR, R. J. & D. M. WHELPDALE. 1977. Spatial distribution of sulfates over eastern Canada during August 1976. Paper No. 77-48.3. *In* Proceedings of the 70th Annual Meeting of the Air Pollution Control Association, Toronto, Canada, June 1977.
15. HUSAR, R. B., N. V. GILLANI, J. D. HUSAR, C. C. PALEY & P. N. TURCU. 1976. Long range transport of pollutants observed through visibility contour maps, weather maps and trajectory analysis. *In* Proceedings, 3rd Symposium on Atmospheric Turbulence, Diffusion and Air Quality, American Meteorological Society, Raleigh, N.C., October 1976: 344.
16. WOLFF, G. T., P. J. LIOY, B. P. LEADERER, D. M. BERNSTEIN & M. T. KLEINMAN. 1979. Characterization of aerosols upwind of New York City. I. Transport. Ann. N.Y. Acad Sci. **322:** 57.
17. HIDY, G. M., P. K. MUELLER & E. Y. TONG. 1978. Spatial and temporal distributions of airborne sulfate in parts of the U.S. Atmos. Environ. **11:** 879.
18. SMITH, F. B. & R. D. HUNT. 1978. Meteorological aspects of the transport of pollution over long distances. Atmos. Environ. **12:** 461.
19. LYONS, W. A., J. C. DOOLEY & K. T. WHITBY. 1978. Satellite detection of long-range pollution transport and sulfate aerosol hazes. Atmos. Environ. **12:** 621.
20. HEFFTER, J. L. & A. D. TAYLOR. 1975. A regional-continental scale transport, diffusion and deposition model. NOAA Technical Memorandum No. ERL ARL 50. U.S. Department of Commerce. Silver Springs, Md.
21. WOLFF, G. T., P. J. LIOY, H. GOLUB & J. S. HAWKINS. 1979. Acid precipitation in the New York metropolitian area: its relationship to meteorological factors. Environ. Sci. Technol. **13:** 209.
22. COGBILL, C. V. & G. E. LIKENS. 1974. Acid precipitation in the northeastern U.S. Water Resour. Res. **10:** 1133.
23. WILLEKE, K. & K. T. WHITBY. 1974. Atmospheric aerosols: size distribution interpretation. J. Air Pollut. Control Assoc. **25:** 529.
24. TANNER, R. L., R. GARBER, W. MARLOW, B. P. LEADERER & M. A. LEYKO. 1979. Chemical composition of sulfate as a function of particle size in New York summer aerosol. Ann. N.Y. Acad. Sci. **322:** 99.

25. MUELLER, P. K., R. W. MOSLEY & L. B. PIERCE. 1972. Carbon in Pasadena aerosol. *In* Aerosols and Atmospheric Chemistry. G. M. Hidy, Ed.: 295. Academic Press, Inc. New York, N.Y.
26. BERNSTEIN, D. M. & K. A. RAHN. 1979. New York Summer Aerosol Study: trace element concentrations as a function of particle size. Ann. N.Y. Acad. Sci. **322:** 87.

TYPES OF TRANSPORT EPISODES IN NORTHERN EUROPE

Cyrill Brosset

Swedish Water and Air Pollution Research Laboratory (IVL)
S-402 24 Gothenburg, Sweden

Introduction

Studies of aerosols—their transport and composition—were started in Scandinavia in the mid-1960s. It was soon established that aerosols have two types of concentration maxima: locally occurring maxima, and those occurring simultaneously over large areas. The latter were also found to be very closely associated with certain directions of upper-level winds. It was later established that this dependence is less marked in certain special cases.

Continued investigations were largely focused on the chemical composition, primarily of the water-soluble part, of the aerosols. The results obtained showed that the aerosols are of different types. Their respective composition and properties are dependent on their origin and on the concentrations of some gases in the air mass in which they reside during transport. This means that, in actual fact, the transport route affects the composition of the aerosols.

Aerosol Episodes

The term "aerosol episode" was coined as early as 1965 when, by coincidence, it was found that soot maxima in different Swedish clean-air areas occurred practically at the same time.[1] An example of such a soot episode is given in Figure 1.

It was later discovered that the soot-containing particles making up the abovementioned aerosols always contain sulfur in the form of sulfate. This is illustrated by Figure 2.

It was then natural to follow up particle episodes with, among other things, measurements of airborne sulfate. This proved to be a fruitful choice, as it was soon found that high concentrations of sulfate do occur without the corresponding increase in soot concentrations.[2]

As a result of these experiences, the term aerosol episode has for us acquired the meaning of a "marked increase" in particle-borne sulfate. Thus, there is no strict definition. However, "episode" can be said to mean an increase in particle-borne $SO_4^{=}$ of up to >100 nmol/m^3 as a daily mean value that remains on this level for at least 24 hours.

The Transport Routes of the Particles

The simultaneous increase in particle concentration over very large areas is no doubt an indication that the responsible sources must be located far from the places of observation. It was therefore investigated whether the episodes were correlated to any

0077-8923/80/0338-0389 $01.75/1 © 1980, NYAS

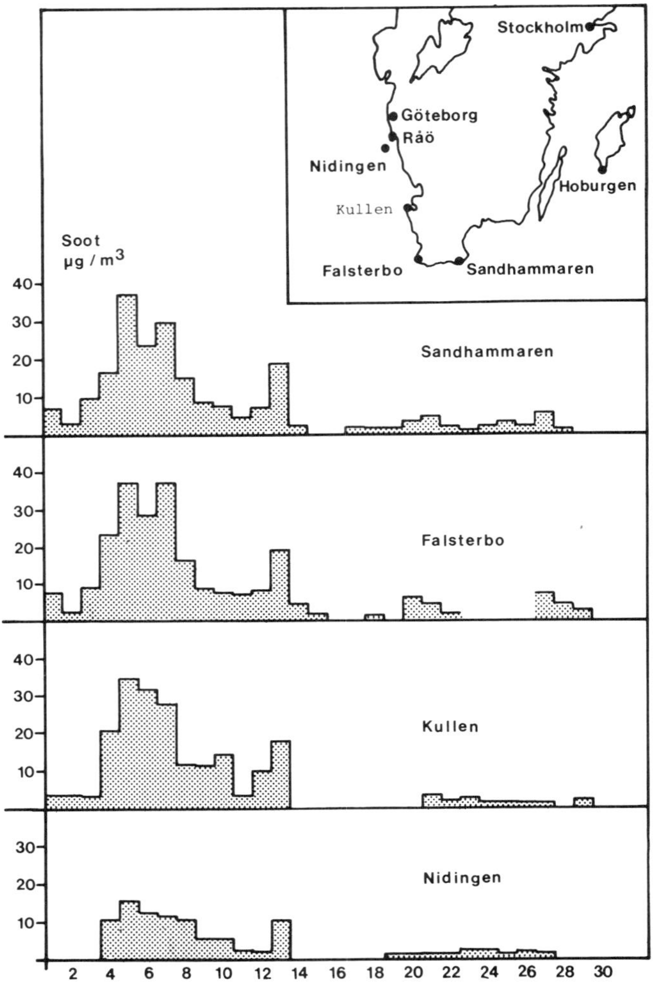

FIGURE 1. Mean daily soot values from lighthouses along the Swedish south and west coasts, February 1968.

special direction of upper-level winds. Rodhe[3] was the first to show that this was the case. As shown by FIGURE 3, high soot concentrations are clearly associated with air movement from SW–SE.

Later, episodes of black particles (so called black episodes) were studied in more detail, and the studies supplemented by trajectory calculations [first provided by the Swedish Meteorological and Hydrological Institute (SMHI) and later by the Norwegian Meteorological Institute and the Norwegian Institute for Air Research]. Examples of transport routes as illustrated by trajectories are given in FIGURE 4. In this case, a high soot (and sulfate) maximum was observed on 22–23 December 1972 all over Sweden, which can be seen from FIGURE 5. It should perhaps be pointed out that black episodes have so far been observed in connection with air transport from the southern sector (European continent) only.

It was mentioned earlier that high sulfate concentrations sometimes occur without high soot concentrations (so called white episodes). A study of the air movements during these episodes indicated, initially, that white episodes are connected to weak

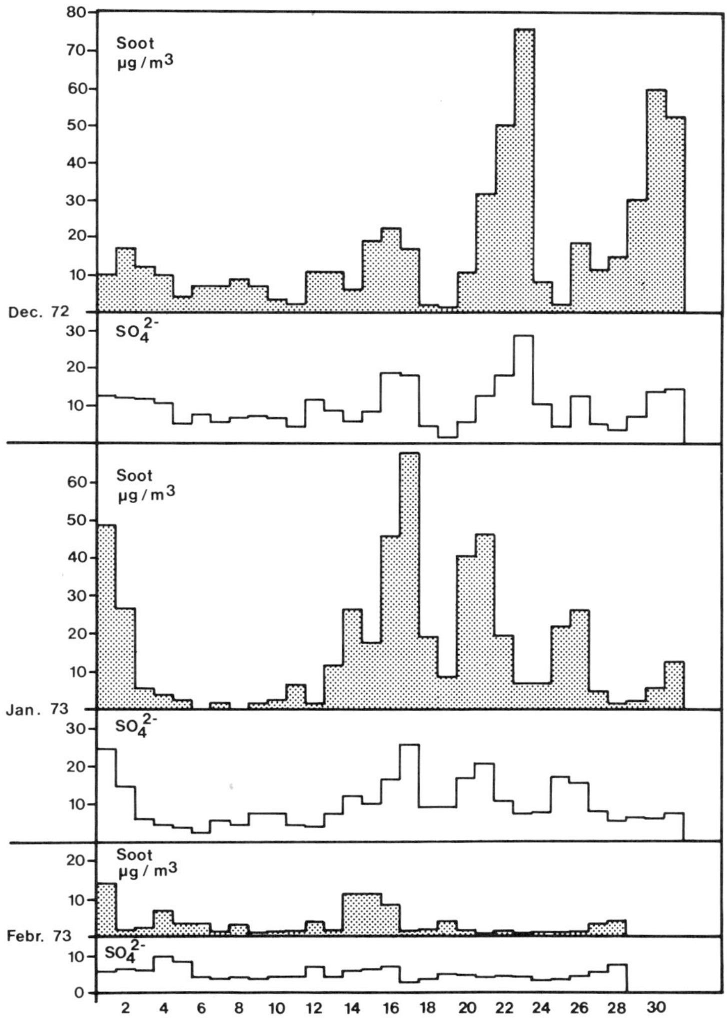

FIGURE 2. Mean daily soot and $SO_4^{=}$ values from Sandhammaren for the period December 1972–February 1973.

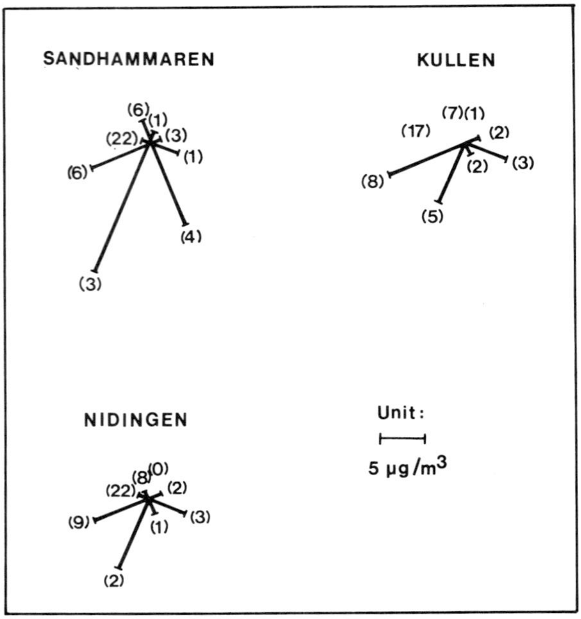

FIGURE 3. Mean values of soot concentration at the coastal stations for different transport directions (45° sectors). The number of days in each class is indicated in brackets. September–December 1969.

westerly winds. However, in February–March 1976, an extensive white episode occurred with largely easterly upper-level winds.[4] This will be dealt with below, in connection with the relevant atmospheric reactions.

THE CHEMICAL COMPOSITION OF THE PARTICLES

Coarse and fine particles have different sources and, consequently, different basic chemical compositions.[5] This has also been shown by investigations carried out on particle samples collected, in most cases, at different wind directions. TABLE 1 lists the equivalent fractions in percent for relevant ions, as observed in such investigations.[6]

As will be seen, the water-soluble part of coarse particles coming in from the north consists almost entirely of NaCl. The fine particles apparently consist of a mixture of NaCl and $(NH_4)_2SO_4$.

In black episodes, the coarse particles are alkaline and contain nitrate. The fine particles also contain nitrate but are weakly acid.

In white episodes, coarse particles are of about the same type as in black episodes, whereas fine particles are markedly acid and low in nitrate.

The differences between fine particles in black and in white episodes are perhaps best illustrated by the mean values compiled in TABLE 2.[6]

Finally, it should be pointed out that, as a rule, black fine particles contain about as much black component (soot) as they do water-soluble salts (ammonium sulfate and nitrate). They also contain metals, some of which are of importance as catalysts in connection with atmospheric reactions. Furthermore, it has been shown[7] that black particles contain a series of polycyclic aromatic hydrocarbons (PAH), the concentrations of which seem to be correlated to the soot concentration. The fine white particles,

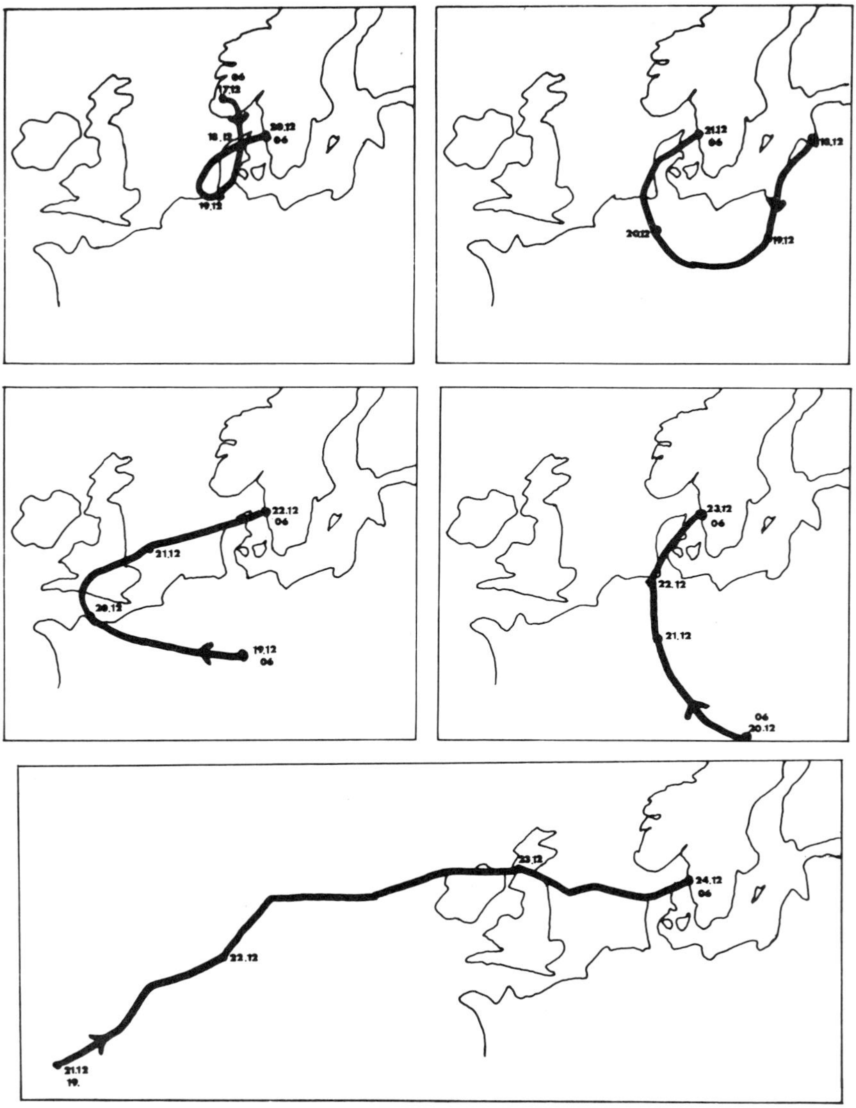

FIGURE 4. Trajectories illustrating the movements of air masses leading to the particle episode recorded on the Swedish west coast on December 22–24, 1972. (After calculations by the Swedish Meteorological and Hydrological Institute.)

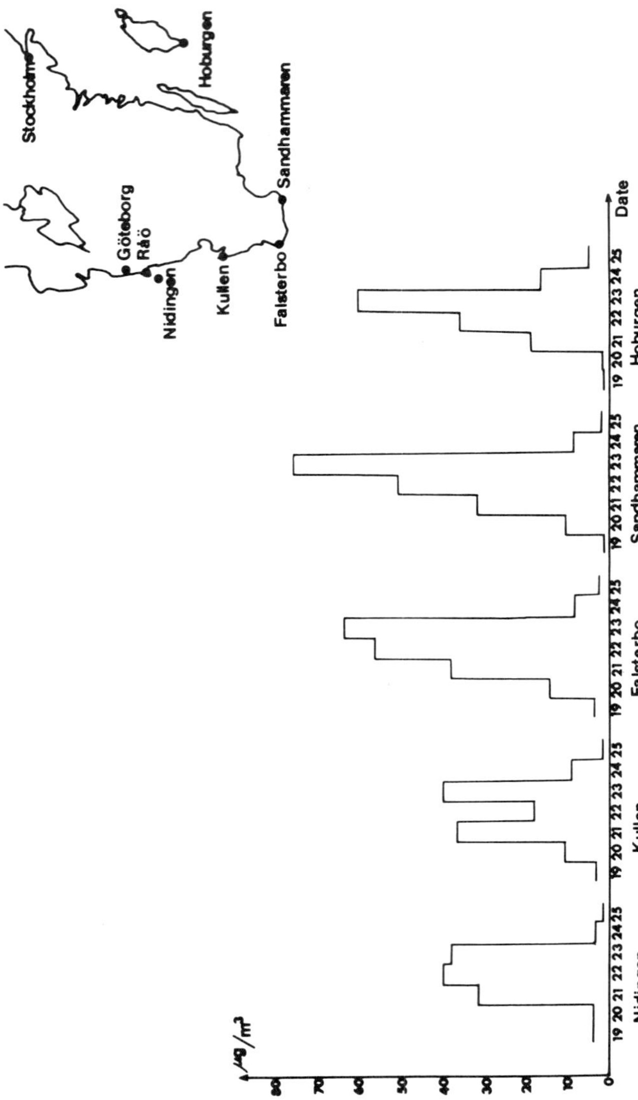

FIGURE 5. Observed concentrations of black airborne particulate matter ($\mu g/m^3$) at Swedish coastal stations during December 1972.

TABLE 1

MEAN ION CONCENTRATIONS IN EQUIVALENT FRACTIONS (%) IN THE WATER-SOLUBLE PART OF "FINE" AND "COARSE" PARTICLES SAMPLED AT Råö* DURING DIFFERENT EPISODE SITUATIONS

Episode	<2 μm (fine particles)						>2 μm (coarse particles)					
Type	SO_4^{2-}	NO_3^-	Cl^-	OH^-	H^+	NH_4^+	SO_4^{2-}	NO_3^-	Cl^-	OH^-	H^+	NH_4^+
None, N–E	57	3	41	0	1	46	7	2	91	0	0	5
None, N–W	82	12	7	0	4	48	45	15	37	3	0	39
Black	75	22	4	0	8	69	29	36	27	8	0	21
White	88	2	10	0	16	65	25	18	52	5?	0	15

*Remote station at the Swedish west coast, FIGURE 1.

TABLE 2

COMPARISON OF PARTICLE COMPOSITION DURING WHITE AND BLACK EPISODES AT Råö, 1975

Episode	n equivalent/m³							Soot
Type	SO_4^{2-}	NO_3^-	$SO_4^{2-} + NO_3^-$	NH_4^+	H^+	$NH_4^+ + H^+$	NH_4^+/H^+	($\mu g/m^3$)
Black	308	33	341	342	7	349	51	23
White	317	<3	~320	233	75	309	3.1	2.3

on the other hand, seem to be made up largely of the phases $(NH_4)_2SO_4$, $(NH_4)_3H(SO_4)_2$, and NH_4HSO_4.

ATMOSPHERIC REACTIONS

The original composition of the water-soluble part of the particles can be regarded, at least at sufficiently high relative humidity, as a solution of the nonvolatile components H_2SO_4 and NaOH (Na represents all cations from basic oxides) in H_2O. Here, the acid will dominate in the fine particles, and the alkali in the coarse ones.

In the atmosphere, this liquid phase is in contact with a gas phase (air) that contains, among other things, the components of interest here, i.e., NH_3, HNO_3, and HCl. In a state of equilibrium, the liquid phase will thus also contain the ions H^+, NH_4^+, NO_3^-, and Cl^-.

We will thus have a system consisting of two phases containing six components. The system consequently has six degrees of freedom. This means that if we choose $c_{SO_4^{2-}}$, c_{Na^+}, and p_{NH_3}, p_{HNO_3} as well as p_{HCl} as independent variables and give them definite values (meaning that we have also determined the variable H_2O), we can, for equilibrium conditions, calculate the concentrations in the liquid phase of c_{H^+}, $c_{NH_4^+}$, $c_{NO_3^-}$ and c_{HCl}. In such case, the following is valid:

$$2c_{SO_4^{2-}} - c_{NA^+} = c_{H^+} + c_{NH_4^+} - c_{NO_3^-} - c_{Cl^-}$$

$$c_{H^+}/c_{NH_4^+} = k_1 \cdot (p_{NH_3})^{-1} = q$$

$$c_{H^+} \cdot c_{NO_3^-} = k_2 \cdot p_{HNO_3} = r$$

$$c_{H^+} \cdot c_{Cl^-} = k_3 \cdot p_{HCl} = s$$

This gives:

$$c_{H^+} = \frac{2c_{SO_4^{2-}} - c_{Na^+}}{2\left(1 + \frac{1}{q}\right)} + \left[\frac{(2c_{SO_4^{2-}} - c_{Na^+})^2}{4\left(1 + \frac{1}{q}\right)^2} + \frac{r + s}{1 + \frac{1}{q}}\right]^{1/2}$$

$$c_{NH_4^+} = \frac{c_{H^+}}{q} \qquad \mathbf{(1)}$$

$$c_{NO_3^-} = \frac{r}{c_{H^+}}$$

$$c_{Cl^-} = \frac{s}{c_{H^+}}$$

The application of Equation 1 can be illustrated by the following case.

Assume that we have two particle populations, corresponding to fine and coarse particles, respectively, in equilibrium with a gas phase. We choose the independent variables (because of previous experience) as shown below:

	Particles (liquid phase mol/l)	
	Fine	Coarse
$2c_{SO_4^{2-}}$	4.00	0.30
c_{Na^+}	0.0	4.00

Gas Phase

$q = 0.2;\ r = 0.2;\ s = 0.1$

By means of Equation 1, the following composition of the water-soluble part of the particles (mol/l) is now obtained:

	Fine	Coarse
c_{H^+}	0.73	0.07
$c_{NH_4^+}$	3.67	0.36
c_{Na^+}	0.0	4.00
$2c_{SO_4^{2-}}$	4.00	0.30
c_{NO_3}	0.27	2.74
c_{Cl^-}	0.14	1.37

This result gives approximately the relative concentrations that are actually observed in the respective particle types.

The above indicates that the particles that make up the two different episode types may be formed as follows:

Coarse, alkaline particles (this applies to both black and white episodes) acquire NO_3^- from a reaction with HNO_3 in the air. In addition, $SO_3^=$ is formed in reaction with sulfur dioxide and is later oxidized to $SO_4^=$. If, in these reactions, an excess of acid is produced, it will react with NH_3, forming NH_4^+.

In white episodes, fine sulfuric acid droplets are formed initially, most likely through photochemical oxidation of SO_2. The droplets will grow and, through turbulent diffusion, reach ground-level air layers, there meeting a flow of NH_3. The degree of neutralization will consequently be dependent on available NH_3.

As long as the white particles remain acid, which seems to be the rule, they will not, of course, take up any large concentration of HNO_3. Regarding their transport routes, it is difficult to pinpoint these for the particles themselves. However, of greater practical importance is the location of the sources that provided the primary material (SO_2) necessary for the white episodes to occur. Such sources seem to be situated both to the west and to the east of Sweden.

The particles of black episodes are considerably more complicated in their formation. The likely process is that initially formed fine soot particles oxidize SO_2 on their surfaces, probably by means of a catalytic reaction. The sulfuric acid formed must be neutralized in order for the catalytic reaction to continue. This takes place partly with the content of alkali, if any, in the soot but mainly by means of NH_3 of the air.

The formation mechanisms indicated above, together with Equation 1, are in part confirmed by the following observations.

At the white episode occurring at easterly upper-level winds in March 1976, particle samples were collected at different places in Sweden. It was then found that the ratio $c_{H^+}/c_{NH_4^+}$ in particles varied systematically with the place of sampling.[4] Thus, this ratio for the same air mass was always greater in samples from a forest area with acid lakes than in samples from an agricultural area. As indicated earlier, $c_{H^+}/c_{NH_4^+}$ is inversely proportional to p_{NH_3}. Consequently, the flow of ammonia from cultivated fields must be noticeably greater than from a coniferous forest. This seems very plausible.

SUMMARY

Long-distance transport of aerosols was studied by means of sampling and analysis of particles and a comparison of the result with the direction of the upper-level winds. The investigations focused mainly on the water-soluble part of particles and their content of the black component.

It was found that coarse particles may be regarded as a mixture of sea salt and alkaline weathering products (or ashes). As a rule, the latter has taken up HNO_3 and SO_2 from the air, resulting in the formation of nitrate and sulfite, which has then been oxidized to sulfate.

Fine particles, formed, among other ways, through combustion of fossil fuel (coal) and which contain soot, are transported in episodes from central to northern Europe. These particles usually contain PAH, certain metals, and quite large amounts of nearly neutral ammonium sulfate. The latter is probably formed through catalytic oxidation, on the particle surface, of SO_2 to H_2SO_4 and neutralization of this acid by the NH_3 of the air.

Sulfuric acid droplets, initially formed through photochemical oxidation of SO_2, are introduced probably through turbulent diffusion into ground-level air layers, where they meet a flow of NH_3 from the ground. Neutralization is usually incomplete, probably due to insufficient NH_3. However, in agricultural areas, the neutralization is more extensive than over coniferous forests.

References

1. Brosset, C. & A. Nyberg. 1971. Investigations of soot and particle-borne sulfur in Sweden. *In* Proceedings of the 2nd International Clean Air Congress, Air Pollution Control Association, Washington, D.C., December, 1970. H. M. Englund & W. T. Beery, Eds.: 481–489. Academic Press, Inc. New York and London.
2. Brosset, C., K. A. Andreasson & M. Ferm. 1975. The nature and possible origin of acid particles observed at the Swedish west coast. Atmos. Environ. **9:** 631.
3. Rodhe, H., C. Persson & O. Åkesson. 1972. An investigation into regional transport of soot and sulfate aerosols. Atmos. Environ. **6:** 675.
4. Brosset, C. 1978. The role of ammonia in the chemistry of atmospheric aerosols. *In* Proceedings of the 175th National Meeting of the American Chemical Society, Anaheim, Calif., March 12–17. (In press.)
5. Whitby, K. T. & B. Cantrell. 1975. Atmospheric aerosols—characteristics and measurement. *In* Proceedings ICESA (International Conference on Environmental Sensing and Assessment), Las Vegas, Nevada, September 14–16. Institute of Electrical and Electronics Engineers, Inc., New York, N.Y., IEEE Catalog **2**(29–1): 1.
6. Brosset, C. 1976. Black and white episodes. Ambio **5**: 157.
7. Björseth, A., G. Lunde & A. Lindskog. 1978. Long-range transport of polycyclic aromatic hydrocarbons. Atmos. Environ. (In press.)

REGIONAL SCALE AIR POLLUTION: SOURCES AND EFFECTS*

R. B. Husar and D. E. Patterson

CAPITA
(Center for Air Pollution Impact and Trend Analysis)
Washington University
St. Louis, Missouri 63130

INTRODUCTION

This report constitutes an attempt to assemble several independent types of information related to the sources and effects of regional scale air pollution. The direct approach involved *in situ* observations of plumes with instrumented aircraft and case studies of individual episodes. Further insight can be obtained by examination of the seasonal and spatial trends and distributions of related parameters. At present, the overall picture is still incomplete, being focused only on coal combustion and sulfur compounds as they relate to optical effects. Nevertheless, the possible relationships outlined below may offer strong clues about the origin and nature of air pollution on a regional scale.

MAN-MADE SOURCES

Man-made air pollutants emitted over the United States are primarily due to the combustion of fossil fuels: coal, oil products, and gas. Additional major emissions result from the smelting of metals. The total U.S. energy consumption in 1978 was about 78 quads [1 quad = 10^{15} Btu/year], of which oil contributed about 40 quads; natural gas, 20; coal, 15; and nuclear power, 3 quads. Energy consumption over the past century has been characterized by a steady increase of the total consumption and by the shifts from one fuel to another in roughly 25-year increments. From 1850 to 1900, wood was the primary energy source. By 1900, and during the first full quarter of this century, the rising energy consumption was matched primarily by increasing use of coal. The depression years of the early 1930s are reflected in the sharp drop of coal consumption, which only recovered by the war years in the mid-1940s. The late 1930s and 1940s began the era of accelerated oil consumption, such that by 1945, the energy supplied by oil exceeded that of coal and maintained its rise up to the current level of the 1970s. The 1973 oil embargo is reflected as a small ripple in the otherwise steadily rising oil consumption. Coal consumption declined to another minimum in 1960, because the increasing energy demands were matched by the much cleaner fuels: natural gas and oil products. The 1950s and 1960s were the era of strongest increase in natural gas consumption. Nuclear energy began to supply a detectable fraction of the total U.S. energy consumption only after 1970. The insert in FIGURE 1

*Supported by the Federal Interagency Energy/Environment Research and Development Program through U.S. Environmental Protection Agency Grant No. R803896.

0077-8923/80/0338-0399 $01.75/1 © 1980, NYAS

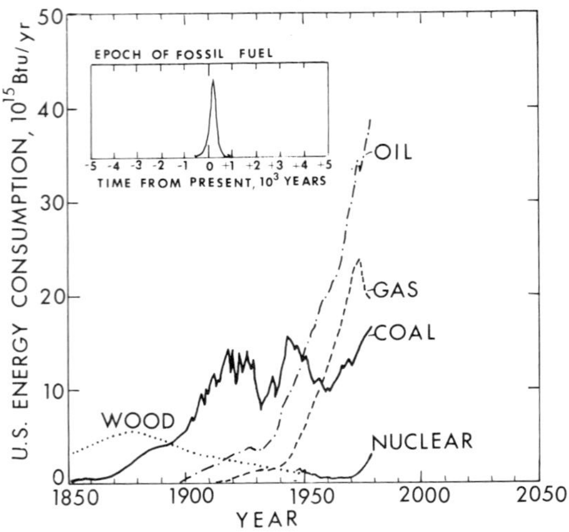

FIGURE 1. Trend of fossil fuel consumption in the U.S. The epoch of fossil fuel exploitation is a mere spike on the time scale of human history (insert). (From Reference 11.)

shows the epoch of fossil fuel consumption, which extends over 10,000 years; the epoch of fossil fuel exploitation is a mere spike on the time scale of human history.

Coal represents over 90% of the total U.S. fossil fuel reserves. Yet it currently supplies only 20% of the energy needs. It is anticipated, however, that coal's importance will grow, considering that domestic oil and gas supplies are limited and declining rapidly; that the U.S. dependence on foreign energy sources continues to

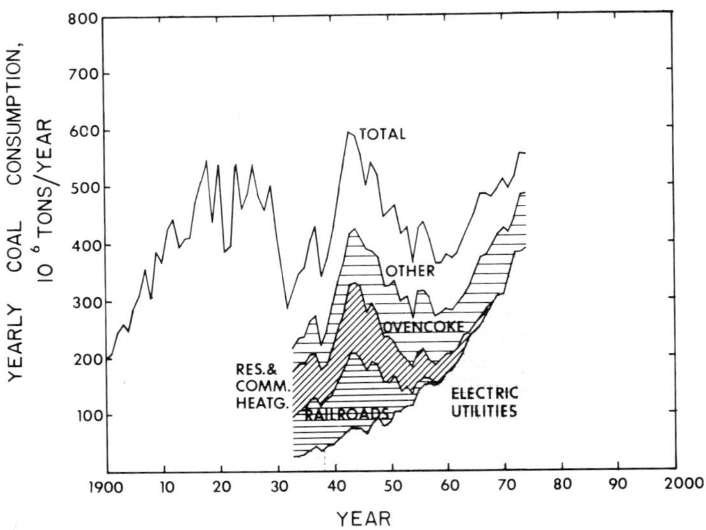

FIGURE 2. Trend of coal consumption in the U.S. The earlier coal demand for railroads and residential and commercial heating is now replaced by electric utility use. The trend of total coal use since 1960 is determined by the electric utility coal use. (Data from Reference 15.)

increase; that the rate of growth of nuclear energy supply is appreciably below the projection; and, finally, that nonconventional energy sources, such as solar and geothermal, are unlikely to contribute significantly to energy supplies for the next decade or two.

In 1975, U.S. coal consumption was about 600 million tons a year, roughly the same as around 1920 and 1943 (FIGURE 2). However, since the 1930s there has been a drastic shift in the economic sectors that consume coal. Before 1940, coal consumption was divided among the economic sectors of railroad, residential and commercial heating, oven coke, and other industrial processes. The railroad demand was particularly high during the war years of the mid-1940s. Within one decade in the 1950s, coal consumption by railroads and by the residential-commercial sector essentially vanished. Currently, electric utilities constitute the main coal-consuming sector, and the trend of total coal use in the United States since 1960 has been determined by the electric utility coal demand.

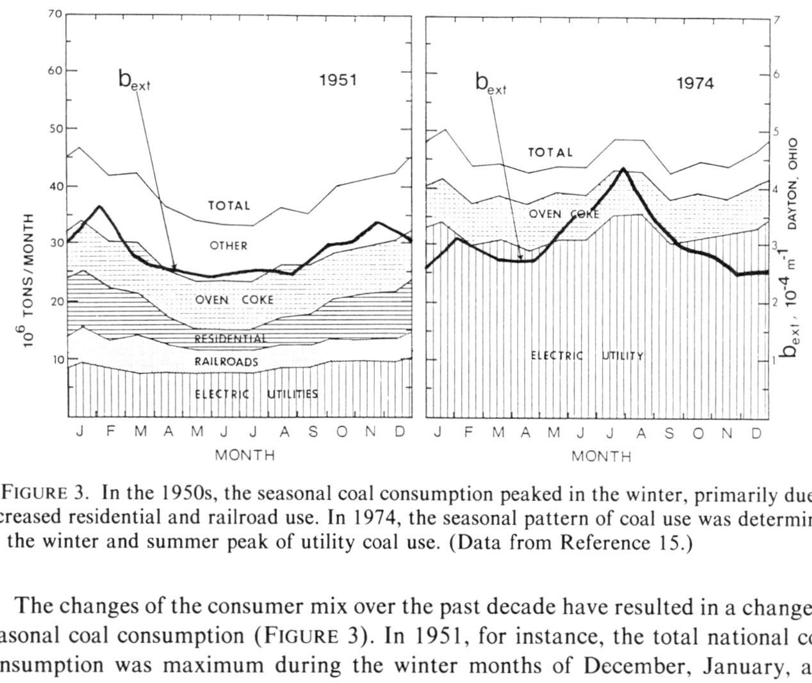

FIGURE 3. In the 1950s, the seasonal coal consumption peaked in the winter, primarily due to increased residential and railroad use. In 1974, the seasonal pattern of coal use was determined by the winter and summer peak of utility coal use. (Data from Reference 15.)

The changes of the consumer mix over the past decade have resulted in a change in seasonal coal consumption (FIGURE 3). In 1951, for instance, the total national coal consumption was maximum during the winter months of December, January, and February, mainly because of the increased residential coal demand. By 1974, the seasonal pattern of coal use was determined by the winter and summer peak of utility coal use. This shift of coal consumption from winter peak to summer peak is reflected in a corresponding shift in the haziness over the eastern United States.

The steady increase in national total coal consumption has not been uniform over the entire U.S. (FIGURE 4). In the Ohio River region (Indiana, Michigan, Ohio, Pennsylvania), coal consumption was high in the 1950s and early 1960s and has further increased since 1960. New England coal consumption has declined since about 1965, reflecting the fuel shift from coal to oil. This has also been the case for the northeastern megalopolis of New York, Maryland, and Virginia. The greatest

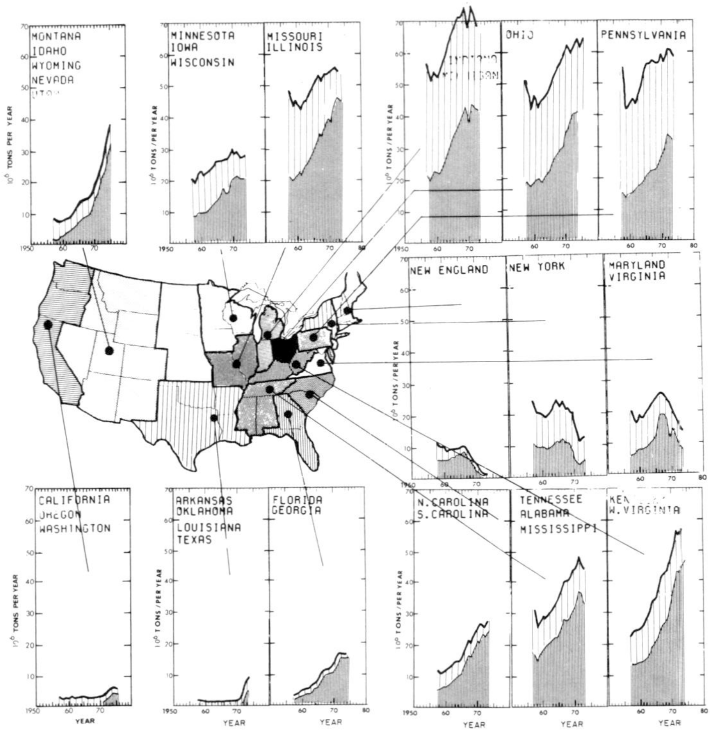

FIGURE 4. The trends of U.S. coal consumption in contiguous areas show strong regional dependence. Electric utility coal consumption is represented by the dark shading. (Data from Reference 15.)

increase in coal consumption has been recorded for the states of Kentucky, West Virginia, North and South Carolina, and Tennessee. In the eastern Sunbelt region, coal consumption has been appreciable only in Georgia and Alabama. In Missouri and Illinois, there has been a moderate increase, again due to the rising demand for electric utility coal. The western states (Montana, Idaho, Wyoming, Nevada, Utah, Colorado, Arizona, New Mexico) have undergone a rapid increase in coal consumption.

The shift of coal consumption from residential, commercial, and railroad to electric utilities has the further consequence of removing the sources from populated areas to rural areas. Currently, most of the electric utilities are located on the banks of major rivers. The increasing size and emission rates of point-source utility boilers and the need for compliance with the ambient short-term and yearly average SO_2 air quality standards have required control strategies for sulfur dioxide. Since the 1960s, point-source stacks have been built increasingly taller to reduce the surface concentrations in the vicinity of the source. For that purpose, the tall stacks admirably utilize the dispersion capability of the atmosphere; there are very few tall-stack point sources that violate either the short-term or long-term air quality standard for SO_2. It is clear, however, that the tall-stack strategy leads only to an effective dilution—not removal—of sulfur oxides. This dispersion and dilution into the upper atmospheric boundary layer inevitably leads to longer residence times and longer transport distances for tall-stack effluents than for those of comparable near-surface sources.

EMISSION RATES OF SO_x AND NO_x

Sulfur dioxide was emitted from man-made sources in the United States at an estimated rate of 30×10^6 tons per year in 1973. Fuel combustion exclusive of transportation accounted for 78%, industrial processes (primary metal industry, petroleum industry, chemical manufacturing, etc.) for 20%, and transportation for 2%. Of the fuels used by utilities and industry, about 65% of the national anthropogenic sulfur oxide came from coal combustion, and 13% from oil combustion.

Some 75% of the sulfur dioxide is emitted east of the Mississippi River, with the highest emission density in the vicinity of the Ohio River Valley (Ohio, Pennsylvania, and Indiana), as shown in FIGURE 5. The emission density for the states in the Ohio River Valley region ranges between 10 and 30 g/m^2 per year of sulfur dioxide. East of the Mississippi, fuel consumption contributes 92% of the total sulfur dioxide emissions. Hence, coal consumption and, to a lesser degree, oil consumption are a proper index of SO_x emissions over the eastern U.S. State-by-state trends of coal consumption are therefore useful indicators of regional SO_x emission trends. About 25% of the national SO_x emissions occur west of the Mississippi River, where fuel consumption contributed only 38%, and industrial and mining processes, 62% of the SO_x emissions in 1973. Electric utility emissions constitute a growing share of the total SO_x emissions. In 1973, utilities contributed about 60% of the SO_x emissions.

The national NO_x emissions of about 24×10^6 tons/year arise in roughly equal

FIGURE 5. Sulfur dioxide emission density is the highest in the Ohio River Valley region. (Data from Reference 16.)

proportions from automobiles, industrial fuel combustion, and electric utility sources. The emission density of NO_x, as seen in FIGURE 6, is highest in the Boston–Washington corridor, followed by the Ohio River Valley states of Indiana, Ohio, and Pennsylvania.

The national trends of SO_x, NO_x, and hydrocarbons (HC) over the past 35 years are illustrated in FIGURE 7. Only primary particulate emissions have exhibited a significant declining trend; since about 1970, effective particulate controls have reduced the emissions by about 30%.

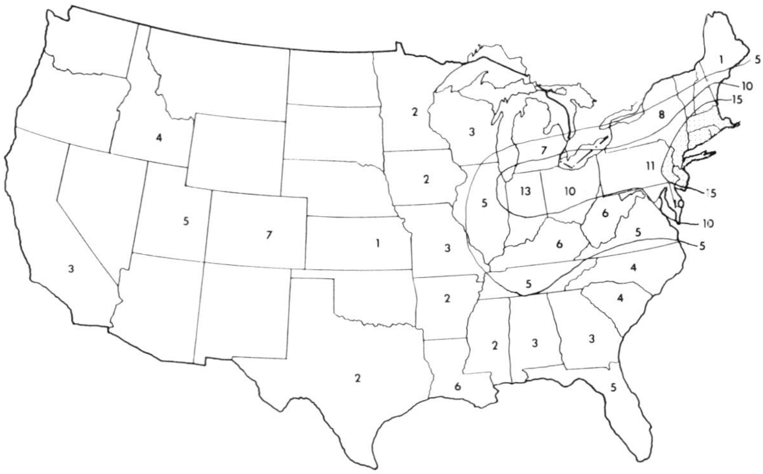

FIGURE 6. The highest NO_x emission density (15 g/m^2 year) is over the northeast metropolitan corridor, Boston–Washington, followed by the industrial states, Indiana, Ohio, and Pennsylvania. (Data from Reference 16.)

AMBIENT CONCENTRATIONS OF SULFUR COMPOUNDS

Altshuller[1] (1976) first noted the anomaly of decreasing urban SO_2 concentrations with increasing rural sulfate trends. The historical data base for SO_2 levels in urbanized areas shows a decline from an annual average of 50 $\mu g/m^3$ in 1964 to one of 26 $\mu g/m^3$ in 1974 (FIGURE 8). This represents an overall decrease by a factor of two, or approximately 6%/year. In 1974, the annual mean sulfur dioxide standard of 80 $\mu g/m^3$ was exceeded at only 3% of the monitoring stations (31 out of 1030). This is a decline from 16% in 1970. The 24-hour SO_2 standard (365 $\mu g/m^3$) was exceeded at less than 5% of the reporting stations (99 out of 2241) in 1974, compared to 11% in 1970. Ambient sulfate particulate concentrations have increased at nonurban sites in the Northeast, however.

Contours of yearly average sulfate concentrations reveal that the maximum concentrations occur in the region of high SO_2 emission density (and of highest coal consumption). In the Ohio River Valley, for instance, the yearly average sulfate concentration exceeds 15 $\mu g/m^3$; and it is between 10 and 15 $\mu g/m^3$ over a large part of the eastern U.S. (FIGURE 9).

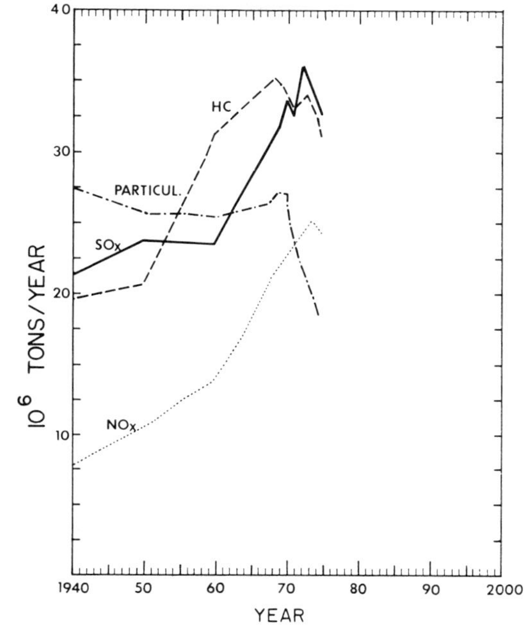

FIGURE 7. The trend of primary emissions for particulates, SO_x, NO_x, and hydrocarbons (HC) shows a dynamic pattern over the past 35 years. From 1950 to 1970, there was a strong upward trend for SO_x, NO_x, and HC; these have leveled off in the 1970s.

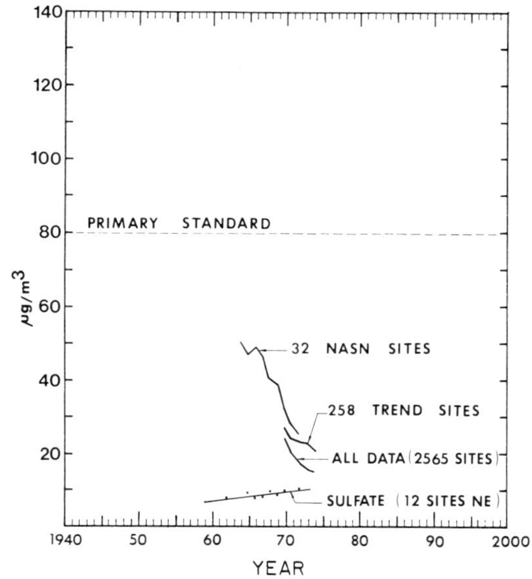

FIGURE 8. The ambient SO_2 concentrations in the U.S. have declined in the past decade, although SO_x emissions have increased. Nonurban sulfate concentration, however, has been increasing in the eastern U.S.

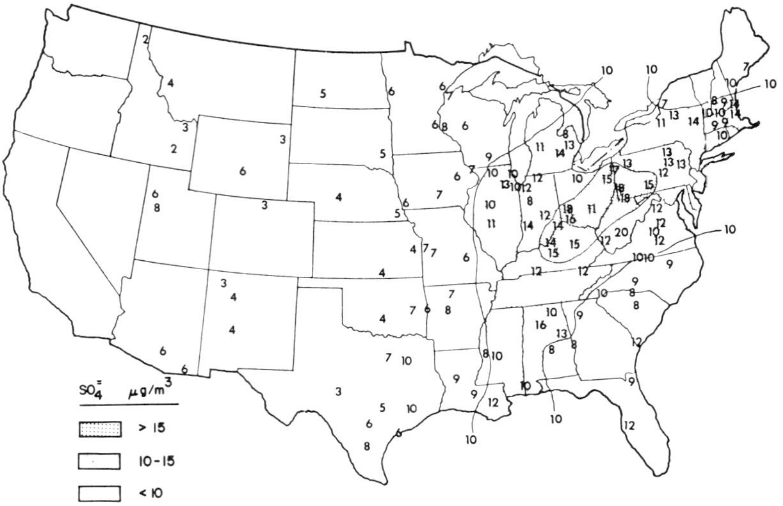

FIGURE 9. Yearly average sulfate concentrations.

The seasonal pattern displays a greater geographical extent of the high sulfate concentrations in the summer than in the winter months (FIGURE 10).

LONG RANGE TRANSPORT

Sulfate aerosol is primarily of secondary origin; it is formed in the atmosphere from reaction of SO_2. The residence time of secondary particles is the sum of the time

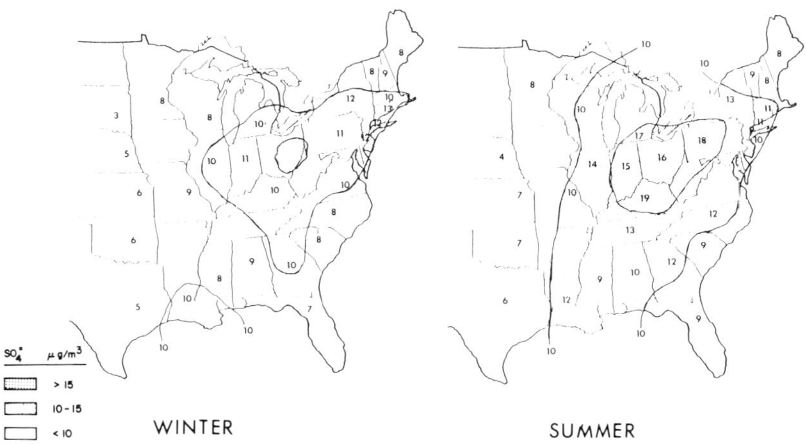

FIGURE 10. Seasonal sulfate concentration patterns.

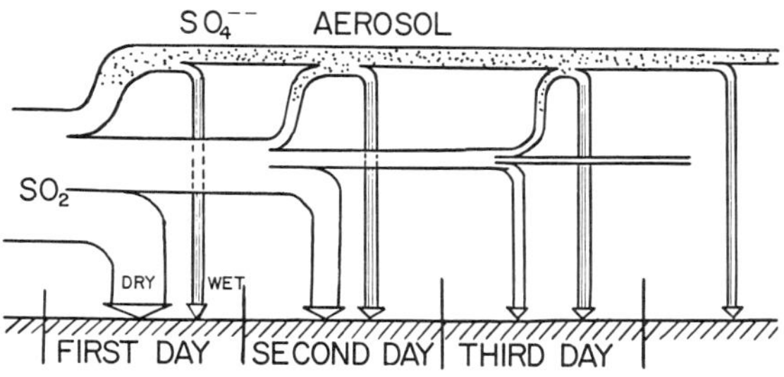

FIGURE 11. Flow diagram of sulfur transmission through the atmosphere. Over half of the SO_2 is removed or transformed to sulfate within the first day of its atmospheric residence. Most of the sulfate, however, may remain for several days of residence in the atmosphere.

required to create them and the subsequent time necessary for removal from the atmosphere. The flow diagram of sulfur transmission through the atmosphere (FIGURE 11) illustrates, for example, that most of the SO_2 emission is removed or transformed to sulfate within the first day of atmospheric residence, but the atmospheric burden of the secondary sulfate may remain relatively high for several days of residence.

It is intriguing to examine the downwind effects of the coal-fired Labadie power plant plume. FIGURE 12 illustrates the plume geometry and the measured sulfur

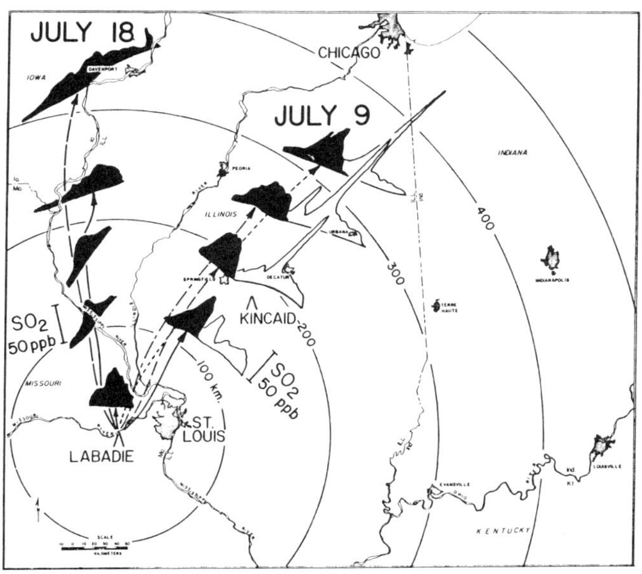

FIGURE 12. Horizontal profiles of SO_2 during selected constant-altitude aircraft traverses on July 9 and July 18, 1976. July 9 traverses are at about 450 m AGL, and July 18 traverses are at about 750 m AGL. The Labadie plume sections are shaded. Also shown are backward trajectories for the Labadie plume.[6]

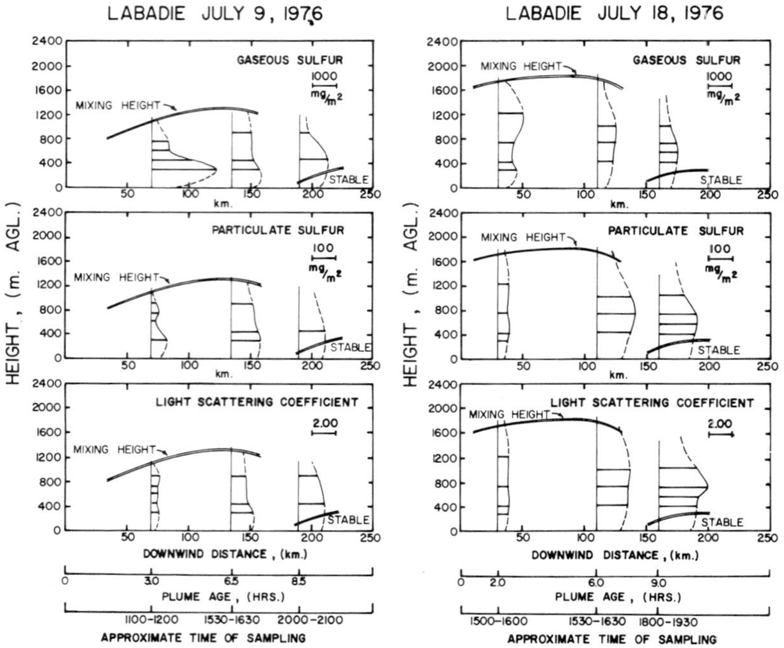

FIGURE 13. Vertical profiles of the crosswind integrals of excess plume concentrations of gaseous sulfur S_g, particulate sulfur S_p, and light scattering coefficient b_{scat} for the transport of the Labadie plume of July 9 and 18, 1976.[6]

FIGURE 14. Schematic illustration for the range of transport of fine sulfate particles. 1500 km is considered a reasonable mean transport distance. The actual transport depends on the synoptic-scale wind field.

dioxide concentrations attributed to the Labadie plume for two long-range-sampling days, July 9 and July 18, 1976. On both days, the plumes were tracked to about 300 km from the source. The optical properties, SO_2, and sulfate crosswind integrals of the aging power plant plume are shown in FIGURE 13. The bottom figure for each date indicates the cross-plume optical depth at different distances from the source, sampled at the indicated altitudes. On both dates, the crosswind plume optical depth at around 50 km distance from the source, sampled during midday, was about $\tau \cong 0.5$.

As the plume aging and light-scattering-aerosol formation continued over the day, the horizontal optical depth increased to about $\tau = 2$ on July 9, and to $\tau \approx 3$ on July 18, 1976. The fivefold increase in the horizontal depth at 150–200 km distance from the source is attributed to the formation of secondary sulfates in the plume, indicating that most of the light scattering aerosol has been formed 100 or more kilometers from the source.

The International Symposium on Sulfur in the Atmosphere[9] concluded that the typical atmospheric residence time of the sulfate aerosol is on the order of 3–5 days,

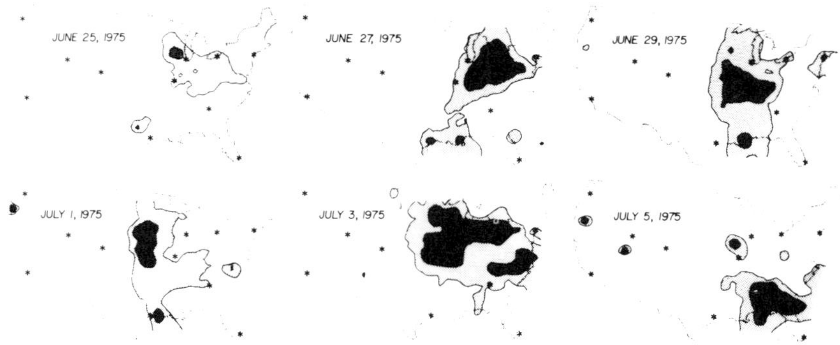

FIGURE 15. Sequential contour maps of noon visibility for June 25–July 5, 1975 illustrate the evolution and transport of a large-scale hazy air mass. Contours correspond to visual range 6.5–10 km (light shade), 5–6.5 km (medium shade), and <5 km (black).[10]

depending on precipitation patterns. Mean winds of 500 km per day are not uncommon in the eastern U.S.,[3] so that a reasonable scale of sulfate impact from an individual SO_2 source is on the order of 1000–2000 km. The implication, as seen in FIGURE 14, is that ambient light scattering sulfate aerosol can result from the additive effects of tens, or even hundreds, of SO_2 plumes of different atmospheric ages.

Since the emission of precursor gases remains relatively constant from day to day, the most important factor governing sulfate concentration is the regional scale dilution of the emissions. Therefore, stagnation of a regional scale air mass can cause high pollutant concentrations and extreme haziness. Such episodes are key contributors to the seasonal and annual average sulfate and haziness.

An example of one such episode over a two-week period in June–July 1975 is presented in FIGURE 15. Inspection of the sequence of contour maps reveals that multistate regions are covered by a haze layer in which noon visibility is less than 10 km ($b_{ext} > 4 \times 10^{-4}$ m^{-1}, outer contours).

The air mass of June 25, 1975, defined by the shaded region of visibility <10 km, was of maritime origin in the Gulf of Mexico and had traversed Louisiana, Illinois, and Indiana. Between June 25 and 27, relative stagnation prevailed in the Great Lakes region, leading to increasing haziness in this high emission density area. The tropical storm "Amy," moving up off the northeast coast, blocked the eastern motion of the hazy air mass as a front approached the Great Lakes from the northwest. Between June 28 and 30, an easterly flow developed, which caused the hazy air mass to drift slowly westward, passing over St. Louis, Missouri, on June 28–29 and continuing across Missouri and Kansas by June 30. From June 30 to July 2, the haze moved in a clockwise pattern up and around to the southern Great Lakes. By July 3, the advancing Canadian front formed the northern border of a massive hazy air mass occupying most of the midwestern and northeastern U.S. The hazy air mass again passed over St. Louis at this time. During July 3–5, the cold front advanced rapidly to the south, pushing the heavily polluted air mass across the Southeast and out to sea.

The apparent motion of the haze was confirmed by geostationary satellite photographs. The region of haziness on July 30 may be clearly seen in Lyons and

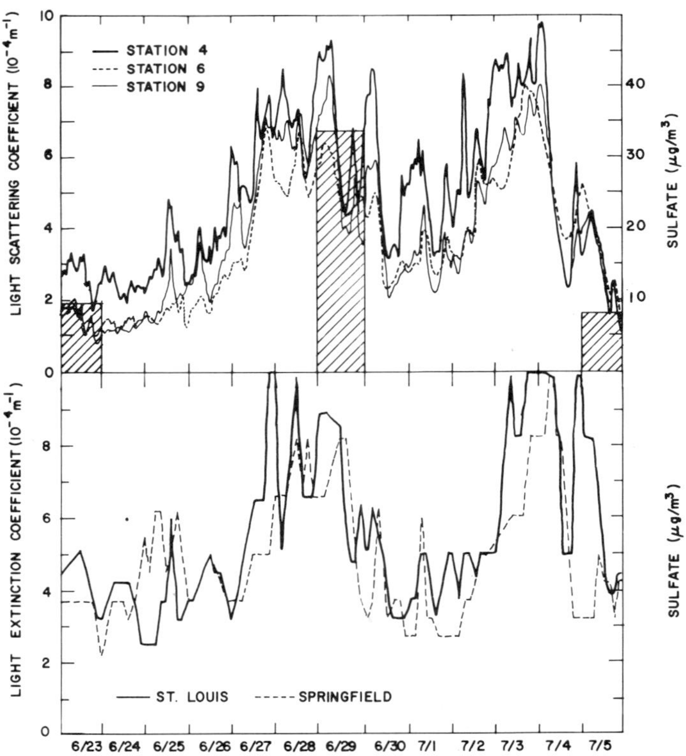

FIGURE 16. Local monitoring data in the St. Louis, Missouri, area during the June–July 1975 haziness episode. (a) Light scattering coefficient (b_{scat}) recorded at three widely spaced locations in the St. Louis metropolitan area, and daily average sulfate concentrations; (b) extinction coefficients (b_{ext}) obtained from visibility observations at St. Louis, Mo., and Springfield, Ill., 150 km apart.[10]

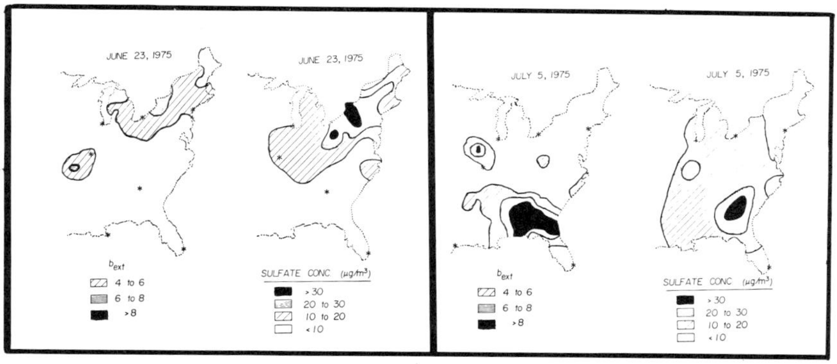

FIGURE 17. Comparison of noon extinction coefficients and daily mean sulfate concentrations on June 23 and July 5, 1975. The regions of highest sulfate concentrations coincide with areas of lowest visibility.[10]

Husar[12] (page 1624, Figure 1), where the turbid air mass is visible over the states of Arkansas, Missouri, Kansas, Iowa, and Minnesota.

Two passages of the hazy air mass over St. Louis, Missouri, resulted in sharp increases of b_{scat} over the entire metropolitan region (FIGURE 16). Sulfate concentration also increased during the haze episode, from about 9 to 33 $\mu g/m^3$. The spatial coherence of the haziness is seen in the correspondence of the extinction coefficients at

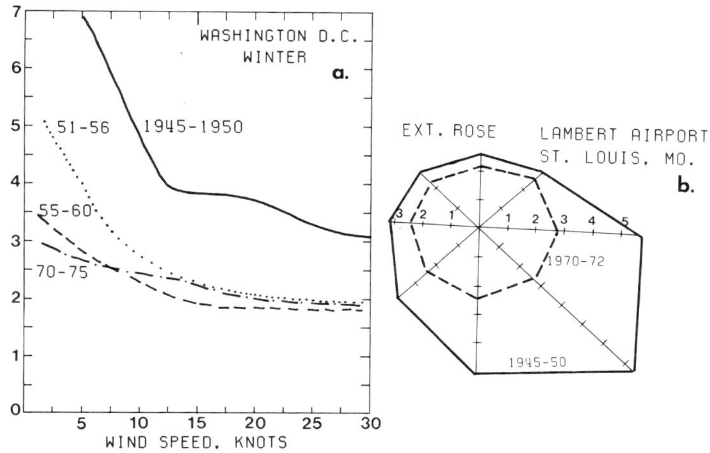

FIGURE 18. (a) In the winters of 1945–50 at Washington, D.C., high extinctions were associated with low wind speeds. By the winters of 1970–75, the level of haziness declined and the wind speed dependence had essentially vanished. The obvious interpretation is that the winter haziness in 1970–75 was due to remote, rather than local, sources. (b) A haziness rose for Lambert Airport in St. Louis, Mo., shows that in 1945–50, the highest extinction levels were observed when the wind was from the city direction; by 1970–72, the haziness was essentially independent of wind direction. Here again, the conclusion is drawn that the haziness is currently due to distant, rather than local, sources.

St. Louis, Missouri, and at Springfield, Illinois (FIGURE 16), again confirming that the observed haziness was primarily due to inflow of the polluted "background" material of the hazy air mass rather than to local contributions.

Sufficient sulfate data were available from the National Aerometric Data Bank for comparison with visibility on two days during the episode period. FIGURE 17 indicates the substantial correspondence between the regions of highest sulfate and those of lowest visibility.

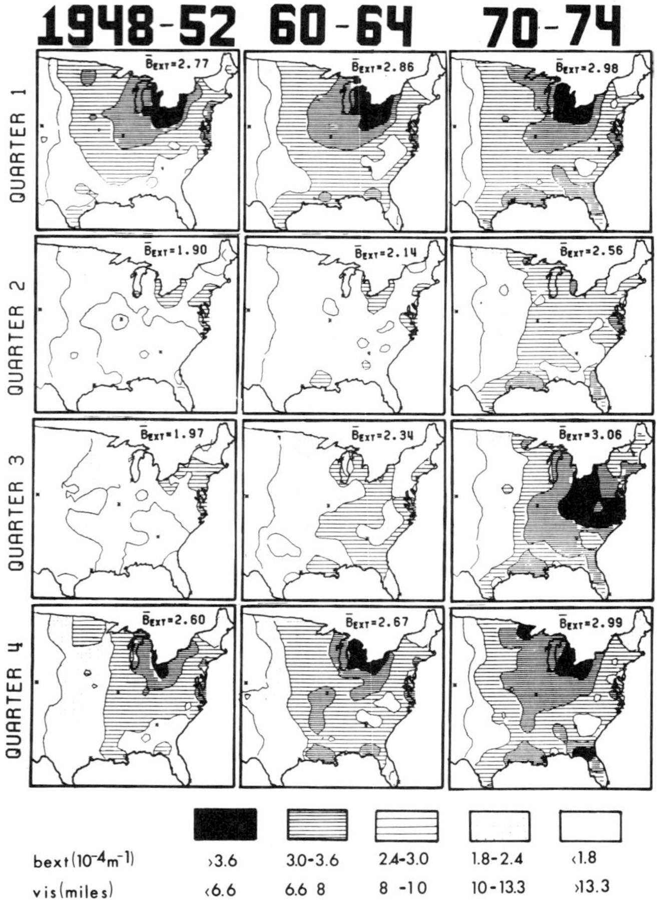

FIGURE 19. The spatial distributions of five-year average extinction coefficients show the drastic increases of Quarter 3 extinction coefficients in the Carolinas, the Ohio River Valley, and the Tennessee–Kentucky area. In the summers of 1948–1952, a 1000-km size multistate region centered around Atlanta, Georgia, had visibility greater than 15 miles, which had declined to less than 8 miles by the 1970s. The spatial trend of winter (Quarter 1) visibility shows improvements in the northeastern megalopolis region and some worsening in the Sunbelt region. Both spring and fall quarters exhibit moderate, but detectable, increase over the entire eastern U.S.

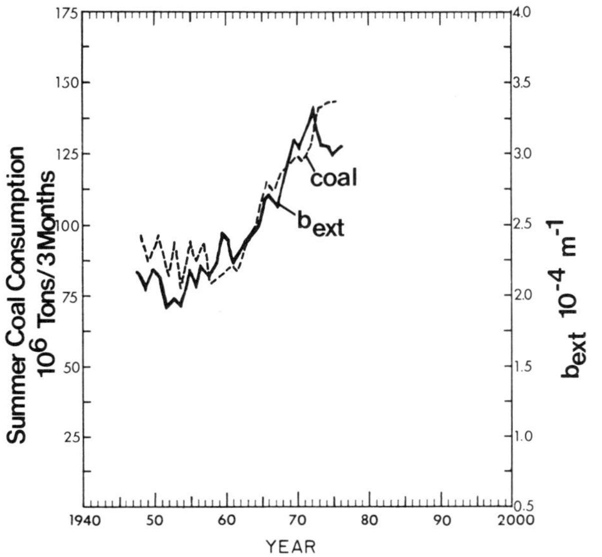

FIGURE 20. Summer trends of U.S. coal consumption (dashed line) and eastern U.S. average extinction coefficient, or haziness (solid line).

The work of other investigators confirms that episodes of regional scale hazy air masses are not rare in the eastern U.S. One of the earliest case studies of transport of large-scale hazy air masses was that of Hall *et al.* (1973).[7] Since about 1975, the evolution and transport of regional scale hazy air masses have received increasing attention by numerous research groups. Detailed case studies of such episodes have been reported by Tong *et al.*,[14] Husar *et al.*,[10] Lyons and Husar,[12] Wolff *et al.*,[18] Samson and Ragland,[13] Vukovich *et al.*,[17] Galvin *et al.*,[5] Hidy *et al.*,[8] and Chung,[4] among others.

VISIBILITY TRENDS

Evidence of the increasing importance of long range transport of pollutants in causing local visibility reduction may be extracted from historical observations taken by National Weather Service sites over the past three decades (FIGURE 18).

The spatial-temporal trend in visibility (as average extinction coefficient) is illustrated in FIGURE 19. The shift from a winter peak towards a summer peak in coal combustion is reflected in the marked worsening of summer visibilities. The Smoky Mountain states, which have undergone the greatest increase in coal use, have also exhibited the strongest increase in summer average haziness.

The comparison of the eastern U.S. summer coal consumption and summer average extinction over the entire eastern U.S. is shown in FIGURE 20. While a statistically high correlation could be extracted from the coal consumption–haze trend data, a cause-effect relationship cannot be established from trend analysis alone.

Turbidity and Solar Radiation

Two independent measures of the relation of sulfates to degradation of the optical environment may be obtained by examination of the recent behavior of turbidity and solar radiation. The seasonal patterns of sulfate and turbidity (Figure 21) are consistent; turbidity is a measure of the vertical integral of aerosol and is thus proportional to the product of concentration and mixing depths, both of which are highest in summer. Evidently, then, the sulfate burden is 2–3 times higher in summer than in winter.

The trend in turbidity over the eastern U.S. (Figure 22) indicates a pronounced summer increase, as was noted for extinction coefficient.

Angell and Korshover[2] (1975) also reported substantial decreases in the total

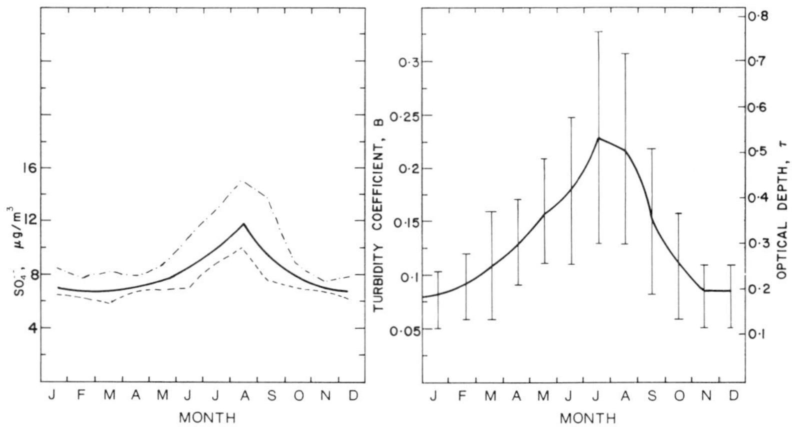

Figure 21. (a) Monthly average sulfate concentration (μ/m^3), 1970–74, for 18 eastern nonurban sites (—). The seasonal pattern at 6 stations in the industrialized Northeast (–·–·) is more pronounced than the seasonal pattern of 12 peripheral stations in the Midwest and Southeast (– – –). (b) The seasonal pattern of turbidity of 26 eastern U.S. sites.

hours of solar radiation over the southeastern U.S. (Figure 23), which could be due to haze "delaying sunrise" and "advancing sunset" (from the point of view of an on-off solar radiation detector). We are aware of the fact that most of these observed optical anomalies could be due to an increase of daytime cloudiness, which may or may not be related to human activities.

Summary and Discussion

The optical effect of a coal-fired power plant plume was measured at 200 km from the stack, where the cross-plume optical depth $\tau > 2$ was associated with sulfate. An episode of extreme haziness covering multistate regions of the eastern U.S. appears to

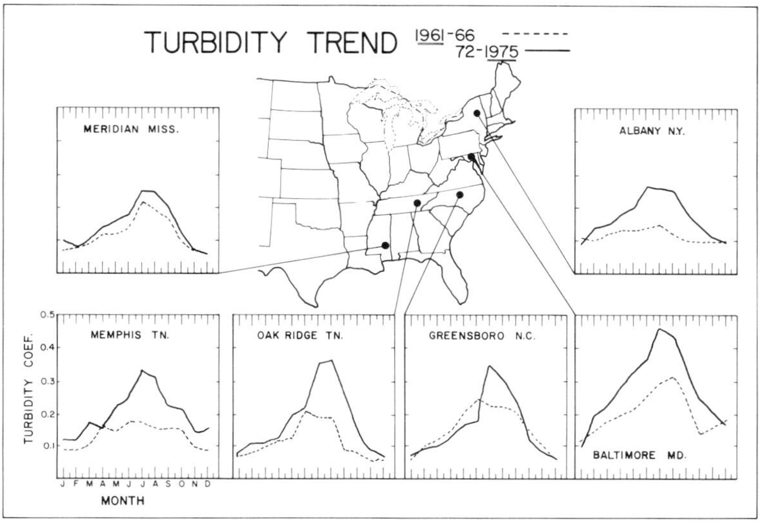

FIGURE 22. The seasonal turbidity trend for eastern U.S. stations is illustrated by comparing the patterns for the period 1961–66 to those of 1972–75. Current *summer* turbidities exceed the 1961–66 values by 10 to 50%.

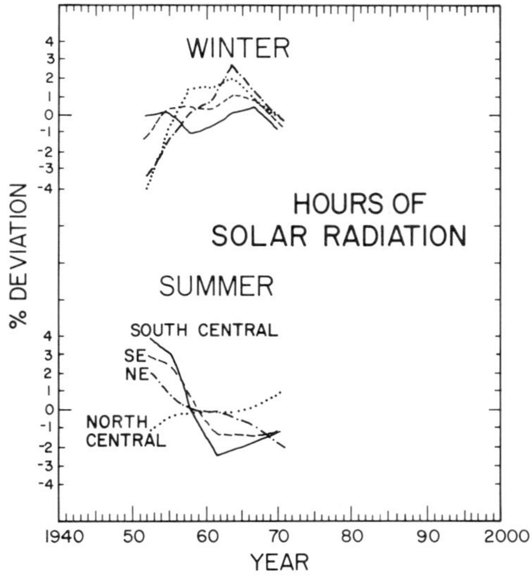

FIGURE 23. The analysis of Angell and Korshover[2] (1975) of the hours of solar radiation shows an *increase* in winter radiation since the 1950s and a *decrease* in summer solar radiation over the eastern U.S.

have been largely due to secondary sulfate aerosol. Finally, the spatial and temporal aerosol trends and distributions of coal use, sulfate, light extinction coefficient, turbidity, and solar radiation all exhibit reasonable internal consistency.

One purpose of this paper was to demonstrate the apparent relationship of patterns of fuel use and effects. We also wished to point out that for proper analysis, the spatial scales of emissions and effects must be matched to the transport distance of sulfate, 1000–2000 km. Finally, it must be emphasized that coal combustion and sulfate are only one factor in the production of optical effects. Until appropriate study of the consumption of gasoline and other oil products is incorporated, no cause-and-effect conclusion can be drawn.

Acknowledgments

This work was conducted as part of projects MISTT (Midwest Interstate Sulfur Transformation and Transport) and STATE (Sulfur Transformation and Transport in the Environment) with the close cooperation of Dr. William E. Wilson, Jr., Project Officer.

References

1. Altshuller, A. P. 1976. Regional transport and transformation of sulfur dioxide to sulfates in the U.S. J. Air Pollut. Control Assoc. **26:** 318.
2. Angell, J. K. & J. Korshover. 1975. Variation in sunshine duration over the contiguous United States between 1950 and 1972. J. Appl. Meteorol. **14:** 1174.
3. Blumenthal, D. L., J. A. Ogren & J. A. Anderson. 1978. Airborne sampling system for plume monitoring. Atmos. Environ. **12:** 613.
4. Chung, Y. S. 1978. The distribution of atmospheric sulfates in Canada and its relationships to long-range transport of air pollutants. Atmos. Environ. **12:** 1471.
5. Galvin, P. J., P. J. Samson, E. C. Peter & D. Romano. 1978. Transport of sulfate to New York State. Environ. Sci. Technol. **12:** 580.
6. Gillani, N. V. 1978 Project MISST: mesoscale plume modeling of the dispersion, transformation and ground removal of SO_2. Atmos. Environ. **12:** 569.
7. Hall, F. P., Jr., C. E. Duchon, L. G. Lee & R. R. Hagan. 1973. Long range transport of air pollution: a case study, August 1970. Mon. Weather Rev. **101:** 404.
8. Hidy, G. M., P. K. Mueller & E. Y. Tong. 1978. Spatial and temporal distributions of airborne sulfate in parts of the United States. Atmos. Environ. **12:** 735.
9. Husar, R. B., J. P. Lodge, Jr. & D. J. Moore, Eds. 1978. Workshop reports. *In* Sulfur in the Atmosphere. Pergamon Press. Oxford, England.
10. Husar, R. B., D. E. Patterson, C. C. Paley & N. V. Gillani. 1976. Ozone in hazy air masses. Paper presented at the International Conference on Photochemical Oxidant and Its Control, Raleigh, N.C., September 12–17.
11. Lapedes, D. M., Ed. 1976. Encyclopedia of Energy. McGraw-Hill. New York, N.Y.
12. Lyons, W. A. & R. B. Husar. 1976. SMS/GOES visible images detect a synoptic-scale air pollution episode. Mon. Weather Rev. **104:** 1623.
13. Samson, P. J. & K. W. Ragland. 1977. Ozone and visibility reduction in the Midwest: evidence for large-scale transport. J. Appl. Meteorol. **16:** 1101.
14. Tong, E. Y., G. M. Hidy, T. F. Lavery & F. Berlandi. 1976. Regional and local aspects of atmospheric sulfates in the northeast quadrant of the U.S. *In* Proceedings, Third Symposium of Turbulence, Diffusion, and Air Quality, American Meteorological Society.

15. U.S. Bureau of Mines. 1933–1974. Minerals Yearbooks.
16. U.S. Environmental Protection Agency. 1976. National Emissions Report (1973): National Emissions Data System (NEDS) of the Aerometric and Emissions Reporting System (AEROS). EPA-450/2-76/007. U.S. Environmental Protection Agency. Research Triangle Park, N.C.
17. VUKOVICH, F. M., W. D. BACH, JR., B. W. CRISSMAN & W. J. KING. 1977. On the relationship between high ozone in the rural surface layer and high pressure systems. Atmos. Environ. **11:** 967.
18. WOLFF, G. T., P. J. LIOY, R. E. MEYERS & R. T. CEDERWALL. 1977. An investigation of long-range transport of ozone across the midwestern and eastern United States. Atmos. Environ. **11:** 979.

EVIDENCE OF TRANSPORT OF HAZY AIR MASSES FROM SATELLITE IMAGERY*

Walter A. Lyons

MESOMET, Incorporated
Chicago, Illinois 60601

INTRODUCTION

Often, a new way of visualizing natural phenomena helps us crystallize our understanding of basic physical processes. The ability of conventional meteorological satellites to observe routinely massive areas of turbidity in the atmosphere—produced by both natural and anthropogenic sources—has shed new perspectives on long range pollution transport. This paper briefly summarizes some of the observations of major aerosol events in the atmosphere and suggests that routine detection and tracking of synoptic scale pollution episodes, along with quantitative measurements of their intensity, are entirely feasible with existing spacecraft and data analysis systems.

Long range transport of certain long-lived atmospheric pollutants was graphically demonstrated three decades ago by the tracking of nuclear bomb debris for, in some cases, several circuits of the globe. However, the Clean Air Act of 1967 and 1970 had as an underlying assumption the notion that the concentrations of various primary pollutants at some unspecified, but not too great, distance downwind from the source would become indistinguishable from the "natural background" due to dilution, wet and dry deposition, and perhaps chemical transformation. During the 1970s, we have gradually come to realize that these chemical transformations produce secondary pollutants of significant concentrations, which become of greater concern at distances typically greater than 100 km.

Measurements by Blumenthal *et al.*[1] showed photochemical oxidants being reported out of the Los Angeles Basin at the rate of 100 tons/hour. Lyons and Cole[2] noted that high ozone values recorded in both urban and rural sections of Wisconsin were the combined result of local emissions, mesoscale transport, and synoptic transport, all commingled in a manner that made it most difficult to separate the fraction resulting from each mechanism. Coffey and Stasiuk,[3] Wolff *et al.*,[4] and Ott and Lyons[5] were among many suggesting that ozone-laden air masses could easily travel hundreds, or even thousands, of kilometers over several days. Periods of elevated ozone in air masses undergoing long range transport were also found to be associated with reduced visibility by Samson and Ragland.[6] As research, such as project MISTT, continued, it became clear that important secondary aerosol-producing mechanisms were present. The conversion of SO_2 gas into aerosols within individual power-plant or urban plumes was documented by many, including White *et al.*[7] The progressive deterioration of regional summer visibility, correlated to increased SO_2 emissions into the upper portion of the boundary layer by the greater use of coal and the use of higher stacks, is reported by Husar.[8]

*Supported primarily by the Regional Field Studies Office of the U.S. Environmental Protection Agency, under Subcontract No. 68-02-3000.

0077-8923/80/0338-0418 $01.75/1 © 1980, NYAS

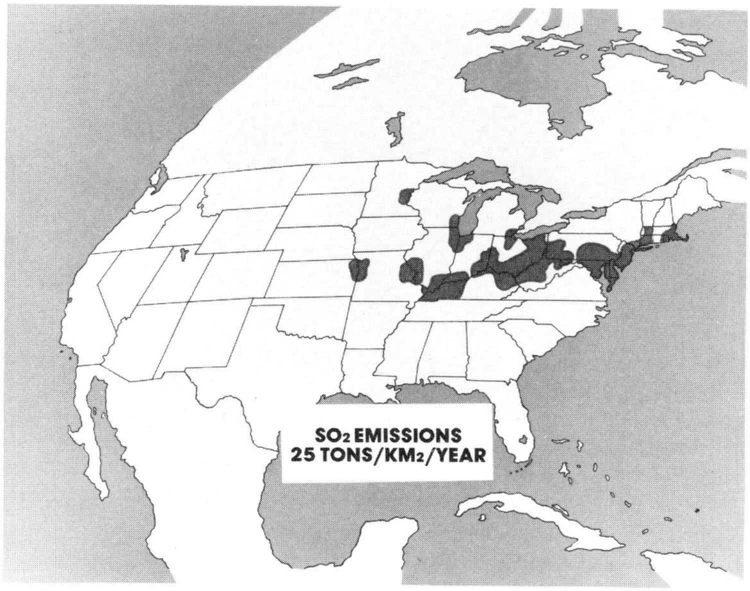

FIGURE 1. Map of regions where SO_2 emission density exceeds 25 tons per square kilometer per year.

Reduced visibility, or visual range, is the result of decreasing contrast of a distant object against a background because of the scattering of light into the intervening volume of air. It is reasonable, then, to assume that measurements of visible-wavelength radiation from an orbital platform would reveal both increased atmospheric brightness and reduced contrast of ground targets, if the aerosol were of the appropriate size and concentration, and if the remote sensing device were of the required sensitivity, resolution, and spectral response.

As early as 1971, Mohr[9] noted anomalous areas of turbid air in ESSA† satellite APT‡ images over Europe and suggested the possibility of using conventional weather satellites to monitor regional air pollution events. In the same year, McLellan[10] had some success in using the digital brightness readings from the geosynchronous ATS§ satellite to detect the smog buildup in the Los Angeles Basin by its enhanced upward light scatter.

The prime contributor to widespread areas of turbidity appears to be the conversion of SO_2 gas to sulfate aerosols, with transformation rates somewhere around one to several percent per hour, according to Wilson *et al.*[11] FIGURE 1 maps the region with the highest SO_2 emission densities in the United States. The contributions of the coal-burning power plants, especially in the Ohio River Valley and surrounding areas, are significant. The isopleths of maximum recorded sulfate levels (FIGURE 2) show a pattern centered near the major source regions, but with values exceeding 20 $\mu g/m^3$

†Environmental Service Science Adminstration.
‡Automatic Picture Transmission.
§Applications Technology Satellite.

possible almost anywhere in the eastern United States. It is in this area that our search for satellite images revealing large-scale pollution episodes will be focused.

SATELLITE DETECTION OF AEROSOLS

Various meteorological satellites have been shown to be capable of detecting a wide variety of aerosol episodes in the atmosphere. Landsat images were analyzed by Lyons,[12] revealing point-source smoke plumes of over 100 km in length over Lake Michigan. Lyons *et al.*[13] also noted apparent indications of high turbidity within entire air masses. Unfortunately, the small areal, and infrequent temporal, coverage of Landsat makes synoptic monitoring impractical. The same could be said of 35-mm photography from manned spacecraft missions, although some individual scenes of note were described by Randerson.[14]

The operational meteorological satellite, while designed to detect bright, high contrast clouds, has been shown to be suitable for mapping large-scale areas of atmospheric dust, haze, and smoke. The SMS/GOES (Synchronous Meteorological Satellite/Geostationary Operational Environmental Satellite) series has been used for this research.[15] Two systems observe the United States from 36,000 km above the equator; one is centered at 75° west longitude, the second at 135° west longitude. Every 30 minutes, 4-km resolution (subpoint) infrared (10.5–12.5 μm) scans of the earth are made from pole to pole. During daylight hours, visible images at 1.0-km subpoint resolution are obtained at the same interval. The visual infrared spin-scan

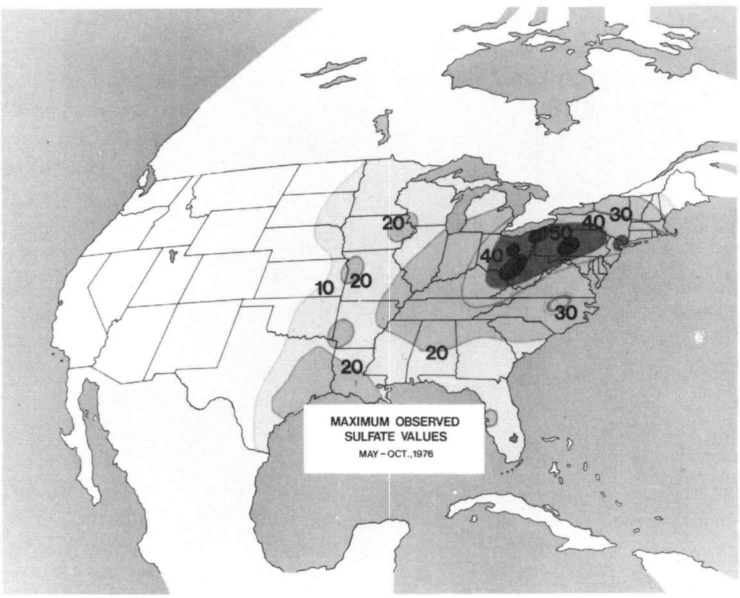

FIGURE 2. Plot of observed maximum sulfate values ($\mu g/m^3$) at NASN sites during the period May through October 1976.

FIGURE 3. SMS/GOES 1.0-km resolution visible image, showing plume of smoke from Hawaiian volcanic eruption drifting several hundred kilometers southwestward, at 0300 GMT, September 16, 1977.

radiometer (VISSR) responds to visible light in the 0.54 to 0.70 μm band. Data are transmitted from satellite at 8-bit resolution. The enormous data volume is managed in many ways, but primarily by conversion to photographic form. However, special computer-based systems do allow investigators to assess and manipulate the data quantitatively.

FIGURE 3 is an example of a major aerosol event in the atmosphere—a massive plume of smoke from the Hawaiian volcanic eruption of September 1977. On subsequent days, the plume drifted in a cohesive manner in the trade winds for many hundreds of miles. Shenk *et al.*[16] were among the first to note the cross-Atlantic transport of large clouds of Saharan dust. In FIGURE 4, a dust cloud reached Cuba after having left the African coast less than five days before, all the while visible in the SMS/GOES visible imagery. During the drought of 1977, several major dust storms occurred in the central U.S. One such storm raged from February 23 to 25, 1977. On February 24, the satellite detected a "plume" of dust stretching from the Oklahoma Panhandle to Georgia (FIGURE 5). It was subsequently seen to spread over the Atlantic, as far east as Bermuda. During the passage, local visibilities dropped to as

FIGURE 4. Full disk SMS/GOES 4.0-km resolution visible image, 1500 GMT, June 18, 1977. Saharan dust cloud had crossed the Atlantic in four days. The position of the leading edge of the dust on June 14 is indicated by an arrow.

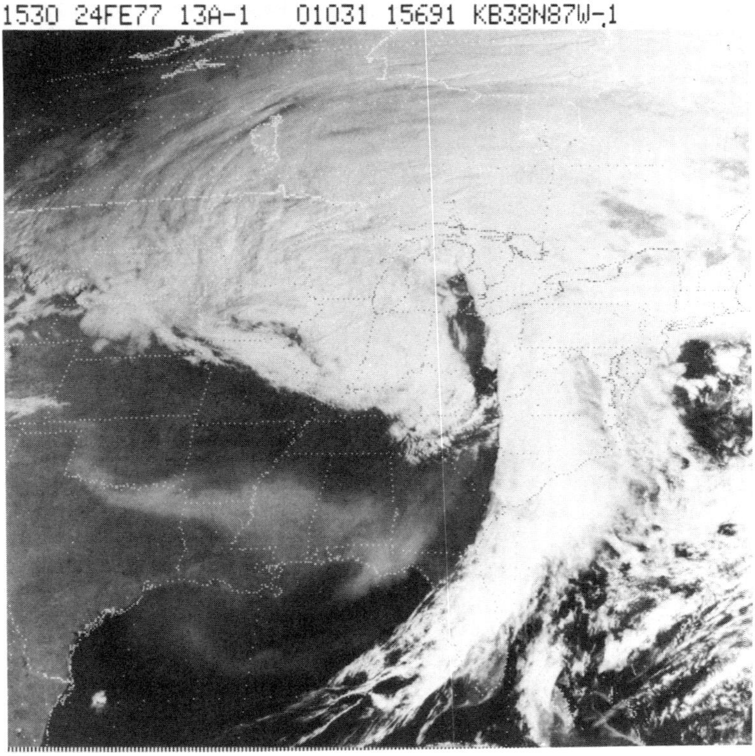

FIGURE 5. SMS/GOES 2.0-km resolution visible image, 1530 GMT, February 24, 1977, shows cloud of dust stretching from Oklahoma Panhandle to Georgia. Dust cloud followed cold front many hundreds of miles into the Atlantic.

FIGURE 6. SMS/GOES 2.0-km resolution visible image, 1700 GMT, April 30, 1975, showing large area of smoke over Yucatan and the western Gulf of Mexico, the result of widespread agricultural slash burning.

FIGURE 7. Blow-up of SMS/GOES 1.0-km resolution visible image, 2200 GMT, September 29, 1976, showing plume of smoke from a small forest fire in northern Minnesota, having drifted about 100 km southeast after about seven hours.

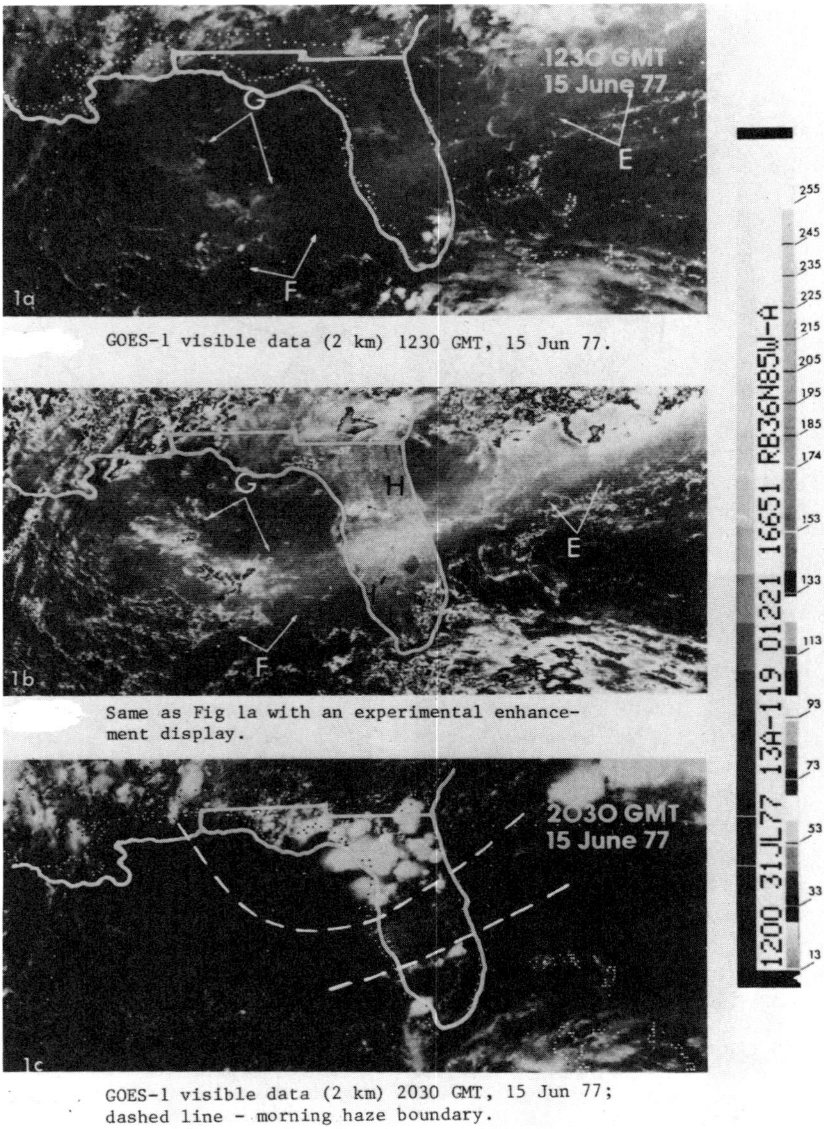

FIGURE 8. (a) Portion of SMS/GOES 2.0-km resolution visible image over Florida, 1230 GMT, June 15, 1977. (b) Same scene, but with imagery digitally enhanced, using experimental technique developed by National Environmental Satellite Service. Band of hazy air now distinct. (c) Same region, but at 2030 GMT, June 15, 1977. Haze was sufficiently dense to apparently suppress afternoon convection present both north and south of haze band.

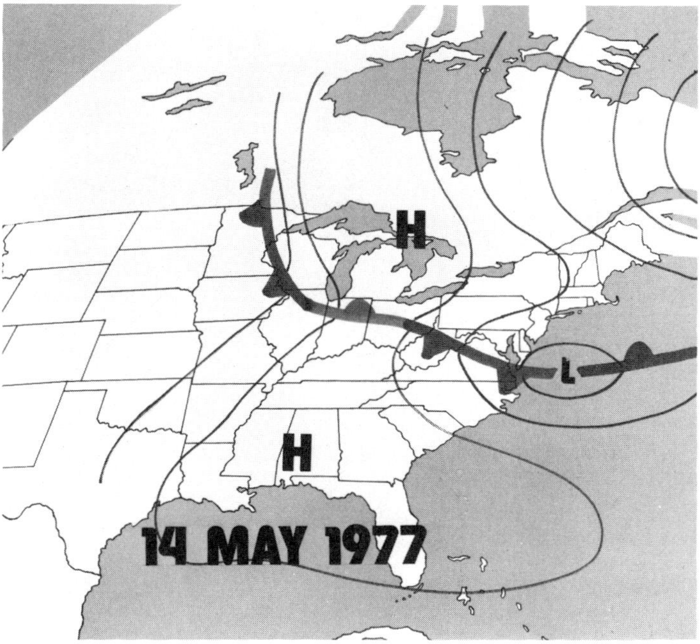

FIGURE 9. Surface weather map, 1200 GMT, May 14, 1977.

low as one to three miles. Hi-vol aerosol measurements taken by the Texas Air Resources Board showed 24-hour averages ranging from around 600 $\mu g/m^3$ near Houston to as high as 2450 $\mu g/m^3$ in northeast Texas.

An often overlooked source of man-made pollution is smoke from agricultural burning. As noted by Parmenter,[17] large areas of slash burning in southern Mexico, Yucatan, and Guatemala during the late winter dry season generate massive smoke palls over much of the Gulf of Mexico. FIGURE 6 shows one of the many such incidents noted in recent years. These smoke-filled air masses frequently reach the southern U.S.

Smoke from giant forest fires in the western U.S. can routinely be monitored on SMS/GOES imagery. However, the plume of smoke from a small northern Minnesota forest fire is shown in its seventh hour (FIGURE 7) after having been tracked at 30-minute intervals from its inception that morning.

DETECTION OF SULFATE AEROSOL HAZES

Lyons and Husar[18] presented an SMS/GOES picture showing a large area of "haze" covering the midwest. Comparisons with other data showed the area to be associated with reduced visibility [6 n mi (nautical mile) or less], high sulfate concentrations, and generally elevated ozone levels. The photograph showed a "typical" episode, on the scale of 1000 km, lasting for several days to nearly two

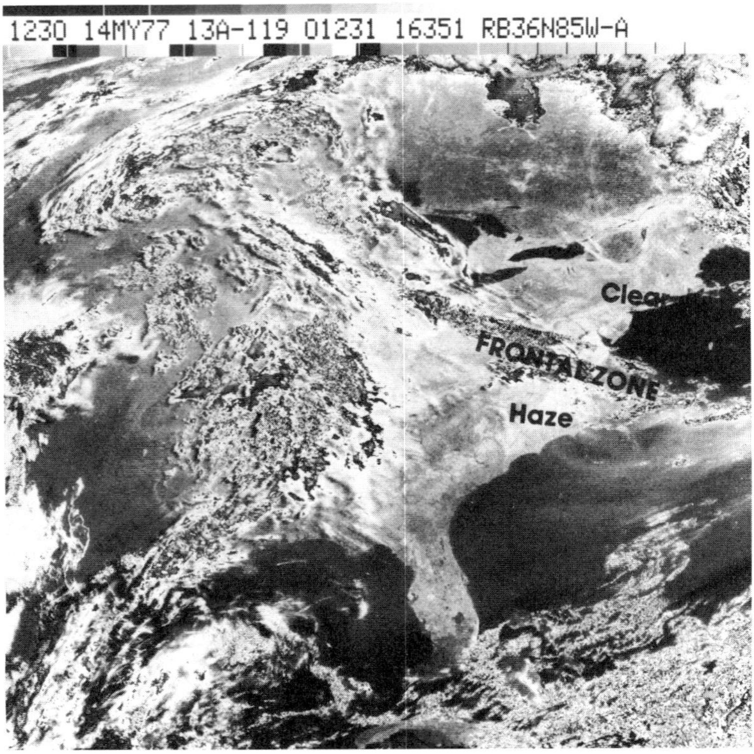

FIGURE 10. SMS/GOES 2.0-km resolution visible image, 1230 GMT, May 14, 1977, specially enhanced by computer processing at NESS. A sharp discontinuity in atmospheric turbidity is clearly found at the front separating the two anticyclones.

weeks, and associated with a warm moist air mass that had significant injections of SO_2 from major source regions. That particular image first needed a certain degree of photographic darkroom enhancement to make the "smog blob" or "hazy blob" or "elevated pollution episode" clearly recognizable.

SMS/GOES image processing is generally geared to highlight high-contrast cloud systems. Haze is not always readily apparent. A technique has been developed by NOAA's¶ National Environmental Satellite Service (NESS), as reported by Parmenter and Anderson,[19] to better delineate the subtly different radiances of polluted versus "clean" air masses. A digital enhancement curve, similar to those applied to infrared imagery, was found very useful. FIGURE 8 shows an example of the NESS technique. The top panel is a segment of a 2-km resolution image over Florida, as would be seen on a standard weather office satellite receiver. Application of digital enhancement makes the band of turbid air across the central part of the peninsula and surrounding waters highly visible (middle panel). Reported surface visibilities were under seven miles within the haze band. Sufficient insolation was reflected to cause

¶National Oceanic and Atmospheric Administration.

significant differential ground heating, as evidenced by the suppression of afternoon convection beneath the haze band (lower panel).

Another example of the NESS technique was produced on May 14, 1977, when a "clean" polar air mass descended into the northeast U.S., displacing southward a stagnant, polluted anticyclone (FIGURE 9). The enhanced image at 1230 GMT (FIGURE 10) shows a clear demarcation of the two air masses, especially over the ocean areas. This effective digital technique, however, is not routinely available from NESS. Thus, researchers studying such phenomena must rely on darkroom enhancement of prints made from negatives or on special processing of "live" or recorded digital satellite data on special image analysis systems. The problem in sulfate haze detection is, therefore, not one of satellite sensor capabilities but one of ground image processing techniques.

CASE STUDIES

The image described by Lyons and Husar[18] was taken on June 30, 1975, the midpoint of an episode lasting from before June 25 until after July 5. It was clearly visible on the prints made from SMS/GOES negatives. FIGURE 11 is a photographi-

FIGURE 11. SMS/GOES 2.0-km resolution visible image, 2200 GMT, June 29, 1975, showing massive "smog blob" west of Appalachians to Kansas and from the Gulf Coast to the upper Great Lakes.

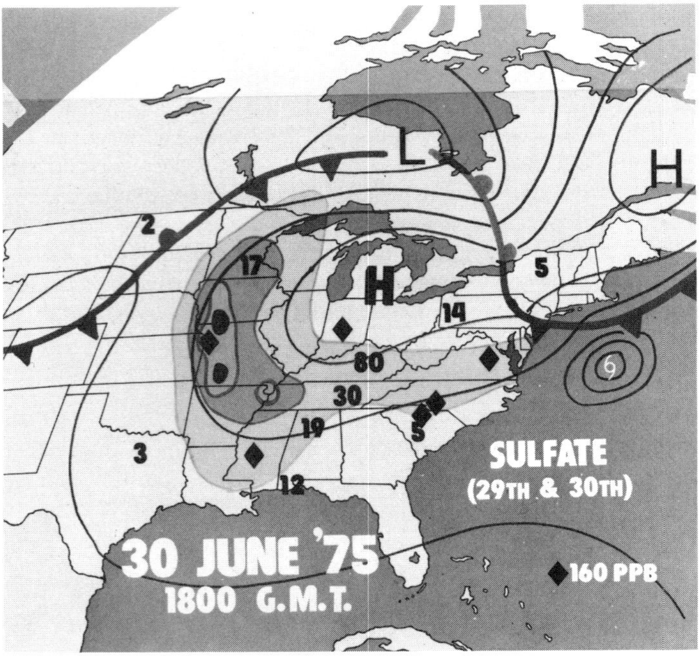

FIGURE 12. Synoptic map, 1800 GMT, June 30, 1975. Shaded areas represent visibility contours of 6, 4, 3, and 2.7 n mi. NASN sulfate values ($\mu g/m^3$) recorded during June 29 and 30th are shown, and diamonds mark oxidant monitors, which recorded maximum hourly values in excess of 160 ppb.

cally enhanced picture, 2-km resolution, June 29, 1975. A persistent high pressure cell over the northeast had caused an accumulation of secondary pollutants over the lower Ohio River Valley and from Louisiana to Minnesota. Trajectories made from 600-m winds confirm this overall flow pattern.[20] Of note in FIGURE 11, in addition to widespread obscuration of ground features in cloud-free areas, are pockets of "reduced turbidity," as suggested by lowered image brightness. Film animation shows that these "pockets" travel with the general low-level winds and retain continuity for a day or more. There is considerable circumstantial evidence relating these to areas of "wash out" by prior thunderstorms. Knowledge of such regions from satellite images would greatly aid in analyzing surface and aircraft aerosol measurements. The volume of air undergoing significant wet removal can also be calculated from such "convective footprints," if correlated to radar data.

FIGURE 12 summarizes some of the measurements made on June 29–30, 1975. The visibility contours show that noontime readings (6, 4, 3, and 2.7 n mi) closely correlate to the haze area seen on the satellite image. Those ozone monitoring sites reporting one-hour averages of 160 ppb or higher were located in, or near, the "blob." Scattered sulfate measurements showed values as high as 80 $\mu g/m^3$ within the haze boundary, and generally 15 $\mu g/m^3$ or less outside.

A major episode occurred from August 16 to 28, 1976. This was discussed in detail by Lyons *et al.*[22] FIGURE 13 shows the "blob" on August 26, 1976. It was particularly "bright" over the Chesapeake Bay area and extended far eastward into the Atlantic.

FIGURE 14 is a map of the 18 GMT, August 22, 1976 weather conditions, with contours of the 5 and 3 n mi visibilities (corrected for precipitation), and with the boundary of the hazy area seen on the satellite picture. This analysis was performed for each 13 days of the episode. In virtually all cases, the 5 n mi visibility contour lay inside the visible boundary of the "hazy blob." It appears that, as regional visual ranges drop below 6 or 7 n mi, the haze becomes "visible" in properly printed SMS/GOES images, especially over water and other low-albedo targets.

FIGURE 15 shows the various parameters for August 28, 1976, overlaid with isopleths of NASN sulfate data. The areas of Pennsylvania and New York State where the visibilities were generally lowest and the image appeared "haziest" were associated within 24-hour sulfate levels of between 30 and 50 $\mu g/m^3$. All sulfate readings made within the hazy area seen on the August 28, 2230 GMT image averaged 22.7 $\mu g/m^3$, but outside the haze area, the average was only 5.7 $\mu g/m^3$. Similarly, a series of sulfate measurements was made in New York City by Brookhaven National Lab[21] during this period (see FIGURE 16). When the "blob," as

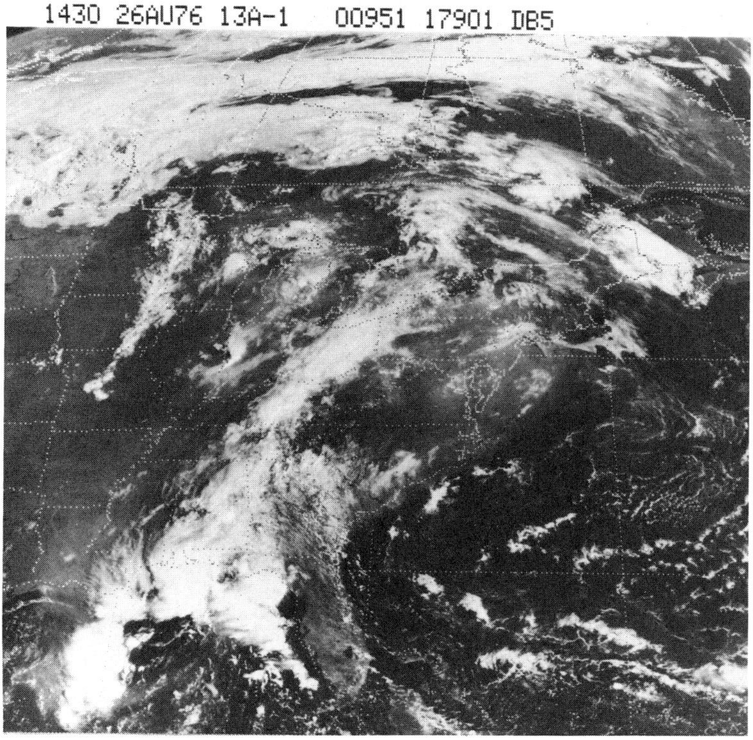

FIGURE 13. SMS/GOES 2.0-km resolution visible image, 1430 GMT, August 26, 1976, showing widespread "smog blob" stretching from the Mississippi River eastward into the Atlantic Ocean.

seen on the satellite prints, covered the area, sulfate values averaged 23.7 $\mu g/m^3$, but were only 7.5 $\mu g/m^3$ when "clean" air covered the area. This suggests overall enhancements of sulfate within satellite-detected "hazy blobs" of 3.98 and 3.16 times for these two cases. Similarly, for each of the 13 episode days, the highest hourly ozone values measured at all stations reported in the National Aerometric Data Bank were plotted. The average for those sites outside the visible "blob" was 69.4 ppb, but inside, the average rose to 90.8 ppb, or 30% higher. This suggests that the correlation of hazy, low-visibility air to sulfates is stronger than it is to ozone levels. This is partly due to the apparently greater spatial variability of ozone, as a result of local NO_x sources, etc.

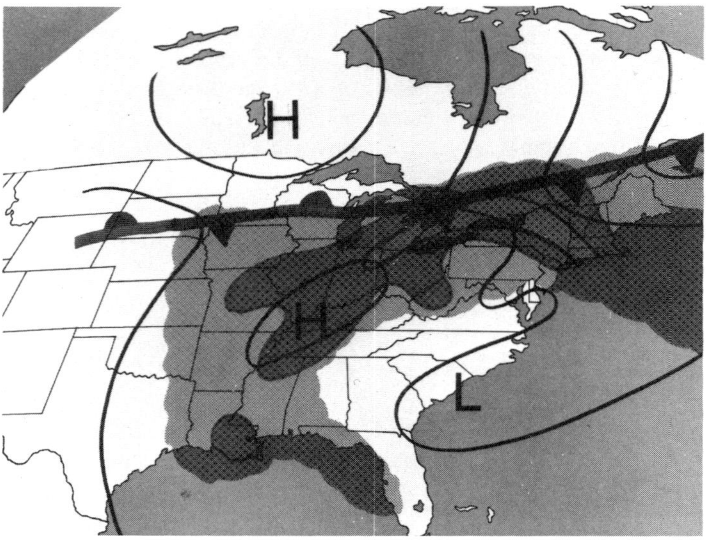

FIGURE 14. Synoptic map, 1800 GMT, August 22, 1976. Stipple represents hazy area as observed on satellite image. Contours of 5 and 3 n mi visibility are also plotted.

As discussed in an earlier paper,[22] the SMS/GOES data can be analyzed quantitatively using such systems as the Man Computer Interactive Data Access System (McIDAS) developed by the University of Wisconsin's Space Science Engineering Center. Digital data tapes from the August 1976 episode were archived. McIDAS allows total pixel-by-pixel manipulations of the imagery. All scenes are navigated and gridded, with animation of sequential images easily accomplished via rapid access from video analog disk. A cursor allows the operator to select any number of image pixels and measure the digital brightness counts (DBC, range 1–256), which can, by application of calibration procedures, be converted into radiances. Over water, with apparently "clean" air masses present (and confirmed by nearby land visibility of seven miles or more), DBCs typically ranged from 40 to 45 units. In hazy, but otherwise cloud-free, areas, values would usually range from 70 to 85 DBCs. Thus, the "signal" of a hazy air mass in the SMS/GOES data was clearly well out of the "noise"

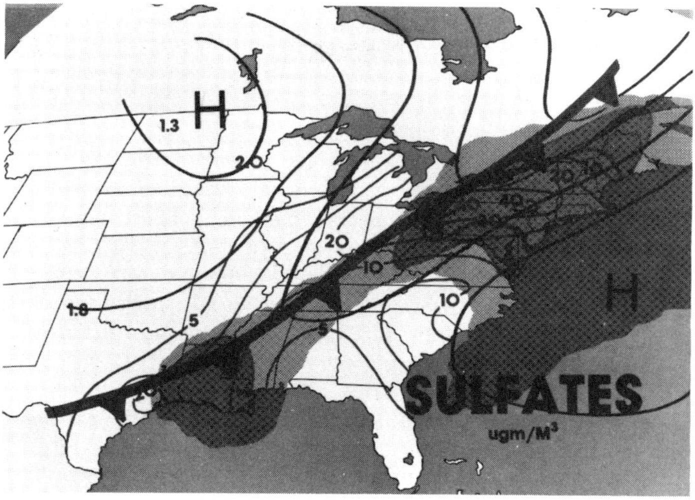

FIGURE 15. Synoptic map, 1800 GMT, August 28, 1976. Overlaid on the hazy-area outline and 3 and 5 n mi visibility contours are the observed NASN sulfate readings ($\mu g/m^3$). Readings in excess of 40 $\mu g/m^3$ were found in low-visibility air in Pennsylvania and New York.

level. Over land, typical values in high-visibility polar air masses during August would be around 60 to 80 DBCs but could be as much as 50% higher when apparent haze covered the target. By careful selection and calibration of ground targets, it appears totally plausible to use digital satellite data to routinely obtain a measure of "atmospheric brightness" and, thus, of turbidity, visual range, and, to lesser degree,

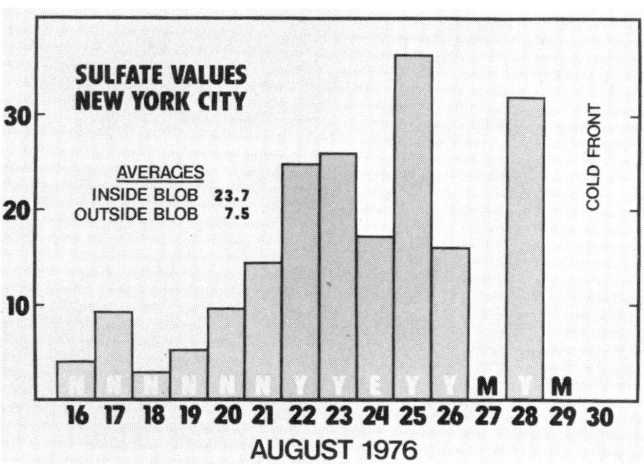

FIGURE 16. Sulfate values, 24-hour averages, recorded in New York City by Brookhaven National Laboratory from August 16 to 28, 1976. N indicates that the smog blob, as seen on the satellite, was not over the area. Y or E indicates that the monitor was in, or on the edge of, the smog blob.

sulfates. What the technique lacks in precision would appear to be compensated for by spatial and temporal coverage—particularly over water, where virtually no routine air quality measurements exist.

CONCLUSION

The current generation of meteorological satellite data, in pictorial, but especially digital, form, permits synoptic monitoring of the formation, growth, and movement of large-scale "hazy blobs" known to be correlated with elevated sulfate episodes. Details of the long range transport and removal process can also be studied. FIGURE 17 shows a "front" of haze moving southwestward along the Atlantic coast on August 25, 1976.

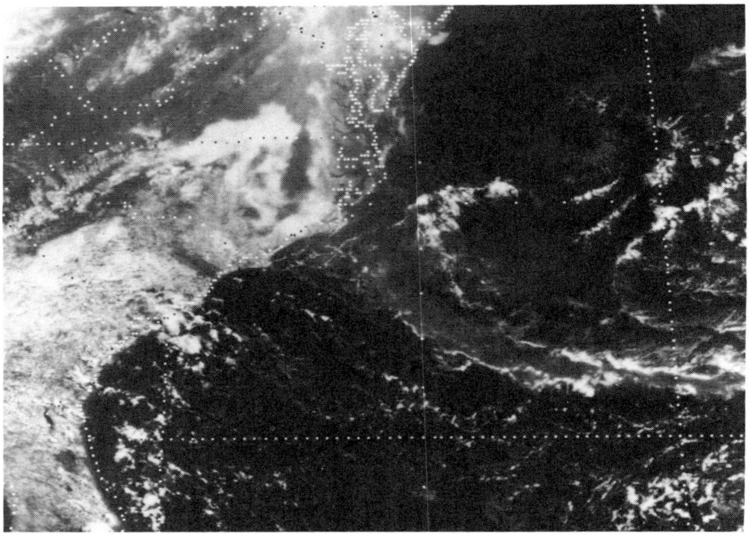

FIGURE 17. Portion of 2.0-km resolution visible image, 2230 GMT, August 25, 1976, showing a "front" of turbid air moving south and west off the U.S. Atlantic coast.

This feature was tracked for three days and finally disappeared on August 26, after being advected into a line of thunderstorms along the Florida peninsula, where its aerosols presumably underwent wash out.

The satellite data allow for rational extrapolation of surface visibility and sulfate measurements, as well as providing a context for the analysis of aircraft data. Tracking well-defined structural haze features provides a possible means for verifying trajectory and transport models. The export of polluted air masses into, and possibly across, the Atlantic can be studied. (The European METEOSAT satellite provides a similar capability for the eastern Atlantic). Actual digital processing of the SMS/GOES data allows for the testing of statistical and physical models relating atmospheric turbidity, visual range, and sulfate aerosol content.

REFERENCES

1. BUMENTHAL, D. L., W. H. WHITE, R. L. PEACE & T. B. SMITH. 1974. Determination of the Feasibility of the Long Range Transport of Ozone or Ozone Precursors. Meteorology Research Inc. Report under EPA Contract 68-02-1462. NTIS Publication No. EPA-450/3-74-0611. U.S. Environmental Protection Agency. Research Triangle Park, S.C.
2. LYONS, W. A. & H. S. COLE. 1976. Photochemical oxidant transport: mesocale lake breeze and synoptic-scale aspects. J. Appl. Meteorol. **15:** 733.
3. COFFEY, P. E. & W. N. STASIUK. 1975. Evidence of atmospheric transport of ozone into urban areas. Environ. Sci. Technol. **9:** 59.
4. WOLFF, G. T., P. J. LIOY, G. D. WRIGHT, R. E. MEYERS & R. CEDERWALL. 1977. An investigation of long-range transport of ozone across the midwestern and eastern United States. Atmos. Environ. **11:** 797.
5. OTT, S. & W. A. LYONS. 1977. Further evidence of long-range photochemical oxidant transport inferred from acoustic sounder data. *In* Proceedings, AMS/APCA Joint Conference on Applications of Air Pollution Meteorology: 33–38.
6. SAMPSON, P. J. & K. W. RAGLAND. 1977. Ozone and visibility reduction in the Midwest: evidence of large-scale transport. J. Appl. Meteorol. **16:** 110l.
7. WHITE, W. H., J. A. ANDERSON, D. L. BLUMENTHAL, R. B. HUSAR, N. V. GILLANI, J. D. HUSAR & W. E. WILSON, JR. 1976. Formation and transport of secondary air pollutants: ozone and aerosols in the St. Louis urban plume. Science **194:** 187.
8. HUSAR, R. B. 1978. Man's impact on the troposphere. *In* Lectures in Tropospheric Chemistry. NASA Reference Publication No. 1022: 319–348. National Aeronautics and Space Administration.
9. MOHR, T. 1971. Picture of the month: air pollution photographed by satellite. Mon. Weather Rev. **99**(8): 653.
10. MCLELLAN, A. 1971. Satellite remote sensing of large scale local atmospheric pollution. *In* Proceedings, Second International Clean Air Congress. H. M. Englund & W. T. Beery, Eds.: 570–574. Academic Press, Inc. New York, N.Y.
11. WILSON, W. E., R. J. CHARLSON, R. B. HUSAR, K. T. WHITBY & D. BLUMENTHAL. 1976. Sulfates in the atmosphere. Paper No. 76-30-06, Preprints, 69th Annual Meeting of the Air Pollution Control Association.
12. LYONS, W. A. & S. R. PEASE. 1973. Detection of particulate air pollution plumes from major point sources using ERTS-1 imagery. Bull. Am. Meteorol. Soc. **54:** 1163.
13. LYONS, W. A., C. S. KEEN & R. NORTHHOUSE. 1974. ERTS-1 satellite observations of mesoscale air pollution dispersion around the Great Lakes. *In* Proceedings, AMS/WMO Symposium on Atmospheric Diffusion and Turbulence. Boston, Mass.: 273–280.
14. RANDERSON, D. 1978. Quantitative analysis of atmospheric pollution phenomena. *In* Skylab Explores the Earth. NASA Publication No. SP 380: 381–400. U. S. Government Printing Office. Washington, D.C.
15. ENSOR, G. J. 1978. User's Guide to the NOAA Geostationary Satellite System. National Environmental Satellite Service, National Oceanic and Atmospheric Administration. U.S. Department of Commerce. Washington, D.C.
16. SHENK, W. E. & R. J. CURRAN. 1974. The detection of dust storms over land and water with visible and infrared measurements. Mon. Weather Rev. **102**(12): 830.
17. PARMENTER, F. C. 1978. Monitoring air quality from satellites. *In* Proceedings, 4th Joint Conference on Sensing of Environmental Pollutants: 254–257. American Chemical Society. Washington, D.C.
18. LYONS, W. A. & R. B. HUSAR. 1976. SMS/GOES visible images detect a synoptic-scale air pollution episode. Mon. Weather Rev. **104**(12): 1623.
19. PARMENTER, F. C. & R. K. ANDERSON. 1977. A satellite overview of inadvertent weather modification. *In* Preprints, Sixth Conference on Inadvertent Weather Modification: 83–86. American Meteorological Society. Boston, Mass.
20. BRAND NIEMANN. Personal communications.
21. LIOY, P. J., G. T. WOLFF & B. P. LEADERER. 1979. A discussion of the New York Summer Aerosol Study, 1976. Ann. N.Y. Acad. Sci. **322:** 153.
22. LYONS, W. A., J. C. DOOLEY, JR. & K. T. WHITBY. 1978. Satellite detection of long range pollution transport and sulfate aerosol hazes. Atmos. Environ. **12:** 621.

THE FATE OF PARTICULATE EMISSIONS FROM AN ISOLATED POWER PLANT IN THE OIL SANDS AREA OF WESTERN CANADA*

Leonard A. Barrie

Atmospheric Environment Service
Downsview, Ontario, Canada M3H 5T4

Introduction

Numerous pathways through the atmosphere can be taken by particulate matter from a point source on its way to a receptor. The route is determined by several factors, including the height of emission, the geographical and meteorological situation, and particle size. Large particles (diameter $> 10\ \mu m$) released near the earth's surface will travel scarcely more than a kilometer before being deposited. On the other hand, small particles (diameter $< 1\ \mu m$) released several hundred meters above ground level may remain airborne for several days before they are removed by precipitation or by dry deposition processes. In recent years, tall stacks have been used to alleviate or avoid severe local pollution problems. Consequently, a larger fraction of gaseous and particulate emissions is exported from the vicinity of the source and contributes to ambient aerosol concentrations hundreds, or even thousands, of kilometers downwind.

In order to assess the impact, locally or regionally, of pollutant releases from a point source, gaseous and particulate deposition as a function of distance and direction from the origin must be known. It is often difficult to distinguish the deposition of material from one source from that of other sources at distances of more than a few kilometers away. From 1976 to 1978, the author had the opportunity to investigate the nature and fate of emissions from an isolated power plant in the Athabasca oil sands area of western Canada (FIGURE 1). Because of low background pollutant levels in that part of the continent, deposition of particulates from the source could be detected up to 40 km away. The following parameters were measured:

1. Particulate elemental concentrations in air 80 km from the source in March 1976 and close to the source in June 1977;
2. Dry deposition patterns of particulate sulfur and heavy metals in June 1977;
3. Total (wet plus dry) deposition patterns of major ions and metals during the winters of 1975/76 and 1977/78.

At the same time, plume chemistry studies were conducted to investigate the oxidation of SO_2 during summer and winter,[1] and fly ash was analyzed for heavy metals.[2]

The Source

The source, an oil sands extraction plant, is located 30 km north of Fort McMurray, Alberta, on the west bank of the Athabasca River. It is situated in a

*Supported by the Alberta Oil Sands Environmental Research Program.

0077-8923/80/0338-0434 $01.75/1 © 1980, NYAS

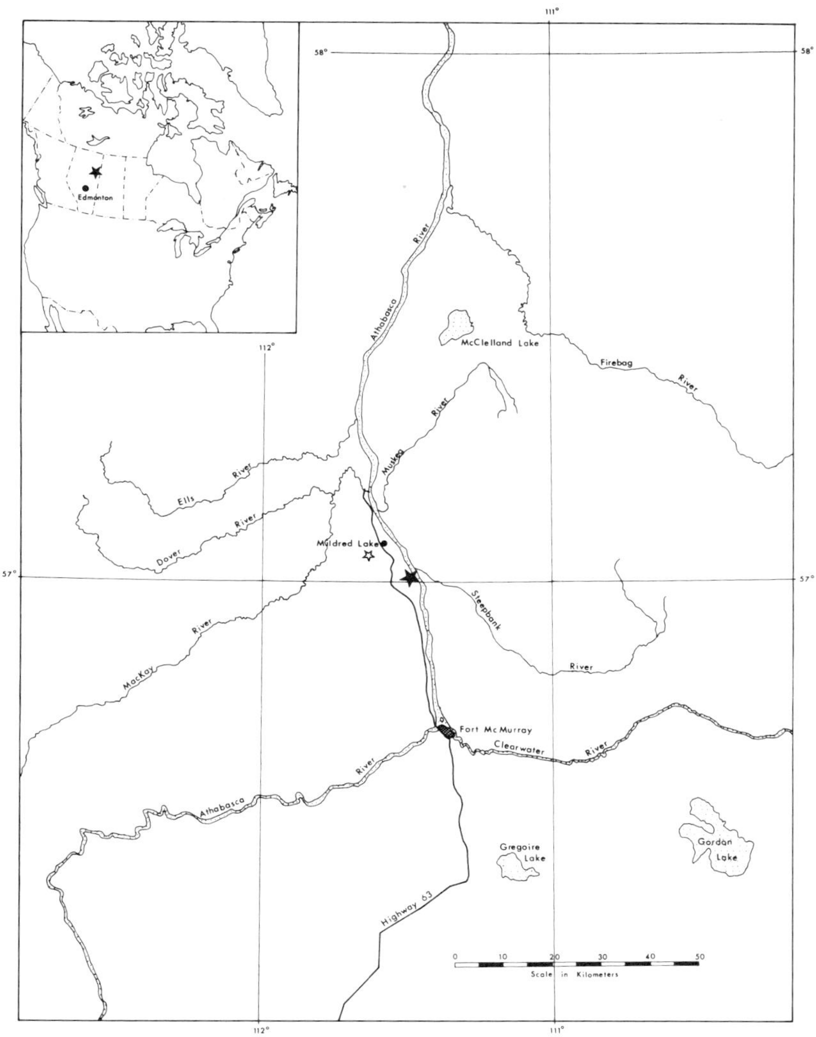

FIGURE 1. The study area. ★ marks the location of the pollutant source. ☆ marks the location of the Syncrude oil extraction plant that was being constructed.

shallow river valley 60 m deep and 2–3 km wide, which cuts through a broad upland plain 50–100 km wide. The bulk of emissions are from a coke-burning power plant, whose 107-m stack emits approximately 200 tonnes† (t) of sulfur dioxide and 27 t of particulate matter into the atmosphere daily. A second, minor source of sulfur dioxide is a 107-m incinerator stack, 260 meters from the main stack, which releases 5–20 t per day. Electrostatic precipitators were not in use.

†tonne = 10^3 kg.

TABLE 1

THE ELEMENTAL COMPOSITION OF THREE FLY-ASH SAMPLES COLLECTED FROM THE POWER PLANT EMISSIONS USING ELECTROSTATIC PRECIPITATION[2]

	Weight Fraction of Sample (%)		
Element	1	2	3
Silica (SiO_2)	34.6	36.3	31.9
C	11.3	10.3	20.7
Al	11.2	11.4	12.3
Fe	4.6	4.5	4.8
V	2.4	2.8	2.4
Ca	2.2	2.2	1.4
Ti	1.9	1.9	1.4
S	1.5	1.6	1.6
Ni	0.99	1.1	0.94
Mg	0.84	0.85	0.71
Mo	0.24	0.25	0.21
Mn	0.090	0.084	0.095

The elemental composition of fly-ash samples collected by electrostatic precipitation from the main stack is summarized in TABLE 1. Three constituents—carbon, aluminum, and silicates—comprise in excess of 50% of the total mass. Six other elements—including vanadium, titanium, and nickel—each account for 1 to 5% of the total mass.

The oil-extraction operation is the main source of oxides of sulfur and nitrogen in the Ft. McMurray area. It is also one of the three particulate sources. The others are windblown dust and motor vehicle emissions. Windblown dust, raised by extraction operations, by traffic on dirt roads, and naturally by the wind from the earth's surface, is predominantly a spring, summer, and fall phenomenon. In winter, snowcover and frozen soil effectively reduce this source to zero. Emission-factor estimates of particulate releases from vehicular activity, centered on the town of Ft. McMurray and along the highway to the north and south, indicate that total particulate emissions equal those from the power plant. Since vehicular emissions are spread out over the region, their contribution to ambient air concentrations and to particulate deposition around the source is probably small and will be neglected.

On a regional basis, the oil sands area is a rather isolated pollutant source. The city of Edmonton (population 500,000), 370 km away, is the nearest large pollution source.

AMBIENT AIR CONCENTRATIONS

Ambient air concentrations of gaseous and particulate sulfur were measured at Mildred Lake 10 km NNW of the source (see FIGURE 1) from June 13–26, 1977. These, and temperature, wind speed, and wind direction, are shown as a function of time in FIGURE 2. Gaseous and particulate sulfur were separated and collected by two filters in series in a device similar to that used by Johnson and Atkins.[3] The filters were then extracted, and the extract analyzed for sulfur. A more detailed description of the sampling technique is available.[4]

Concurrently with the sulfur measurements, three 3-day aerosol samples were collected on Whatman 41, 47 mm filters (sampling periods A, B, and C in FIGURE 2) and analyzed for their elemental content by instrumental neutron activation analysis. As indicated by ambient-sulfur concentrations, emissions from the source reached the sampling site throughout most of period B, during part of period A, and not at all

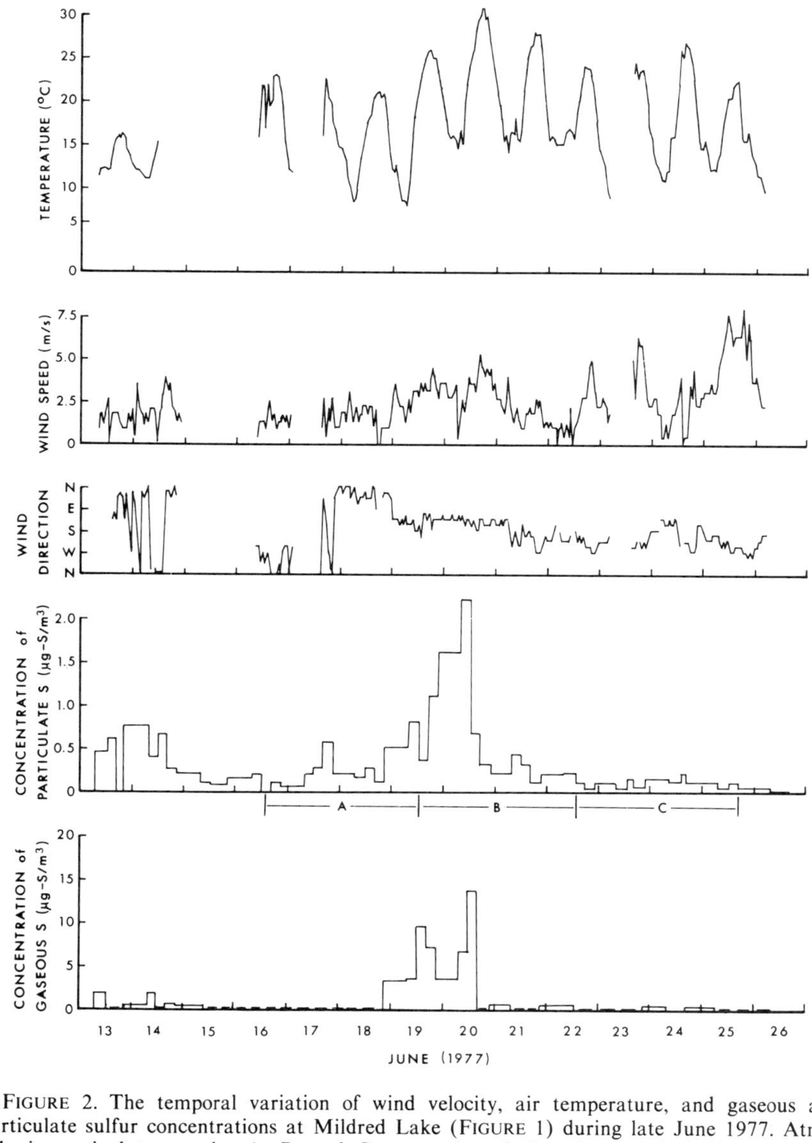

FIGURE 2. The temporal variation of wind velocity, air temperature, and gaseous and particulate sulfur concentrations at Mildred Lake (FIGURE 1) during late June 1977. Atmospheric particulate samples A, B, and C were taken during the periods indicated on the abscissa.

during period C. In TABLE 2, the concentrations of airborne particulate elements during these periods are compared to those obtained at Birch Mountain, 80 km NNW of the source, under clean air conditions (i.e., northerly winds) in March 1976.

Atmospheric particulate matter at Mildred Lake can be subdivided into three components:

1. A windblown dust component originating from particles raised from the earth's surface as the result of natural processes and of man's activities;
2. An anthropogenic component originating from particulate emissions;
3. A background component that prevails in the absence of the other two sources and owes its existence to a variety of complex processes that contribute to the residual particulate content of an air mass remote from large sources.[5]

From the ambient concentrations reported in TABLE 2, it is possible to estimate quantitatively the relative magnitude of the source components for each element during polluted period B at Mildred Lake. The background component can be represented by the average concentration of samples 1, 2, and 3 from Birch Mountain, which were taken when the land was snow covered (no windblown dust) and when winds were persistently from the northern sector. The windblown dust component is the difference between the period C concentration and the background concentration. As noted above, sample C was taken during summer when anthropogenic emissions were absent, as indicated by low sulfur concentrations.

The anthropogenic component is represented by the difference between concentrations during the polluted and unpolluted periods, B and C, respectively. The relative magnitude of windblown, anthropogenic, and background aerosol components during a polluted period near the source were thus calculated (TABLE 3). Even during a pollution episode, most elements have a substantial windblown dust component. Sulfur and vanadium were the only exception. Therefore, they are the best tracers of anthropogenic emissions. In winter, the windblown component is low. Consequently, such elements as Al, Mn, Sc, Ti, K, and, to a lesser degree, Ca join S and V in the list

TABLE 2

AMBIENT AIR CONCENTRATIONS OF ELEMENTS IN PARTICULATE MATTER AT MILDRED LAKE 10 km NNW OF THE SOURCE AND AT BIRCH MOUNTAIN 80 km NNW OF THE SOURCE (ng m^{-3})

	Mildred Lake, June 1977			Birch Mt., March 1976		
	A	B	C	1	2	3
Element	16–19	19–22	22–25	3–7	7–13	13–17
Al	1500	2300	1500	41	56	70
As	0.68	0.87	0.37	0.37	1.1	0.44
Ca	2200	2800	2700	19	41	40
Fe	3000	2400	890	—	—	—
K	350	650	430	32	25	35
Mg	260	870	780	31	23	26
Mn	28	49	38	0.66	0.93	7.0
Sc	0.45	0.41	0.16	—	0.014	0.017
Ti	110	210	82	—	6.7	7.9
V	80	110	4.2	0.52	4.5	4.7

TABLE 3

THE RELATIVE MAGNITUDE OF PARTICULATE SOURCE COMPONENTS IN AMBIENT AIR AT MILDRED LAKE DURING A POLLUTED EPISODE (PERIOD B FIGURE 2)

	Anthropogenic	:	Windblown	:	Background
Al	20	:	28	:	1
As	0.6	:	0.0	:	1
Ca	3	:	80	:	1
K	7	:	13	:	1
Mg	3.3	:	28	:	1
Mn	13	:	43	:	1
Sc	16	:	9.3	:	1
Ti	18	:	10	:	1
V	32	:	0.3	:	1
S	5.9	:	1	:	1

of useful anthropogenic tracers. One would expect, therefore, that near the source in winter, the deposition patterns of S and V would be similar to those of the other elements but would be quite different in summer. A more detailed discussion of the relative contribution of various sources to deposition is given in the following section.

DRY DEPOSITION

Dry deposition of particulates occurs by sedimentation, impaction, interception, and diffusion to objects at the earth's surface. Measuring dry deposition fluxes representatively is very difficult. Recently, the development of a fast-response monitor for particles of 0.05–0.1 μm diameter has enabled micrometeorologists to use eddy correlation flux techniques.[6] This approach demands elaborate and costly instrumentation, uniform surfaces, and quasi-steady state conditions, which makes its use in a dense network near a source impractical. Alternatively, artificial collectors can be used.

A common drawback of artificial dry deposition collectors is our poor understanding of their representativeness. A collector developed at Harwell, England, by Cawse[7] was chosen for this study (FIGURE 3), because its collection efficiency relative to a natural surface (rye grass) has been experimentally determined as a function of particle size.[8] The collector consists of a 25 x 20 cm Whatman 541 filter paper enclosed in a polyethylene filter holder and exposed horizontally, face up, 12 cm beneath an acrylic rain protector (1 m x 1 m) and 1.5 m above the ground. The collector's collection efficiency relative to rye grass is 100% in the particle size range 2–10 μm (where the main mass of fly-ash particles occurs) and is less than 100% for smaller particles. The poorest collection efficiency is 10% for particles 0.5 μm diameter. At present, the relationship between deposition to the Harwell collector and to other natural surfaces, such as a forest canopy, is not known and is likely to be the subject of much research in the next decade. Nevertheless, useful information about the rates and pattern of coarse-particle deposition in an area can be obtained using this collector.

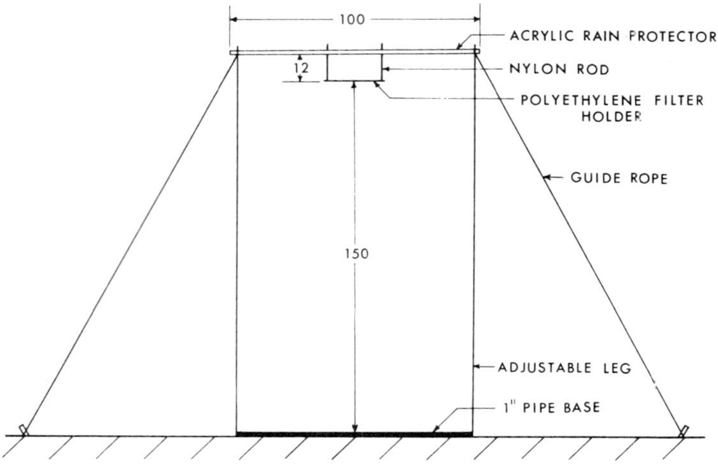

FIGURE 3. The Harwell particulate dry deposition collector.

The Harwell collector was chosen not only because it has been investigated experimentally, but also because it has several favorable characteristics:

1. The filter collection medium is easily handled and analyzed; it is also SO_2 inert;
2. The simple geometry of the collector lends itself to theoretical considerations;
3. Recent tests by the author at the Atmospheric Environment Service, Downsview, Ontario, Canada, using six collectors simultaneously, have shown that the reproducibility of a deposition measurement is better than 25% for all elements.

For the last two weeks in June 1977, a 15-station network of Harwell collectors was used to measure particulate deposition patterns about the source (FIGURE 4). After two weeks of exposure, seven elements were detected above field blank levels in exposed filters by neutron activation analysis, namely, Al, Ca, Mg, Mn, Ti, and V. Sulfur was also measured by isotope dilution analysis after hot water extraction.

Deposition patterns for sulfur and vanadium (FIGURE 5) were quite different than those for other elements (FIGURE 6). There is an easterly bulge in the deposition patterns of S and V that is absent in the patterns of the other elements. The bulge corresponds to deposition occurring when surface winds were from the west to southwesterly sector during the sampling period. The elements Al, Ca, Mg, Mn, and Ti are most heavily deposited on the west side of the river near the Syncrude oil extraction plant that was under construction and along the dirt road north of the source.

A principal component analysis conducted independently on the data also grouped the elements in the same configuration suggested by the deposition patterns. Three statistical components were identified that best explained the variance in the data set when used in a linear regression analysis. They were (1) Al, Ca, Mg, Mn, and Ti; (2) S; and (3) V. Sulfur and vanadium were not well correlated, even though ambient air data at Mildred Lake suggest that both are predominantly of anthropogenic origin during a pollution period (TABLE 3). The reason for this becomes clear when one calculates what fraction of total deposition of each element is anthropogenic in origin.

The relative magnitude of the various source components of an element in deposition depends on: the relative contribution of each source to the airborne particulate during pollution and nonpollution events; the deposition velocity of each component; and the frequency of occurrence of anthropogenic emissions at a site. Consider a site that receives polluted air p percent of the time in which the ratio of

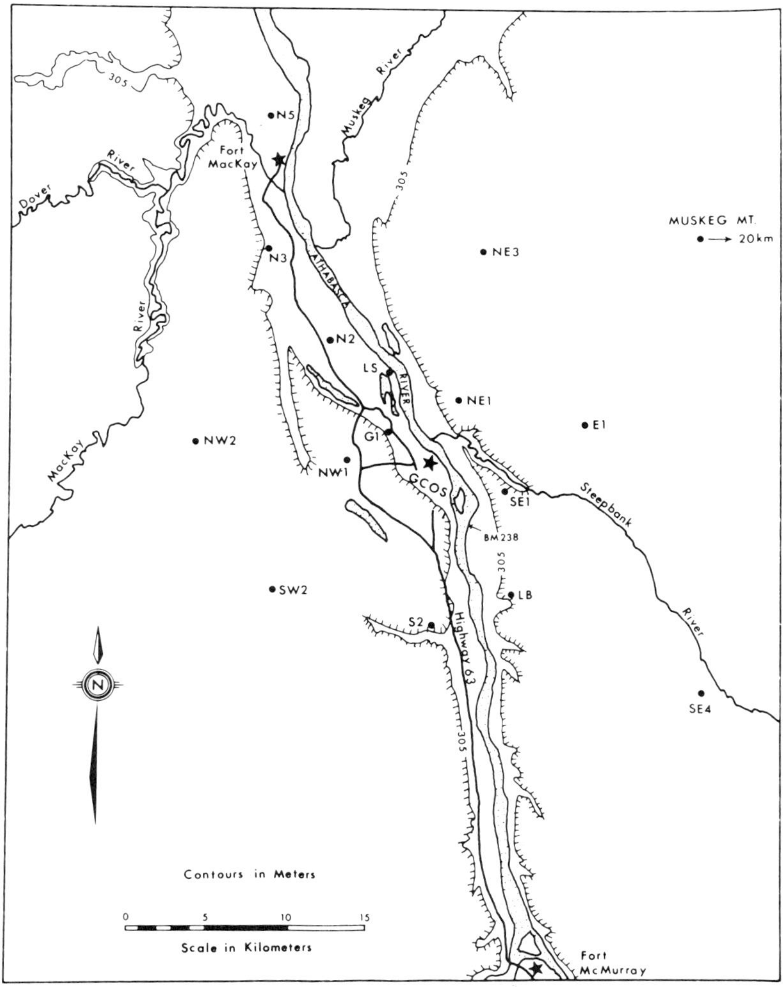

FIGURE 4. Sites at which particulate dry deposition was collected between June 10 and 26, 1977, using the Harwell collector.

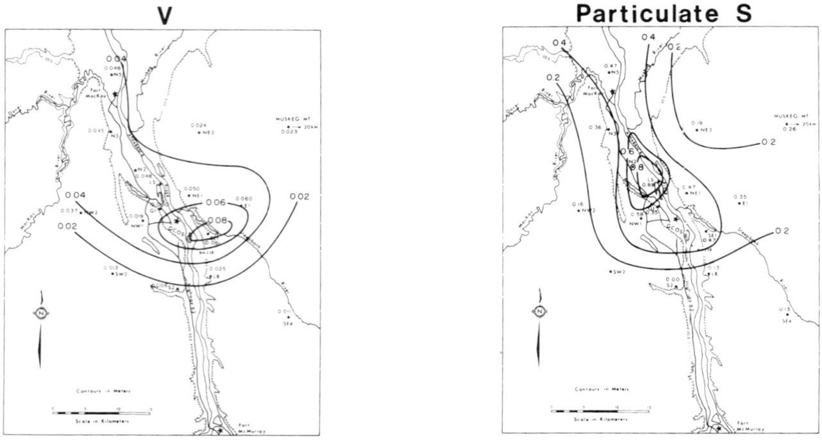

FIGURE 5. Dry deposition patterns of particulate vanadium and sulfur (in kg hr^{-1} yr^{-1} = 3.2 ng m^{-2} s^{-1}) during June 10–26, 1977.

anthropogenic : windblown dust : background components of a particular element is $a : d : b$. Then it can be shown that in deposited particles, the relative proportion of the element from each source is approximately:

$$0.01\, p \frac{V_a}{V_b} a : \frac{V_d}{V_b} d : b \qquad \mathbf{(1)}$$

where V_a, V_d, and V_b are representative deposition velocities for each of the source components. For particles larger than 0.5 μm, deposition velocity increases with increasing particle size. Consequently, coarse windblown dust particles have higher deposition rates than smaller particles. In the derivation of Equation 1, it was assumed that windblown and background concentrations averaged over several days do not vary significantly between polluted and unpolluted events.

At Mildred Lake, p was approximately 20% during the study period (inferred from gaseous S in FIGURE 2). The relative magnitude of source components for sulfur and vanadium in deposition (TABLE 4) were calculated using Equation 1, the airborne concentration ratios in TABLE 3, and representative deposition velocity ratios. Anthropogenic vanadium from this source comprised about 80% of the total amount deposited, while only 10% of deposited sulfur was related to the source. It is therefore not surprising that vanadium and sulfur were not well correlated in deposited matter. In areas with less windblown dust than the Mildred Lake site and with more frequent pollution episodes (viz., the area to the east of the source), the anthropogenic component would be higher; while in remote areas where p is low, anthropogenic particles may make up an even smaller fraction of the deposition than at Mildred Lake.

A mass balance of anthropogenic sulfur and vanadium was done, using the measured dry deposition patterns (FIGURES 5 and 6). During the study period, 5500 kg of particulate sulfur were released, and 584 kg were deposited within 25 km. Bearing the above considerations in mind, only about 10% of that deposited originated from

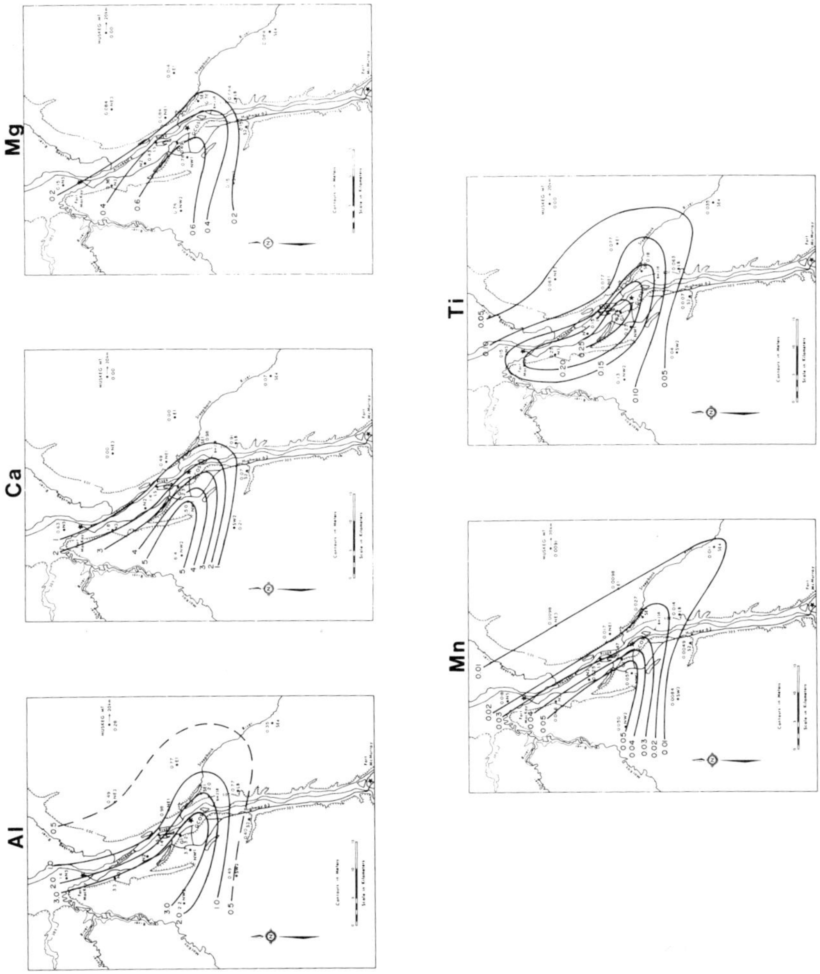

FIGURE 6. Dry deposition patterns of particulate aluminum, calcium, magnesium, manganese, and titanium (in kg hr^{-1} yr^{-1} $\equiv$ 3.2 ng m^{-2} s^{-1}) during June 10–26, 1977.

TABLE 4

ESTIMATED RELATIVE MAGNITUDE OF PARTICULATE SOURCE COMPONENTS IN DRY DEPOSITION NEAR MILDRED LAKE IN LATE JUNE 1977*

Element	$a : d : b$	V_a/V_b	V_d/V_b
S	1.2 : 10 : 1	1	10
V	6.4 : 0.6 : 1	1	2

*a, d, and b are the anthropogenic, windblown dust, and background components, respectively. V_a, V_d, and V_b are the corresponding deposition velocities.

the source. Therefore, approximately 1% of the particulate sulfur released by the source was dry deposited within 25 km.

A similar estimate was made for vanadium; 9200 kg were released by the source; 70.6 kg were dry deposited, of which about 80% was anthropogenic. Consequently, about 0.6% of the man-made vanadium released was dry deposited within 25 km.

TOTAL DEPOSITION

In winter, a persistent snow cover can be used as a natural collector of material removed from the atmosphere by wet and dry deposition processes. Precautions must be taken to sample snow that has not been subjected to a prolonged thaw, since as much as 40% of a snowpack's pollutant content can be leached away by the first 10% of meltwater.[9]

In March 1976, a snow chemistry survey was conducted, with light winds and generally stable conditions, in which the deposition patterns of sulfur and hydrogen ion were measured.[10] It was found that 0.14% of the total sulfur released by the source in a three-week period was deposited within 25 km. An average dry deposition velocity of 0.2–0.4 cm/s for SO_2 was deduced from the deposition measurements on the snow and from measured or calculated ambient air concentrations.[10]

In late January 1978, a second snow chemistry survey was conducted to examine deposition not only of sulfur, but also of major ions and metals. The sampling and analysis techniques employed are reported in detail elsewhere.[11] However, for the sake of completeness, a brief description will be given here.

Snow was sampled at 60 sites within 140 km of the source (FIGURE 7). An acrylic plastic snow corer was used. One set of three cores was melted and analyzed for major ions and pH, using standard wet chemical techniques.

Another set of cores was melted and filtered through a 0.45-micron Sartorius cellulose acetate paper, and analyzed for soluble and insoluble metals using neutron activation analysis and atomic absorption spectroscopy. To avoid metal contamination or loss, care was taken to store filtered snowmelt at a pH of 1.5 in polyethylene containers presoaked in nitric acid. Only those results relevant to the theme of this paper—namely, mass budgets—are presented here; the full data set can be found elsewhere.[11]

The snowpack had accumulated pollutants without experiencing a melt for the 70-day period November 18, 1977 to January 26, 1978. Snowpack loading of insoluble aluminum, vanadium, and manganese, as well as soluble aluminum, vanadium, sulfate, and nitrate, was determined (TABLE 5). At all but a few sites, less than 10% of vanadium and 1% of aluminum in deposited matter were soluble. For all intents and purposes, the insoluble fraction of these metals therefore represents the total amount deposited. (In the absence of soluble data for manganese, it will be assumed in mass balance calculations that most of it is insoluble.) Total deposition of sulfur and nitrate is accurately characterized by that of their soluble fractions.

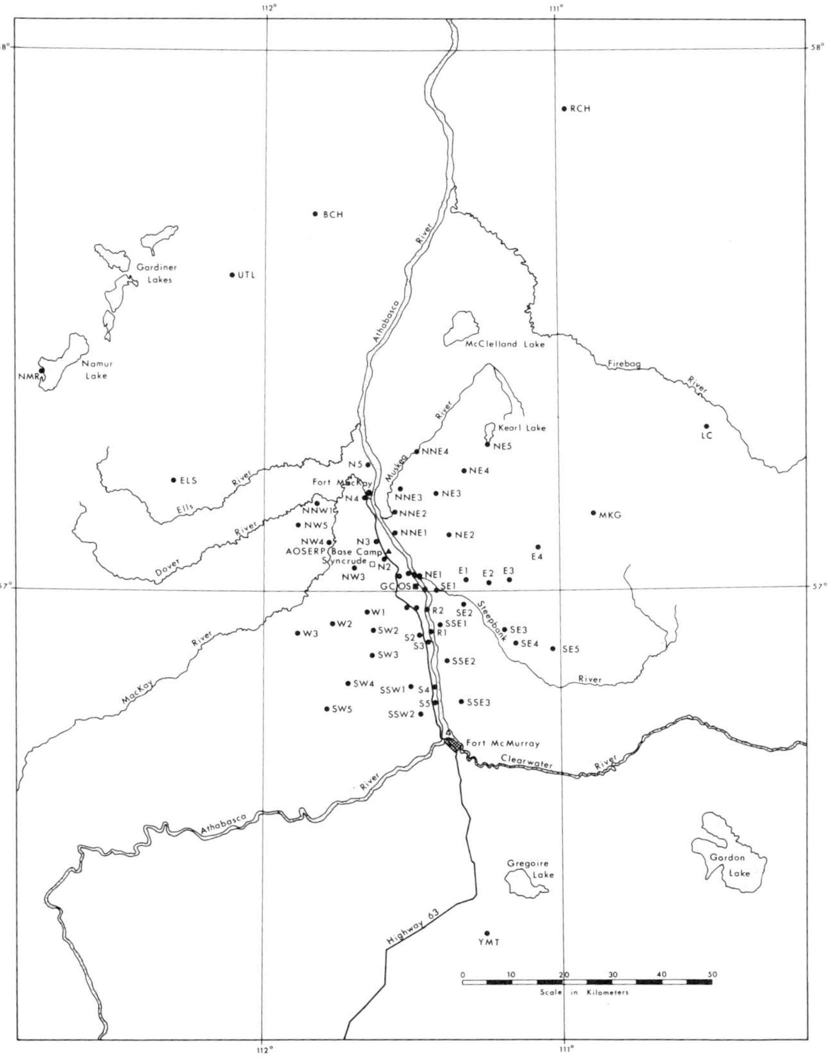

FIGURE 7. Location of snow sampling sites for the January 1978 snow chemistry survey.

TABLE 5

SNOWPACK CHEMISTRY DATA FOR SITES IN FIGURE 7*†

SITE	SNOWPACK LOADINGS ($mg\ m^{-2}$) Al SOL	Al INS	V SOL	V INS	Mn INS	$SO_4^=$-S	$NO_3^=$-N	pH
NNE1	0.7466	M	1.7919	M	M	16.8	8.08	4.9
NNE2	0.7127	100.0	1.1912	17.20	1.00	17.1	7.26	4.7
NNE3	0.4757	53.0	0.6554	8.70	0.30	11.5	6.45	5.0
NNE4	0.3268	43.0	0.5065	6.80	0.20	9.5	5.71	5.0
NE1	0.1958	380.0	2.5653	57.30	2.90	20.6	7.39	6.0
NE2	0.4554	46.0	D.L.	7.60	0.30	7.3	3.93	5.0
NE3	0.3941	44.0	0.6418	6.90	0.30	12.5	7.06	4.9
NE4	0.3006	44.0	0.3006	7.10	0.30	10.0	4.74	5.0
NE5	0.2444	44.0	0.3239	5.90	D.L.	21.0	7.63	6.6
E1	0.3141	130.0	0.9676	17.20	0.80	28.4	9.70	5.1
E2	0.4100	81.0	0.6355	12.60	0.60	8.4	5.87	5.0
E3	0.2333	46.0	0.3966	7.20	0.30	7.4	4.78	5.0
E4	0.5586	49.0	0.3072	6.10	0.20	10.0	4.83	5.0
MA	0.1300	7.0	0.2275	0.90	D.L.	3.2	2.34	5.2
MB	D.L.	6.0	0.1960	0.80	D.L.	M	2.22	5.0
MC	0.3903	9.0	0.1774	0.90	D.L.	3.8	3.85	5.0
MD	0.1857	9.0	0.2105	1.00	D.L.	3.4	3.21	5.0
ME	0.1257	7.0	0.2388	0.80	D.L.	3.0	2.64	5.0
SE1	0.7125	370.0	2.7731	57.30	2.40	20.2	6.84	5.4
SE2	0.5439	240.0	1.4006	33.20	1.10	13.5	6.07	5.1
SE3	0.1452	44.0	0.3242	5.60	0.30	7.4	4.36	6.2
SE4	0.1376	30.0	0.2843	4.20	0.20	12.6	4.81	5.4
SE5	0.0901	26.0	0.1126	3.70	0.10	14.1	2.60	6.6
SSE1	0.5799	110.0	1.1856	17.20	0.80	15.4	6.38	5.0
SSE2	0.1510	60.0	0.6241	3.30	D.L.	17.4	4.85	6.8
SSE3	0.3250	48.0	0.2990	6.30	0.30	M	2.10	5.9
S1	0.1985	1000.0	6.0465	149.00	8.40	51.1	10.98	6.1
S2	0.3428	460.0	2.9070	66.50	3.60	32.3	11.35	5.4
S3	0.3354	140.0	1.6366	21.80	1.10	31.6	1.84	4.9
S4	0.3210	130.0	0.6035	18.30	0.90	24.4	3.05	6.6
S5	0.1605	110.0	0.4173	12.60	0.80	24.1	8.42	7.1
SSW1	0.9117	210.0	2.1881	33.20	1.50	16.0	7.83	5.4
SSW2	0.5549	100.0	1.1300	17.20	0.70	15.9	6.78	5.1
SW1	D.L.	640.0	1.8954	56.20	3.80	23.8	11.61	6.6
SW2	D.L.	92.0	0.4076	9.70	0.90	17.1	6.72	M
SW3	D.L.	80.0	0.4095	10.80	0.60	7.8	6.35	6.4
SW4	D.L.	36.0	0.3462	4.50	0.20	16.4	8.28	5.2
SW5A	D.L.	6.0	0.1703	1.00	D.L.	8.5	6.36	5.4
SW5B	D.L.	5.0	0.0949	0.80	D.L.	M	9.78	5.3
SW5C	0.0986	D.L.	0.1183	0.60	D.L.	8.2	6.32	5.1
SW5D	0.1078	6.0	0.0808	0.70	D.L.	8.0	6.20	5.1
SW5E	0.1101	5.0	0.1596	0.80	D.L.	7.3	5.82	5.8
W1	0.3594	49.0	0.5895	6.50	0.30	10.7	7.23	5.0
W2	D.L.	18.0	0.2122	2.10	1.00	5.8	3.59	7.3
W3	0.1548	17.0	M	2.70	D.L.	M	5.53	5.0
NW3	0.3347	64.0	0.3681	5.00	0.40	8.0	5.99	4.9
NW4	0.2169	32.0	0.2440	4.40	0.20	7.4	6.27	5.1
NW5	0.0894	23.0	0.1476	3.30	0.20	20.1	5.23	7.7
NNW1	M	M	M	M	M	7.3	5.49	5.3
N1A	0.3451	100.0	1.7452	14.90	0.80	16.7	7.44	6.0
N1B	M	100.0	2.1265	13.80	0.70	16.4	7.87	5.8
N1C	0.2763	190.0	1.9010	25.20	1.30	17.4	7.71	5.7
N1D	0.3302	110.0	1.7831	16.10	1.00	15.4	7.49	6.2
N1E	0.3440	110.0	2.0294	14.90	0.80	17.5	7.55	6.1
N2	0.1445	370.0	1.2231	27.50	2.20	16.3	8.60	6.9
N3	0.2678	110.0	0.8972	14.90	2.50	15.2	8.48	6.7
N4	D.L.	69.0	0.6654	8.50	0.50	11.2	7.59	5.1
N5	0.5237	40.0	0.8118	5.90	0.20	11.4	7.31	5.1
BM	0.0973	9.0	D.L.	0.30	D.L.	6.8	1.70	5.1
G1	0.1331	540.0	5.3245	75.70	4.10	24.5	11.48	6.1
G5	0.1926	780.0	1.7047	61.90	6.40	39.0	9.89	6.9
R1	0.2761	150.0	1.5540	24.10	1.40	18.2	7.67	6.7
R2	0.1784	480.0	1.5788	35.50	4.00	19.5	10.74	6.2
R3	0.0864	510.0	3.6049	79.10	3.40	23.3	6.91	5.6
R4A	0.1651	160.0	3.0488	18.30	8.90	25.1	9.02	6.7
R4B	0.1617	110.0	2.9853	14.90	0.70	23.5	6.82	6.4
R4C	0.2110	150.0	2.8691	79.10	1.00	26.5	8.05	6.5
R4D	0.2155	220.0	3.2548	31.00	1.30	29.1	8.48	6.5
R4E	0.2752	180.0	3.3790	26.40	1.30	32.4	9.07	6.5
ELS	0.1428	8.0	0.1357	1.10	0.60	10.0	M	M
RHS	0.1368	50.0	D.L.	9.00	0.70	M	8.89	4.9
NMR	0.0896	M	D.L.	M	M	6.3	3.49	4.8
SMT	0.4465	12.0	0.2143	1.00	0.50	23.2	7.86	4.8
LC	0.1980	M	0.1188	M	M	M	6.27	4.9
UT	0.1404	2.0	D.L.	0.20	D.L.	M	6.67	4.9
LS	0.2408	140.0	2.7978	18.40	1.00	41.6	19.45	6.3

*The snowpack was exposed without experiencing a melt from November 18, 1977 to January 26, 1978.

†Analyses of five samples, labeled A–E, are given for sites M, SW5, N1, and R4.

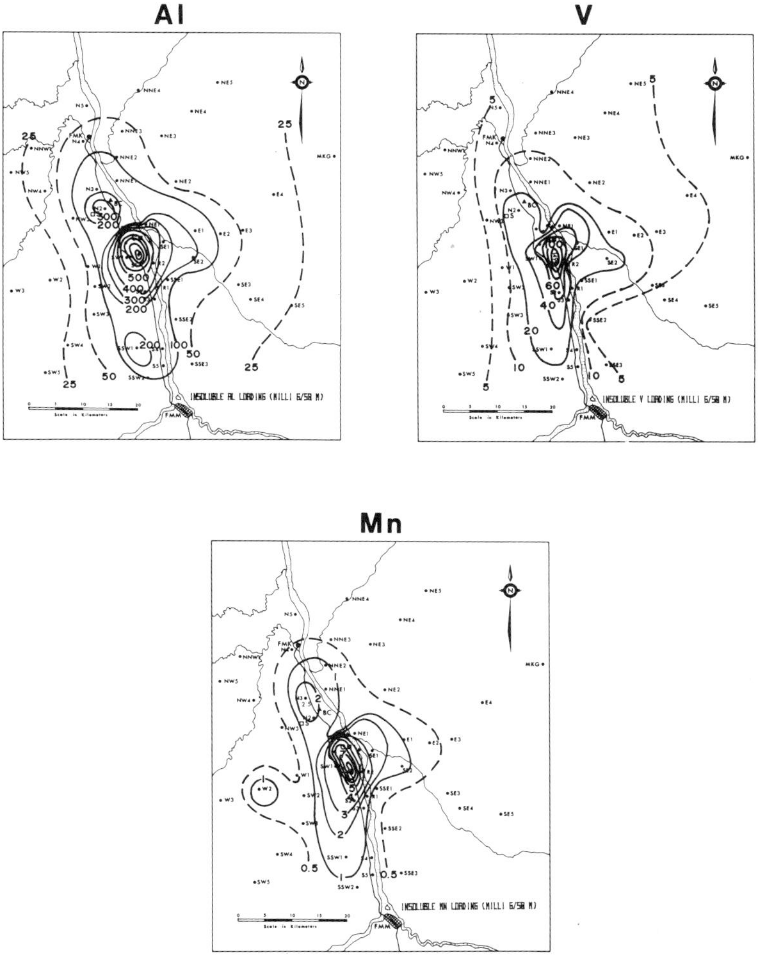

FIGURE 8. Deposition patterns of insoluble aluminum, vanadium, and manganese for the period November 18, 1977 to January 26, 1978, as determined from snowpack sampling (see TABLE 5).

Deposition patterns are shown in FIGURES 8 and 9. Aluminum and manganese deposition was similar to that of vanadium and sulfate. This was not the case in summer, as was shown in the previous section. Pollutants were deposited north and south of the source in an elongated band centered on the river valley. A secondary maximum in deposition occurred immediately to the east of the plant. The deposition pattern is correlated to the surface wind distribution at Mildred Lake during the snowpack's lifetime (FIGURE 10). Measurable winds occurred 75% of the time. They blew along the valley in a northerly or southerly direction 52% of the time, and from the west or southwest 13% of the time.

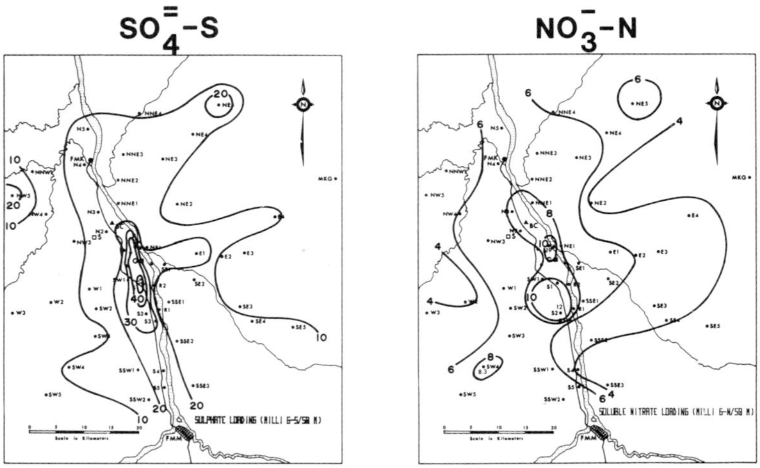

FIGURE 9. Deposition patterns of sulfur and nitrate for the period November 18, 1977 to January 26, 1978, as determined from snowpack sampling (see TABLE 5).

Deposition gradients near the source were much greater for the metals than for sulfate and nitrate. The ratio of loadings between site S1, 5 km south of the source, and site SSW2, 25 km south, was about 10 : 1 for the metals and less than 3:1 for sulfate and nitrate. This contrast can be largely attributed to the preferential removal of large metal-bearing fly-ash particles over smaller sulfate and nitrate particles or over gaseous sulfur and nitrogen compounds. Plume measurements with an Anderson cascade impactor showed that the mass median diameter of Al- and Mn-bearing particles was 3 μm, somewhat larger than that of vanadium (2.5 μm). The size of sulfate- and nitrate-bearing particles is generally well below this. The mass median diameter of pollutant particulate sulfur at ground level 5 km north of the source was found to be 0.4 μm.

A mass budget was done for each of the elements. The total amount that was deposited within 25 km was calculated from the deposition patterns. To obtain total anthropogenic deposition, a background amount based on loadings at remote sites was subtracted. The correction was less than 2% for metals and about 20% for sulfate and nitrate. In TABLE 6, the anthropogenic amount deposited is compared to the total amount released by the source.

A large fraction of substances that are predominantly in particulate form, namely Al, Mn, and V, was deposited within 25 km. On the other hand, over 98% of the more volatile oxides of sulfur and nitrogen was transported out of the area. Of the total sulfur released by the source, 0.3% was removed within 25 km. This is twice as high as the 0.14% found to be deposited in snow during the three-week period ending March 2, 1976.[10] These results are consistent with the results of a study at a sour gas plant in central Alberta, where it was concluded that less than 2% of the sulfur released was deposited within 40 km of the source during winter.[12]

Of particulate vanadium, 58% was removed by wet and dry processes within 25 km of the oil extraction plant. In the previous section, summertime dry deposition measurements were reported, in which approximately 0.6% of particulate vanadium was dry deposited within 25 km. If this conclusion is applicable in the winter, then most of the vanadium was brought down to the earth's surface by precipitation.

In-cloud and below-cloud precipitation scavenging would have had to have occurred at least 58% of the time for this to be the case. The actual frequency of precipitation was only 2%. Therefore, either much more than 0.6% was dry deposited and/or other wet removal mechanisms were effectively removing pollutants. In fact, another, rather interesting type of wet scavenging, "plume snow-out," occurred approximately 25% of the time. Whenever the air temperature was below −25°C, pollutants were incorporated into ice particles, which formed in the moist plume a few hundred meters from the source, grew, and fell to the ground as snow. The concentration of soluble vanadium in one such snowfall was 0.25 mg l^{-1}, 4–15 times higher than that measured in the snowpack. In the author's opinion, plume snow-out is a very effective scavenging mechanism, especially at low temperatures under light wind conditions.

The spatial distribution of pH around the source on a regional scale is shown in FIGURE 11. Near the source, pH is a maximum of between 6 and 7. It then falls off,

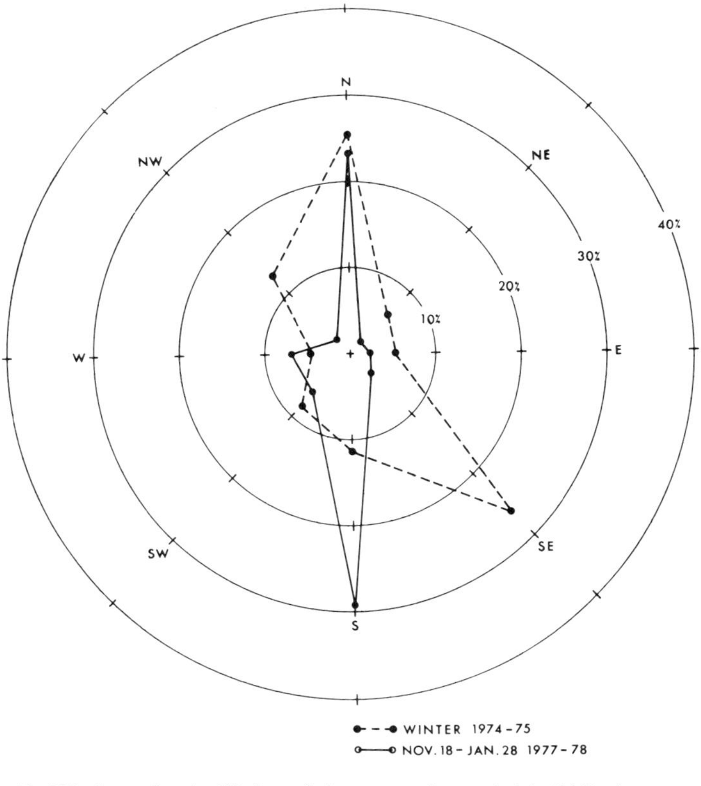

FIGURE 10. Wind rose for the lifetime of the snowpack sampled (solid line) compared to the average wind rose of January–March 1974 and 1975 (dashed line).

TABLE 6

THE MASS BUDGET OF VARIOUS SUBSTANCES RELEASED BY THE SOURCE AND DEPOSITED WITHIN 25 km FROM NOVEMBER 18, 1977 TO JANUARY 26, 1978

	Amount ($\times 10^6$ g)		Fraction Deposited within 25 km (%)
	Released	Deposited within 25 km	
Gaseous/particulate			
$SO_2/SO_4^{=}$–S	7744 ± 700	23.3	0.30
NO_x/NO_3^-–N	509 ± 50	10.2	2.0
Particulate			
V	49 ± 7	28.4	58
Al	224 ± 27	208.0	93
Mn	1.68 ± .2	1.66	99

rapidly to the east and west, and less rapidly north and south along the river valley, to values between 4.7 and 5. Ion balances in snowmelt taken near the source show that in order of decreasing equivalents, Ca^{++}, Mg^{++}, Na^+, NH_4^+, H^+ and $SO_4^{=}$, NO_3^-, and Cl^- are the predominant ions. In outlying areas having pH 4.7–5.1, H^+, $SO_4^{=}$, and NO_3^- are the dominant ions. High pH near the source is likely due to calcium and magnesium oxides in fly-ash particles. In outlying areas, alkaline particles are absent, leaving acidic oxides of sulfur and nitrogen to dominate the deposition. Similar observations of a relatively alkaline source region ringed by a maximum in acidity in outlying areas have been made around cities in Scandinavia.[13,14]

CONCLUSIONS

Many elements in particulate matter deposited around the oil extraction plant originate primarily from a different source in summer than in winter. In particular, the elements Al, Ca, Mg, Mn, Ti, and S are predominantly from windblown dust sources in summer, and from power plant emissions in the winter when the soil is snow covered. Vanadium deposition is mainly of anthropogenic origin in both summer and winter. In late June 1977, dry deposition processes removed about 1% of the primary particulate sulfur and vanadium within 25 km of the source.

Wet and dry removal processes during winter essentially separate alkaline fly-ash particles from the more acidic emissions and deposit them within 25 km of the source. In winter 1977–78, a unique pollutant removal process, plume snow-out, was found to be active at temperatures below about −25°C. Near the source, deposition is more alkaline (snowpack pH 6–7) than in outlying areas (snowpack pH 4.7–5), probably due to the soluble oxides of calcium and magnesium in fly ash. Results strongly suggest that wet and dry deposition of acidic oxides of sulfur and nitrogen from the power plant emissions is the main source of snowpack acidification in downwind areas. One cannot, however, exclude entirely the possibility that pollutants transported into the region from very remote sources are also contributing to the low pH values observed at sites off the atmospheric pathways of the oil-extraction plant emissions.

Mass balances of the oxides of sulfur and nitrogen demonstrate clearly that acidic compounds can be transported over long distances before being removed. Only 0.3 and

2% of total sulfur and nitrogen oxide emissions were deposited within 25 km of the source, respectively.

SUMMARY

The transport, transformation, and removal of atmospheric particulate matter originating from a thermal generating station in a remote area are discussed in the light of measurements made using low volume samplers, a network of artificial, dry deposition collectors, and snow chemistry surveys. Ambient concentrations, deposition

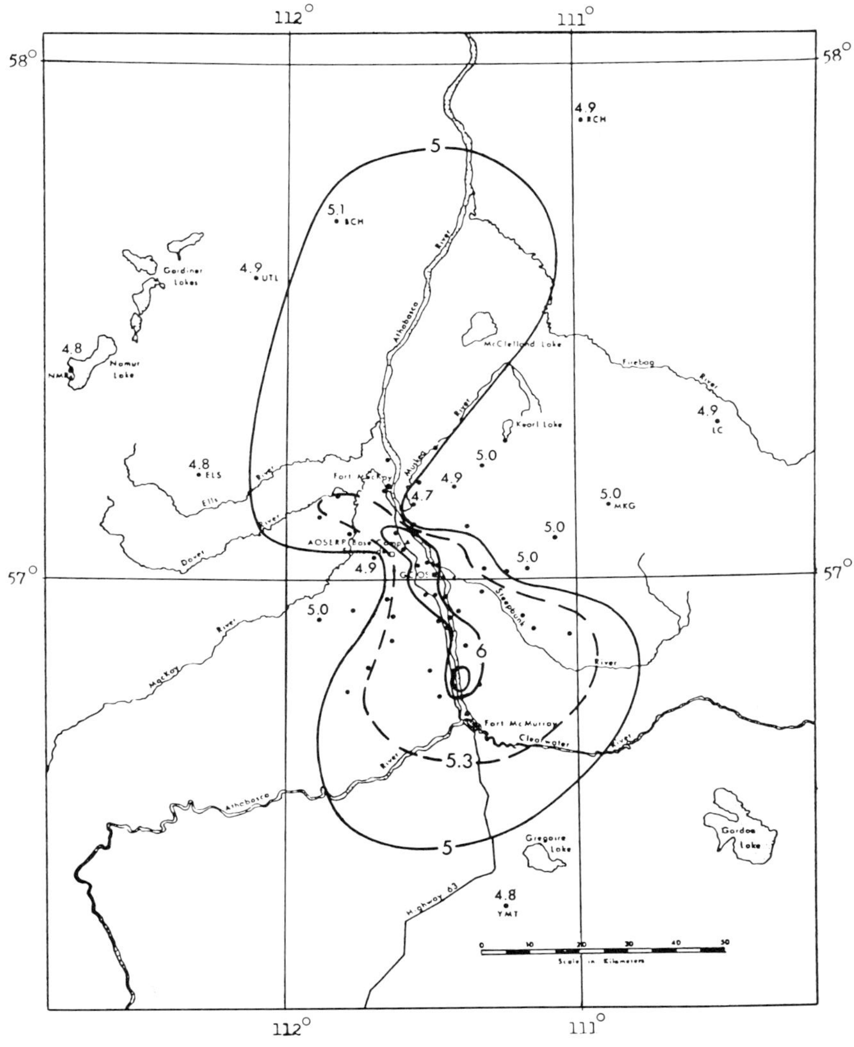

FIGURE 11. The spatial distribution of pH in the snowpack on January 26, 1978.

patterns, and deposition rates of several elements, including V, Al, Mn, S, Ti, Ca, and Mg, are reported for distances up to 100 km from the source. For each element, the fraction of total emissions that is transported out of the region is estimated from a mass balance.

ACKNOWLEDGMENTS

The author thanks J. Kovalick for his assistance in preparing this paper, and D. Whelpdale, K. Puckett, and E. Finnegan for their helpful comments and assistance in data analysis.

REFERENCES

1. LUSIS, M. A., K. G. ANLAUF, L. A. BARRIE & H. A. WIEBE. 1979. Plume chemistry studies at a northern Alberta power plant. Atmos. Environ. **12**(12): 2429.
2. SNC Tottrup. 1977. An Inventory System for Atmospheric Emissions in the AOSERP Study Area. Report No. 29 Me 2.2. Alberta Oil Sand Environmental Research Program. Edmonton, Alberta, Canada.
3. JOHNSON, D. A. & D. H. F. ATKINS. 1975. An airborne system for the sampling and analysis of sulphur dioxide and atmospheric aerosols. Atmos. Environ. **9:** 825.
4. LUSIS, M. A., L. A. BARRIE, H. A. WIEBE & K. G. ANLAUF. 1977. Development and Testing of a Filter Pack for Simultaneous Sampling of Gaseous and Particulate Sulphur Compounds in the Atmosphere. Report No. ARQA 49-77. Atmospheric Environment Service. Downsview, Ontario, Canada.
5. LUDWICK, J. D., T. D. FOX & S. R. GARCIA. 1977. Elemental concentrations of northern hemispheric air at Quillayute, Washington. Atmos. Environ. **11:** 1083.
6. WESELY, M. L., B. B. HICKS, W. P. DANNEVIK, S. FRISELLA & R. B. HUSAR. 1977. An eddy correlation measurement of particulate deposition from the atmosphere. Atmos. Environ. **11:** 561.
7. CAWSE, P. A. 1974. A Survey of Atmospheric Trace Elements in the U.K. (1972–73). AERE Report No. R-7669. Atomic Energy Research Establishment. Harwell, England.
8. CLOUGH, W. S. 1973. Transport of particles to surfaces. Aerosol Sci. **4:** 227.
9. JOHANNESSEN, M. & A. HENRIKSEN. 1978. Chemistry of snow meltwater: changes in concentration during melting. Water Resour. Res. **14**(4): 615.
10. BARRIE, L. A. & J. L. WALMSLEY. 1978. A study of sulphur dioxide deposition velocities to snow in northern Canada. Atmos. Environ. **12:** 2321.
11. BARRIE, L. A. 1979. A Wintertime Investigation of the Deposition of Pollutants around an Isolated Power Plant in Northern Alberta. Report No. ARQT-4-78. Atmospheric Environment Service. Downsview, Ontario, Canada.
12. SUMMERS, P. W. & B. HITCHON. 1973. Source and budget of sulphate in precipitation from central Alberta, Canada. J. Air Pollut. Control Assoc. **23:** 194.
13. OVERREIN, L. N. 1972. Sulphur pollution patterns observed; leaching of calcium in forest soil determined. Ambio **1**(4): 145.
14. HOGSTROM, U. 1974. Wet fallout of sulfurous pollutants emitted from a city during rain or snow. Atmos. Environ. **8:** 1291.

OECD PROGRAM ON LONG RANGE TRANSPORT OF AIR POLLUTANTS—MEASUREMENTS FROM AIRCRAFT

Y. Gotaas

Norwegian Institute for Air Research
N-2001 Lillestrøm, Norway

Introduction

The Organization for Economic Cooperation and Development (OECD) started, in 1972, a "Cooperative Technical Program to Measure the Long Range Transport of Air Pollutants (LRTAP)." The Norwegian Institute for Air Research (NILU) was responsible for the coordination and also for the preparation of the final report.

During the first measurement phase, completed in January 1974, it became clear that long range transport of sulfur dioxide and particulate sulfate was of considerable importance in the acidification of precipitation. The importance of measurements above the ground level also became evident. The second, and last, measurement phase was completed in March 1975. In all, 11 European countries participated in the program, and a total of 76 ground-level sampling stations were established.

Measurements and findings of the LRTAP program have been presented by OECD[1] and were also presented at the International Symposium on Sulfur in the Atmosphere, in Dubrovnik, 1977.[2] The present author was responsible for the collection and presentation of the aircraft measurements in the final report.[1] The following presentation, including tables and figures, is a summary.

Measurement Techniques

At the beginning of the LRTAP program, sulfur dioxide was absorbed in wash bottles preceded by a filter to collect sulfate particles. The true detection limit of sulfur dioxide was not better than 5 $\mu g/m^3$, partly due to unavoidable contamination problems.[3] A much improved method was introduced by the British—using impregnated filters to collect sulfur dioxide.[4] The air is drawn through the filter holder, which contains two filters in series. The first filter collects the aerosol particles. The second filter is impregnated with potassium hydroxide and absorbs the sulfur dioxide from the air sample. In the LRTAP program, the filters were analyzed either by the barium perchlorate thorin method, by isotopic dilution analyses, or by x-ray fluorescence spectroscopy.

Recording instruments are needed for detailed information during sampling flights. A coulometric cell designed by Novak[5] was installed in the NILU aircraft for continuous recording of sulfur dioxide. The detection limit was originally 10 $\mu g/m^3$, but was gradually brought down to 1–2 $\mu g/m^3$ with a time constant of seconds. The readings gave valuable information on the heights to be selected for filter sampling, and discrete plumes were easily detected. However, readings are affected by changes

0077–8923/80/0338–0453 $01.75/1 © 1980, NYAS

in air density, and drift of the zero line does occur. Relative values, rather than absolute values, were therefore obtained.

To estimate particle concentrations, different types of optical instruments were used. Integrating nephelometers were installed in the Swedish and Norwegian aircraft. These instruments give a continuous recording of the light scattering from illuminated particles in the size range of 0.3 μm to 1 μm.[6]

From the readings, the mass concentration of aerosol sulfate may be estimated.[7] But readings are highly affected by the relative humidity. Concentration of Aitken nuclei was measured in the NILU aircraft by a small, hand-operated particle counter. An automated version was used in the Danish aircraft. In these instruments, the number of fog droplets formed in an expansion chamber is measured electrooptically.

Air temperatures and visual weather observations are available from most flights. For various technical reasons, information about wind direction and wind speeds was generally not available.

TABLE 1 gives a list of participating countries, sampling techniques, and number of flights. The geographical distribution of the aircraft measurements of sulfur compounds is shown in FIGURE 1. The aircraft sampling was mainly limited to situations where high concentrations of sulfur dioxide or sulfate aerosols were forecast. The total number of sampling flights was 111 on 92 different days, providing 611 single observations of either sulfur dioxide or sulfate concentration or both.

Intercalibration took place between the Norwegian and German aircraft during parallel flights. The German measurements (wet bubblers) gave a mean value (6

TABLE 1
PARTICIPATING COUNTRIES AND SAMPLING TECHNIQUES

Country	Type of Aircraft	Method	Number of Flights
Denmark	DC-3	Filters in series—IDA*	3
Federal Republic of Germany	Beechcraft Queen Air	Wet absorption—TCS†	8
France	Cessna 206	Filters in series—BPT‡	5
Norway	Piper Aztec	Filters in series—BPT	37
Sweden	Beechcraft Travel Air	Before March 1974: Wet absorption, peroxide solution (SO_2), glass fiber filter (SO_4)	35
		After March 1974: Filters in series—BPT	
United Kingdom	Varsity Hercules	Filters in series—BPT	14 (+6 in 1971) 1
	Hastings	SO_2: TCS	8
		SO_4: Filters—BPT	

*IDA = Isotope dilution analysis.
†TCS = Tetrachloromercurate solution (SO_2)—West and Gaeke.
‡BPT = Barium perchlorate thorin.

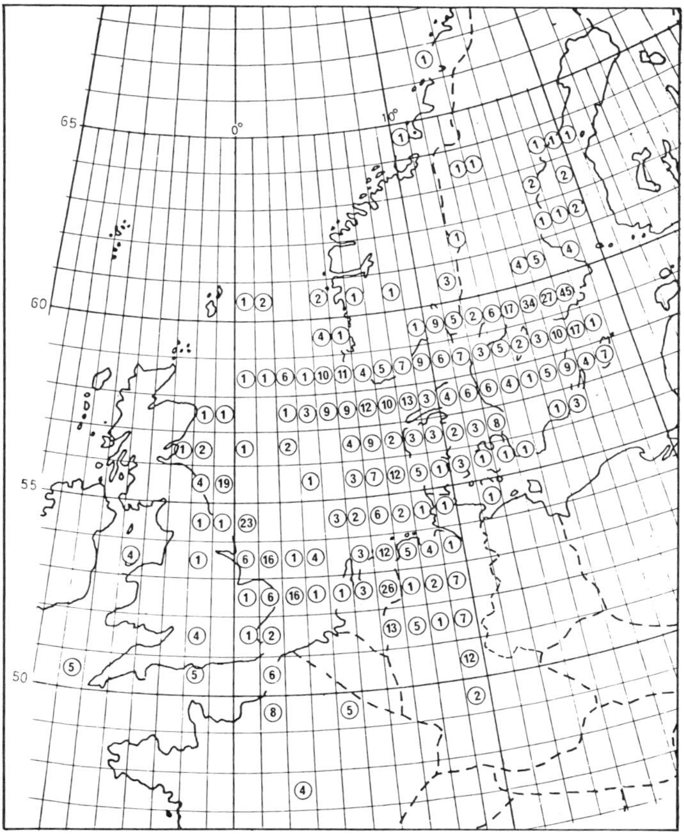

FIGURE 1. Number and geographical distribution of aircraft measurements of sulfur compounds (including 26 observations made prior to the LRTAP program by the British Meteorological Office on 6 flights during 1971).

$\mu g/m^3$) that was 50% higher than the Norwegian measurements (filters). A similar comparison flight between the Swedish and the Norwegian aircraft gave a much better agreement between mean values but showed individual scatter, reflecting the uncertainties involved in measuring low concentrations.

STATISTICAL RESULTS

A complete listing of the sampling data from each flight is given in a separate report.[8] Results of selected flights have been published elsewhere.[7,9,10] In the following will be given some statistical results, including all data and also some United Kingdom measurements from 1971. Results from two individual NILU flights are shown as examples.

TABLE 2

MEAN SO_2 AND SO_4 CONCENTRATIONS FROM AIRCRAFT MEASUREMENTS AT DIFFERENT ALTITUDES, COMPARED WITH DISPERSION MODEL ESTIMATES AND 24-HOUR GROUND OBSERVATIONS

Height (m)	SO_2 ($\mu g/m^3$)			SO_4 ($\mu g/m^3$)		
	Observed	Estimated	Ground	Observed	Estimated	Ground
<500	20 (245)*	22 (245)	16 (228)	11 (243)	6 (243)	9 (230)
501–1000	13 (233)	18 (233)	—	9 (227)	6 (227)	—
1001–1500	7 (132)	—	—	8 (127)	—	—
>1500	6 (77)	—	—	5 (75)	—	—

*Number of data points is given in parentheses.

Flights were planned to take place in situations with high concentrations, but, for various reasons, this was not always the case. Flights were made during a great variety of weather conditions, some in fair weather, but mostly below, between, or above cloud layers. Clouds and precipitation were generally avoided. However, a few samples obtained within thin clouds or in light precipitation show the concentration of sulfur compounds to be little affected.

In order to compare them with model estimates, the aircraft measurements were separated into four height intervals (TABLE 2). The main part of the sulfur dioxide is seen to be contained in a layer below 1000 m, which is the assumed mixing height in the models. The mean values of the sulfur dioxide concentrations measured by aircraft are lower than the corresponding mean values estimated from the models, while the

TABLE 3

MEAN SO_2 AND SO_4 VALUES, STANDARD DEVIATIONS (σ), AND CORRELATION COEFFICIENTS (r) FOR CORRESPONDING DATA FROM AIRCRAFT SAMPLING (AIR), GROUND-LEVEL SAMPLING (GRO), AND DISPERSION MODEL ESTIMATES (CAL)

	All Observations below 1000 m		North Sea		Sweden		Continent (French + German)	
	SO_2	SO_4	SO_2	SO_4	SO_2	SO_4	SO_2	SO_4
AIR*	16	10	18	9	5	7	26	17
σ*	22	9	24	8	6	8	30	10
GRO*	15	8	15	8	9	5	25	11
σ*	14	6	12	5	5	5	23	8
r	.56	.37	.53	.18	.55	.79	.46	.24
n†	448	445	141	134	123	131	71	60
AIR*	17	10	18	9	7	8	29	17
σ*	23	9	24	8	9	9	34	10
CAL*	20	6	20	5	12	7	31	4
σ*	16	9	14	5	11	9	17	3
r	.37	.17	.32	.12	.28	.31	.26	.36
n†	478	470	147	139	137	143	74	63

*Values in $\mu g/m^3$.
†n = number of observation pairs.

observed mean sulfate concentrations are higher. For the total sulfur content, there is a relatively good agreement.

One might expect little agreement between surface observations made over a 24-hour period and those obtained by aircraft over a much shorter time period and at higher altitude. However, TABLE 2 shows that the 24-hour surface concentrations of both sulfur dioxide and sulfate are generally similar to, although slightly lower than, the average of the aircraft measurements made below 500 meters. The discrepancy could be explained by dry deposition at the surface.

One must be careful not to draw conclusions from this apparent relationship, due to the wide scatter and poor correlations of the individual data. TABLE 3 shows correlation coefficients of corresponding data from aircraft sampling, ground-level sampling, and model estimates. Only in the case of sulfate aerosols over Sweden is the

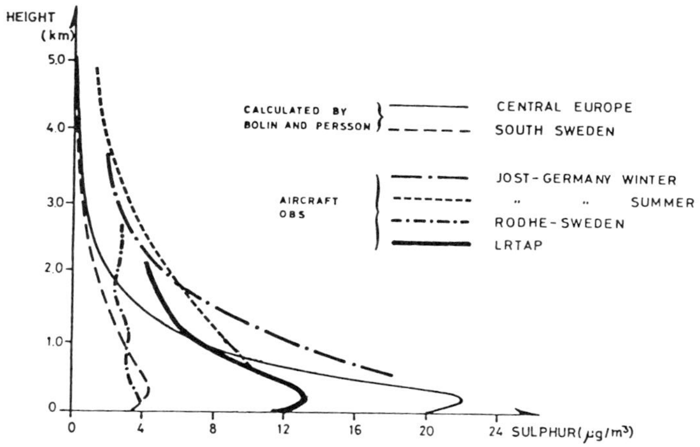

FIGURE 2. Calculated and observed mean vertical concentration profiles for sulfur (from Bolin and Persson[13]). Mean of LRTAP observations shown as the heavy line.

correlation sufficiently high (0.79) to confirm that individual observations at the surface and in the air are related.

The observed average vertical concentration distributions are subject to the selection of weather situations and sampling levels, and to the fact that most flights were conducted to obtain horizontal distribution over large areas. However, FIGURE 2 shows that the average vertical concentration distribution is not much different from those reported elsewhere[11,12] and that a low-level maximum concentration agrees with the calculations of Bolin and Persson.[13]

The mixing height may be defined as the top of a surface-based layer of good vertical mixing and is therefore often associated with an inversion base. During nighttime, in winter, and over the sea in summer, this definition may not be appropriate. Therefore, a mixing height that is defined by using meteorological observations alone sometimes may be difficult to apply. Based on observed vertical distributions, Garland and Branson[14] defined the mixing height (H) as the ratio of the content of a column of the atmosphere of unit cross section to the concentration at ground level (C_0):

$$H = \int_0^\infty C(z)\,dz / C_0 \qquad (1)$$

where C is the concentration at height z. This definition makes mixing height rather sensitive to the value of C_0. The sensitivity may be reduced by using an alternative definition. In the idealized case of a constant concentration below the height H, the standard deviation σ_z becomes $H/\sqrt{3}$. Including this limiting case, the mixing height may therefore be defined as:

$$H = \sqrt{3}\,\sigma_z = \sqrt{3\int_0^\infty z^2 C(z)\,dz \Big/ \int_0^\infty C(z)\,dz} \qquad (2)$$

In both formulae, the integration has to be performed to a height where the concentration has fallen to zero or to a representative background value. The resulting

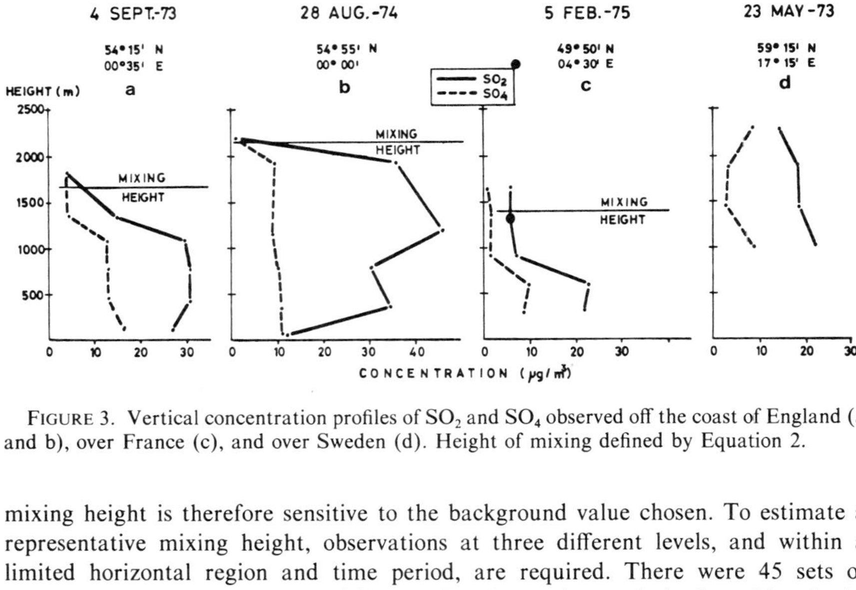

FIGURE 3. Vertical concentration profiles of SO_2 and SO_4 observed off the coast of England (a and b), over France (c), and over Sweden (d). Height of mixing defined by Equation 2.

mixing height is therefore sensitive to the background value chosen. To estimate a representative mixing height, observations at three different levels, and within a limited horizontal region and time period, are required. There were 45 sets of observations with the horizontal difference less than 2 degrees latitude and longitude, with time difference less than two hours, and with at least one value of sulfur dioxide equal to or greater than 10 $\mu g/m^3$.

Some profiles obtained are shown in FIGURE 3. Case **d** illustrates a profile that does not allow an estimate of the mixing height. TABLE 4 shows the results for 23 sets of data for sulfur dioxide and 18 sets for sulfate aerosols. The mean mixing heights for sulfur dioxide, computed from Equations 1 and 2, are nearly equal. However, the standard deviation for sulfur dioxide is much lower for Equation 2 (550 m versus 1170 m). For sulfate particles, standard deviations are about 550 m in both cases. The

TABLE 4

MIXING HEIGHT AND HEIGHT TO THE CENTER OF GRAVITY

	$H = \int_0^\infty C(z)\mathrm{d}z/C_0$ (1)*		$H = \sqrt{3} \cdot \sigma_z$ (2)
Mixing height—SO_2†	1200 m (1170)§		1285 m (550)
Mixing height—SO_4‡	1235 m (520)		1530 m (585)
Center of gravity—SO_2†		620 m (280)	
Center of gravity—SO_4‡		755 m (285)	

*One vertical profile was excluded when using Equation 1, because a very low value of C_0 gave an unreasonably large mixing height.

†Mean of 23 profiles.

‡Mean of 18 profiles.

§Standard deviations are given in parentheses.

height of the center of gravity is also given in TABLE 4, indicating the level of the bulk of material.

A few aircraft measurements can be used to estimate conversion rates of sulfur dioxide to sulfate. Smith and Jeffrey[9] found that the fraction converted to sulfate depended quite sensitively on relative humidity.

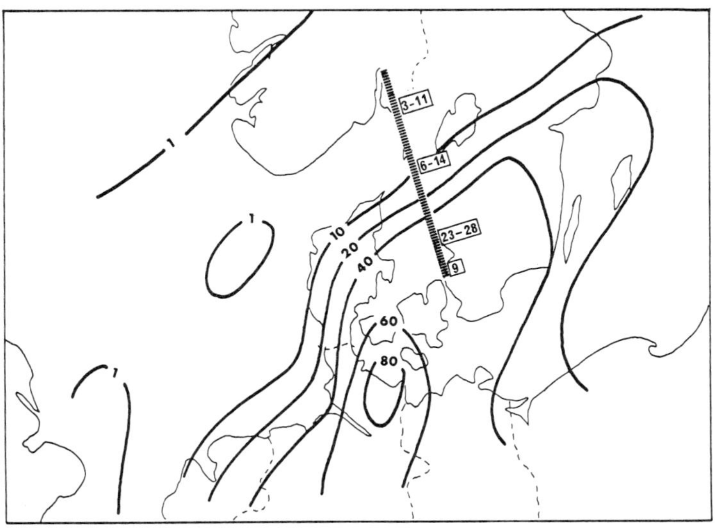

FIGURE 4. SO_4 concentrations ($\mu g/m^3$) from aircraft sampling (in squares) and Lagrangian model estimates (isolines) on August 26, 1974. Leftmost aircraft data are from the southward flight at 300 m, rightmost from the return flight at 700 m.

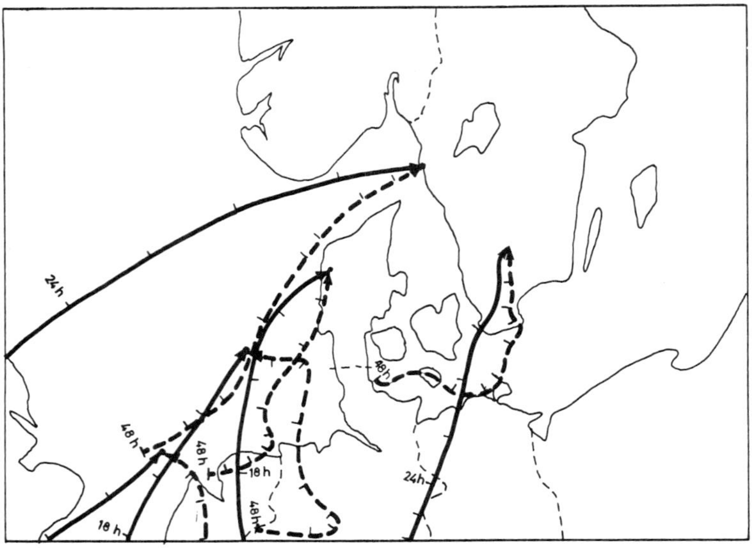

FIGURE 5. Trajectories on September 12, 1974 computed from observed 850 mb winds (full lines) and surface winds (broken lines).

RESULTS FROM SELECTED FLIGHTS

Two examples of sampling flights will be presented. The first covers a situation with transport of polluted air towards Scandinavia and is illustrated in FIGURES 4, 5, and 6. Measurements were carried out at different levels. The vertical distribution of sulfur dioxide in the cross section along the flight is shown in FIGURE 4. The distribution of sulfate is similar, only the concentration values are somewhat lower.

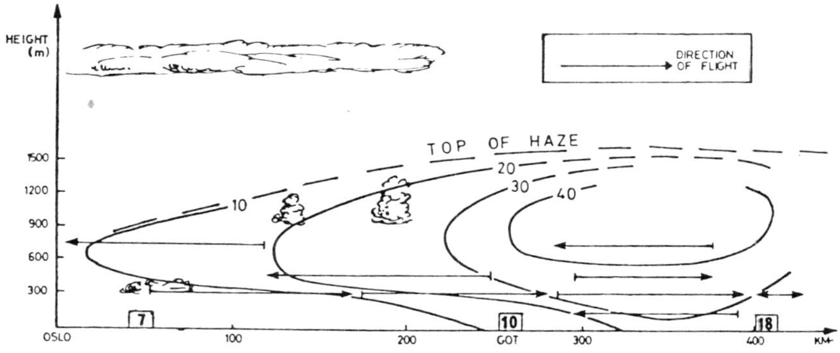

FIGURE 6. Vertical cross section of SO_2 concentrations ($\mu g/m^3$) along the flight track on August 26, 1974 (see FIGURE 4). Numbers in squares are 24-hour ground-level measurements. (GOT: Gothenburg.)

Pollutants were all contained within a well-defined haze layer. FIGURE 5 shows air trajectories.

The cross-sectional transport into Sweden, from 55° N to the Norwegian border, is estimated to be 130 t sulfur/hour. FIGURE 5 shows a northwest/southeast gradient in good agreement with both model estimates and 24-hour ground-level measurements. However, there is a marked disagreement for the southernmost aircraft observations, which show low values for both sulfur dioxide and sulfate. The possibility of a sampling or analysis error cannot be overlooked, but the aircraft measurements are supported by an increase in the visibility.

FIGURE 7 shows results from an NILU flight, including some vertical ascents. In this case, concentrations were considerably higher. Over the North Sea, a thick layer

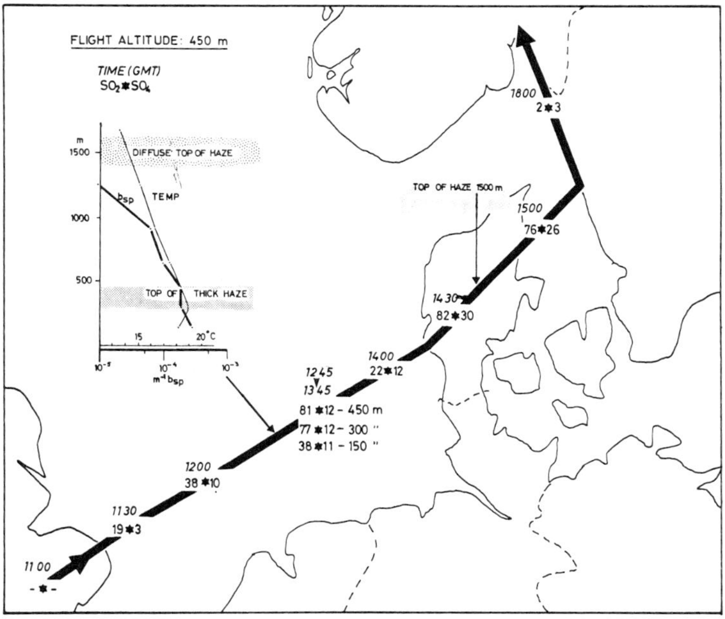

FIGURE 7. SO_2 and SO_4 concentrations from aircraft sampling on September 12, 1974. Insert: Vertical profiles of scattering coefficient (b_{sp}) and temperature.

of haze extended to 500 m. Below 400 m, there was an inversion. Above 400 m, the temperature decreased steadily with height. A thin layer of haze extended to about 1500 m. The nephelometer indicated a general decrease in the amount of aerosols with height, without any abrupt change at the top of the haze layers as most often observed. Similar behavior near haze layers has also been observed on other flights. Thick haze layers are not always capped by an inversion. As in the previous example, the maximum sulfur concentrations were observed at heights of several hundred meters.

Conclusions

The aircraft sampling program was rather ambitious and could only be partially implemented. For different reasons, standby aircraft equipped with the required instruments could not be maintained in all the participating countries. Considering these practical difficulties, there was a reasonably large number of flights.

The quality of the aircraft data equals or exceeds that of the ground-level network. Most of the data from sampling flights relate to constant levels over long distances, for comparison with model calculations. Individually observed values do not agree particularly well either with calculated values or with ground-level observations. But this is to be expected because of the differences in time and space between the three sets of data. However, the agreement is better in the case of mean values.

In the model calculations, a constant vertical distribution has to be assumed, although the aircraft flights have shown this not to be the case. In the examples of individual sampling flights described, maximum concentrations were observed at heights of several hundred meters. However, the mixing height (1000 m) assumed in model calculations was in quite good agreement with the mean mixing height (1200–1500 m) obtained from the 23 sulfur dioxide profile measurements. The assumption of a distinct mixing height seems somewhat less justified for sulfate and other aerosol concentrations.

References

1. OECD. 1977. The OECD Programme on Long Range Transport of Air Pollutants. Organization for Economic Cooperation and Development. Paris, France.
2. Husar, R. B., J. P. Lodge, Jr. & D. J. Moore, Eds. 1978. Sulfur in the Atmosphere. Atmos. Environ. **12**(1–3).
3. Semb, A. 1978. Aircraft Sampling. Methods for the Determination of Sulphur Dioxide and Sulphate Aerosol Using a Light Aircraft. Scandinavian Council for Applied Research (NORDFORSK). Helsinki, Finland.
4. Johnson, D. A. & D. H. F. Atkins. 1975. An airborne system for the sampling and analysis of sulphur dioxide and atmospheric aerosols. Atmos. Environ. **9:** 825.
5. Lahman, E. & K. E. Prescher. 1971. Automatische Schwefeldioxid Bestimmungen mit einem Polarographisch-Coulometrisch arbeitenden Gerät. Wasser Luft Betr. **15:** 366.
6. Charlson, R. J., N. C. Ahlquist & H. Howarth. 1968. On the generality of correlation of atmospheric aerosol near concentration and light scatter. Atmos. Environ. **2:** 455.
7. Waggoner, A. P., A. J. Vanderpol, R. J. Charlson, S. Larsen, L. Granat & C. Trägårdh. 1976. Sulphate–light scattering ratio as an index of the role of sulphur in tropospheric optics. Nature **261:** 120.
8. Gotaas, Y. 1975. Aircraft Measurements. Data Collected within the LRTAP Project. LRTAP Report No. 17/75. Norwegian Institute for Air Research. Lillestrøm, Norway.
9. Smith, F. B. & G. M. Jeffrey. 1975. Airborne transport of sulphur dioxide from the U.K. Atmos. Environ. **9:** 643.
10. Heidam, N. Z. 1977. Airborne measurements of air pollution over Denmark and the adjacent seas. Tellus **24:** 345.
11. Rodhe, H. 1972. Measurements of sulfur in the free atmosphere over Sweden 1969–1970. J. Geophys. Res. **77:** 4494.
12. Jost, D. 1974. Aerological studies on the atmospheric sulphur budget. Tellus **26:** 206.
13. Bolin, B. & C. Persson. 1975. Regional dispersion and deposition of atmospheric pollutants with particular application for sulphur pollution over western Europe. Tellus **27:** 281.
14. Garland, J. A. & J. R. Branson. 1976. The mixing height and mass balance of SO_2 in the atmosphere above Great Britain. Atmos. Environ. **10:** 353.

THE OCCURRENCE OF ATMOSPHERIC AEROSOLS IN THE NORTHEASTERN UNITED STATES*

P. K. Mueller, G. M. Hidy, K. Warren, T. F. Lavery, and R. L. Baskett

Environmental Research & Technology, Inc.
Westlake Village, California 91361

Concerns with the origins and effects of the occasionally widespread occurrence of dense hazes over the northeastern United States[1] have led to major regional-scale studies.[2-4] Early results from these studies, involving data primarily from the summer months, indicate that the regional hazes occur in multiday episodes (2–6 days) that involve two types of synoptic-scale weather regimes:[5]

1. Stagnation of high pressure systems moving eastward across the eastern and midwestern United States. Such air masses appear to maintain themselves over periods of 1 to 4 days. The air is characteristically warm and moist, with light winds around the southern and western parts of a high pressure area. Under these conditions, zones of polluted air tend to collect within 100 to 300 km of high sulfur oxide emission sources.

2. Channeling of air flow between the west side of the Appalachian Mountains and cold fronts over the Great Lakes moving southeastward provides a means for transporting aerosols that have accumulated on the western side of a Bermuda high pressure system. The accumulation is associated with limited vertical mixing due to subsidence. Such conditions have been observed for up to 6 days and appear to be conducive to long range (1500 km) transport of polluted air.

Sulfate is a major component of the particles in these hazes[6] and, therefore, continues to be of considerable interest. However, to fully understand the genesis and potentials for control of the hazes, it is important to obtain a complete knowledge of the aerosol.

The Sulfate Regional Experiment (SURE), sponsored by the Electric Power Research Institute, is therefore in the process of establishing an extensive data base on the mass concentration, size distribution, and chemical composition of the atmospheric particulate matter for several seasons. Samples and data have been acquired from a 54-ground-station network in the nonurban eastern United States from August 1977 through October 1978, and from aircraft during six days of flying during one month each season (a total of more than 200 flights). The area covered is approximately 2400 × 1840 km—from eastern Kansas to the Atlantic coast and from mid-Alabama to southeastern Canada. The resolution of phenomena of interest is set by a system of 80 km × 80 km grids. The purpose of this paper is to summarize some of the preliminary findings of SURE.

*Sponsored in part by the Electric Power Research Institute, Palo Alto, California, under RP861/862.

0077–8923/80/0338–0463 $01.75/1 © 1980, NYAS

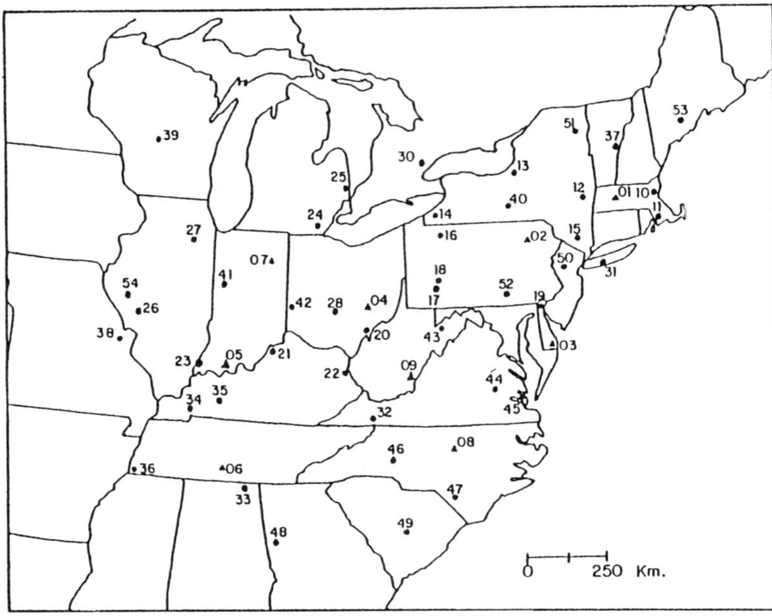

FIGURE 1. SURE region showing locations and numbers of the ground measurement stations. Triangles are Class I, circles are Class II stations.

MEASUREMENTS

The SURE ground-station network is shown in FIGURE 1. Nine of the stations are instrumented to measure the air quality and meteorological variables listed in TABLE 1. These Class I sites (shown by triangles) have been operating continuously since July 1977, and are scheduled to continue until July 1, 1980. The sequential filter samplers were operated during the intensive sampling periods of August and October 1977 and mid-January to mid-February, April, July, and October 1978. The 45 Class II stations (shown by dots) were instrumented with a continuous SO_2 analyzer and a high-volume filter sampler, and operated only during these intensive sampling months. All particulate matter was collected on Teflon®-coated glass fiber filters. These filters, specified as TX40HI20(ERT), are made to our specifications and subjected to quality control tests at ERT. The particles collected on the filters were analyzed for mass, sulfates, nitrates, ammonium, chloride, carbon, and some metals. The stations were located according to criteria designed to measure rural air quality. Distances and directions from nearby point and area emission sources, and practical facility siting and exposure requirements, were considered. These criteria are described in more detail elsewhere.[4]

Two SURE aircraft were flown simultaneously with up to three other aircraft sponsored by the Department of Energy (DOE) MAP3S program for six days during each of the six seasonal intensive study periods. The flight plans called for vertical and horizontal measurements at various altitudes up to 3,000 meters mean sea level in the area of Class I ground-station sites, and horizontal measurements over the SURE region using flight paths coordinated among the aircraft. All aircraft were equipped for continuous measurements of meteorological and position parameters, SO_2, O_3, NO/NO_x, b_{scat}, and condensation nuclei. Sulfur dioxide was also sampled in 30-minute composites with impregnated filter packs. Suspended particles were collected in 10- and 60-minute composites. The 30-minute collections were analyzed for mass by weighing, and all collections were analyzed for sulfates and nitrates. The 10-minute collections provided resolution every vertical 500 meters, and the 30-minute collections provided resolution below and above 1,500 meters.

The vertical profiles were flown over two selected Class I stations at least 300 miles apart. At each location, three vertical profiles were flown 20 to 50 miles apart at locations that were directly above, upwind, and downwind of the station. Two flights were completed each day; one early in the morning beginning approximately at sunrise, and the other at midday. The sampling periods and locations for each flight are summarized in TABLE 2. The data collection times have been published.[6]

To assess and document the quality of the SURE data base, considerable effort was put into quality control, which is described in further detail elsewhere.[4] Several activities were implemented:

1. Establishment of criteria for Class I and Class II site selection, including tests to verify the extent to which these criteria were met.
2. Achievement of uniform, or equivalent, Class I and Class II station designs, construction, and installation.
3. Verification that existing Class II station designs, configurations, and operations were consistent with the SURE objectives and quality goals.

TABLE 1

INSTRUMENTATION FOR CLASS I STATIONS

Parameter	Instrument	Data Base Content
Gases		
Sulfur dioxide	Meloy 285	1-hour average
Ozone	Monitor Labs 8410	1-hour average
Nitric oxide/Nitrogen dioxide	Monitor Labs 8440	1-hour average
Particles		
Total suspended particles, <50 μm AMMD*	High volume filter sampler using Teflon-coated glass fiber filters (TGF)	24-hour average
Total suspended particles, <15 μm AMMD	ERT sequential filter sampler using TGF	3-hour average
Refined suspended particles, <3 μm	ERT sequential filter sampler using TGF	3-hour average
Meteorology		
Dew point/temperature	EG&G Model 220	1-hour average
Surface winds	Various	1-hour average

*AMMD = aerometric mass median diameter.

TABLE 2

PERIODS AND LOCATIONS OF AIRCRAFT FLIGHTS DURING SURE

Period	Aircraft and Location
August 1977	MRI — Rockport
	RTI — Duncan Falls/Zanesville
	MAP3S — Support with horizontal traverses
October 1977	MRI — Duncan Falls/Zanesville
	RTI — Scranton
January 1978	MRI — Lewisburg
	RTI — Zanesville/Duncan Falls
	MAP3S — Limited support in northeast
March 1978	MRI — Duncan Falls/Zanesville
	RTI — Raleigh/Durham
July 1978	MRI — Duncan Falls/Zanesville
	RTI — Scranton
	MAP3S — Horizontal flights for major haze episodes
October 1978	MRI — Lewisburg
	RTI — Giles County
	MAP3S — Horizontal flights depending on haze occurrence

4. Preparation of standard operating procedures and manuals for Class I and Class II stations.

5. Control of specifications for equipment and materials, both at vendor facilities and upon delivery.

6. Establishment and verification of quality control processes within the central laboratories. Such laboratory activities include:

- Preinstallation instrument performance verification and postinstallation problem solving by way of laboratory simulations;
- Preparation and performance verification of various types of sampling media;
- Establishment and implementation of transfer and shipping procedures for sampling media, to and from the sampling sites and among operational units;
- Establishment of sample receiving, custody, and validation of procedures;
- Certification of calibration standards;
- Documentation of sample analysis, data processing, and data reporting procedures;
- Documentation and implementation of laboratory quality control procedures, and documentation of performance; and
- Conducting interlaboratory performance comparison studies.

7. Simultaneous calibration of the measurements systems on the two SURE aircraft and at one nearby Class I ground station. Calibrations were also cross checked in-flight between the aircraft and the nearby SURE Class I ground station. Calibration checks were performed on the nonflight day during the 13-day aircraft operation in each intensive sampling period.

8. Establishment, continual revision, and documentation of data validation guidelines and criteria, and data processing procedures.

9. Establishment and documentation of data base management procedures and reporting formats.

10. Verification of data treatment and analysis outputs.

11. Initial and periodic training of staff throughout SURE, both formally at vendor and ERT facilities and informally in-service.

12. Periodic recalibration in the laboratories and at the stations throughout the course of the program.

13. Maintenance of a performance data record to enable retrospective estimation of the actual precision and accuracy of the various components of the data base.

Observations

An initial examination of the data taken during the August 1977 through October 1978 intensive sampling periods at the ground stations and from the aircraft has been completed. These data are now available for distribution in either hard copy or on tape.[6] The data from SURE are being released to the DOE/MAP3S data bank at the Brookhaven National Laboratory, Upton, N.Y., and to the U. S. EPA's Environmental Science and Research Laboratory at Research Triangle Park, N.C. ERT releases data to other interested parties upon request. Because we are still in the process of condensing the information in this large data base into formats that are suitable for presentation within the confines of this type of paper, the general nature of the information is illustrated here by way of examples.

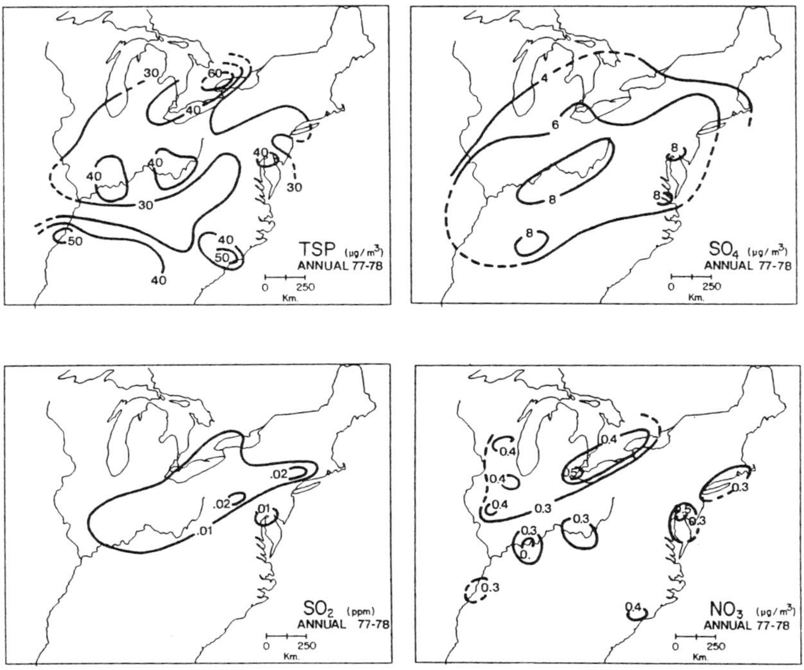

FIGURE 2. Geographic distribution of annual geometric means for 24-hour mass, sulfate, and nitrate and 1-hour SO_2 values.

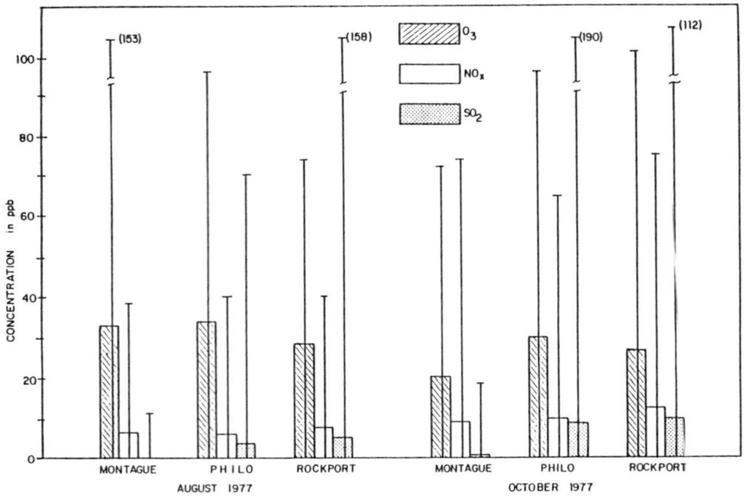

FIGURE 3. Monthly and extreme hourly averages of gas concentrations for August and October 1977 at three sites in the SURE region.

The geographic distributions of the annual geometric means, based on data from the intensive sampling months, for 24-hour Hi-vol suspended particulate matter (TSP), sulfate (SO_4), nitrate (NO_3), and sulfur dioxide (SO_2) are shown in FIGURE 2. This annual distribution includes two summer months, August 1977 and July 1978. Distinct differences are seen in the geographic distribution of these four parameters. The TSP, SO_2, and nitrate maxima tended to be localized in terms of a few hundred km or less. Concentrations of SO_2 in excess of 10 ppb occurred along and to the northeast of the Ohio River. The highest annual averages were observed near Charleston, W.Va., and Boston, Mass. Sulfate concentrations in excess of 8 $\mu g/m^3$ occurred in the areas of central Tennessee; the Ohio River Valley; Richmond Va.; and Baltimore–Philadelphia. Annual average sulfate concentrations of 6 to 8 $\mu g/m^3$ were observed over an area about 750 km wide and 1800 km long from the southwest to the northeast. Annual average nitrate concentrations greater than 0.3 $\mu g/m^3$ occurred primarily in the vicinity of urban areas in the southeast, the Ohio Valley, the midwest, the vicinity of Lake Erie, the Washington–Philadelphia complex, and the New York–Boston complex.

Pollutant Gases

Monthly averages and the extremes of hourly pollutant gas concentrations at three Class I stations are shown in FIGURE 3. The station numbers refer to those shown in FIGURE 1. Included are the periods of August 1977 and October 1977. The three stations are located near Rockport, Ind. (03) in the lower Ohio Valley; Philo (04)† in the central Ohio Valley; and near Montague (01) in western Massachusetts.

†Philo = Duncan Falls.

The ozone (O_3) levels were variable from day to day, with monthly means largest in August when maximum hourly averages in excess of 100 ppb occurred.

The nitrogen oxides ($NO + NO_2 = NO_x$) were generally less than 10 ppb. The NO_x concentrations were higher in October than in August at all three stations, when maximum hourly averages above 70 ppb occurred. Based on this condensation of the data, there appears to be no significant difference in NO_x occurrences among these three sites. However at some sites, especially at Research Triangle Park (08) and Giles County (06), the diurnal NO_x patterns are consistent with influences from motor vehicle emissions.

The SO_2 levels shown in FIGURE 3 are low—a few ppb. The monthly 1-hour averages vary from about 5 to 40 $\mu g\ SO_2/m^3$. There appear to be significant differences among the sites. Philo and Rockport have concentrations higher than those at Montague. The measurements taken in October were higher than those in August.

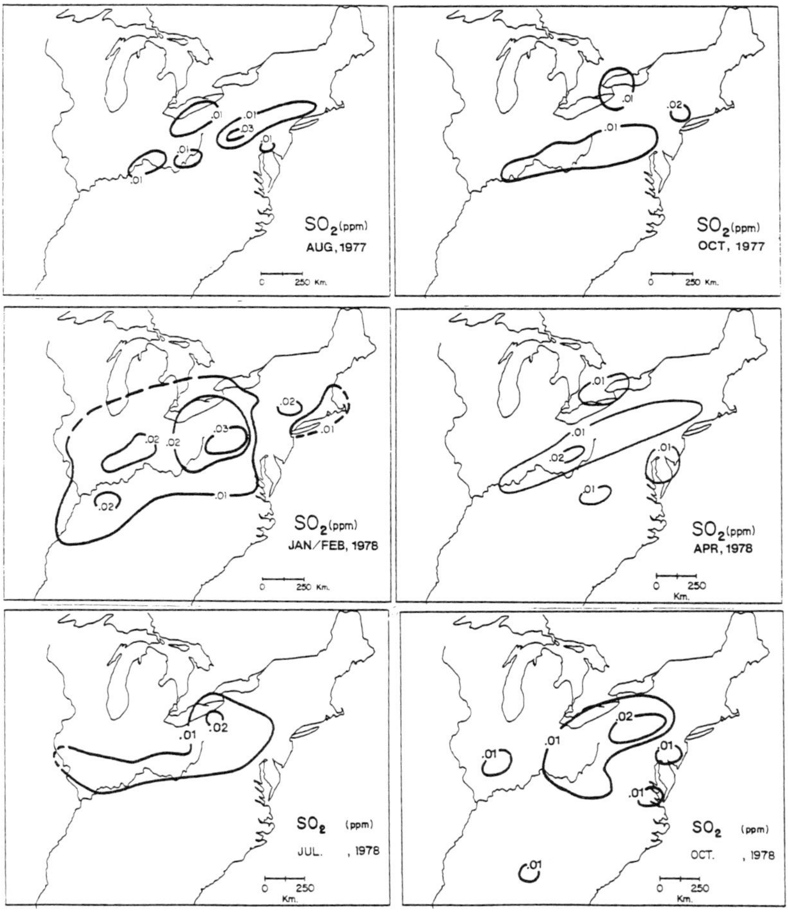

FIGURE 4. Geographic distribution of monthly arithmetic means for SO_2.

The maximum 1-hour averages in August at Philo and Rockport were in the range of 200 to 420 $\mu g\ SO_2/m^3$, and 30 $\mu g\ SO_2/m^3$ at Montague. The maximum 1-hour averages in October at Philo and Rockport were in the range of 300 to 500 $\mu g\ SO_2/m^3$, and 50 $\mu g\ SO_2/m^3$ at Montague.

Diurnally, O_3 tended to peak in the midafternoon, NO_x in midmorning, and SO_2 at midday on high (>50 ppb) ozone days. The data at some sites show diurnal fluctuations of O_3 and NO_x that appear to be characteristic of influences from motor vehicle emissions. This is a subject for a separate analysis.

The monthly geographic distribution of 1-hour SO_2 concentrations is shown for the central months of six quarters in FIGURE 4. Covering of the largest area by greater than 10 ppb occurred in January–February. The area appears to be located west of the Appalachian Mountains; south to Richmond, Va.; and in the New York to Boston corridor. The maximum concentrations were observed in the vicinity of the Ohio

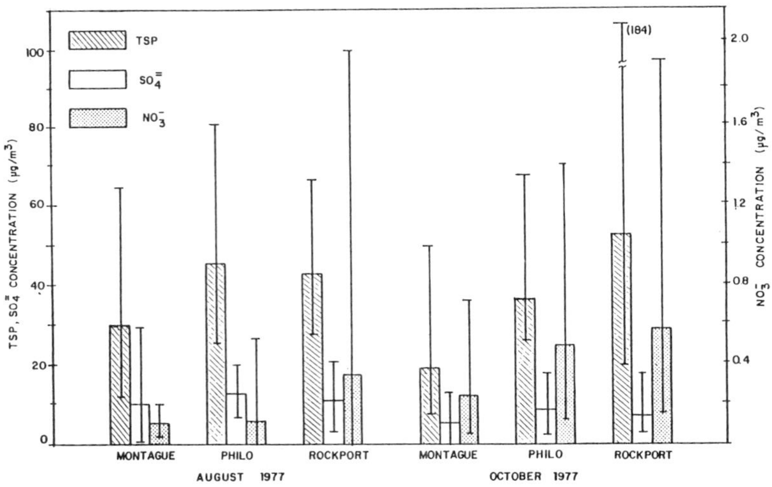

FIGURE 5. Monthly mean and extremes of 24-hour average particulate data for August and October 1977 at three sites in the SURE region.

Valley and western Pennsylvania. This pattern of maxima appears to be maintained for all the intensive sampling months. SO_2 maxima are generally apparent in the urban areas of Richmond, Va., Philadelphia, New York, and Boston during one or more of the months. However, during July 1978, no monthly hourly averages of greater than 10 ppb SO_2 were observed along the Atlantic seaboard. The maximum monthly average was 20 ppb, just south of Lake Erie. In August 1977, the SO_2 distribution was somewhat different, with a maximum monthly average of 30 ppb being observed in southwestern Pennsylvania. The generally lower SO_2 values during April and October are associated with frequent frontal passages through the area during spring and fall. Since the utility emissions of SO_2 did not vary greatly seasonally, the lower SO_2 values during the two summer months as compared to January–February could be explained by increased low-level emissions in winter, more rapid deposition in summer, more rapid conversion to sulfate in summer, or more intense inversions during the winter.

Particulate Matter

Some 24-hour particulate data are shown in FIGURE 5 for the Montague, Philo, and Rockport stations. With the exception of Rockport, the total particulate mass (TSP) was higher during August than in October 1977. During August, averages ranged from 30 to 45 $\mu g/m^3$, while extremes ranged from 10 to 80 $\mu g/m^3$. In October, the range was similar, except that a 24-hour value of 184 $\mu g/m^3$ was recorded at Rockport.

The diurnal fluctuations in particulate matter at these sites did not show a regular pattern, even though we have observed, during haze episodes, a tendency for sulfate to peak diurnally in the early afternoon. The complete data set shows diurnal peaks occurring early morning, midday, late afternoon, or at night. The 3-hour diurnal concentration changes are usually in the range of 3 to 5 times the daily minimum. Factors of up to 10 have occurred.

The particulate size distribution data indicate that during August, the fraction of particles with an aerodynamic mass median diameter (AMMD) of less than 3 μm was in the range of from 32 to 98% of the total particulate mass as measured with high volume samplers at all three sites for all days in August. The small particle mass fraction tended to be greater at Montague than at the other two sites. During October, the small particle mass fraction was also highly variable, ranging from about 20 to 85%. Again, the small particle mass fraction tended to be greater at Montague.

The 24-hour sulfate (SO_4) concentrations tended to be greater during August than October. The average monthly 24-hour concentration for August was about 10 μg SO_4/m^3, with daily values fluctuating from less than 1 to about 30 $\mu g/m^3$. In October, the average concentration was about 5 μg SO_4/m^3 at Montague and 8 $\mu g/m^3$ at Philo and Rockport. The daily values ranged from less than 1 to about 18 $\mu g/m^3$. For August, 79–93% of the sulfate mass occurred in particles less than 3 μm AMMD at these three sites. During October, 74–85% of the sulfate occurred in the small particle size fraction.

The monthly 24-hour nitrate (NO_3) concentrations ranged from 0.11 to 0.35 μg NO_3/m^3 at Montague, Philo, and Rockport in August, and from 0.22 to 0.58 $\mu g/m^3$ during October. Thus, the sulfate concentrations were 30 to 100 times greater than nitrate during August, and about 10 to 20 times greater than nitrate during October. The observed nitrate concentrations were substantially lower than those reported by the National Air Sampling Network (NASN) for urban sites. These stations reported values in the range of 1 to 5 $\mu g/m^3$. For rural or remote sites, NASN reported 0.1 to 3 $\mu g/m^3$ for summer and fall months.[7] Aside from sampling location and time differences, the differences in filter media may account for the observed discrepancies. The glass fiber filter material used in NASN adsorbs more gas phase nitrates than the Teflon-coated glass fiber (TGF) used in this study. The problem of what fraction of the nitrates recovered from the filters occurred in the atmosphere as particles or gases has not yet been resolved. The TGF material used in this study does not adsorb significant amounts of NO_x, but it may adsorb some of the atmospheric nitric acid.

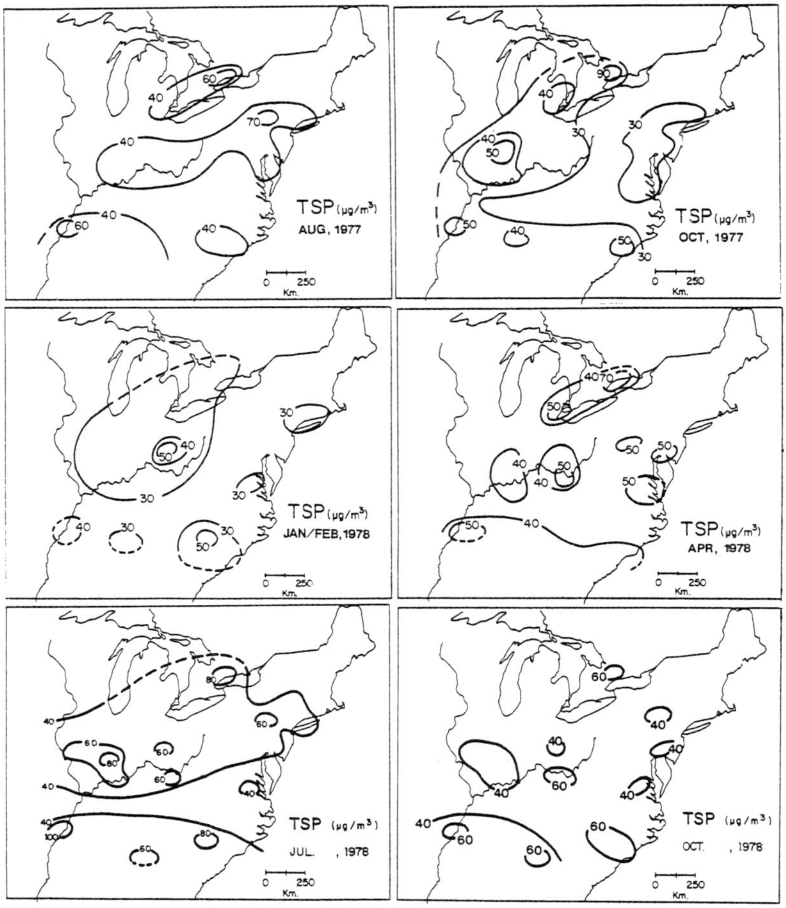

FIGURE 6. Geographic distribution of monthly geometric mean 24-hour Hi-vol TSP values.

The geographic distributions observed for monthly mean 24-hour Hi-vol TSP samples from all sites reporting data are shown for the intensive sampling months in FIGURE 6. Discernible geographic distributions appear to be related to source areas. The highest observed monthly mean value was 90 $\mu g/m^3$, during October 1977 in the vicinity of Toronto. The 24-hour Hi-vol TSP mass values do not correlate well with the TSP content of sulfates and nitrates.

The geographic distributions of monthly mean sulfates observed in 24-hour Hi-vol samples are shown in FIGURE 7. The highest and most widespread monthly average concentrations—12 $\mu g\ SO_4/m^3$ and greater—were observed in July 1978. Monthly means of 8 $\mu g\ SO_4/m^3$ or greater were observed to extend from the Ohio Valley to the Atlantic seaboard also in August 1978. While no really regional distribution of sulfates is apparent in the January–February data, subregional occurrences of 8 $\mu g/m^3$ were observed ranging from Memphis, Tennessee, northeast to northern Ohio.

In April 1978, occurrences of 8 $\mu g/m^3$ and greater were localized in the greater Philadelphia area and south from central Tennessee to north of Toledo and Detroit. October 1977 and 1978 showed the lowest sulfate occurrences. There is a tendency for sulfate maxima to shift to the southern part of the SURE region during these two months.

The geographic distributions of monthly mean nitrates observed in 24-hour Hi-vol samples are shown in FIGURE 8. The ranking of the months with respect to decreasing nitrate concentrations is January–February 1978 > October 1978 > April 1978 = October 1977 > August 1977 = July 1978. Thus, from this set of observations of two summers, two falls, and one winter and one spring month, it appears that nitrates peak seasonally during winter, and sulfates during the summer and winter. This differential

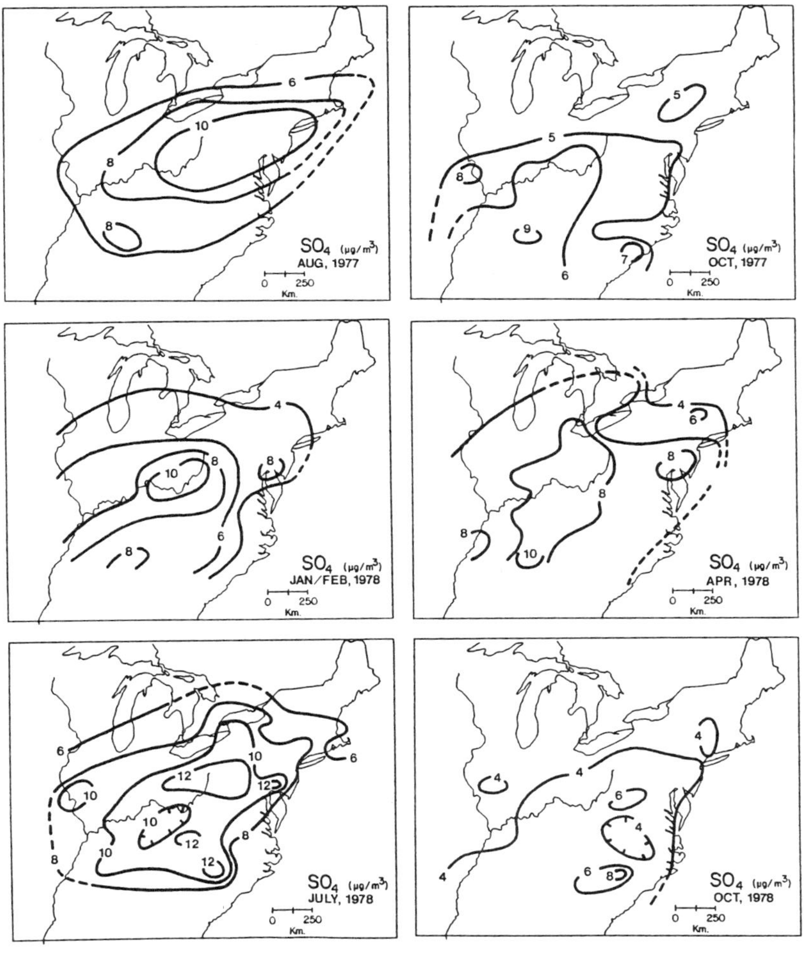

FIGURE 7. Geographic distribution of monthly geometric mean 24-hour Hi-vol sulfate values.

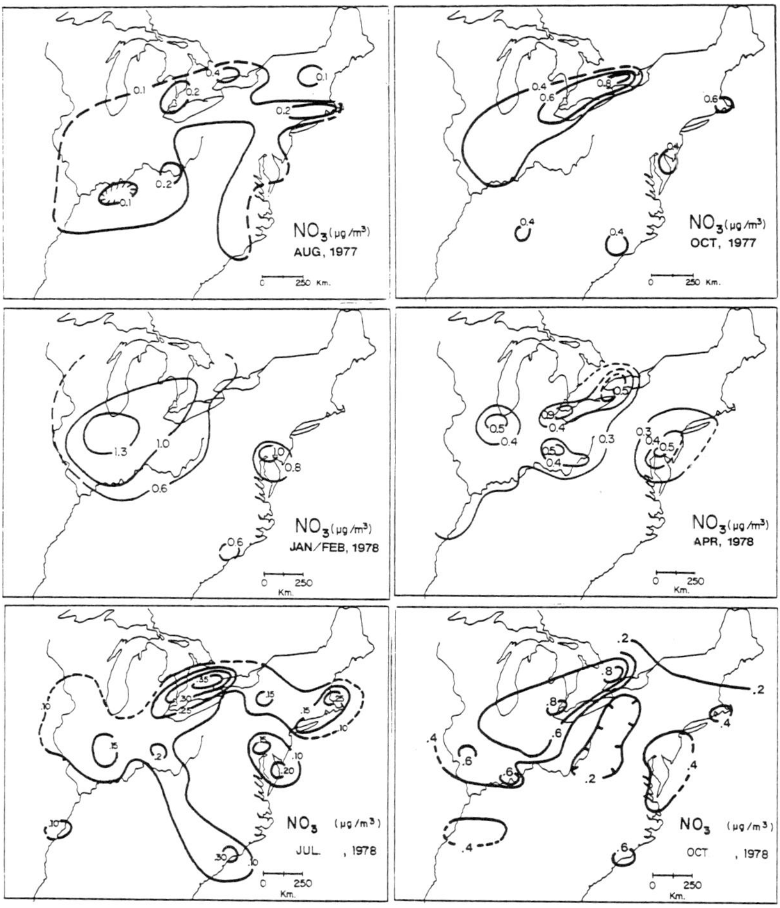

FIGURE 8. Geographic distribution of monthly geometric mean 24-hour Hi-vol nitrate values.

seasonal occurrence for sulfates and nitrates confirms our previous analysis of quarterly NASN data.[8]

The locations of the nitrate monthly maximum values are distinctly different from those for sulfates. Generally, the nitrate maxima were observed north of the Ohio River Valley; south of Lake Michigan; in the vicinity of Lake Erie, and in the vicinity of urban areas in the south, Baltimore–Philadelphia, and New York–Boston. Thus, the nitrates appear not to be associated with the utility sources concentrated in the Ohio River Valley.

Aircraft Observations

Very few data exist that characterize vertical differences in particulate matter on a regional scale. Data of this kind were obtained in the SURE aircraft sampling program operated by Meteorology Research, Inc. and the Research Triangle Institute.

The average levels of sulfate and nitrate for six morning and afternoon flights in early August were measured within two height intervals between approximately surface to 1.5 km above mean sea level (MSL), and 1.5 to 3.0 km above MSL. The results show sulfate and nitrate concentrations were higher in the bottom than in the top layer. On the average, the sulfate levels in the bottom layer appeared to be similar to those observed at the ground. But the nitrate levels aloft appeared to be significantly greater than those at the ground.

The variability in sulfate and nitrate aloft for August 1977 and January 1978 is illustrated for the Philo (04) area in FIGURE 9. The data shown are for the height range between surface and 1.5 km. In this particular location, there was a systematic difference between morning and afternoon sulfate data in August, possibly reflecting differences in day-night formation rates or in boundary layer dynamics. The sulfate levels observed were higher during August. The nitrate levels in August were less than half the sulfate levels, as was the case in January. Also, the January levels of nitrate were less than in summer, in contrast to the ground data in some locations. Again, the nitrate values aloft in January were substantially greater than those observed at the ground. However, the negative concentration gradient to the surface was considerably smaller in January than in August.

The findings of higher nitrate concentrations in January than in August are consistent with the higher nitrate concentrations reported for winter months for NASN.[8] The higher ground-level nitrates observed in winter may be due to a greater proportion of atmospheric nitrates residing in the particle phase than in the gas phase during winter than during summer, or due to higher concentrations of nitric acid near the ground in winter because of lower deposition rates. The concentration gradients to

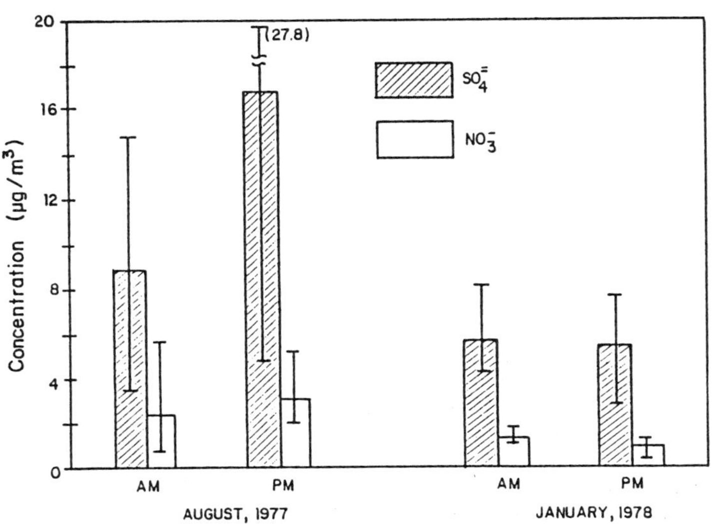

FIGURE 9. Comparison of summer and winter sulfates and nitrates near Philo, Ohio, between 0.3 and 1.5 km aloft during 6 days of sampling. The means and ranges are shown.

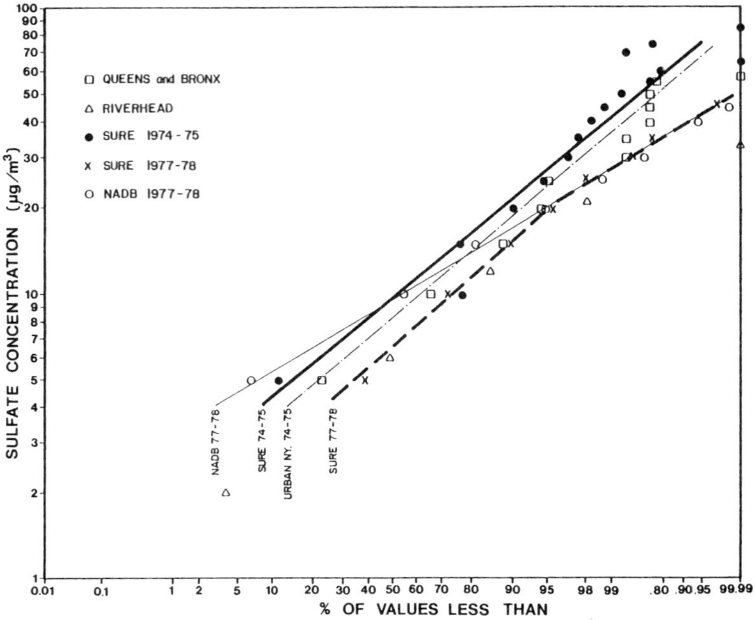

FIGURE 10. Frequency of occurrence of sulfate concentrations in the SURE region as shown by various data sets.

the surface suggest that a substantial portion of the nitrate measured may have been generated in the atmosphere. All of these possibilities remain to be investigated.

TEMPORAL AND SPATIAL DISTRIBUTIONS

Techniques are being devised to depict the temporal and spatial distributions of the aerosol properties for the entire SURE region. Preliminary approaches are presented here to describe the regional occurrence of the 24-hour particulate sulfate data.

Cumulative distributions of 24-hour sulfate concentrations are shown in FIGURE 10. Four sets of data are included:

1. 1974–75 daily data from the Bronx and Queens, N.Y., obtained as part of the Community Health Air Monitoring Program (CHAMP) operated by the U.S. EPA. The daily data from the Riverhead, N.Y., rural site from CHAMP are also plotted. The samples were collected on glass fiber filters.

2. 1974–75 daily data from 12 rural sites in the northeastern U.S. from SURE. These samples were collected on glass fiber filters.

3. 1977–78 every-sixth-day data from governmental sampling networks in the eastern U.S. for urban and rural sites reporting into the National Aerometric Data Bank (NADB). These samples were collected on glass fiber filters.

4. 1977–78 daily data from 9 Class I SURE sites. These samples were collected on Teflon-coated glass fiber filters.

The logarithmic frequency plot of these data leads to some interesting conclusions as follows:

1. The 1974–75 SURE and urban N.Y. data followed approximately the same distribution.
2. The 1977–78 NADB data show a smaller geometric standard deviation than the 1974–75 SURE and urban N.Y. data.
3. The 1977–78 NADB data appear to be log-normally distributed, whereas the 1977–78 SURE data are not. Above 20 $\mu g/m^3$, these two sets of data appear to be similar. Fifty percent of the SURE data was less than 6.5 $\mu g\ SO_4/m^3$, while 50% of the NADB data was less than 10 $\mu g/m^3$.
4. A number of factors could explain the differences between the NADB and SURE data sets for 1977–78:

- The use of glass fiber filter media could have contributed a greater relative amount of sulfate artifact from SO_2 at sulfate concentrations below 20 $\mu g/m^3$.
- The urban sites in the NADB data may have experienced generally higher sulfate concentrations.
- The sampling frequency of every sixth day tends to miss the extremes in sulfate concentrations observed with the SURE and CHAMP daily samplings.

Thus, the question arises of whether the actual distribution of sulfate concentrations is log-normal. The histogram of the distribution is shown in FIGURE 11. About 6.6% of the samples had concentrations greater than 20 $\mu g/m^3$. While the data sets shown in FIGURE 10 suggest a widespread, uniform occurrence of sulfate in the greater

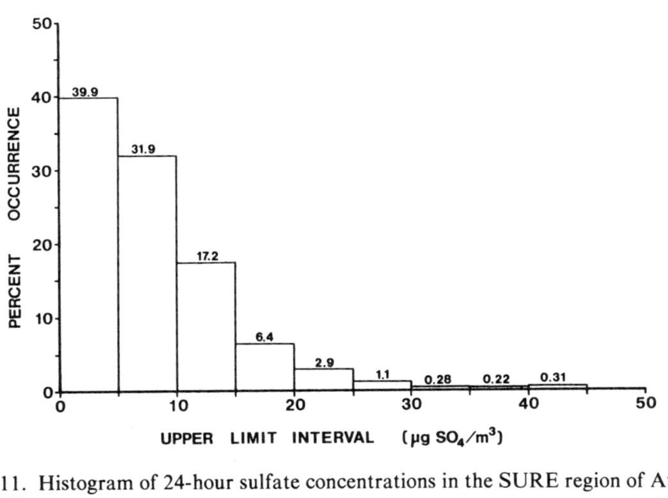

FIGURE 11. Histogram of 24-hour sulfate concentrations in the SURE region of August 1977 through 1978.

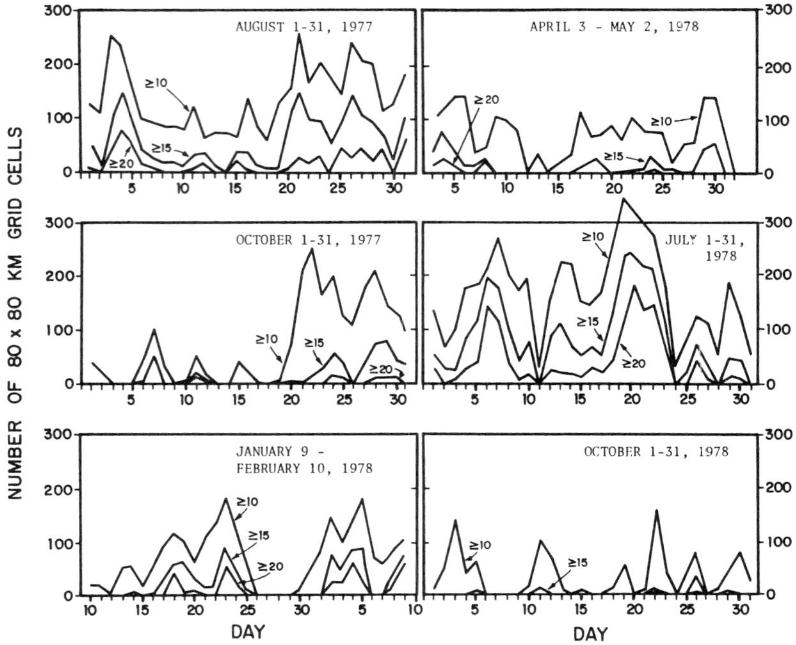

FIGURE 12. Calculated number of grid cells in the SURE region where 24-hour sulfate concentrations exceed 10, 15, and 20 $\mu g/m^3$ during intensive sampling months in 1977–78.

Northeast, they do not reveal the spatial and temporal characteristics of sulfate patterns.

For the purpose of comparing spatial distributions, the greater northeastern region shown in FIGURE 1 has been segmented into approximately 80 × 80 km grids—23 from south-to-north and 30 from west-to-east. An approximate estimation of average concentration in each 80 × 80 km grid of the SURE area was made by extrapolating the 54 station observations to other grid squares, using an inverse distance squared interpolation procedure. The extrapolation from any one site was limited to a distance of three grids from the site. The approximate estimates in this paper include extrapolation into areas outside the boundary of the SURE station network. This introduces errors, because we know little about the actual sulfate concentrations outside this boundary. It is planned to use a calculation routine that includes only the interpolation within the boundary of the sites. However, the more approximate estimates given in this paper provide some initial insights about the temporal and spatial distributions.

Within the above limitation, the extent of the SURE region experiencing concentrations above 10, 15, and 20 $\mu g\ SO_4/m^3$ is plotted in FIGURE 12 by day for August and October 1977, and mid-January to mid-Feburary, April, July, and October 1978. The episodic nature of the regional behavior of sulfate, based on nonurban data, shows up clearly. For instance, during August 1977, there was a 3-day regional sulfate episode that peaked on about August 4. Toward the end of August, a

3-day episode beginning on August 21 was followed by a 5-day episode beginning on August 25. During July 1978, five apparently distinct regional sulfate episodes occurred, one following the other. However, as shown in FIGURE 7, the spatial extent varies geographically among episodes. This variability is currently being analyzed, and methods for describing it succinctly are being developed. In October 1977, virtually no levels above 20 $\mu g/m^3$ were seen until the 20th. After this, an increase in regional sulfate concentrations occurred. Despite intense cold and stormy weather in the January–February 1978 period, two cases of elevated sulfate levels appeared—one around January 23 and one in early February. During April 1978, the levels and extent of sulfate aerosol occurrences were less than those in the summer and winter of 1977 and 1978. Large-scale regional sulfate hazes did not develop during October 1978.

An impression of the spatial extent of the occurrence of the elevated sulfate

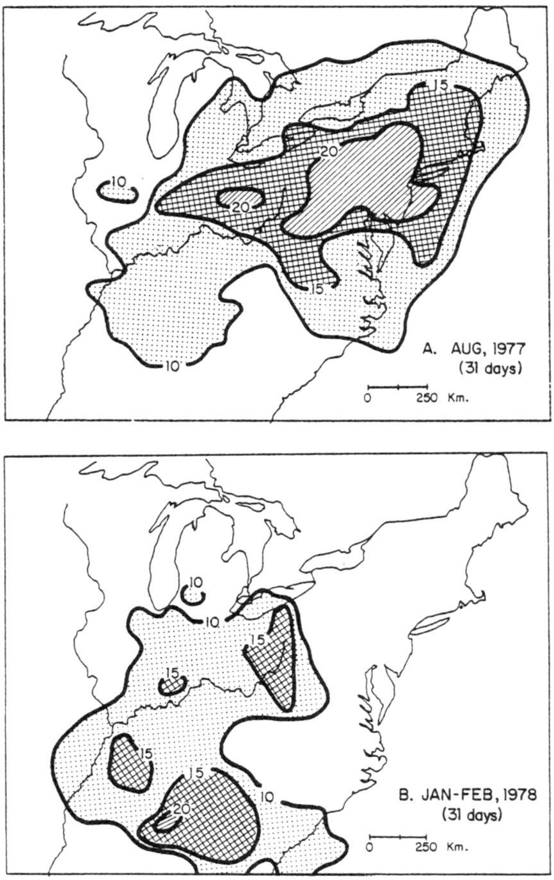

FIGURE 13. Geographic distribution of numbers of days per month where sulfate ≧ 10 $\mu g/m^3$ using r^{-2} extrapolation of station data: (A) August 1977; (B) January–February 1978.

concentrations during each season can be obtained by summing for each grid the estimated number of days exceeding 10 or 20 $\mu g/m^3$. This is illustrated, for August 1977 and January–February 1978, in FIGURE 13. During August (FIGURE 13A), sulfate concentrations in excess of 10 $\mu g/m^3$ were observed during 10 or more days over most of the eastern United States. In the Pennsylvania–West Virginia area, these concentrations occurred on 20 or more days. Interestingly, the January data also show a widespread geographic exposure of 10 or more days to 10 $\mu g/m^3$ (FIGURE 13B). However, this occurred in the southwest-southcentral portion of the SURE region. The zones of >10 $\mu g/m^3$ for half the month (or more) were located in the central Ohio Valley and in the Tennessee–Alabama–Georgia areas. This shift in occurrence of high sulfate southward during January may be related to penetration of intense cold fronts, which thereby prevents the warm, moist air from flowing northward over the regions of high emission density. These results suggest major differences in seasonal sulfate behavior. Similar depictions of other aerosol data are expected to reveal the characteristics of interactions among aerosol components and emission sources on temporal and spatial scales.

Summary and Conclusions

To understand the genesis and potentials for control of regional atmospheric hazes, it is important to obtain a complete knowledge of the aerosol. As part of SURE, an extensive data base on the mass concentration, size distribution, and chemical composition is being established for the atmospheric aerosol occurring in the greater northeastern United States. The design and implementation of the SURE ground-station and aircraft observations have been briefly described. Some observations made from August 1977 through October 1978 during quarterly intensive sampling months have been presented. Procedures of accessing the data base have been indicated. Analysis and interpretation of data are in progress. Highlights of the SURE findings so far indicate:

1. Hi-vol mass (TSP), sulfur dioxide, and nitrate maxima tended to be localized on scales of about 100 km on an annual average basis.

2. TSP tended to be highest near source areas and showed no systematic month-to-month differences. The highest monthly average of about 70 $\mu g/m^3$ was observed during July 1978.

3. The highest annual rural SO_2 concentrations occurred along the Ohio River Valley. The maximum annual averages ranged from about 50 to 80 $\mu g/m^3$. The highest monthly average SO_2 concentrations of about 80 $\mu g/m^3$ were observed during January–February of 1978.

4. The highest annual rural nitrate concentrations occurred in the vicinity of urban areas. The maximum annual averages ranged from about 0.30 to 0.50 $\mu g/m^3$. The highest monthly nitrate values of about 1.3 $\mu g/m^3$ were observed during January–February of 1978.

5. Nitric acid may be an important component of the total atmospheric nitrates.

6. Nitrate concentrations were higher aloft up to 1.5 km above mean sea level (MSL) than at the ground and in the altitude range of 1.5 to 3.0 km above MSL. A strong, positive, up-to-tenfold, concentration gradient was observed for nitrates from ground level to about 1.5 km above MSL. This gradient was greater during August 1977 than during January 1978.

7. Sulfates appeared to be regionally distributed, with annual maxima located in central Tennessee, the Ohio River Valley, and in the vicinity of the urban areas along the northeastern Atlantic seaboard from Richmond, Va., to New York, N.Y. The highest monthly average concentrations in the range of 10 to 12 $\mu g/m^3$ were observed during August 1977, July 1978, and January–February of 1978.

8. Sulfate concentrations were relatively uniform from the ground to 1.5 km above MSL. From 1.5 to 3.0 km above MSL, the sulfate concentrations were less than in the lower layer.

9. The particle size distributions of the TSP mass were highly variable. The small particle mass (<3 μm AMMD) ranged from about 20 to 98% of the TSP mass.

10. The particle size distributions of the TSP sulfate were fairly uniform. About 75 to 95% of the total sulfate occurred in small particles.

11. Diurnal patterns of mass and sulfates were highly variable. Maximum diurnal fluctuations in 3-hour average concentrations ranged from 3 to 5 times the daily minimum 3-hour value. During persistent haze conditions, there was a tendency for diurnal sulfate maxima to occur in the early afternoon.

12. The 24-hour sulfate data from rural and urban sites in the northeastern United States, obtained by sampling on glass fiber filters in 1974–75 and 1977–78, fit log-normal distributions with geometric means in the range of 8 to 10 $\mu g/m^3$. The 1977–78 SURE 24-hour sulfate data from rural sites, obtained by sampling on Teflon-coated glass fiber filters, are not log-normally distributed. About 50% of the SURE data have values less than 6.5 $\mu g/m^3$. About 6.6% of the SURE data have values greater than 20 $\mu g/m^3$.

13. Monthly time-area displays have been devised to chart the regional and episodic distribution of sulfates. The longest and regionally most extensive occurrences of sulfate concentrations greater than 20 $\mu g/m^3$ were observed in the summer quarters during August 1977 and July 1978, and in the winter quarter during January–February 1978.

14. Monthly time-concentration maps have been devised to illustrate the regional monthly dosage distribution of sulfates. The data show a tendency for the zones of maximum sulfate dosage to shift from the northeast during the summer months (August 1977 and July 1978) to the southeast during fall and winter months (October 1977, 1978 and January–February, 1978.)

Acknowledgments

The U.S. Environmental Protection Agency has supplied some of the instrumentation for the ground-station network. We remain indebted to the dedicated work of a large number of unnamed individuals who have been working on the many aspects of SURE.

References

1. Husar, R. B., N. V. Gillani, J. D. Husar, C. C. Paley & P. N. Turco. 1976. Long range transport of pollutants. *In* Proceedings, 3rd Symposium on Atmospheric Turbulence, Diffusion, and Air Quality, American Meteorological Society: 344–347.
2. Perhac, R. M. 1978. Sulfate Regional Experiment in northeastern United States: The SURE Program. Atmos. Environ. **12:** 641.
3. MacCracken, M. C. 1978. MAP3S: An investigation of atmospheric energy related pollutants in the northeastern United States. Atmos. Environ. **12:** 649.
4. Hidy, G. M., P. K. Mueller, V. Deyo & K. C. Detore. 1979. Design and implementation of the Sulfate Regional Experiment. *In* Proceedings, 4th Symposium on Turbulence, Diffusion, and Air Pollution, American Meteorological Society: 314–321.
5. Lavery, T. F., G. M. Hidy, R. L. Baskett & J. Thrasher. 1979. Occurrence of long range transport of sulfur oxides in the northeastern United States. *In* Proceedings, 4th Symposium on Turbulence, Diffusion, and Air Pollution, American Meteorological Society: 330–336.
6. Warren, K. & P. K. Mueller. 1979. Data Available from the EPRI/SURE Data Bank. ERT Document No. P-5042DB2. Environmental Research & Technology, Inc. Westlake Village, Calif.
7. Mueller, P. K., G. M. Hidy, T. F. Lavery, K. Warren & R. L. Baskett. 1979. Some early results from the Sulfate Regional Experiment (SURE). *In* Proceedings, 4th Symposium on Turbulence, Diffusion, and Air Pollution. American Meteorological Society: 314–321.
8. Hidy, G. M., E. Y. Tong & P. K. Mueller. 1976. Design of the Sulfate Regional Experiment. Vol. 1, Report EC–125, NTIS *PB–274947*. Electric Power Research Institute. Palo Alto, Calif.
9. Hidy, G. M. *et al.* 1977. Preliminary Air Quality Analysis of Urban and Regional Sulfate Distributions. Report 4295. American Petroleum Institute. Washington, D.C.

MESOSCALE AND SYNOPTIC SCALE TRANSPORT: GENERAL DISCUSSION*

George T. Wolff, *Moderator*

Environmental Science Department
General Motors Research Laboratories
Warren, Michigan 48090

J. W. Winchester (*Florida State University, Tallahassee, Fla.*): Dr. Wolff, do you think that the transport of the sulfate is primarily as sulfate particles or as the SO_2 gas precursor? On this scale that you just talked about.

G. T. Wolff: Our observations on this have only been as far as High Point to New York City, and in that range, it's definitely sulfates that we're seeing. I think that your question will probably be answered in Dr. Mueller's presentation today. We have looked at visibility—transported visibility blobs—the same way that Dr. Husar has, and based on that, I think that it is the sulfate.

Unidentified Speaker: Just a quick comment, Dr. Wolff. I guess it's a statement of the obvious, but certainly there are conditions, with no reaction activators, no fog, no heavy ultraviolet radiation, and very low accumulation of other pollutants, where the sulfur dioxide is moving and just keeps on moving because there's nothing to change it or stop it. Have you observed this, or shall we wait for Dr. Mueller's paper on that?

G. T. Wolff: We've never noticed an important contribution to the SO_2 in New York. I don't know about other areas.

W. R. Pierson (*Ford Motor Company, Dearborn, Mich.*): Dr. Brosset, I was interested in your comment about the lack of correlation between manganese and iron in the black episodes. We're also finding it in the fine particles, and people here are reporting it in the coarse particles. It occurred to me that we didn't get any size information in the Allegheny experiment that I described, at least as far as manganese and iron are concerned. However, our findings were in contrast to your results.

Now, we're dealing with what you would call a white episode, okay? We were getting a correlation coefficient of 0.85 between manganese and iron. However, the manganese to iron ratio was very low—only about 0.028 on the average. I just wondered if maybe the pattern that we're seeing is that in the black episodes, we have one manganese story; but perhaps if we would look at the manganese in the white episodes, we might find that it was in the large particle regime.

C. Brosset (*Swedish Water and Air Pollution Laboratory, Gothenburg, Sweden*): I think you are quite correct. We have, in fact, analyzed a number of fly ashes for heavy metals, and there are a lot of differences. But those that we suspect are coming from central Europe contain a lot of manganese in the fine particles.

L. Newman (*Brookhaven National Laboratories, Upton, N.Y.*): Dr. Brosset, in your last slide, showing the chemical composition of black particles, you showed

*This is a compilation of the discussions held after the presentation of each of the papers in this section.

essentially all of it as ammonium sulfate and gave data of the mass of soot. I made a quick mental calculation of the very highest number, and it appeared that essentially all of the soot could be accounted for by the mass of ammonium sulfate. Is that true, or is that generally the case?

C. BROSSET: I think that is true.

K. A. RAHN (*University of Rhode Island, Kingston, R.I.*): Dr. Brosset, at the risk of prolonging the discussion of manganese and iron, I'd just like to say—from my own experience in the northeastern United States and a little bit in Europe—that there seems to be quite a spectrum of the preferred particle sizes of the various chemical elements. On the one hand, you have aluminum and silicon, which are essentially coarse particles, and on the other hand, you have things like lead and sulfur and vanadium, which are exclusively fine particles.

By contrast, manganese and iron are sort of in the middle. What I have seen is that in both Europe and United States, a reasonable fraction of the mass of each of these two elements is in the fine particles as well as in the coarse particles. This is not always recognized for iron, but it seems to be the case. Manganese seems to be shifted a little bit more towards the fine particles, so there's a systematically higher percentage of manganese than of iron in fine particles, but still nowhere near the extent of lead, sulfate, and vanadium.

I also have been looking around lately and have observed that there seems to be a lot more manganese in European aerosol characteristically than in northeastern United States aerosol; more by a factor of 3, or 4, or 5.

C. BROSSET: That is quite correct. I had the privilege of analyzing a number of fly ashes from the United States, and your observation is quite correct. I would like to mention one thing that we had discussed together earlier. Those two episode types that I have mentioned were synoptic scale episodes—they are identified very well in my account here. There may be another type—a global episode. We have two observations up to now that are indicating the possibility of a global episode, but as they are only two times observed, I don't wish to discuss it more today.

D. S. LANG (*University of Minnesota, St. Paul, Minn.*): Dr. Husar, for the areas in which you showed average sulfate concentrations of 25 micrograms per cubic meter, would you care to speculate on what a background sulfate concentration might be. And second, for the areas where we are seeing 15 micrograms per cubic meter now, would you also care to speculate as to how long it has taken for that kind of a concentration to develop.

R. B. HUSAR (*Washington University, St. Louis, Mo.*): I would think that two or three micrograms per cubic meter would be an upper limit on background—the annual average concentration for sulfate over the eastern U.S. if you remove everybody and let everybody go back home to Europe and wherever. I didn't quite understand the second question. How long does it take to develop? These are yearly averages.

D. S. LANG: How far would we have to go back—to the 1800s? How long has this condition been prevalent in this area?

R. B. HUSAR: Using the industrialization as a rough guide, it would be possible to have a good estimate. However, that estimate could be improved by plotting the trend of coal consumption again as a qualitative index that has been rising; in 1900, it was about 10% of the current consumption, so 1850 would certainly be a very safe time to go back to, to make an estimate of the sulfate concentrations.

L. NEWMAN: Dr. Barrie, I was intrigued with the fact that the deposition of the sulfur was about one-tenth the deposition of nitrogen. Do you have any understanding as to why that's the case?

L. A. BARRIE (*Atmospheric Environment Service, Downsview, Ont., Canada*): I can only guess, but there are two things. One is that particulate nitrates tend to be in slightly larger particles. However, since most of the NO_x [nitrogen oxides] and SO_x [sulfur oxides] are gases, it probably lies in the reactivity of SO_2 and nitric acid.

L. NEWMAN: I guess, in my naive view, I don't think the SO_2 is any more reactive than the NO_x. While the SO_2 is more reactive than the NO, it is probably not more reactive than NO_2. I would guess that the deposition velocity, if you will, for these two gases should be about comparable. And, on balance, I would have thought you would have gotten more deposition of sulfur than of nitrogen. I'm kind of surprised to see that inversion there.

L. A. BARRIE: What about nitric acid, though?

L. NEWMAN: I don't think you're forming that much nitric acid at 25 kilometers, in the time scale that you're following.

G. M. KLAUBER (*Environmental Measurements, Inc., Arnold, Md.*): I'd just like to ask Dr. Lyons, if he had his druthers, what he would like to have NASA put in the next satellite.

W. A. LYONS (*MESOMET, Inc., Chicago, Ill.*): There have actually been several reports. John Seinfeld is on one of the committees that have made recommendations to NASA. NASA is asking the basic question, What shall we do for the next decade? I think there is a growing need not only to utilize the information we have right now, but also to develop a new generation of sensors. These, I think, should have two capabilities. One is to measure at least a portion of the tropospheric aerosol mass for an estimate of the total mass loading. That is probably going to be some type of laser technique.

The second device we need is one that measures lower tropospheric ozone loading. This flying around with aircraft may be a lot of fun for some people, but it's also very expensive; and you a have a sampling problem that you can't solve unless you have the entire SAC air force fleet at your disposal. We need some sort of remote sensing capability to give us the lower 1500 meters of ozone. I know there's a lot of concern about what's going on up in the stratosphere, but I think for the immediate health effects, we need that lower tropospheric sensor. So that's being considered, but it's probably still at least five years off.

S. DATTNER (*Texas Air Control Board, Austin, Tex.*): Dr. Lyons, earlier in your talk, you showed photographs of the winds going counterclockwise and coming off the gulf coast and having these blobs traveling in opposite directions than usual. We had done extensive studies on some of our highest ozone levels in the Houston-gulf area, and we're finding that they're coming off the gulf. We've gone ahead and tried to track these situations, to make sure that we're not getting a day-night sea breeze effect. We're seeing very much the same kind of thing—the patterns are very different than normally expected.

W. A. LYONS: There's a response to that. In the movies of that August episode, you see one situation where a huge mass of haze, which I'm sure is loaded with ozone, comes all the way around the east coast, down the Atlantic coast about a thousand miles offshore, crosses Florida where some of it is washed in a thunderstorm, goes through the gulf and then right into Houston.

ON THE ORIGIN AND TRANSPORT OF THE WINTER ARCTIC AEROSOL*

Kenneth A. Rahn and Richard J. McCaffrey

Graduate School of Oceanography
University of Rhode Island
Kingston, Rhode Island 02881

Midlatitude Pollution Aerosol in the Arctic

The study of Arctic air chemistry is still in its infancy. Until very recently, only fragmentary data on trace gases and aerosols from the Arctic were available. During the last 2–3 years, however, a small but systematic study of the Barrow, Alaska aerosol has been undertaken at the University of Rhode Island. The results have shown, contrary to all expectations, that the atmosphere of Arctic Alaska contains an abundance of pollution-derived aerosol, particularly during the winter half-year. This is in great contrast to the Antarctic, which has one of the cleanest atmospheres yet studied. In this article we first summarize the new Arctic data, which have totally altered our understanding of the Arctic atmosphere, and then attempt to demonstrate that they are quantitatively consistent with a midlatitude origin followed by rather direct transport to the Arctic. We conclude with a series of arguments that Europe is the main source region for the winter Arctic aerosol.

Since September 1976 we have been continuously sampling the Barrow aerosol, in conjunction with the University of Alaska and the National Oceanic and Atmospheric Administration. High-volume samples are taken for 3–7 days on 20 × 25-cm (8 × 10-inch) Whatman No. 41 cellulose filters, which are then analyzed chemically for a variety of natural and anthropogenic constitutents. In this article we concern ourselves with the results of three constituents—V (more properly excess, or noncrustal, V), sulfate (more properly excess, or nonmarine, sulfate), and ^{210}Pb. Noncrustal V is used as an indicator of midlatitude pollution emissions. Nonmarine sulfate, a major constituent of the Barrow aerosol, has a variety of sources, but the sulfate reaching Barrow during winter seems to have predominantly anthropogenic sources. ^{210}Pb is a natural daughter product of the radioactive decay of gaseous ^{222}Rn emitted from continents.

Barrow is a typical Arctic village. It has a population of about 2,500 and is situated at 71° north latitude on the shore of the Arctic Ocean. The Barrow area is snow covered for roughly nine months each year, September through May. Mean temperatures are about −25°C in winter and about 3–5°C in summer. The nearest town of any size is Prudhoe Bay, about 300 km to the east-southeast.

The first two years of data from Barrow have shown that its aerosol changes drastically from summer to winter in a reproducible fashion. The Arctic winter aerosol is at least an order of magnitude more concentrated than the summer aerosol. The precise winter-summer ratio depends on the site and chemical species; at Barrow it is

*Supported by the Office of Naval Research (Contract N00014-76-C-0435) and by the National Science Foundation (Grant DPP77-27242).

0077–8923/80/0338–0486 $01.75/1 © 1980, NYAS

about 15 for ^{210}Pb, 20 for sulfate, and nearly 50 for V. Monthly mean concentrations of sulfate and V for 1976–77 are shown in FIGURE 1. The overall pattern is one of rapid increase during fall to a general winter maximum, followed by a similarly rapid decrease in late spring. Other Arctic sites, for which we have less detailed information, such as Spitsbergen, Bear Island, and Greenland, also show winter maxima, which indicates that this phenomenon is Arctic-wide. Where systematic, multielemental data are available, as at Barrow, further details within the broad winter maximum may be resolved. The strikingly similar patterns of sulfate and V, each with an early winter plateau and spring maximum in March, are, we believe, typical of the Arctic winter aerosol as a whole, at least at this site.

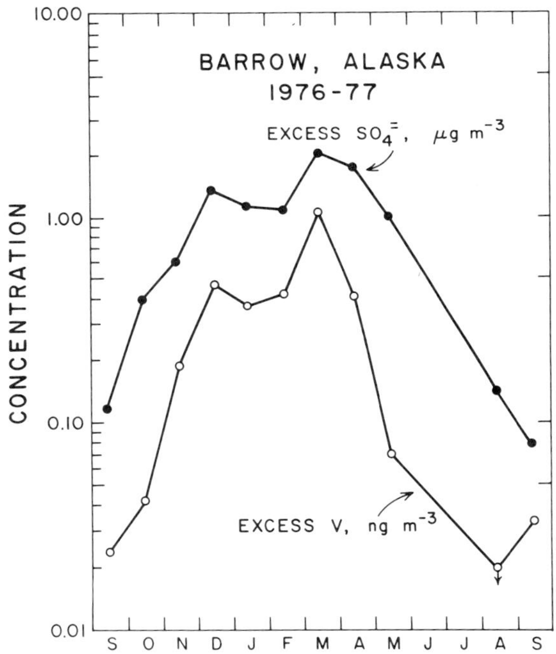

FIGURE 1. Monthly mean concentrations of excess sulfate and vanadium at Barrow, Alaska, 1976–77.

The winter-summer contrast of the Barrow aerosol manifests itself in other ways as well. Unpublished data from J. Rosen and D. Hofmann of the University of Wyoming show that the concentrations of all particles larger than radius 0.15 μm increase by roughly an order of magnitude throughout the depth of the Arctic troposphere during winter, which is a valuable extension of our surface data. The color of the Arctic aerosol changes in winter—summer filters are typically colorless, whereas winter filters are gray. Winter is also the season of Arctic haze, but in summer the Arctic skies are extremely clear, as one might intuitively expect for the entire year.

The winter-summer contrast of the Arctic aerosol is much greater than in

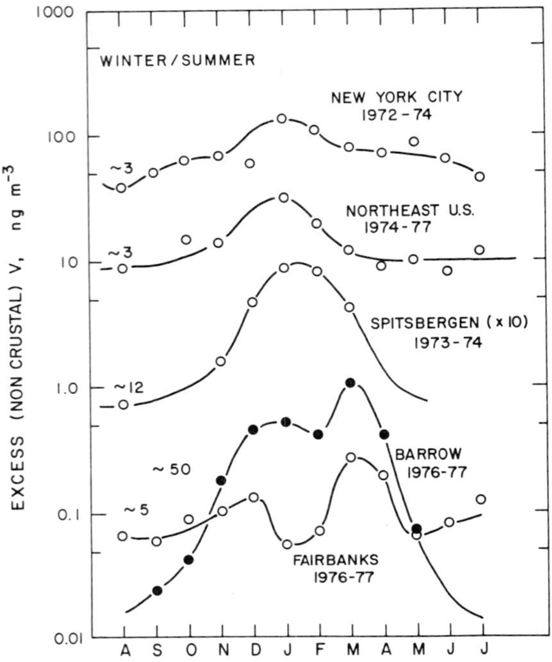

FIGURE 2. Monthly mean concentrations of excess vanadium at five Arctic and non-Arctic sites.

midlatitudes. This is shown for V in FIGURE 2, where seasonal patterns from New York City, the Northeast United States, Spitsbergen, Barrow, and Fairbanks are given. Variations at the midlatitude sites (including Fairbanks) are factors of 3–5, whereas the two Arctic sites have factors of 10–50. This tells us that transport from midlatitudes to the Arctic (which we show below must occur) is more effective in winter than in summer. Another interesting feature of this plot is the nature of the Fairbanks aerosol, which, in spite of being only 800 km south of Barrow, appears to be much more midlatitude than Arctic in character.

The winter Barrow aerosol appears to be much more pollution-derived than its summer aerosol. This is shown in FIGURE 3, where aerosol–crust enrichment factors of V and Mn are seen to have a strong maximum in winter. Interestingly, the enrichment factors of both elements are 2–3 times lower in Barrow during winter than they are in the suspected midlatitude source regions, a change which seems to indicate that the aerosol picked up some crustal material en route.

The midlatitude origin of the Barrow aerosol is attested to by the presence of excess V. This element is a most useful one in atmospheric chemistry, because its pollution component (as measured at Barrow) comes nearly exclusively from combustion of residual oil,[1] which is too viscous to be easily handled in cold climates and so is burned nearly exclusively in temperate latitudes.

The pollution character of the Barrow aerosol is also indicated by the high absolute concentrations of sulfate. Each spring the monthly mean concentrations of sulfate reach or exceed 2 $\mu g\ m^{-3}$, a remarkably high value. By comparison, mean winter sulfates in the northeastern United States are only about 5–6 $\mu g\ m^{-3}$. Similarly high sulfate concentrations are also observed at Spitsbergen.

Another interesting feature of the Barrow aerosol is its highly secondary nature. This is shown here by the sulfate/V ratio, which, as shown in FIGURE 4, is one order of magnitude higher in the Arctic than in typical midlatitude source regions. Interestingly, the values for the intermediately located northern Norway seem to be also intermediate. In our opinion, the source of the excess sulfate of the Arctic must be oxidation of the gaseous SO_2 of midlatitude polluted air masses. This oxidation can occur along the path to the Arctic, and shows that the Arctic aerosol is an aged version of the midlatitude aerosol. The factor of increase in the sulfate/V ratio is also reasonable, because midlatitude polluted air masses have several times more SO_2-S than sulfate-S. This point will be explored in more detail in the second half of this article.

In general, ^{210}Pb collected on our Barrow filters has three components: a "supported" fraction, which is in equilibrium with nuclides of the ^{238}U series of the soil fraction of the Barrow aerosol; an "unsupported," or excess, component, which was also present in those same parent soils and was simply carried to Barrow with them; and an excess component formed from ^{222}Rn en route. Because the specific activity of total ^{210}Pb that we measure in the Barrow aerosol (15×10^4 dpm g^{-1} of soil aerosol) is much larger than the specific activity of supported ^{210}Pb in soils (about 1 dpm g^{-1}) or

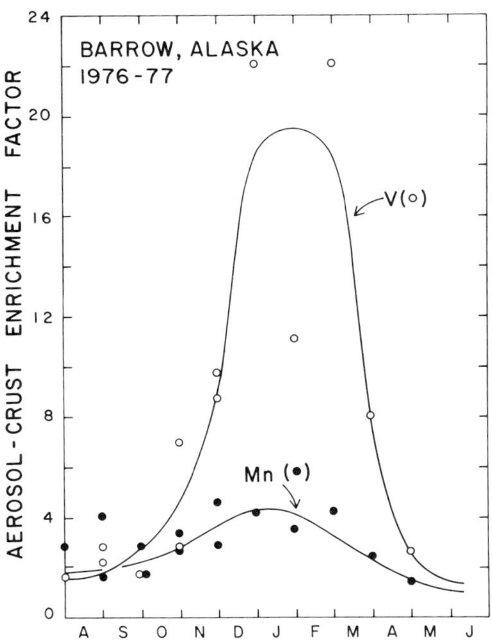

FIGURE 3. Monthly mean aerosol–crust enrichment factors for V and Mn at Barrow, Alaska, 1976–78.

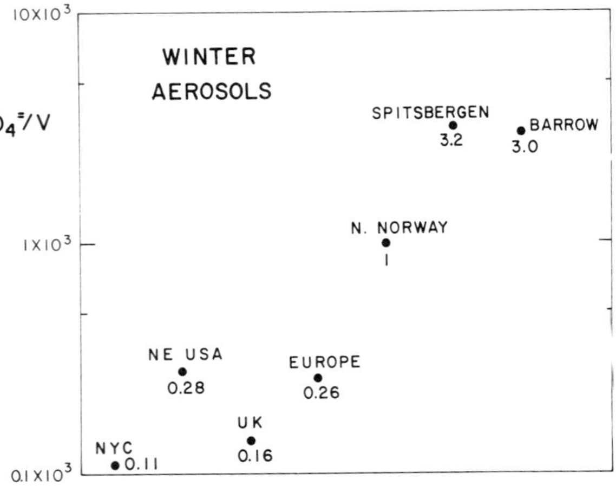

FIGURE 4. Winter $SO_4^{=}$/V ratios at Arctic and non-Arctic sites.

typical excess ^{210}Pb in soils (≦100 dpm g^{-1}),[2] we believe that essentially all of the total ^{210}Pb that we measure has been formed by decay of ^{222}Rn en route.

FIGURE 5 shows seasonal patterns of total ^{210}Pb at New York City and Barrow.[3] At New York, the mean ^{210}Pb concentration is about 15 fCi m^{-3}, with no strong seasonal variation. At Barrow, its annual mean is lower; but during winter the ^{210}Pb

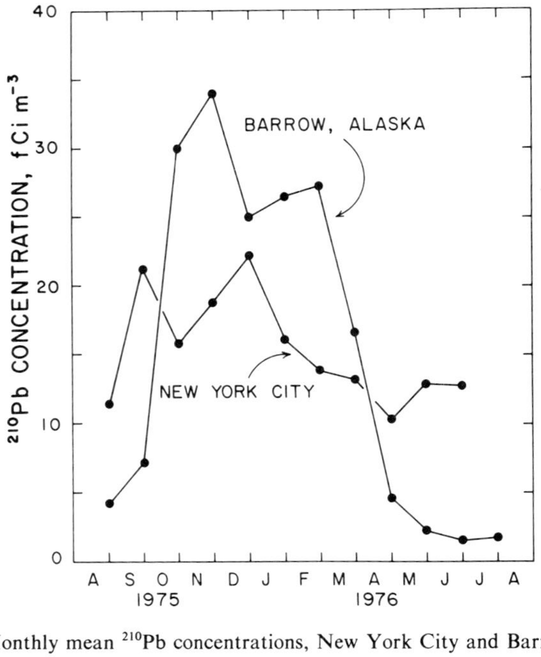

FIGURE 5. Monthly mean ^{210}Pb concentrations, New York City and Barrow, Alaska.[3]

concentrations are nearly double the New York values, a completely unexpected result. We suggest that the most likely explanation for the high ^{210}Pb concentration at Barrow is an unusually long residence time for the Arctic aerosol, so that ^{210}Pb produced in the Arctic (from parent ^{222}Rn, which must be about an order of magnitude less concentrated there than near New York) accumulates with little loss due to precipitation. We return to this point later.

One can immediately see the parallels between the SO_2-sulfate and ^{222}Rn-^{210}Pb systems: both have gaseous precursors with residence times of a few days that are emitted over continents, and both are converted into submicron secondary aerosols. Both daughter products are enriched in the Barrow winter aerosol by an order of magnitude relative to what was expected. The parallelism between ^{210}Pb and sulfate at Barrow is even greater: they are extremely well correlated (FIGURE 6), with a

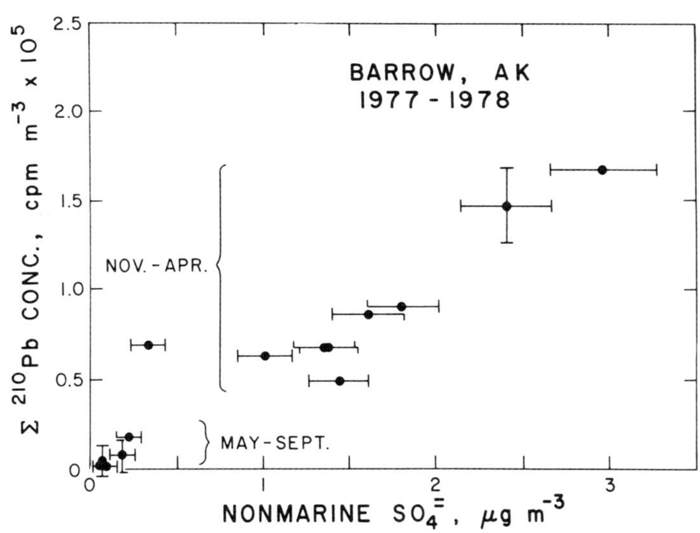

FIGURE 6. ^{210}Pb vs. excess sulfate, Barrow, Alaska.

correlation coefficient of 0.95 or greater. This strongly suggests that sulfate, itself a major fraction of the Arctic aerosol, is serving as a substrate for the ^{210}Pb. But we still do not understand why the correlation is so good.

In summary, then, the Barrow winter aerosol seems to be derived from midlatitude polluted air masses that are efficiently transported to the Arctic. Primary and secondary aerosol components are both observed, with the secondary components being enriched relative to the primary ones by an order of magnitude compared to midlatitudes. Sulfate is the principal secondary component measured so far. Because it may represent most of the mass of the aerosol, the Arctic winter aerosol may—to a first approximation—be considered secondary, that is, derived from midlatitude pollution gases.

TRANSPORT OF AEROSOL TO THE ARCTIC DURING WINTER

In the second half of this article we propose a simple transport model for the winter Arctic aerosol and its precursor gases. Because the study of the Arctic aerosol is yet so new, too few data exist to warrant the development of a detailed model. We believe, however, that the available data do place reasonable constraints on the origin and pathway of the Arctic winter aerosol and gases. The spirit of the following section is then to extract the maximum from the present data and to use these first semiquantitative conclusions as a guide for future studies.

In general, the concentration of a constituent of the Arctic aerosol will depend on its concentration in the midlatitude source region, the speed of transport to the Arctic, and the rate of removal along the way. Each of the above quantities will have seasonal variations. As has been seen in FIGURE 2, the fact that the seasonal variation of V at Barrow is a factor of 50, as compared to a factor of only 3–4 at the source, implies that the seasonal variation of transport plus removal is more than an order of magnitude, i.e., several times greater than the source variations. Thus, transport and removal are important in determining the concentration of the aerosol at Barrow. We will now examine these two processes in more detail, in particular for the winter half-year.

Consider first the speed of transport from midlatitudes to the Arctic. The best indicator of this parameter is the concentration of ^{222}Rn in the Arctic. We are aware of data from Spitsbergen,[4] and from Kodiak and Wales, Alaska,[5] but have none from Barrow itself. These data all show that ^{222}Rn concentrations are several times higher in winter than in summer, the actual factors being 10 for Spitsbergen and 4 for the Alaskan sites. This suggests that the speed of transport from continental regions to the Arctic is faster in winter than in summer. On the basis of this and the timing of the onset and disappearance of the winter maximum of aerosol at Barrow, a very general air-mass picture of transport to the Arctic was proposed,[6] which has been called the coupling-decoupling hypothesis.

This picture considers the Northern Hemisphere atmosphere to be comprised of polar and tropical air masses, which are separated by the polar front. Air flow within each air mass is less restricted than air flow across the polar front, which by its very nature represents a barrier to exchange of air. In summer, when the polar front is located in its most northerly position, it is between the midlatitude source areas and Barrow, so that Barrow and the rest of the Arctic are cut off, or decoupled, from the midlatitudes. Air reaching the Arctic from midlatitudes is very aged and cleansed of contaminants. In winter, on the other hand, the polar front lies just south of the midlatitude source areas, so that they and the Arctic are essentially in the same air mass. Air flow between the two areas is much freer than in the summer, and Arctic aerosol concentrations are consequently much higher.

It is interesting that the polar front does not migrate smoothly between its summer and winter positions. Rather, its motion is quite sudden, during roughly October–November and May–June. These are just the times when the Arctic aerosol changes its character.

Obviously, such a simple, broad picture of the basic forces controlling aerosol transport to the Arctic cannot supply any details of the specific sources or of the mechanics of transport, nor is it intended to. But it is useful as a first picture of the seasonal influences on transport.

Concerning the effect of removal rates, we have already seen that the high concentration of ^{210}Pb at Barrow during winter almost certainly has to be caused by longer residence times in the Arctic than in midlatitudes. The same argument applies to the sulfate-SO_2 system. The main points about residence times in the midlatitudes vs. those of the Arctic can be derived, we feel, from FIGURE 7, which gives monthly mean precipitation rates for a typical midlatitude site (Providence, Rhode Island) and Barrow. (We assume that precipitation controls the residence time of atmospheric aerosols.)

FIGURE 7 has two main features. First, Providence has 10 times the annual precipitation that Barrow does. For this reason, we believe that the residence time of aerosol near Barrow ought to be approximately 10 times longer than in midlatitudes. Second, the seasonal patterns of precipitation at the two sites are quite different. Providence has its precipitation distributed quite evenly over the year, with only gentle

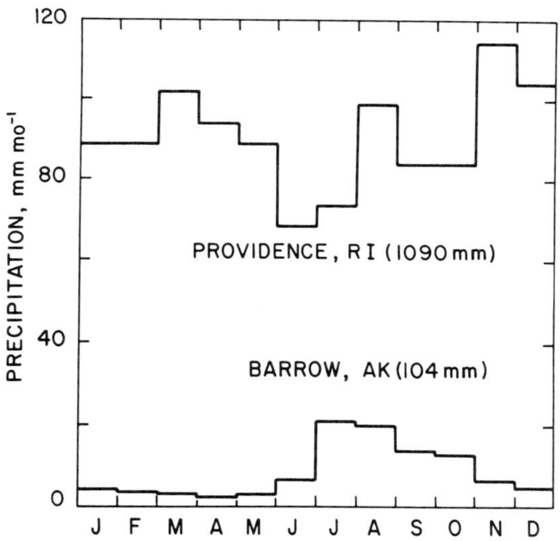

FIGURE 7. Monthly mean precipitation, Providence, R.I., and Barrow, Alaska.

maxima in spring and fall. Barrow, by contrast, has a broad winter and spring minimum followed by a strong summer maximum. The difference between spring and summer precipitation rates is about an order of magnitude. According to FIGURE 7, winter Arctic residence times ought to be roughly 20 times higher than in the midlatitudes, if precipitation is indeed the controlling factor. In summer, on the other hand, Arctic residence times ought to be only 4 times higher than those of midlatitudes.

An equally important question is the midlatitude source area for Arctic aerosol. Regions to be considered include Japan/Korea/China, the northeastern United States, Europe/UK, and possibly central Siberia. Likely pathways from these sources, based on major air flow patterns, are shown in FIGURE 8. The pathway from eastern Asia corresponds to that derived for an episode of Asian dust at Barrow in the spring

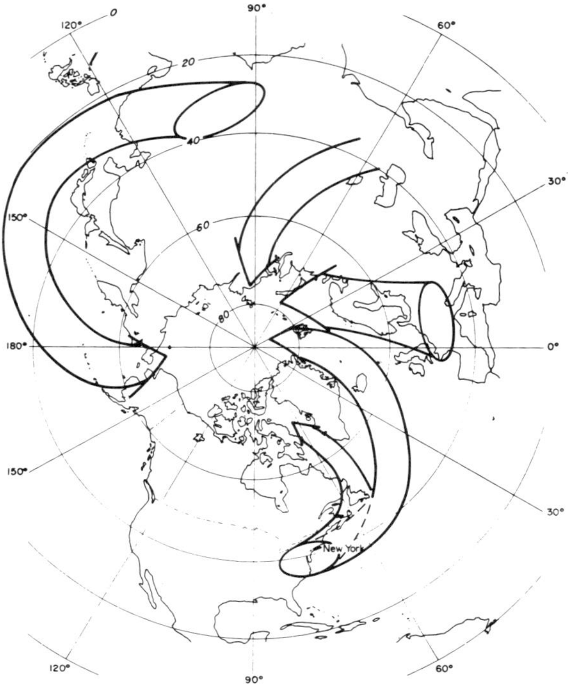

FIGURE 8. Possible main pathways of aerosol transport to the Arctic.

of 1976,[7] but would probably be about the same for transport of pollution aerosol. The pathway from the northeastern United States basically follows the jet stream through the Icelandic low, with a branch around the western side of Greenland corresponding to periods when the Baffin Island low is strong. The significance of this latter branch is not yet certain. The pathway northward from central Europe is roughly that followed by pollution episodes observed throughout Scandinavia. Lastly, the hypothesized pathway from central Asia is based on motions observed around the Siberian high. This path is also highly uncertain.

Until recently, we had no clear indication of which source or combination of sources might be most important for the Arctic during winter. Within the past few months, however, several pieces of data have emerged that seem to be narrowing the field. First, we see no evidence that central Siberia can be an important source, in spite of the rapid industrialization of the region. Population densities are simply too low there. Second, it now appears that air masses coming up to Barrow and Fairbanks from the south, which have come from eastern Asia via the Pacific, are very clean compared to Arctic air coming from the north (see episode during mid-April 1978 in FIGURE 9). We interpret this to mean that Japan/Korea/China is not an important source region, at least during winter. This is reasonable in light of the fact that air from this region has traveled about 7,000 km over the stormy Pacific Ocean, where removal of aerosol must be rapid.

We are thus left with the northeastern United States and Europe, the two sources that from the beginning seemed most likely. Very recent evidence suggests that the United States may be a relatively weak source compared to Europe. For one thing, transport from this source to the Arctic via Iceland and Spitsbergen is quite analogous to the Asia/Pacific case. If the cleansing effect of the Icelandic low is at all similar to that of the Aleutian low, air from North America ought to be very clean by the time it reaches the Arctic. This indeed seems to be the case. Recent data for northern and southern Greenland during the winter of 1977–78 (obtained in cooperation with the Danish Meteorological Institute) show that mean concentrations of V at both sites were several times lower than in Spitsbergen (data obtained in cooperation with the Norwegian Institute for Air Research) and Barrow (FIGURE 9). Thus, the source of V at these latter sites does not seem to be North America. Another piece of evidence is the proportions of V and Mn. At Spitsbergen and Barrow, Mn is more abundant than V, as it is in the European aerosol. In the northeastern United States, on the other hand, V is several times more abundant than Mn. But perhaps the most powerful evidence for a European source is a series of calculations of the evolution of the composition and concentration of the aerosol as it moves from midlatitudes to the Arctic, our transport model, which is presented below.

These calculations were performed to see whether the actual concentrations of V,

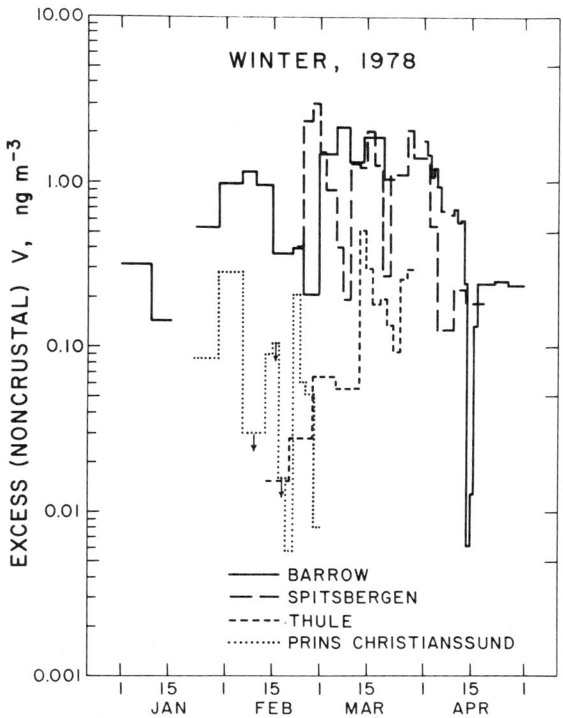

FIGURE 9. Concentrations of excess vanadium at four northern sites, 1977–78.

sulfate, and ^{210}Pb at Barrow could be accounted for by reasonable source concentrations, transport times, oxidation and decay rates, removal rates, and dilutions. The results showed that this was indeed possible. Before presenting these calculations, we wish to make it clear that they are quite elementary and will doubtless be revised. No vertical motions are incorporated, and many of the parameters are imprecisely known. Yet even this simple approach yields essentially only one "solution," i.e., the combination of source area, transport time, pathway, and rate constants that effectively reproduces the Barrow data. All parts of this solution are physically reasonable. Further, there exist sufficient constraints to argue effectively against other solutions that we have tried.

Our model starts with an air mass whose bulk chemical characteristics are derived from surface measurements in a particular midlatitude polluted region, normally either the northeastern United States or western Europe. This air mass is then allowed to age for a period of time designed to simulate transport to the Arctic. During this time, gases are allowed to decay and/or react, particles and gases are removed, and the air mass is steadily diluted by external air having no aerosol or trace gases. For ease of calculation, discrete one-day time steps are used. The chemical characteristics of the air mass after a preset number of days are compared with the observed winter means for Barrow.

The controlling equations are:

$$\mathrm{V}_{n+1} = \mathrm{V}_n e^{-(k_\mathrm{p})_n \Delta t}(1/D_{n+1})$$

$$(\mathrm{SO_4}^{=})_{n+1} = e^{-(k_\mathrm{p})_n\Delta t}\left[(\mathrm{SO_4}^{=})_n + \frac{1.5(k_\mathrm{ox})_n(\mathrm{SO_2})_n}{A_n}(1 - e^{-A_n\Delta t})\right](1/D_{n+1})$$

$$(^{210}\mathrm{Pb})_{n+1} = e^{-(k_\mathrm{p})_n\Delta t}\left\{(^{210}\mathrm{Pb})_n + \frac{8.41\times 10^{-5}(\mathrm{Rn})_n}{[0.1824 - (k_\mathrm{p})_n]}[1 - e^{-(0.1824-(k_\mathrm{p})_n\Delta t)}]\right\}(1/D_{n+1})$$

$$(\mathrm{SO_2})_{n+1} = (\mathrm{SO_2})_n e^{-(k_\mathrm{wd}+k_\mathrm{ox})_n\Delta t}(1/D_{n+1})$$

$$(\mathrm{Rn})_{n+1} = (\mathrm{Rn})_n e^{-k_\mathrm{Rn}\Delta t}(1/D_{n+1})$$

where:

$$A_n = (k_\mathrm{wd})_n + (k_\mathrm{ox})_n - (k_\mathrm{p})_n$$

and:

k_wd = rate constant for combined wet and dry removal of SO_2;
k_ox = rate constant for dry oxidation of SO_2;
k_p = rate constant for removal of (submicron) aerosol;
Δt = time step of 1 day;
D_n = dilution factor ($D_n \geqq 1$).

The subscripts n, $n + 1$, etc., refer to the number of days of transport.

Calculations are executed as follows: Initial values for sulfate, SO_2, V, ^{222}Rn, and ^{210}Pb are entered, together with initial and final values for k_p, k_ox, k_wd, and D_n ($D_0 = 1$). The k's may decrease exponentially between initial and final values or remain constant, and D_n may increase exponentially or remain constant. First, D_{n+1} is calculated as follows:

$$D_{n+1} = \sqrt[N]{1/D_n}$$

where N is the number of days of transport. Next, V_{n+1}, $(SO_4^{=})_{n+1}$, and $(^{210}Pb)_{n+1}$ are calculated, followed by $(SO_2)_{n+1}$, and $(Rn)_{n+1}$. Finally, values of k_{n+1} are calculated using one of the following equations:

$$k_{n+1} = k_n - \frac{(k_0 - k_N)}{N} \Delta t \quad \text{(if linear decay)}$$

$$k_{n+1} = k_n e^{-\{[\ln (k_0/k_N)]\Delta t/N\}} = k_n e^{-[(\ln k_0 - \ln k_N)\Delta t/N]} \quad \text{(if exponential decay).}$$

This sequence is repeated N times.

We now present a series of typical results of these calculations, designed to illustrate in a simplified way how European, but not North American, air masses can account for the Barrow winter aerosol. For all these calculations we have used a value of 20 days for the transport time to Barrow—a somewhat arbitrary figure. At present, the best that can be said is that the travel time is somewhere between 10 and 30 days. The best paths that we can construct from both the United States and Europe to Barrow seem each to be about 13,000 km long (see below for a discussion of the best European path). For mean wind speeds of 7 and 10 m s^{-1}, this corresponds to travel times of 21 and 15 days, respectively. Most importantly, though, it will be seen below that the total travel time to the Arctic is not an important variable, because the composition of the aerosol, i.e., the ratio between primary and secondary aerosol, becomes fixed after about 10–15 days of transport.

FIGURE 10 shows results for air masses of the northeastern United States. This first calculation is particularly simple, with the three k's constant at 0.2 day^{-1} (corresponding to residence times of 5 days for particle removal, oxidation of SO_2, and removal of SO_2) and no dilution during transport. These assumptions are of course not justified, because at least the two k's that depend on precipitation and turbulence (k_p and k_{wd}) will decrease during transport, and there is sure to be a dilution as well. Initial conditions were taken as ^{222}Rn = 170 pCi m^{-3} (winter mean for Chester, N.J.);[8] ^{210}Pb = 15 fCi m^{-3} (winter mean for New York City);[3] SO_2 = 16 $\mu g\ m^{-3}$ (survey of available literature); $SO_4^{=}$ = 6 $\mu g\ m^{-3}$ (also available literature); and V = 22 ng m^{-3} (our own data).

FIGURE 10 shows that V is slightly underestimated (0.4 ng m^{-3} instead of about 0.6 ng m^{-3}), but that sulfate and ^{210}Pb are seriously underestimated. To remedy this, one can incorporate exponentially decreasing k's (initial and final residence times of 5 and 50 days for k_{ox} and k_{wd}, and 4 and 50 days for k_p) into a second series of calculations, starting with the same initial conditions.

The Barrow aerosol cannot be derived from midlatitude aerosols by using constant aerosol residence times. The residence time required to reproduce V and $SO_4^{=}$ is about 10 days, which is too long to be defended for midlatitudes. Further, use of this value produces a $^{210}Pb/SO_4^{=}$ ratio that is too low. Testing a series of possible constant residence times has shown that no value can produce both the correct concentrations of V, sulfate, and ^{210}Pb as well as their correct ratios. On the other hand, residence

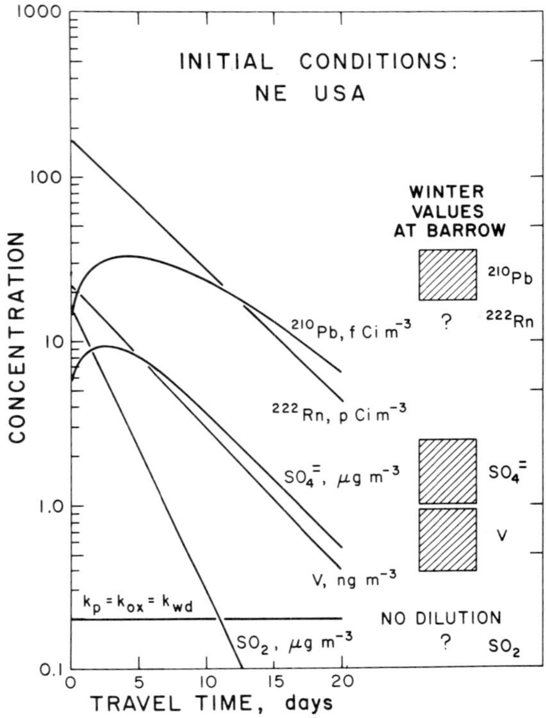

FIGURE 10. Modeled transformation of a polluted air mass as it travels from the northeastern United States to the Arctic.

times that increase with travel time satisfy these conditions rather easily, as will be seen below. This approach also makes good physical sense: (a) precipitation at Barrow is much less than in midlatitudes; (b) our preliminary calculations of residence times for the Barrow aerosol, based on trace elements in aerosol and snow there, yield residence times of roughly 30–100 days; and (c) calculations of deposition of ^{210}Pb in the Arctic and in midlatitudes show that its residence times are about an order of magnitude longer in the Arctic.

In order to adjust the final V value to 0.6 ng m^{-3}, a dilution factor, or D_N, of 6 can be used. These calculations, displayed in FIGURE 11, show that sulfate and ^{210}Pb are still much too low. Sensitivity calculations have shown that the results of this kind of calculation are very insensitive to the values of k_{ox} and only moderately sensitive to k_{wd}. We have therefore concluded that the only way that the Barrow values for sulfate and ^{210}Pb can be reached is by altering the source conditions, i.e., starting with an air mass with more SO_2 and ^{222}Rn. (Initial values of sulfate and ^{210}Pb are unimportant to these calculations, because they are diminished by factors of 20–30, as is V.)

To match the sulfate at Barrow, or more properly stated, to match the $SO_4^{=}/V$ ratio at Barrow, initial SO_2 has to be doubled to about 32 $\mu g\ m^{-3}$. This is very similar to European conditions, where area-wide SO_2 is roughly 30 $\mu g\ m^{-3}$ during winter.[9]

Indeed, taking European initial conditions (^{222}Rn estimated at 100 pCi m^{-3}; ^{210}Pb estimated at 15 fCi m^{-3}; SO_2 = 32 μg m^{-3}; $SO_4^=$ = 6 μg m^{-3}) and letting all 3 *k*'s decrease exponentially from 0.2 day^{-1} to 0.02 day^{-1} after 20 days, and keeping the dilution factor at 6, as before, both the V and sulfate of Barrow are reproduced. ^{210}Pb, however, is still several times too low.

In order to account for the actual ^{210}Pb found at Barrow, its initial European value would have to be set at about 600 pCi m^{-3}, a value that is unacceptably high. The problem is intensified when one realizes that Europe, being essentially a broad peninsula, probably has ^{222}Rn concentrations considerably lower than the 100 pCi m^{-3} used in FIGURE 12. We found, however, that a prolonged source of ^{222}Rn (keeping its initial value constant for the first few days), or still better, a source that increased the ^{222}Rn during the first few days, greatly increased the final ^{210}Pb concentrations at Barrow. One somewhat arbitrary system that worked is shown in FIGURE 13. Here, an initial ^{222}Rn concentration of 50 pCi m^{-3} was taken for the first 5 days, followed by 150 pCi m^{-3} for the next 5 days. After this, the ^{222}Rn decayed normally. This system reproduced concentrations of V, $SO_4^=$, and ^{210}Pb, as well as their ratios.

This suggests to us that the polluted European air mass passes over a much less polluted continent on its way to the Arctic. The physical interpretation is obvious from

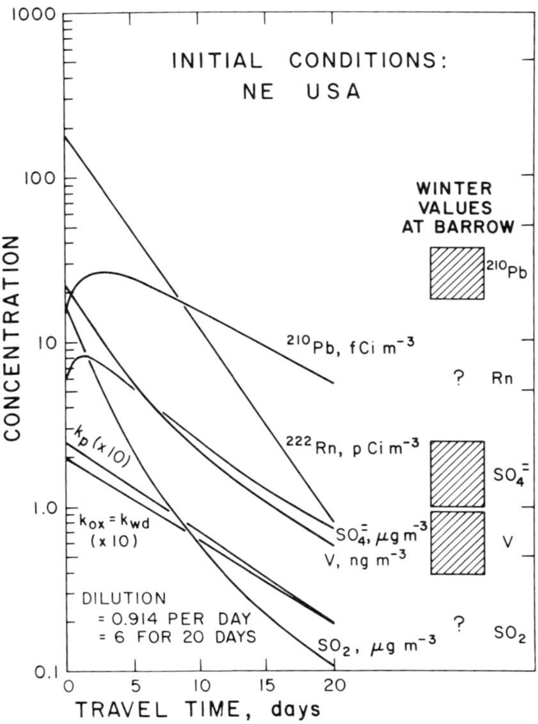

FIGURE 11. Modeled transformation of a polluted air mass as it travels from the northeastern United States to the Arctic.

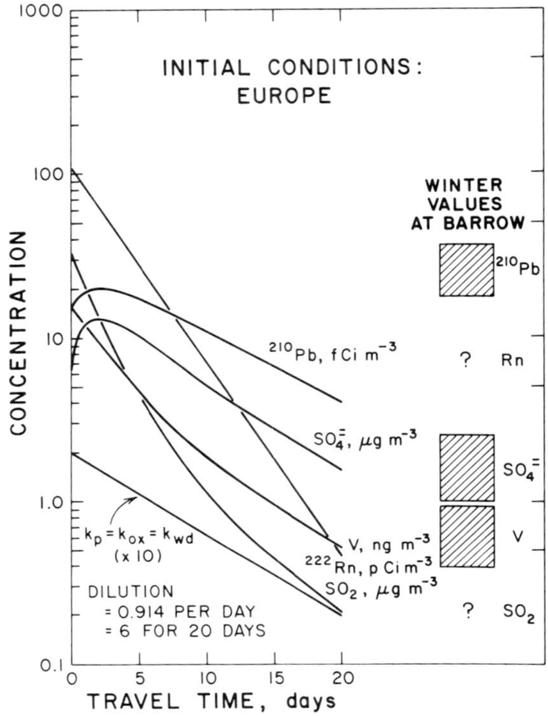

FIGURE 12. Modeled transformation of a polluted air mass as it travels from Europe to the Arctic.

a map and from knowledge of mean air flows over Europe—the air mass moves to the northeast, into European Russia (where population density is a few times less than in central Europe), then at some point turns northward to the Arctic.

Several pieces of independent evidence seem to confirm this view. First, the mean European winter 850-mb wind, as calculated by Bolin and Persson[10] for air over southern England, is indeed to the east-northeast. For air leaving central Europe, this would correspond to a mean path somewhere between Moscow and Leningrad. For evidence that the air flow after this should curve more to the north, we invoke the blocking influence of the winter Siberian high, which diverts this air almost straight to the north (see Figure 3.30 in Trewartha[11]). Thus, we envision this aged European air leaving the Eurasian landmass somewhere between northern Scandinavia and the Taimyr Peninsula, i.e., in the general vicinity of Novaya Zemlya.

After this, we can only guess about the trajectory. According to FIGURE 14, which shows the annual mean surface pressure distribution for the Arctic (after Dunbar and Wittman[12]), surface flow in the Arctic is broken into two distinct regions—the Greenland-Spitsbergen-Novaya Zemlya region, which is dominated by the Icelandic low and where cyclonic flow is the rule, and the Alaskan side, where the Arctic high dominates and anticyclonic flow is the rule. It is easy to imagine European air passing over Novaya Zemlya, then splitting (in the mean) into two branches, one that flows southward along the east coast of Greenland and another that flows westward along northern Greenland and the Canadian Arctic Islands, eventually reaching Barrow.

Here we emphasize again that the above picture must be regarded with extreme caution. It is a first approach only, and is intended only to test whether it is possible, using the best available information, to synthesize a mean pathway from Europe to Barrow that makes good physical sense and that satisfies the aerosol data at both places. The truth is undoubtedly much more complicated. For example, the European air probably rises during its passage northward, and by a height of about 700 mb (3 km), the Arctic high has disappeared, having been replaced by an Arctic-wide low. At this level, European air will come to Barrow from the west, not the east.

In summary, then, we have attempted to demonstrate that observed winter concentrations of V, sulfate, and ^{210}Pb are indeed consistent with a midlatitude pollution source and transport for roughly 20 days with reasonable values of transformation rate constants. Considering the relatively small number of measurements available, there seems to be a remarkable set of constraints imposed upon the system, such as the need for increased residence times in the Arctic and, perhaps most

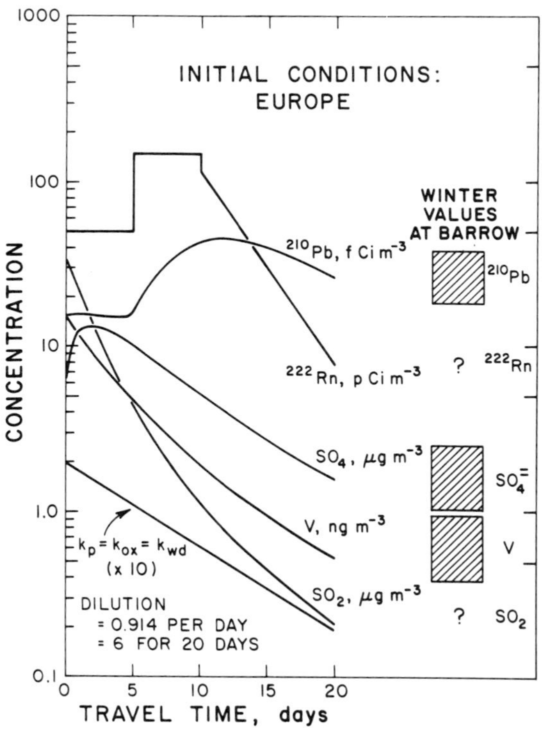

FIGURE 13. Modeled transformation of a polluted air mass as it travels from Europe to the Arctic.

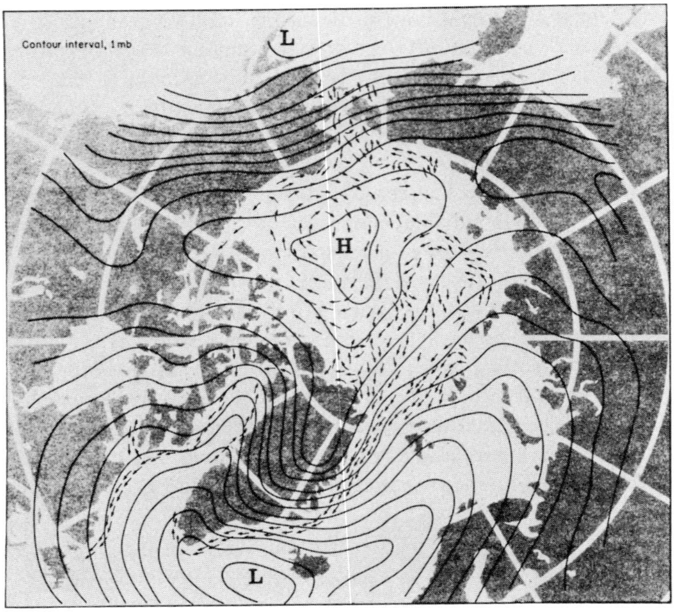

FIGURE 14. Annual mean surface pressure distribution for the Arctic (after Dunbar and Wittman[12]).

important, the apparent identification of Europe rather than the United States as the prime source, with a mean winter transport path over the European part of the Soviet Union.

So far, we have been unable to find any other realistic solution to the equations of transport than the source and path just proposed. We will continue to search for others, however, and at the same time will be carefully studying the model for its sensitivity to variations in the different parameters. These results already suggest further experiments that should be carried out to test our hypothesis.

SUMMARY

Recent research on the Arctic aerosol has revealed that it is an order of magnitude more concentrated in winter than in summer throughout most of the troposphere, is gray in winter, but colorless in summer (on filters), and is much more enriched in pollution-derived constituents like V, Mn, and sulfate in winter than in summer. The winter Arctic aerosol also appears to be highly secondary in nature, containing unusually high concentrations of gas-to-particle conversion products, such as sulfate and ^{210}Pb. On this basis, then, the Arctic aerosol in winter seems to be strongly pollution-derived. A simple transport model is presented, from which it is shown that the actual concentrations of V, sulfate, and ^{210}Pb at Barrow, Alaska are quantitatively consistent with polluted European air masses as the source, followed by transport over European Russia and then to the north.

ACKNOWLEDGMENTS

Barrow aerosol samples were collected as part of a cooperative program with the NOAA Geophysical Monitoring for Climatic Change (GMCC) program, using equipment designed by R. D. Borys. The cooperation of staff members of the GMCC Clean-Air Observatory and the Naval Arctic Research Laboratory in Barrow, Alaska is gratefully acknowledged. Samples were chemically analyzed at the Rhode Island Nuclear Science Center, Narragansett, Rhode Island, with assistance from R. D. Borys and T. J. Conway. Many helpful discussions with G. E. Shaw are acknowledged.

REFERENCES

1. ZOLLER, W. H., G. E. GORDON, E. S. GLADNEY & A. G. JONES. 1973. The sources and distribution of vanadium in the atmosphere. *In* Trace Elements in the Environment. Adv. Chem. Ser. No. 123: 31.
2. NOZAKI, Y., D. J. DEMASTER, D. M. LEWIS & K. K. TRUEKIAN. 1978. Atmospheric ^{210}Pb fluxes determined from soil profiles. J. Geophys. Res. **83:** 4047.
3. Environmental Measurements Laboratory. 1978. Appendix to Environmental Quarterly. Environmental Measurements Laboratory Report EML-344, UC-11. U.S. Department of Energy. New York, N.Y.
4. WILKNISS, P. E., J. W. SWINNERTON, D. J. BRESSAN, R. A. LAMONTAGNE & R. E. LARSON. 1975. CO, CCl_4, Freon-11, CH_4 and Rn-222 concentrations at low altitude over the Arctic Ocean in January 1974. J. Atmos. Sci. **32:** 158.
5. LOCKHART, L. B., JR. 1961. Atmospheric Radioactivity at Kodiak and Wales, Alaska. NRL Report 5658. U.S. Naval Research Laboratory. Washington, D.C.
6. RAHN, K. A., R. D. BORYS & G. E. SHAW. Particulate air pollution in the Arctic: Large-scale occurrence and meteorological controls. *In* Proceedings of the 9th International Conference on Atmospheric Aerosols, Condensation and Ice Nuclei, Galway, Ireland, September 21–27, 1977. (In press.)
7. RAHN, K. A., R. D. BORYS & G. E. SHAW. 1977. The Asian source of Arctic haze bands. Nature **268:** 713.
8. Environmental Measurements Laboratory. 1978. Regional Baseline Station, Chester, New Jersey. Environmental Measurements Laboratory Report EML-347, UC-11. U.S. Department of Energy. New York, N.Y.
9. OTTAR, B. 1978. An assessment of the OECD study on long range transport of air pollutants (LRTAP). Atmos. Environ. **12:** 445.
10. BOLIN, B. & C. PERSSON. 1975. Regional dispersion and deposition of atmospheric pollutants with particular application to sulfur pollution over Western Europe. Tellus **27:** 281.
11. TREWARTHA, G. T. 1968. An Introduction to Climate. 4th edit. McGraw-Hill Book Company. New York, N.Y.
12. DUNBAR, M. & W. WITTMAN. 1963. Some features of ice movement in the Arctic basin. *In* Proceedings of the Arctic Basin Symposium, October 1962: 90. Arctic Institute of North America. Washington, D.C.

THE LONG RANGE TRANSPORT OF SULFUROUS AEROSOL TO SCANDINAVIA

Brynjulf Ottar

Norwegian Institute for Air Research
N-2001 Lillestrøm, Norway

INTRODUCTION

On the basis of precipitation chemistry data collected since 1955–56 at a number of stations in Europe, Odén[1] in 1968 related the observed acidification of rivers and lakes in Scandinavia to an increase in the acidity of the precipitation over the whole area. Separately, incidents of grey snow had been noted in remote areas far from pollution sources.[2] Chemical analysis of the polluted precipitation showed a high content of sulfuric acid, soot, and fly ash. The lead content caused the health authorities to warn the population against drinking the water in cisterns.

At the time, this was believed to be a relatively new development, and the observations caused much alarm. In 1969, OECD*—after having examined the available evidence on a request from the Scandinavian countries—recommended a full investigation. After a few years of preparatory studies primarily by the Nordic countries and coordinated through NORDFORSK, the Scandinavian Council for Applied Research, the OECD program "Long Range Transport of Air Pollutants" (LRTAP) was launched in 1972.[3] Eleven western European countries participated.

THE LRTAP STUDY

The design of the LRTAP cooperative research program has been described elsewhere.[4] Basically, one-layer atmospheric dispersion models, based on trajectories and on an emission survey for sulfur dioxide, were used to calculate concentration and deposition fields for sulfur dioxide and sulfate in a system of 32 × 32 grid squares—each measuring 127 km on a side—covering the northwestern part of Europe. Deposition rates and parameters for the chemical transformation of sulfur dioxide to sulfate were adjusted by fitting the model to observations from aircraft and from more than 70 ground-level stations. The measurement program was completed in 1975, and the final report was published by OECD in 1977.[5]

The results from this study showed clearly that the acidification of the precipitation in Europe is due to the increasing use of sulfur-containing fossil fuels and to the increased emissions of nitrogen oxides. The total emissions of sulfur dioxide in Europe from coal and oil during the last hundred years are shown in FIGURE 1.[6] Considering the highly episodic character of the long range transport of air pollutants at some distance from the major source areas, and the fact that the concentration of sulfur dioxide and sulfate aerosols in episodes may increase 10–20 times above the annual mean value, incidents with polluted air and acid precipitation probably have been taking place for a long period of time without being noticed.

*Organization for Economic Cooperation and Development.

0077–8923/80/0338–0504 $01.75/1 © 1980, NYAS

The OECD study was centered on the year 1974. The mean annual concentration of sulfate aerosol for 1974 is shown in FIGURE 2. The highest concentrations are observed over the major emission areas in England and central Europe. The annual mean concentration of sulfur dioxide is similar, but tails off more rapidly with distance from the source areas.

The dry deposition of sulfur dioxide and sulfate was calculated by multiplying the concentrations with mean deposition velocities of 0.8 cm/s and 0.2 cm/s, respectively. The total deposition of sulfate, shown in FIGURE 3, reflects the large variations in the precipitation pattern. For the whole area, it is estimated that dry deposition accounts

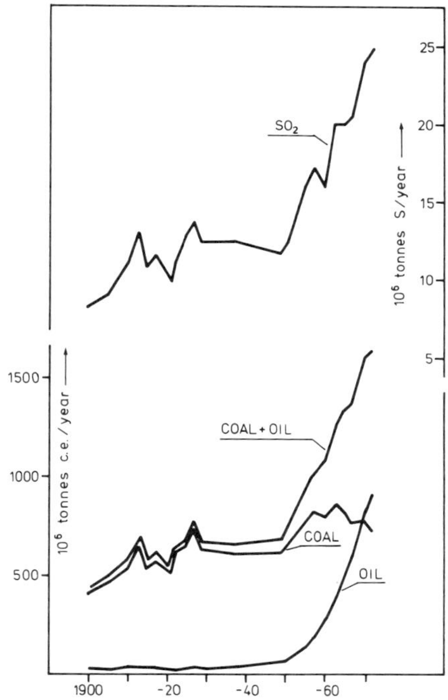

FIGURE 1. Fossil fuel consumption and estimated anthropogenic SO_2 emissions in Europe 1910–1972.

for removal of 50% and precipitation for removal of 30% of the emitted sulfur. In areas such as southern Norway, comparatively far from major emissions and with high rates of precipitation, the estimated dry deposition is only about 30% of the total deposition.[7]

COMPOSITION OF THE SULFATE AEROSOL

The main sources of air pollution in Europe are the fossil fuels used for heat and power generation, motor vehicles, and various industrial activities. The composition of

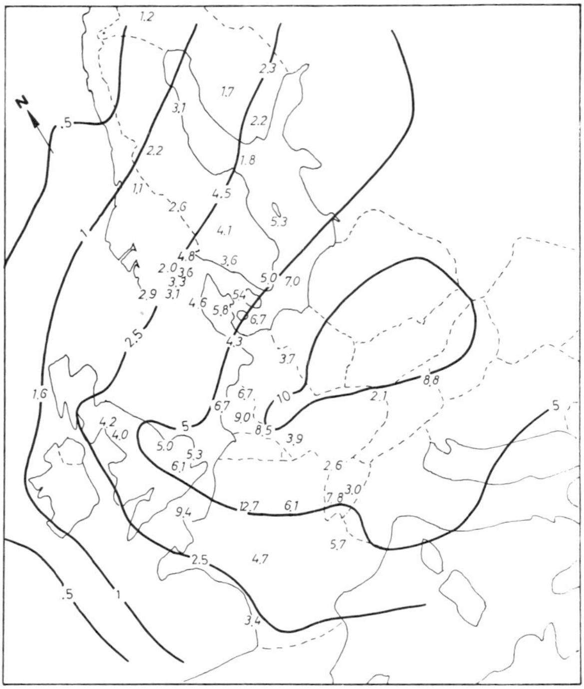

FIGURE 2. Estimated mean concentration field for particulate sulfate for 1974. Observed mean concentrations given by italic numbers. Unit: $\mu g\ SO_4/m^3$.

the aerosol reaching Scandinavia represents a mixture of components from all these sources. Due to the time of transport, one or two days, a substantial part of the sulfur dioxide has been transformed to sulfate.

Normally, the aerosol filter samples are black, due to the soot content, and contain considerable amounts of trace elements, such as Pb, Cd, Zn, and others, as well as polyaromatic hydrocarbons and other organic components. About 40% of the aerosol mass consists of more-or-less-acid ammonium sulfate. Most of this material is found in the accumulation mode below 2–3 μm.

Under special weather conditions, particularly in the summer, so-called white aerosol episodes may occur. This phenomenon, which was first described by Brosset *et al.*[8] on the basis of data collected at a station near Gothenburg on the Swedish west coast, has also been observed in southern Norway. The white aerosol contains highly acid ammonium sulfates, and crystals of NH_4HSO_4, $(NH_4)_3H(SO_4)_2$, and $(NH_4)_2SO_4$ have been identified by x-ray analysis.

Due to the continuous exchange between particles and gaseous components in the atmosphere, details of the aerosol composition depend on the route taken by the polluted air masses. When passing over agricultural areas, ammonia is picked up by the air masses and acid particles are neutralized. When neutralized particles pass over the highly acid soils of forested areas in Scandinavia, they may give off ammonia and become more acid. Nitric acid is largely present in the gaseous phase and not in the acid particles. The nitric acid is, however, easily picked up by cloud droplets and precipitation. When the polluted air masses pass over the sea surface, there is no contribution of ammonia. Therefore, the catalytic oxidation of sulfur dioxide by absorption on the particles is slowed down as the particles become increasingly acid. The relative importance of photochemical oxidation processes increases, and new sulfate-particle nuclei are formed. Under these circumstances, the acidity of the precipitation may occasionally reach extreme values. FIGURE 4 shows a case from April 1974 when high sulfate-aerosol concentrations were formed over the Atlantic in polluted air from Europe. When these air masses later entered the coastal areas of Norway and Scotland, pH values down to 2.5–2.7 were measured in precipitation. A few days earlier, a case of pH 3.5 in 10 mm precipitation had been observed at Rjupnahaed, Iceland.

The larger aerosol particles consist of sea salts, soil dust, and finely divided materials from various mechanical processes. Obviously, samples of the smaller particles obtained with a size segregation sampler would be more directly related to long-range-transported aerosols. However, in most cases, total samples are collected. Reactions then take place on the filter between the large and the small particles. The

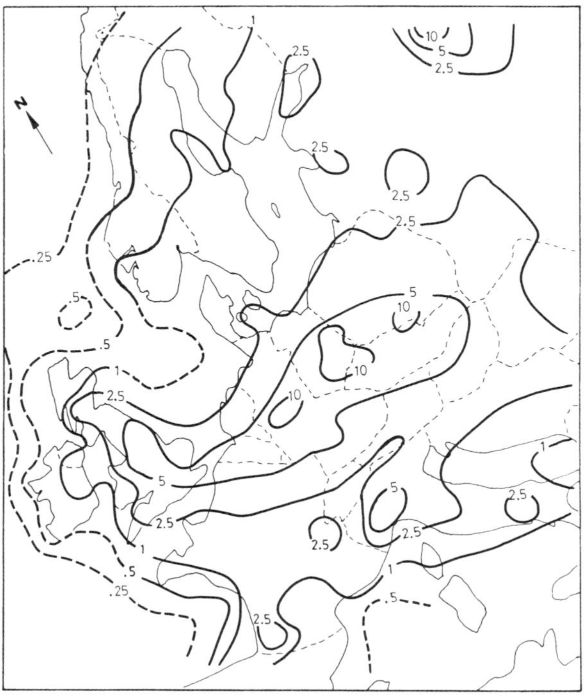

FIGURE 3. Estimated total (dry and wet) sulfur deposition pattern for 1974. Unit: g S/m^2.

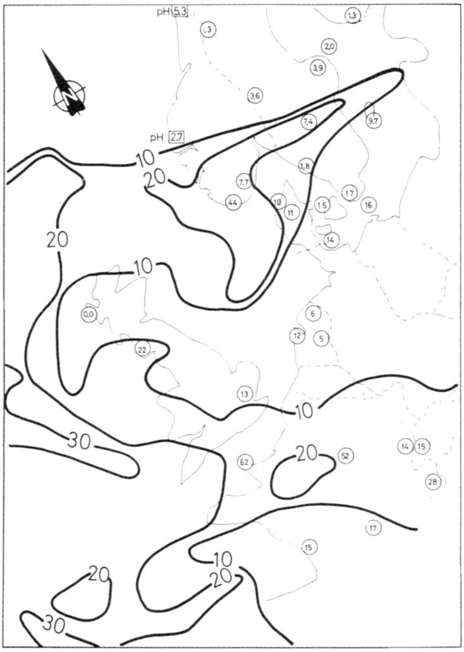

FIGURE 4. Concentration field for aerosol sulfate, April 8, 1974, and ground-level observations ($\mu g/m^3$).

type of reaction depends on the composition. The chloride in the large sea-salt particles will be removed due to reactions with the small, acid sulfate particles. The same will happen with nitrate, which may be present in the larger particles. If the alkalinity of the larger particles dominates the mixture, ammonia may be expelled from the smaller particles. In Scandinavia, this is probably a rare situation.

In order to obtain more meaningful information on the chemistry of aerosols and on the chemical reactions and equilibria between gaseous atmospheric constituents and particles (or liquid droplets), more sophisticated sampling must be applied. At the same time, the methods must be sufficiently simple to be utilized in a measurement network of wide regional coverage. Routine methods and sampling equipment for this purpose are being developed and tested as part of the cooperative research program of the Nordic countries.

As the polluted air masses pass farther north, the reactive gaseous components gradually disappear due to chemical reactions and dry deposition. The aerosol particles, particularly those in the submicron size range, have a considerably longer atmospheric lifetime than do reactive gases, such as SO_2. High-volume sampling and neutron activation analysis have been used to trace anthropogenic aerosols as far north as northern Norway and Spitsbergen.[9,10] FIGURE 5 shows how the concentration of some of these components is reduced in passing from Denmark up to northern Norway. The episodic character of the pollution transport is retained, however. FIGURE 6 shows the variations in concentration of total aerosol mass at Skoganvarre in northern Norway.[11] Apart from an unusual forest fire in the USSR, the occurrence of elevated concentrations is directly related to strong south-to-north transport of air masses.

Deposition

The main sink for atmospheric aerosols is precipitation. The annual deposition of excess sulfate by precipitation in Norway[12] is shown in FIGURE 7. The maxima in the southern part of the country are due to a combination of simultaneous intrusion of polluted air masses and orographic precipitation. The sulfate aerosol is accompanied by minor amounts of various metals and organic components,[13] and the deposition patterns for trace elements, such as lead and cadmium, obtained by analysis of moss and lichen samples are similar to the deposition of excess sulfate.[14,15]

In FIGURE 8, monthly average chemical compositions of precipitation and aerosol filter samples for the station at Birkenes in southern Norway are compared. (With the units used, the concentration in precipitation corresponds to dissolution of the aerosol phase in 1 g/m^3 of H_2O.)

From the relative proportions of H^+, NH_4^+, SO_4^-, and NO_3^- in aerosol and in precipitation, it appears that a significant fraction of the sulfate and most of the

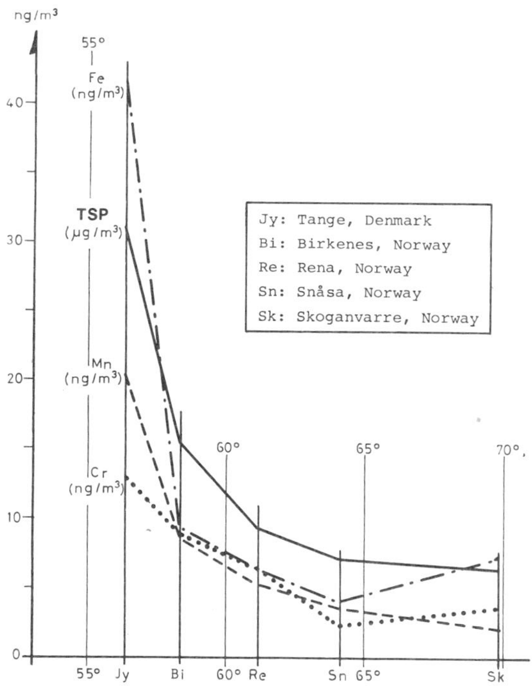

FIGURE 5. Mean concentrations of total suspended particulates (TSP) and the elements Fe, Mn, and Cu at background stations in Denmark and Norway at different latitudes.

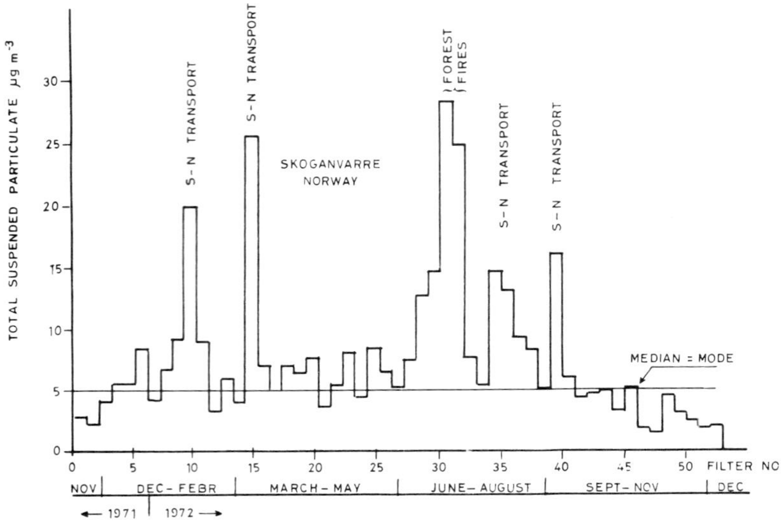

FIGURE 6. Weekly mean concentrations of total suspended particulates in air at Skoganvarre, Norway.

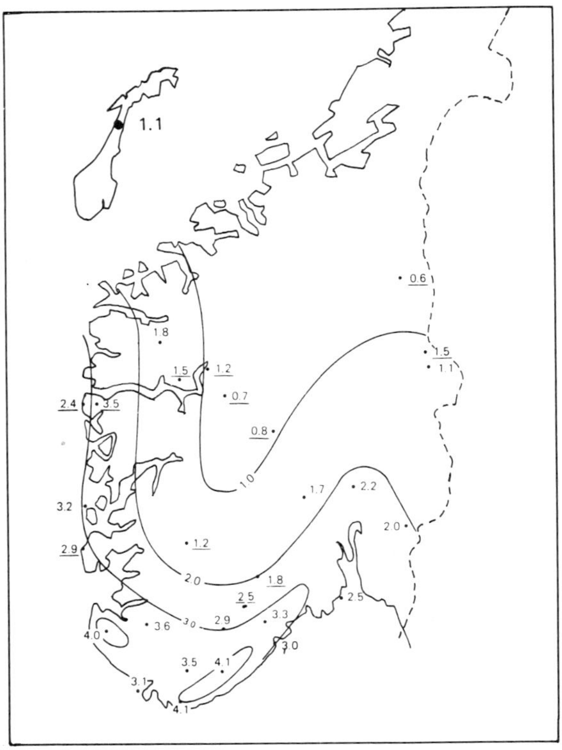

FIGURE 7. Annual deposition of excess sulfate from precipitation (g SO_4/m^2) based on data from July 1972 to June 1975. Numbers underlined refer to shorter sampling periods.

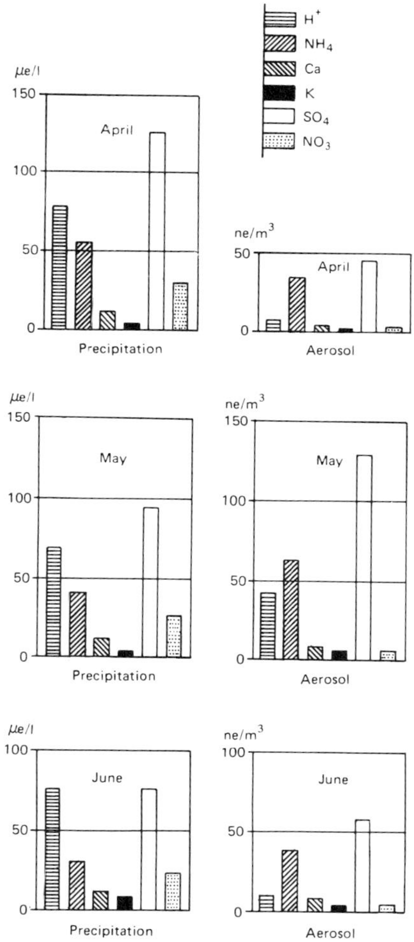

FIGURE 8. Concentrations of water-soluble compounds in precipitation and aerosols at Birkenes, Norway, April–June 1975.

nitrate in precipitation are due to removal of sulfur dioxide and gaseous HNO_3 by cloud formation and precipitation processes.

From the ratio between the concentrations in air and in precipitation, it is also apparent that the removal efficiency for aerosols is high.

Dry deposition is difficult to measure and is highly dependent on surface structure and weather conditions.[16,17] The simplified procedures used in the model calculations of the OECD project indicate that dry deposition of aerosols contributes only a few percent to the total sulfate deposition in Norway. Intuitively, the open structure and large surface area of coniferous forests suggest a greater aerosol particle deposition velocity, and measurements of precipitation washout from forest canopies indicate a 30% increase of the deposition of sulfate under trees as compared to precipitation in the open. However, more direct studies of the impaction of particles, e.g., in wind

TABLE 1

ESTIMATED BUDGET FOR DRY-PLUS-WET DEPOSITION OF SULFUR FOR 1974*

Emitters / Receivers	Austria	Belgium	Denmark	Federal Republic of Germany	Finland	France	The Netherlands	Norway	Sweden	Switzerland	United Kingdom and Ireland	Czechoslovakia	German Democratic Republic	Italy	Poland	Other Areas	Undecided	Sum	Annual Emission
Austria	60	6	0	40	0	20	2	0	0	5	20	20	20	30	7	20	30	300	221
Belgium	0	100	0	20	0	30	5	0	0	0	30	1	4	0	0	1	10	200	499
Denmark	0	1	60	6	0	3	1	0	2	0	10	1	6	0	2	2	10	100	312
Federal Republic of Germany	8	60	7	700	0	100	40	0	2	7	100	20	80	7	10	10	90	1300	1964
Finland	0	2	8	10	100	4	2	2	30	0	10	7	30	0	20	80	70	400	274
France	2	40	1	50	0	600	10	0	0	6	100	5	20	30	2	30	150	1000	1616
The Netherlands	0	10	1	10	0	10	60	0	0	0	30	1	4	0	1	0	10	150	391
Norway	0	4	8	10	1	9	4	30	9	0	60	3	10	0	5	4	100	250	91
Sweden	0	7	30	30	10	10	6	6	100	0	40	8	50	0	20	30	100	500	415
Switzerland	1	2	0	7	0	20	1	0	0	30	10	2	1	6	1	2	20	100	76
United Kingdom and Ireland	0	8	2	10	0	20	4	0	0	0	800	2	9	0	2	1	100	1000	2883†
Czechoslovakia, German Democratic Republic, Italy, Poland and other areas	60	60	80	400	40	200	40	9	50	10	600	900	1300	900	1000	4500	1000	11000	—
Sum	100	300	200	1300	150	1000	200	40	200	60	1800	1000	1500	1000	1100	4600	1900	17000	—

*Unit: 10^3 tonnes S. Numbers rounded to one significant figure and accurate to within about ± 50%. The sums are calculated from unrounded figures and thereafter rounded separately.
†Including 80×10^3 tonnes S from Ireland.

tunnels, are required to give quantitative data on the deposition velocity as a function of surface structure, particle size, and wind velocity.

The model calculations were used to calculate the exchange of sulfur pollutants between the European countries. The result for 1974 is given in TABLE 1. It shows that the Scandinavian area receives about two times as much sulfate from outside sources as from indigenous ones. However, corresponding calculations for 1975 and 1976, based on the same emission data, show that the contributions from one country to another may change by a factor of two or more from year to year, because of changes in the annual weather pattern. The total deposition in Scandinavia was, however, remarkably constant for these years.

CONCLUSION

The LRTAP program and other studies have shown that considerable long range transport of air pollutants takes place from the United Kingdom and central Europe to Scandinavia. The sulfate deposition pattern in Scandinavia is largely governed by precipitation and wind direction. In the southern part of Scandinavia, maximum values of 3–5 tonnes† of SO_4/km^2 are found. Farther north, the contribution from foreign sources to the sulfate deposition in Scandinavia becomes insignificant—except for the northernmost part, which receives a contribution with air masses from the south (up the Gulf of Bothnia). The contribution from precipitation outweighs the dry deposition, particularly in the southern part of Norway. The dry deposition over forested areas may be more significant than the present data indicate.

About 20% of the total sulfur emissions in Europe leaves the area and contributes to the general background pollution of the atmosphere. A part of this material has been traced all the way into the arctic region.[18]

The experience of the LRTAP program and other studies indicates that in order to obtain better estimates for the long range transport of air pollutants, a better understanding of the chemical reactions in the atmosphere and of the deposition processes is required. Efforts in this direction are at present being made in connection with the new European Monitoring and Evaluation Program under the United Nations Economic Commission for Europe.

REFERENCES

1. ODÉN, S. 1968. Nederbördens och luftens försurning dess orsaker, förlopp och verkan i olika miljöer. Statens Naturvetensk. Forskningsråd, Ekologikomm. Bull. No. 1. Stockholm.
2. ELGMORK, K., A. HAGEN & A. LANGELAND. 1973. Polluted snow in Southern Norway during the winters 1968–1971. Environ. Pollut. **4:** 41.
3. OECD. 1972. Co-operative Technical Programme to Measure the Long Range Transport of Air Pollutants. Organization for Economic Cooperation and Development. Paris, France.
4. OTTAR, B. 1977. An assessment of the OECD study on long range transport of air pollutants. Atmos. Environ. **12:** 445.

†One tonne = 10^3 kg.

5. OECD. 1977. The OECD Programme on Long Range Transport of Air Pollutants. Measurements and Findings. Organization for Economic Cooperation and Development. Paris, France.
6. SEMB, A. 1977. Sulphur emissions in Europe. Atmos. Environ. **12:** 455.
7. ELIASSEN, A. 1977. The OECD study of long range transport of air pollutants. Long range transport modelling. Atmos. Environ. **12:** 479.
8. BROSSET, C., K. ANDREASSON & M. FERM. 1975. The nature and possible origin of acid particles observed at the Swedish west coast. Atmos. Environ. **9:** 631.
9. LARSSEN, S. & K. RAHN. 1977. Elemental concentrations of total suspended particulate matter in background areas in Scandinavia as a function of northern latitude. Discussion Paper No. R.3/COM.5, presented at the Economic Commission of Europe Seminar on Fine Particulates, Villach, Austria, October 17–22.
10. Workshop Report. 1977. Sources and Significance of Natural and Man-made Aerosols in the Arctic. U.S. Office of Naval Research and Norwegian Institute for Air Research.
11. RAHN, K. A. 1972. Study of National Air Pollution by Combustion. Progress Report. Institute of Nuclear Sciences. University of Ghent. Belgium.
12. DOVLAND, H., E. JORANGER & A. SEMB. 1976. Deposition of Air Pollutants in Norway. Research Report No. 6/76. The SNSF Project. Institute of Forest Research. ÅS-NLH, Norway.
13. SEMB, A. 1978. Deposition of Trace Elements from the Atmosphere in Norway. Research Report No. 13/78. The SNSF Project. Institute of Forest Research. ÅS-NLH, Norway.
14. RÜHLING, Å. & G. TYLER. 1973. Heavy metal deposition in Scandinavia. Water Air Soil Pollut. **2:** 445.
15. STEINNES, E. 1977. Atmospheric Deposition of Trace Elements in Norway Studied by Means of Moss Analysis. Report No. KR-154. Institute for Atomenergy. Kjeller, Norway.
16. DOVLAND, H. & A. ELIASSEN. 1976. Dry deposition on a snow surface. Atmos. Environ. **10:** 783.
17. SLINN, W. G. N. 1974. Dry deposition and resuspension of aerosol particles—A new look at some old problems. *In* Atmosphere–Surface. Exchange of Particulate and Gaseous Pollutants: 1–40. Proceedings of a symposium held in Richland, Wash., 1974.
18. RAHN, K. A. & G. E. SHAW. 1978. Briefing on Arctic Haze and Arctic Aerosol. Bureau of Oceans and International Environmental and Scientific Affairs. Department of State. Washington, D.C.

LONG RANGE TRANSPORT OF DESERT DUST WITH SPECIAL EMPHASIS ON THE SAHARA*

Lothar Schütz†

Max-Planck-Institut für Chemie
D-6500 Mainz, Federal Republic of Germany

INTRODUCTION

In the last few years, much attention has been given to the arid and semiarid regions of the earth, which cover approximately 30% of the continental surface. There are several reasons for this attention. During the Sahelian drought, it was proposed that some of the great deserts may extend their areas. Thus, for instance, the ecology of these areas is strongly affected. Climatic effects caused by an increased production of airborne mineral dust were discussed.

Mineral dust contributes to a considerable extent to the atmospheric aerosol. Therefore, it might affect both the radiation balance and precipitation processes. Mineral dust in the atmosphere can be important for soil formation. The classic examples of dust deposition on the continents are the loess deposits of the glacial age. Depending on the extension of the deserts and the meteorological transport conditions, airborne mineral dust also contributes to the formation of the deep-sea sediments. This clearly shows how important the atmospheric transport of desert-derived aerosol is for atmospheric and geological processes.

At present, mineral dust transport from the Sahara Desert is fairly well investigated and is the subject of discussion in the following section. Furthermore, mineral dust transport in other regions will be described; and, finally, the global source strength of mineral dust will be estimated.

MINERAL DUST TRANSPORT FROM THE SAHARA

The Sahara Desert is one of the major source areas for windblown dust of the Northern Hemisphere. Large amounts of dust are produced by weathering of rocks and soils at the surface and are emitted into the atmosphere by strong surface winds and distributed by vertical turbulent mixing that is caused by intensive insolation. Thus, a more or less continuous vertical flux of mineral particles builds up a dust reservoir several kilometers thick within the atmosphere. This dust is fed into the major wind systems and transported across the desert's margins. Large amounts of mineral dust are transported across the northern equatorial Atlantic Ocean to the Caribbean Sea and, in a southwesterly direction, towards the coast of Guinea. Under suitable flow conditions, Saharan dust is also transported across the Mediterranean to Europe and the Near East. The transport of dust across the Atlantic Ocean is a more

*Supported by the Deutsche Forschungsgemeinschaft through its "Sonderforschungsbereich 73" (atmospheric trace compounds).

†Present affiliation: Institute for Meteorology, University of Mainz, D-6500 Mainz, Federal Republic of Germany.

0077–8923/80/0338–0515 $01.75/1 © 1980, NYAS

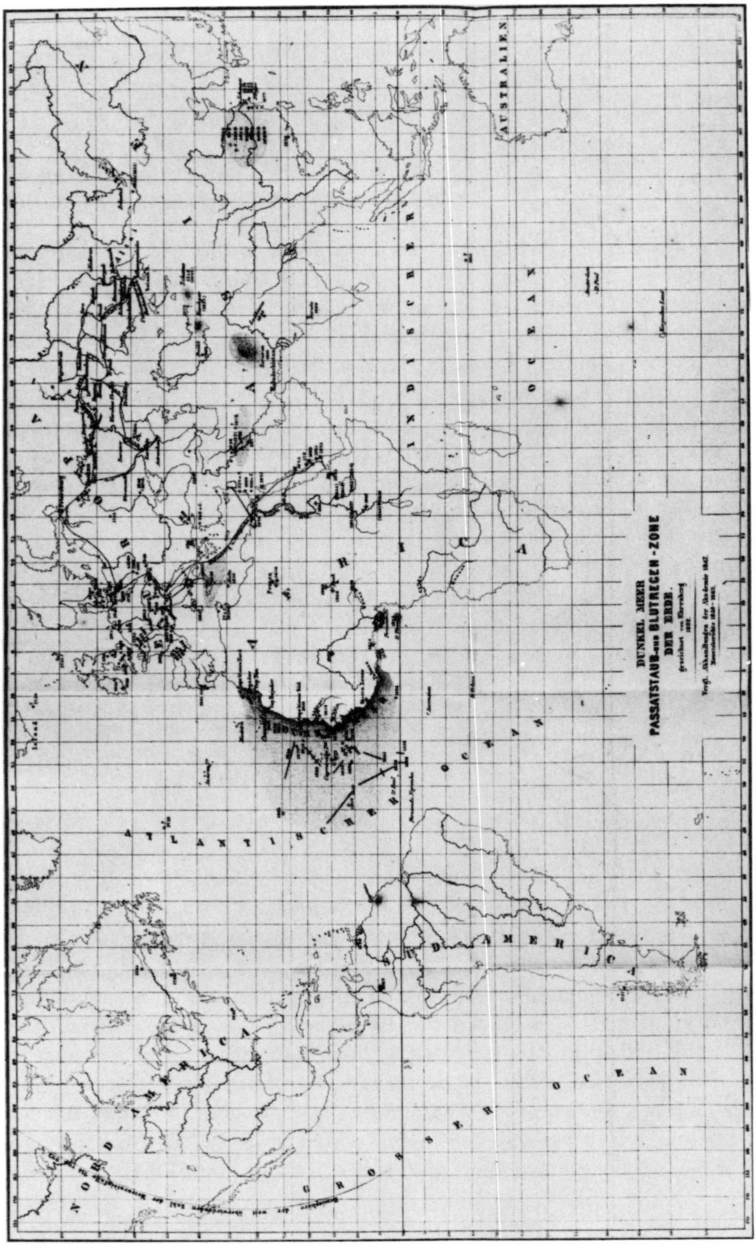

FIGURE 1. Historical map of dustfall records around the world (compiled by Ehrenberg in the year 1862). Dotted areas indicate regions of heavy dustfalls.

or less continuous flow, upon which perturbations caused by heavy sandstorms occasionally are superimposed. Transport to the south is a seasonal phenomenon that must be seen in connection with the north-south migration of the intertropical convergence zone (ITCZ). Europe and the Near East usually are affected by Saharan dust transport during spring, when polar air masses penetrate into the northern Sahara for a short period of time, generating strong low pressure systems.

Transport over the North Atlantic Ocean

Dust transport and fallout over the northern equatorial Atlantic have been observed from ships for centuries. Ehrenberg[1] published a map of the global distribution of dust fallout based on historical records (FIGURE 1). An area close to northwest Africa was called the dark sea ("Dunkelmeer") due to considerable reduction of horizontal visibility. This area is found between 10°–25° north latitude extending up to about 35° west longitude. Semmelhack[2] investigated the seasonal variation of the dust fallout region and found a shift of the fallout area to the south during winter, whereas during summer, it is shifted into a northerly position. The maxima of fallout activity were observed around January and August. Both effects he attributed to the seasonal shift of the ITCZ. Semmelhack[2] further claimed that the main transport of dust takes place above the trade-wind inversion within the Harmattan wind system.

During the last 10 years, a number of comprehensive studies dealing with the physical and chemical properties of Saharan mineral aerosol have been carried out. The transport characteristics and vertical and areal distribution of dust have been reported by a number of authors.[3–10] Their results can be summarized as follows:

Strong insolation over the desert creates a very deep isentropic mixing layer, filled with dust particles up to altitudes of about 6 km. The dry and dust-laden air masses leave Africa and move westward within a very persistent[11,12] wind system towards the Caribbean Sea. This dust flow takes place in the so-called Saharan air layer above the trade-wind inversion (1.5 km above sea level). As the Saharan air masses emerge from the continent, they are undercut by the relatively cool and moist northeast trade winds that originally are free of dust. This sublayer receives its dust content by vertical downward mixing and by fallout of larger particles from the overlying Saharan air layer, which can be considered as a large reservoir for mineral particles. Vertical soundings showed that the dust layer ranges from about 1 km to 6 km altitude near the African coast and converges to a layer between 1.5 and 3 km over the Caribbean Sea.

The lower limit of the Saharan air layer is always well defined by the trade-wind inversion. The upper limit, however, is determined by several factors, such as the vertical thickness of the original dust layer over the desert, the vertical wind profile, and the subsidence within the subtropical airmasses. These parameters show a seasonal dependence, and so does the transport. For instance, the mean vertical wind profile over the Atlantic in FIGURE 2 (average wind data[13] for Sal, Cape Verde Islands, for the years 1951–1960) shows favorable conditions for westward transport in the summer between May and October. During that time, easterly winds are dominant throughout altitudes up to more than 6 km (500 mb). During the winter season, however, easterly winds are weaker, and they are restricted to a very shallow layer of about 1.5 km (850 mb). Above 3 km (700 mb) during that time of the year, westerly winds ("counter trade winds") do prevail. Hence, the transported mass of dust reaches a maximum during summer and a minimum in winter. This seasonal trend is well documented by a series of turbidity measurements for Sal (Cape Verde Islands) as compiled by Jaenicke *et al.*[10] and by a 10-year record of the mineral dust content over the Barbados (Caribbean Sea) by Prospero *et al.*[14]

The latitudinal extent of the dust layer ranges from about 5° N to 25° N. A seasonal variation is observed that can be explained by the annual migration of the ITCZ. During summer, when the ITCZ moves northward, the dust layer is usually confined to a latitudinal band between 15° N and 25° N (see FIGURE 3). In the winter season, the layer over the Atlantic Ocean extends more in a southwestern direction and is located at the West African coast between about 5° and 15° north latitude (FIGURE 3). The seasonal shift of the dust layer correlates well with the findings made

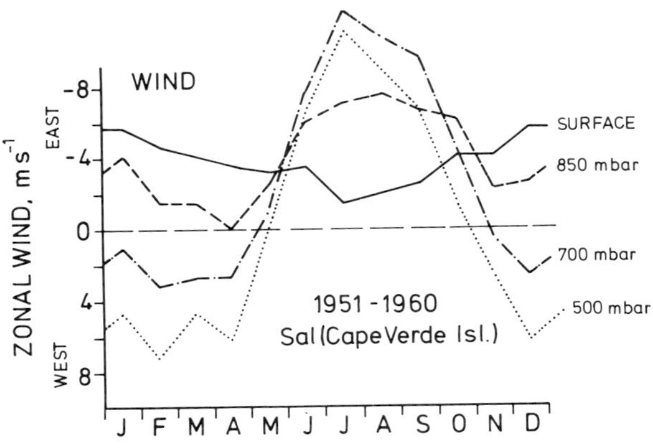

FIGURE 2. Vertical zonal wind field over Sal (Cape Verde Islands). The vertical wind profile can be assumed as being representative for the North Atlantic trade-wind region.

earlier[2] on the annual geographical distribution of the dust fallout area, as can be seen from FIGURE 3. It also explains the winter minimum of mineral dust concentrations over the Caribbean Sea, as mentioned above.

The turbidity measurements over the Atlantic Ocean indicate that the dust transport is a rather continuous phenomenon. However, short-term variations occasionally are observed, too. For instance, wave disturbances—called "easterly waves"—that travel westward interrupt the zonal dust flow over the ocean. On the other hand, a number of heavy dust outbreaks caused by large sandstorms over the desert are superimposed on the continuous dust flow. The path of such individual dust outbreaks has been observed very often on satellite photographs. Dust clouds with horizontal extensions up to more than 1000 km^2 have been observed over West Africa. Their subsequent movement over the North Atlantic in the vicinity of Africa has been observed fairly well, especially when the fronts of the clouds were sharply defined. Usually, these clouds reach the Caribbean Sea within 5–7 days.

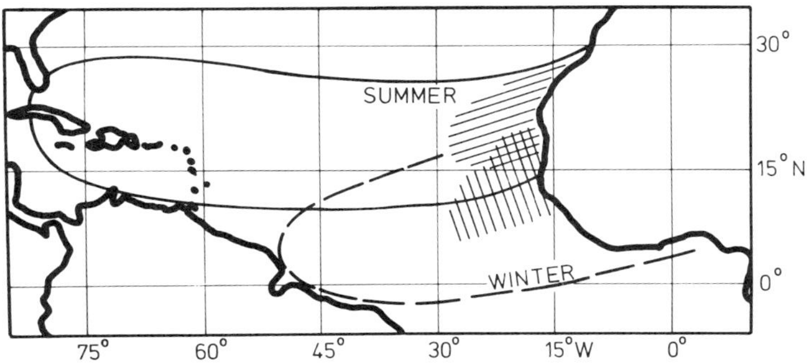

FIGURE 3. Areal extension of the Saharan dust plume and dust fallout areas over the North Atlantic Ocean during summer and winter. Hatched areas close to the West African coast represent the mean extension of dust fallout regions (after Semmelhack[2]). Delineated areas show the mean extension of the dust plume as derived from turbidity measurements by Prospero *et al.*[6]

Measuring campaigns carried out in the past by several authors[5,7,8,14,15] give quite good estimates of the mass concentration ranges to be expected when Saharan air masses travel from the source to the Caribbean Sea. TABLE 1 shows surface air data representing the summer situation (for Barbados, winter season data are also available). The data clearly show a decrease in the mass concentration of dust as a function of the distance from the source area. The upper values represent sandstorm or dust outbreak conditions. The strong concentration decrease between the source and the Cape Verde Islands is due to fallout of the particle fraction with radii $r > 1$ μm, containing the majority of the mineral dust mass. As the dust plume reaches Barbados, the concentration decrease is comparatively smaller, because a considerable portion of the removed mineral mass consists of submicron particles transported downward only by slow mixing. These characteristic features can be explained by a transport model coupled with surface air measurements across the North Atlantic by Schütz.[7]

A steady state model in two dimensions considers the same basic transport processes as proposed by Carlson *et al.*,[4] i.e., horizontal transport of dust in an

TABLE 1

AVERAGE MASS CONCENTRATIONS OF SAHARA DUST IN SURFACE AIR

Location	Concentration (μg/m^3)	Source
Sahara (Libya, Mauretania)	2×10^2–1.8×10^5*	Schütz[15]
		Lepple[5]
Sal, Cape Verde Islands	20–200*	Schütz[7]
		Savoie *et al.*[8]
Barbados	5–80*	Savoie *et al.*[8]
	5†	Prospero *et al.*[14]

*Summer data.
†Winter data.

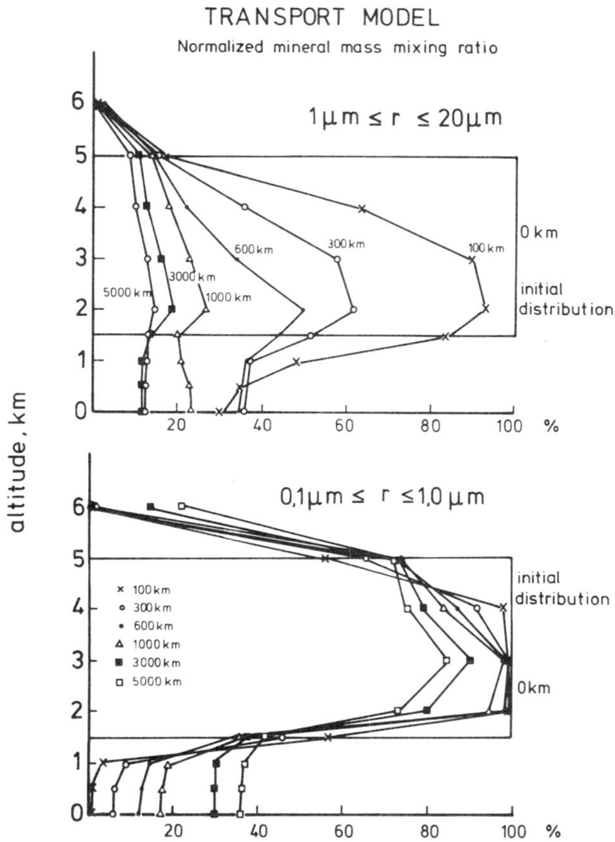

FIGURE 4. Vertical distributions of mineral dust for selected transport distances from the source (model calculation). Vertical mass mixing ratios are plotted for mineral particles in the submicron range (0.1 μm $\leqq r \leqq$ 1 μm) and the giant particle range (1.0 μm $\leqq r \leqq$ 20 μm) normalized to the initial mixing ratio at the source (x = 0).

east-west direction and vertical transport by turbulent diffusion and sedimentation of the dust particles. The trade-wind inversion is also incorporated into the model by using small vertical diffusion coefficients calculated from vertical wind and temperature profiles. The geographic starting point of the model is fixed at the West African coast. A dust layer with a constant vertical mixing ratio is assumed between 1.5 km and 5 km in altitude. The trade-wind layer (0 to 1.5 km) is regarded as originally free of mineral dust. Particles with radii between 0.1 μm and 20 μm are considered. In order to study size distributions as a function of the transport distance, an initial distribution at the West African coast is assumed. According to measurements of the mineral aerosol size distribution in the center of the Sahara (Libya) by Schütz *et al.*,[16] an initial power function distribution with an exponent of $\nu^* = 2$ is assumed.‡ Only if

‡Approximations of aerosol size distributions for particles with radii $r > 0.1$ μm can be obtained by a power function $n^*(r) = n^*(r_0) \cdot (r/r_0)^{-\nu^*}$.

such an initial distribution was assumed did the results of the model agree with surface air measurements across the Atlantic.

FIGURE 4 shows how the Saharan dust plume develops during transport and how the mass mixing ratio of the mineral particles changes. The values of the mass mixing ratio are normalized to their initial ratio. Submicron particles ($0.1\ \mu m \leqq r \leqq 1\ \mu m$), which mainly are causing atmospheric turbidity, are only inappreciably removed from the dust layer aloft. Their mixing ratio within the plume is reduced to only about 70% of the initial value. Simultaneously, at sea level, their mixing ratio increases as the dust moves westward. Vertical turbulent mixing transfers these particles only slowly into the trade-wind layer. Submicron-size particles are therefore subject to long range transport. In contrast, the particle fraction $1\ \mu m \leqq r \leqq 20\ \mu m$, representing most of the mineral mass, shows much faster variations with increasing transport distance. Because of the large sedimentation velocity of particles with $r > 1\ \mu m$, most of the

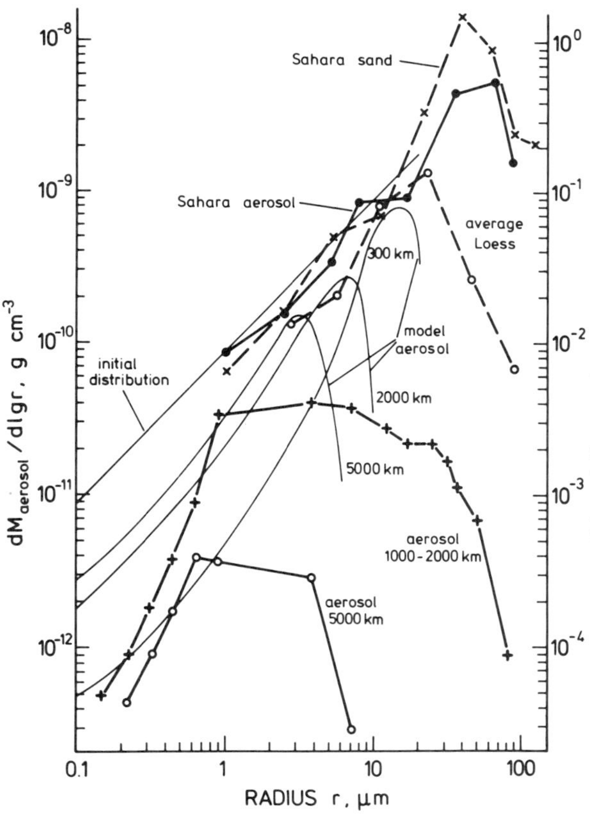

FIGURE 5. Mineral mass distributions at sea level for soil material and mineral aerosol for various distances from the source. Dotted lines represent mean Sahara sand from Libya (after Schütz *et al.*[15]) and average loess (after Junge[18]). Heavy lines represent Sahara aerosol over the source and mineral aerosol for selected transport distances over the North Atlantic. Thin lines represent model calculations assuming an initial distribution at the source.

mass is deposited within the first 1000 km. At a distance of approximately 1000 km, a nearly constant vertical mass mixing ratio is achieved. Within that distance, dust falls[2] usually occur.

These characteristic features predicted by the model agree well with size distribution measurements carried out by Savoie *et al.*[17] during aircraft flights in the Cape Verde Islands area. Savoie *et al.*[17] observed that the concentrations of giant particles ($r > 1$ μm) within the plume varied much more than those of the large particles ($r < 1$ μm). They also found that, because of rapid fallout, the concentration of particles with $r > 10$ μm was greatly reduced within the plume and correspondingly enhanced in the trade-wind layer. The different behavior of the large and giant particle ranges was also observed during surface air measurements at the Cape Verde Islands by Jaenicke *et al.*[10] When a dust cloud passed the sampling site, it was observed that particles with $r > 20$ μm reached their concentration peak 12 to 24 hours earlier than did the submicron fraction.

A property that is useful in characterizing aerosols is their mass distribution. Most of the measurements dealing with the Saharan dust had been performed in surface air across the Atlantic. In these, Savoie *et al.*[17] and Schütz[7] found very characteristic changes of the mineral aerosol mass distribution as the plume moves westward. These can be explained qualitatively by the model.

In FIGURE 5, calculated, as well as measured, mass distributions of particles are plotted as a function of the distance from the Sahara. FIGURE 5 also includes mass distributions of desert soil and aerosol and a mean distribution of loess deposits,[18] which can be taken as fallout material close to a source. According to FIGURE 5, one can observe a general feature when soil dust is injected into the atmosphere and is transported long distances. It can be easily seen that both the maximum of the distributions and the absolute concentrations decrease with increasing distance from the source. The distributions of soil and sandstorm aerosol are similar for particles with $r < 100$ μm. During short distance transport, particles with $r > 50$ μm are deposited very rapidly, forming loess deposits. If the desert aerosol emerges from the source area, the maximum of the mass distribution will be observed at approximately 30 μm radius and subsequently shifts to 5 μm and 1 μm for transport distances of about 1500 km and 5000 km, respectively.

The annual mineral mass budget of the Saharan dust plume can be estimated using the transport model (for details, see Schütz *et al.*[19]). For this calculation, it was assumed that the main transport across the North Atlantic takes place within a transport channel 1000 km wide (15°–25° N) and 3.5 km high (1.5 km to 5 km altitude). It turned out that approximately $260 \cdot 10^6$ tonnes§ of mineral dust per year leave the desert towards the west. This figure compares well with an estimate by Lepple[5] on the order of $290 \cdot 10^6$ tonnes/year for the same latitudinal band. About 70% of the emitted mass will be removed during the first 2000 km of transport, due to rapid fallout of giant particles (see also FIGURE 4). Only small variations are found for the subsequent westward movement; thus, roughly $50 \cdot 10^6$ tonnes/yr reach the Caribbean area. This figure also agrees reasonably well with an estimate of semiannual flux of about $37 \cdot 10^6$ tonnes/yr by Prospero *et al.*[3]

§One tonne = 10^3 kg.

Transport toward the Gulf of Guinea

The dust transport over the southern border of the Sahara is a strong seasonal phenomenon, with the migration of ITCZ as an important factor. The region of northeasterly winds with dry and dust-laden air is restricted during summer to the middle and northern Sahara, when the ITCZ has its most northern position (Griffiths *et al.*[20]). Southern parts and the equatorial region, however, are influenced by dust-free maritime air masses blowing from southwesterly directions (southwest monsoon). As the ITCZ moves to the south during winter, the northeast winds carrying dust from the Sahara can reach the Gulf of Guinea. During heavy dust outbreaks at Abidjan (Ivory Coast), mineral dust concentrations up to 900 $\mu g/m^3$ were reported (Bertrand *et al.*[21]). Mineral dust penetrating into that region is transported by northeast winds at higher altitudes, because the dry and hot air masses are undercut by the relatively cool and humid air masses of the southwest monsoon.[21] The boundary layer between both air masses is the ITCZ, which is a strongly inclined layer (see GARP,[22] Figures 4.3 and 4.4). It is therefore remarkable that the dust is transferred from a layer above the ITCZ down to the surface, as is the case for the Saharan air layer above the trade-wind inversion over the North Atlantic. The amount of mineral dust transported to the south has not yet been estimated.

Transport to Europe and the Near East

Dust transport across the northern borders of the Sahara into the Mediterranean area and to Europe occurs only in short-term outbreaks mainly during the spring season. No continuous dust flow has been observed. Usually in early spring, polar air masses penetrate into the northern Sahara, generating strong low pressure systems. Associated with large vertical motions in the vicinity of the cold front, huge amounts of dust are released into the atmosphere.[23,24,15] Very often, a trough is situated over western Europe so that, within strong southerly winds above the 850 mb level, dust is transported to central Europe. Under these conditions, hundreds of thousands of tonnes of mineral dust are deposited over the continent, causing "red snow" in the Alps.[25–27]

Recently, a case of a strong Saharan dust outbreak and its transport across the Alps has been well documented by an air monitoring station at the Wank (1780 m above sea level, Garmisch-P., FRG). TABLE 2 (after Reiter *et al.*[28]) shows how the aerosols' chemical matrix changes as a dust cloud passes the station. For crust-derived compounds (Fe_2O_3, SiO_2, and Al_2O_3), an increase in the concentrations of up to a factor of 10–20 can be observed, whereas "noncrustal" compounds (NH_4^+, $SO_4^=$, and PbO) are only negligibly affected by the dust impact.

The region of the Near East is affected by springtime dust storms that are created by strong low pressure systems over the Libyan and Egyptian desert.[23] These storms, called "Khamsin depressions," travel into the eastern Mediterranean and produce considerable dust deposits over northern Egypt and Israel. Estimates of Salam *et al.*[29] and Yaalon *et al.*[30] lead to deposition rates of about 0.5–2.5 $tonnes/km^2$ hr for average Khamsin storms of a mean duration of 1 day. Roughly 15 Khamsin episodes can be expected in that area each spring. The annual dust production of the Khamsin storms over the Near East has been estimated by Josef *et al.*[31] (based on turbidity measurements over Israel) and by Yaalon *et al.*[30] to be on the order of 3–30 · 10^6 tonnes. These figures suggest that the output of the Sahara across its northern borders is small, compared to the transport to the west.

Source Areas for Airborne Saharan Dust

Distinct source areas for the dust emerging from the desert are not well known. Satellite photographs showing dust after sandstorms in the atmosphere cannot be used to determine small scale source regions. Usually, there are only a few surface observations from weather stations available to determine local sources.

For the western half of the Sahara, sandstorm statistics (Dubief[32]) are available that indicate that the area of southwest Algeria, Morocco, and Mauritania can be assumed to be a very frequent source region. This area is obviously the source of dust that is transported over the North Atlantic. The source region of dust transported during winter towards the Gulf of Guinea is supposed to be the Niger and Chad area.[21]

TABLE 2

CONCENTRATIONS OF VARIOUS AEROSOL COMPOUNDS DURING PASSAGE OF A SAHARAN DUST CLOUD AT WANK PEAK (1780 m, FRG) IN 1977*

Chemical Compound	Concentration (ng/m^3)		
	May 16–18	May 18–21†	May 21–25
NH_4^+	1290	561	592
SO_4^{2-}	3584	2182	2799
PbO	19	9.9	19
Fe_2O_3	122	1351	74
SiO_2	411	8027	422
Al_2O_3	305	3109	129

*Data derived from Reiter *et al.*[28]
†Passage of dust cloud.

For dust transported across the northern margins, no specific regions have been mentioned in the literature.

The generation of airborne desert dust as a function of meteorological conditions and several soil parameters has been studied only for North American semiarid regions (for details, see Gillette[33]). In the case of the Sahara, however, little information is available on physical and chemical parameters of both airborne dust and soil material for different geomorphological regions. Size distribution studies of various desert soils and aerosols in Libya by Schütz *et al.*[16] indicate that dune areas are relatively unproductive for the generation of airborne dust particles. Within dune material, the size range below 20–30 μm radius, which is subject to atmospheric transport, was strongly depleted. Thus, the question of which surface formations are the most productive is still unsettled. For instance, Yaalon *et al.*[30] assume that desert dust originates from alluvial deposits in desert mountain valleys. Hagedorn[34] believes that flood plains around mountain ranges and dried out river beds (wadis) are most probably sources for airborne dust.

Contributions of Mineral Dust to the Atmosphere and Oceans on a Global Scale

The input of mineral dust from arid and semiarid regions and its distribution within the atmosphere have not yet been studied thoroughly on a global basis. The mean distribution of dust in the atmosphere and its pathways of transport can be described only in the case of the Sahara. For other deserts of the world, little information is presently available. However, downwind of these areas, dust transport must be expected. The distribution of the frequency of dry haze over the oceans and the distribution of clay minerals in deep-sea sediments can serve as evidence for mineral dust transport. Their distribution patterns can be considered as results of the areal distribution of arid and semiarid regions and the effect of major wind systems. Both dry haze (caused by particles with about 0.2 $\mu m \leq r \leq 2$ μm) and clay minerals (particles with $r < 1$ μm) consist of particles that are subject to long range atmospheric transport, as mentioned above.

The distribution of the frequency of dry haze was published by McDonald.[35] These maps were compiled from visual reports by navigators. Such observations can be biased by misinterpretations, but it can be assumed that the occurrence of dry haze downwind of deserts is mainly caused by the presence of mineral dust particles in the atmosphere.

More than 50% of the carbonate-free mass of ocean sediments consists of clay minerals that are, to a certain extent, of eolian origin (Griffin *et al.*[36]). For instance, the distribution of two species, illite and kaolinite, is related in part to atmospheric transport. Both kaolinite and illite also have strong riverborne inputs in the vicinity of the continents. But farther from the continents, these inputs become less important; thus the distribution pattern of these minerals can serve as an indicator of atmospheric contributions.

Tropical North Atlantic

The distribution of dry haze, like that of the clay minerals, reflects the strong influence of the Saharan dust plume.

For dry haze, one finds the greatest frequency near the Sahara Desert. A seasonal shift of the haze plume is observed, also, which corresponds well with the findings displayed in FIGURE 3. The maximum of the haze frequency is recorded at the Gulf of Guinea area during the winter season. This indicates that, obviously, a considerable dust transport towards the south occurs that has not yet been investigated.

The kaolinite distribution shows a maximum similar to the haze distribution between 5° and 15° N. Although at lower latitudes, kaolinite is formed by soil weathering; in the North Atlantic, it probably is derived mostly from eolian dust, because large river inputs are confined toward equatorial regions. Close to the West African coast, the maximum of the distribution extends in a southwesterly direction; this can be explained by a southern shift of the fallout material of the dust plume due to southerly coastal winds and deep-sea currents. The pattern as a whole extends well over the Atlantic into the Caribbean Sea, indicating the extension of the Saharan dust plume. The illite pattern across the subtropical North Atlantic also supports the idea of eolian dust contributions into the sediments, except for the area close to the South

American coast, where strong riverborne inputs are clearly delineated. These distribution patterns of kaolinite and illite are confirmed by mineralogical investigations of airborne dusts[37–39] that resemble the deep-sea sediments.

Further evidence for eolian dust transport is given by the distribution of quartz particles in the deep-sea sediments. According to Beltagy *et al.*,[40] quartz is found in maximum abundance in a region centered near the West African coast. Decrease of mean particle radius occurs toward the Middle Atlantic Ridge (MAR). However, west of the MAR, increasing particle radius is observed, indicating other sources in addition to dust from the Saharan plume. These findings compare well with Rb/Sr and $^{87}Sr/^{86}Sr$ isotope ratios of both airborne dust and sediments by Biscaye *et al.*[41] Sediments east of the MAR showed a strong Saharan dust origin, whereas west of the MAR, a subsequent dilution with other material can be observed.

South Atlantic

The haze distribution at the coast of southwest Africa shows an increased frequency during fall, which can be attributed to airborne dust from the Namib and Kalahari deserts. Its pattern is oriented in northwesterly directions due to the southeasterly trades, which obviously carry the dust towards the sea. In contrast, the kaolinite and illite distributions of the deep-sea sediments do not indicate a similar pattern. Thus, a considerable eolian transport can be excluded. In fact, data reported by Chester *et al.*[38] and Aston *et al.*[37] confirm this assumption, because they showed at that area an increase in dust concentrations of only a factor of 2 compared to undisturbed maritime air. Isotope investigations[41] of both Kalahari dust and sediment also exclude this desert region as a significant source for airborne dust.

Pacific Ocean

The haze distribution over the open ocean shows only in some rare cases an appearance of dry haze, suggesting that no major dust inputs are observed. A closer look at the surrounding coastline, however, shows that downwind of some desert areas, dry haze occurs over the ocean. For instance, during spring, an increased haze activity is reported near the Mexican highlands, the South American coastal deserts, and over the Sea of Japan close to the great Asian deserts.

In the case of the eastern Pacific haze close to Mexico and South America, only the investigation by Prospero *et al.*[42] is available. This study supports the idea that soil-derived material is transported into this ocean area. According to the kaolinite and illite pattern, however, no major inputs can be observed.

Dry haze over the Sea of Japan can be fairly well attributed to mineral dust transport from the Asian continent. Recent studies of arctic haze bands (Rahn *et al.*[43]) have given evidence of major dust transport across the North Pacific into the Arctic Basin and even across the pole into the Spitsbergen area. From air mass trajectories and surface observations, the pathway of transport was traced back over a distance of about 12,000 km into the area of the Gobi Desert.

In fact, transport of airborne dust across the Pacific has been mentioned in the literature in connection with studies of the deep-sea sediments and Hawaiian soils. For instance, the illite distribution shows a strong band across the North Pacific extending between 20° and 40° north latitude, which is attributed by Griffin *et al.*[36] to westerly "jetstream transport" of soil material from the Asian mainland. A band of quartz particles having a similar distribution is also reported by Rex *et al.*,[44] suggesting airborne dust transport. According to $^{18}O/^{16}O$ studies of quartz particles in the ocean sediments and Hawaiian soils by Clayton *et al.*,[45] an eolian origin can be confirmed. Quartz both in sediments over a wide area of the ocean and in the soils was found to have the same isotope composition, one that is distinctly different from that of underlying material of the Hawaiian Islands. The grain size distribution of the quartz particles also was found to be similar in Pacific sediments and Hawaiian soils, with a maximum around 3 μm radius, supporting the argument for eolian transport.

Any conclusions concerning dust transport into the South Pacific based on the distribution of dry haze and that of kaolinite are rather vague. Dry haze will be observed only in very rare cases, and the clay minerals do not show a distinct pattern that can be related to continental sources. But it can be assumed that the large Australian desert contributes to a certain extent to the formation of the ocean sediments. Kaolinite and illite are abundant in the eastern desert soils, and some oxygen isotope data from quartz particles of Australian and New Zealand soils and South Pacific sediments[45] also support the idea of eolian dust transport.

Indian Ocean

Dry haze is observed very frequently over the Arabian Sea during summer. The surrounding continents between North Africa and the Indian peninsula are covered with deserts, which are most probably sources for airborne dust over the northern part of the Indian Ocean. The haze maximum in summer coincides with the presence of the strong southwestern monsoon by which, most probably, the main dust transport takes place.

In contrast to the haze distribution, the distribution of kaolinite and illite in the northern Indian Ocean does not reflect any strong eolian contribution to the ocean sediments. This is surprising, because large areas of the northern Arabian Sea are usually covered by a dust layer up to 8000 m in altitude and even up to 10,000 m over the Rajasthan desert.[46] In that area, however, the Indus River contributes the major part of the ocean sedimentary material. According to Goldberg *et al.*,[47] the riverborne input of illite, for instance, is approximately 80% of the total sedimentary input.

Another dust source for the southeastern part of the Indian Ocean is the Australian desert. With respect to the haze distribution, an increased transport of dust within the southeastern trades can be expected between September and February. In fact, this transport appears to be rather significant, because the kaolinite pattern shows a distinct plume extending in a northwesterly direction. Eolian transport seems to be the most important source in that area, because no major river inputs are observed and the desert soils of western Australia are very rich in kaolinite.[36]

Global Source Strength Estimates for Atmospheric Mineral Dust

Global estimates of the production rate for atmospheric mineral dust have been made in the past. For instance in the SMIC report,[48] a global mineral dust production on the order of 100–500 $\cdot$ 10^6 tonnes/yr was estimated. However, if one considers the more recent figures for the Sahara Desert alone, it becomes obvious that the older estimates should be scaled up.

In the following section, the global production rates will be estimated on the basis of (a) mean sedimentary accumulation rates, (b) vertical dust flux into the atmosphere during sandstorms, and (c) average mineral dust concentrations in the atmosphere.

a. Average ocean sedimentation rates are reported[36,49] to be on the order of 0.1 cm/1000 yr. This figure, representing carbonate-free material, can be considered to be the result of atmospheric dust deposition.[36] On the basis of a packing density for sedimentary deposits[49] of 2 g/cm^3, one can estimate a global production rate of 10^{15} g/yr of mineral dust. With an average residence time for tropospheric aerosols on the order of $4 \cdot 10^{-2}$ yr, a total atmospheric mineral dust load of $4.1 \cdot 10^{13}$ g is calculated. This is equivalent to a mass concentration of about 10 μg/m^3 in the atmosphere, which seems to be reasonable for a global average. However, if one considers that these

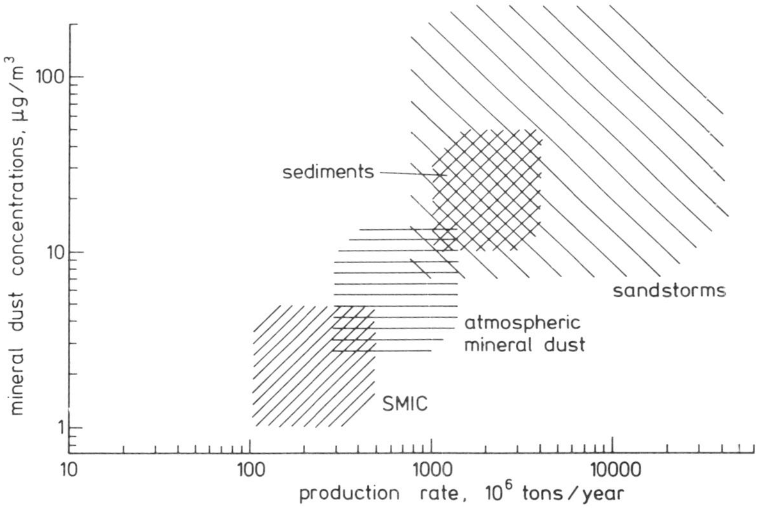

FIGURE 6. Production rates and concentrations of atmospheric mineral dust for various estimates. Hatched areas represent the range of uncertainty. (ton = tonne = 10^3 kg.)

figures are based on an average sedimentation that is made up preferentially by particles being transported over long distances, a larger rate for the total mineral dust production should be assumed. Referring to the Saharan dust transport, the long range transport fraction is found to be about 20% of the total production. Thus, a total annual mineral dust production rate on the order of $5000 \cdot 10^6$ tonnes/yr can be estimated.

b. Jackson *et al.*[50] estimate vertical fluxes of soil-derived aerosols during sandstorms in the range of $7 \cdot 10^{-2}$ up to 21 μg/cm^{2}s. Assuming that one sandstorm per year with a mean duration of 12 hours is generated for 5% of the arid and semiarid areas of the world ($2.5 \cdot 10^{17}$ cm^2), a total mineral dust production of between $760 \cdot 10^6$ and $230{,}000 \cdot 10^6$ tonnes/yr can be calculated. The lower value seems to be much closer to reality than the upper one, if one considers the Saharan output of about $300 \cdot 10^6$ tonnes/yr and the previous estimate of the global production rate of about $5{,}000 \cdot 10^6$ tonnes/yr.

c. Finally, an estimate will be made on the basis of mean mineral mass concentrations in the atmosphere. Elemental concentrations of Al, Fe, and other crustal elements measured in background aerosol studies (compiled by Junge[18]) can be used to estimate the mineral mass load of the atmosphere, but this technique yields only average values because of the lack of data for large areas of the ocean. Based on these data, one can distinguish between three air masses; desert, continental, and oceanic air masses with average mineral concentrations of 50, 5, and 0.5 $\mu g/m^3$, respectively. Assuming a height of the homogenous dust atmosphere of $H = 4$ km, 2 km, and 1 km for these air masses, and surface areas of $0.5 \cdot 10^{18}$ cm^2, 10^{18} cm^2, and $3.6 \cdot 10^{18}$ cm^2, one can calculate mineral mass loads of $10 \cdot 10^{12}$ g, 10^{12} g, and $0.18 \cdot 10^{12}$ g. With an average residence time of $4 \cdot 10^{-2}$ yr, a mean production rate—including all three air masses—of about $280 \cdot 10^6$ tonnes/yr is found. Assuming, as before, that this value represents only the fraction undergoing long range transport, one can estimate a total production rate for mineral dust released from continental surfaces on the order of $1{,}400 \cdot 10^6$ tonnes/yr.

The results of these estimates, summarized in FIGURE 6, show a wide spread; but they seem to indicate that the total global mineral dust production is on the order of a thousand million tonnes per year.

CONCLUSIONS

The mechanism of mineral dust transport into the atmosphere, its worldwide distribution, and its possible effect on the climate are still not well understood. First, there is a lack of information about the areal and vertical distribution of mineral dust over most of our deserts on the earth. Our present knowledge is derived mainly from studies of selected areas, thus it is not easy to extrapolate this information on a more general basis.

A wider range of information is available about the Sahara Desert than about any other major dust-producing locality. The Sahara studies have clearly shown that the largest portion of mineral dust generated over the desert is deposited in the vicinity of the source area within transport distances of less than 2000 km. Only about 20% of the total production is available for long range transport. The mass distribution of the mineral particles shows characteristic changes during transport. As the dust plume emerges from the source area, the maximum of the distribution is shifted towards smaller particle sizes, and the total concentrations decrease. After a transport distance of about 5000 km, most of the mineral airborne mass consists of submicron particles. Based on this information, it is possible to explain the distribution of dry haze close to various deserts and the distribution of larger areas of deep-sea sediment deposits, which obviously reflect the amount and areal distribution of mineral dust within the atmosphere.

The budget calculation for the mineral mass transported across the Atlantic and estimates of the global dust production rate based on sediment data, vertical dust fluxes over the desert, and average dust concentration within the atmosphere clearly show that previous estimates for the world must be revised upward.

Acknowledgments

I would like to thank Prof. R. Jaenicke, Institute for Meteorology, Mainz, and Prof. E. Perry, Northern Illinois University, for valuable suggestions and discussions.

References

1. Ehrenberg, C. G. 1862. Erläuterungen eines neuen wirklichen Passatstaubes aus dem atlantischen Dunkelmeere vom 29. Oct. 1861. Monatsber. Kgl. Preuss. Akad. Wiss. Berlin: 202.
2. Semmelhack, W. 1934. Die Staubfälle im nordwest-afrikanischen Gebiet des Atlantischen Ozeans. Ann. Hydrol. **62**(7): 273.
3. Prospero, J. M. & T. N. Carlson. 1972. Vertical and areal distribution of Saharan dust over the Western Equatorial North Atlantic Ocean. J. Geophys. Res. **77**(27): 5255.
4. Carlson, T. N. & J. M. Prospero. 1972. The large-scale movement of Saharan Air Outbreaks over the Northern Equatorial Atlantic. J. Appl. Meteorol. **11**(2): 283.
5. Lepple, F. K. 1975. Eolian Dust over the North Atlantic Ocean. Ph.D. Thesis. University of Delaware. Newark, Del.
6. Prospero, J. M., D. L. Savoie, T. N. Carlson & R. T. Nees. 1979. Monitoring Saharan aerosol transport by means of turbidity measurements. *In* Saharan Dust: Mobilisation, Transport, Deposition. Scope 14. C. Morales, Ed.: 171–186. John Wiley and Sons. Chichester, United Kingdom.
7. Schütz, L. 1977. Die Saharastaub-Komponente über dem subtropischen Nord-Atlantik. Ph.D. Thesis. University of Mainz. Mainz, Federal Republic of Germany.
8. Savoie, D. L. & J. M. Prospero. 1977. Aerosol concentration statistics for the northern tropical Atlantic. J. Geophys. Res. **82**(37): 5954.
9. Gringel, W. & R. Mühleisen. 1978. Sahara dust concentration in the troposphere over the North Atlantic derived from measurements of air conductivity. Beitr. Phys. Atmos. **51**: 121.
10. Jaenicke, R. & L. Schütz. 1978. A comprehensive study of physical and chemical properties of the surface aerosols in the Cape Verde Islands region. J. Geophys. Res. **83**(C7): 3585.
11. Newell, R. E., J. W. Kidson, D. G. Vincent & G. J. Boer. 1972. The General Circulation of the Tropical Atmosphere and Interactions with Extratropical Latitudes **1**. MIT Press. Cambridge, Mass.
12. Lamb, P. J. 1977. On the surface climatology of the Tropical Atlantic. Arch. Meteorol. Geophys. Bioklimatol. Ser. B **25**: 2.
13. Climatologia Aeronautica de Portugal. 1968. Fasciculo II, Aeroporto do Sal: 1–160. Servico Meteorologico National. Lisboa, Portugal.
14. Prospero, J. M. & R. T. Nees. 1977. Dust concentration in the atmosphere of the equatorial North Atlantic: possible relationship to the Sahelian drought. Science **196**: 1196.
15. Schütz, L. 1971. Messung der Grössenverteilung von luftgetragenem Staub und Bodenmaterial in der Libyschen Sahara. Master's Thesis. University of Mainz. Mainz, Federal Republic of Germany.
16. Schütz, L. & R. Jaenicke. 1974. Particle number and mass distribution above 10^{-4} cm radius in sand and aerosol of the Sahara desert. J. Appl. Meteorol. **13**(2): 863.
17. Savoie, D. L. & J. M. Prospero. 1976. Saharan aerosol transport across the Atlantic Ocean: characteristics of the input and the output. Bull. Am. Meteorol. Soc. **57**(1): 145.
18. Junge, C. 1979. The importance of mineral dust as an atmospheric constituent. *In* Saharan Dust: Mobilisation, Transport, Deposition. Scope 14. C. Morales, Ed.: 49–60. John Wiley and Sons. Chichester, United Kingdom.

19. SCHÜTZ, L., R. JAENICKE & H. PIETREK. 1980. Saharan dust transport over the North Atlantic Ocean. *In* Proceedings of the AAAS Symposium, Desert Dust: Origin, Characteristics and Effect on Man, Denver, Colorado, February 1977. (In press.)
20. GRIFFITHS, J. F. & K. H. SOLIMAN. 1972. Climates of Africa. *In* World Survey of Climatology. H. E. Landsberg, Ed. **10**: 75–131. Elsevier Publishing Company. Amsterdam, The Netherlands.
21. BERTRAND, J., J. BAUDET & A. DROCHON. 1974. Importance des aerosols naturels en Afrique de l Ouest. J. Rech. Atmos. **8**(3–4): 845.
22. Global Atmospheric Research Program. 1978. The West African Monsoon Experiment (WAMEX). GARP Publication Series No. 21: 1–88. World Meteorological Organization. Geneva, Switzerland.
23. ELFANDY, M. G. 1940. The formation of depressions of the Khamsin type. Q. J. R. Meteorol. Soc. **66**: 323.
24. SCHMIDT, G. 1942. Schirokko-Lagen. Herausgegeben von der Wetterwarte beim Oberbefehlshaber Süd: 1–7. Deutscher Wetterdienst. Offenbach, Federal Republic of Germany.
25. GLAWION, H. 1939. Staub und Staubfälle in Arosa. Gerlands Beitr. Geophys. freien Atmos. **25**: 1.
26. BRANDTNER, E. 1947. Der Staubfall in Westeuropa am 29. März 1947. Meteorol. Rundsch. **1**: 222.
27. NEUWIRTH, R. 1962. Der Staubfall vom 17.4.1962. Staub **22**(10): 412.
28. REITER, R., R. SLADKOVIC & K. PÖTZL. 1978. Chemische Komponenten des Reinluftaerosols in Abhängigkeit von Luftmassencharakter und meteorologischen Bedingungen. Ber. Bunsenges. Phys. Chem. **82**: 488.
29. SALAM, M. S. A. & M. A. SOWELIM. 1967. Dustfall caused by the spring Khamsin storms in Cairo: A preliminary report. Atmos. Environ. **1**: 221.
30. YAALON, D. H. & E. GANOR. 1979. East Mediterranean trajectories of dustcarrying storms from the Sahara and Sinai. *In* Saharan Dust: Mobilisation, Transport, Deposition. Scope 14. C. Morales, Ed.: 187–193. John Wiley and Sons. Chichester, United Kingdom.
31. JOSEPH, J. H., A. MANES & D. ASHBEL. 1973. Desert aerosols transported by Khamsinic depressions and their climatic effects. J. Appl. Meteorol. **12**: 792.
32. DUBIEF, J. 1953. Les vents de sable au Sahara Francais. Colloq. Int. CNRS **35**: 45.
33. GILLETTE, DALE. 1980. Major contributions of natural primary continental aerosols: source mechanisms. Ann. N.Y. Acad. Sci. (This volume.)
34. HAGEDORN, H. 1978. Personal communication.
35. MCDONALD, W. F. 1938. Atlas of Climatic Charts of the Oceans. U.S. Department of Agriculture, Weather Bureau. Washington, D.C.
36. GRIFFIN, J. J., H. WINDOM & E. D. GOLDBERG. 1968. The distribution of clay minerals in the world ocean. Deep-Sea Res. **15**: 433.
37. ASTON, S. R., R. CHESTER, L. R. JOHNSON & R. C. PADGHAM. 1973. Eolian dust from the lower atmosphere of the Eastern Atlantic and Indian Oceans, China Sea and Sea of Japan. Mar. Geol. **14**: 15.
38. CHESTER, R., H. ELDERFIELD, J. J. GRIFFIN, L. R. JOHNSON & R. C. PADGHAM. 1972. Eolian dust along the eastern margins of the Atlantic Ocean. Mar. Geol. **13**: 91.
39. DELANY, A. C., A. C. DELANY, D. W. PARKIN, J. J. GRIFFIN, E. D. GOLDBERG & B. E. F. REIMANN. 1967. Airborne dust collected at Barbados. Geochim. Cosmochim. Acta **31**: 885.
40. BELTAGY, A. I., R. CHESTER & R. C. PADGHAM. 1972. The particle size distribution of quartz in some North Atlantic deep-sea sediments. Mar. Geol. **13**: 297.
41. BISCAYE, P. E., R. CHESSELET & J. M. PROSPERO. 1974. Rb-Sr, $^{87}Sr/^{86}Sr$ isotope system as an index of the provenance of continental dusts in the open Atlantic Ocean. J. Rech. Atmos. **8**(3–4): 819.
42. PROSPERO, J. M. & E. BONATTI. 1969. Continental dust in the atmosphere of the eastern equatorial Pacific. J. Geophys. Res. **74**(13): 3362.
43. RAHN, K. A., R. D. BORYS & G. E. SHAW. 1977. The Asian source of arctic haze bands. Nature **268**(5622): 713.

44. REX, R. W. & E. D. GOLDBERG. 1958. Quartz contents of pelagic sediments of the Pacific Ocean. Tellus **10**(1): 153.
45. CLAYTON, R. N., R. W. REX, J. K. SYERS & M. L. JACKSON. 1972. Oxygen isotope abundance in quartz from Pacific pelagic sediments. J. Geophys. Res. **77**(21): 3907.
46. BRYSON, R. A. & D. A. BAERREIS. 1967. Possibilities of major climatic modification and their implications: northwest India, a case for study. Bull. Am. Meteorol. Soc. **48**(3): 136.
47. GOLDBERG, E. D. & J. J. GRIFFIN. 1970. The sediments of the northern Indian Ocean. Deep-Sea Res. **17**: 513.
48. Study of Man's Impact on Climate (SMIC). 1971. Inadvertent Climate Modification. MIT Press. Cambridge, Mass.
49. GOLDBERG, E. D. 1971. Atmospheric dust, the sedimentary cycle and man. Comments Earth Sci. Geophys. **1**: 117.
50. JACKSON, M. L., D. A. GILLETTE, E. F. DANIELSEN, I. H. BLIFFORD, R. A. BRYSON & J. K. SYERS. 1973. Global dustfall during the Quarternary as related to environments. Soil Sci. **116**(3): 135.

ARCTIC HAZE: PERTURBATION OF THE POLAR RADIATION BUDGET*

Glenn E. Shaw and Knut Stamnes

Geophysical Institute
University of Alaska
Fairbanks, Alaska 99701

Introduction

Arctic haze refers to turbid air in the northern polar regions. It was first noticed and commented on in 1956 by Murray Mitchell, Jr.,[1] who more or less described it as being an amorphous haze extending from the northern Alaskan coasts to the highest polar latitudes. In the period 1971–75, Shaw and Wendler,[2] Shaw,[3] and Holmgren *et al.*[4] measured anomalously high values of optical extinction through the atmosphere at McCall Glacier in remote northeastern Alaska and at Point Barrow on the northern tip of Alaska. The high values of extinction (haze optical depth = 0.1–0.2 at 500 nm wavelength) were extremely puzzling, especially since the atmospheric optical extinction underwent a seasonal variation in the opposite direction from what one would expect: the values of optical extinction were large during the months March–April, when the oceans and surface are covered with ice and snow, but were smaller in summer. This behavior is distinctly anomalous, since turbidity at most locations is at its greatest in summer[5,6,7] and at its minimum in middle or late winter.

Since 1975, arctic haze has been studied extensively in Alaska by Rahn *et al.*[8] Its composition, size distribution, and other information indicate that it may not be natural but may be associated with pollution emitted at the middle latitudes. Chemical studies indicate that sulfates and carbonaceous particles are present in the haze. Vanadium and other elements indicating a pollution source are strongly enriched, for example. There are other indications that the haze is pollution-derived (see Rahn and McCaffrey[9] for an account of the properties and probable origin of the haze).

The subject of this paper is a somewhat specialized topic pertaining to arctic haze—we will consider the perturbation that arctic haze makes on the polar radiation balance. It will be shown that the presence of haze at the time of the spring equinox has a dramatic effect on the arctic radiation balance.

The Radiation Balance Calculation

The objective of the study was to theoretically model the response of the atmospheric and earth–atmosphere radiation budget when a homogeneous layer of haze is introduced into the air poleward of the Arctic Circle. We used a complete

*Sponsored by National Science Foundation Grant No. DPP-77-27242 and by Office of Naval Research Grant No. N-00014-76-C-0435.

0077–8923/80/0338–0533 $01.75/1 © 1980, NYAS

solution to the equations of transfer that accounted for high-order scattering and for the reflection of light from the surface ice and snow. The reported solutions apply only to visible-band radiation; the effects of arctic haze on infrared heat balance will later be shown to be small.

The solutions reported here were obtained for a monochromatic wavelength band centered at 500 nm. We made the assumption that 500 nm would be a representative wavelength to use for the visible-spectrum radiation balance. This is an approximate technique, but one that will be rectified in the future by weighting multiwavelength calculations against the solar spectrum. The numbers will change slightly when a more exact treatment is given, but the basic conclusions reached will remain the same.

Computational Scheme

The transfer of radiation through an absorbing/scattering atmosphere is described by the radiative transfer equation; in this case, for a plane-parallel, horizontally homogeneous geometry. An azimuthally integrated version of the transfer equation is:

$$\mu \frac{dI(\tau, \mu)}{d\tau} = I(\tau, \mu) - \frac{\omega(\tau)}{2} \int_{-1}^{1} \bar{P}(\tau; \mu, \mu') I(\tau, \mu') \, d\mu' \tag{1}$$

where I is the azimuthally averaged intensity, μ is the cosine of the zenith angle, ω is the single scattering albedo, and τ is the optical thickness. The azimuthally integrated phase function is:

$$\bar{P}(\tau, \mu, \mu') = \frac{1}{2\pi} \int_{0}^{2\pi} P[(\mu\mu' + (1 - \mu^2)^{1/2}(1 - \mu'^2)^{1/2} \cos \phi)] d\phi$$

where P is the phase function satisfying the usual normalization:

$$\frac{1}{2} \int_{0}^{\pi} P(\cos \theta) \sin \theta \, d\theta = \frac{1}{2} \int_{-1}^{1} \bar{P}(\mu, \mu') \, d\mu' = 1 \tag{2}$$

In general, the single scattering albedo and the phase function depend on the optical depth, τ. For the purpose of this preliminary study, we assume a constant mixing ratio of aerosols to molecules. This approximation has the effect of making ω and P independent of τ. Herman and Browning[10] showed that the vertical inhomogeneity has very little effect on the reflected flux. We conjecture that if altitude profiles of the energy deposition (heating/cooling rate profiles) are not required, the vertical distribution of the aerosols will not be important. Thus, our model is adequate for determining the integrated quantities: albedo, transmissivity, and total absorption within the atmosphere.

The discrete ordinate approximation[11,12] consists of replacing the integral term in (**1**) by a numerical quadrature containing $2n$ terms. This leads to a system of $2n$ coupled differential equations for which eigen solutions can be obtained. As shown by Stamnes,[11] this system of equations can be written in matrix form, and the symmetry properties of the phase function can be used to reduce the order of the resulting algebraic eigenvalue problem by a factor of 2.

The numerical implementation is accomplished by using algebraic eigenvalue routines. The program is rapid and reliable.

Model Atmosphere

To estimate the effect of adding aerosols of given optical thicknesses and absorbing/scattering properties to a Rayleigh-scattering atmosphere, we make the following assumptions:

1. Constant mixing ratio of aerosols to molecules (see discussion above).
2. The ground is a "Lambert reflector" with albedo A_s.
3. The aerosols scattered according to a Henyey-Greenstein phase function:

$$P_A(\cos\theta) = P_{HG}(\cos\theta) = \frac{1 - g^2}{[1 + g^2 - 2g\cos\theta]^{3/2}} \tag{3}$$

with asymmetry factor:

$$g = \langle\cos\theta\rangle = \frac{1}{2}\int_0^{\pi} \cos\theta \, P(\omega s\,\theta)\sin\theta \, d\theta \tag{4}$$

Denoting the Rayleigh scattering optical thickness τ_R and the aerosol optical thickness $\tau_A = \tau_{A,S} + \tau_{A,A}$, where $\tau_{A,S}$ and $\tau_{A,A}$ refer to scattering and absorption, respectively, the single scattering albedo is:

$$\omega = \frac{\tau_R + \tau_{A,S}}{\tau_R + \tau_A} \tag{5}$$

Similarly, the phase function becomes:

$$P(\cos\theta) = (1 - b)P_A(\cos\theta) + bP_R(\cos\theta) \tag{6}$$

where P_R $(\cos\theta)$ is the Rayleigh scattering phase function and:

$$b = \frac{\tau_R}{\tau_R + \tau_{A,S}} \tag{7}$$

Clearly, $b = 1$ and $\omega = 1$ refer to a nonabsorbing Rayleigh-scattering atmosphere.

Determination of the Optical Parameters of Arctic Haze

The necessary parameters to solve the transfer equation for arctic haze conditions are: τ, the optical thickness; ω, the albedo of single scattering; $\langle\cos(\theta)\rangle$, the asymmetry factor; and A_s, the underlying surface albedo. These four parameters were estimated from multiple-wavelength optical measurements during haze episodes in March–April 1978 at Fairbanks and Barrow. The following experimental measurements were made: (1) the intensity of direct solar radiaion at the surface with a precision sun photometer; (2) the sky radiance along the solar almucantar with a photoelectric coronameter and directional-scanning spectrometer; (3) the monochromatic downwelling and upwelling diffuse radiation fluxes with specially developed flux sensors; and (4) reflectivity of exposed millipore filters.

Optical thickness was derived according to standard sun photometry methods.[13] The albedo of single scattering, ω, was determined in two ways: (1) by use of Kubla-Monk theory applied to the measurements of exposed filters,[14,15] and (2) by comparing measured downwelling diffuse radiation with that calculated by radiative transfer theory for a series of trial values of $\tilde{\omega}$.[16] The asymmetry factor was determined from the scattering phase function, which itself was derived from the scans of sky intensity along the solar almucantar.[17] There are uncertainties in the derived optical parameter, because some of the techniques were approximate and the optical parameters fluctuated with time. Nevertheless, the values are accurate enough to do preliminary modeling.

During times of moderately heavy haze, the optical parameters were: τ (λ = 500 nm) = 0.1–0.2; $\tilde{\omega}$, 0.5–0.8; $\langle \cos(\theta) \rangle$, 0.6–0.8; A_s, 0.6–0.8. The low values of $\tilde{\omega}$ indicate

TABLE 1

VISIBLE-BAND RADIATION BALANCE AT 65 AND 75° NORTH LATITUDE DURING VERNAL EQUINOX*†

	Net Radiation Absorbed (kcal month^{-1} cm^{-2})					
	65° N			75° N		
	CLR‡	NAH§	AH¶	CLR	NAH	AH
Earth–Atmosphere	2.13	1.98	3.68	1.22	1.13	2.24
Atmosphere	0	0	1.94	0	0	1.32
Surface	2.13	1.98	1.75	1.22	1.13	0.91+

*Solar Constant: 1.96 cal cm^{-2} min^{-1} (8.7 kcal cm^{-2} month^{-1}) = 136 mw cm^{-2}.

†Haze optical thickness, 0.2; albedo of single scattering, 0.8; surface reflection, 0.8; haze scattering asymmetry factor, 0.75.

‡CLR, uncontaminated atmosphere.

§NAH, atmosphere charged with nonabsorbing haze.

¶AH, atmosphere charged with absorbing haze.

that the haze is an absorbing haze, as is evident also from the gray color of the exposed filters and from the presence of what appeared to be carbonaceous particles when the filters were inspected with a scanning electron microscope.

EVALUATION OF THE MODEL

The radiative transfer model was evaluated for latitudes at 65° and 75° N, assuming the haze to be homogeneous with $\tau_H = 0.2$, $\tilde{\omega} = 0.8$, $A_s = 0.8$, and $\langle \cos(\theta) \rangle = 0.75$. Comparative calculations were also made for a pure Rayleigh-scattering atmosphere. Results are tabulated in TABLE 1.

Arctic haze heats the atmosphere and forces a slight cooling at the surface. The earth–atmosphere system radiative heating increases (when haze is added) by 73% at latitude 65° north and 84% at latitude 75° north. The dramatic change in the amount of visible-band radiation absorbed by the arctic regions, by simply introducing an optically thin haze layer, can be understood by realizing that (a) the actual amount of radiation absorbed by the Arctic during haze-free periods is small to begin with, for the pack ice is highly reflecting and the sun angle is low, and (b) the long atmospheric ray paths and the multiple reflections between ice and haze both enhance the amount of radiation absorbed by the atmosphere. The net effect is to heat the atmosphere by an amount quite comparable with heating of ice by the sun in spring. At the same time, net radiation absorbed at the surface of the Arctic is decreased (by 22% and 32% at 65 and 75° N, respectively) when haze is introduced. The deficit is mostly absorbed in the atmosphere.

So far, nothing has been said about the interaction between haze and emitted thermal radiation. The optical thickness of arctic haze decreases with wavelength in the visible spectrum as λ^{-2}; by extrapolation, this would suggest a haze optical thickness of only 5×10^{-4} at $\lambda = 10\ \mu$, or 400 times smaller than the optical thickness in the visible region of the spectrum. If the haze is entirely absorbing in the long wave infrared region (as it probably is), the atmosphere would gain 0.01 kcal cm^{-2} $month^{-1}$ (for a radiating temperature of 250°K)—a negligible amount in comparison with the energy deposited by absorption of visible-band radiation.

Discussion

The radiation term represents only a small fraction of the arctic heat budget except for about two months centered around summer solstice (during which time there is no haze). Essentially, arctic heat transfer is a balance between advected heat and emitted thermal radiation. The distribution of radiant fluxes is, however, obviously disturbed by clouds.

It is difficult to state exactly what the influence of arctic haze is, or what it could be, on climate. There are secondary effects that must be considered: (a) The redistribution of radiant heating terms in the atmosphere would increase dynamic stability, thereby decreasing cloudiness. (b) If the haze is confined to the Arctic only, the model predicts that the energy balance contrast (which drives the general circulation) between high and low latitudes† would decrease; this would probably also act to reduce cloudiness because of smaller amounts of heat and moisture entering by advection. On the other hand, the net increase in the meridional gradient of radiation balance in the Arctic itself may increase horizontal scale dynamical turbulence. In summary, cloudiness would be predicted to *decrease* in the Arctic during haze episodes, and circulation patterns might alter, though in ways entirely unknown at this stage of the study.

Heating of the hazy atmosphere from 1.94 kcal cm^{-2} $month^{-1}$ of absorbed radiation (see TABLE 1) in a 2-km-thick layer would occur at a rate of 1.35°C day^{-1}. If the haze were uniformly mixed through the atmosphere, the heating rates would have to be reduced by a factor of four, but these are still substantial heating rates.

The most important effect on climate of the heating of the atmosphere probably would be to decrease cloudiness. Decreased cloudiness, lowered surface temperatures

†It is not actually necessary to suppose that the haze is maintained only in the Arctic. The heating becomes smaller, or even turns to cooling, when arctic haze is placed over temperate latitudes, which in general have lower surface reflectances.

(because of loss of infrared radiation), and dryer air masses would be expected to be *caused* by arctic haze. Clear skies, lower than normal surface temperatures, and lower than normal air humidities are exactly what were found at Barrow during arctic haze episodes in April 1978. The correlations suggest that arctic haze could have substantial influences on arctic climate.

The actual climatic alterations caused by arctic haze will have to be evaluated with numerical grid climate models. The parameters that would have to go into the models are, however, known only poorly. It is only recently that arctic air has been suspected of being contaminated. This first study indicates that significant climatic alterations may be brought about by the haze.

Summary

Air in the northern polar regions is sometimes charged with an enigmatic scattering-absorbing haze that cuts visibility and alters the amount and distribution of radiant energy flowing through the atmosphere and the absorption at the terrestrial surface. The perturbation of the radiation energy balance caused by the haze over the reflecting pack ice has been estimated by solving the equation of transfer with a discrete ordinate method. By using what are believed to be representative values for the optical parameters of arctic haze (albedo of single scattering = 0.8, optical thickness in visible spectrum = 0.20, and an asymmetry ratio of 0.75), we calculate that in spring the haze will introduce a heating of the earth–atmosphere system of 1–1.5 kcal cm^{-2} $month^{-1}$, a heating of the atmosphere of 1.3–1.9 kcal cm^{-2} month, and a cooling at the surface of 0.3–0.4 kcal cm^{-2} $month^{-1}$. The changes in the radiation budget will tend to introduce air subsidence and decrease cloudiness. Arctic haze may therefore have a substantial influence on arctic climate.

References

1. Mitchell, M., Jr. 1956. Visual range in the polar regions with particular reference to the Alaskan Arctic. J. Atmos. Terr. Phys. (Supplement): 195.
2. Shaw, G. E. & G. Wendler. 1972. Atmospheric turbidity measurements at McCall Glacier in northern Alaska. *In* Proceedings of the Atmospheric Radiation Conference, Ft. Collins, Colorado, August 1972. University of Colorado Press. Ft. Collins, Colo.
3. Shaw, G. E. 1975. The vertical distribution of atmospheric aerosols at Barrow, Alaska. Tellus **27**(1): 39.
4. Holmgren, B., G. E. Shaw & G. Weller. 1974. Turbidity in the Arctic atmosphere. *In* AIDJEX Bulletin **27:** 135–148.
5. Flowers, E. C., R. A. McCormick & K. R. Kurfis. 1969. Atmospheric turbidity over the United States, 1961–66. J. Appl. Meteorol. **8**(6): 955.
6. Volz, F. E. 1978. Atmospheric Turbidity in Europe, 1963–1969. Report AFGL-TR-78-0108, Environmental Research Papers No. 631, Air Force Geophysics Laboratory. Hanscom Air Force Base, Mass.
7. Laulainen, N. S., B. J. Taylor & P. W. Hodge. 1977. Analysis of atmospheric extinction obtained by astronomers. II. Seasonal variations in astronomical extinction. Atmos. Environ. **11:** 21.
8. Rahn, K., R. Borys & G. E. Shaw. 1977. The Asian source of Arctic haze bands. Nature **268:** 713.

9. Rahn, K. A. & R. J. McCaffrey. 1980. On the origin and transport of the winter arctic aerosol. Ann. N.Y. Acad. Sci. (This volume.)
10. Herman, B. M. & S. R. Browning. 1975. The effect of Aerosols on the earth–atmosphere albedo. J. Atmos. Sci. **32:** 1430.
11. Stamnes, K. 1978. A new look at the discrete ordinate method for radiative transfer calculations in anisotropically scattering atmospheres. I. Theory. (Submitted for publication.)
12. Stamnes, K. & R. A. Swansen. 1978. A new look at the discrete ordinate method for radiative transfer calculations in anisotropically scattering atmospheres. II. Results. (Submitted for publication.)
13. Shaw, G. E., J. A. Reagan & B. M. Herman. 1973. Investigations of atmospheric extinction using direct solar radiation measurements made with a multiple wavelength radiometer. J. Appl. Meteorol. **12**(2): 374.
14. Lindberg, J. D. 1975. Absorption coefficient for atmospheric dust and other strongly absorbing powders: an improvement on the method of measurement. Appl. Opt. **14**(12): 2813.
15. Lin, C. I., M. Baker & R. J. Charlson. 1973. Absorption coefficient of atmospheric aerosol: a method of measurement. Appl. Opt. **12**(6): 1356.
16. Herman, B. M., R. S. Browning & J. J. DeLuisi. 1975. Determination of the effective imaginary term of the complex index of refraction of atmospheric dust. J. Atmos. Sci. **32**(5): 918.
17. Deirmendjian, D. 1970. Use of Scattering Techniques in Cloud Microphysics Research. I. The Aureole Method. RAND Report R-590-PR.

ACID PRECIPITATION AND ITS EFFECTS ON TERRESTRIAL AND AQUATIC ECOSYSTEMS

Ellis B. Cowling

School of Forest Resources
North Carolina State University
Raleigh, North Carolina 27650

INTRODUCTION

Acid rain is becoming a dominant feature of man-induced change in the chemical climate of the earth. But the increasing acidity of rain is only part of the changing chemistry of precipitation, dry particulate matter, aerosols, and gases. Recent research has demonstrated that (1) atmospheric deposition contains both beneficial nutrients and injurious substances; (2) plants, animals, and ecosystems vary greatly in susceptibility, tolerance, and adaptability to changes in atmospheric deposition; (3) injury is most likely when rapid changes in the chemical climate coincide with the presence of a vulnerable life form or life stage; (4) increasing acidity of precipitation has caused extinction of fish and other changes in species composition at all trophic levels in oligotrophic lakes in northern Europe and eastern North America; and (5) simulated acid rain has caused leaching of nutrients from soil and both direct and indirect injury to terrestrial vegetation. With rare exceptions, however, economic damage to crops by naturally occurring acid rains has not been demonstrated to date.

These results show why a National Atmospheric Deposition Program is currently being developed to determine spacial and temporal trends in the amount, chemical form, and geographical distribution of atmospheric deposition and its effects on vegetation, soils, and aquatic ecosystems.

ACID PRECIPITATION AS PART OF A GENERAL PHENOMENON OF ATMOSPHERIC DEPOSITION

Terrestrial vegetation, soils, and surface waters are the primary deposition sites for all types of substances dispersed in the atmosphere. Most plants take up dissolved nutrients directly from precipitation as well as from the soil solution.[44,49] Fish and other aquatic organisms live in water derived from precipitation but filtered through the soil. Uptake of nutrients from atmospheric sources is especially important in natural ecosystems, such as lakes, estuaries, wetlands, forests, and rangelands, where nutrients from other sources are scarce and where fertilization is not a normal management procedure. But this uptake process also increases the vulnerability of terrestrial and aquatic organisms to injury by acid precipitation and toxic aerosols and gases.[13]

Precipitation formed in an atmosphere relatively free from natural or anthropogenic sources of contamination would have a pH of about 5.6 due to dissolution of gaseous CO_2. Thus, acid precipitation has been defined operationally as rain or snow

0077–8923/80/0338–0540 $01.75/1 © 1980, NYAS

with a pH value below 5.6. But acidity in precipitation must be understood as a reflection not only of the amounts of substances yielding hydrogen ions (such as sulfuric, nitric, hydrochloric, and organic acids) but also of the total ionic balance between all the cations in precipitation.[34,40]

Airborne substances that influence terrestrial plants and animals include: sea spray from oceans and large lakes; dust resulting from wind erosion of soil as well as from volcanic and cosmic sources; gases, such as CO_2, NH_3, SO_2, H_2S, CH_4, H_2, etc., released from decomposing organic matter and volcanoes; biogenic particles, such as spores, hyphal fragments, bacteria, and pollen; particulate matter, aerosols, and gases produced by wild fires and controlled burning of agricultural, forest, and urban wastes as well as from industrial, agricultural, residential, and commercial heating and transportation operations.[3,44]

Combustion of fossil fuels is the largest anthropogenic source of airborne substances.[25] Some of these substances include beneficial nutrient elements; others are injurious to terrestrial and aquatic ecosystems.[39,44,46,49]

Most of the substances dispersed in the atmosphere are at least partially soluble in water. These dissolved substances can be absorbed or assimilated by plants when precipitation is intercepted by vegetation. Since some of the substances dissolved in precipitation are beneficial while others are injurious to plants and animals, the net effect of atmospheric deposition can be beneficial or injurious depending on the chemical composition of the deposited matter, the duration and intensity of deposition episodes, the species and genetic characteristics of the organisms on which the substances are deposited, and the physiological condition, structure, phenology, and stage of maturity of the organisms.[13]

Direct injury to living organisms by acid precipitation is most likely when a particularly vulnerable life form (for example, lichens or fish) is exposed at a particularly vulnerable life stage (for example, the early stages of reproduction), and is growing in a poorly buffered environment (for example, sandy soils or oligotrophic lakes and bogs) during a season of the year when acid precipitation is most likely (for example, spring rains).[13,20]

Rain and snow change in chemical composition within, as well as between, precipitation events. In cold climates, acid substances accumulate in the snowpack, where they are released in concentrated form with the first meltwater and thus cause very sudden increases in acidity of soils, vegetation, and surface waters.[35,36] Thus, a given organism or ecosystem may be subject to beneficial atmospheric influences at one time, and to negative influences at another time, within a given day, month, growing season, or the years of its development in the case of perennial plants and animals.[46] Even a given molecular species, such as SO_2 or NO_2, may be absorbed and utilized as a beneficial nutrient at one concentration; while at another, higher concentration, even on the same day, the same substance may be absorbed and found to be toxic or even injurious to the very same plant, animal, or microorganism.

Herbaceous and perennial plants are subject to changes in atmospheric deposition within their annual cycles of growth. Trees and fish, on the other hand, persist for many years (or even decades in the case of trees); they are subject to long-term, as well as seasonal, changes in the chemistry of the atmosphere and precipitation.

For all of the above reasons, it is difficult to assess the effects of acid substances in precipitation in isolation from the general chemistry of the atmosphere. The effects of acidic precipitation on terrestrial and aquatic ecosystems are only one facet of the much larger subject of surface interactions of plants with the atmosphere and the soil. The effects of a given "acid rain" or a prevailing condition of "acid rains" are very complex, variable in time, and involve significant interactions that are only partially understood.[20]

Direct and Indirect Effects of Acid Precipitation on Vegetation

A partial list of theoretical effects of acid precipitation on vegetation was developed earlier by Tamm and Cowling[46] and is reproduced in Table 1. The effects are classified as either direct or indirect, although most direct effects will have many indirect consequences as well. A decreased rate of growth would be the expected consequence of most of the effects postulated in Table 1, but unequivocal evidence of significant growth effects has yet to be demonstrated in agricultural, forest, or range ecosystems. Specific biological effects that have been proven to occur in at least one experimental plant are marked with a dot (•) next to the numbers in Table 1.

The Availability of Essential and Injurious Elements from the Atmosphere

Plant life and animal life as we know it would be impossible without atmospheric sources of carbon for photosynthesis, nitrogen for biological fixation and protein synthesis, sulfur for protein synthesis, oxygen for respiration and synthesis of carbohydrates, and water for transpiration within plants. Many essential elements and injurious substances are present in the gaseous, particulate, and aerosol emissions from natural and anthropogenic sources. Essentially all of these diverse emissions into the atmosphere return to the surface of the earth where they are deposited into ecosystems by one of four major processes: (1) wet fallout of substances dissolved or suspended in raindrops or adsorbed on snowflakes; (2) gravitational settling of windblown dust and particulate matter; (3) impaction of aerosols; and (4) absorption of gases. Processes 2, 3, and 4 are component parts of so-called dry deposition.

Many natural ecosystems (both terrestrial and aquatic) are markedly influenced by deposition of nutrient and toxic elements from the atmosphere. Among the essential elements, carbon, hydrogen, and oxygen are rarely limiting. Nitrogen, phosphorus, and potassium are often limiting and, consequently, are added as major fertilizer elements. Sulfur, calcium, and magnesium are sometimes limiting, as are the so-called micronutrients—iron, copper, zinc, manganese, molybdenum, and boron.

Some elements are both essential and injurious to organisms. For example, small amounts of sulfur and large amounts of nitrogen are needed for synthesis of proteins, nucleic acids, and other substances in plants and animals; but gaseous oxides, as well as acids, of sulfur and nitrogen are injurious to many organisms at very low concentrations. Similarly, excess amounts of the micronutrient elements and heavy metals can be injurious. Atmospheric fluoride is toxic to plants at 25–50 ppm. Aluminum is the most abundant toxic metal in the earth's crust. The availability and toxicity of aluminum and other metal ions are influenced greatly by the acidity of

TABLE 1

POTENTIAL EFFECTS OF ACID PRECIPITATION ON VEGETATION*

Direct Effects

●1. *Damage to protective surface structures such as cuticle.*

Damage to surface structures may occur due to accelerated erosion of the cuticular layer that protects most foliar organs. It also could result from direct injury to surface cells by high concentrations of sulfuric acid and other harmful substances that are concentrated by evaporation or adherence of soot particles on plant surfaces.

●2. *Interference with normal functioning of guard cells.*

Malfunction of guard cells will lead to loss of control of stomata and thus to altered rates of transpiration and gas-exchange processes and possibly increased susceptibility to penetration by foliar plant pathogens.

●3. *Poisoning of plant cells after diffusion of acidic substances through stomata or cuticle.*

This could lead to development of deep necrotic or senescent spots on foliar organs including leaves, flowers, twigs, and branches.

●4. *Disturbance of normal metabolism or growth processes without necrosis of plant cells.*

Such disturbances may lead to decreased photosynthetic efficiency, altered respiratory patterns and intermediary metabolism, as well as abnormal development or premature senescence of leaves or other organs.

●5. *Alteration of leaf- and root-exudation processes.*

Such alterations may lead to changes in populations of foliar-surface and root-surface microflora and microfauna, including nitrogen-fixing organisms.

●6. *Interference with reproduction processes.*

Such interference may be achieved by decreasing the viability of pollen, interference with fertilization, decreased fruit or seed production, decreased germinability of seeds, etc.

●7. *Synergistic interaction with other environmental stress factors.*

Such reinforcing interactions may occur with gaseous sulfur dioxide, ozone, fluoride, soot particles, and other air pollutants, as well as drought, flooding, etc.

Indirect Effects

●1. *Accelerated leaching of substances from foliar organs.*

Damage to cuticle and surface cells may lead to accelerated leaching of mineral elements and organic substances from leaves, twigs, branches, and stems.

2. *Increased susceptibility to drought and other environmental stress factors.*

Erosion of cuticle, interference with normal functioning of guard cells, and direct injury to surface cells may lead to increased evapotranspiration from foliar organs and vulnerability to drought, air pollutants, and other environmental stress factors.

●3. *Alteration of symbiotic associations.*

Changes in leaf- and root-exudation processes and accelerated leaching of organic and inorganic substances from plants may affect the formation, development, balance, and function of symbiotic associations, such as mycorrhizae, nitrogen-fixing organisms, lichens, etc.

●4. *Alteration of host-parasite interactions.*

Resistance and/or susceptibility to biotic pathogens, parasites, and insects may be altered by subjecting plants to any environmental stress. Acid precipitation may increase the susceptibility of plants to these injurious agents, alter their capacity to tolerate disease or injury, or alter the virulence of pathogens. The effects of acid precipitation may vary with the following: the nature of the pathogen involved (whether a fungus, bacterium, mycoplasma, virus, nematode, parasitic seed plant, insect, or multiple-pathogen complex); the species, age, and physiological status of the host; and the stage in the disease cycle in which the acidic stress is applied. For example, acid rain might decrease the infective capacity of bacteria before infection and increase the susceptibility of the host to disease development after infection.

*From Tamm and Cowling.[46]

soils, which in turn is influenced by the abundance of acid precipitation. Aluminum ions are toxic to many organisms at concentrations below 1 ppm. Estimates of the threshold concentrations for detectable injury or economic damage by air pollutants, including NO_2, SO_2, O_3, toxic metal ions, and acid precipitation, require substantial additional research with a great variety of organisms and ecosystems.

In earlier decades, it was believed that essential nutrients were taken up by plants almost entirely from the soil. Now, it is recognized that airborne gases, particulate matter, and aerosols significantly augment the supply of both essential and injurious elements.[44] In fact, all of the nutrient elements essential for growth of plants can be taken up readily through foliar organs as well as by absorption by roots from the soil solution.[39,44,49]

Most cultivated crops require large amounts of essential nutrients. For this reason, more nutrients are required for sustained-yield agriculture than for sustained-yield range management or forestry. This is true for rangelands because biomass yields are very low, and for forests because the parts of trees that are usually harvested—the wood and bark of tree stems—contain very much less of most essential elements than do the seeds and fruits that are commonly harvested in agriculture. This is a major reason why soil fertilization is so common in agriculture and so rare in forestry and range management. In some forests and natural grasslands, atmospheric deposition is adequate to permit sustained-yield cropping of the land without fertilization. This is much less likely to remain so, as more and more of the nutrient-rich branches, foliage, and roots of trees are being harvested in so-called whole-tree chipping and other modern forestry practices.

Many scientists believe that acid rain and snow are deposited only onto soils, where acid substances could be neutralized in well-buffered soils or by applications of lime. This is true for some agricultural lands, especially after harvesting of crops. But it is not true for forests, rangelands, or even for most agricultural lands during the growing season. Most raindrops are intercepted by the foliage of plants, where substances dissolved in the rain can induce various physiological changes before reaching the soil. These early effects include erosion of epicuticular waxes and leaching of essential elements and various metabolites from foliar organs. In a mature forest, for example, rain will wash over three tiers of foliage before reaching the soil. Most forests and rangelands in the United States are rarely fertilized and have almost never been limed to neutralize soil acidity. Furthermore, forests cover about half the total land area of North America. Forest trees develop very large canopies of leaves and branches that extend high into the air. Thus, forests provide a very large surface for deposition and assimilation of all the beneficial and injurious substances dispersed in the atmosphere and deposited as precipitation.[44,46,49]

Sources, Amounts, and Distance of Transport of Beneficial and Injurious Substances in the Atmosphere

There are three major reasons why agricultural, range, forest, and aquatic biologists are becoming increasingly concerned about atmospheric transport and deposition of both nutritionally beneficial and potentially injurious substances: (1) vegetation, soils, and surface waters are the primary deposition sites for precipitation and airborne particulate matter of all types; (2) atmospheric deposition constitutes an important source of nutrients and potentially injurious substances that affect the productivity and stability of agricultural, forest, and aquatic ecosystems;[45] and (3) human activities are steadily increasing the amounts and variety of substances dispersed in the atmosphere.[3,21,25,33]

Previously, it was believed that most anthropogenic emissions were removed from the atmosphere near the site of emission. Now it is recognized, particularly with increased use of tall stacks in power plants, that atmospheric processes can lead to extensive mixing, which leads to both chemical and physical interactions and transformations of atmospheric particles, aerosols, and gases. Furthermore, these substances and their reaction products are dispersed by meteorological processes and finally enter the biosphere in fields of deposition that extend hundreds or even thousands of kilometers from the original sources of emission. The fallout of radioactive materials all over the Northern Hemisphere since the early 1950s as the result of atomic bomb tests provides a dramatic and continuing reminder of the long-distance transport and deposition of pollutants.

TABLE 2

BIOGEOCHEMICAL AND TECHNOLOGICAL FORCES IN THE BIOSPHERE OF THE EARTH*

Biosphere Components	Tons per Year
Biogeochemical Processes:	
Yield of photomass	1×10^{10}
Cycle of inorganic elements	1×10^{10}
River discharges:	
Dissolved substances	3×10^{9}
Suspended substances	2×10^{10}
Anthropogenic Sources:	
Output of fertilizers	3×10^{8}
Industrial dust	3×10^{8}
Garbage, urban wastes and by-products	2×10^{10}
Mine refuse	5×10^{9}
Aerosols and gas discharges	1×10^{9}

*Data of Kovda.[25]

The amounts of various substances introduced deliberately or inadvertently by man into the biosphere of industrialized regions of the earth are becoming so large that man is becoming a major force in the biogeochemistry of these regions.[25] TABLE 2 contains a tabulation of data on annual output of fertilizers, industrial dusts, garbage and other urban wastes and by-products, mine refuse, and discharges of aerosols and gases mainly from combustion of fossil fuels. All of these categories of matter are becoming comparable in magnitude with the discharges of dissolved and suspended substances in all the rivers of the world, the annual yield of photosynthetic products, or the cycling of inorganic elements in the earth as a whole. Anthropogenic emissions into the atmosphere are also very large, as shown in TABLE 3. Most gases, carbon oxides, and aerosols result from combustion of fossil fuels. A very large fraction of these global emissions was produced in the United States.

If the United States is to so greatly augment the amounts of substances dispersed in the atmosphere and deposited into the biosphere of the earth, it is essential that we should measure the amount and chemical form of the deposited matter and

TABLE 3

ANTHROPOGENIC EMISSIONS INTO THE ATMOSPHERE*

Type of Emission	Tons per Year†
Dust	2.5×10^8
Gases (mainly SO_2, HC, and NO_x)	6.5×10^8
Carbon oxides ($CO + CO_2$)	2.0×10^9
Aerosols	1.0×10^9

*Data of Kovda.[25]
†Doubling about every 7–10 years.

understand the biological consequences of that deposition in our country and abroad.

Regrettably, our understanding of these processes in North America is very fragmentary. Fortunately, however, more extensive measurements of atmospheric deposition and its biological consequences have been made in Europe, where an atmospheric deposition network has been maintained since the late 1940s.[33]

The European Air Chemistry Network began in Sweden in 1952 and has gradually spread to include most of western Europe and parts of eastern Europe, including Poland and the Soviet Union.* Since the mid 1950s, a network of about 100 stations has provided monthly measurements of changes in the chemistry of precipitation and of the air at ground level. The substances analyzed at most of these stations include the following major cations and anions: NH_4, Na, Ca, K, Mg, SO_4, NO_3, HCO_3, and Cl, as well as pH, conductivity, and, more recently, titratable acidity and alkalinity. These data have shown long-term trends in the chemistry of precipitation. For example, the amount of nitrate nitrogen in precipitation (an important fertilizer element) increased markedly in many parts of Europe during the 15 years between 1955 and 1970.[34] Nitrate nitrogen helps plants grow. Thus, the nitrogen added in precipitation probably increased yields of agricultural and forest crops. A similar trend between the 1920s and the 1970s has been measured at Ithaca, New York.[26]

But not all the substances detected in precipitation are beneficial. Long-term trends of injurious sulfate and hydrogen ions also have been detected.[21] These changes are attributed to strong acids formed in the atmosphere, mainly from oxides of sulfur and nitrogen produced during combustion of fossil fuels. More recent data show that these trends of increasing acidity are continuing,[2,33,34] although the relative contribution of sulfuric and nitric acids is changing.[26]

SPATIAL AND TEMPORAL CHANGES IN THE CHEMISTRY OF PRECIPITATION IN NORTH AMERICA

Some monitoring of the chemistry of precipitation has also been carried on in the United States.[10,14,27] Some of these studies provide excellent and reliable information about spatial variability in the acidity of precipitation. But most studies in this country

*A list of publications based on the European Air Chemistry Network can be obtained from the International Meteorological Institute, Stockholm, Sweden.

have suffered from three major shortcomings: (1) the data were collected for a limited land area—typically, only a single point or a few points in one or two states;[14] (2) the data were collected for very limited periods of time—typically, only 1 or 2 years; and (3) very few direct measurements of pH were made. There is only one location in the United States—the Hubbard Brook Experiment Forest in New Hampshire—where the acidity of rain has been measured directly and consistently for more than 10 years. The longest-term national monitoring program was operated by the National Center for Atmospheric Research for 6 years from 1960 to 1966.[27]

Using fragmentary bits of information, obtained indirectly and in limited areas and periods of time, Cogbill and Likens[5] managed to calculate the probable changes in average acidity of rainfall in various parts of the eastern United States from 1955–1973. As shown in FIGURE 1,[26] precipitation in a large portion of the eastern United States was less than pH 5.6 in 1955–56; the region of greatest acidity (lowest pH) was generally consistent with the region where sulfur emissions were high—parts of Ohio, Pennsylvania, West Virginia, New York, and New England. By 1972–1973, however, the area with an average rain pH of below 4.5 had extended to include parts of Missouri, Arkansas, Mississippi, Alabama, Georgia, South Carolina, North Carolina, Virginia, Kentucky, Illinois, Michigan, and farther north into New England and Canada. Essentially, it embraced most of the area east of the Mississippi River. Individual rainstorms with pH values below 3.6 have been reported in New York, Illinois, Indiana, New Hampshire, Massachusetts, and North Carolina—in some cases, many hundreds of kilometers from major sources of air pollution.[26]

The relative contribution of sulfate and nitrate to the total acidity of precipitation apparently has changed markedly during the years since 1964–65. At Hubbard Brook, New Hampshire, the ratio of sulfate to nitrate ions changed from 5.5:1 in 1964 to 2.2:1

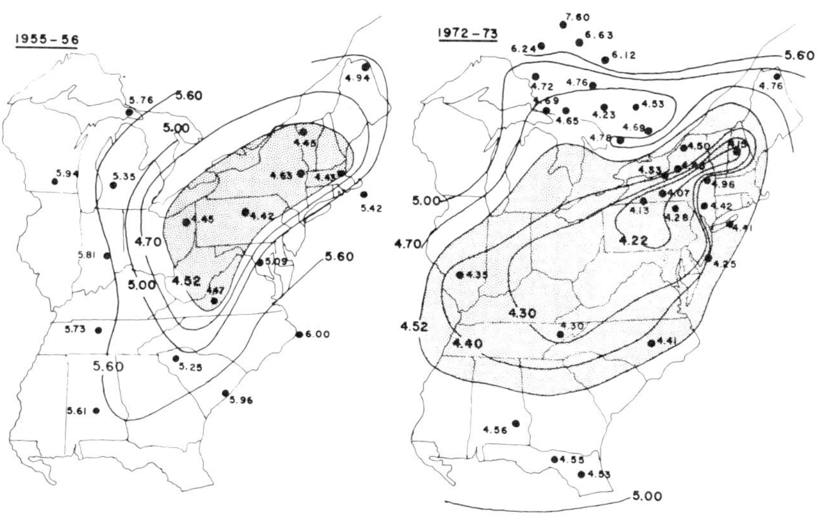

FIGURE 1. Changes in the weighted annual average pH of precipitation in the eastern United States in 1955–56 and 1972–73. Adapted from Likens[26] and used by permission of *Chemical and Engineering News*.

in 1974. During this same decade, the total input of hydrogen ions increased by 36%. Thus, most of this increase appears to be due to increased amounts of nitric acid.[26]

EFFECTS OF ACID PRECIPITATION ON ECOSYSTEMS

Research on the effects of change in the chemistry of precipitation on terrestrial and aquatic ecosystems has been conducted in Europe since about 1955 and in North America mainly since about 1966. Several publications are worthy of special note in this connection. The pioneering and outstanding research by Eville Gorham—first in England, later in Canada, and most recently in the United States—dealt with effects on lake waters, aquatic vegetation, terrestrial vegetation and human health.[15,16] Svante Odén published his famous Ecological Committee Report in 1968.[33] In 1971, Bolin and coworkers completed a case study about sulfur deposition and acid precipitation in Sweden, as a contribution to the United Nations Conference on the Human Environment.[3] In 1972, the Norwegian government established a special multiinstitute research project called Acid Precipitation: Effects on Forests and Fish, with an annual budget of about 10 million kroner (about $2 million).[43] This huge project was initiated with two specific objectives:

1. To establish as precisely as possible the effects of acid precipitation on forests and freshwater fish; and
2. To investigate the effects of air pollutants on soil, vegetation, and water, to the extent required to support the primary objective.

The First International Symposium on Acid Precipitation and the Forest Ecosystem was held at The Ohio State University at Columbus in May of 1975.[8] In June of 1976, the International Conference on Effects of Acid Precipitation was held at Telemark, Norway, and the major papers assembled for this meeting were published by Braekke *et al.*[4] and in a special issue of *Ambio.*[2] In November 1976, Likens published his summary report in *Chemical and Engineering News.*[26] In 1977, the Organization for Economic Cooperation and Development[37] published their comprehensive findings regarding the international exchange of sulfur compounds among the various nations of Europe. In May 1978, a NATO Advanced Research Institute on Ecological Effects of Acid Precipitation was held at Toronto, Canada. In September 1978, the Central Electricity Generating Board in England hosted an international symposium on the biological effects of acid precipitation. The proceedings of these last two meetings are scheduled for publication in 1979 and 1980, respectively.

Effects on Aquatic Ecosystems

The effects of acid precipitation on fish populations and aquatic ecosystems have been very pronounced in various regions of northern Europe and eastern North America.[7] The most important of these effects were summarized at the First International Symposium on Acid Precipitation and the Forest Ecosystem[8,9] and at the International Congress on the Effects of Acid Precipitation held at Telemark, Norway, in June 1976.[2] The latter report reads in part:

> Water quality has changed during the last decades in numerous lakes and rivers in southern Scandinavia and eastern North America, pH often falling below 5, with sulfate

becoming the most important anion. There is strong evidence that this change is due to acid precipitation. It is associated with the loss of buffering capacity, and the occurrence of additional short-term decreases in pH, related to meltwater from snow or episodic inputs of acid precipitation from polluted air masses.

The acidification of freshwater ecosystems leads to many changes, most of which involve decreases in biological activity and important changes in nutrient cycling. For example, decomposer organisms are less active in acid waters, resulting in increased accumulations of organic matter. When the pH drops below 6, numbers of species in several groups of organisms (phytoplankton and zooplankton, bottom fauna and several other groups of invertebrates), decrease considerably, thus affecting the variety of food for fish and other animals depending on freshwater ecosystems. Shifts have been observed from higher aquatic plants toward mosses, and this will influence not only the bottom fauna but also nutrient exchange with the sediments.

High acidity (pH <5.5) seriously affects fish populations, particularly when occurring in waters of low ionic strength. Rapid extinction rates of fish populations inhabiting acidified waters have been observed during the past few decades in southern Scandinavia as well as in parts of eastern North America.

Case studies of several fish populations and experiments clearly indicate that the elimination of fish is often a result of chronic reproductive failure in acid conditions and of damage done to sensitive stages, especially the newly hatched larvae. Such a process is insidious and not readily evident in terms of fishery yield until extinction is imminent.

In lakes and streams with soft waters, acid stress has also been shown both experimentally and in field studies to cause mortality among adult fish as a result of interference with physiological mechanisms regulating active ion exchange across gill membranes. In this case factors such as size, age, acclimation history, genetic background and ionic strength of the water interact in complex ways to determine the relative acid tolerance of the fish.

There is strong evidence that the increased acidity of precipitation is now the main cause of these extensive losses of salmonid fish stocks as well as other populations of economic importance both in southern Scandinavia and the northeastern part of the United States and parts of southeastern Canada.

Sensitivity is related to the tolerance of the differing species (reflecting their genetic diversity), to the timing of episodic acid precipitation in relation to the stage in the life cycle, and to the influence of the geological environment, e.g., lakes and rivers on bedrock, overburden and soils highly resistant to chemical weathering. This may also be the case in geological areas of this type elsewhere in the world. Special consideration should be given to the preservation of unique gene pools and habitats.

There may be complex interactions with other environmental factors, including some organic compounds. In waters of low pH the adverse effects of heavy metals on fish and other organisms can be enhanced, whereas at pH values nearer neutrality these substances would have been tolerated.[2]

Effects on Terrestrial Ecosystems

Effects on terrestrial ecosystems generally have been less well documented than those on populations of freshwater fish and other aquatic organisms. Nevertheless, certain definite effects have been reported. The most striking of these effects was development of peat moss (*Sphagnum* sp.) as a submarine, rather than a terrestrial, plant in acidified lakes and streams in Sweden.[17] Dense mats of *Sphagnum* and an apparently parasitic aquatic fungus develop on the bottom of these lakes in water as deep as 18 meters. This growth induces oligotrophication (opposite of eutrophication)—a self-accelerating process that leads to a substantial nutrient impoverishment of lake waters.

Analyses of forest growth in southern Sweden from 1896 to 1965 showed a 2 to 7% decrease in growth between 1950 and 1965. Johnsson and Sundberg[23,45] "found no good reason for attributing [this] reduction in growth to any cause other than acidification." Similar attempts to quantify possible effects on growth of forests in the United States have been inconclusive.

Other reports given at the meetings in Columbus, Ohio, and Telemark, Norway, indicated that:

1. Forest canopies filter out sulfur and hydrogen ions from precipitation;[24,29]
2. Acidity of bark and development of cuticular features, such as frequency and size of stomata, can be used as biological indicators of sulfur pollution and acidification;[41] for example, pH values of 2.2 to 4.7 were reported in the bark of several species of trees in Poland;[8]
3. Vegetation developing in recently formed sandy soils or glacial outwash areas is more vulnerable to the effects of acid precipitation than is vegetation growing in older, well-buffered soils of high clay content and consequently large base exchange capacity;[12,47]
4. Before reaching the soil, acidic substances in precipitation can induce changes in the physiology of foliar organs;[46]
5. After reaching the soil, these substances also can induce changes in root function and in the availability of essential cations;[12]
6. Ammonia and other cations combine with sulfate ions in the atmosphere and tend to neutralize and thus hide atmospheric acidity—when the cations are absorbed by plants, however, additional hydrogen ions are released by the plants and contribute to the further acidification of agricultural and forest ecosystems;[34]
7. Acid rains increase the leaching of nutrient cations from the root zone of plants;[32,38]
8. Acid precipitation also increases the solubility (and therefore the toxicity) of metals, such as Al, Pb, and Hg, in soil.[30]

Simulated "rain" acidified with sulfuric acid has been reported to:

1. Induce direct injury to foliage of pines, birch, and mosses;[42,47,51]
2. Induce symptoms similar to those caused by certain other pollutants and biotic pathogens;[50]
3. Induce poorer germination of spruce seeds;[1]
4. Accelerate leaching of nutrients from foliage[1,51] and from soil;[1,28,48]
5. Inhibit decomposition of leaf litter from the forest floor;[1]
6. Increase erosion of epicuticular waxes on oak and bean leaves;[42]
7. Decrease uptake of nitrogen by endomycorrhizae of sweetgum seedlings;[19]
8. Inhibit reproduction of root-knot nematodes;[42]
9. Inhibit development of bean rust and production of telia by the oak-leaf-rust fungus, *Cronartium fusiforme*;[42]
10. Inhibit or stimulate development of halo blight in bean seedlings, depending on the time in the disease cycle during which the simulated "rain" was applied;[42] and
11. Inhibit nodulation and fixation of nitrogen by *Rhizobium* in bean and soybean seedlings.[42]

As yet unknown constituents of precipitation have been shown to induce "russetting"—brown blemishes that decrease the market quality and price of yellow delicious apples.[52]

With the exception of this latter effect on apples, the possible economic consequences of the various biological effects listed above are not known. Nevertheless, the variety of biological influences that have been observed suggests that (1) much more research is needed to evaluate the ecological and possible economic influences of changes in the acidity and other chemical properties of precipitation; and (2) a well-coordinated network to measure long-term changes in the chemistry of air and precipitation is needed in rural and urban areas throughout North America.

Based on a general analysis of effects reported to date, Heck *et al.*[20] offered the following generalizations about the relative vulnerability of various organisms and ecosystems to acid precipitation:

1. Freshwater fish are more vulnerable to acute acid shock than is freshwater vegetation;
2. Aquatic vegetation is more vulnerable than terrestrial vegetation;
3. Herbaceous vegetation is more vulnerable than woody vegetation;
4. Plants growing in recently formed sandy soils or glacial outwash areas are more vulnerable than plants growing in well-buffered soils of high clay content and consequently large base exchange capacity;
5. Reproductive stages are more vulnerable than vegetative stages;
6. Plants growing in natural ecosystems are more vulnerable than plants growing in agroecosystems where liming and other fertilization practices are common;
7. Coniferous forests are more vulnerable than hardwood forests; and
8. Plants growing in regions of high rainfall are more vulnerable than desert plants.

Development of a National Atmospheric Deposition Program for the United States

In 1975, the National Academy of Sciences Committee on Atmospheric Sciences published its report on *Atmospheric Chemistry: Problems and Scope.*[31] In May of 1975, a workshop to define future needs for research on acid precipitation was held immediately following the First International Symposium on Acid Precipitation and the Forest Ecosystem. The recommendations of this workshop were published by Dochinger and Seliga.[9] In May 1977, the director of the Norwegian Special Project on Acid Precipitation provided an updated analysis of research needs on acid precipitation as perceived from the standpoint of knowledge of acid precipitation effects in Scandinavia.[43] In September 1978, the Federal Interagency Advisory Committee on Water Data completed its report on *Research and Monitoring of Chemistry in the United States: Present Status and Future Needs.*[22] In September 1977, the congressionally mandated "Committee of Nine," responsible for regional research in the agricultural experiment stations of the United States, approved plans for establishment of a long-term interagency research project designated NC-141, Chemical Changes in Atmospheric Deposition and Effects on Agricultural and Forested Land Surface Waters in the United States.[6] In March of 1978, the Fish and Wildlife Service published their extensive analysis of impacts of coal-fired power plants (including acid precipitation) on fish, wildlife, and their habitats.[11] In November 1978, the name of the NC-141 program was changed to the National Atmospheric Deposition Program (NADP); and in December 1978, a subcommittee of NADP completed a report to the President's Council on Environmental Quality entitled *A National Program to Assess the Problem of Atmospheric Deposition (Acid Rain).*[13]

In each of these major studies, the need for action to increase research on the changing chemistry of precipitation and its ecological effects was described. A remarkable degree of similarity in research recommendations is evident in all of these reports: (1) establish a permanent network to continously monitor changes in the chemistry of precipitation, and (2) coordinate this long-term monitoring program with research to determine the sources, transport, and transformation of pollutants, the meteorology of their dispersal and deposition, and the nature and magnitude of their ecological effects on agricultural land, forested land, rangeland, urban lands, and aquatic ecosystems.

An adequate program of monitoring and research ideally should include the various state agricultural experiment stations, the regional forest experiment stations of the Forest Service, and the Agricultural Research Division of the Department of Agriculture, as well as pertinent units of the United States Geological Survey, the Department of Energy, the National Oceanic and Atmospheric Administration, the Environmental Protection Agency, the Tennessee Valley Authority, the National Center for Atmospheric Research, the National Science Foundation, and the National Aeronautics and Space Administration. An effective program will also require the interest and expertise of many scientists in a diverse array of publicly and privately supported colleges and universities in the United States and Canada.

The design for such an integrated, multinational, multidisciplinary, and multiagency program of research and monitoring has been developed[13] and endorsed by President Carter in his Second Environmental Message.[53] This Presidential Initiative on Acid Rain calls for a research investment of $10 million per year for 10 years and will include each of the following elements:

1. Quantitative analysis of emissions, chemical transformations, and transport and deposition processes;
2. Measurement of spatial and temporal trends in atmospheric deposition in various regions of North America;
3. Analysis of the effects of atmospheric deposition on water quality, vegetation, soils, fish, microorganisms, and simple synthetic ecosystems;
4. Calibrated watershed studies to provide an understanding of the chemical linkages between the atmosphere and both terrestrial and aquatic natural ecosystems; and
5. Economic assessment and criteria development.

As Galloway and his colleagues[13] have emphasized: "The time has come to stop talking and to implement a comprehensive and well-coordinated program of research and monitoring. The United States already lags well behind other nations (including Canada, Norway, and Sweden) despite the severe threat posed by acid precipitation to large areas of this country."

Let's get on with the job!

References

1. Abrahamsen, G., K. Bjor, R. Horntvedt & B. Tveite. 1976. Effects of acid precipitation on coniferous forest. *In* Research Report FR-6, SNSF Project. F. H. Braekke, Ed.:37–63. Norwegian Institute for Forestry Research (NISK). Ås, Norway.
2. Ambio. 1976. Report of the International Conference on the Effects of Acid Precipitation in Telemark, Norway. Ambio **5:** 200.
3. Bolin, B., L. Granat, L. Ingelstam, M. Johannesson, E. Mattsson, S. Odén, H. Rodhe & C. O. Tamm. 1971. Air Pollution across National Boundaries. The Impact of Sulfur in Air and Precipitation. Sweden's Case Study for the United Nations Conference on the Human Environment. P. A. Norstedt and Sons. Stockholm, Sweden.
4. Braekke, F. H., Ed. 1976. Impact of Acid Precipitation on Forest and Freshwater Ecosystems in Norway. Research Report #FR-6, SNSF Project. Norwegian Institute for Forestry Research. Ås, Norway.
5. Cogbill, C. V. & G. E. Likens. 1974. Acid precipitation in northeastern United States. Water Resour. Res. **10:** 1133.
6. Cooperative State Research Service. 1976. Regional Research Project on Atmospheric Deposition: Chemical Changes in Atmospheric Deposition and Effects on Agricultural and Forested Land and Surface Waters in the United States. Cooperative State Research Service, U.S. Department of Agriculture. Washington, D.C.
7. Dickson, W. 1978. Some effects of the acidification of Swedish lakes. Verh. Int. Ver. Limnol. **20:** 851.
8. Dochinger, L. S. & T. A. Seliga, Eds. 1976. Proceedings, First International Symposium on Acid Precipitation and the Forest Ecosystem. USDA Forest Service General Technical Report NE-23. Northeastern Forest Experimental Station. Upper Derby, Penn.
9. Dochinger, L. S. & T. A. Seliga, Eds. 1976. Workshop Report on Acid Precipitation and the Forest Ecosystem. USDA Forest Service General Technical Report NE-26. Northeastern Forest Experimental Station. Upper Derby, Penn.
10. Feth, J. H., S. M. Rogers & C. E. Roberson. 1964. Chemical Composition of Snow in the Northern Sierra Nevada and Other Areas. Water Supply Paper 1535-J. U.S. Geological Survey. Washington, D.C.
11. U.S. Fish and Wildlife Service. 1978. Impacts of Coal-Fired Power Plants on Fish, Wildlife, and Their Habitats. Report FWS/OBS-78/29. National Power Plant Team, U.S. Department of the Interior. Ann Arbor, Michigan.
12. Frink, C. R. & G. K. Voigt. 1976. Potential effects of acid precipitation on soils in the humid temperate zone. *In* Proceedings, First International Symposium on Acid Precipitation and the Forest Ecosystem: 685–710. (See Reference 8.)
13. Galloway, J. N., E. B. Cowling, E. Gorham & W. W. McFee. 1978. A National Program for Assessing the Problem of Atmospheric Deposition (Acid Rain). Unnumbered Report. Council on Environmental Quality. Washington, D.C.
14. Gambell, A. W. & D. W. Fisher. 1966. Chemical Composition of Rainfall, Eastern North Carolina and Southeastern Virginia. Water Supply Paper 1535-K. U.S. Geological Survey. Washington, D.C.
15. Gorham, E. 1958. The influence and importance of daily weather conditions in the supply of chloride, sulphate and other ions to fresh waters from atmospheric precipitation. Philos. Trans. Roy. Soc. London Ser. B **241:** 147.
16. Gorham, E. 1976. Acid precipitation and its influence upon aquatic ecosystems—an overview. *In* Proceedings, First International Symposium on Acid Precipitation and the Forest Ecosystem: 425–458. (See Reference 8.)
17. Grahn, O., H. Hultberg & L. Landner. 1974. Oligotrophication—a self accelerating process in lakes subjected to excessive supply of acid substances. Ambio **3:** 93.
18. Grodzinska, K. 1976. Acidity of tree bark as an indicator of forest pollution in southern Poland. *In* Proceedings, First International Symposium on Acid Precipitation and the Forest Ecosystem: 905–912. (See Reference 8.)
19. Haines, B. & R. G. Best. 1976. The influence of an endomycorrhizal symbiosis of nitrogen movement through soil columns under regimes of artificial throughfall and artificial acid rain. *In* Proceedings, First International Symposium on Acid Precipitation and the Forest Ecosystem: 951–962. (See Reference 8.)

20. HECK, W. W., A. S. HEAGLE & E. B. COWLING. 1977. Air pollution: impact on plants. Proc. Soil Conservation Soc. Am.: 193–202.
21. HUSAR, R. B., J. P. LODGE, JR. & D. J. MOORE. 1978. Sulfur in the Atmosphere. Pergamon Press, Inc. New York, N.Y.
22. Interagency Advisory Committee on Water Data. 1978. Research and Monitoring of Precipitation Chemistry in the United States—Present Status and Future Needs. Office of Water Data Coordination, U.S. Geological Survey. Reston, Va.
23. JOHNSSON, B. & R. SUNDBERG. 1972. Has the Acidification by Atmospheric Pollution Caused a Growth Reduction in Swedish Forests? Research Note No. 20. Department of Forest Yield Research, Royal College of Forestry. Stockholm, Sweden.
24. KNABE, W. & K. H. GÜNTHER. 1976. Investigation on effects of the forest canopy on acid and sulfur precipitation in the Ruhr district, Germany. *In* Proceedings, First International Symposium on Acid Precipitation and the Forest Ecosystem: 895–911. (See Reference 8.)
25. KOVDA, V. A. 1975. Biogeochemical Cycles in Nature, Their Disturbance and Study. Kuaka Publishing House. Moscow, USSR. (In Russian.)
26. LIKENS, G. E. 1976. Acid rain. Chem. Eng. News **54**(48): 29.
27. LODGE, J. P., J. B. PATE, W. BASBERGILL, G. S. SWANSON, K. C. HILL, E. LORANGE & A. L. LAZRUS. 1968. Chemistry of United States Precipitation. Final report on the National Precipitation Sampling Network. National Center for Atmospheric Research. Boulder, Colo.
28. MALMER, N. 1976. Acid precipitation: chemical changes in the soil. Ambio **5:** 231.
29. MAYER, R. & B. ULRICH. 1976. Acidity of precipitation as influenced by the filtering of atmospheric sulfur and nitrogen compounds—Its role in the element balance and effect on soil. *In* Proceedings, First International Symposium on Acid Precipitation and the Forest Ecosystem: 737–744. (See Reference 8.)
30. MCFEE, W. W., J. M. KELLY & R. H. BECK. 1976. Acid precipitation effects on soil pH and base saturation of exchange sites. *In* Proceedings, First International Symposium on Acid Precipitation and the Forest Ecosystem: 725–736. (See Reference 8.)
31. National Academy of Sciences. 1975. Atmospheric Chemistry: Problems and Scope. National Academy of Sciences. Washington, D.C.
32. NORTON, S. A. 1976. Changes in chemical processes in soils caused by acid precipitation. *In* Proceedings, First International Symposium on Acid Precipitation and the Forest Ecosystem: 711–724. (See Reference 8.)
33. ODÉN, S. 1968. The Acidification of Air and Precipitation and Its Consequences in the Natural Environment. Ecology Committee Bulletin No. 1. The State National Science Research Council. Stockholm, Sweden. (Available through Translation Consultants, Ltd. Arlington, Va.)
34. ODÉN, S. 1976. The acidity problem—an outline of concepts. *In* Proceedings, First International Symposium on Acid Precipitation and the Forest Ecosystem: 1–36. (See Reference 8.)
35. ODÉN, S. & T. AHL. 1970. Försurningen au skandinaviska vatten. Ymer, Arsbok 1970. Stockholm, Sweden.
36. ODÉN, S. & J. BERGHOLM. 1971. The Chemistry of Snow. I. Ionic Separation during Melting. Unnumbered report. Department of Soil Science, Agricultural College. Uppsala, Sweden.
37. Organization for Economic Cooperation and Development. 1977. The OECD Programme on Long Range Transport of Air Pollutants: Measurements and Findings. Unnumbered Report. OECD. Paris, France.
38. OVERREIN, L. N. 1972. Sulphur pollution patterns observed; leaching of calcium in forest soil determined. Ambio **1:** 145.
39. RENNIE, P. J. 1955. The uptake of nutrients by mature forest growth. Plant Soil **7:** 49.
40. REUSS, J. O. 1975. Chemical/Biological Relationships Relevant to Ecological Effects of Acid Rainfall. Report No. EPA-660/3-75-032. U.S. Environmental Protection Agency. Corvallis, Oreg.
41. SHARMA, G. K. 1976. Cuticular features as indicators of environmental pollution. *In* Proceedings, First International Symposium on Acid Precipitation and the Forest Ecosystem: 927–932. (See Reference 8.)

42. SHRINER, D. S. 1976. Effects of simulated rain acidified with sulfuric acid on host-parasite interactions. *In* Proceedings, First International Symposium on Acid Precipitation and the Forest Ecosystem: 919–925. (See Reference 8.)
43. SNSF Project. 1977. SNSF Project—Condensed Program 1977–79. Norwegian Institute for Forestry Research. Ås, Norway.
44. TAMM, C. O. 1958. The atmosphere. Handb. Pflanzenphysiologie **6:** 233.
45. TAMM, C. O. 1976. Acid precipitation: biological effects in soil and on forest vegetation. Ambio **5:** 235.
46. TAMM, C. O. & E. B. COWLING. 1976. Acidic precipitation and forest vegetation. *In* Proceedings, First International Symposium on Acid Precipitation and the Forest Ecosystem: 845–885. (See Reference 8.)
47. TAMM, C. O., G. WIKLANDER & B. POPOVIC. 1976. Effects of applications of sulfuric acid to poor pine forests. *In* Proceedings, First International Symposium on Acid Precipitation and the Forest Ecosystem: 1011–1024. (See Reference 8.)
48. WIKLANDER, L. 1973/74. The acidification of soil by acid precipitation. Grundförbättring **26:** 155.
49. WITTWER, S. H. & M. J. BUKOVAC. 1969. The uptake of nutrients through leaf surfaces. *In* Handbuch der Pflanzenernährung und Dungung. K. Scharrer & H. Linser, Eds.: 235–261. Springer-Verlag. New York, N.Y.
50. WOOD, F. A. & S. P. PENNYPACKER. 1976. Etiological study of the short needle disease of pine. *In* Proceedings, First International Symposium on Acid Precipitation and the Forest Ecosystem: 977–978. (See Reference 8.)
51. WOOD, T. & F. H. BORMANN. 1974. The effects of an artificial acid mist upon the growth of *Betula alleghaniensis* Britt. Environ. Pollut. **7:** 259.
52. KENDER, W. J. 1977. Personal communication.
53. Reference to this Presidential Initiative was added in proof.

LONG RANGE TRANSPORT: GENERAL DISCUSSION*

Kenneth A. Rahn, *Moderator*

Graduate School of Oceanography
University of Rhode Island
Kingston, Rhode Island 02881

R. JAENICKE (*Max-Planck-Institut für Chemie, Mainz, Federal Republic of Germany*): Dr. Rahn, you showed that you obviously have some points of origin for your long range transport, and I wonder if you ever checked the following possibility. The vanadium that you have calculated as excess vanadium—that means that you compared it to the aluminum and that the excess vanadium is calculated from the average crossed ratio of the vanadium to aluminum. Is that correct?

K. A. RAHN: That's right.

R. JAENICKE: So if you now have different starting points for your transport, is it also possible that the soil in these areas is so different—I mean the vanadium to aluminum ratio is so different—that it could explain the excess vanadium?

K. A. RAHN: I suppose, in principle, it's possible; but we don't really think that's happening, for a number of reasons. If you look, for example, at average soil as opposed to average crust, you find a vanadium to aluminum ratio very similar to the one we were using here. That's one point. We've seen some data, for example, from U.S. soils, that confirm that average quite nicely. Dr. Schütz, who will be speaking this afternoon, conducted, while at our laboratory, a study of vanadium, aluminum, and a large number of other elements in various, individual desert soils, and the values are very close to the mean processor ratio that we have been using independent of particle size.

I feel as confident of the vanadium to aluminum ratio in the crust as I do about any pair of elements in the crust studies seen so far. Anyhow, we are a factor of ten to twenty higher than that in Barrow during the winter, and I don't see any possible explanation for this on the basis of soil alone.

R. B. HUSAR (*Washington University, St. Louis, Mo.*): Dr. Rahn, it is my recollection that the second pause, that is, the time interval between the pulses of the Chinese bombs, is about 10 days to two weeks. Monitoring of the radioactivity at the surface indicates that this is about the time required for the air to go around. So it would seem to me then that 20 days is a long time, longer than would actually be the requirement. Thus, you wouldn't really need to use such a low—evidently a somewhat unrealistically low—rate constant for the removal, as you have been doing. Meaning that other European or North American areas could very well make an impact like that.

K. A. RAHN: We have wondered about the "20" figure, too. Although we can't prove it, I have a feeling that most of the transport of aerosol to the Arctic is at rather low levels and that transport speeds may be slower than transport of bomb debris, which ought to climb considerably higher in the troposphere. At least the evidence for

*This is a compilation of the discussions held after the presentation of each of the papers in this section.

0077–8923/80/0338–0556 $01.75/1 © 1980, NYAS

European transport as far north as northern Norway shows that this material is found within 1½ to 2 kilometers of the surface most of the time. Perhaps Dr. Ottar can comment on this in the course of his talk.

I don't think that our rate constants are all that remarkably low. For example, a residence time of five days in midlatitudes is not unusual. We've also done a number of test calculations for the Arctic—independent calculations suggesting that the residence time ought to be on the order of one to two months up there; that is an order of magnitude longer than in the midlatitudes.

If you like, we could discuss this afterwards, but we've felt fairly confident about this. I admit that there is a great deal of room for play and I'm presenting you with only the first results of calculations that we've done recently. But the main thing is that, quantitatively, we seem to have a compatible source versus receptor area here.

R. JAENICKE: Dr. Ottar, you showed us a map with the large distribution from Europe. There are several major cities within Scandinavia—Stockholm, Gothenburg, Copenhagen, Oslo—that didn't show up on the graph. Is that due to the large group you have? Can you comment about that.

B. OTTAR (*Norwegian Institute for Air Research, Lillestrøm, Norway*): Well, when we look at just the sulfate concentrations or the sulfur dioxide concentration, it is definitely due to the large system used, because in the calculation, you have to assume that everything is completely mixed within this large grid. And most of this analysis was done for areas far away from the sources.

Of course, I could add that in Europe, there was a tremendous interest in how much is coming from one country to another, and discussing this is really difficult. For instance, you have Greenland, which is a part of Denmark. Well, should we divide our emissions between Denmark and Greenland, or not? Similar things apply to all these countries; there is a major deposition in some areas, while other areas do not receive much.

R. B. HUSAR: In passing, Dr. Ottar, you mentioned that the increase in acidity recently has been attributed to the consumption of sulfur-containing fuel. Of course, there is debate in this country—and it has already started in Europe—about the relative role of nitrogen oxides, nitric acid specifically. I wondered if you could reconfirm now whether the feeling is that sulfur oxides are the key contributors to the acidity or whether there is still a debate about it.

B. OTTAR: I can say that we are analyzing the last of the precipitation samples for nitrate and sulfate. Of course, the nitrates and sulfates are mixed; you can't say which was the acid one. But if you assume that this is due to nitric acid and sulfuric acid, then the nitric acid contributes from 20 to 50% of the total acidity.

R. J. MCCAFFREY (*University of Rhode Island, Kingston, R.I.*): Looking at the maps of the distribution of dust, Dr. Schütz, a question occurred to me. How good are the deposition maps when you consider that there's considerable ocean transport of the dust after it's deposited on the surface of the ocean and before it gets to the bottom? It seems to me that this could be a considerably delayed transport and that it might severely modify the deposition patterns. Therefore, the patterns may not imply simply the deposition by aerosol directly, but deposition plus transport while suspended in the ocean.

L. SCHÜTZ (*Max-Planck-Institut für Chemie, Mainz, Federal Republic of*

Germany): I think you are definitely right. There is an additional transport within the ocean by the ocean currents, and there is a very good case of this on one of the slides. A plume off the coast of western Africa is a little bit more extended to the south. This is due to the fact that the deep-sea current is extending in a southerly direction. So this also affects this distribution pattern.

D. GILLETTE (*National Center for Atmospheric Research, Boulder, Colo.*): Dr. Schütz, I'd like to expand a bit on what I think is possibly a reason for the low estimates of the annual dust production based upon sea sediments. If we do as Goldberg does, that is, base our global-scale sedimentation rates on the sediments in the middle of the ocean, we have to connect these with the production of dust, and we have to have a mechanism for long-scale transport.

Now, I think that in desert situations, in cases of very large negative Richardson numbers, we've obviously seen cases where we have generation of large quantities of dust lacking the potential for long-scale transport. The dust is then produced and kept more or less in the area without having the benefit of traveling to the oceans.

Smaller Richardson numbers would require higher winds—which happen in over smaller times here—rather than large heating with smaller gradient winds. We would then expect that we have a very large production of dust—much larger than is indicated by the dust that is carried thousands of miles into the middle of oceans. So I think that your point here is well taken, and I gather that your estimate of a thousand million tonnes is taken from the production in the Saharan wind strips.

L. SCHÜTZ: If you'll excuse me, I'd like to make a mention of something. I have assumed that all these data from sediments, etc., are produced by 20% of the dust based on the Sahara dust model. We have seen that only about 20% is undergoing long range transport. I assume all the material falling into the ocean and producing sediments is only 20% of the total production, though I had scaled up everything by a factor of five.

D. GILLETTE: I see. Another point I'd like to make is that we have made some estimates of production in large-scale dust storms in the southwestern United States. In these storms, we've flown an airplane within the dust cloud and have made estimates of the volume of the dust storm using radiosonde sensor and plotting out potential temperatures, etc. The volume and concentrations then gave us certain totals of dust produced in the southwestern United States that had the potential for long range transport. This is in the range of about five million tons per year, which is a very, very small number compared to the Sahara. However, I'd have to say that there is a huge amount of dust produced in that area that is not transported those great distances.

R. JAENICKE: Dr. Rahn,† you showed a very nice picture on the turbidity, comparing New York City—or say the continental U.S.—with values from Alaska. The values came very close together in wintertime, but the turbidity spread out very much in summertime.

In one of your talks on another occasion, you indicated the idea that the polluted episodes in Alaska might be caused by the fact that in wintertime, some parts of the continental U.S. might be included in the polar air mass. I would consider this picture

†Dr. Rahn presented the paper by G. E. Shaw and K. Stamnes, as the authors could not be present at the conference.

as an indication for your theory. Are you still following up on this idea? Because I haven't heard it mentioned.

K. A. RAHN: Yes, we still do believe in it. What Dr. Jaenicke is referring to is our general concept that in the winter, transport is faster between the midlatitudes and the Arctic, because midlatitude sources and the Arctic essentially are in one and the same air mass, namely, some degree of modified polar air.

In the summer, however, midlatitude sources basically are in a tropical air mass separated from the receptor Arctic by a jet stream/polar front combination, and transport between the two is considerably slower. There is no evidence to the contrary, as far as I can see. So yes, we definitely still believe that.

T. V. LARSON (*University of Washington, Seattle, Wash.*): Dr. Rahn, do you have any estimates of how much of the total extinction is due to the absorption in these aerosols, based on those measurements you made?

K. A. RAHN: I don't think so quite yet. The only thing I can say is that the absorption by the aerosol seems to be higher than was expected. There is a lot of carbon in that aerosol, but we have no quantitative measurements on the carbon yet; this is something we propose to do within the next year or two.

S. O. WILSON (*New York State Energy Research Development Authority, Albany, N.Y.*): Dr. Rahn, have you had a look at satellite imagery and the visibility as mentioned by Dr. Lyons?

K. A. RAHN: No, but this is something that is on our list of things to do in the future. I think it could be quite useful when done properly.

S. O. WILSON: Having spent some time in the Arctic, it strikes me that one reason why the haze may be less perceptible from the ground is that there is, in fact, a tremendous amount of reradiation. That is, when the ground is covered with snow or ice, a single particle would appear to be illuminated from below as well as from above. That's a very simplistic idea of the matter. However, I asked that first question with respect to the idea that open water would seem to permit your seeing haze a little better than if you were standing on a white surface at ground level.

R. JAENICKE: Dr. Cowling, you gave us an idea about the development of acidity during the last ten—or, for some cases, even more—years. What would be your outlook for the next ten years?

E. B. COWLING (*North Carolina State University, Raleigh, N.C.*): There undoubtedly will be factors that will moderate the extent and rate of change. We can say that sulfur concentrations have leveled off pretty much, pehaps in part due to control technology. But we have nitrogen oxides continuing to increase in abundance, and my expectation is that, with the increasing dependence upon coal in this country at least, we will further contaminate the atmosphere. Now, whether or not that will include enough alkaline material to allow for a change in the trend, I don't know. But that is why we need a network; we need to know what we're doing in this country and around the world.

I hope that the major lesson you would take home from this is that if the United States and western Europe are going to so significantly augment the supply of substances in the atmosphere, it is prudent that we invest a small sum in finding out what is the chemical nature of the substances that fall into ecosystems and what are the biological effects.

Our view is that this network will provide a rational data base that the protagonists on all sides of these environmental issues can use. I really hope that, collectively, we can adjust the aspirations of mankind to allow for a decent life within the capacity of this planet.

G. M. KLAUBER (*Environmental Measurements, Inc., Arnold, Md.*): Dr. Cowling, yesterday Dr. Gillani was talking about the correlations between sulfates and ozone. Perhaps you might want to say a word or two about any effects the two of them together would have on the ecosystem.

E. B. COWLING: They interact in a synergistically detrimental way, and that's true of most of the pollutants. If you combine sulfur oxides with nitrogen oxides, they are more injurious together than the sum of their individual injuriousness. That's also true of acids in relation to ozone. There are also correlations in time between ozone and acidity, as detected in the few data that are accumulated near Cape Kennedy.

D. GILLETTE: Dr. Cowling, some of the models of carbon dioxide increase in the atmosphere have shown oceans, which have sufficient isolation from limestone (calcium carbonate) bases, to have their pH significantly decreased from its present value of 8.4 or so to some acid level. It seems that these models show the primary influence on this to be an increase of the carbon dioxide and, consequently, of the temperature. Would you have any comments about the effects that the decrease of pH in oceans might have on the fishing industry and biology?

E. B. COWLING: Except in places like the Sargasso Sea, where, as I understand it, atmospheric nutrients are a major source of supply for that region of the ocean, it would not be my expectation that there would be much change. I'm sure that an oceanographer could tell us about the buffering capacity of ocean water, but I have no experience with which to respond to your question.

SOME FUTURE PROBLEMS ASSOCIATED WITH ENERGY: INTRODUCTORY REMARKS

Thomas R. Glenn

Interstate Sanitation Commission
New York, New York 10019

In this session, we will move beyond the discussions related to the identification and characterization of the atmospheric aerosols, and examine the problems and possible consequences associated with the continued use of fossil fuels and other interactions of significance to the environment and/or public health.

In light of recent national energy policy decisions, there will be an acceleration towards the reintroduction of coal and towards greater use of diesel fuels as energy sources. As a result of these and other industrial emissions, there is a distinct probability that (1) there will be increases in the transportable and inhalable atmospheric aerosol loading; (2) the documented effects of acid rain, which have been observed in the northeastern U.S. and Europe, will intensify; and (3) climatological effects, such as local or urban weather modification, will become more apparent. We hope that this session can yield information on these issues.

Since 1974, the Interstate Sanitation Commission has assisted its compact states of New York, New Jersey, and Connecticut in the investigation of the photochemical oxidant problem in the the northeastern U.S. The results have shown that, at times, long range transport of ozone from places as far away as Texas, and more often from the Midwest, contributes significantly to the northeastern problem. Subsequent to, and concurrently with, the investigations by the Commission and other organizations, prevailing ozone control strategies came under serious question. These are presently being reevaluated from a number of perspectives by state and federal agencies. In all cases, however, determination of (1) the controls necessary on hydrocarbon and nitrogen oxide emissions for the reduction of ozone in an urban area and (2) the extent to which the same source area is accountable for precursors that yield downwind (transport) contributions is an important issue.

While the ozone studies were being done, the Commission also participated in sulfate and total suspended particulate aerosol investigations. The results have shown that in summer smog episodes, as much as 73% of the sulfate and 35% of the total suspended particulates are transported into the New York metropolitan area.

The general nature of the contributions is illustrated in TABLE 1, which shows the range of ozone, sulfate, and total suspended particulate concentrations found during 1976 transport episodes. It is clear that the severity of an episode is quite variable. From detailed wind trajectory analyses, it has been determined that the levels accumulated are dependent on the path traveled by the high pressure system and on the persistence of the same system associated with an episode. Periodically, the sulfate concentrations observed in the photochemical smog exceeded the anticipated 24-hour health standard of 25 $\mu g/m^3$; and in most, if not all, cases, the transported total suspended particulates were in the inhalable particulate size range.

Under the Clean Air Act, the Environmental Protection Agency is required to

TABLE 1
TYPICAL VALUES OF OZONE, SULFATE, AND TOTAL SUSPENDED PARTICULATES (TSP) OBSERVED DURING TRANSPORT EPISODES

	Ozone (ppb)	Sulfate ($\mu g/m^3$)	TSP ($\mu g/m^3$)
Total	<80 to >200	2.5 to 32	50 to 194
Transport Fraction	<80 to 150	1.9 to 26.7	43 to 141

*Observed during the New York Summer Aerosol Study, 1976.[1]

revise all air quality criteria documents and reconsider existing standards by 1980. As part of this effort, a sulfate standard and an inhalable particulate matter standard are also being considered. Therefore, on the basis of the preceding results, it is inconceivable that the scientific and regulatory issues of local production and subsequent transport of aerosol will be any less difficult to resolve than those of ozone. In addition, the greater use of coal and diesels will probably increase aerosol loadings. Thus, the quantitative significance of increases in fine particles and sulfate precursors must be accurately projected.

The recent actions of the OPEC nations point out another problem that may have environmental consequences. The unpredictability of national and world oil economics may, at some future time, require the use of "dirty" fuels. Although the magnitude of these emissions cannot be estimated, they could yield a significant adverse contribution to the atmospheric aerosol.

The results of the discussions from the preceding sessions, this background information, and the presentations to follow should stimulate many questions. At a minimum, it could be expected that observations can be made on the types of information required for the continued protection of the environment and health and on the development of effective control strategies. The consequences of other effects, such as climate alterations and waste heat generation, may not be as tangible as some of the above. However, the net effect must also be discussed.

REFERENCES

1. KNEIP, T. J. & M. LIPPMANN, Eds. 1979. The New York Summer Aerosol Study, 1976. Annals **322.** New York Academy of Sciences. New York, N.Y.

POTENTIAL EFFECTS OF THE PROJECTED INCREASE IN COAL USE

Arthur M. Squires

Department of Chemical Engineering
Virginia Polytechnic Institute
and State University
Blacksburg, Virginia 24061

In the competition with cheap gas and oil after World War II, coal in the United States held its own only for the generation of electricity and the manufacture of steel and cement. Small-scale uses of coal nearly disappeared—for example, fewer than 1% of our homes are now heated with coal—and systems for distributing and marketing coal to the small user have atrophied.

Few people have wept for coal, since its removal from cities has greatly improved the urban scene. We have heard a good deal in the previous sessions about St. Louis, but I can say from personal experience that this city has nothing now to compare with the black winter days of the 1920s and the 1930s. Pittsburgh, too, is nothing like its smoky past.

Is this just to be a temporary reprieve? Must we return to smoky St. Louis and Pittsburgh? I have been asked to speak on the effects of an expected return to the use of coal for many of our energy needs that are now met by oil and gas.

The recent discoveries of oil in Mexico and the heralded "glut" of natural gas following deregulation may postpone for a few years the "crisis" in the supply of these fuels; but no one who has read Hubbert or Warman or other dispassionate students of oil and gas geology can doubt that the supplies of oil and gas will flag in the working lifetime of the younger attendees at this conference.

Prediction is difficult but fun, and most predictions are soon forgotten, and so I will hazard a few concerning some consequences of an inability of the world's oil producers to meet the demand for their product.

First, let me dispose of a few easy predictions. Limiting my consideration simply to emission of pollutants into the atmosphere from new equipment, I do not believe the electricity industry can be faulted much longer. A number of approaches to the clean generation of electricity from coal are in the works, but I will speak of only one, which is probably the furthest along.

Last summer, Ruhrchemie in West Germany operated equipment for gasifying coal with steam and oxygen to produce a clean fuel gas suitable for firing a gas turbine. Texaco supplied Ruhrchemie with the design of this equipment, which was developed at Texaco's research laboratory at Montebello, California. Texaco's approach to coal gasification is a spin-off from its commercial success, going back 25 years, in the gasifying of heavy residual oils at elevated pressure. Coal is mixed with water, and the mixture is pumped and handled almost as if it were oil. Gasification is accomplished by a partial combustion with oxygen in a flame at more than 1,500°C contained in a refractory-lined pressure vessel. Control of coal gasification uses intruments and an approach already thoroughly proven in operations on oil, and tests

0077–8923/80/0338–0563 $01.75/1 © 1980, NYAS

have shown that the Texaco coal gasifier has excellent load-following capabilities. Most impressive was the fact that the Ruhrchemie test operated smoothly just seven months after it was authorized—a sharp contrast with some recent experiences by others.

Encouraged by the Ruhrchemie experience, the Electric Power Research Institute, Southern California Edison, and Texaco are now designing a commercial-scale test facility for construction at Southern California's Cool Water station near Barstow. Ruhrchemie operated at a scale of 150 tons of coal per day; Cool Water will be designed for 1,000 tons/day and will supply fuel gas at high pressure for generation of 100 megawatts of electricity by a gas turbine followed by a waste-heat steam turbine, i.e., a "combined cycle."

The attraction of this approach is that cleaning out both gaseous and particulate matter from a fuel gas at high pressure is relatively simple and inexpensive. Most important is the fact that increments in efficiency of removal can be bought by reasonable additions of equipment. The high-pressure gasification approach can probably meet almost any reasonable atmospheric emission standards that might be set, even for trace elements. It is easy to conceive of gasifiers that ought to be better than the Texaco design, and a success at Cool Water will surely produce a rapid development of coal gasification technology. In the end, this development will give us power-generating equipment affording sharply higher efficiencies: with gas-turbine designs now under vigorous development, combined-cycle electricity-generating efficiencies of beyond 50% from coal are in view. By the way, nitrogen oxide emissions from combined-cycle equipment can probably be held at extremely low levels.

If I now consider the probable rate of growth of the electricity industry, and the probable extent to which electricity may replace oil and gas as these fuels dwindle, my crystal ball gets pretty cloudy. Electricity-generating equipment has become mighty expensive, costing something like five to ten times what it did just ten years ago. The marginal cost of electricity is much larger than the average price customers are now paying, and it is not clear just how the industry will weather the political storms that will accompany its efforts to put the price structure for electricity into shape to allow for expansion. If I were pressed, I would have to say that I am not optimistic that the historical rate of electricity growth will be maintained, and I will elaborate on some of my reasons for this view later in this paper.

Will synthetic fuels from coal take the place of oil and gas in their various uses? Again, I am not optimistic. The capital equipment for conversion of coal to synthetic petroleum or synthetic natural gas is much too costly for a coal conversion industry to get under way very quickly.

As an aside, let me comment that the pollution problems of a coal conversion plant making either gas or oil are not dissimilar to the problems of a large oil refinery. Just as in the case of a refinery, siting will not be easy, but the capabilities are there to limit atmospheric and water emissions to acceptable levels. No great amount of research is needed. I deplore the attention being given to paper studies of the pollution problems of a large number of coal conversion processes in development, even processes yet lacking pilot-plant experience. Many of the processes obviously will not make it to the marketplace, and some were poorly conceived from the standpoint of selection of process steps that inherently eliminate or reduce pollution problems.

Consider what will happen in the marketplace when the world supply of oil flags.

By the verb "flag," I mean a loss of ability to expand the production of oil to meet the increase in demand for petroleum products. Note that I am not speaking of the topping out of world oil production. I am speaking of a loss of supply at the margin. If the world oil market were free, loss of supply at the margin would trigger a sharp rise in price. Even with controlled prices, there will be great pressure to maintain growth in petroleum consumption for the "higher uses," i.e., transport fuels and petrochemical feedstocks. These account today for roughly one-half of the world's oil. When world oil flags, it will be much easier and cheaper to build plants to convert fuel oils to transport fuels and chemical feedstocks than to build plants to convert coal to synthetic petroleum. Fuel-oil conversion equipment can be located in existing refineries, right where the feedstock is available; it requires the opening of no mines and the locating of no new plant sites; and it uses conversion processes already proven. In short, it will be easier to get some more gasoline by denying oil to a boiler than by making gasoline from coal.

Accordingly, I predict a rapid drying up of the supply of fuel oils in the decade after world oil supply just begins to fall short of world demand.

By far the cheapest and quickest way to replace oil with coal will be to return to the combustion of coal, and I believe this will happen even for relatively small-scale heating needs.

It is instructive to consider how quickly the rise in the price of oil in late 1973 brought about a revival of wood burning. Last year, there were 175,000 wood-burning stoves sold in the United States, but there were also 40,000 *reported* chimney fires, indicating that a good many people began to burn wood without much knowledge or experience. It is disturbing to see so many advertisements for wood-burning stoves asserting that the equipment can also burn coal, when many of these stoves would not even burn wood without copious emissions of tarry smoke. It will be important for the United States not to remain too long in ignorance of the principles of design of a good coal furnace. We do not *have* to go back to smoky Pittsburgh and St. Louis, but there is danger that we *will* if the return to coal combustion is without some sense and order.

To acquire some assurance that we do not have to go back to smoke and smog, one has merely to visit Bradford or Leeds in northern England, where coal is still used for space heat, and see the dramatic effects of the British Clean Air Act of 1956 in these cities. The smoke of St. Louis, Pittsburgh, Bradford, and Leeds in the "bad old days" came from the combustion of cheap, bituminous coals in simple furnaces. In both countries, rich folk burned the more expensive, smokeless fuels—anthracites and cokes. The British Clean Air Act required everyone to use smokeless fuels—or to burn bituminous fuel in a smokeless manner, as I shall describe shortly—and the change was dramatic. I personally experienced "killer" smogs in London in 1958 and 1959, but these are now history. The citizens of Bradford and Leeds have cleaned their black buildings, revealing the beautiful pink granite preferred by the Victorian builders of these cities, and the buildings remain clean.

The householder burned bituminous coal in simple furnaces in the old days because nothing better was available. The larger furnaces used by industry were often clean simply because the heat value present in smoke represented a serious thermal loss. Power engineers all the way back to James Watt understood perfectly well how to get rid of smoke and recover this heat value. They provided a zone of intense secondary combustion, above the fuel bed, to burn up the smoke. For example, in the underfeed stoker, coal is pushed from below into a bed of burning coke particles. Combustion air is supplied from holes in a hollow casting surrounding the bed of coke. As a coal particle rises toward the hot zone, it carbonizes and releases volatile matter, which must pass through the hot zone to escape from the upper surface of the coke bed. Secondary combustion air is played upon the upper coke surface, burning up any of the volatiles that escape from the coke bed. Accumulated ash matter is pushed to one side from the top of the coke bed as additional coal is supplied, and can usually be removed as a clinker in one or two lumps.

As competition for the U.S. coal industry arose from oil and gas in the 1920s, underfeed stokers were developed for domestic heating. Their attraction was their automatic operation, as well as their cleanliness. Late models of these furnaces, developed in the 1940s (some members of the audience will remember the Iron Fireman), included features for automatic removal of ash and worked almost without attention through an entire heating season. As late as 1950, their annual sales were about one-tenth that of domestic oil furnaces.

The American designs were too large for most of the British market, and after the passage of the British Clean Air Act, the National Coal Board took up development of "smoke-eaters" to allow the small householder to burn bituminous coal. Several designs are now marketed at ratings below 10 kilowatts thermal, and with smoke emissions generally below a few grams per hour. The Act would allow 5 grams per hour from these sizes. Smoke-eaters are coming into use in New Zealand and South Africa, as well as in England.

The Iron Fireman underfeed stoker and the British smoke-eaters were developed with a supply of "stoker coal" as a given. That is to say, coal that had been screened twice—hence, having a top size and a bottom size—was readily purchasable throughout both countries. Such coal is still available in England, but the supply has contracted to a very small amount in the United States. I understand that an expansion of the supply of stoker coal here might not be easy.

Although stoker coal has never been available in New Zealand, a few Iron Firemen were made there, and were used in spite of difficulties arising from an accumulation of fine coal at the outer curve of a horn through which coal is pushed upward into the burning coke bed. After the crisis of late 1973, a strong demand developed for coal-burning equipment in New Zealand, especially from greenhouses and other agricultural activities. Taylors Ltd. of Christchurch was able to revise the underfeed stoker design to accommodate the presence of fines in the coal. The new design pushes coal from opposite directions into the bottom of the vertical shaft for upward movement of the coal into the coke bed. The coal pushes against itself, as it changes direction from horizontal to vertical motion, instead of pushing against a curved metal surface. Taylors offers designs for 0.35 to 3.5 megawatts thermal. They may find a place here for burning low-sulfur coals, such as those available from western Virginia, southern West Virginia, and eastern Kentucky and Tennessee, as well as from the West. It will be important to obtain emissions data for Taylors' design on U.S. coals.

Fluidized-bed combustion is an approach that should permit use of high-sulfur coals at relatively small scales, perhaps down to about 3.5 megawatts thermal. In Germany in 1922, Fritz Winkler invented a new technique for gasifying coal. His idea was to increase the rate of flow of gas upward through a bed of coal particles to the point at which each particle in the bed was buoyed by the rising gas, and then to increase the gas flow well beyond that point. When the pull of gravity downward upon each particle was just canceled by the upward drag of the gas, the particles became free of one another, and the mass of particles took on much of the character of a liquid, finding its own level and possessing hydrostatic head. As the gas velocity was increased further, much of the gas rose through the "fluidized" mass of coal particles in the form of bubbles, and the bed took on the character of a vigorously boiling liquid.

In fluidized-bed combustion of coal, the bed consists almost entirely of a noncombustible solid; and there are just a few particles of coal floating around in the bubbling mass. The solid serves two major functions. It provides a reservoir of heat that almost instantaneously ignites a coal particle added to the bed. The solid also carries the heat away from the narrow region of intense heat, just at the burning coal particle, and conveys the heat to the surface of metal tubes that contain water to receive the heat.

Wayne McCurdy, of the Department of Energy, and the firm Pope, Evans and Robbins deserve enormous credit for having picked up the British lead in fluidized bed combustion in the mid-1960s. John Bishop and Ernie Robison of Pope Evans quickly proposed that a third function be imposed upon the noncombustible solid of the fluidized bed. They conducted experiments using limestone in the bed, instead of the sand used by the British, and showed that the stone has the power to absorb sulfur dioxide generated during the combustion.

The evidence is not yet in, but I believe that fluidized-bed furnaces will need high-efficiency dust collection, probably fabric filters, to keep emissions of fine particulate matter below acceptable limits.

A number of relatively small fluidized-bed furnaces are in early stages of commercial operation. One of the most interesting is a 25-megawatt furnace in the District Heating Scheme of Enköping, Sweden. The unit is perhaps the most flexible, since it is capable of burning high-sulfur residual oil (using limestone to capture sulfur), coal, peat, wood and wood wastes, and municipal wastes. There is a bag filter to capture dust. The furnace provides for combustion in two stages: (1) primary combustion in the fluidized bed; and (2) secondary combustion in an unusually tall space above the bed. A portion of the combustion products is recycled to the primary air, thus returning part of the heat to the bed. These features lead to negligible emissions of nitrogen oxides and to the capability of burning fuels of low and varying heating value, such as wet industrial or municipal wastes.

Let me tell you just a little about district heating, which I consider a major opportunity for expanding the use of coal. Just since 1970, Enköping, Sweden, has installed hot water pipes to furnish about 90% of its space heating needs. Enköping measures 1.5 by 2.5 miles and has 20,000 people. The current load is approaching 100 megawatts thermal. The development of this scheme was strictly a commercial proposition: it can give cheaper heat. This is largely a matter of thermal efficiency. A household furnace is seldom better than 50% efficient—almost never better than 60%. Enköping's scheme is running at 82 to 83%. There is a stack loss of about 10 or 11%, and about 7% loss of heat in transmission. The rest of the heat is paid for by customers according to the reading of their heat meters.

How does district heat compare with synthetic gas and with electricity as a market for coal? For starters, the efficiency of district heat is roughly twice as good. For finishers, the capital cost, even for retrofit, appears to be well below half. The piping in Enköping could be replaced for about $70 per kilowatt thermal. A unit for coal combustion at Enköping's needs of 100 megawatts thermal would cost about $100 per kilowatt, including environmental controls and coal receiving, handling, and stockpiling.

This cost comparison will serve to illustrate my earlier remark that a return to coal combustion will be cheaper than synthetic fuels and electricity. Denmark, indeed, is showing the way; and I understand that all of the electricity generation and much of the space heat there this winter is coming from coal. Odense, Denmark, has converted a large combined station for electricity and heat from oil to coal, and it is striking that the reported cost of the conversion is only $100 per kilowatt electric. U.S. boilermakers have stated that such a conversion is "impossible," but I suspect that economic necessity will cause them to change their tune. By the way, the Danes tell me that they can buy coal on a world coal market at about two-thirds the cost of oil.

There is much to be done. We need to test a wide range of furnace types on U.S. coals, including measurement of emissions. We need to plan the development of coal supplies and dealerships, furnace manufacturers and dealers, etc. Will it be profitable to revive supplies of smokeless fuels? The anthracite mines are mostly under water, but perhaps cokes can be furnished by processes that "skim some cream" from coal for conversion to synthetic fuels. Perhaps this is the best path to a small supply of synthetics. Is there any hope of reviving coal combustion in individual homes for space heat? I am pessimistic on this question, but I do not believe we should close out this option arbitrarily.

Some early progress could be made by targeting a few heat needs for prompt conversion to coal. My favorite target is the chicken farmer. Three-quarters of the growers of chickens now heat with propane. We need to study the best way to replace propane with coal.

METEOROLOGICAL EFFECTS OF REJECTED HEAT

H. E. Landsberg

Institute for Physical Science and Technology
University of Maryland
College Park, Maryland 20742

Man-caused heat rejection in the modern world is a complex phenomenon. It has a number of components that are not easily disentangled. One of these stems simply from man-made changes of the natural surface. This causes a different partitioning of solar energy than occurs over undisturbed ground with natural vegetation. Then there is the heat dissipated into the atmosphere from all combustion processes, be they from engines, industrial activities, power production, or space heating. Coupled with these activities is invariably a production of various aerosols and water vapor. The latter is one of the end products of fossil fuel consumption and is augmented by vaporization from cooling processes that use water as a coolant. In addition, combustion of organic materials, principally fossil fuels, adds carbon dioxide (CO_2) to the atmosphere. All these components interact in the atmosphere and have a joint effect on meteorological conditions and events on various scales.

It is axiomatic that all energy used is ultimately degraded to heat and ends up in the atmosphere. The effects of this heat will depend on how concentrated the release is in time and space. Presently, the main concern is, of course, heat effects in small spaces, such as metropolitan areas and even individual energy installations.

Before attempting to disentangle the anthropogenic heat rejection, it is necessary to take a glance at the effects of man-made surface changes. These are most noticeable in urbanized areas. Such areas absorb and store more solar heat because of the physical characteristics of stone, concrete, asphalt, and other building materials, which differ strikingly from those of natural, principally vegetated, surfaces. The end result is the formation of the well-known urban heat island. In larger metropolitan areas, this can raise the annual mean air temperature by 1 to 2°C and, on calm clear nights, maximally by 6 to 10°C. In addition, the change in energy partitioning can lead to cloud and shower formation downwind. These effects have been documented in many localities, so that we may safely refer to the pertinent literature (Landsberg, 1970;[1] Hosler and Landsberg, 1977;[2] Changnon, 1978[3]).

The heat added by the various man-made energy transformations in urban areas is generally only a small fraction of the solar radiation. However, in some densely settled local areas, especially in higher latitudes during winter, the rejected heat can far exceed the contemporary energy income from the sun. By and large in urban areas, the heat directly added from fuel use is only a fraction of the net energy added to the atmosphere, except in polar regions during the long polar night. However, in the middle and higher latitudes, there is net energy added to the atmosphere by the greater absorption and storage of solar energy compared with natural surfaces. The fraction of direct heat generation varies with latitude, season, and degree of cloudiness. In most midlatitude urban areas, it usually does not exceed 30% of the net radiation between sun, surface, and atmosphere.

The situation is quite different for heat rejection in very small areas. This refers to

0077–8923/80/0338–0569 $01.75/1 © 1980, NYAS

such installations as refineries or power stations. There, a major fraction of the energy used is released locally and continuously. For example, in electric power production, whether by use of fossil or nuclear fuels, only 30 to 40% of the fuel energy is converted to electricity and used elsewhere. The remainder, 60 to 70%, is dissipated locally. Hence for a 10 megawatt (MW) power installation, about 15 to 17 MW must be dissipated locally through direct heat release from stacks or by latent heat release through evaporation of cooling water.

Some of the effects are clear enough. They become obvious through the formation of clouds. If the meteorological conditions bring an air mass that is already fairly moist and has the right temperature distribution into an area of excessive heat release, clouds can grow rapidly and become persistent. There are well-documented cases in the literature (Auer, 1976[4]). Such cloud formation is well known for urban areas where, in summer, convective clouds may form one or two hours before they show up in the surrounding rural environs.

Where stack plumes and cooling tower plumes interact, cloud formation is even more apparent. These man-made clouds are more extended and more persistent where the natural moisture content of the local atmosphere is higher (Kramer *et al.*, 1976;[5] Smith *et al.*[6]). When the air is close to saturation for water vapor, visible plumes many miles long have been observed. At times, these merge with natural cloud formations, and Kramer *et al.*[5] conclude that "there is no doubt that cooling tower plumes can modify natural cloud formations." Low-level effluents from steel mills and mechanical draft towers have been observed to produce fog or to reinforce natural fog, especially in hilly terrain (Hosler and Landsberg, 1977[2]).

Marked heat injection into the atmosphere on a sustained basis can lead to showers from convective clouds. This has been documented in a number of cases. Urban heat islands can induce such precipitation in summer even if natural conditions would not have been sufficient to initiate rain (Harnack and Landsberg, 1975[7]). The effect of the urban heat island plume has also been effectively demonstrated in the increased summer rainfall in, and downwind from, metropolitan areas (Changnon, 1978[3]). This increased precipitation is not insignificant. Values of 10 to 20% augmentation have been indicated as plausible. Whether or not this is indeed an undisguised blessing for the farmers downwind of cities remains to be seen. The washout effect of urban aerosols will not only lead to acidification but will also bring down a wide variety of trace elements, which may or may not be beneficial. Increases in thunderstorm and hail frequency have also been predicted.

The question most frequently asked is, How does all this affect the global climate? In order to arrive at an answer, a look at the energy-density flux over various areas is useful. TABLE 1 shows this flux for the mid-1970s in regions that differ in size and in industrialization, with an estimate of how this compares with the net natural radiation energy flux for these same areas. A variety of sources have been used and updated (Bach, 1976;[8] Flohn, 1970;[9] Flohn, 1973;[10] Kellogg, 1974;[11] Kellogg, 1974;[12] Kutzbach, 1974[13]).

It is immediately clear that the fluxes from current energy sources on a global, continental, and even regional scale are far too small, compared with natural fluxes, to have any appreciable effect on weather and climate. This is corroborated by a number of modeling experiments. These use well-known mathematical models of the atmosphere and add the known concentrated local heat sources to the natural forces driving the general atmospheric circulation (Llewellyn and Washington, 1977;[14] Washington and Chervin, 1979[15]). These authors have also made similar calculations for a heat release of 90 watts m^{-2} for a postulated megalopolis in the eastern United States from Boston to Washington. The model calculations showed temperature rises at the ground of 12°C in January and 3°C in July. The model also showed increases in precipitation over the region. However, the experiment also showed that this type of energy release remains regional and has no provable effects on a global basis.

Before leaving the topic of local heat effects, a look at other consequences is needed. There is already clear evidence that the local heat island in winter turns snow into rain, a result that would certainly benefit traffic and reduce snow removal costs. This beneficial aspect, however, usually occurs with light snowfalls. In summer, however, the extra heat is most unwanted. It increases energy demand for air conditioning and is physiologically detrimental for the inhabitants. Death rates in metropolitan areas always soar during heat waves, and anthropogenic additions to the heat aggravate this cause of mortality (Hodge, 1978[16]).

TABLE 1

ENERGY FLUX FROM MAN-MADE HEAT REJECTION IN VARIOUS AREAS*

Locality	Size of Area (km^2)	Heat Rejection ($W\ m^{-2}$)	Heat Rejection in % of Natural Net Energy Exchange
1.5-GW Power Plant	0.2	10^4	10^4
Manhattan	60	600	650
West Berlin	234	22	40
St. Louis	250	65	65
Moscow	880	130	310
Chicago	1800	30	35
Los Angeles	3500	25	23
Rhine-Ruhr Industrial Area	10×10^3	10	20
Washington-Boston Corridor	90×10^3	4.5	5
Japan	370×10^3	0.8	1.2
Central & Western Europe	156×10^3	0.7	1.2
European Russia	4×10^6	0.25	0.6
U.S. less Alaska	7.8×10^6	0.26	0.3
Africa	30×10^6	0.004	0.004
All Continents	136×10^6	0.047	0.05
Globe	510×10^6	0.016	0.02

*Mid-1970s estimate.

Yet, it must be concluded that even with a vastly increased energy consumption projected, global effects will stay subliminal within the large natural climatic fluctuations. The local and regional effects in urbanized and industrial areas will remain confined to their areas and will stay within acceptable limits. There is, however, one important exception. This is power production in very restricted spaces. Popular pressure and public policies have forced utilities to restrict power production by fossil and nuclear fuels to highly concentrated and somewhat isolated premises. This has led to planning and implementation of power parks. Such installations, producing 10 to 20 MW electricity, are already—or soon will be—reality. Depending on the efficiency of the plants, 1½ to 2 times as much energy will be released continuously into the atmosphere. This energy—flowing essentially from a spot—is equivalent to that of a large urban area sprawled over many square kilometers. Some have calculated the energy flux density by using the dimensions of the power park, some tens or hundreds of square kilometers. This is misleading, because the energy is released from a few stacks and cooling towers that are usually in close proximity. Only if the heat is dissipated by cooling ponds or from mixing with a water body is it justified to spread the heat flux over a substantial surface area.

What, then, are the potential meteorological consequences of rejecting thousands and even ten thousands of watts per square meter into the atmosphere? This question has been discussed in the literature (Koenig and Bhumralkar, 1974;[17] Hanna and Gifford, 1975;[18] Hosler and Landsberg, 1977;[2] Laurman, 1978[19]). There is no doubt that this energy release is equivalent to that of severe local storms. The problem, as always in studies of man-made effects on the atmosphere, is the capability of the atmosphere to resist, or its readiness to respond to, a local stimulus of this kind. In a very-stably stratified, cool air mass, the effect of such heat release may be nil. In a less stable air mass, the only result may be local cloud formation. But in an unstable, moist air mass, the extra heat—both in the form of sensible heat from stack flux and as latent heat release from cooling towers—may be all that is needed to trigger a severe thunderstorm or even to organize existing convection into a tornado. The effects of these will be felt downstream. The environmental impact of such severe storms may far outweigh any imagined environmental benefits contemplated in the concentration and segregation of power production.

It remains for us to explore briefly the interaction of man-made heat flux with particulate pollution. The two generally go hand in hand.

Man-made particulates have a measurable influence on the energy budget of the earth's surface. Although on a global scale, they cannot rival the amounts of dust produced by nature, they have significant effects near their sources. The natural dusts originate principally in deserts and over the oceans as salt from evaporated seawater spray. Volcanic eruptions also contribute sporadically to the aerosol, as do lightning-induced forest fires. The man-made sources include agriculture—especially the slash-and-burn methods used in the tropics, industry, traffic, and space heating. Urban areas again stand out as areas with a high dust load.

Most of these particulates stay close to the surface. They do scatter a small amount of the incoming solar energy back to space, but their major influence is as interceptors of long-wave radiation from the surface. This will heat up the layers in which the particulates are situated, and some of this energy is reradiated back to the surface (Riches *et al.*, 1975;[20] Bühlen and Eiden, 1977;[21] Idso *et al.*, 1978[22]). Thus, the particulates reinforce the urban heat island effect. Their influence is greatest at night. In contrast to these particulates, man-made aerosols in the stratosphere will generally effectively cause cooling by increasing the earth's albedo. How much of the stratospheric aerosol is anthropogenic is presently unknown, but, both in quantity and effectiveness, it is enormously outweighed by the tropospheric dust loads. These, as stated, are mostly close to the densely settled areas. Various atmospheric processes, such as fall-out, rain-out, wash-out, and dispersion, reduce this concentration rapidly from the source. Hence, their meteorological effects remain essentially local or, at worst, regional. The efforts to control particulates may keep their influence down, in contrast to that of heat rejection, which is bound to grow with increasing energy rejection.

For the long-range future, neither energy rejection nor particulates are apt to affect global climatic conditions. This is in contrast to the projected increases in atmospheric CO_2. If fossil fuel use increases, this must indeed be viewed as a threat to the present global climatic equilibrium. It would far exceed any effects of heat rejection in terms of global warming. The heat rejection will have local repercussions that are only likely to have serious consequences in the case of massed power production. Therefore, the development of power parks with tens of gigawatt output should be resisted.

References

1. Landsberg, H. E. 1970. Man-made climatic changes. Science **170:** 1265.
2. Hosler, C. L. & H. E. Landsberg. 1977. The effect of localized man-made heat and moisture sources in mesoscale weather modification. *In* Energy and Climate: 96–105. National Academy of Sciences. Washington, D.C.
3. Changnon, S. A., Jr., Ed. 1978. METROMEX studies. J. Appl. Meteorol. **17:** 565–715.
4. Auer, A. H., Jr. 1976. Observations of an industrial cumulus. J. Appl. Meteorol. **15:** 406.
5. Kramer, M. L., M. E. Smith, M. J. Butler, D. E. Seymour & T. J. Frankenberg. 1976. Cooling towers and the environment. J. Air Pollut. Control Assoc. **26:** 582.
6. Smith, M. E., M. L. Kramer, D. Seymour & T. T. Frankenberg. (No date.) Cooling Towers and the Environment. American Electric Power Service Corporation. New York, N.Y.
7. Harnack, R. P. & H. E. Landsberg. 1975. Selected cases of convective precipitation caused by the metropolitan area of Washington, D.C. J. Appl. Meteorol. **14:** 1050.
8. Bach, W. 1976. Global air pollution and climatic change. Rev. Geophys. Space Phys. **14:** 429.
9. Flohn, H. 1970. Produzieren wir unser eigenes Klima? Meteorol. Rundsch. **23:** 161.
10. Flohn, H. 1973. Natürliche und anthropogene Klimamodifikationen. Ann. Meteorol. N.F. (6): 59.
11. Kellogg, W. W. 1974. Long Range Influences of Mankind on The Climate. World Conference "Toward a Plan of Action for Mankind: Needs and Resources—Methods of Forecasting." Institut de la Vie. Paris, France.
12. Kellogg, W. W. 1974. Mankind as a factor in climate change. *In* The Energy Question, An International Failure of Policy. E. W. Erickson & L. Waverman, Eds. **1:** 241–255. University of Toronto Press. Toronto, Ont., Canada.
13. Kutzbach, J. E. 1974. Possible impact of man's energy generation on climate. Presented at the International Council of Scientific Union/World Meteorological Organization International Study Conference on the Physical Basis of Climate and Climate Modelling, Stockholm, Sweden.
14. Llewellyn, R. A. & W. M. Washington. 1977. Regional and global aspects. *In* Energy and Climate: 106–118. National Academy of Sciences. Washington, D.C.
15. Washington, W. M. & R. M. Chervin. 1979. Regional climatic effects of large-scale thermal pollution: simulation studies with the NCAR General Circulation Model. J. Appl. Meteorol. **18.** (In press.)
16. Hodge, W. 1978. Weather and mortality. EDIS (September): 12.
17. Koenig, L. R. & C. M. Bhumralkar. 1974. On Possible Undesirable Atmospheric Effects of Heat Rejection from Large Electric Power Centers. Report No. R-1628-RC. RAND Corporation. Santa Monica, Calif.
18. Hanna, S. R. & F. A. Gifford. 1975. Meteorological effects of energy dissipation at large power parks. Bull. Am. Meteorol. Soc. **52:** 1069.

19. Laurman, J. 1978. Modification of Local Weather by Power Plant Operation. Technical Planning Study No. 76-660. Electric Power Research Institute. Palo Alto, Calif.
20. Riches, M. R., J. T. Peterson & E. C. Flowers. Effects of Atmospheric Aerosols on Infrared Irradiance at the Earth's Surface in a Nonurban Environment. Report No. EPA-650/4-75-017, Environmental Monitoring Series. U.S. Environmental Protection Agency. Research Triangle Park, S.C.
21. Bühlen, T. & R. Eiden. 1977. Vergleichende Studie zum Einfluss des Aerosols auf den Energiehaushalt der ruralen and urbanen Atmosphäre. Ann. Meteorol. (12): 222.
22. Idso, S. B. & A. J. Brazel. 1978. Climatological effects of atmospheric particulate pollution. Nature **274:** 781.

CLIMATIC EFFECTS OF ATMOSPHERIC AEROSOLS

James E. Hansen, Andrew A. Lacis, Pauthon Lee, and Wei-Chyung Wang

Goddard Institute for Space Studies
National Aeronautics and Space Administration (NASA)
New York, New York 10025

INTRODUCTION

It is worthwhile to briefly consider how ideas about the effect of aerosols on climate have developed over the years. We mention only the contributions of Franklin in 1784,[1] Humphreys in 1940,[2] and Rasool and Schneider in 1971.[3] Many other important contributions were made during that period, but perhaps these few will serve as an adequate sample.

In May 1784, Benjamin Franklin wrote:

> During several of the summer months of the year 1783, when the effects of the sun's rays to heat the earth in these northern regions should have been the greatest, there existed a constant fog over all Europe, and great part of North America. This fog was of a permanent nature; it was dry, and the rays of the sun seemed to have little effect towards dissipating it, as they easily do a moist fog, arising from the water. They were, indeed, rendered so faint in passing through it, that, when collected in the focus of a burning-glass, they would scarce kindle brown paper. Of course, their summer effect in heating the earth was exceedingly diminished.
>
> Hence, the surface was early frozen.
>
> Hence, the first snows remained on it unmelted, and received continual additions.
>
> Hence perhaps the winter of 1783–4, was more severe than any that happened for many years.
>
> The cause of this universal fog is not yet ascertained. Whether it was adventitious to this earth, and merely a smoke proceeding from the consumption by fire of some of those great burning balls or globes which we happen to meet with in our course round the sun, and which are sometimes seen to kindle and be destroyed in passing our atmosphere, and whose smoke might be attracted and retained in our earth; or whether it was the vast quantity of smoke, long continuing to issue during the summer from Hecla, in Iceland, and that other volcano which arose out of the sea near that island, which smoke might be spread by various winds over the northern part of the world, is yet uncertain.[1]

Franklin was uncertain about whether the aerosols originated from comets or volcanoes, but it is clear that the idea that aerosols cause climatic cooling is at least two centuries old. Franklin's theory was well stated but not very quantitative, and he did not give any data on the aerosols. With knowledge of the diameter of his burning glass and the type of brown paper he had available, it might be possible to estimate the optical thickness of the aerosols. However, since the climate data for that period are also sparse, it probably would be more fruitful to investigate more recent volcanic events.

Humphreys performed a detailed quantitative analysis of the effect of volcanic aerosols in his book *Physics of the Air,* first published in 1920 and revised in 1940.[2] By considering the energy balance at the earth's surface, he estimated that volcanic aerosol injections, such as those by Krakatoa in 1883, could cool the surface by 6–7°K

0077–8923/80/0338–0575 $01.75/1 © 1980, NYAS

due to their interception of solar radiation. He recognized the counteractive warming by the aerosols due to their "greenhouse" blanketing of thermal radiation from the earth's surface; but he argued that because of the small size of the aerosols, their opacity to thermal radiation would lead to only a negligible warming of a few tenths of a degree. Humphreys overestimated the cooling effect of the aerosols, in part because he thought the aerosols must reflect most of the radiation they intercept back out of the atmosphere, while in reality they scatter most of it into the forward hemisphere. Also, Humphreys' simple assumption of an energy balance at the earth's surface must be replaced by a proper radiative transfer calculation, and the effect of the thermal inertia of the ocean in damping the radiative response of the earth must be included, to yield a realistic magnitude for the temperature perturbation, since the thermal forcing due to the volcanic aerosols is too short-lived to reach an equilibrium.

Rasool and Schneider[3] focused attention on the aerosol/climate problem in 1971, when they made radiative transfer calculations that indicated that an increase in atmospheric aerosols by a factor of 4 could cool the earth's surface temperature by about 4°K. Subsequently, a large number of further investigations of this problem have been made, and a fairly detailed understanding of the radiative aspects of the problem has been developed.

At the present time, the two major factors that prevent reliable prediction of the future effect of aerosols on climate are (1) large uncertainties about aerosol optical properties, particularly anthropogenic aerosols, and lack of knowledge of how these will change in the future; and (2) the primitive state of the models for the complete climate system into which the aerosols' radiative perturbations must be inserted.

In the next section of this paper, we illustrate quantitatively how the radiative climatic forcing by aerosols depends upon their physical properties. This type of information permits one to estimate the nature of the mean climatic response to a presumed change in aerosol amount or aerosol properties under the (naive) assumption that feedbacks and interactions with other parts of the climate system are small or understood.

In the following section, we then consider the special case of atmospheric aerosols produced by volcanic explosions, a case that provides an excellent opportunity to test not only our theories about the climatic effect of aerosols, but also climatic modeling capabilities in a rather general sense. We present evidence that even the simple climate models available today may be able to capture some of the basic effects of aerosols on global climate, and we also emphasize the enormous potential scientific value of the natural climatic experiment that could be provided by future large volcanic eruptions.

Finally, we briefly consider the possible effects of anthropogenic aerosols on global climate.

Radiative Effect of Aerosols

The direct radiative effect of aerosols on atmospheric temperatures can be accurately computed with available radiative transfer techniques. Such calculations have limited significance because they exclude important feedback effects in the climate system as well as effects of the aerosols on processes such as cloud formation. In fact, most of the estimates of aerosol impact on climate have been based on calculations of only the direct radiative effect, and it is logical to first try to understand this before proceeding to additional complications.

To illustrate the degree to which aerosols can impact global temperature through their direct radiative effect, including an indication of how the results depend upon various aerosol parameters, we could collect results from the literature. However, to provide a uniform basis for comparison, we have chosen instead to make new computations with a single model. The one-dimensional radiative-convective model we employ is described briefly by Hansen, Wang, and Lacis[4] and in detail by Wang and Lacis.[5] It is basically similar to the model first employed by Manabe and Wetherald[6] more than 10 years ago. The model we employ computes global average temperature under the assumption that relative humidity, cloud cover, and cloud height do not change in the climate perturbation experiment.

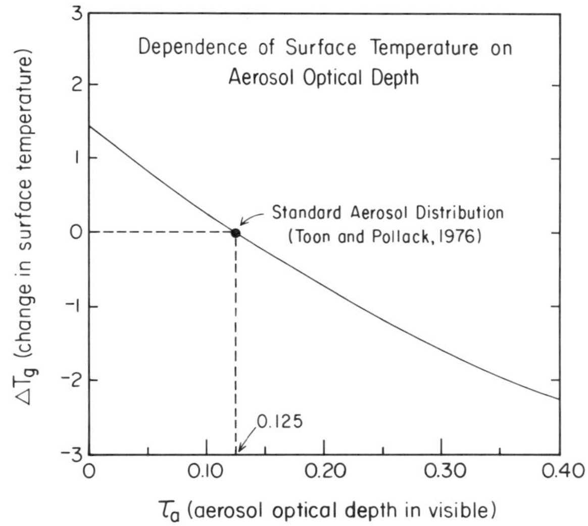

FIGURE 1. Change in the surface temperature as the aerosol amount is increased, compared to the standard aerosol distribution of Toon and Pollack.[7]

FIGURE 1 shows the global average temperature perturbation that results from changing the amount of atmospheric aerosols while keeping everything else constant. The global average aerosol distribution of Toon and Pollack[7] is employed as a standard for comparison. This distribution has a total optical depth of 0.125; it contains sea salt, basalt, and sulfates in the lowest three kilometers of the atmosphere, where the effective particle radius of the size distribution is about 2 μm; it contains sulfates and basalt between 3 and 12 kilometers, with an effective radius of about 0.5 μm; it has sulfuric acid particles in the stratosphere, with an optical depth of 0.005 and an effective particle radius of 0.2 μm. The important point to note is that, for these standard aerosols, increasing the amount of aerosols by a factor of 2 would tend to cause a cooling of about 1.2°K. The effect is essentially linear within the range of optical depths of interest. We emphasize again that these are computations of direct radiative response for a simple one-dimensional climate model and all of the caveats that should be attached to such computations are in effect.

FIGURE 2 illustrates how the surface temperature would change if the single scattering albedo of the aerosols were changed, with all other parameters left unchanged. It shows that making the aerosols perfectly conservative (nonabsorbing) would result in a further cooling of only 0.1°K. On the other hand, by introducing substantial absorption in the aerosols, it is possible to cause a very large heating. The graph may be somewhat misleading in the sense that most materials tend to fall close to the conservative scattering end of the diagram, but it is appropriate from the theoretical standpoint, since the temperature effect is almost linear with change in single scattering albedo.

This strong dependence of the surface temperature on the single scattering albedo of the aerosols is a major difficulty when it comes to estimating how additional aerosols will affect the temperature. In practice, it means that we need to know not only the primary composition of the aerosols, but also the composition of any impurities that may affect the single scattering albedo. However, we do note that what we call the "critical albedo"—the single scattering albedo for which a change in aerosol amount would neither heat nor cool the surface—is relatively small (~0.85). Thus, most aerosols tend to have a cooling effect. There may be some significant exceptions, such as some desert and soil aerosols and man-made aerosols containing a substantial amount of carbon.

FIGURE 3 shows that the temperature impact of aerosols also depends upon the altitude at which they are located. In this case, the standard aerosol distribution has been employed, but an additional layer of aerosols has been added at the height h_a. The results are more complicated than in the previous figures, because there are

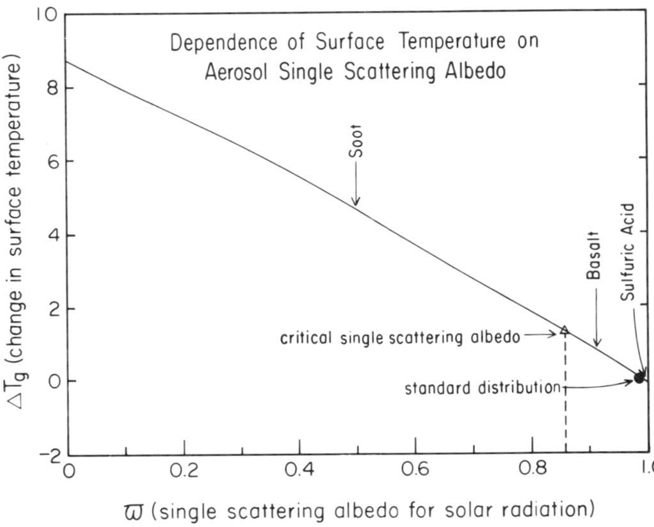

FIGURE 2. Change in surface temperature as the single scattering albedo of the aerosols is changed, compared to the standard aerosol distribution of Toon and Pollack.[7] The "critical albedo" is that value that would eliminate the cooling effect of the aerosols.

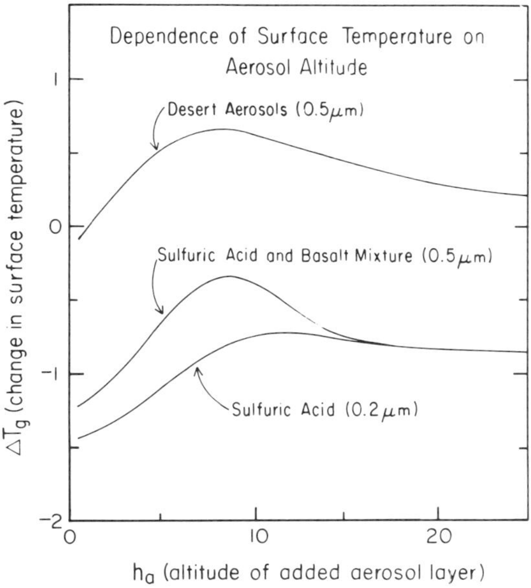

FIGURE 3. Change in surface temperature as the altitude of a layer of aerosols is varied. The atmosphere contains the standard aerosol distribution of Toon and Pollack[7] plus a layer of either (1) sulfuric acid, (2) 60% sulfuric acid, 40% basalt mixture, or (3) desert aerosols. The optical thickness of the added layer is 0.1, except in the case of the desert aerosols for which it is 0.025. The optical properties of the aerosols are from Toon and Pollack,[7] except for the desert aerosols which are from Patterson *et al.*[17] and Volz.[18]

competing effects in the solar and thermal parts of the spectrum. The cooling effect due to the increase of the earth's albedo becomes somewhat stronger when the aerosols are at greater altitude, where they are above the other radiative constituents. On the other hand, the greenhouse warming in the thermal infrared also increases as the altitude increases within the troposphere, because the aerosols radiate at a colder temperature. Aerosols of the type in Toon and Pollack's standard model cool the surface for all aerosol altitudes, with maximum cooling for low level aerosols. However, it should also be emphasized that if the aerosols are either (1) more absorbing or (2) larger, the effect will tend to be more in the direction of heating. This is illustrated in FIGURE 3 for desert aerosols with substantial absorption.

FIGURE 4 illustrates the effect of changing the aerosol optical thickness in the thermal infrared window region. In the computations for this figure, all other quantities except the infrared optical depth were held fixed, using the values computed for the standard distribution of Toon and Pollack.[7] The infrared optical thickness is very sensitive to the largest particles in the distribution, i.e., to the large particle tail of the distribution. Therefore, it is clear that we must have accurate knowledge of the aerosol size distribution in order to reliably state the direct radiative effect of the aerosols.

There are, of course, several other parameters that influence the radiative effect of the aerosols, in addition to those we chose to illustrate in FIGURES 1–4. One parameter that has received considerable attention is the ground albedo; however, it is within the

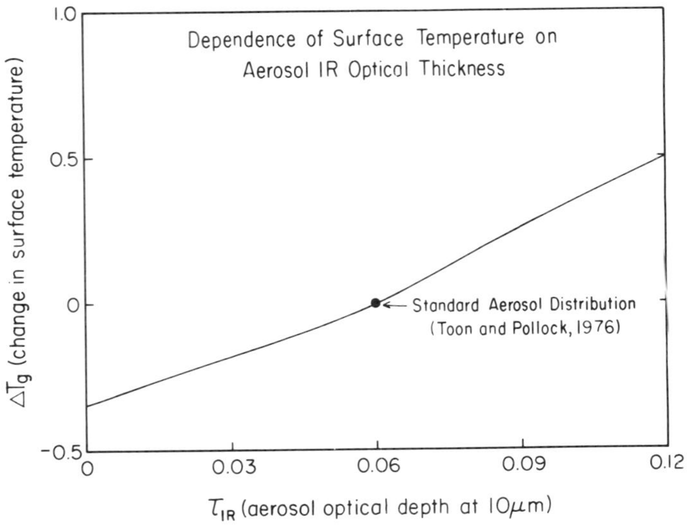

FIGURE 4. Change in surface temperature as the optical thickness of the aerosols in the thermal infrared window changes with all other parameters fixed.

range 5–25% over most of the earth, and this does not significantly impact the above results computed for a 10% ground albedo. Another is the asymmetry of the single scattering phase function of the aerosols, but this is included implicitly in our results through the Mie scattering computations for different size distributions.

The parameters we have illustrated, i.e., the aerosol optical thickness, the single scattering albedo, the altitude of the aerosols, and the infrared optical thickness (or particle size distribution), are perhaps the most critical parameters. In any case, they are sufficient to indicate that a rather comprehensive knowledge of the aerosol properties is necessary in order to understand what their radiative and climatic effect will be.

These results should not leave us pessimistic about the prospects for getting a good handle on the climatic effect of atmospheric aerosols. On the contrary, we will argue that there are good reasons for being very optimistic about understanding future aerosol effects *if* proper observations are made. But it is, of course, necessary that we try our best to recognize what the complications are, and then take them into account in future studies and observational programs.

VOLCANIC AEROSOLS AS A TEST CASE OF CLIMATE PERTURBATIONS

A large volcanic eruption provides an excellent opportunity for a case study of the climatic response to a change in atmospheric aerosols. Volcanic aerosols injected into the stratosphere tend to be spread globally by stratospheric winds, so if their optical thickness is substantial, they should be capable of producing a measurable climatic response.

The effect of volcanic aerosols on climate has been considered by a large number of different investigators. In addition to those discussed in the introduction, we note especially the contributions of Lamb,[8] Pollack *et al.*,[9,10] and Mass and Schneider,[11] where the reader may also find references to many other papers. Lamb[8] has made the most comprehensive study, at least from an empirical point of view. He tabulates a chronology of volcanic eruptions and a "dust veil index" for the period subsequent to 1500 A.D., and, in general, he does find some correlations of volcanic events and climate perturbations. It is difficult to make a more quantitative statement than that because of the impossibility of obtaining accurate knowledge of the amount and properties of atmospheric aerosols produced by these past eruptions. In fact, Lamb's dust veil index is based, in part, on observed temperature data because of the lack of direct knowledge about the aerosols.

Pollack *et al.*[9,10] focus on the past century, 1870–1970, for which it is possible to crudely estimate the aerosol optical thickness from measurements of the transmission through the atmosphere for solar radiation. The measurements do not cover the full period; they are made with several different techniques, and they exist for only a small number of locations. However, for periods of substantial volcanic contributions, the estimates are probably reliable within about a factor of two. With the assumption that the measured opacity was due to stratospheric aerosols composed of sulfuric acid, Pollack *et al.* made radiative energy balance calculations to obtain the Northern Hemisphere temperature perturbation, as shown in FIGURE 5. They obtained an increase in temperature of about 0.5°K between 1890 and 1920, which is in rather good agreement with observations. However, between 1940 and 1963, the observed temperature declined by 0.2–0.3°K, while the model showed an increase because of an absence of substantial volcanic eruptions and a continuing CO_2 buildup.

It is clear that there are other mechanisms besides volcanic aerosols and CO_2 affecting the mean global or hemispheric temperature. This makes it difficult to compare observations with model results over a long time period and draw a firm conclusion about the role of aerosols in climate change. One way to try to minimize the effect of the simultaneity of different mechanisms is to make a statistical analysis of the temperature for the few years before and the few years after a large number of different volcanic explosions. This has, in fact, been done by Mass and Schneider,[11]

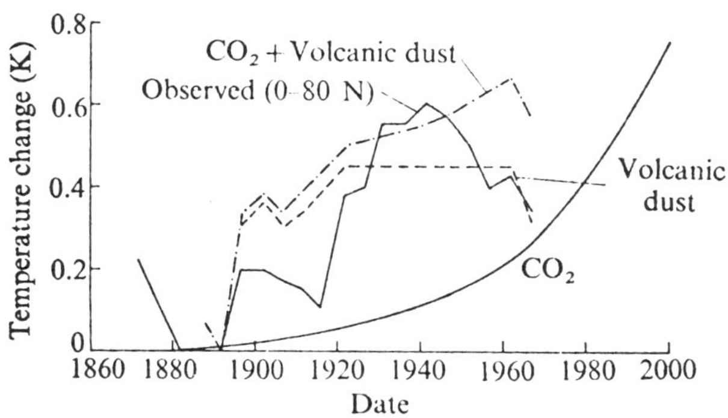

FIGURE 5. Observed Northern Hemisphere surface temperature changes and model results of Pollack *et al.*[10] The figure is from Pollack *et al.*,[10] who took the observed temperature curve from Mitchell.[16]

who find a small, but possibly significant, cooling after volcanic events, with the magnitude on the order of 0.1°K.

A different approach is to pick only the largest and best-documented volcanic eruptions and examine them in detail. This is the approach used by Hansen, Wang, and Lacis[4] for the eruption of Mount Agung on the island of Bali in 1963. This eruption occurred at a time when tropospheric temperatures were being measured routinely, accurate measurements of the aerosol optical depth were being made at several astronomical observatories, and there was direct sampling of the composition of stratospheric aerosols. Using the measured time history of the aerosol optical depths and a one-dimensional radiative-convective climate model, Hansen *et al.*[4] computed the expected effect of the aerosols on stratospheric and tropospheric temperatures. In their model, they employed a thermal inertia at the surface for 70 meters of ocean.

FIGURE 6 shows the computed and observed temperatures in the stratosphere. The observations[12] were made over Port Hedland, Australia, and are smoothed with a 3-month running mean. At altitudes near 20 km (pressures ~50 mb), the amplitude of the temperature increase was 4 to 8°C throughout the region from 10° N to 30° S. Part of the observed temperature increase may be due to the quasi-biennial oscillation, but it seems clear that the aerosols did result in stratospheric warming of a few degrees, with a time scale for the increase on the order of several weeks.[13]

FIGURE 7 shows the computed and observed[14] temperatures in the troposphere for the latitude range 30° N to 30° S. A more recent analysis[15] of the observations is in substantial agreement with the illustrated observations. It appears that after the Agung eruption, the average tropospheric temperatures did indeed decrease by a few tenths of a degree with a time scale on the order of 1 year, in agreement with the theoretical result.

Based on the computations in the preceding section, the physical explanation for the computed effects is straightforward. The overall effect of the added aerosols on the

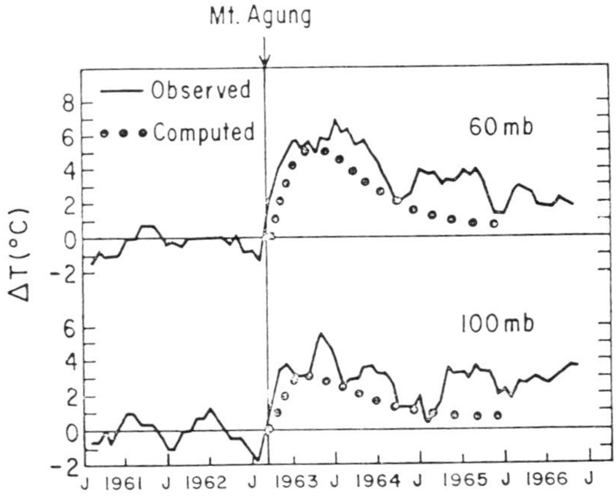

FIGURE 6. Observed stratospheric temperatures over Australia[12] and computed temperatures after the eruption of Mount Agung, assuming that the added stratospheric aerosols are sulfuric acid and that the average depth of the mixed layer of the ocean is 70 m. (Copyright 1978 by the American Association for the Advancement of Science. Reprinted from Hansen *et al.*[4] with the permission of the publisher.)

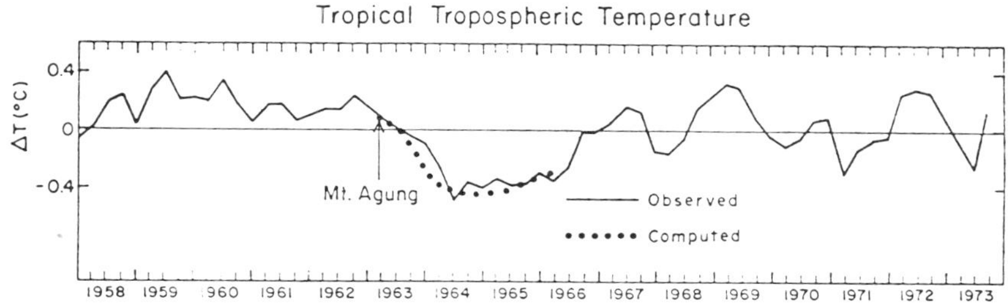

FIGURE 7. Observed tropospheric temperatures between 30° N and 30° S[14] and computed temperatures after the eruption of Mount Agung, assuming that the added stratospheric aerosols are sulfuric acid and that the average depth of the mixed layer of the ocean is 70 m. (Copyright 1978 by the American Association for the Advancement of Science. Reprinted from Hansen *et al.*[4] with the permission of the publisher.)

bulk of the atmosphere and the surface is cooling, because sulfuric acid is highly reflective to solar radiation and thus tends to decrease the amount of solar radiation absorbed by the earth-atmosphere system. The aerosols also interact with the thermal radiation and, in fact, warm the earth-atmosphere system through a "greenhouse" blocking of radiation from the surface and troposphere; however, aerosols of the assumed size have too small an optical thickness in the infrared for the greenhouse effect to exceed the albedo effect. The effect of the aerosols on the local (stratospheric) temperature is heating, because (1) they absorb thermal radiation from the warmer atmosphere and surface below more effectively than they release heat to space, and (2) they absorb a small amount of solar radiation because they have broad absorption bands in the near infrared. The amount of energy involved in the stratospheric heating is small, but the local temperature is substantially influenced because of the low density of the atmosphere in that region.

From the results in the preceding section, we also know that the thermal effect of the aerosols must depend on their composition and size distribution, properties that are not very accurately known for the first few weeks after the volcanic eruption. At that time, the aerosols may have included a substantial fraction of absorbing silicate (basaltic) dust. To examine the potential effect of aerosol composition, computations were made for two extreme cases—pure basaltic material and pure sulfuric acid (75% concentration by weight in water). TABLE 1 shows that the basaltic dust results in a substantial increase in the stratospheric heating and some decrease in tropospheric cooling.

The model results also depend on the assumed thickness of the ocean mixed layer and on restriction of the model domain to latitudes between 30° N and 30° S, as indicated in TABLE 1, but in ways that are easily understood. With a 100-m mixed layer instead of a 70-m one, the maximum tropospheric cooling is reduced, and there is no significant change in the stratospheric warming. The date of maximum tropospheric cooling is little changed because, in either case, the time constant for cooling the mixed layer is longer than that for changing the aerosol optical thickness. The stratospheric temperature change depends mainly on the assumed stratospheric aerosol properties, but also on tropospheric temperature, since the effect is basically due to absorption of upwelling thermal radiation. If global average properties are used

TABLE 1

TEMPERATURE CHANGE IN THE TROPOSPHERE AND STRATOSPHERE COMPUTED WITH THE ONE-DIMENSIONAL RADIATIVE-CONVECTIVE MODEL*

	Troposphere				Stratosphere (55 mb)			
Mixed Layer Depth (m):	70	100	70	70	70	100	70	70
Aerosol Composition:	Sulfate	Sulfate	Sulfate	Silicate	Sulfate	Sulfate	Sulfate	Silicate
Model Domain:†	Tropical	Tropical	Global	Tropical	Tropical	Tropical	Global	Tropical
Δt Months								
1	−0.02	−0.02	−0.01	−0.02	1.25	1.25	0.38	5.5
3	−0.08	−0.06	−0.03	−0.05	3.65	3.65	1.50	10.8
6	−0.25	−0.18	−0.11	−0.13	5.29	5.31	2.13	12.7
9	−0.40	−0.28	−0.17	−0.18	5.01	5.05	2.02	12.3
12	−0.50	−0.36	−0.22	−0.21	3.72	3.77	1.48	10.2
18	−0.57	−0.43	−0.24	−0.22	1.86	1.91	0.80	7.2
24	−0.56	−0.44	−0.24	−0.20	0.75	0.80	0.32	4.9
30	−0.52	−0.42	−0.22	−0.17	0.16	0.18	0.03	3.2
36	−0.47	−0.39	−0.21	−0.14	−0.13	−0.10	−0.03	2.2

*After Hansen *et al.*[4]

†The tropical model is limited to latitudes 30° N–30° S.

instead of those for latitudes 30° N to 30° S, the tropospheric cooling is reduced to about half of that illustrated in FIGURE 7, primarily because of the reduction of the average aerosol optical thickness. With a one-dimensional model, perhaps the most that can be said is that the best estimate is somewhere between the results computed for the global average and those computed for the isolated tropical cases. Despite this uncertainty, the results are in excellent agreement with the observations. We believe that the observations are fairly conclusive in indicating tropospheric cooling by a few tenths of a degree 1 or 2 years after Agung; but the very close fit to the theoretical curve in FIGURE 7 is almost certainly fortuitous, particularly in view of the noisy appearance of the observed temperatures in earlier and later years.

The greatest weakness in these model computations is probably the absence of interactions of the computed heating with the atmospheric dynamics. A related defect of the computations is the omission of potential cloud-cover feedbacks; the radiative perturbation due to aerosols could conceivably induce a change in the clouds that would counteract (or reinforce) the direct climatic effect of the aerosols. Proper analysis will require a model in which the cloud cover is computed from first principles and is allowed to interact realistically with the radiation field and the atmospheric dynamics. However, the extent to which the observed climatic effect agrees with that obtained from a simple radiative model in fact provides some evidence that such potential feedbacks do not overwhelm the direct radiative effect.

Finally, we emphasize the enormous potential scientific value of the natural climatic experiment provided by a large volcanic eruption. One very useful strategy for investigating the global climatic system would be to make detailed observations and analyses of the next large volcanic explosion. The observations should include global monitoring of the spread, growth, and decay of the aerosols, sampling of the aerosol properties *in situ,* and accurate global monitoring of the climatic response. These data would permit testing of global climate models and would aid in-depth analyses of radiative, chemical, and dynamic processes, which are essential for obtaining an improved understanding of the physical mechanisms and interactions involved.

DISCUSSION

The preceding sections suggest that aerosols can significantly influence climate. If additional evidence were needed, it could certainly be found in data obtained from recent spacecraft missions to our neighboring planets, Venus and Mars. Venus is found to be covered by a smog of sulfuric acid aerosols that veils the lower atmosphere from view. This planet-wide aerosol layer is apparently similar to the Junge layer on Earth in terms of its composition and the atmospheric pressure level at which it occurs. But the Venusian layer is much thicker, and the particles are larger. According to Pollack and Young,[19] the larger particle size causes the Venusian aerosols to act as a greenhouse material rather than to cool the planet. In the case of Mars, the atmospheric dynamics seem to be intimately tied in with the presence of aerosols. The coupling apparently works through the radiation, as a result of the fact that the aerosols are primary atmospheric absorbers of both solar and thermal radiation. Study of the aerosol processes on these planets should help us to understand the possible role of aerosols in the terrestrial climate.

In the case of the Earth, perhaps the main question about aerosols is: How is man modifying the nature and amount of atmospheric aerosols, and how will this affect the climate? We don't believe that we have the evidence to confidently answer even the first half of the question, let alone the second half. It is obvious that man is influencing aerosols, but we do it in many different ways, and it is not clear how significant this is on a global basis. The claim has often been made that the global cooling since about 1940 is a result of man-made aerosols, but there are not adequate observations of aerosols or adequate climate models to prove or disprove this.

In order to begin to answer these questions, the first thing that we will need is better observations. And these should be on a global basis, because aerosols are very heterogeneous and there are many factors that can influence their properties and spatial distributions in different ways. The second thing that is needed, and which should be emphasized, is an improved understanding of aerosol processes, e.g., the physics that governs the formation and life cycle of aerosols as well as the influence of aerosols on other climate processes, such as cloud formation. A third thing needed is, of course, the construction of more realistic climate models, and the verification of the models in some more or less well-defined cases.

We have tried to emphasize that one fruitful approach may be to focus attention on a particular problem or problems, such as the volcanic aerosol problem. By doing this, it may be practical to handle the complete range of tasks: getting the global observations, doing the necessary process studies, and building and testing the climate models. If such an effort were successfully carried out, it could give us considerable confidence in our ability to understand other climate perturbations—not only aerosols, but other radiative perturbations, such as that due to changing CO_2.

This approach should not be taken at the expense of other aerosol research. In fact, as we have also emphasized, a lot can be learned by pursuing other problems, and we gave the Venus and Mars aerosols as an example. But, at the same time, there is a good rationale for focusing an effort on carrying through a full investigation in one case, and we think that the next large volcanic eruption would provide a remarkably good opportunity to do that.

Summary

We briefly review past ideas and calculations concerning the effect of atmospheric aerosols on climate. Current modeling capabilities permit an accurate assessment of the direct radiative effects of aerosols, provided that the physical properties of the aerosols are known. We illustrate that this climatic forcing by aerosols depends substantially upon the aerosol composition, size distribution, and altitude distribution, as well as upon the aerosol amount. Modeling capabilities for the complete climatic system are at a more primitive state, but we argue that the radiative forcing due to aerosols produced by volcanic explosions provides an excellent opportunity for testing climate modeling capabilities. Examination of the best-documented large volcanic eruption—of Mount Agung in 1963—provides encouraging evidence with regard to our ability to understand and model the climatic effect of aerosols. Finally, we discuss the possible climatic impact of anthropogenic aerosols.

References

1. Sparks, "Life of Benjamin Franklin," Vol. 6, 455–457 (as cited in Proc. Am. Phil. Soc., 45; 127, 1906).
2. Humphreys, W. J. 1940. Physics of the Air. McGraw-Hill Book Co. New York, N.Y.

3. RASOOL, S. I. & S. H. SCHNEIDER. 1971. Atmospheric carbon dioxide and aerosols: effects of large increases on global climate. Science **173:** 138.
4. HANSEN, J. E., W. C. WANG & A. A. LACIS. 1978. Mount Agung eruption provides test of a global climate perturbation. Science **199:** 1065.
5. WANG, W. C. & A. A. LACIS. 1979. A one-dimensional radiative-convective model of the earth's atmosphere for climate applications. (In preparation.)
6. MANABE, S. & R. T. WETHERALD. 1967. Thermal equilibrium of the atmosphere with a given distribution of relative humidity. J. Atmos. Sci. **21:** 241.
7. TOON, O. B. & J. B. POLLACK. 1976. A global average model of atmospheric aerosols for radiative transfer calculations. J. Appl. Meteorol. **15:** 225.
8. LAMB, H. H. 1970. Volcanic dust in the atmosphere; with a chronology and assessment of its meterological significance. Philos. Trans. R. Soc. London Ser. A **266:** 425.
9. POLLACK, J. B., O. B. TOON, C. SAGAN, A. SUMMERS, B. BALDWIN & W. VAN CAMP. 1976. Volcanic explosions and climatic change: a theoretical assessment J. Geophys. Res. **81:** 1071.
10. BALDWIN, B., J. B. POLLACK, A. SUMMERS & O. B. TOON. 1976. Stratospheric aerosols and climatic change. Nature **263:** 551.
11. MASS, C. & S. H. SCHNEIDER. 1977. Statistical evidence on the influence of sunspots and volcanic dust on long-term temperature records. J. Atmos. Sci. **34:** 1995.
12. NEWELL, R. E. 1970. Stratospheric temperature change from the Mt. Agung volcanic eruption of 1963. J. Atmos. Sci. **27:** 977.
13. MCINTURFF, R. M., A. J. MILLER, J. K. ANGELL & J. KORSHOVER. 1971. Possible effects on the stratosphere of the 1963 Mt. Agung volcanic eruption. J. Atmos. Sci. **28:** 1304.
14. NEWELL, R. E. & B. C. WEARE. 1970. Factors governing tropospheric mean temperature. Science **194:** 1413.
15. ANGELL, J. K. & J. KORSHOVER. 1977. Estimate of the global change in temperature, surface to 100 mb, between 1958 and 1975. Mon. Weather Rev. **105**(4): 375.
16. MITCHELL, J. M. 1970. A reassessment of atmospheric pollution as a cause of long-term changes of global temperature. *In* Global Effects of Environmental Pollution. F. Singer, Ed.: 139. Springer Verlag. New York, N.Y.
17. PATTERSON, E. M., D. A. GILLETTE & B. H. STOCKTON. 1977. Complex index of refraction between 300 and 700 nm for Saharan aerosols. J. Geophys. Res. **82:** 3153.
18. VOLZ, F. E. 1972. The complex index of refraction of aerosols from 0.2 to 40 μm. *In* Conference on Atmospheric Radiation, American Meteorological Society, August 7–9, 1972: 114–115.
19. POLLACK, J. B. & R. YOUNG. 1975. Calculations of the radiative and dynamical state of the Venus atmosphere. J. Atmos. Sci. **32:** 1025.

FUTURE AEROSOLS OF THE SOUTHWEST: IMPLICATIONS FOR FUNDAMENTAL AEROSOL RESEARCH*

S. K. Friedlander

Department of Chemical, Nuclear, and Thermal Engineering
University of California
Los Angeles, California 90024

Substantial increases in coal utilization in the United States are anticipated as a result of shortages in the supply of gas and oil. (At least temporarily, however, there is a gas surplus, and the Department of Energy has been encouraging industry and utilities to switch from oil to gas.)[1] The National Energy Plan (NEP), proposed by the administration in April 1977, is probably the best known projection for increased coal utilization. The environmental implications of NEP have been reviewed in a report prepared by the Congressional Research Service,[2] based on studies by outside contractors.

According to NEP, the amount of coal burned in industrial boilers and electrical utilities in the U.S. will almost double from 1975 to 1985 and about triple by 2000. Without NEP, with "business as usual," coal combustion is expected to increase about 2½ times by 2000.

The most striking trend in the projections of emissions is the substantial increases in emissions of particulate matter, SO_x, and NO_x for the part of the U.S. west of the Mississippi (TABLE 1). Both hydrocarbons and CO are expected to decrease.

The largest increases in emissions are projected for Air Quality Control Region 6, which includes Texas and New Mexico. From 1975–85, a 60 to 70% increase in emissions of both SO_x and NO_x is anticipated for this region. The increase in SO_x emissions will result from the anticipated switch from gas to coal combustion; increased NO_x emissions will be due in part to coal combustion and in part to trucking activities. The total *mass* of particulate emissions in Region 6 is projected to remain the same from 1975–85. However, as discussed below, a shift will probably take place in the primary particulate emissions from coarse to submicron aerosols.

Thus, the West can expect substantial increases in emissions of the aerosol precursor gases, SO_x and NO_x; emissions of hydrocarbons are expected to decrease. The other important aerosol precursor, NH_3, has both natural and man-made sources. The relative amounts are not known.

NATURE OF THE SUBMICRON AEROSOL IN THE SOUTHWEST

The visibility-reducing aerosol in the Southwest has been studied as part of the EPA-sponsored VISTTA Project.[3] Flights were made over Arizona, Utah, Colorado, and New Mexico on October 5 and 9, 1977. The total flight path was over 1,000 km, and each day's flight covered the same path but in reverse directions. Measurements

*Supported in part by Environmental Protection Agency Grant No. R805736.

0077–8923/80/0338–0588 $01.75/1

were made of the light scattering and chemical composition for the same aerosol samples. The measured light scattering on the days of observation was in the range of long-term averages for the region. While the results are of limited statistical significance, they provide a useful, detailed picture of the nature of the background submicron aerosol on these two days.

Many of the chemical species present in the submicron aerosol have been identified, providing important clues to the origins of this material. The element present in highest concentration was silicon, with sulfur levels a little lower. The mass concentration of sulfur in combined form as sulfate is somewhat higher than that of silicon as silicon dioxide. The presence of silicon in the submicron aerosol, if confirmed, indicates a significant component of primary fine particles. The high concentration of silicon is surprising; it has generally been assumed that the submicron aerosol is largely of secondary origin, resulting from the conversion of SO_x, NO_x, and organic compounds to aerosols. [**Note added in proof:** More recent data indicate lower silicon concentrations than originally reported.]

A breakdown of the submicron aerosol according to chemical composition for

TABLE 1
ANNUAL EMISSIONS OF CRITERIA AIR POLLUTANTS FOR THE WEST[2]

	1975*	2000* (NEP Projected)
Particulates	3.6	4.7
Sulfur oxides	5.8	8.6
Nitrogen oxides	5.4	10.9
Hydrocarbons	6.4	4.7
Carbon monoxide	34.7	19.5

*Millions of tons.

these two days is shown in TABLE 2. Of the identified components, more than half are of primary origin. However, the components that have not been measured are mostly of secondary origin.

What will be the effect of the increased emissions projected under NEP on submicron aerosols? Quantitative estimates of the fractional conversion of SO_2 to sulfate starting from first principles will be difficult. If the increased emissions of NO_x, SO_x, particulate matter, and hydrocarbons were in the same proportion as in existing sources, a reasonable assumption would be a linear scale-up with SO_2 emissions. This would hold best for a collection of noninteracting sources in similar meteorological regimes.

As noted above, however, SO_x and NO_x emissions in Region 6 will rise proportionally but from somewhat different sources, and hydrocarbon emissions will decrease. The photochemical conversion of SO_2 to sulfate depends in a complex way on the levels of NO_x and hydrocarbons, as well as on SO_2 concentrations.[4] Thus, a linear scale-up of sulfate levels with SO_2 emissions cannot be justified on fundamental grounds.

Reliable data on ammonium nitrate concentrations in ambient air in the Southwest are not available. In addition, the equilibrium between particulate-phase ammonium nitrate and gas-phase ammonia and nitric acid is very sensitive to temperature. Thus, the distribution of ammonium nitrate between gas and aerosol phases may change markedly during the course of the day. This complicates attempts to scale up ammonium nitrate levels with emissions. It also indicates that great care must be exercised in sampling for nitrate. Equilibrium effects for ammonium nitrate are discussed below.

The total mass of particulate emissions is projected to remain about the same for Region 6. This will be accomplished by tightening the new-source performance standard for the total mass of particulate emissions. However, most of the particulate mass in the postcombustion gases from pulverized coal combustion is in the size range larger than 1 μm.[5] The collection efficiency of electrical precipitators is greatest for such large particles, passing through a minimum in the size range between 0.1 and 1 μm.[6] Thus, the tightened new-source performance standard for *total* mass will probably be met by preferential removal of coarse particles. The result will be an increase in the mass of primary submicron particles.

TABLE 2
SUBMICRON AEROSOL CHEMICAL COMPOSITION: SOUTHWEST BACKGROUND AEROSOL ($\mu g/m^3$)

	1977[3]	Projected (see text)
Secondary		
SO_4^{2-}	2.04	increase
NH_4^+ (with sulfate)	0.76	increase
NO_3^-	unknown	increase (perhaps large)
NH_4^+ (with nitrate)	unknown	increase
Organics	unknown	decrease or unchanged
Primary		
SiO_2	1.54	increase
Other (as oxides)	0.93	increase
Water	unknown	increase
Total	>5.27	≫5.27

In summary, quantitative estimates of the levels of the various components of the Southwest aerosol resulting from increasing emissions cannot be made with confidence. However, qualitative estimates of the trends (TABLE 2) indicate that a substantial increase can be expected in the mass of the submicron aerosol.

In the following sections, we discuss recent research by our group on aerosol behavior as it relates to secondary aerosol formation in the Southwest. The problems considered are the distribution of sulfur (as sulfate) with respect to particle size, new particle formation, and the ammonium nitrate equilibrium between gas and aerosol phases.

Distribution of Sulfur with Respect to Particle Size

The development of a low pressure impactor capable of fractionating aerosol particles larger than about 500 Å[7,8] has made it possible to measure the distribution of sulfur and other chemical species over most of the particle size range of interest. The new impactor has eight single-jet stages and a flow rate of 1 liter per minute. The aerodynamic cutoff diameters corresponding to a particle collected with an efficiency of 50% are 4.0, 2.0, 1.0, 0.5, 0.25, 0.11, 0.075, and 0.05 μm. The cutoff diameters were determined by calibration using monodisperse polystyrene latex and uranine aerosols.

Particles larger than 0.5 μm are fractionated at near-atmospheric pressure over the first four stages of the impactor. The last four stages operate at pressures from 150 mm down to 8 mm Hg (absolute) to size segregate the smaller particles. Material deposited at each stage can be analyzed for sulfur-containing compounds at the nanogram level by flash volatilization followed by flame photometric analysis (Roberts and Friedlander[9]). Enough material for chemical analysis can be collected by sampling ambient air with sulfate levels of 10 μg m^{-3} for 15 minutes. The impactor has been checked with ammonium sulfate and sulfuric acid aerosols by comparison with an electrical mobility analyzer. Good agreement between the two instruments was found, indicating that the evaporation of these compounds in the impactor was not important. A further check was carried out by using both the impactor and a filter—simultaneously and in parallel—to sample an ammonium sulfate aerosol. Excellent agreement was obtained.

Measurements have been made of the sulfur mass distribution of the Los Angeles aerosol with the low pressure impactor. The results of many measurements indicate that the sulfur mass distribution usually peaks in the size range around 0.5 μm (Figure 1). This is the size range that corresponds to the wavelength of the visible light and is strongly light scattering.

We have found quite different distributions of sulfur with respect to particle size in desert areas. Figure 2 shows the results of measurements made near Trona, California, about 170 miles northeast of Los Angeles, west of Death Valley. On the day the measurements were made, wind speeds averaged about 2 miles/hour, and the aerosol could be considered to be developing in an almost stationary air mass. As shown in Figure 2, the distributions tend to peak in the size range between 0.1 and 0.2 μm. Particles in this size range are much less efficient (on a mass basis) in scattering light than are the aerosols observed in the Los Angeles area. Measurements in other arid areas have shown similar size distributions.

What is the cause of the marked difference in sulfur size distributions between the coastal (type "C") and desert (type "D") regions? It is possible that in a coastal area, such as Los Angeles, much of the SO_2 conversion takes place in fogs in droplet phase and that this leads to the accumulation of sulfates in the larger particle sizes typical of type "C" aerosols. Other possible explanations are discussed by Friedlander.[10] The

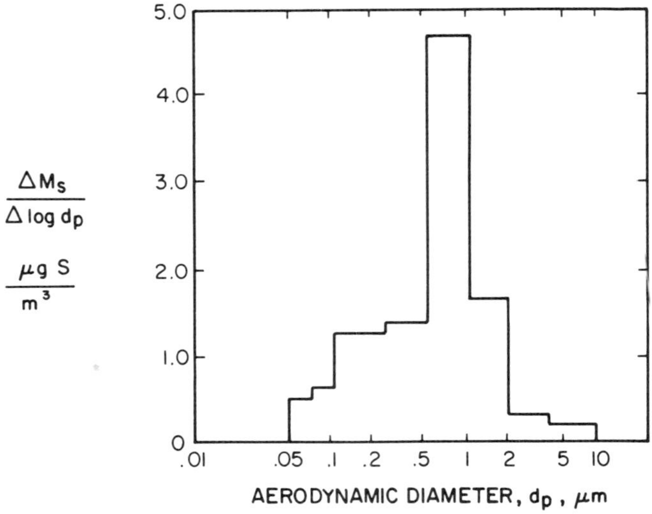

FIGURE 1. Aerosol sulfur mass distribution, Pasadena, California, December 26, 1978. Average of two samples from 1428–1600 PST. Total mass loading of sulfur was 10.6 $\mu g/m^3$. (Courtesy Mr. J. Ouimette, California Institute of Technology.)

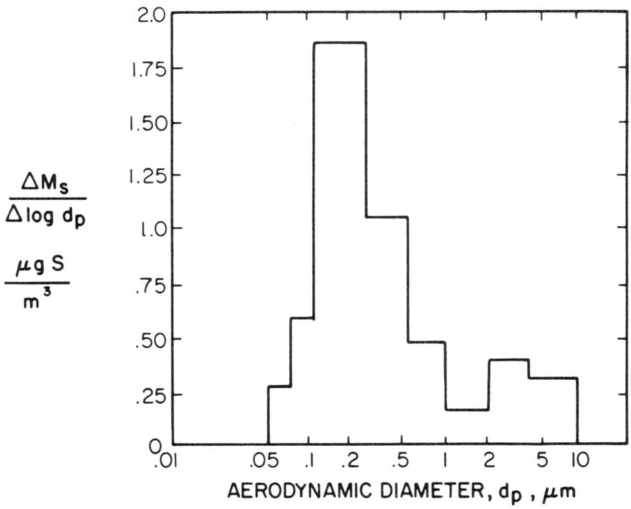

FIGURE 2. Aerosol sulfur mass distribution, Trona, California, May 13, 1978, from 0946–1145 PDT. Total mass loading of sulfur was 1.42 $\mu g/m^3$. (Courtesy Mr. J. Ouimette, California Institute of Technology.)

type "D" aerosols almost certainly result from homogeneous gas-phase reactions involving the conversion of SO_2 to sulfate; this type of process has been studied in balloon reactor experiments, which confirm that a sulfur mass peak develops in the smaller particle size range.[10] Further measurements are needed to verify the existence of these two types of sulfate-containing aerosols and to establish their origins.

New Particle Formation

Atmospheric measurements of condensation nuclei in rural sites have been made by Went,[11] Hogan,[12] and Lopez, Servant, and Fontan.[13] All of these investigators concluded that new particles were formed in the atmosphere during daytime hours. White *et al.*[14] measured concentrations of condensation nuclei in the plume of the Labadie power plant in St. Louis and observed peaks in the nuclei counts at the edges of the plume in some cases. They concluded that new particles were forming even in the presence of the particles already in the plume.

McMurry[15] studied new particle formation in the presence of the high particle concentrations present in the Los Angeles atmosphere. From his measurements, he concluded that new particle formation did take place, particularly during periods of intense solar radiation. Thus, new particle formation has been observed in rural areas, at the edges of power plant plumes, and even in particle-laden urban atmospheres.

When an aerosol is present in a nucleating gas, classical nucleation theory must be significantly modified because of scavenging of clusters by the aerosol.[18] The process is analogous to the scavenging of ions—which form continuously in the atmosphere—by aerosol. While a complete theory has not been worked out for cluster scavenging, a similitude analysis gives interesting results.

Let n_g be the concentration of clusters containing g monomer molecules. In the steady state, the cluster balance equation takes the following form:[18]

$$I_g = I_{g+1} + n_g \left(\frac{kT}{2\pi m_g}\right)^{1/2} A \quad \textbf{(1)}$$

where I_g, the particle droplet current, is the net rate of formation of particles of size g from clusters of size $g - 1$. The net rate of loss of particles of size g to the range $g + 1$ by growth processes is I_{g+1}. Collision and coagulation among the clusters is neglected.

Clusters deposit on the aerosol at a rate proportional to the aerosol surface area if the particles are smaller than the mean free path of the carrier gas. (It is assumed that the aerosol particles are uniformly dispersed on a scale comparable to that on which the chemical reactions are taking place.) The mean velocity of the clusters is given by $(kT/2\pi m_g)$, where k = Boltzmann's constant, T = absolute temperature, m_g = mass of a cluster of size g, and A = surface area of preexisting aerosol per unit volume of gas. We consider only the case of a single nucleating species for which the particle current can be written as follows:[16,17]

$$I_g = \beta n_{g-1} s_{g-1} - \alpha_g n_g s_g \quad \textbf{(2)}$$

where β, the monomer flux, is given by:

$$\beta = n_1 \left(\frac{kT}{2\pi m_1} \right)^{1/2} \tag{3}$$

and α_g, the evaporative flux, is given by:

$$\alpha_g = \beta \frac{s_{g-1}}{s_g} \exp\left[\frac{4\sigma v_1}{d_p kT} - \ln S \right] \tag{4}$$

where s_g = surface area of a cluster, σ = surface tension, v_1 = volume of monomer molecule, d_p = cluster diameter, and S = saturation ratio.

The general nature of the solution to (**1**) can be examined by a similitude analysis: substituting (**2**) in (**1**), the difference equation for the cluster size distribution is given by:

$$\beta n_{g-1} s_{g-1} - \alpha_g n_g s_g - \beta n_g s_g + \alpha_{g+1} n_{g+1} s_{g+1} = n_g \left(\frac{kT}{2\pi m_g} \right)^{1/2} A \tag{5}$$

This is a second order difference equation, and two boundary conditions are needed. These are taken to be the same as in classical nucleation theory. For $g = 1$, $n_g = n_1$; and for $g \to \infty$, $n_g \to 0$. The ratio α_g/β can also be expressed as follows, using (**4**):

$$\frac{\alpha_g}{\beta} = \frac{s_{g-1}}{s_g} \exp\left[\left(\frac{d_p^*}{d_p} - 1 \right) \ln S \right] \tag{6}$$

$$= \frac{s_{g-1}}{s_g} S^{[(d_p^*/d_p)-1]} \tag{6b}$$

$$= \left(\frac{g-1}{g} \right)^{2/3} S^{[(g^*/g)^{1/3}-1]} \tag{6c}$$

where the critical particle diameter is given by:

$$d_p^* = \frac{4\sigma v_1}{kT \ln S} \tag{7}$$

Rearranging (**5**) and substituting (**6c**), the result for the dimensionless cluster concentration $\tilde{n}_g = n_g/n_1$ is:

$$\tilde{n}_{g-1} \left(\frac{g-1}{g} \right)^{2/3} - \tilde{n}_g \left\{ \left(\frac{g-1}{g} \right)^{2/3} S^{[(g^*/g)^{1/3}-1]} + 1 \right\} + \tilde{n}_{g+1} S^{\{[g^*/(g+1)]^{1/3}-1\}} - \frac{\tilde{n}_g}{g^{7/6}} \frac{A}{n_1 s_1} = 0 \tag{8}$$

with the boundary conditions:

$$g = 1, \quad \tilde{n}_g = 1$$

$$g \to \infty, \quad \tilde{n}_g \to 0$$

Three dimensionless groups appear in (**8**): the saturation ratio S, the critical cluster size g^*, and the area ratio $A/n_1 s_1$. For the steady state, the dynamics of new particle formation are controlled by these dimensionless groups.

The limiting cases can be distinguished; when the concentration of preexisting aerosol vanishes, $A/n_1 s_1 \rightarrow 0$. This corresponds to the case of constant particle current, well known in classical nucleation theory:[16,17]

$$\frac{I}{\beta n_1 s_1} = \left(\frac{\pi}{6}\right)^{1/2} (g^*)^{1/6} (\ln S)^{1/2} S^{-g^*/2} \tag{9}$$

The dimensionless particle current depends only on g^* and S and not on $A/n_1 s_1$.

For large values of $A/n_1 s_1$, a solution for the dimensionless particle current can be found as follows. Setting $I_g = 0$ in (**2**) gives the equilibrium size distribution for the clusters:

$$\frac{\alpha_g}{\beta} = \left(\frac{n_{g-1} s_{g-1}}{n_g s_g}\right)_{\text{eq}} \tag{10}$$

Substituting (**10**) in (**2**) and rearranging gives:

$$I_g = \beta n_{g-1} s_{g-1} (1 - r_{g-1}) \tag{11}$$

where:

$$r_{g-1} = \frac{(n_{g-1} s_{g-1}/n_g s_g)_{\text{eq}}}{n_{g-1} s_{g-1}/n_g s_g}$$

When $A/n_1 s_1$ is very large, the clusters are rapidly scavenged by the foreign aerosol. The cluster size distribution drops sharply with increasing g, much more rapidly than the equilibrium distribution. Thus $r_{g-1} \ll 1$ and (**11**) becomes:

$$\frac{n_{g-1} s_{g-1}}{n_g s_g} \approx \frac{1}{g^{7/6}} \left[\frac{A}{n_1 s_1}\right] \tag{12}$$

The particle current is no longer independent of g and is given by:[18]

$$\frac{I_{g+1}}{\beta n_1 s_1} = \frac{[g!]^{7/6}}{[A/n_1 s_1]^{g-1}} \tag{13}$$

The dimensionless particle current depends only on $A/n_1 s_1$ (and on g).

The dependence of the dimensionless particle current $I_g/\beta n_1 s_1$ on the area ratio $A/n_1 s_1$ is shown in FIGURE 3. For small values of $A/n_1 s_1$, classical nucleation theory applies and $I_g/\beta n_1 s_1$ is independent of $A/n_1 s_1$. For large values of $A/n_1 s_1$, $I_g/\beta n_1 s_1$ depends only on the area ratio.

A more complete formulation than (**2**) would take into account the motion of the cluster in the collision process. This would lead to different limiting results for the particle current, but the same dimensionless groups are still sufficient to describe the system.

New particle formation in the presence of an aerosol has been studied under controlled experimental conditions by McMurry.[15] In smog chamber experiments, he measured the rate of formation of particles larger than 100 Å using an electrical aerosol analyzer. He found that the rate of new particle formation was strongly

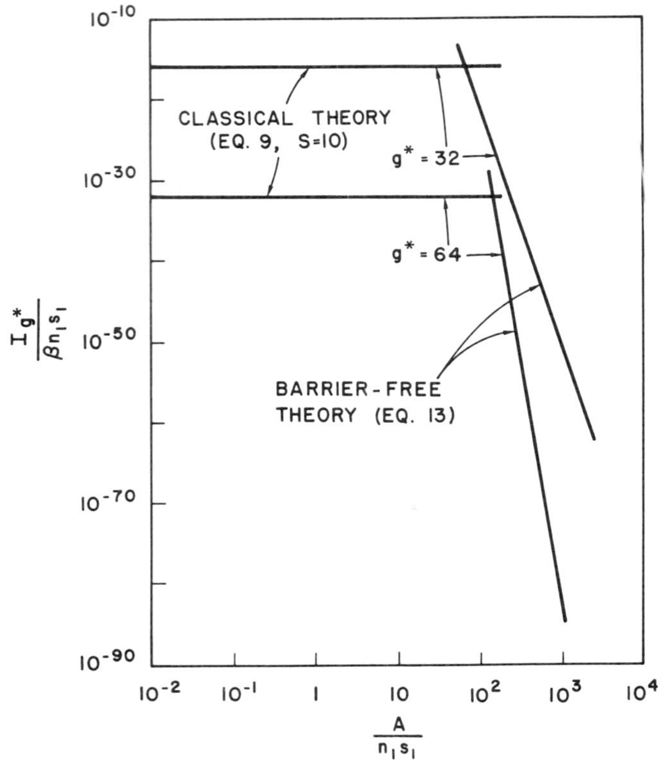

FIGURE 3. Variation of the dimensionless particle current with the ratio, A/n_1s_1 for $S = 10$ and $g^* = 32$ and 64. For small values of A/n_1s_1, the particle current depends only on g^* and S (Equation 9). For large values of A/n_1s_1, the particle current is independent of g^* and S but drops sharply with increasing A/n_1s_1 (Equation 13).

affected by the amount of foreign aerosol present. He was able to correlate his results by a theory similar to the analysis developed here for large values of A/n_1s_1. In his analysis, he included the rate of formation of condensable species by gas-phase chemical reaction.

No data are available on the rate of new particle formation in the arid regions of the Southwest. Such measurements should be made, and these new concepts tested.

The Ammonium Nitrate Equilibrium

Substantial amounts of gas phase NH_3 and HNO_3 exist in equilibrium with solid ammonium nitrate at ambient temperatures.[19] The equilibrium relation can be written:

$$NH_4NO_3(s) \rightleftharpoons NH_3(g) + HNO_3(g) \quad \textbf{(14)}$$

and the partial pressure product, based on an extrapolation of experimental data, is given by:

$$\ln K_c = 62.296 - \frac{21510}{T} \quad \textbf{(15)}$$

where $K_c = c_{NH_3}\, c_{HNO_3}$ is the equilibrium constant expressed in ppm^2 and T is the absolute temperature in °K. This relationship holds for a pure solid NH_4NO_3 phase even if mixed with other solid salts. It must be modified if the NH_4NO_3 is present as part of a solid solution.

The distribution of ammonium nitrate between the gas and aerosol phases is very sensitive to temperature. Over the range from 20 to 30°C, the equilibrium mass concentration in the gas phase of NH_3 and HNO_3 (equimolar) increases from 11.0 to 38.4 $\mu g/m^3$. This is in the range of mass concentrations observed for NH_4NO_3 in the aerosol phase in Los Angeles.

If the partial pressure product of NH_3 and HNO_3 is less than the equilibrium value at a given temperature, there can be no pure ammonium nitrate present in the aerosol phase. Once the gas is saturated with ammonium nitrate, the addition of more NH_3 or HNO_3 results in the precipitation of solid until the partial pressure product reaches the equilibrium value.

It may be that the atmosphere in the Southwest already contains substantial amounts of NH_3 and HNO_3, but at concentrations less than the equilibrium partial pressure product. The formation of additional HNO_3 from increased emissions of NO_x would then result in the formation of aerosol ammonium nitrate if K_c were exceeded.

The high volatility of ammonium nitrate has important implications for sampling aerosol nitrate. Measurements by Smith *et al.*[20] of nitrate on filters indicate that volatilization occurs rapidly for filters stored in the laboratory at room temperature and open to the atmosphere.

Summary

Increased coal utilization is projected to result in a 50% increase in emissions of sulfur oxides and a doubling of nitrogen oxide emissions in the western U.S. by the year 2000. The conversion of these gases to particulate form will increase ambient levels of aerosol sulfates and nitrates. Submicron primary particle emissions probably will rise also, so that a large increase in submicron aerosol mass is likely.

Some work has been done on the distribution of sulfur compounds with respect to particle size in the Southwest. Two types of sulfate size distributions have been observed, corresponding to coastal and desert sites. Desert sulfates peak in a smaller particle size range, probably because they result from homogeneous gas-phase reactions.

The total aerosol number concentration will probably increase as a result of increased emissions of aerosol precursor gases. The rate of new particle formation depends in a complex way on the concentration of the preexisting aerosol when condensable species form by chemical reaction. Limiting cases for the general theory are discussed.

Little is known about the nitrate component of the Southwest aerosol. The situation is complicated by the high vapor pressure of ammonia and nitric acid over solid ammonium nitrate.

REFERENCES

1. Anonymous. 1978. The pressure shifts back to natural gas. Bus. Week (December 4): 36.
2. TRUMBULE, R. E., J. E. BINIEK & J. E. BLODGETT. 1977. Environmental Challenges of the President's Energy Plan: Implications for Research and Development. Report prepared for the Subcommittee on the Environment and the Atmosphere of the Committee on Science and Technology. U.S. Government Printing Office. Washington, D.C.
3. MACIAS, E. S., D. L. BLUMENTHAL, J. A. ANDERSON & B. K. CANTRELL. 1980. Size and composition of visibility-reducing aerosols in southwestern plumes. Ann. N.Y. Acad. Sci. (This volume.)
4. SANDER, S. P. & J. H. SEINFELD. 1976. Environ. Sci. Technol. **10:** 1114.
5. FLAGAN, R. C. & S. K. FRIEDLANDER. 1978. Particle formation in pulverized coal combustion—a review. *In* Recent Developments in Aerosol Science. D. T. Shaw, Ed. **2:** 25–29. John Wiley & Sons, Inc. New York, N.Y.
6. MCCAIN, J. D., J. P. GOOCH & W. B. SMITH. 1975. J. Air Pollut. Control Assoc. **25:** 117.
7. HERING, S. V., R. C. FLAGAN & S. K. FRIEDLANDER. 1978. Environ. Sci. Technol. **12:** 667.
8. HERING, S. V., S. K. FRIEDLANDER, J. J. COLLINS & L. W. RICHARDS. 1979. Environ. Sci. Technol. (In press.)
9. ROBERTS, P. T. & S. K. FRIEDLANDER. 1976. Atmos. Environ. **10:** 403.
10. FRIEDLANDER, S. K. 1978. Atmos. Environ. **12:** 187.
11. WENT, F. W. 1964. Proc. Nat. Acad. Sci. **51:** 1259.
12. HOGAN, A. W. 1968. Atmos. Environ. **2:** 599.
13. LOPEZ, A., J. SERVANT & J. FONTAN. 1974. Atmos. Environ. **8:** 733.
14. WHITE, W. H., J. A. ANDERSON, W. R. KNUTH, D. L. BLUMENTHAL, J. C. HSUING & R. B. HUSAR. 1976. Mapping Large Pollutant Plumes by Instrumented Aircraft: Support for Project MISTT, 1974. Report MRI 76 FR-1414 to EPA, Contract No. 68-02-1919. Meteorology Research Inc. Altadena, Calif.
15. MCMURRY, P. H. 1977. On the Relationship between Aerosol Dynamics and the Rate of Gas-to-Particle Conversion. Ph.D. dissertation. Environmental Engineering Science, California Institute of Technology. Pasadena, Calif.
16. FRIEDLANDER, S. K. 1977. Smoke, Dust and Haze. Wiley-Interscience. New York, N.Y.
17. FRENKEL, J. 1955. Kinetic Theory of Liquids. Dover Publications, Inc. New York, N.Y.
18. FRIEDLANDER, S. K. 1978. A note on new particle formation in the presence of an aerosol. J. Colloid Interface Sci. **67:** 387.
19. STELSON, A. W., S. K. FRIEDLANDER & J. H. SEINFELD. 1978. A note on the equilibrium relationship between ammonia and nitric acid and particulate ammonium nitrate. Atmos. Environ. **13:** 369.
20. SMITH, J. P., D. GROSJEAN & J. N. PITTS, JR. 1978. J. Air Pollut. Control Assoc. **28:** 930.

OVERVIEW OF AEROSOL RESEARCH

Merril Eisenbud

Institute of Environmental Medicine
New York University Medical Center
New York, New York 10016

We of course have been concerned here only with the physical and chemical aspects of the atmosphere, and there have been only passing references to the biological effects. The impression I have—concerning which, I realize, there is not likely to be general agreement—is that during the past decade, the effects of air pollution on people have proven to be less severe than many predicted would be the case. This, however, has been offset by findings that indicate that the effects on the physical environment are more severe than had been appreciated formerly.

The epidemiologists have evidence of air pollution effects on human health. But they are what I, from the point of view of a public health officer, would describe as second-order effects. These probably, of themselves, are not worth the $10 to $15 billion a year that is now being spent to control just two types of sources, namely, power-plant and tail-pipe emissions.

On the other hand, I think the past ten years have shown that there are geophysical and geochemical effects that were almost unexpected ten years ago and that are proving to be far more serious than even the most pessimistic of the estimates about health effects.

I don't know what to conclude about haze. If we want to live in a comfortable society with plenty of energy and lots of leisure time, and be productive enough so that we have the money we need to enjoy our leisure time, I think we can tolerate some haze. This, of course, would be contrary to the views of many people. However, it may be a price that we have to pay in some parts of the country for enjoying the benefits of energy use.

On the other hand, when one considers the implications of the greenhouse effect, we have—for the first time in my opinion—an environmental impact that is totally unacceptable. Civilization could be destroyed almost totally by the effects of the global geophysical changes that may be within our power to trigger.

Although Humboldt and Arrhenius pointed out the possibility of a greenhouse effect a century ago, it wasn't really until after World War II that the first of the important papers were published on this subject.

I am sure that we can assemble all the scientific information we need for the models required to predict the consequences of increasing carbon dioxide concentrations, but this process will certainly take 25 to 50 years. However, I am not sure we will have the time needed to develop the institutional mechanisms that will be required on a global scale to control CO_2 emissions, should that prove to be necessary.

A hundred years seems to be a long time when one is young, but as I look around the room, I think there are four of us whose professional experience totals just about 200 years. From the point of view of a mature scientist, a hundred years is not a long time, and in terms of history, it's only a point in time.

I know I may have shocked some of you by saying that the health implications

0077–8923/80/0338–0599 $01.75/1 © 1980, NYAS

have turned out to be less in my opinion than many people thought they were. I realize there are exceptions. Ray Wanta, who is here, and I worked on an air pollution problem that was killing people 31 years ago. We all know about Donora, and you know about the London smog. But those are fairly rare cases, and I think the changes that have taken place in the last 20 or 25 years minimize the likelihood of repeated such examples.

On the other hand, it does bother me when I think back about ten years to the widely publicized statement—made by the then President of the United States when he signed one of the Clean Air Acts—that the act had come just in time, that the

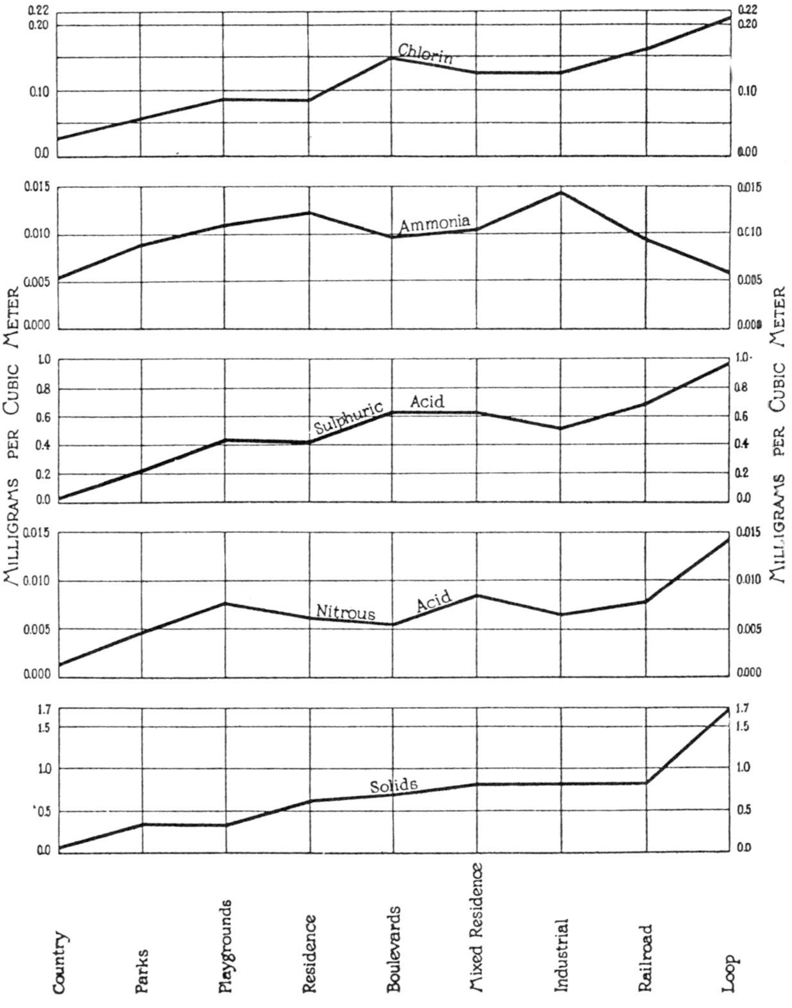

FIGURE 1. Average results of air analyses, City of Chicago, circa 1915. (Courtesy of Dr. Ben Linsky.)

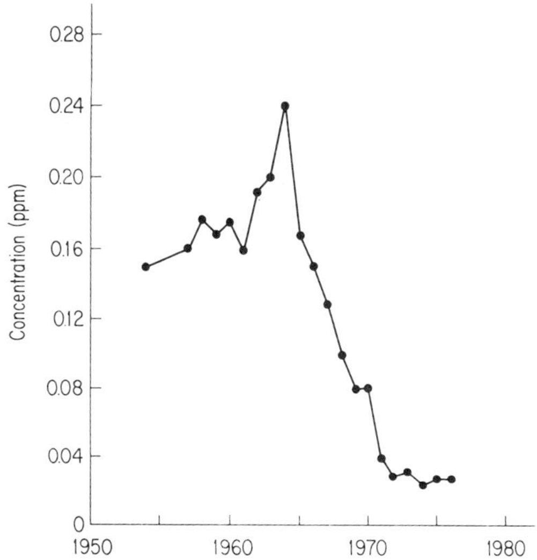

FIGURE 2. Citywide annual average SO_2 concentration in New York (1954–1976).

country must get on with the job of cleaning up the air before it becomes unbreathable. I'd like to show you a few things that might put the matter in perspective.

I am indebted to Dr. Ben Linsky for FIGURE 1. It is taken from his new edition of the 1915 Chicago study, and gives the concentrations of various pollutants that were observed as one moved out into the Chicago suburbs.[1] The concentration of sulfuric acid in downtown Chicago was about 1 mg/m^3, and suspended solids were 1.7 mg/m^3.

FIGURE 2 shows that substantial improvement in the annual average concentration of SO_2 has occurred in New York City. New York City instituted controls in 1964, which resulted in a spectacular decrease in the concentration of SO_2. The job had very largely been accomplished by the time the Environmental Protection Agency was formed in 1970.

I have not seen similar data for St. Louis, Pittsburgh, or other cities, but I'm sure they exist, and I hope that someone will find it possible to summarize them for the historical record.

Now, the question that FIGURE 2 suggests is the very basic one of, How clean is clean? Until the late 1960s, we had no standards and simply required what seemed to be reasonable controls. However, by 1969, a standard for SO_2 seemed desirable, and an advisory council appointed by the Health Department suggested that we set it at 0.1 ppm temporarily and then do some epidemiological work to see whether we needed to go any further. We were considering only health effects, not the effects on materials or visibility. I might say that during that period, Dr. James McCarroll, who was then with the Cornell Medical Center, was undertaking an epidemiological study in Manhattan, New York. This study was discontinued for various reasons in 1965, the beginning of a very important period. He had been keeping records of the incidence and severity of respiratory disease in a Manhattan population, and correlating the effects with air pollution data. From the time Dr. McCarroll's studies were discontinued in 1965, there were no further studies conducted until 1970, when the CHESS[3] studies began. As can be seen from FIGURE 2, they had only a very narrow gradient within which to conduct their studies.

It would have been very valuable to the nation if epidemiological studies could have been performed in New York City during the period 1964–1970, but that is an opportunity lost forever. The federal government set the standard at 0.03 ppm, and that's where it is today. But the difference between 0.1 ppm and 0.03 ppm to New York City alone is about $200 million a year, or nearly $2 billion during the past 10 years. That is in 1970 dollars. I remind you that this is in a city where the school playgrounds were closed last summer because the city did not have enough money to provide supervisors.

The reduction in settled dust that has taken place in New York City during the past 40 years is shown in FIGURE 3. Much of the improvement resulted from the transition from coal to oil and gas for space heating and power generation. There has been a remarkable reduction, and I'd like to say that there is one resulting benefit that I can personally vouch for. It was a daily occurrence in this city, as in many other coal-burning cities, to get at least one cinder in your eye. This has become a relatively rare occurrence. New York is unquestionably much cleaner, and we're all grateful for that. The same is true for Pittsburgh, where, I remember, it was just impossible to wear a white shirt back in the 30s.

Studies of total suspended particulates (TSP) in New York City go back to about 1930. The first studies were conducted over a period of several months by Arthur Stern's group and by Schrenk and Yant. It can be seen from FIGURE 4 that the TSP was then about 0.5 mg/m^3, which is about one-third of the concentration reported in the Chicago Loop. The concentrations of TSP have since diminished rapidly. Perhaps

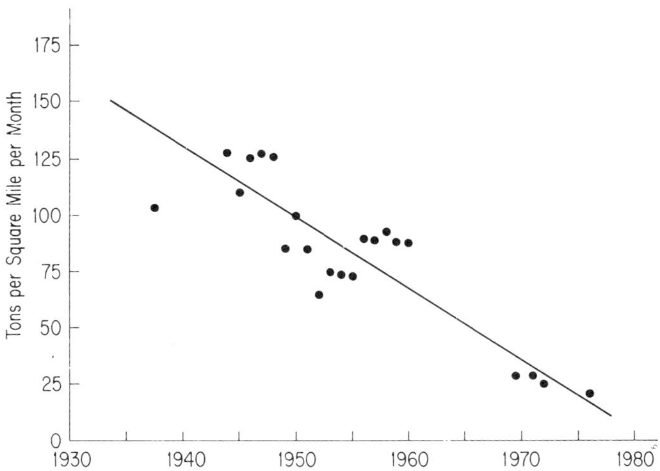

FIGURE 3. Settled dust in Manhattan (1936–1976). (Sources of data: 1936, Reference 2; 1944–60, Reference 3; 1969–76, Reference 4.)

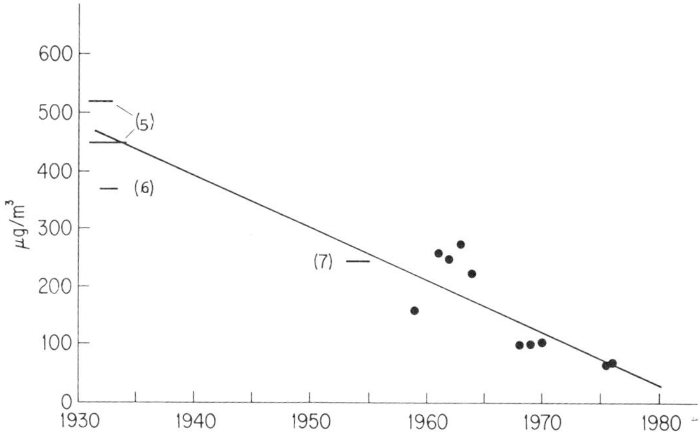

FIGURE 4. Total suspended particulates in New York City (1930–1976). (Numbers in parenthesis refer to References. Data after 1955 are from reports of the New York City Department of Air Resources; the sampling locations change from year to year.)

the principal effect has been to control the larger particles, but I can't help but think that there's been some improvement in the smaller particles as well.

A question that must be asked as we examine these geophysical problems is whether the kinds of standards that we're developing here are going to be adopted world-wide. Remember that we are but 6% of the world's population, yet we use about 30% of the energy now being consumed in the world. Sometime in the next century, when the developing countries emerge to the level of energy use characterized now by the western countries and Japan, they will be consuming enormous quantities of energy but may not have the economic means to put in the kinds of controls that will be needed.

We are already seeing that problem in connection with water pollution control. As these countries develop, one of the first things they've got to do is install water supply systems of good quality, and they must learn how to treat their sewage. A 200-million-gallon sewage treatment plant nowadays may cost a billion dollars, and a plant of this size may service no more than a relatively small city in Southeast Asia. A country that can't afford the billion dollars will simply discharge the wastes into its bays and rivers. How are we going to convince the developing countries to install flue gas scrubbers, which may increase the cost of a 1000 megawatt plant by about $150 million?

The developing countries will be existing on a marginal economy for some years. They will need help from the outside; and either the developed countries must provide the money for such controls, which in turn will benefit us, or it's not going to be done.

Dr. Moss has commented on the need for research, and I would like to spend just a moment emphasizing that point. I think it's a great tragedy that we spend billions of dollars for hardware to control emissions, and yet are reluctant as a nation to spend relatively small quantities of money to do the research necessary to develop the controls in a rational way.

My overview message is optimistic with respect to health effects. I know there are many people who don't agree with me, but we have had a decade now of epidemiological study after epidemiological study, and have found at most only second-order health effects. This is minor compared to other things that public health officers must be concerned about. However, on the geophysical side, the potential for cataclysmic climatological effects is very real. It is my belief that, in the decades ahead, there will be far greater concern with these effects than with the effects on public health.

REFERENCES

1. Chicago Association of Commerce Committee. 1915. The Chicago Report on Smoke Abatement: A Landmark Survey of the Technology and History of Air Pollution Control. (Appraisal by Benjamin Linsky.) Chicago Association of Commerce. Chicago, Ill. (Reprinted in 1971 by Maxwell Reprint Co., Elmsford, N.Y.)
2. SIEGEL, J. & B. FEINER. 1945. Sootfall studies for New York City. Part II of air pollution survey report 1945. Heat./Piping/Air Cond. **17:** 495.
3. Community Health and Environmental Surveillance System. 1974. Health Consequences of Sulfur Oxides: A Report from CHESS, 1970–1971. Report No. EPA-650/1-74-004. U.S. Environmental Protection Agency. Research Triangle Park, N.C.
4. Department of Air Resources. 1976. Annual Report. City of New York.
5. IVES, J. E., R. H. BRITTEN, D. W. ARMSTRONG, *et al.* 1936. Atmospheric Pollution of American Cities for the Years 1931 to 1933. Public Health Bulletin No. 224, U.S. Public Health Service. U.S. Government Printing Office. Washington, D.C.
6. YANT, W. P. *et al.* 1941. Carbon Monoxide and Particulate Matter in Air of Holland Tunnel and Metropolitan New York. Report of Investigations, U.S. Department of Interior, Bureau of Mines. Washington, D.C.
7. MEGONNEL, W. H. 1957. Historical Review. Prepared for the Study of Air Pollution in the New York-Metropolitan Area conducted by Interstate Sanitation Commission. New York, N.Y.

SOME FUTURE PROBLEMS ASSOCIATED WITH ENERGY: PANEL DISCUSSION

Moderator: Theodore J. Kneip

Panel Members: Merril Eisenbud, Sheldon K. Friedlander, James E. Hansen, Helmut E. Landsberg, Paul J. Lioy, Arthur M. Squires, William N. Stasiuk, and Thomas Moss

T. V. LARSON (*University of Washington, Seattle, Wash.*): I wanted to ask a somewhat technical question of Dr. Friedlander. I was intrigued by the speculation on the different reaction mechanisms for the different sulfate distributions that you observed. I'm wondering if you considered and ruled out the possibility of coagulation moving particles into the larger size ranges. Is that a difficult phase?

S. K. FRIEDLANDER (*University of California, Los Angeles, Calif.*): Yes, that was the first mechanism that we considered, actually. The question was, What are the mechanisms that might explain why over the arid regions, there seems to be a tendency for the sulfates to accumulate in the 0.1 to 0.2 micron size range, whereas in the very large mass of data for Los Angeles, the sulfates peak in the 0.6 to 1 micron range? We suggested that one difference could be droplet-phase reactions for the coastal site and homogeneous gas-phase reactions for the desert area.

We did some calculations for coagulation. If the original mechanism is a homogeneous gas-phase reaction, it would take about a day for the aerosol to grow from the 0.1–0.2 micron range to the 0.6–1 micron range. At least a day, and it just didn't seem to us that there was that much time, unless there's a lot of sloshing around with the aerosol blowing out during the night and coming back again pretty much the same.

But that's a possible alternative, and it's discussed in a review paper that I wrote in *Atmospheric Environment*—part of the Dubrovnik conference that Jim Lodge and Rudy Husar edited in the last year.

R. JAENICKE (*Max-Planck-Institut für Chemie, Mainz, Federal Republic of Germany*): I would like to make some comments to Dr. Hansen. The effect you described—from a report by Benjamin Franklin about the year 1783—was a very exceptional volcanic eruption in Iceland. Usually, the aerosol from volcanic eruptions goes far into the stratosphere. The effect of such an eruption as you have convincingly shown lasts roughly two years. What is the removal time of aerosol particles from these layers? In 1783, the effect was gone by the fall. That shows that this eruption only went into the troposphere; it never penetrated into the stratosphere. So all the particles were removed with the tropospheric residence time for aerosol particles—roughly on the order of, say, weeks or months. So the effect was gone in the fall of that year, but the climatological effect for that year took a smaller period of time. This raises the possibility of another comment.

I wonder that you did not refer to Budyko's values about the average temperatures of the Northern Hemisphere, because if you compare Budyko's values with turbidity values from the beginning of the century on, you can see a very good correlation of the turbidity of the atmosphere and the temperature, the average temperature of the Northern Hemisphere.

J. E. HANSEN (*Goddard Institute for Space Studies, New York, N.Y.*): Yes, there

0077–8923/80/0338–0605 $01.75/1

are two things you brought out in regard to the 1783 volcanic eruption. I did not attempt to show the correlation between the climate and the aerosols just for the reason that you really don't have adequate data on either, particularly on the climate. The observing stations at that time were concentrated to a small part of the globe. I agree with you that the available evidence indicates that most of the volcanic aerosols in that case were tropospheric, but it doesn't affect the arguments that I made.

In regard to your second point on the correlations between aerosols and climate, empirical calculations that Budyko and Bryson and others have made show that there is a correlation, but again, the problem is the measurements. The measurements are very limited to particular geographic areas, but it's a global problem that we are talking about. So I don't think you can convince the scientific community that you have an explanation for the global climate trends based on the meager observations that exist.

R. JAENICKE: May I add something? I don't know if you are aware about two works in this field. One is by Hammer, published in *Nature,* who showed sulfur layers in the Greenland ice cap that strongly correlated with volcanic eruptions. The second is, as we have shown, that you can easily calculate the turbidity of the atmosphere from measurements with the circumference sunshine recorder. This has not been done in the past, but it's very easy. We have shown that, and it's published.

So with this instrument, which has been around since the turn of the century, one could easily have a worldwide network of turbidity data and correlate that to, for instance, the surface temperature.

T. J. KNEIP (*New York University Medical Center, New York, N.Y.*): Dr. Hansen, it seems to me that during the conference, we've heard that we're creating a haze both in Europe and the United States in the summer that consists of white or light-colored particles in the troposphere and that in the winter, we have a potential for conversion to fuels that might give us more dark particles. If I interpret all of this correctly, it sounds like we get a cooling in the summer and a heating in the winter. Is it possible that, rather than just affecting the overall global average, we are affecting things differently in different seasons?

J. E. HANSEN: I think that's quite possible, given the fact that tropospheric aerosols have a relatively short lifetime. But again, I emphasize that if you just pick a piece of the data, then it's hard to have much confidence in any predictions. So you really would need to look at it on a global basis, I think.

M. EISENBUD (*New York University Medical Center, New York, N.Y.*): I'd like to hear a little discussion about the stratosphere. I have a kind of queasy feeling that there may be some sleepers ahead of us. After all, it was only ten years ago or so, less maybe, that the SST controversy arose, and this stimulated interest in some of the photochemical reactions that would take place in the stratosphere. Then a few years after that, maybe 5 or 6 years ago, we began to get reports of the possible effects of the fluorocarbons. Now we hear talk about even agricultural chemicals that may infiltrate the stratosphere and produce photochemical reactions that interfere with the ozone layer. I wonder how many more are ahead of us that we haven't thought of and what we should do to try to take stock of those possibilities, those possible anthropogenic effects on stratospheric ozone. I'd like to hear some discussion.

A. M. SQUIRES (*Virginia Polytechnic Institute and State University, Blacksburg,*

Va.): I wonder if I could raise the question of whether we're talking about particulates—aerosols—or whether we're not. Because that's outside the subject.

There are a number of things that *are* related to aerosols that I would like to see discussed. Some of them were aroused by the comments in your overview, Dr. Eisenbud, and so I'd like to address them to someone else. You referred to the 1915 Chicago study, which had a very nice, mobile air pollution sampling laboratory with 1912 license plates. The study was mounted because there had been a feeling in the city of Chicago that it was too filthy, too stinky, too dirty, too hazy, and they proposed the electrification of the railways and the railway terminal. They'd gotten just so far, had gotten ready to go, ready to be made operational by the Board of Aldermen when in order to stop it—in order to stall it, in other words—the Chicago Association of Commerce said, Well, now there are some questions. (Just raising a question is the way to stop anything, almost). So they mounted a three- or four-year study. Meantime World War I came along and stopped everything, as it did most of the cities in the country.

But it was the undesirability of the air in those days—they didn't like it being murky, stinky, hazy, and dirty. I've observed in the 30 years I've been in this field that there is continuing rising expectation for cleaner air among the people who were there and certainly among the younger people coming along—an expectation reinforced, fortunately, by the fact that periodically the winds blow clear and people can see what they're missing.

So I wonder, and I'd like to get some feeling from the others of the panel, would you settle for a hazy, stinky world? Now Bob McNamara told the U.N. Conference, speaking for the World Bank in 1972 in Stockholm, that putting a clean steel mill into a steel mill town would add 2% to the cost. Would you settle for a stinky, hazy world and also ignore the as yet unmeasured health impairment effect, the nonspecific health stress, rather than spend the extra 2%?

T. J. Kneip: I think you're asking the wrong group. If we could get a group here from Harlem, or the South Bronx, or the poverty stricken areas of India and ask them if they would accept a hazy, stinky world if they could get enough to eat and a warm place to sleep at night, I think they'd answer that they'll take the haze if that's what it costs to get them some food.

A. M. Squires: I wonder whether they would if they knew that we're only talking about 2 or 3% of the GNP or of the operation, especially for the big dirties and the small dirties. We're only talking 2 or 3% to avoid allowing a messy world to develop around these new industries and new industry towns like Gary, Indiana—one of the first planned towns in the country.

Unidentified Speaker: Yes, but who wants to go back to those days?

A. M. Squires: That's exactly my point. We are in an era of rising expectation for cleaner air, and yet we anticipate a continuing increase in the use of fuels.

W. N. Stasiuk (*New York State Department of Health, Albany, N.Y.*): Is Tom Graedel still in the audience? He had some comments on NO_2 from power plants that may be pertinent to Dr. Eisenbud's question about the stratospheric issues. Does anybody know anything about that and want to bring it up?

T. V. Larson: I'd just like to make a general comment. Since no one is volunteering as an expert on the stratosphere, I'll venture some statements. The particles in the stratosphere typically are in the lower stratosphere as the classical sulfate haze (Junge layer), whereas the concern over ozone depletion from the fluorocarbons involves chemistries in the upper stratosphere. The stratosphere is so named—as opposed to the troposphere—because it is so stratified, and horizontal motions are much greater than the vertical ones.

The point I wanted to make was that concerns over consequences of reactions in the upper stratosphere are seemingly much more ominous than are effects on the lower stratosphere because, if we're doing something now that may have consequences in the upper stratosphere, we might not see those consequences for another 20 or 30 years. So our foresight must be much greater in that case.

P. J. LIOY (*New York University Medical Center, New York, N.Y.*): Dr. Eisenbud, in his overview, missed one effect that—aside from health effects—is important. I guess since we are summarizing, we should reemphasize the fact that we still have a continuing problem with the acidity of the rain. This problem is not going to go away, and the environmental or ecological impact may get more severe. Maybe we will be cleaning up the fuel and it will not have large particle emissions, but we'll have fine particle emissions from coal-burning plants that may be much more acidic in nature. Added SO_2 emissions, more than likely from tall stacks, may also be converted to acidic sulfates. Therefore, we have to concern ourselves with all the implications of this problem.

It is interesting that recently manufactured automobiles have a warranty disclaimer in their owner's package that states that the surface of the car can no longer be guaranteed or warranteed in the eastern United States because environmental factors may cause a degradation of the paint material. Now this is not, I think, just a minor problem. I think it is an indication of something that will be pervasive, and we have to concern ourselves with the degradation of materials and other secondary effects, as well as potential health effects.

S. K. FRIEDLANDER: I agree that we have to emphasize here the technical aspect. I think that's probably what most of us feel that we're best at. I think, though, that it's important that—when we can—we participate in the political processes that really decide how the funds are allocated—what fraction of the cost of capital investment goes into control, and in fact, on the national scale, how much goes into say the space program or military spending. All these are political issues at base, and ours is just as much a political issue. I think we have to be prepared to argue in support of what we're doing, but I don't think this is the place to do it.

I felt encouraged about some of the technical developments that were reported here. There was a lot of discussion of the chemical element balance method, and I was very encouraged by the advances made by all those retreaded nuclear chemists. I think they're really doing an excellent job, and it seems to me that they are developing a very powerful tool. When we develop methods of that kind, we should make it possible to resolve sources based on the contributions from different types of pollution sources. That, in the long run, will reduce the cost of pollution control because we can, in a more rational way, make decisions on what kinds of sources to control in order to get the most control for our money. It seems to me that that type of work is very important.

I was reasonably encouraged also by some of the results of predictive models, although I think that's still pretty far from being a really reliable tool for the kind of problem I was talking about today, where we like to estimate the effects of new sources on air quality say in the Southwest. I don't see that we're going to be able to do much more than parameterize the data, if we want to estimate the amounts of sulfates and nitrates that are going to be present in the atmosphere.

I talked before about the importance of carbon and organic balances. I still feel that that's an important next step in this procedure, as far as doing source resolution and also in accounting for agents in the environment. I think the comment that Dr. Newman made on the sulfur conversion was an important one. For example, one of the figures on conversion, which somebody showed in a table, was 1 to 13%; but that represented a high range and was especially chosen. We chose those days to get very high rates of conversion, and I imagine that quite a few of the other reports on conversion also reflected high rates or conditions that would lead to high rates. So we probably ought to indicate in the tables some measure of the severity of the conditions producing sulfates.

For example, it might be something like the ozone bulge at some characteristic distance downwind from the source. I don't know if that's a good measure, but there should be some measure of the severity of conversion conditions.

Finally, with all due respect to your wisdom, Dr. Eisenbud, I don't know whether you've ever lived in Los Angeles for any period of time. I don't know whether you count eye irritation and difficulty in breathing as health effects, but we still have episodes of that kind, though not as frequently as in the past. I think most people would be willing to pay a substantial amount for controls, if they're cost effective. I don't think people just want to throw money away, but I think a lot of us would be willing to pay more. In fact, it's claimed that property values reflect this; property values on the west side of town are increasing much more rapidly than in the area in which I live, unfortunately.

M. EISENBUD: I agree completely with what you say, Dr. Friedlander, but I have never seen a justification for saddling the rest of the country with the cost of the administrative responsibilities of solving your problem. You can't rouse your own community out there, so we have got to pay for, in essence, four cars per family. Why should we worry about it in other parts of the country?

We had to look at this problem, because New York State had to take a position on automobile emission. The State Advisory Council flatly took a position that we do not have an automobile emission problem of significant health importance in the state. Yet the national policy, which is geared pretty much to your problem out there I think, is burdening us with about a billion-dollars-a-year cost.

B. LINSKY (*Morgantown, W. Va.*): Starting in 1954, residents of Washington, D. C., began to have observable eye irritation. They began to measure the oxidant there in 1959 and '60, and that is one of the major reasons that we were able to get adequate support all over the country.

A. M. SQUIRES: I just wanted to say a few words about acid rain and something that's puzzled me. It struck me as kind of curious that Odén's maps show the explosion of acid rain over northwest Europe in just exactly the years when Europe abandoned coal for small-scale use—small-scale energy needs—and went to oil. Bill Stasiuk pointed out, in one short sentence or two in his talk, that oil fires make more SO_3 than coal fires do, and SO_3 of course comes out as a sulfuric acid aerosol from the stack.

What's more, the sulfur trioxides produced in an oil fire don't track the sulfur level of the oil. Here in New York City—going from 2½% sulfur in oil, roughly in the late 1960s, down to the present level of about 0.3%—the meager data in the literature would suggest that the sulfur trioxide production in New York City may have come down by about a half. The data do not really permit a very good estimate of that; it could have come down only 20 or 30%, Dr. Eisenbud, or it might have come down by a factor of three, but that's sort of the level of the uncertainty.

I would suggest that the sulfur conversion to SO_3 at 0.3% sulfur in oil may be as high as about 10% in an oil fire. Again, I don't know if it's that high; the data are too meager. It's certainly more than 5%, and quite possibly 10%

I heard estimates of conversion in the urban plume for some kind of average situation (a gentleman from Brookhaven threw this out) as being something under 1%. We saw the much higher figures in summer days when it was interesting to study the high conversion rates and do the modeling. But if it's less than 1% on the average, then the sulfur conversion in the small oil fires that are heating New York City apartment buildings may be the equivalent of 10 to 15 hours of downwind travel of the plume. I would suggest that this may be interesting to the people who are studying urban plumes.

T. J. KNEIP: Dr. Moss, do you have any last comment?

T. MOSS (*Office of Rep. George E. Brown, Jr., Washington, D.C.*): I was just going to attempt to justify my seat at the table by saying something about these public policy questions that have been mentioned.

I think whether we like it or not in many cases, there obviously are certain fields of science that—unfortunately or fortunately—have been directly coupled into these decision-making processes, and there's probably excessive faith in the functioning of the mechanism by which that occurs. I was thinking about it here, and I thought there were two ways that seem to be operating. One is the idea that it can work in an adversary fashion. The whole line of thinking related to science courts is an example. The press releases from the environmental groups, or the ads that are placed in the newspapers, etc., this is part of the adversary process in dealing with the coupling of the data to public policy. And that may be a good way to do it, and it may not be.

Then another is the arbitration line of thought, which is probably related to the National Academy of Sciences being asked to do things. But it seems to me that both mechanisms really imply an organizational need in the science community that is far beyond what we have ever done in the past and that breaks a lot of traditions in science. There is a healthy anarchy in science that keeps up competitiveness, that allows us to go on our own ways and not talk about it until we spring it at a meeting, and all that kind of thing. All those traditions—those anarchical traditions—make it very difficult to couple either the adversary process or the arbitration process into the public policy. And that may be a whole new challenge.

The CO_2 problem, I think, is a very good example. Here we have perhaps 20 years to begin to very systematically accumulate monitoring and other kinds of data on CO_2 effects before the science community goes to the political community and says, Look, there's really a problem and you guys have to do something and make some sacrifices. Yet it's not at all clear that we will organize to systematically build the case that will need to be presented.

I've seen a whole range of challenges—organizational challenges. I think, on the simplest end, it's a decision of an individual person who has some data to call up a colleague that he knows is working in a similar area and say, Let's publish a joint paper on this; we both have some new findings, instead of just doing it individually, which is the normal, prideful way to do it. Then that decision to do the joint paper may involve three or four groups in exchanging manuscripts and may involve a lot of clumsy work through the mail, etc. It's not our normal way, and yet it may make a great difference in the credibility to the public of that scientific case.

Then the other extreme comes to societies jointly doing, planning, and hosting topical conferences and other efforts. But I think one of the real challenges the scientific community has to face is whether it's willing to put up with that organizational stress in order to play this role in public policy.

T. J. KNEIP: Thank you, Dr. Moss.

M. GRIFFEL (*University of Iowa, Iowa City, Iowa*): I think the last comment was an extremely important one. I'd like to call attention to the fact that in medical research—clinical trials, for example, where it's been necessary to involve a good number of major medical centers in the investigation of a particular procedure or a particular drug—this type of cooperation is now customary and very frequent, and may involve as many as a hundred investigators. The resulting paper is published not by individuals, but by units from different universities.

So I think it is possible, it is very possible.

W. N. STASIUK: I'd just like to add something that I believe is very important. Dr. Eisenbud called for it at the end of his talk, and that's a collection of good, continuous baseline data using the same techniques for a long period of time, be it epidemiological, be it atmospheric, or be it source. There's a tremendous shortfall here. The analogy that we go in spurts of activity and then stop that activity at the crucial time almost seems to be the rule rather than the exception.

All the good work on transport was not developed by theorists, it was developed by people who observed data and said, Hey, something is wrong. Traditional air pollution cause and effect doesn't explain this. I add to yesterday's comments by Dr. Cowling, who was looking for the support to establish a network. I don't know if his network is the one that has the highest priority for funding or not, but I think there is a need that the federal government should recognize—that somebody should sit on top of a data-gathering group and make sure that there's a continuity to the effort so that there will be a baseline that can serve as a bench mark for other studies.

J. P. LODGE (*Consultant in Atmospheric Chemistry, Boulder, Colo.*): Considered from the standpoint of a former air pollution control official, this conference has set forth an interesting dilemma with regard to standards for total suspended particulate matter.

We have learned that, on the average, half of the mass of TSP lies below about 2.5 micrometers, and half above. Of the former category, roughly 80% is a condensation aerosol from gaseous contaminants emitted at substantial distances from the point of measurement. Roughly 80% of the coarse aerosol consists of soil particles which, at least in the dry climate where I live, arise from such inherently uncontrollable activities as agriculture.

Faced with these figures, if we are unable to meet ambient air quality standards, whom can we sue?

It has been noted, both here and elsewhere, that the essence of success in obtaining research funding is a good acronym. I therefore propose a study aimed specifically at determining the real feasibility of achieving air quality goals when a large fraction of the emissions is either natural or originates outside the governing jurisdiction. A suitable name for the project might well be Spontaneous Injections of Naturally Elevated Contaminants at Unprecedented Rates in the Environment.

P. J. LIOY: I'd like to continue what Dr. Friedlander was saying a few minutes ago, and talk about the things that I was encouraged by with respect to aerosol transport.

There was mention of understanding the summertime urban plume reaction rates a little bit more thoroughly. Also, the presentations yesterday by Dr. Husar and by Dr. Lyons gave us a good visual and perceptual understanding of the meteorological conditions associated with transport and hazy air masses.

However, we're now at the stage where we have to go to the next step, which is the toughest. We have obtained a lot of data; and if we're to define the direct contribution of transport to pollution in major regional areas, we have to go beyond looking at the qualitative aspects of it—drawing isopleths. I think this is the next big step, and I hope we can continue it in a consistent pattern.

INTERPRETIVE DISCUSSION

Paul J. Lioy and Theo. J. Kneip

Institute of Environmental Medicine
New York University Medical Center
New York, New York 10016

The papers presented in this conference have covered a series of interrelated topics in the field of aerosol science. Several major relationships between sources and receptors of atmospheric aerosols were discussed, particularly in relation to potentially serious effects on climates and ecosystems. This discussion paper presents an integrated and concise view of the major conclusions supported by the papers in the conference.

The traditional approach to identifying the atmospheric aerosols has been in terms of local contributions from sources, such as industries, space heating, utilities, and automobiles. Institutionally, emission controls have been implemented for these types of sources in order to obtain compliance with minimum air quality standards. Many studies have been conducted on urban and rural aerosols in order to determine composition and size and mass distributions. Most of these studies were designed as intensive, short-term programs. Their findings indicated large-scale effects to be present. Those findings are now supported by the results of regional and global-scale programs discussed during this meeting.

As was first noted in Sweden, atmospheric aerosols should not be considered only in terms of local sources, particle size, and composition, but also in terms of the degree to which they are formed and transported in the atmosphere. In other words, at times, local excursions in particle size and mass concentrations can be related to the movement of material by the upper air wind and its eventual distribution over a wide area. Separation of the local and regional components is an intriguing problem and requires the simultaneous use of both meteorologic and air pollution data.

A thorough understanding of these regional and global-scale phenomena has not been achieved as yet; however, the hypotheses developed thus far will be tested in ongoing or future research programs. All research workers should consider these factors in sampling site selections and research project designs in order to achieve a more rapid understanding of the large-scale problems.

Much of the success of the more recent programs can be attributed to the development of instrumentation that provides direct, continuous measurements of optical properties or size distributions of the aerosol. Recent developments in methods for the measurement of various chemical species in the aerosol have also been of critical importance.

The earlier recognition of the trimodal size distribution of aerosols and its physical significance, i.e., nucleation, accumulation, and coarse modes, has spurred development of improved instrumentation. Sophisticated and reliable instruments are now available for continuous measurement of particle size distributions or for collection of size-stratified samples for analysis. These instruments have been used in condensation and particle growth studies, and in observations of the characteristics of fresh and aged aerosols in both the rural and urban environments.

0077-8923/80/0338-0613 $01.75/1 © 1980, NYAS

Knowledge of aerosol composition has been enhanced by the use over the last five years of multielement techniques, such as instrumental neutron activation analysis (INAA), x-ray fluorescence (XRF), and proton-induced x-ray emission analysis (PIXE). In addition, techniques that characterize, or fingerprint, individual particles—such as microprobe analysis—have been used on a limited basis in a number of studies. The introduction of new and more precise chemical measurement of sulfur and nitrogen species, plus methods for hydrogen ion determination, has helped to improve our understanding of the secondary aerosol. Methods for measurement of organic compounds have become increasingly more reliable and sophisticated. The further development and application of all of these techniques should become even more important as society turns to the use of more coal, new sources of oil, and diesel fuel for energy.

The sources of the organic fractions will have to be determined in cases where adverse effects may exist. Thus, mathematical techniques that have been reasonably successful in defining the contribution of major sources to the atmospheric particle loading will increase in importance. The techniques include the chemical element balance method (CEB), factor analysis (FA), multiple regression, and other multivariate mathematical techniques. These methods have been successfully used to assign the sources of the many elements present in aerosols and to estimate the sources of the aerosol mass.

A major difficulty that still has to be addressed in mathematical source assignment is the frequent lack of a means of distinguishing sources of aerosols of similar composition. Two particularly difficult cases are the crustal soil type and secondary aerosol fractions. These may have a multiplicity of sources and, therefore, could actually be assigned to several sources. Application of meteorological data, further development of pattern recognition methods, and the use of newer instrumentation, such as the electron microprobe, may be helpful in solving this problem. As a result of the discussions at the conference, it appears that enough overlap exists between the various mathematical techniques to allow for a more integrated use of their respective assets in the further development of apportionment models.

As a result of major plume and regional studies, we are becoming more aware of a potential for accumulation of secondary particulate matter in a plume, and its transport to receptors downwind. Sulfur-containing contaminants appear to have become increasingly important in the eastern U.S. and in northern Europe in recent years. Studies have not only examined the growth and transportation of particles within a plume, but also the influence of the character of the upwind or background air on the transformation rates of SO_2 to sulfate. Summertime plume studies have shown that the condition of the background air with respect to nonmethane hydrocarbons, nitrogen oxides, and free radicals may significantly influence the rate of ozone production and subsequently the formation of aerosol particles.

For the particular cases discussed at the conference—the Milwaukee plume, the Labadie power plant plume, and the St. Louis plume—formation of high $SO_4^=$ concentrations occurred when the plume was embedded in a smoggy air mass. Because of the limited number of meteorological variables covered by these and other plume studies, the foregoing statements must be qualified and research on the reactions and aerosol formation rates that occur is necessary. In addition, plumes must be investigated for types of transformations and accumulation patterns that can occur at night and during seasons other than summer.

The various studies of anthropogenic aerosols have been paralleled by the study of the emission and transport characteristics of natural aerosols. The natural aerosol has various origins, including stratospheric, maritime, and continental. The particles are generated by resuspension both as coarse and accumulation mode particles and, through physical and chemical transformations of organic, sulfur, and nitrogen gases or vapors, as accumulation and nucleation particles.

The lifetimes of aerosol particles indicate that natural aerosol particles $< 1.0\ \mu m$ can be transported distances in excess of 1,000 km. This phenomenon has been identified on a number of occasions dating back to the 19th century for emissions from the Sahara and Gobi deserts, volcanic eruptions, and forest fires. Satellite images of both natural and man-made events were discussed during the conference. Obviously, the primary uncertainty associated with major natural emission events is the unpredictability of the time of occurrence.

It is curious that despite the very early observation of significant events involving the atmospheric transport of naturally occurring materials, it has only been within the last decade that our concepts on the movement of anthropogenic emissions have changed. With the evolution of the industrialized countries of the world, the production of man-made contaminant emissions has increased, and it is now clear that this has resulted in periodic episodes of regionally distributed aerosols in the eastern U.S. and in Europe. Since the 1960s, the summer visibility in the eastern U.S. has been degrading progressively, and this appears to be related, statistically at least, to the increase in summer coal consumption by electrical utilities. Through visibility analyses, ozone isopleth analyses, and satellite haze photographs, the degradation has been found to occur periodically during the summer and is associated with the movement of high pressure systems through the eastern U.S.

The visibility-reducing aerosol is comprised of a number of species; however, $SO_4^=$ appears to be a dominant influence. The work of the Organization for Economic Cooperation and Development (OECD) in Europe supports this conclusion. However, it was reached in an environment where high-sulfur oil is the major source of utility energy. As the demand for energy increases in the underdeveloped countries of the world, the visibility degradation problem will probably be intensified. This is presently being observed in the "clean areas" of the southwestern U.S. as a result of increased industrialization and urbanization.

On a broader scale, arctic haze has been increasing in the wintertime. The problem appears to result from the presence of secondary $SO_4^=$ and has its origin in the midlatitude industrialized countries. It has been hypothesized that in the winter, the restrictive boundary of the polar cold front is below the midlatitude regions, which allows free movement of man-made pollution to the Arctic. In the summer, however, the cold front is above the midlatitudes, which reduces mixing of clean and polluted air. This case indicates that even without local industrialization, degradation can be observed in remote areas.

The conclusions from this conference can be summed up in terms of current knowledge of interactions of changing fuel quality, atmospheric chemistry, long range transport, and effects. Because of the desire of underdeveloped countries to grow, and because of the anticipated changes in fuels dictated by the economics of oil pricing, it appears that energy-related gaseous and particulate emissions will increase. As steps are taken to lessen local impacts, the indications are that the resulting aerosols will be distributed periodically over very wide areas and may contribute to climatic changes and potentially adverse effects on the ecosystem. With these existing possibilities, it is necessary to continue collection of atmospheric aerosol data, and to extend the analyses of the existing data in order to make recommendations as to the effects that new or changing sources could have on the atmosphere and on mankind.

Index of Contributors*

Alkezweeny, A. J. **219–232**
Allen, J. **86–92**
Anderson, J. A. **233–257**

Barrie, L. A. 190–197, 377–378, **434–452,** 483–485
Baskett, R. L. **463–482**
Beyak, R. A. **174–189**
Blanchard, D. C. **330–347,** 377–378
Blumenthal, D. L. **233–257**
Brachaczek, W. W. **145–173**
Brosset, C. 190–197, **389–398,** 483–485
Butler, J. W. **145–173**

Cantrell, B. K. **233–257**
Cowling, E. B. **540–555,** 556–560
Cunningham, P. T. 190–197

Daisey, J. M. **50–69,** 86–92
Dattner, S. 86–92, 190–197, 483–485
Dzubay, T. G. **126–144,** 190–197, 309–316

Eisenbud, M. **599–604,** 605–612
Ellestad, T. G. **202–218,** 309–316

Fein, R. S. 86–92
Friedlander, S. K. 86–92, 309–316, 377–378, **588–598,** 605–612

Gillani, N. V. **276–296,** 309–316
Gillette, D. 309–316, **348–358,** 377–378, 556–560
Glenn, T. R. **561–562**
Gordon, G. E. 86–92, **93–102,** 190–197
Gotaas, Y. **453–462**
Graedel, T. E. 309–316, 377–378
Griffel, M. 605–612

Hahn, J. **359–376,** 377–378
Hansen, J. E. **575–587,** 605–612
Hidy, G. M. **463–482**
Hopke, P. K. **103–115,** 190–197
Husar, R. B. **399–417,** 483–485, 556–560

Jaenicke, R. 86–92, 309–316, **317–329,** 377–378, 556–560, 605–612

Klauber, G. M. 86–92, 309–316, 483–485, 556–560
Kleinman, M. T. **116–125,** 190–197
Kneip, T. J. **ix,** 190–197, 309–316, 605–612, **613–616**
Korniski, T. J. **145–173**

Lacis, A. A. **575–587**
Landsberg, H. E. **569–574**
Lang, D. S. 483–485
Larson, T. V. **26–38,** 86–92, 556–560, 605–612
Lavery, T. F. **463–482**
Leaderer, B. P. **70–85,** 86–92
Lee, P. **575–587**
Linsky, B. 605–612
Lioy, P. J. **ix,** 190–197, 605–612, **613–616**
Lippmann, M. **1–12,** 86–92, 190–197
Lodge, J. P. 605–612
Lyons, W. A. **418–433,** 483–485

Macias, E. S. **233–257,** 309–316
McCaffrey, R. J. **486–503,** 556–560
Miller, D. F. **219–232,** 309–316
Moss, T. 605–612
Mueller, P. K. 86–92, 309–316, 377–378, **463–482**

Newman, L. 309–316, 483–485

Ottar, B. **504–514,** 556–560

Patterson, D. E. **399–417**
Peterson, T. W. 86–92, **174–189,** 190–197
Pierson, W. R. 86–92, **145–173,** 190–197, 483–485

Rahn, K. A. 190–197, 483–485, **486–503,** 556–560

Schiermeier, F. A. **198–201**
Schütz, L. **515–532,** 556–560
Sears, E. 86–92
Shaw, G. E. **533–539**

*The darker numbers refer to papers. The lighter numbers refer to comments made in discussion sessions.

Shaw, R. W. **13–25**
Squires, A. M. **563–568,** 605–612
Stamnes, K. **533–539**
Stasiuk, W. N. 605–612
Stevens, R. K. **13–25,** 86–92
Stolwijk, J. A. J. **70–85**
Swift, D. L. 86–92

Tanner, R. L. **39–49,** 86–92, 309–316
Truex, T. J. **145–173**
Tsoukalas, S. N. 377–378

Wang, W.-C. **575–587**
Warren, K. **463–482**
Whitby, K. T. 86–92, 190–197, **258–275,** 309–316, 377–378
Wilson, S. O. 556–560
Wilson, W. E. **276–296**
Winchester, J. W. 190–197, **297–308,** 309–316, 483–485
Wolff, G. T. 86–92, 309–316, **379–388,** 483–485
Woodcock, A. H. **330–347**